Industrial Water Quality

Industrial Water Quality

W. Wesley Eckenfelder, Jr.

Davis L. Ford

Andrew J. Englande, Jr.

Fourth Edition

New York Chicago San Francisco
Lisbon London Madrid Mexico City
Milan New Delhi San Juan
Seoul Singapore Sydney Toronto

The McGraw·Hill Companies

Copyright © 2009, 2000, 1989, 1967 by The McGraw-Hill Companies, Inc. All rights reserved. Printed in the United States of America. Except as permitted under the United States Copyright Act of 1976, no part of this publication may be reproduced or distributed in any form or by any means, or stored in a data base or retrieval system, without the prior written permission of the publisher.

1 2 3 4 5 6 7 8 9 0 DOC/DOC 0 1 4 3 2 1 0 9 8

ISBN 978-0-07-154866-3
MHID 0-07-154866-1

Sponsoring Editor
Larry S. Hager

Production Supervisor
Richard C. Ruzycka

Editing Supervisor
Stephen M. Smith

Project Manager
Aparna Shukla, International Typesetting and Composition

Copy Editor
Megha RC, International Typesetting and Composition

Proofreader
Upendra Prasad

Indexer
Broccoli Information Management

Art Director, Cover
Jeff Weeks

Composition
International Typesetting and Composition

Printed and bound by RR Donnelley.

McGraw-Hill books are available at special quantity discounts to use as premiums and sales promotions, or for use in corporate training programs. To contact a special sales representative, please visit the Contact Us page at www.mhprofessional.com.

This book is printed on acid-free paper.

Information contained in this work has been obtained by The McGraw-Hill Companies, Inc. ("McGraw-Hill") from sources believed to be reliable. However, neither McGraw-Hill nor its authors guarantee the accuracy or completeness of any information published herein, and neither McGraw-Hill nor its authors shall be responsible for any errors, omissions, or damages arising out of use of this information. This work is published with the understanding that McGraw-Hill and its authors are supplying information but are not attempting to render engineering or other professional services. If such services are required, the assistance of an appropriate professional should be sought.

About the Authors

W. Wesley Eckenfelder, Jr., is one of the world's foremost authorities on wastewater treatment and industrial water quality management. He provides technical direction and management of wastewater-related projects at AquAeTer. Dr. Eckenfelder has written scores of books and more than 200 scientific and technical papers.

Davis L. Ford, president of Davis L. Ford & Associates, is a multinational expert and consultant in wastewater treatment and environmental engineering. Dr. Ford has consulted with over 200 companies in his career and co-authored or contributed to seven books.

Andrew J. Englande, Jr., is a professor in the Department of Environmental Health Sciences at Tulane University's School of Public Health and Tropical Medicine. Dr. Englande has authored over 150 scientific papers and has consulted for numerous industries, governments, and consulting firms.

Contents

Preface xix
Acknowledgments xxi

1 Source and Characteristics of Industrial Wastewaters 1
   1.1 Introduction 1
   1.2 Undesirable Wastewater Characteristics ... 2
   1.3 Partial List of Regulations Which Affect Wastewater Treatment Requirements within the United States 6
        Air 6
        Liquid 6
   1.4 Sources and Characteristics of Wastewaters ... 7
   1.5 Industrial Waste Survey 8
   1.6 Waste Characteristics—Estimating the Organic Content 18
   1.7 Measuring Effluent Toxicity 32
        Toxicity Identification of Effluent Fractionation 36
        Source Analysis and Sorting 39
   1.8 In-Plant Waste Control and Water Reuse ... 40
        Waste Minimization 40
   1.9 Problems 46
        References 48

2 Wastewater Treatment Processes 49
   2.1 Introduction 49
   2.2 Process Selection 52

3 Pre- and Primary Treatment 65
   3.1 Introduction 65
   3.2 Equalization 67
   3.3 Neutralization 83
        Types of Processes 83
        System 87
        Control of Process 87
   3.4 Sedimentation 94
        Discrete Settling 94
        Flocculent Settling 98
        Zone Settling 106

		Laboratory Evaluation of Zone Settling and Calculation of Solids Flux	106
		Clarifiers	107
	3.5	Oil Separation	110
	3.6	Oil Processing in a Typical Petroleum Refinery	116
	3.7	Sour Water Strippers	119
	3.8	Flotation	120
		Air Solubility and Release	120
	3.9	Problems	131
		References	134
4	**Coagulation, Precipitation, and Metals Removal**		**137**
	4.1	Introduction	137
	4.2	Coagulation	137
		Zeta Potential	139
		Mechanism of Coagulation	141
		Properties of Coagulants	143
		Coagulant Aids	144
		Laboratory Control of Coagulation	146
		Coagulation Equipment	147
		Coagulation of Industrial Wastes	148
	4.3	Heavy Metals Removal	151
		Arsenic	157
		Barium	159
		Cadmium	159
		Chromium	159
		Copper	166
		Fluorides	166
		Iron	167
		Lead	169
		Manganese	169
		Mercury	169
		Nickel	170
		Selenium	171
		Silver	171
		Zinc	172
	4.4	Summary	172
	4.5	Problem	177
		References	177
5	**Aeration and Mass Transfer**		**179**
	5.1	Introduction	179
	5.2	Mechanism of Oxygen Transfer	179

	5.3	Aeration Equipment	195
		Diffused Aeration Equipment	197
		Turbine Aeration Equipment	201
		Surface-Aeration Equipment	202
		Measurement of Oxygen Transfer Efficiency	205
		Other Measuring Techniques	210
	5.4	Air Stripping of Volatile Organic Compounds	211
		Packed Towers	213
	5.5	Problems	221
		References	223
6	**Principles of Aerobic Biological Oxidation**		**225**
	6.1	Introduction	225
	6.2	Organics Removal Mechanisms	225
		Sorption	227
		Stripping	228
		Sorbability	229
		Biodegradation	232
	6.3	Mechanisms of Organic Removal by Biooxidation	233
		Sludge Yield and Oxygen Utilization	237
		Oxygen Requirements	246
		Nutrient Requirements	253
		Mathematical Relationships of Organic Removal	256
		Specific Organic Compounds	278
	6.4	Effect of Temperature	284
		Effect of pH	293
		Toxicity	294
	6.5	Sludge Quality Considerations	299
		Filamentous Bulking Control	307
		Biological Selectors	308
		Design of Aerobic Selectors	308
	6.6	Soluble Microbial Product Formation	316
	6.7	Bioinhibition of the Activated Sludge Process	318
		OECD Method 209	321
		Fed Batch Reactor	322
		Glucose Inhibition Test	323
	6.8	Stripping of Volatile Organics	326
		Treatment of VOC Emissions	332
	6.9	Nitrification and Denitrification	334
		Nitrification	334

x Contents

		Nitrification Kinetics	335
		Nitrification of High-Strength Wastewaters	342
		Inhibition of Nitrification	343
		Batch Activated Sludge Nitrification	355
		Fed Batch Reactor Nitrification Test	357
		Denitrification	358
		Nitrification and Denitrification Systems	368
		Nitrification Design Procedure	369
		Denitrification Design Procedure	372
	6.10	Phosphorus Removal	375
		Chemical Phosphorus Removal	375
		Biological Phosphorus Removal	380
		Mechanism for Biological Phosphorus Removal	381
		Glycogen Accumulating Organisms	383
		Biological Phosphorus Removal Design Considerations	383
		Other Mechanisms of Phosphorus Removal	385
		Membrane Bioreactors (MBRs)	385
	6.11	Laboratory and Pilot Plant Procedures for Development of Process Design Criteria	386
		Wastewater Characterization	386
		Reactor Operation	388
		Volatile Organic Carbon	389
		Reduction of Aquatic Toxicity	392
	6.12	Problems	396
		References	399
7	**Biological Wastewater Treatment Processes**		**403**
	7.1	Introduction	403
	7.2	Lagoons and Stabilization Basins	403
		Type I. Facultative Ponds	404
		Type II. Anaerobic Ponds	404
		Type III. Aerated Lagoons	405
		Lagoon Applications	405
	7.3	Aerated Lagoons	414
		Aerobic Lagoons	415
		Facultative Lagoons	418
		Temperature Effects in Aerated Lagoons	420
		Aerated Lagoon Systems	422
	7.4	Activated Sludge Processes	434
		Plug Flow Activated Sludge	435

		Complete Mix Activated Sludge	436
		Extended Aeration	437
		Oxidation Ditch Systems	439
		Sequencing Batch Reactor	440
		Batch-Activated Sludge	445
		Oxygen-Activated Sludge	448
		Deep-Shaft-Activated Sludge	451
		Biohoch Process	451
		Integrated Fixed Film Activated Sludge	453
		Thermophilic Aerobic Activated Sludge	454
		Final Clarification	454
		Flocculation and Hydraulic Problems	462
		Treatment of Industrial Wastewaters in Municipal-Activated Sludge Plants	462
		Effluent Suspended Solids Control	465
	7.5	Trickling Filtration	474
		Theory	474
		Oxygen Transfer and Utilization	479
		Effect of Temperature	482
		Trickling Filter Applications	482
		Tertiary Nitrification	485
	7.6	Rotating Biological Contactors	488
	7.7	Anaerobic Treatment Processes	494
		Process Alternatives	495
		Mechanism of Anaerobic Fermentation	499
		Biodegradation of Organic Compounds under Anaerobic Conditions	505
		Factors Affecting Process Operation	507
	7.8	Laboratory Evaluation of Anaerobic Treatment	511
	7.9	Problems	518
		References	521
8	**Adsorption**		**525**
	8.1	Introduction	525
	8.2	Theory of Adsorption	525
		Formulation of Adsorption	526
	8.3	Properties of Activated Carbon	530
		Laboratory Evaluation of Adsorption	532
		Continuous Carbon Filters	532
		Carbon Regeneration	534
		Adsorption System Design	536
		GAC Small Column Tests	550

Contents

		Performance of Activated Carbon Systems	553
	8.4	The PACT Process	554
	8.5	Problems	562
		References	563
9	**Ion Exchange**		**565**
	9.1	Introduction	565
	9.2	Theory of Ion Exchange	565
		Experimental Procedure	570
	9.3	Plating Waste Treatment	572
	9.4	Problem	576
		References	576
10	**Chemical Oxidation**		**577**
	10.1	Introduction	577
	10.2	Stoichiometry	578
	10.3	Applicability	581
	10.4	Ozone	582
	10.5	Hydrogen Peroxide	586
	10.6	Chlorine	590
	10.7	Potassium Permanganate	596
	10.8	Oxidation Overview	596
	10.9	Hydrothermal Processes	596
	10.10	Problem	600
		References	601
11	**Sludge Handling and Disposal**		**603**
	11.1	Introduction	603
	11.2	Characteristics of Sludges for Disposal	606
		Leaching Tests to Characterize Residuals	606
	11.3	Aerobic Digestion	609
	11.4	Gravity Thickening	616
	11.5	Flotation Thickening	621
	11.6	Rotary Drum Screen	625
	11.7	Gravity Belt Thickener	625
	11.8	Disk Centrifuge	626
	11.9	Basket Centrifuge	626
	11.10	Specific Resistance	628
		Laboratory Procedures	631
		Capillary Suction Time Test	634
	11.11	Centrifugation	635
	11.12	Vacuum Filtration	638
	11.13	Pressure Filtration	642

11.14	Belt Filter Press	645
11.15	Screw Press	647
11.16	Sand Bed Drying	648
11.17	Factors Affecting Dewatering Performance	649
11.18	Land Disposal of Sludges	650
11.19	Incineration	657
11.20	Problems	658
	References	661

12 Miscellaneous Treatment Processes 663

12.1	Introduction		663
12.2	Land Treatment		663
	Irrigation		663
	Rapid Infiltration		666
	Overland Flow		667
	Waste Characteristics		667
	Design of Irrigation Systems		668
	Performance of Land Application Systems		671
12.3	Deep-Well Disposal		672
12.4	Membrane Processes		679
	Pressure		685
	Temperature		686
	Membrane Packing Density		686
	Flux		686
	Recovery Factor		686
	Salt Rejection		686
	Membrane Life		686
	pH		686
	Turbidity		687
	Feedwater Stream Velocity		687
	Power Utilization		687
	Pretreatment		687
	Cleaning		687
	Applications		688
12.5	Membrane Bioreactors		693
	Types of Membranes and Reactor Configuration		693
	Benefits of Membranes Compared to Conventional Technology		694
	Membrane Issues		695
	Application		697
	Case Study A		698
	Case Study B		699

	12.6	Granular Media Filtration	700
	12.7	Microscreen	706
		References	708

13 Treatment: Oil/Gas Exploration/Production Residuals ... 711

13.1	Introduction and Background Information	711
	Introduction	711
	Background Information	712
13.2	Regulations	717
	Introduction	717
	Federal Regulations	717
	Exempt and Nonexempt E&P Wastes	720
	State Regulations	724
	Local Regulations	725
	Lease Agreements and Miscellaneous Issues	725
13.3	E&P-Related Fluid Characterization	726
13.4	E&P Treatment Processes, Waste Sources, and Residual Reuse/Disposal	729
	Water Sources Description	734
	E&P Waste Residuals and Treatment Options	735
13.5	Problems	744
	References	745

14 Chlorinated Compounds, VOCs, and Odor Control ... 747

14.1	Introduction	747
14.2	Polychlorinated Biphenyls	748
	Introduction	748
	Environmental Impacts	749
	Regulatory History of PCBs	750
	Treatment Methodologies	751
14.3	Chlorinated Solvents	755
	Introduction and Historical Perspective	755
	Treatment Methodologies	758
14.4	Chlorinated Pesticides	766
	Introduction	766
	Regulatory History	766
	Pesticide Characterization	767
	Treatment Methodologies	775
14.5	Perchlorates	775
	Introduction	775
	Treatment Technologies	775

14.6	Other Miscellaneous Chlorinated Organics	777
14.7	Chlorinated By-Products, Other VOCs, and Odor Control	779
	Odor Control	780
14.8	Problems	786
	References	787

15 Waste Minimization and Water Reuse 791

15.1	Introduction	791
15.2	Water Recycle and Reuse	791
	Limits of Water Reuse	792
	Reuse or Wastewater Treatment Plant Effluents	793
	The Decision-Making Process	794
	Case Histories	795
	Benchmarking	802
15.3	Waste Minimization—RCRA Hazardous Waste Issues	806
15.4	Zero Effluent Discharge and Economic Concepts	809
15.5	Case History	810
	Formosa Plastics, Point Comfort, Texas	810
	History of Zero Discharge Technologies	811
	Industry Applications of Zero Discharge Technology	813
	Formosa Studies Zero Discharge Options	813
	Zero Effluent Discharge Technologies Appropriate for Formosa	815
	Initial Evaluation Results	817
	Recycle Effects on Effluent Toxicity Testing	821
15.6	Summary	821
15.7	Problem	822
	References	823

16 Allocation of Superfund Disposal Site Response Costs 825

16.1	Introduction and Literature Review	825
16.2	Cost Allocation Principles	828
	Volume, Weight, Operating Time History	828

	16.3 Contaminant Selection, Mobility, Toxicity, Persistence, Content, and Physical State—Multiple Off-Site Contributors	830
	16.4 Allocation Methods for Same-Site Multiple Contributors	837
	16.5 Summary	842
	16.6 Problem	843
	References	845
17	**Industrial Pretreatment**	**847**
	17.1 Introduction	847
	17.2 National Categorical Pretreatment Standards and Local Limits Development	848
	17.3 Pretreatment Compliance Monitoring for Industrial Users	849
	17.4 POTW Ordinance Guidelines for Industrial Users (EPA Model)	851
	17.5 User Charge Rates and POTW Cost Recovery	852
	17.6 Case Histories	854
	City of Chicago, Illinois	854
	City of Indianapolis, Indiana	855
	City of San Diego, California	855
	City of Shreveport, Louisiana	855
	City of Austin, Texas	858
	Barceloneta POTW, Puerto Rico, and Pharmaceutical Pretreatment	863
	Determining Limits for Pollutants Regulated under PSES	866
	Determining Compliance Monitoring for PSES Pollutants	871
	Final Limits as They Would Appear in a Permit for Facility B	872
	Pretreatment of Leachate Discharges	873
	References	874
18	**Environmental Economics**	**875**
	PART 1 INDUSTRIAL ENVIRONMENTAL ECONOMICS	875
	18.1 Introduction	875
	18.2 Industrial Environmental Economics and Regulatory Compliance Metrics	875
	Evolution of Environmental Laws and Regulations in the United States	876
	18.3 Capital and Operational Economic Planning	878

18.4	Economics of New Facility Siting Analysis and Planning	880
18.5	Environmental Compliance	883
18.6	CERCLA, Superfund, and Joint and Several and Retroactive Economic Exposure	895
18.7	Environmental Litigation Exposure	896
18.8	Industrial Environmental Governance	901
	Sarbanes-Oxley Act (SOX) (Publicly Owned Companies)	902
	ISO 14001 Certification	902
	Creation of Internal Environmental Cost Accounting System	902
	Annual Reports for Publicly Owned Industrial Complexes	903
	Utilization of Third-Party Consulting Specialists, Advisory Groups, and Regulation Experts	903
	Organizational Structure and Participatory Policies	903
PART 2	ENVIRONMENTAL ECONOMICS FROM THE CONSULTANT AND CLIENT PERSPECTIVES	904
18.9	Introduction	904
18.10	SUM Concept	905
	Basic Equation	905
	Parameter Sensitivity	907
18.11	Consulting Internals—Salary to Expense Ratio and Utilization	910
18.12	Consulting Externals: Multiplier and Pricing	916
18.13	Project Economics	918
	References	920
Index		**921**

Preface

The three editions of *Industrial Water Pollution Control* were published primarily as a graduate engineering text. Due to changes in many graduate curricula, there was less need for a fourth edition specifically written as a graduate text. However, the authors thought that a comprehensive volume on industrial water quality would be a useful reference. Therefore, this volume, *Industrial Water Quality*, is a significant expansion and update of the third edition of *Industrial Water Pollution Control* that emphasizes practical applications and provides examples and information on current technology and water quality management. It will be particularly useful for industrial managers, consulting engineers, regulators, and academics addressing industrial water quality issues in the twenty-first century.

Acknowledgments

The authors would like to acknowledge the assistance of several individuals in the development of *Industrial Water Quality*. We are most appreciative of the support and help of our wives, Agnes Eckenfelder, Gwen Ford, and Bonnie Englande, throughout the process of writing and editing this book. The book's reviewers, Dr. Lial Tischler (Tischler/Kocurek Consultants) and Mr. Dave Marrs (Valero Energy Corporation), both busy and highly competent practitioners, were instrumental in providing excellent technical review and input. Additionally, the authors appreciate the review and comments from Dr. James Barnard from Black and Veach, who is an acknowledged expert on nutrient control, and Mr. Fred Gaines, of Enviroquip, who currently chairs the membrane committee for the Water Environment Federation.

We also appreciate the input of Dr. Cecil Lue-Hing, formerly of the Metropolitan Water District of Greater Chicago, Mr. Raj Bhattari, City of Austin, Texas, and Mr. Brian Flynn, MRE Associates, all of whom made significant contributions to Chaps. 17 and 18.

We also would like to acknowledge Mr. Larry Hager and his associates at McGraw-Hill, who provided extraordinary assistance to the authors as this book was developed. The authors recognize the efforts of Ms. Pam Arthur, who coordinated and electronically recorded major portions of this manuscript.

Industrial Water Quality

CHAPTER 1
Source and Characteristics of Industrial Wastewaters

1.1 Introduction

This book is designed to impart fundamental concepts of water pollution control technologies to the practicing environmental engineer. Emphasis is placed on the technical feasibility and application of these techniques to meet predetermined water quality criteria. A holistic perspective of waste management is stressed.

Wastewater treatment is defined as the cost-effective stabilization of wastewaters and residuals so as to produce minimum adverse effects on the environment and public health and foster sustainability of resources. Its objectives include:

1. The selection, engineering, and management of a sequence of unit operations to allow for cost-effective physical, chemical, and biological conversion and stabilization to occur under optimally controlled conditions
2. To promote and design for conservation, recycle, and reuse opportunities, and
3. To outline possible industrial approaches in enhancing compliance metrics

Its ultimate goal is to protect the environment and human health. Sustainable development and production are common to this goal. This book is written to assist and facilitate compliance to this goal. This chapter presents the first step of the waste management process—determination of sources and characteristics of industrial wastewaters.

1.2 Undesirable Wastewater Characteristics

Depending on the nature of the industry and the projected uses of the waters of the receiving stream, various waste constituents may have to be removed before discharge. These may be summarized as follows:

1. Soluble organics causing depletion of dissolved oxygen—Since most receiving waters require maintenance of minimum dissolved oxygen, the quantity of soluble organics is correspondingly restricted to the capacity of the receiving waters for assimilation or by specified effluent limitations.

2. Suspended solids—Deposition of solids in quiescent stretches of a stream will impair the normal aquatic life of the stream. Sludge blankets containing organic solids will undergo progressive decomposition resulting in oxygen depletion and the production of noxious gases.

3. Priority pollutants such as phenol and other organics discharged in industrial wastes will cause tastes and odors in the water and in some cases are carcenogenic. If these contaminants are not removed before discharge, additional water treatment will be required.

4. Heavy metals, cyanide, and toxic organics—In 1977, as per Section 307 of the EPA Clean Water Act, a list of 126 chemicals has been designated as priority pollutants. These are presented in Table 1.1. This list has been modified only slightly based on pollutant toxicity, persistence, and degradability and ecological effects on aquatic life. Specific limitations are imposed in relevant permits. The U.S. Environmental Protection Agency (EPA) has defined a list of toxic organic and inorganic chemicals that now appear as specific limitations in most permits. The identified priority pollutants are listed in Table 1.1.

5. Color and turbidity—These present aesthetic problems even though they may not be particularly deleterious for most water uses. In some industries, such as pulp and paper, color removal can be difficult and expensive.

6. Nitrogen and phosphorus—When effluents are discharged to lakes, ponds, and other recreational areas, the presence of nitrogen and phosphorus is particularly undesirable since it

Compound Name	Compound Name
1. Acenaphthene*	Dichlorobenzidine*
2. Acrolein*	28. 3,39-Dichlorobenzidine
3. Acrylonitrile*	
4. Benzene*	Dichloroethylenes* (1,1-dichloroethylene and 1,2-dichloroethylene)
5. Benzidine*	
6. Carbon tetrachloride* (tetrachloromethane)	29. 1,1-Dichloroethylene
	30. 1,2-*trans*-Dichloroethylene
Chlorinated benzenes (other than dichlorobenzenes)	31. 2,4-Dichlorophenol*
7. Chlorobenzene	Dichloropropane and dichloropropene†
8. 1,2,4-Trichlorobenzene	32. 1,2-Dichloropropane
9. Hexachlorobenzene	33. 1,2-Dichloropropylene (1,2-dichloropropene)
Chlorinated ethanes* (including 1,2-dichloroethane, 1,1,1-trichloroethane, and hexachloroethane)	34. 2,4-Dimethylphenol*
	Dinitrotoluene*
10. 1,2-Dichloroethane	35. 2,4-Dinitrotoluene
11. 1,1,1-Trichloroethane	36. 2,6-Dinitrotoluene
12. Hexachloroethane	37. 1,2-Diphenylhydrazine*
13. 1,1-Dichloroethane	38. Ethylbenzene*
14. 1,1,2-Trichloroethane	39. Fluoranthene*
15. 1,1,2,2-Tetrachloroethane	
16. Chloroethane (ethyl chloride)	Haloethers* (other than those listed elsewhere)
Chloroalkyl ethers* (chloromethyl, chloroethyl, and mixed ethers)	40. 4-Chlorophenyl phenyl ether
	41. 4-Bromophenyl phenyl ether
17. Bis(chloromethyl) ether	42. Bis(2-chloroisopropyl) ether
18. Bis(2-chloroethyl) ether	43. Bis(2-chloroethoxy) methane
19. 2-Chloroethyl vinyl ether (mixed)	
	Halomethanes* (other than those listed elsewhere)
Chlorinated naphthalene*	44. Methylene chloride (dichloromethane)
20. 2-Chloronaphthalene	45. Methyl chloride (chloromethane)
Chlorinated phenols* (other than those listed elsewhere; includes trichlorophenols and chlorinated cresols)	46. Methyl bromide (bromomethane)
	47. Bromoform (tribromomethane)
21. 2,4,6-Trichlorophenol	48. Dichlorobromomethane
22. *para*-Chloro-*meta*-cresol	49. Trichlorofluoromethane
23. Chloroform (trichloromethane)*	50. Dichlorodifluoromethane
24. 2-Chlorophenol*	51. Chlorodibromomethane
	52. Hexachlorobutadiene*
Dichlorobenzenes*	53. Hexachlorocyclopentadiene*
25. 1,2-Dichlorobenzene	54. Isophorone*
26. 1,3-Dichlorobenzene	55. Naphthalene*
27. 1,4-Dichlorobenzene	56. Nitrobenzene*

TABLE 1.1 EPA List of Organic Priority Pollutants

Compound Name	Compound Name
Nitrophenols* (including 2,4-dinitrophenol and dinitrocresol) 57. 2-Nitrophenol 58. 4-Nitrophenol 59. 2,4-Dinitrophenol* 60. 4,6-Dinitro-o-cresol	85. Tetrachloroethylene* 86. Toluene* 87. Trichloroethylene* 88. Vinyl chloride* (chloroethylene)
Nitrosamines* 61. N-Nitrosodimethylamine 62. N-Nitrosodiphenylimine 63. N-Nitrosodi-n-propylamine 64. Pentachlorophenol* 65. Phenol*	Pesticides and metabolites 89. Aldrin* 90. Dieldrin* 91. Chlordane* (technical mixture and metabolites)
Phthalate esters* 66. Bis(2-ethylhexyl) phthalate 67. Butyl benzyl phthalate 68. Di-n-butyl phthalate 69. Di-n-octyl phthalate 70. Diethyl phthalate 71. Dimethyl phthalate	DDT and metabolites* 92. 4,4′-DDT 93. 4,4′-DDE 94. 4,4′-DDD
Polynuclear aromatic hydrocarbons (PAH)* 72. Benzo(a)anthracene (1,2-benzanthracene) 73. Benzo(a)pyrene (3,4-benzopyrene) 74. 3,4-Benzofluoranthene 75. Benzo(k)fluoranthene (11,12-benzofluoranthene) 76. Chrysene 77. Acenaphthylene 78. Anthracene 79. Benzo(ghi)perylene (1,12-benzoperylene) 80. Fluorene 81. Phenanthrene 82. Dibenzo(a,h)anthracene (1,2,5,6-dibenzanthracene) 83. Indeno (1,2,3-cd) pyrene (2,3-o-phenylenepyrene) 84. Pyrene	Endosulfan and metabolites* 95. α-Endosulfan-alpha 96. β-Endosulfan-beta 97. Endosulfan sulfate
	Endrin and metabolites* 98. Endrin 99. Endrin aldehyde
	Heptachlor and metabolites* 100. Heptachlor 101. Heptachlor epoxide
	Hexachlorocyclohexane (all isomers)* 102. α-BHC-alpha 103. β-BHC-beta 104. γ-BHC (lindane)-gamma 105. δ-BHC-delta

TABLE 1.1 EPA List of Organic Priority Pollutants

Source and Characteristics of Industrial Wastewaters

Compound Name	Compound Name
Polychlorinated biphenyls (PCB)*	111. PCB-1260 (Arochlor 1260)
106. PCB-1242 (Arochlor 1242)	112. PCB-1016 (Arochlor 1016)
107. PCB-1254 (Arochlor 1254)	113. Toxaphene*
108. PCB-1221 (Arochlor 1221)	114. 2,3,7,8-Tetrachlorodibenzo-*p*-dioxin (TCDD)*
109. PCB-1232 (Arochlor 1232)	
110. PCB-1248 (Arochlor 1248)	

*Specific compounds and chemical classes as listed in the consent degree.

TABLE 1.1 (*Continued*)

enhances eutrophication and stimulates undesirable algae growth.

7. Refractory substances resistant to biodegradation—These may be undesirable for certain water-quality requirements. Refractory nitrogen compounds are found in the textile industry. Some refractory organics are toxic to aquatic life.
8. Oil and floating material—These produce unsightly conditions and in most cases are restricted by regulations.
9. Volatile materials—Hydrogen sulfide and volatile organics will create air-pollution problems and are usually restricted by regulation.
10. Aquatic toxicity—Substances present in the effluent that are toxic to aquatic species and are restricted by regulation.
11. Persistent organic pollutants (POPs). These are persistent toxic chemicals which adversely affect human health globally and can accumulate in the food chain. The Stockholm Convention Treaty has listed the "Dirty Dozen" to be eliminated or reduced. These include PCBs, dioxins, furans, and various pesticides, including DDT, which are discussed in Chap. 14.
12. Emerging pollutants. These represent a group of contaminants which may represent an environmental and public health issue and for which additional research information is needed. Among these compounds of concern include Pharmaceuticals and Personal Care Products (PPCPs), Endocrine Disrupting Chemicals (EDCs), brominated flame retardants, phthalate esters, and others.

1.3 Partial List of Regulations Which Affect Wastewater Treatment Requirements within the United States

It is not the intent of this book to discuss federal and state regulations, but a brief summary of present regulatory requirements relative to industrial water pollution control will serve as guidance to the reader. Details of these regulations can be found in the cited Code of the Federal Register (CFR) as noted. A more detailed description of pertinent regulations is presented in corresponding chapters.

Air

Air National Emission Standards for Hazardous Air Pollutants, NESHAP

- NESHAPs are emissions standards set by EPA for air pollutants not covered by NAAQS that may cause an increase in fatalities or serious, irreversible, or incapacitating illness. Standards are set for 186 hazardous air pollutants based on Maximum Achievable Control Technology (MACT). Authorization is by Section 112 of the Clean Air Act and regulations are published in 40 CFR, parts 61 and 63.

NESHAP (40 CFR, part 61)

- NESHAP (40 CFR, part 61) Regulates 33 designated hazardous air pollutants in terms of mass loadings and concentration limits. Off-gas capture and treatment is required until a specified effluent level is achieved. For example, those streams with a 10 percent (or greater) water content must be considered in calculating total annual benzene (TAB). If the source is 10 Mg/yr or greater, then all wastes with 10 ppmw benzene on a flow-weighted annual average basis will need to be treated and controlled regardless of water content.

Occupational Safety and Health Administration (OSHA) Standards

- Regulates hydrogen sulfide and contaminants which pose exposure risks.

Liquid

Federal Industry Point Source Category Limits (40 CFR, part 405-471)

- Mass-based for raw material processing, e.g., pulp and paper, and concentration-based for synthetic chemicals and pharmaceuticals for conventional pollutants.

Source and Characteristics of Industrial Wastewaters

- Concentration-based and/or mass-based limits for nonconventional pollutants (metals and priority pollutants).

Regional Initiatives (e.g., Great Lakes Initiative)

- For example, concentration-based limits for total phosphorus.

State Water Quality Standards

- Limits for pollutants based on design receiving stream low flow (i.e., 7–Q10, the average 7-day low flow every 10 years) for the use classification.

Local Pretreatment Limits (USEPA, PB92-129188, December 1987)

- Those regulated under point source categories, plus those required to ensure POTW (publicly owned treatment works) effluent compliance.

1.4 Sources and Characteristics of Wastewaters

The volume and strength of industrial wastewaters are usually defined in terms of units of production (e.g., gallons per ton* of pulp or cubic meters per tonne† of pulp and pounds of [BOD biochemical oxygen demand] per ton of pulp or kilograms of BOD per tonne of pulp for a pulp-and-paper-mill waste) and the variation in characteristic as described by a statistical distribution. In any one plant there will be a statistical variation in wasteflow characteristics. The magnitude of this variation will depend on the diversity of products manufactured and of process operations contributing waste, and on whether the operations are batch or continuous. Good housekeeping procedures to minimize dumps and spills will reduce the statistical variation. Plots showing the variation in flow resulting from a sequence of batch processes are shown in Fig. 1.1. Variation in waste flow and characteristics within a single plant are shown in Fig. 1.2.

Wide variation in waste flow and characteristics will also appear among similar industries, e.g., the paperboard industry. This is a result of differences in housekeeping and water reuse as well as of variations in the production processes. Very few industries are identical in their sequence of process operations; as a result, an industrial waste survey is usually required to establish waste loadings and their variations. Variations for several industries are shown in Table 1.2. The variation in suspended solids and BOD discharge from 11 paperboard mills is shown in Fig. 1.3. Probability plots are generally more representative using log-probability analysis since this reflects the nature of the data. However, arithmetic-probability plots may also work as well as demonstrated in Fig. 1.1.

---

*ton = 2000 lb.
†tonne = 1000 kg.

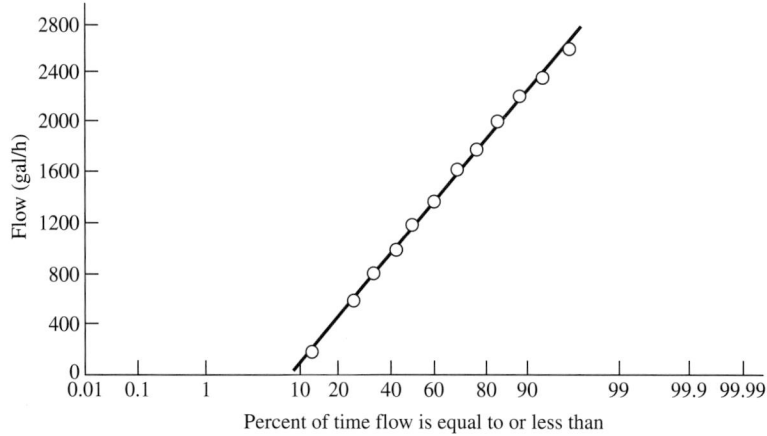

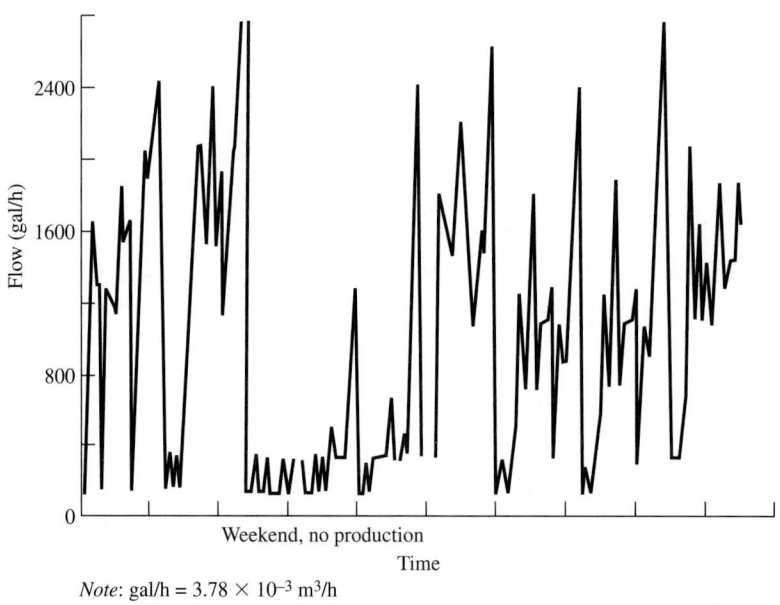

Note: gal/h = 3.78 × 10⁻³ m³/h

FIGURE 1.1 Variation in flow from a batch operation.

1.5 Industrial Waste Survey

The industrial waste survey involves a procedure designed to develop a flow-and-material balance of all processes using water and producing wastes and to establish the variation in waste characteristics from specific process operations as well as from the plant as a whole. The results of the survey should establish possibilities for water conservation, reuse, and source treatment. It should result in reduction of flow,

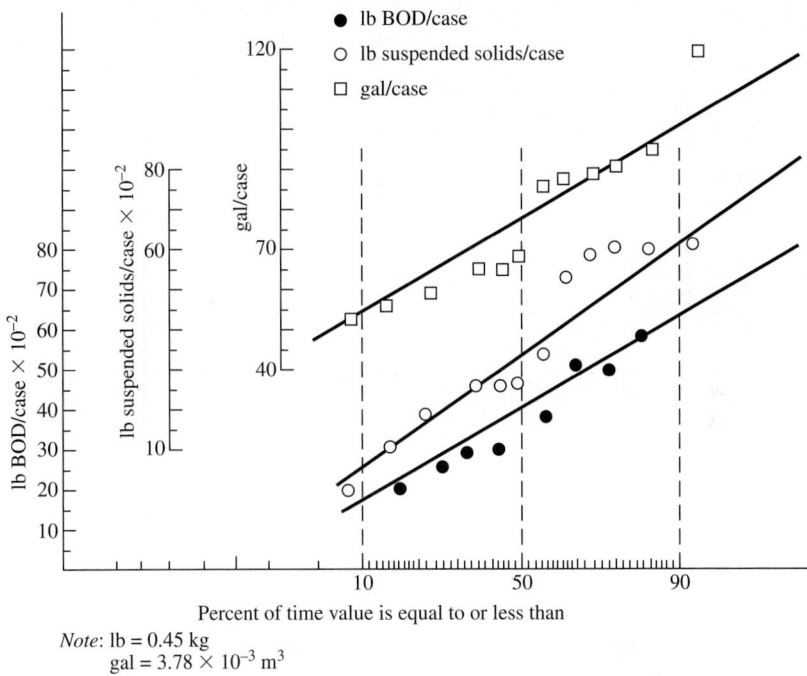

FIGURE 1.2 Daily variation in flow and characteristics: tomato waste.

contaminant loadings, and variations to the wastewater treatment system.

The basic components of the survey include: identify the problem; evaluate the problem and determine most feasible solution; implement abatement measures; and monitor to evaluate effectiveness of these measures. A wastewater audit protocol schematic is presented in Fig. 1.4.

The selected method of flow measurement will usually be contingent on the physical location to be sampled. When the waste flows through a sewer, it is frequently possible to measure the velocity of flow and the depth of water in the sewer and calculate the flow from the continuity equation. Since $Q = AV$, the area in a partially filled circular sewer can be determined given the depth from Fig. 1.5. This method applies only for partially filled sewers of constant cross section. The average velocity of flow can be estimated as 0.8 of the surface velocity timed from a floating object between manholes. More accurate measurements can be obtained by the use of a current meter. In gutters or channels, either a small weir can be constructed or the flow can be estimated as above by measuring the velocity and depth of flow in the channel. In some cases the flow can be obtained from the pumping rate and the duration of pumping of a wastestream. Total waste flow from an industrial plant can be measured by use of a weir or

Waste	Flow (gal/Production Unit) % Frequency			BOD (lb/Production Unit) % Frequency			Suspended Solids (lb/Production Unit) % Frequency		
	10	50	90	10	50	90	10	50	90
Pulp and paper*	11,000	43,000	74,000	17.0	58.0	110.0	26.0	105.0	400.0
Paperboard*	7,500	11,000	27,500	10	28	46	25	48	66
Slaughterhouse†	165	800	4,300	3.8	13.0	44	3.0	9.8	31.0
Brewery‡	130	370	600	0.8	2.0	44	0.25	1.2	2.45
Tannery§	4.2	9.0	13.6	575¶	975	1400	600¶	1900	3200

*Tons paper production.
†1000 lb live weight kill.
‡bbl beer.
§Pounds of hides; sulfides as S vary from 260 (10%) to 1230 mg/L (90%).
¶As mg/L.
Note: gal = 3.78×10^{-3} m³
lb = 0.45 kg
ton = 907 kg
bbl beer = 0.164 m³

TABLE 1.2 Variation in Flow and Waste Characteristics for Some Representative Industrial Wastes

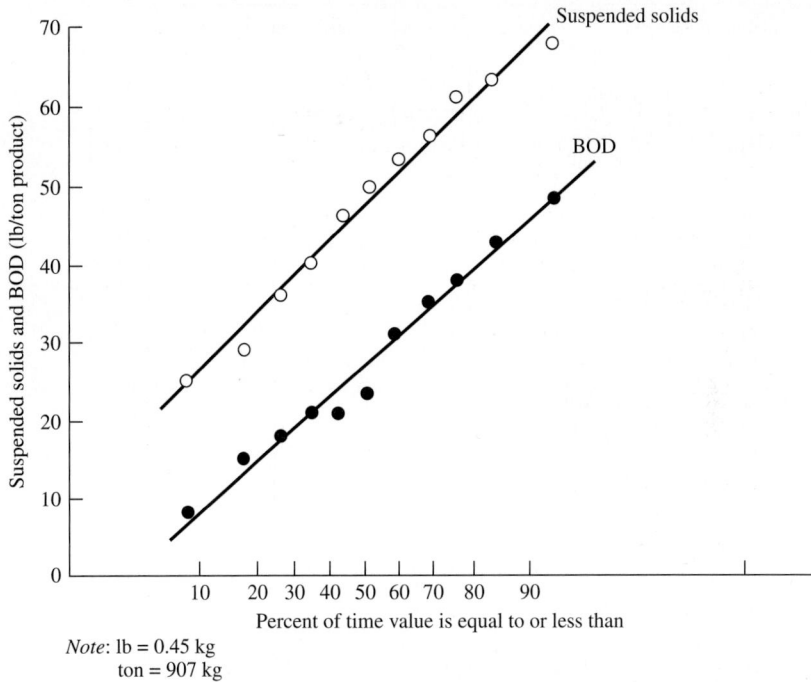

Figure 1.3 Variations in suspended solids and BOD from 11 paperboard mills.

other suitable measuring device. In certain instances the daily waste flow can be estimated from water consumption records.

The general procedure to be followed in developing the necessary information with a minimum of effort can be summarized in four steps:

1. Develop a sewer map from consultation with the plant engineer and an inspection of the various process operations. This map should indicate possible sampling stations and a rough order of magnitude of the anticipated flow.

2. Establish sampling and analysis schedules. Continuous samples with composites weighted according to flow are the most desirable, but these either are not always possible or do not lend themselves to the physical sampling location. The period of sample composite and the frequency of sampling must be established according to the nature of the process being investigated. Some continuous processes can be sampled hourly and composited on an 8-, 12-, or even 24-h basis, but those that exhibit a high degree of fluctuation may require a 1- or 2-h composite and analysis. Where source treatment is to be considered frequent compositing is needed. In other cases more

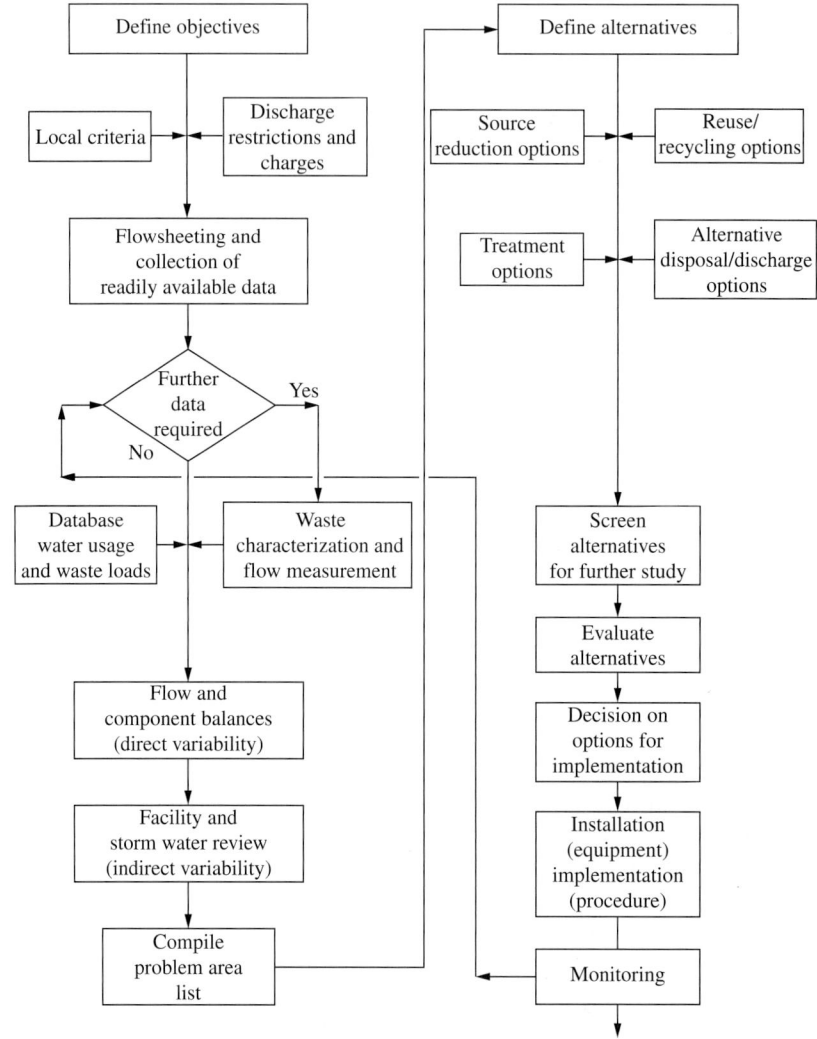

FIGURE 1.4 A wastewater audit protocol.

frequent samples are rarely required, since most industrial waste-treatment processes have a degree of built-in equalization and storage capacity. Batch processes should be composited during the course of the batch dump.

3. Develop a flow-and-material-balance diagram. After the survey data are collected and the samples analyzed, a flow-and-material balance diagram should be developed that considers all significant sources of waste discharge. How closely the summation of the individual sources checks the measured total effluent provides a check on the accuracy of the survey.

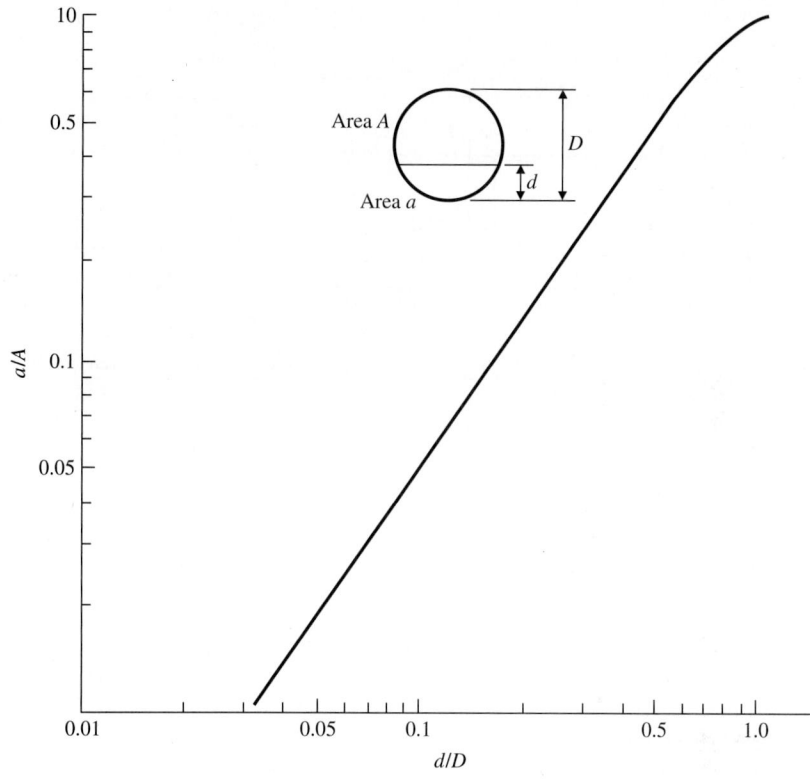

FIGURE 1.5 Determination of waste flow in partially filled sewers.

A typical flow-and-material-balance diagram for a corn-processing plant is shown in Fig. 1.6.

4. Establish statistical variation in significant waste characteristics. As was previously shown, the variability of certain waste characteristics is significant for waste-treatment plant design. These data should be prepared as a probability plot showing frequency of occurrence.

The analyses to be run on the samples depend both on the characteristics of the samples and on the ultimate purpose of the analysis. For example, pH must be run on grab samples, since it is possible in some cases for compositing to result in neutralization of highly acidic and basic wastes, and this would yield highly misleading information for subsequent design. Variations in BOD loading may require 8-h or shorter composites for certain biological-treatment designs involving short detention periods, while 24-h composites will usually suffice for aerated lagoons with many days' retention under completely mixed conditions. Where constituents such as nitrogen or phosphorus are to

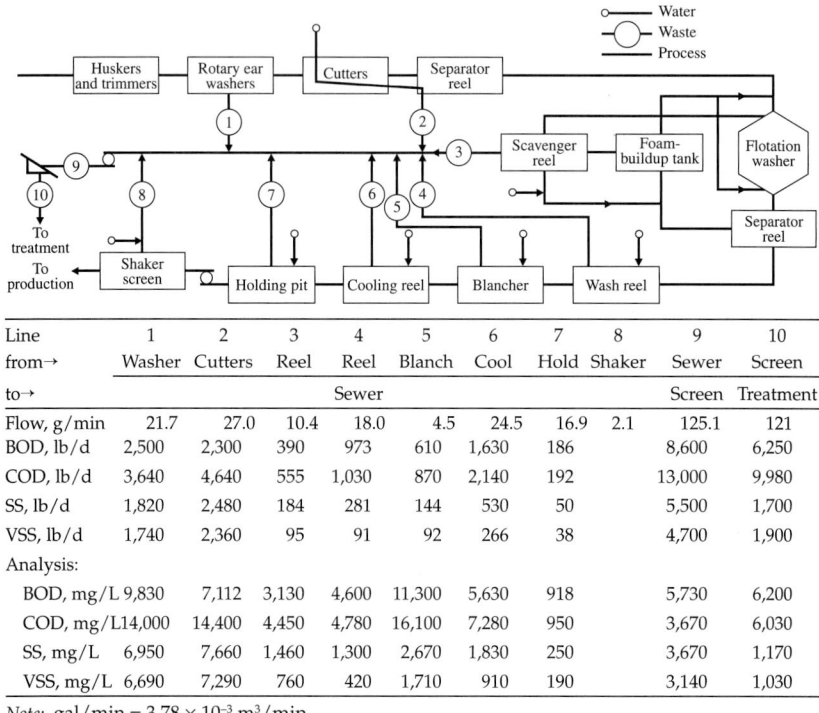

FIGURE 1.6 Waste flow diagram and material balance at a corn plant.

be measured to determine required nutrient addition for biological treatment, 24-h composites are sufficient since the biological system possesses a degree of buffer capacity. One exception is the presence of toxic discharges to a biological system. Since a one-shot dose of certain toxic materials can completely upset a biological-treatment process, continuous monitoring of such materials is required if they are known to exist. It is obvious that the presence of such materials would require separate consideration in the waste-treatment design. Other waste-treatment processes may require similar considerations in sampling schedules.

Data from industrial waste surveys are highly variable and are usually susceptible to statistical analysis. Statistical analysis of variable data provides the basis for process design. The data are reported in terms of frequency of occurrence of a particular characteristic, which is that value of the characteristic that may be expected to be equaled or not exceeded 10, 50, or 90 percent of the time. The 50 percent chance value is approximately equal to the median. Correlation in this manner will linearize variable data, as shown in Fig. 1.7. The probability of occurrence of any value, such as flow, BOD, or suspended solids,

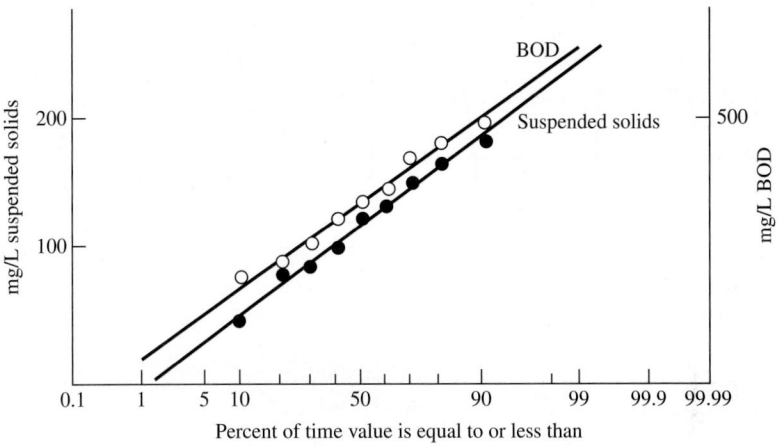

FIGURE 1.7   Probability of occurrence of BOD and suspended solids in raw waste.

may be determined as shown in the probability plot. This can also be determined by a standard computer program.

The suspended solids and BOD values are each arranged in order of increasing magnitude. n is the total number of solids or BOD values and m is the assigned serial number from 1 to n. The plotting positions $m/(n + 1)$ are equivalent to the percent occurrence of the value. The actual values are then plotted against the percent occurrences on probability paper, as indicated in Fig. 1.7. A smooth curve of best fit may usually be drawn by eye or, if desirable, it may be calculated by standard statistical procedures. The probability of occurrence of any values can be obtained. The statistical calculation is shown in Example 1.1. In order to extrapolate the results of an industrial waste survey to future production, it is desirable to relate waste flow and loading to production schedules. Since some effluent-producing operations do not vary directly with production increase or decrease, the scale-up is not always linear. This is true of the cannery operation shown in Fig. 1.8, in which six process operations were independent of the number of washing and cleaning rigs in operation. Log-probability or arithmetic-probability plots can be used depending on the best fit of the data.

Example 1.1.   For small amounts of industrial waste survey data (i.e., less than 20 datum points), the statistical correlation procedure is as follows:

1. Arrange the data in increasing order of magnitude (first column of Table 1.3).
2. In the second column of Table 1.3, m is the assigned serial number from 1 to n where n is the total number of values.

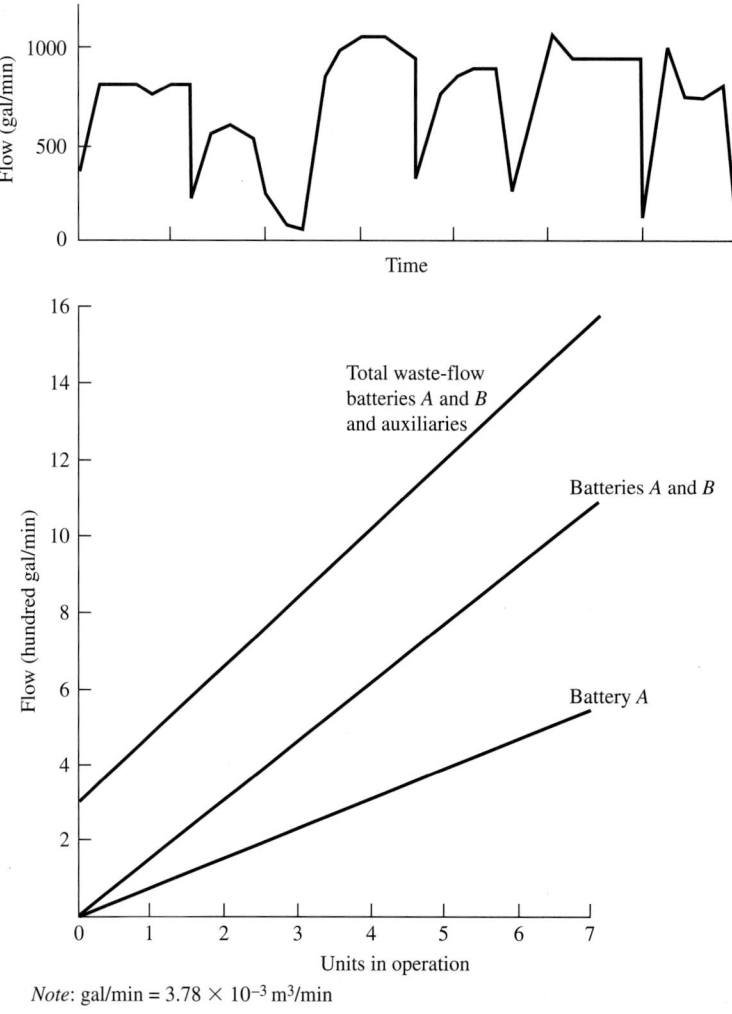

FIGURE 1.8 Variation in plant waste flow from unit operations.

3. The plotting position is determined by dividing the total number of samples into 100 and assigning the first value as one-half this number (third column of Table 1.3):

$$\text{Plotting position} = \frac{100}{n} + \text{previous probability}$$

For $m = 5$:

$$\text{Plotting position} = \frac{100}{9} + 38.85$$

$$= 49.95$$

BOD (mg/L)	m	Plotting Position
200	1	5.55
225	2	16.65
260	3	27.75
315	4	38.85
350	5	49.95
365	6	61.05
430	7	72.15
460	8	83.75
490	9	94.35

TABLE 1.3 Statistical Correlation of BOD Data

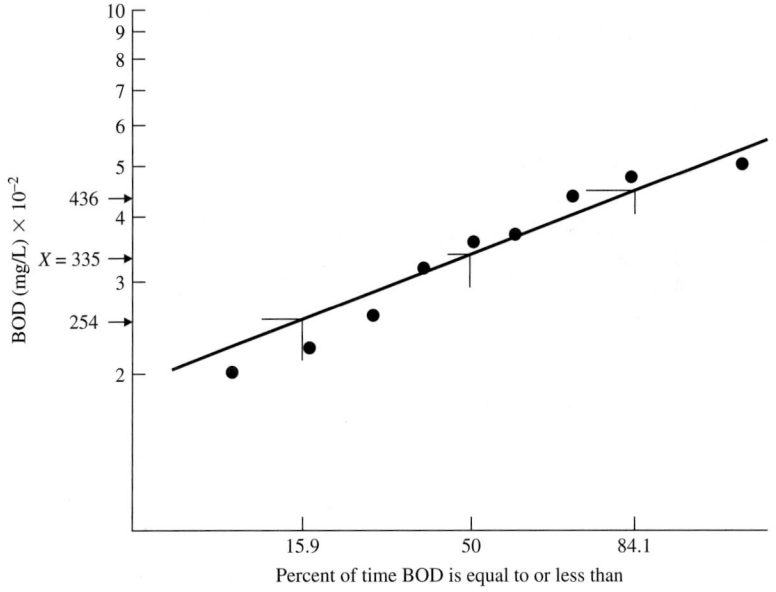

FIGURE 1.9 Statistical correlation of waste survey data.

4. These data are illustrated in Fig. 1.9. The standard deviation of these data (S) is calculated by

$$S = \frac{X_{84.1\%} - X_{15.9\%}}{2}$$

From Fig. 1.9:

$$S = \frac{436 - 254}{2}$$

$$= 91 \text{ mg/L}$$

and median:

$$\overline{X} = X_{50.0\%}$$
$$= 335 \text{ mg/L}$$

When large numbers of data are to be analyzed it is convenient to group the data for plotting, for example, 0 to 50, 51 to 100, 101 to 150, and so on. The plotting position is determined as $m/(n + 1)$, where m is the cumulative number of points and n is the total number of observations. The statistical distribution of data serves several important functions in developing the industrial waste management program. This methodology can be used to determine the reduction in both the loading and variability of the pollutants of concern.

1.6 Waste Characteristics—Estimating the Organic Content

Although the interpretation of most of the waste characteristics is straightforward and definitive, special consideration must be given to the organic content. The organic content of the waste can be estimated by the BOD, COD (chemical oxygen demand), TOC (total organic carbon), or TOD (total oxygen demand). Considerable caution should be exercised in interpreting these results:

1. The BOD_5 test measures the biodegradable organic carbon and, under certain conditions, the oxidizable nitrogen present in the waste. Nitrification may be suppressed so that only carbonaceous oxidation is recorded as $CBOD_5$.

2. The COD test measures the total organic carbon with the exception of certain aromatics, such as benzene, which are not completely oxidized in the reaction. The COD test is an oxidation-reduction, so other reduced substances, such as sulfides, sulfites, and ferrous iron, will also be oxidized and reported as COD. NH_3^-N will not be oxidized in the COD test.

3. The TOC test measures all carbon as CO_2, and hence the inorganic carbon (CO_2, HCO_3^-, and so on) present in the wastewater must be removed prior to the analysis or corrected for in the calculation.

4. The TOD test measures organic carbon and unoxidized nitrogen and sulfur.

Remember to exercise considerable caution in interpreting the test results and in correlating the results of one test with another. Correlations between BOD and COD or TOC should usually be made of filtered samples (soluble organics) to avoid the disproportionate relationship of volatile suspended solids in the respective tests.

The BOD by definition is the quantity of oxygen required for the stabilization of the oxidizable organic matter present over 5 days of incubation at 20°C. The BOD is conventionally formulated as a first-order reaction:

$$\frac{dL}{dt} = -kL \tag{1.1}$$

which integrates to

$$L = L_o e^{-kt} \tag{1.2}$$

Since L, the amount of oxygen demand remaining at any time, is not known, Eq. (1.2) is re-expressed as

$$y = L_o(1 - e^{-kt})$$

where y is the amount of BOD exerted at time t:

$$y = L_o - L$$

or

$$y = L_o(1 - 10^{-kt}) \tag{1.3}$$

By definition, L_o is the oxygen required to stabilize the total quantity of biologically oxidizable organic matter present; if k is known, L_5 is a fixed percentage of L_o. In order to interpret the BOD_5 obtained on industrial wastes, certain important factors must be considered.

It must be recognized that the oxygen consumed in the BOD test is the sum of (1) oxygen used for synthesis of new microbial cells using the organic matter present and (2) endogenous respiration of the microbial cells as shown in Fig. 1.10. The rate of oxygen utilization during phase 1 is 10 to 20 times that during phase 2. In most readily degradable substrates, phase 1 is complete in 24 to 36 h.

In wastes containing readily oxidizable substrates, e.g., sugars, there will be a high oxygen demand for the first day, as the substrate is rapidly utilized, followed by a slower endogenous rate over the subsequent days of incubation. When these data are fitted to a first-order curve over a 5-day period, a high k value will result because of the high initial slope of the curve. Conversely, a well-oxidized effluent will contain very little available substrate, and for the most part only endogenous respiration will occur over the 5-day incubation period. Since this rate of oxygen utilization is only a fraction of the rate obtained in the presence of available substrate, the resulting k rate will be correspondingly lower. Schroepfer[1] showed this by comparing the k_{10} rates of a well-treated sewage effluent and raw sewage containing a large quantity of available substrate. The average rate

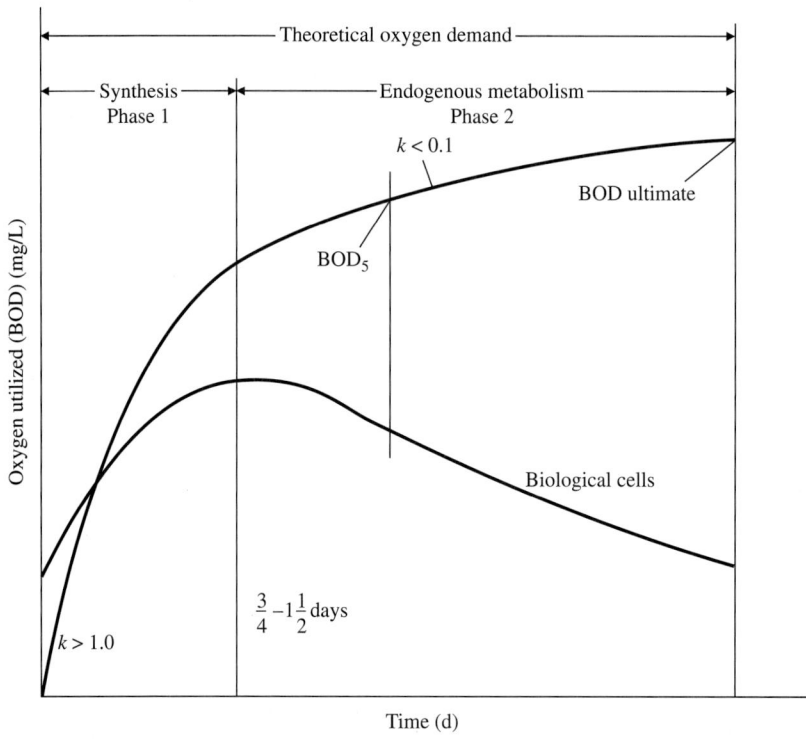

Figure 1.10 Reactions occurring in the BOD bottle.

was 0.17 per day for the sewage and 0.10 per day for the effluent. It is obvious that under these conditions a direct comparison of 5-day BODs is not valid. Typical rate constants are shown in Table 1.4.

Many industrial wastes are difficult to oxidize; they require a bacterial seed acclimated to the specific waste, or a lag period may occur which yields an erroneous interpretation of the 5-day BOD values. Stack[2] showed that the 5-day BOD of synthetic organic chemicals varied markedly depending on the acclimation of the seed used. Some typical BOD curves are shown in Fig. 1.11. Curve A is normal exertion of BOD. Curve B is representative of what might be expected from sewage which slowly acclimated to the waste. Curves C and D are

Substance	k_{10} (day^{-1})
Untreated wastewater	0.15–0.28
High-rate filters and anaerobic contact	0.12–0.22
High-degree biotreatment effluent	0.06–0.10
Rivers with low pollution	0.04–0.08

Table 1.4 Average BOD Rate Constants at 20°C

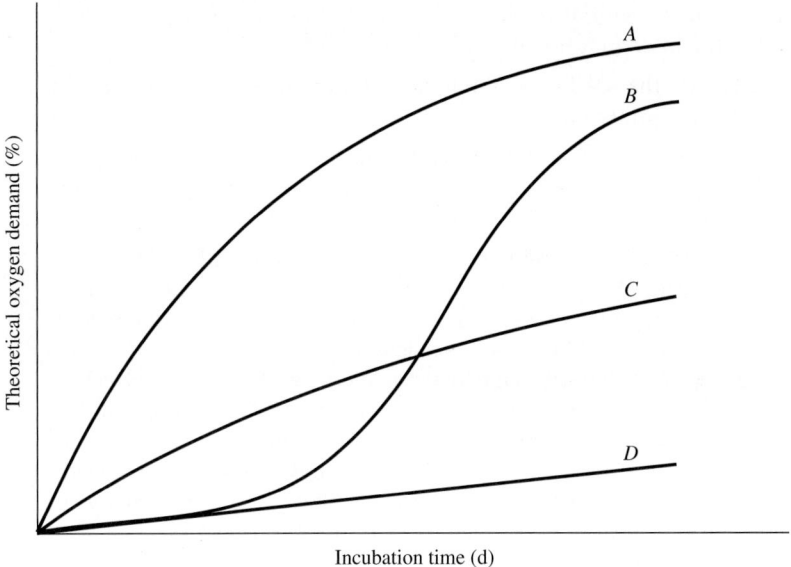

Figure 1.11 Characteristic BOD curves.

characteristic of nonacclimated seed or a inhibitory wastewater. Acclimation of microorganisms to organics is shown in Table 1.5. In some cases, the 1-day BOD may provide a good control test for treatment plant performance.

Although modifications of the BOD procedure such as the short-term test proposed by Busch[3] may eliminate some of the errors resulting from the first-order assumption and the variation in k_{10} due to substrate level, these procedures have not found broad application in

1. Nontoxic aliphatic compounds containing carboxyl, ester, or hydroxyl groups readily acclimate (<4 days acclimation).
2. Toxic compounds with carbonyl groups or double bonds, 7–10 days acclimation; toxic to unacclimated acetate cultures.
3. Amino functional groups difficult to acclimate and slow degradation.
4. Seeds for dicarboxylic groups longer to acclimate compared to one for carboxylic group.
5. Position of functional group affects lag period for acclimation. Primary butanol   4 days Secondary butanol   14 days Tertiary butanol   Not acclimated

Table 1.5 Effect of Structural Characteristics on Bio-Acclimation

industry. It is essential, therefore, that the following factors be considered in the interpretation of the BOD for an industrial waste:

1. That the seed is acclimated to the waste and all lag periods are eliminated.
2. That long-term BOD tests establish the magnitude of k_{10} on both the raw waste and treated effluent. In the case of acidic wastes, all samples should be neutralized before incubation.

Toxicity in a waste is usually evidenced by so-called sliding BOD values, i.e., an increasing calculated BOD with increasing dilution. If this situation exists, it is necessary to determine the dilution value below which the computed BODs are consistent.

The COD test measures the total organic content of a waste which is oxidized by dichromate in acid solution. When a silver sulfate catalyst is used, the recovery for most organic compounds is greater than 92 percent. However, some aromatics such as toluene are only partially oxidized. Since the COD will report virtually all organic compounds, many of which are either partially biodegradable or nonbiodegradable, it is proportional to the BOD only for readily assimilable substances, e.g., sugars. Such a case is shown in Fig. 1.12 for a readily assimilable chemical and refinery waste. Tables 1.6 and 1.8 show the BOD and COD characteristics for a variety of industrial effluents.

Because the 5-day BOD will represent a different proportion of the total oxygen demand for raw wastes than for effluents, the

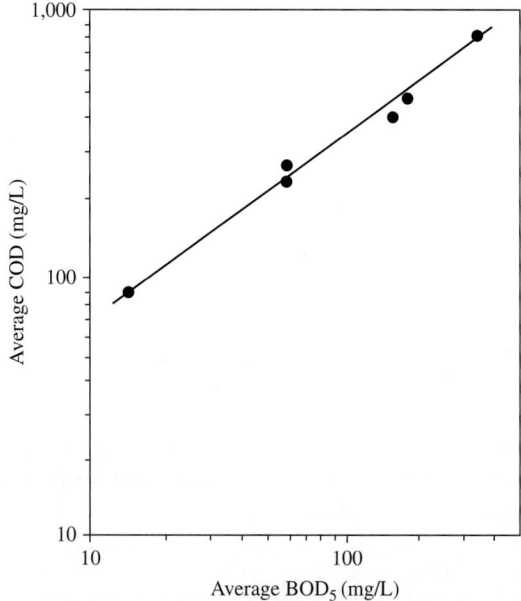

Figure 1.12 Relationship between BOD and COD for a chemical and refinery wastewater.

Waste	BOD$_5$ (mg/L)	COD (mg/L)	TOC (mg/L)	BOD/ TOC	COD/ TOC
Chemical*	—	4,260	640	—	6.65
Chemical*	—	2,410	370	—	6.60
Chemical*	—	2,690	420	—	6.40
Chemical		576	122	—	4.72
Chemical	24,000	41,300	9,500	2.53	4.35
Chemical—refinery	—	580	160	—	3.62
Petrochemical	—	3,340	900	—	3.32
Chemical	850	1,900	580	1.47	3.28
Chemical	700	1,400	450	1.55	3.12
Chemical	8,000	17,500	5,800	1.38	3.02
Chemical	60,700	78,000	26,000	2.34	3.00
Chemical	62,000	143,000	48,140	1.28	2.96
Chemical	—	165,000	58,000	—	2.84
Chemical	9,700	15,000	5,500	1.76	2.72
Nylon polymer	—	23,400	8,800	—	2.70
Petrochemical	—	—	—	—	2.70
Nylon polymer	—	112,600	44,000	—	2.50
Olefin processing	—	321	133	—	2.40
Butadiene processing	—	359	156	—	2.30
Chemical	—	350,000	160,000	—	2.19
Synthetic rubber	—	192	110	—	1.75

* High concentration of sulfides and thiosulfates.
Source: Adapted from Ford, 1968.[4]

TABLE 1.6 Oxygen Demand and Organic Carbon of Industrial Wastewaters

BOD/COD ratio will frequently vary for effluents as compared to untreated wastes. There will be no correlation between BOD and COD when organic suspended solids in the waste are only slowly biodegradable in the BOD bottle, and, therefore, filtered or soluble samples should always be used. Pulp and fiber in a paper mill waste are an example. There will also be no correlation between BOD and COD in complex waste effluents containing refractory substances such as ABS. For this reason, treated effluents may exert virtually no BOD and yet exhibit a substantial COD.

Total organic carbon (TOC) has become a common and popular method of analysis due to its simplicity of measurement. There are presently several carbon analyzers on the market.

When considering routine plant control or investigational programs, the BOD is not a useful test because of the long incubation time. It is therefore useful to develop correlations between BOD and COD or TOC. The changes in ratios of these parameters may indicate variability in composition of wastestreams.

The theoretical oxygen demand (THOD) of a wastewater containing identified organic compounds can be calculated as the oxygen required to oxidize the organics to end products; e.g., for glucose:

$$C_6H_{12}O_6 + 6O_2 \rightarrow 6CO_2 + 6H_2O$$

$$\text{THOD} = \frac{6M_{O_2}}{M_{C_6H_{12}O_6}} = 1.07 \; \frac{\text{mg COD}}{\text{mg organic}}$$

For most organics, with the exception of some aromatics and nitrogen-containing compounds, the COD will equal the THOD. For readily degradable wastewaters, such as from a dairy, the COD will equal the $BOD_{ult}/0.92$. When the wastewater also contains nondegradable organics, the difference between the total COD and the $BOD_{ult}/0.92$ will represent the nondegradable content.

It has been found that some nondegradable organics will accumulate during biooxidation by oxidation by-products of the organics in the wastewater and by-products of endogenous metabolism. These are defined as soluble microbial products (SMP). Hence the effluent COD through biological treatment will increase over the influent nondegradable COD.

When the compounds are identified, the TOC can be related to COD through a carbon-oxygen balance:

$$C_6H_{12}O_6 + 6O_2 \rightarrow 6CO_2 + 6H_2O$$

$$\frac{\text{COD}}{\text{TOC}} = \frac{6M_{O_2}}{6M_C} = 2.66 \; \frac{\text{mg COD}}{\text{mg organic carbon}}$$

Depending on the organic in question, the COD/TOC ratio may vary from zero when the organic material is resistant to dichromate oxidation to 5.33 for methane. Since the organic content changes during biological oxidation, it can be expected that the COD/TOC ratio will also change. This same rationale also applies to the BOD/TOC ratio. Values of BOD and COD for a variety of organics are shown in Table 1.7. Since only biodegradable organics are removed in the activated sludge process, the COD remaining in the effluent will consist of the nondegradable organics present in the influent wastewater [$(SCOD_{nd})_i$]

Source and Characteristics of Industrial Wastewaters

Chemical Group	THOD	Measured COD (mg/mg)	Measured BOD$_5$ (mg/mg)	COD THOD (%)	BOD$_5$ COD (%)
Aliphatics					
Methanol	1.50	1.05	0.92	70	88
Ethanol	2.08	2.11	1.58	100	75
Ethylene glycol	1.26	1.21	0.39	96	32
Isopropanol	2.39	2.12	0.16	89	8
Maleic acid	0.83	0.80	0.64	96	80
Acetone	2.20	2.07	0.81	94	39
Methyl ethyl ketone	2.44	2.20	1.81	90	82
Ethyl acetate	1.82	1.54	1.24	85	81
Oxalic acid	0.18	0.18	0.16	100	89
Group average				91	64
Aromatics					
Toluene	3.13	1.41	0.86	45	61
Benzaldehyde	2.42	1.98	1.62	80	82
Benzoic acid	1.96	1.95	1.45	100	74
Hydroquinone	1.89	1.83	1.00	100	55
O-Cresol	2.52	2.38	1.76	95	74
Group average				84	69
Nitrogenous organics					
Monoethanolamine	2.49	1.27	0.83	51	65
Acrylonitrile	3.17	1.39	nil	44	—
Aniline	3.18	2.34	1.42	74	61
Group average				58	63
Refractory					
Tertiary butanol	2.59	2.18	0	84	—
Diethylene glycol	1.51	1.06	0.15	70	—
Pyridine	3.13	0.05	0.06	2	—
Group average				52	—

Source: Adapted from Busch, 1961.

TABLE 1.7 Comparison of Measured COD and BOD$_5$ with the Theoretical Oxygen Demand of Selected Organic Compounds

and residual degradable organics (as characterized by the soluble BOD) and soluble microbial products generated in the treatment process. The SMP are not biodegradable (designated as SMP_{nd}) and, thus, exert a soluble COD (or TOC) but no BOD. Data indicate that the SMP_{nd} is 2 to 10 percent of the influent degradable COD. The actual percentage depends on the type of wastewater and the operating solids retention time (SRT) of the biological process. COD, BOD, and SMP relationships for industrial wastewaters are presented in Table 1.8, where the SMP_{nd} are assumed as 5 percent of the influent degradable SCOD.

A schematic illustrating COD and TSS composition for influent and effluent is presented in Fig. 1.13. The effluent total COD ($TCOD_e$) can be calculated as the sum of the degradable plus nondegradable soluble COD ($SCOD_d + SCOD_{nd}$) plus the "particulate" COD due to the effluent suspended solids (TSS_e). If the effluent solids are primarily activated sludge floc carryover, their COD can be estimated as $1.4\,TSS_e$. This is expressed as follows:

$$TCOD_e = (SCOD_{nd})_e + (SCOD_d)_e + 1.4\,TSS_e \tag{1.4}$$

$$(SCOD_{nd})_e = SMP_{nd} + (SCOD_{nd})_i \tag{1.5}$$

$$(SCOD_{nd})_i = SCOD_i - (SCOD_d)_i \tag{1.6}$$

$$(TCOD)_e = SCOD_i - (SCOD_d)_i + SMP_{nd} + (SCOD_d)_e + 1.4\,TSS_e \tag{1.7}$$

The degradable SCOD of the influent or effluent wastewater can be estimated from the ratio of BOD_5 to ultimate BOD (BOD_u) (designated as f_i or f_e). Assuming that $BOD_u = 0.92\,SCOD$, the degradable SCOD in the influent (i) or effluent (e) can be estimated as:

$$(SCOD_d)_{i/e} = \frac{(BOD_5)_{i/e}}{f_{i/e} \cdot 0.92} \tag{1.8}$$

The effluent TCOD can then be estimated by combining Eqs. (1.4) through (1.8).

$$(TCOD)_e = SCOD_i - \left[\frac{(BOD_5)_i}{f_i \cdot 0.92}\right] + SMP_{nd} + \left[\frac{(BOD_5)_e}{f_e \cdot 0.92}\right] + 1.4\,TSS_e \tag{1.9}$$

The calculations relative to BOD, COD, and TOC are illustrated in Example 1.2.

The BOD, COD, and TOC tests are gross measures of organic content and as such do not reflect the response of the wastewater to various types of biological treatment technologies. It is therefore desirable to

	Influent		Effluent				
Wastewater	BOD (mg/L)	COD (mg/L)	BOD (mg/L)	COD (mg/L)	SMP_{nd}* (mg/L)	$(COD_{nd})e$[†] (mg/L)	BOD_5/COD_{deg}[‡]
Pharmaceutical	3,290	5,780	23	561	261	526	0.60
Diversified chemical	725	1,487	6	257	62	248	0.56
Cellulose	1,250	3,455	58	1,015	122	926	0.47
Tannery	1,160	4,360	54	561	190	478	0.28
Alkylamine	893	1,289	12	47	62	29	0.69
Alkyl benzene sulfonate	1,070	4,560	68	510	202	405	0.25
Viscose rayon	478	904	36	215	35	160	0.61
Polyester fibers	208	559	4	71	24	65	0.40
Protein process	3,178	5,355	5	245	256	237	0.59
Tobacco	2,420	4,270	139	546	186	332	0.59
Propylene oxide	532	1,124	49	289	42	214	0.56
Paper mill	380	686	7	75	31	64	0.58

*$0.05 (COD_{deg})_i$;
[†]$(COD_{nd})_e = SCOD_e − [BOD_5)_e]/0.65]$
[‡]$(COD_d)_i = COD_i − (COD_{nd})_e + SMP_{nd}$

TABLE 1.8 COD, BOD, and SMP Relationships for Industrial Wastewaters

	Influent		Effluent				
Wastewater	BOD (mg/L)	COD (mg/L)	BOD (mg/L)	COD (mg/L)	SMP_{nd}* (mg/L)	$(COD_{nd})e$[†] (mg/L)	BOD_5/COD_{deg}[‡]
Vegetable oil	3,474	6,302	76	332	298	215	0.55
Vegetable tannery	2,396	11,663	92	1,578	504	1,436	0.22
Hardboard	3,725	5,827	58	643	259	554	0.67
Saline organic chemical	3,171	8,597	82	3,311	264	3,185	0.56
Coke	1,618	2,291	52	434	93	354	0.79
Coal liquid	2,070	3,160	12	378	139	360	0.70
Textile dye	393	951	20	261	35	230	0.53
Kraft paper mill	308	1,153	7	575	29	564	0.50

*$0.05 (COD_{deg})_i$
[†]$(COD_{nd})_e = SCOD_e - [(BOD_5)_e/0.65]$
[‡]$(COD_d)_i = COD_i - (COD_{nd})_e + SMP_{nd}$

TABLE 1.8 COD, BOD, and SMP Relationships for Industrial Wastewaters *(Continued)*

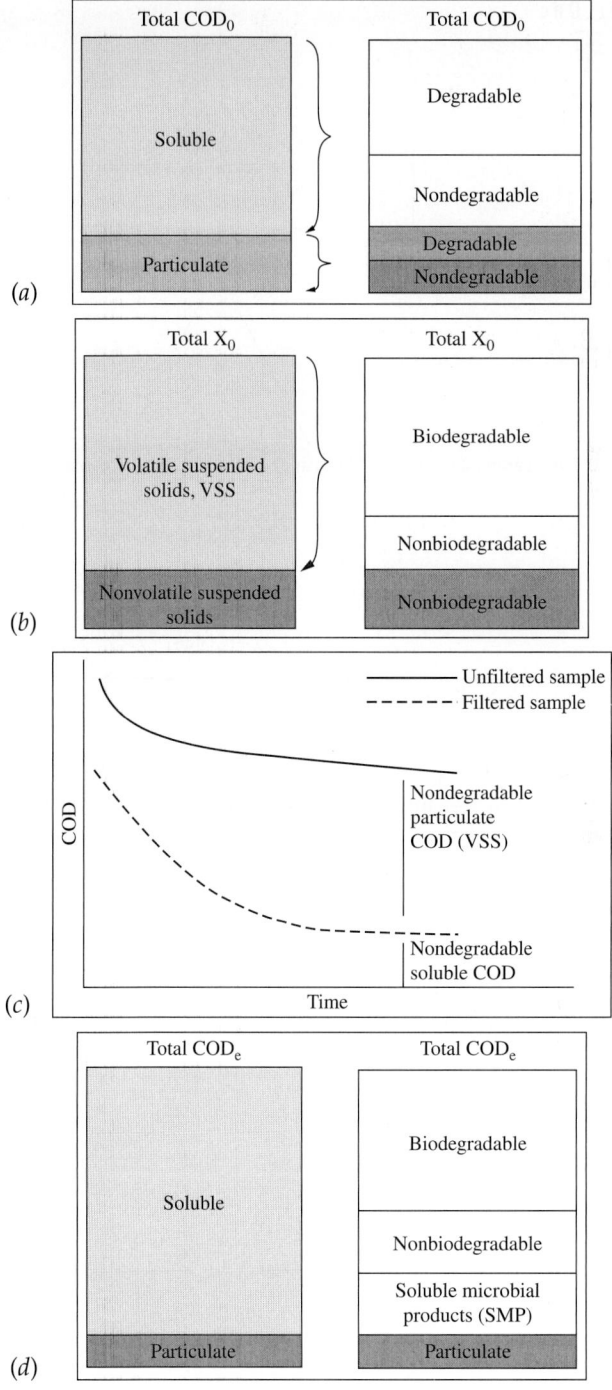

FIGURE 1.13 COD and TSS composition for influent and effluent. (a) Influent chemical oxygen demand, COD_0; (b) Influent total suspended solids, X_0; (c) Determination of degradable on non-degradable influent COD; (d) Effluent COD, COD_e.

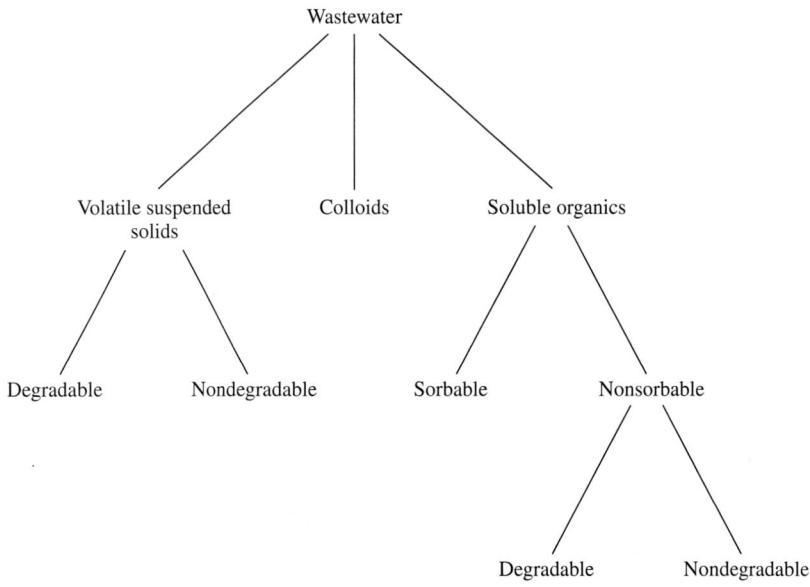

Figure 1.14 Partition of organic constituents of a wastewater.

partition the wastewater into several categories, as shown in Fig. 1.14. The distinction between sorbable and nonsorbable BOD is important in selecting the appropriate process configuration for the control of sludge quality.

Example 1.2. A wastewater contains the following:
150 mg/L ethylene glycol
100 mg/L phenol
40 mg/L sulfide (S^{2-})
125 mg/L ethylene diamine hydrate (ethylene diamine is essentially nonbiodegradable)
(a) Compute the COD and TOC.
(b) Compute the BOD_5 if the k_{10} is 0.2/day.
(c) After treatment, the BOD_5 is 25 mg/L. Estimate the COD ($k_{10} = 0.1$/day).

Solution

(a) The COD is computed:
Ethylene glycol

$$C_2H_6O_2 + 2.5O_2 \rightarrow 2CO_2 + 3H_2O$$

$$COD = \frac{2.5(32)}{62} \times 150 \text{ mg/L} = 194 \text{ mg/L}$$

Source and Characteristics of Industrial Wastewaters

Phenol

$$C_6H_6O + 7O_2 \rightarrow 6CO_2 + 3H_2O$$

$$COD = \frac{7(32)}{94} \times 100 \text{ mg/L} = 238 \text{ mg/L}$$

Ethylene diamine hydrate

$$C_2H_{10}N_2O + 2.5O_2 \rightarrow 2CO_2 + 2H_2O + 2NH_3$$

$$COD = \frac{2.5(32)}{78} \times 125 \text{ mg/L} = 128 \text{ mg/L}$$

Sulfide

$$S^{2-} = 2O_2 \rightarrow SO_4^{2-}$$

$$COD = \frac{2(32)}{32} \times 40 \text{ mg/L} = 80 \text{ mg/L}$$

The total COD is 640 mg/L.
  The TOC is computed:
Ethylene glycol

$$\frac{24}{62} \times 150 \text{ mg/L} = 58 \text{ mg/L}$$

Phenol

$$\frac{72}{94} \times 100 \text{ mg/L} = 77 \text{ mg/L}$$

Ethylene diamine

$$\frac{24}{78} \times 125 \text{ mg/L} = 39 \text{ mg/L}$$

The total TOC is 174 mg/L.

(b) The ultimate BOD can be estimated:

$$COD \times 0.92 = BOD_{ult}$$

$$(194 \text{ mg/L} + 238 \text{ mg/L} + 80 \text{ mg/L}) \times 0.92 = 471 \text{ mg/L}$$

Thus

$$\frac{BOD_5}{BOD_{ult}} = (1 - 10^{-(5 \times 0.2)}) = 0.9$$

BOD_5 is 471 mg/L × 0.9 = 424 mg/L.

(c) The BOD_{ult} of the effluent is

$$\frac{25 \text{ mg/L}}{1 - 10^{-(5 \times 0.1)}} = \frac{25 \text{ mg/L}}{0.7} = 36 \text{ mg/L}$$

The COD is 36/0.92 = 39 mg/L. Therefore the COD will be 128 mg/L + 39 mg/L + residual by-products.

1.7 Measuring Effluent Toxicity

The standard technique for determining the toxicity of a wastewater is the bioassay, which estimates a substance's effect on a living organism. The two most common types of bioassay tests are the chronic and the acute. The chronic bioassay estimates longer-term effects that influence the ability of an organism to reproduce, grow, or behave normally; the acute bioassay estimates short-term effects, including mortality.

The acute bioassay exposes a selected test organism, such as the fathead minnow or *Mysidopsis bahia* (mysid shrimp), to a known concentration of sample for a specified time (typically 48 or 96 h, but occasionally as short as 24 h). The acute toxicity of the sample is generally expressed as the concentration lethal to 50 percent of the organisms, denoted by the term LC_{50}. The chronic bioassay exposes a selected test organism to a known concentration of sample for longer periods of time than the acute bioassay, usually 7 days with daily renewal. The toxicity of the sample is currently expressed as the IC_{25} value, which represents the sample concentration that produces 25 percent inhibition to a chronic characteristic of the test species (e.g., growth weight or reproduction). NOEC is the concentration at which there was no observed effect.

The LC_{50} and IC_{25} values are determined through statistical analysis of mortality-time data or weight- or reproduction-time data, respectively. The lower the LC_{50} or IC_{25} values, the more toxic the wastewater. The bioassay data can be expressed as a concentration of a specific compound (e.g., mg/L), or in the case of whole-effluent (i.e., overall effluent toxicity), as a dilution percentage or as toxic units. Toxic units can be calculated as 100 times the inverse of the dilution percentage expression of whole-effluent toxicity; a whole-effluent toxicity of 25 percent is equal to 100/25 or 4 toxic units. This results in a more logical measure where increasing values indicate increasing toxicity. The toxic unit expression of the data is applicable to acute or chronic tests using any organism. It is simply a mathematical expression.

Various organisms are used to measure toxicity. The organism and life stage (e.g., adult or juvenile) selected depends on the receiving stream salinity, the stability and nature of the expected contaminants, and the relative sensitivity of different species to the effluent. Different organisms exhibit different toxicity thresholds to the same compound (Table 1.9), and there is a relatively large variability in toxicity for a single compound and any one test species as a result of biological factors. This is shown in Fig. 1.15 for a petroleum refinery effluent. It should also be noted that variability in a plant effluent will result in highly variable effluent toxicity as shown in Fig. 1.16. In addition, results for multiple tests vary because of factors such as the species of organism, test conditions, and the number of replicates (i.e., duplicates)

	Units	Fathead Minnow	*Daphnia Magna*	Rainbow Trout
Organics				
Benzene	mg/L	42	35	38
1,4-Dichlorobenzene	mg/L	3.72	3.46	2.89
2,4-Dinitrophenol	mg/L	5.81	5.35	4.56
Methylene chloride	mg/L	326	249	325
Phenol	mg/L	39	33	35
2,4,6-Trichlorophenol	mg/L	5.91	5.45	4.62
Metals				
Cadmium	µg/L	38	0.29	0.04
Copper	µg/L	3.29	0.43	1.02
Nickel	µg/L	440	54	—

TABLE 1.9 Acute Toxicity of Selected Compounds (96-h LC_{50})

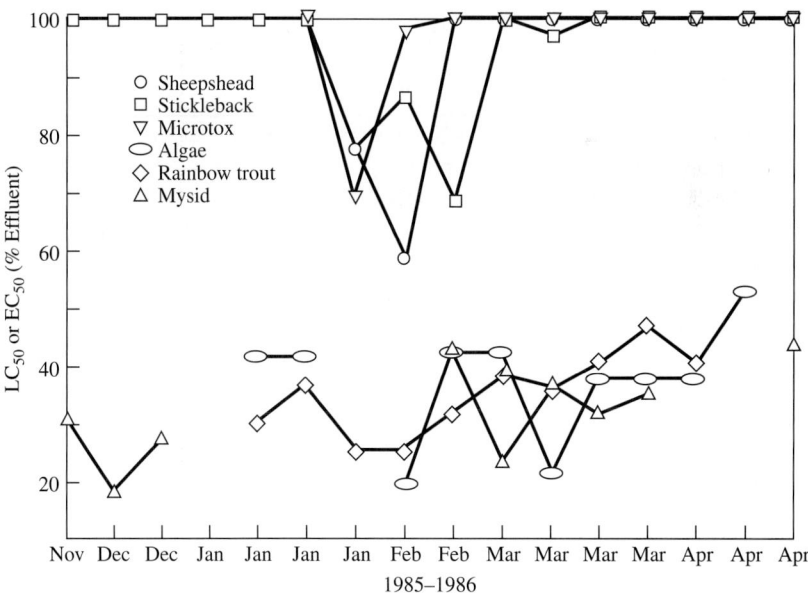

FIGURE 1.15 Acute toxicity of six species to refinery effluent. (*Adapted from Dorn, 1992.*[5])

and organisms used, as well as the laboratory conducting the test (with greater variability observed when more than one lab is involved).

The precision of toxicity test results decreases significantly as the actual toxicity decreases. For example, in one series of bioassays conducted with *Mysidopsis bahia*, for an LC_{50} of 10 percent (10 toxic units), the 95 percent confidence level was about 7 to 15 percent, whereas for

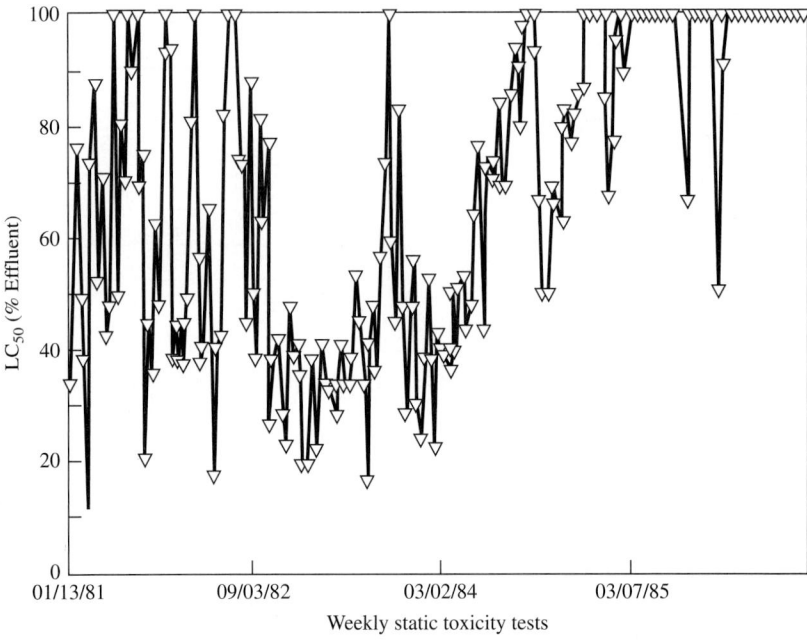

Figure 1.16 Acute toxicity of refinery effluent to stickleback. (*Adapted from Dorn, 1992.*[5])

an LC_{50} of 50 percent (2 toxic units), it was 33 to 73 percent. This variation is a result of the statistical nature of the test. At high LC_{50} values, there is a low mortality rate, which, if the test is done with only a few organisms, can result in a wide range of actual LC_{50} concentrations. If, conversely, there is a high mortality, the results are more precise, since a greater percentage of the organisms was affected by the sample, thus giving a more statistically precise estimate of actual toxicity. It must be kept in mind that any such test gives only estimates of the actual toxicity and should be accorded the precision of an estimate.

Because of the wide variability in bioassay test results, numerous confirmatory data points are needed to be confident that the true extent of the toxicity problem has been estimated. No single data point should be relied upon for any conclusion. This may require long-term operation of bench, pilot-scale, or full-scale systems to determine the effectiveness of treatment.

A thorough toxicity reduction program will involve a large number of tests on treated and untreated samples. For the initial screening stages, one should consider employing either a shortened, simplified bioassay technique (or surrogate test) such as a 24-h or 48-h version of the required test, or a rapid aquatic toxicity test such as Microtox, IQ, or Ceriofast to determine whether there is a correlation between the surrogate test and the required test. If a reasonably strong correlation exists, the surrogate test allows much quicker data turnaround, usually at a much lower cost.

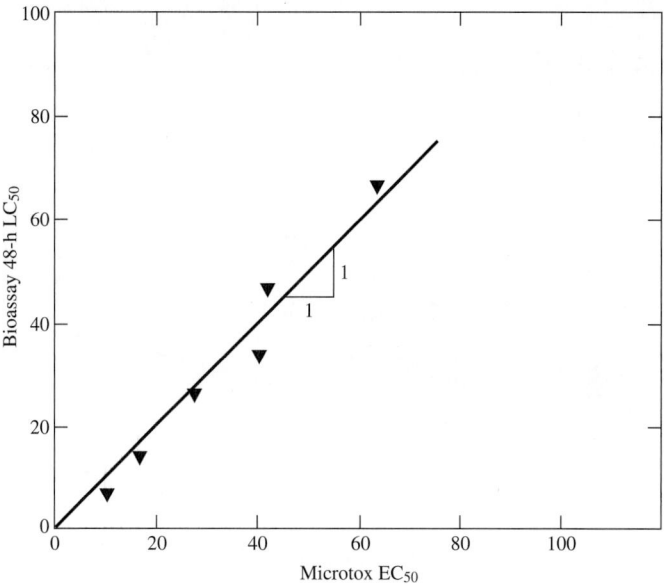

Figure 1.17 Bioassay/Microtox comparison—carbon-treated effluent.

Microtox (Microbics Corporation, Carlsbad, California) uses a freeze-dried marine luminescent microorganism *Vibrio fischeri* incubated at constant temperature in a high-salt, low-nutrient growth medium. Light outputs are measured to determine the effect of the wastewater on the luminescence of the organisms. Microtox has been found to correlate reasonably well in many domestic wastewater cases and in some relatively simple industrial cases. Figure 1.17 shows a correlation for a carbon-treated chemical plant effluent.

IQ (Aqua Survey, Flemington, New Jersey) uses less than 24-h-old Cladoceran species *(Daphnia magna, Daphnia pulex)* hatched from the eggs provided with the test kit. The test organisms are exposed to serial dilutions of the samples for 1 h. Then, the organisms are fed a fluorometrically tagged sugar substrate for 15 min. The intensity of the fluorescence of the control organisms is compared with the fluorescence of the test organisms to determine the effect of the wastewater on the capacity of the organisms to digest the substrate. Figure 1.18 shows a correlation for a pharmaceutical plant effluent.

Ceriofast (Department of Environmental Engineering Sciences, University of Florida, Gainesville, Florida) uses an in-house culture of *Ceriodaphnia dubia*. The 24- to 48-h test organisms are exposed to dilutions of the samples for 40 min. Then, the test organisms and the controls are fed for 20 min with a yeast substrate that contains a nontoxic fluorescent stain. The presence of fluorescence in the intestinal tract of the control organisms is compared with the presence or absence of fluorescence in the test organisms to determine the effect of the wastewater on the feeding ability of the organisms.

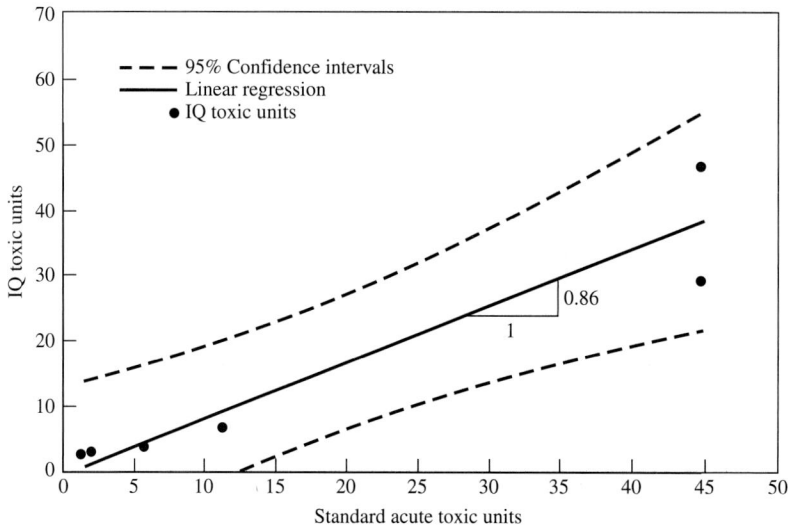

Figure 1.18 IQ toxic units versus standard acute toxic units for a pharmaceutical effluent.

Toxicity Identification of Effluent Fractionation

Toxicity identification of effluent fractionation investigations determine, generically or specifically, the cause of effluent toxicity. Either the actual effluent is fractionated by chemical and/or physical means, or a sample that simulates the known toxic effect is synthesized. The goal is to measure the toxicity of each suspected key component in the absence of other toxic components, but with identical or equivalent nontoxic background components.

The procedure generally involves sample manipulation to eliminate toxicity associated with specific chemical groups. Results of toxicity tests on the treated samples are compared with tests on the unmanipulated effluent. Any difference found indicates that the constituent, or class of constituents, removed is likely to be responsible for the toxicity.

The specific components to be isolated, and the means for doing this, are quite varied and may require extensive investigation. Some techniques are now fairly standard and reliable; others may need to be further developed specifically for the subject wastewater. However, this evaluation can result in significant savings of time and money if properly executed.

Before beginning the fractionation, a thorough plan needs to be developed, one that must address the specific conditions encountered at the facility. Not all separation options (summarized in Fig. 1.19) are employed in every fractionation, and treatments vary from case to case.

The first step is to study the process flowsheet, as well as any long-term data on the chemical makeup of the discharge and plant production. This analysis may provide clues to the source of the toxicity.

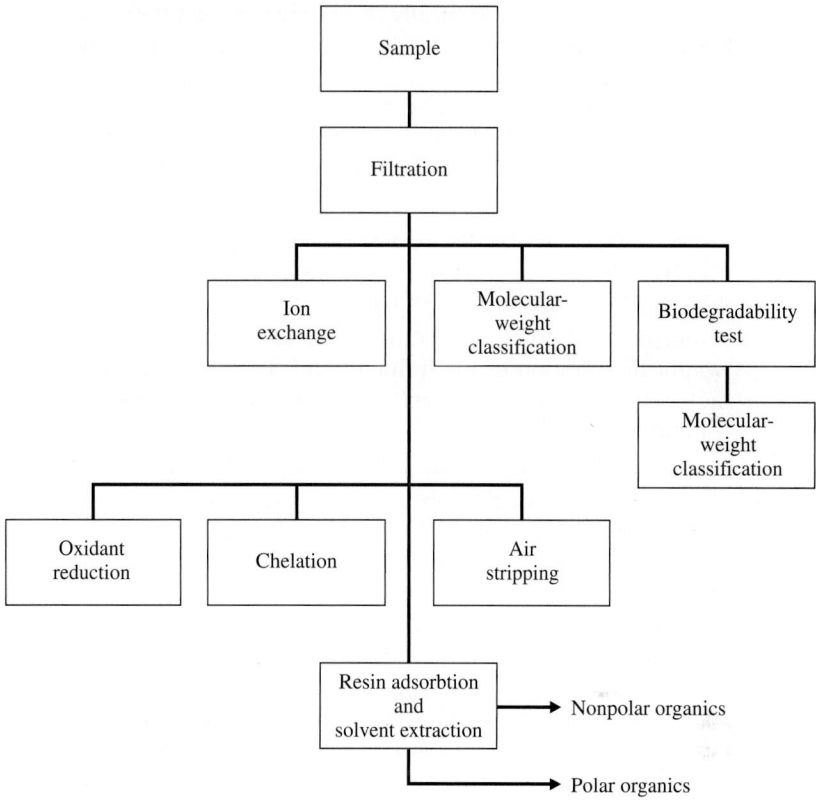

FIGURE 1.19 Various techniques can be employed to fractionate a wastewater sample.

(However, these are just clues, since only the response of the organism in the actual sample is evidence of a toxic effect. In some cases, a compound whose concentration is above the reported toxic level is not harmful in the actual wastewater stream because it is unavailable to the organisms. For example, many heavy metals are toxic at very low levels in soft water but not in hard water.)

If nothing conclusive results from the process analysis (which will often be the case), a literature search should be performed to obtain information on toxicity of similar wastewaters and toxicity of the toxic compounds known to be present. These data will aid in later investigations.

The next step is to actually fractionate the discharge sample. In all cases, blanks (i.e., control samples) should be analyzed to make certain that toxicity is not being introduced by the fractionation or testing procedure. The following manipulations may be considered:

- **Filtration** Filtration is generally performed first to determine whether the toxicity is related to the soluble or insoluble phase of the sample. Typically, 1-μm glass fiber filters that have been prewashed with ultrapure water are used. The insoluble phase should be resuspended in control water to be sure that filtration, not adsorption on the filter medium, removed the toxicity. A 0.45-μm filter should be used in order to determine if the colloidal fraction is responsible for the toxicity.

- **Ion exchange** Inorganic toxicity can be studied by using cationic and anionic exchange resins to remove potentially toxic inorganic compounds or ions.

- **Molecular weight classification** Evaluating the molecular weight distribution of the influent, and the toxicity of each molecular-weight range, can often narrow the list of suspected contaminants.

- **Biodegradability test** Controlled biological treatment of effluent samples in the lab can result in almost complete oxidation of the biodegradable portion of organics. Bioassay analysis can then quantify the toxicity associated with the nonbiodegradable components, as well as the reduction in toxicity attainable by biological treatment.

- **Oxidant reduction** Residual chemical oxidants carried over from a process (e.g., chlorine and chloramines used for disinfection, or ozone and hydrogen peroxide used in treatment) can be toxic to most organisms. A simple batch reduction of these oxidants at various concentrations, using an agent such as sodium thiosulfate, will indicate the toxicity of any remaining oxidants.

- **Metal chelation** The toxicity of the sum of all cationic metals (with the exception of mercury) can be determined by chelation of samples, using varying concentrations of ethylenediaminetetraacetic acid (EDTA) and evaluating the change in toxicity.

- **Air stripping** Batch air stripping at acid, neutral, and basic pH can remove essentially all volatile organics. At a basic pH, ammonia is removed as well. Thus, if both volatile organics and ammonia are suspected toxicants, an alternative ammonia removal technique, such as a zeolite exchange, should be used. (Note that ammonia is toxic in the nonionized form, so ammonia toxicity is very pH dependent.)

- **Resin adsorption and solvent extraction** Specific nonpolar organics can sometimes be identified as toxics, using a resin-adsorption/solvent-extraction process. A sample is adsorbed on a long-chain organic resin, the organics are reextracted from the resin with a solvent (e.g., methanol), and the sample's toxicity is determined via a bioassay.

Source Analysis and Sorting

In a typical wastewater collection system, multiple streams from various sources throughout the plant combine into fewer and fewer streams, ultimately forming a single discharge stream or treatment plant influent. To identify exactly where in the process the toxicity originates, one performs source analysis and sorting. This procedure starts with the influent to the treatment plant and proceeds upstream to the various points of wastestream combination, until the sources of key toxicants have been identified.

Treatment characteristics for each source are evaluated to determine whether the stream can be detoxified by the existing end-of-pipe technology at the facility, usually a biological treatment system. During this evaluation, the relative contribution of each source to effluent toxicity can be defined. (Methods for reducing or eliminating contributions through source treatment at the individual processing unit will be addressed in a later phase of the program.)

Gathering and interpreting this information for a large number of sources requires a well-planned and organized program; a procedure such as the one outlined in Fig. 1.19 can be used. Sources are first classified (before treatment) according to the following criteria:

- Bioassay toxicity, in terms of mg/L of the key chemical component
- Flow, as a percentage of the total effluent
- Concentration of the key chemical component (e.g., organic material expressed as total organic carbon [TOC])
- Biodegradability

Relatively nonbiodegradable wastestreams are the most likely to induce toxic effects in the discharge and should be evaluated carefully. Some may require physical or chemical treatment because they are relatively nonbiodegradable. Others may have relatively high biodegradation rates but result in high residual levels after biode-gradation. These will require additional testing to assess whether they remain significantly toxic following biological treatment.

Wastestreams found to be highly biodegradable have low probabilities for inducing toxic effects. However, their actual impact on effluent toxicity or inhibition must be confirmed. This can be done by feeding a composite wastestream, of all (or most of) the streams to a continuous-flow biological reactor, and then determining the reactor effluent's toxicity.

It is also necessary to determine whether interactions are occurring between streams. To do so, the individual wastestream toxicities are compared with the toxicity of a biological-reactor effluent after treating a blend of the streams. If the toxicity (expressed in toxic units) of a blend of several samples is exactly additive, then no synergistic or

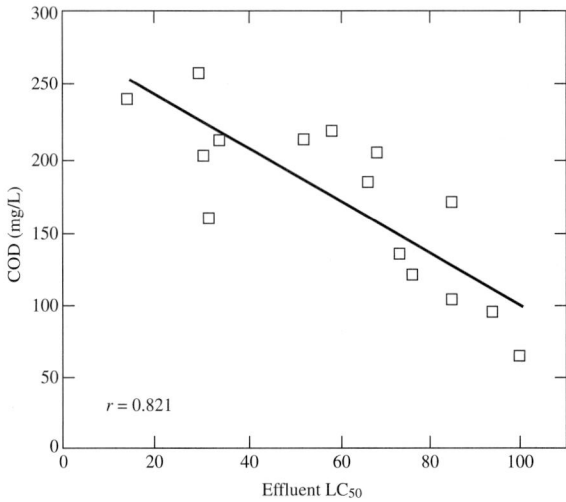

Figure 1.20 Toxicity/COD correlation for wastewater treatment plant effluent.

antagonist effects are taking place. If the measured toxic units value for the composite stream is lower than the calculated value (i.e., the composite sample is nontoxic), the streams are antagonistic; if higher, they are synergistic.

In many cases the cause of effluent toxicity cannot be isolated, and a correlation must be made on total effluent COD as shown in Fig. 1.20 for a petroleum refinery effluent.

1.8 In-Plant Waste Control and Water Reuse

Waste Minimization

Before end-of-pipe wastewater treatment or modifications to existing wastewater treatment facilities to meet new effluent criteria are considered, a program of waste minimization should be initiated.

Reduction and recycling of waste are inevitably site- and plant-specific, but a number of generic approaches and techniques have been used successfully across the country to reduce many kinds of industrial wastes.

Generally, waste minimization techniques can be grouped into four major categories: inventory management and improved operations, modification of equipment, production process changes, and recycling and reuse. Such techniques can have applications across a range of industries and manufacturing processes, and can apply to hazardous as well as nonhazardous wastes.

Many of these techniques involve source reduction—the preferred option on EPA's hierarchy of waste management. Others deal with on- and off-site recycling. The best way to determine how these general approaches can fit a particular company's needs is to conduct

Inventory management and improved operations
• Inventory and trace all raw materials
• Purchase fewer toxic and more nontoxic production materials
• Implement employee training and management feedback
• Improve material receiving, storage, and handling practices
Modification of equipment
• Install equipment that produces minimal or no waste
• Modify equipment to enhance recovery or recycling options
• Redesign equipment or production lines to produce less waste
• Improve operating efficiency of equipment
• Maintain strict preventive maintenance program
Production process changes
• Substitute nonhazardous for hazardous raw materials
• Segregate wastes by type for recovery
• Eliminate sources of leaks and spills
• Separate hazardous from nonhazardous wastes
• Redesign or reformulate end products to be less hazardous
• Optimize reactions and raw material use
Recycling and reuse
• Install closed-loop systems
• Recycle on-site for reuse
• Recycle off site for reuse
• Exchange wastes

TABLE 1.10 Waste Minimization Approaches and Techniques

a waste minimization assessment, as discussed below. In practice, waste minimization opportunities are limited only by the ingenuity of the generator. In the end, a company looking carefully at bottom-line returns may conclude that the most feasible strategy would be a combination of source reduction and recycling projects.

Waste minimization approaches as developed by the EPA are shown in Table 1.10. In order to implement the program, an audit needs to be made as described in Table 1.11. Case studies from three industries following rigorous source management and control are listed in Table 1.12. Pollution reduction can be directly achieved in several ways.

1. *Recirculation.* In the paperboard industry, white water from a paper machine can be put through a save-all to remove the pulp and fiber and recycled to various points in the paper making process.

2. *Segregation.* In a soap and detergent case, clean streams were separated for direct discharge. Concentrated or toxic streams may be separated for separate treatment.

Phase I—Preassessment
• Audit focus and preparation
• Identify unit operations and processes
• Prepare process flow diagrams
Phase II—Mass balance
• Determine raw material inputs
• Record water usage
• Assess present practice and procedures
• Quantify process outputs
• Account for emissions:
To atmosphere
To wastewater
To off-site disposal
• Assemble input and output information
• Derive a preliminary mass balance
• Evaluate and refine the mass balance
Phase III—Synthesis
• Identify options
Identify opportunities
Target problem areas
Confirm options
• Evaluate options
Technical
Environmental
Economic
• Prepare action plan
Waste reduction plan
Production efficiency plan
Training

TABLE 1.11 Source Management and Control

3. *Disposal.* In many cases concentrated wastes can be removed in a semidry state. In the production of ketchup, after cooking and preparation of the product, residue in the kettle bottoms is usually flushed to the sewer. The total discharge BOD and suspended solids can be markedly reduced by removal of this residue in a semidry state for disposal. In breweries the secondary storage units have a sludge in the bottom of the vats which contain both BOD and suspended solids. Removal of this as a sludge rather than flushing to the sewer will reduce the organic and solids load to treatment.

4. *Reduction.* It is a common practice in many industries, such as breweries and dairies, to have hoses continuously running

Case Studies	Before	After
1. Chemical industry		
• Volume (m³/day)	5000	2700
• COD (t/day)	21	13
2. Hide and skin industry		
• Volume (m³/day)	2600	1800
• BOD (t/day)	3–6	2–6
• TDS (t/day)	20	10
• SS (t/day)	4–83	3–7
3. Metal preparation and finishing		
• Volume (m³/day)	450	270
• Chromium (kg/day)	50	5
• TTM (kg/day)	180	85

Note: t = tonne = 1000 kg.

TABLE 1.12 Source Management and Control

for cleanup purposes. The use of automatic cutoffs can substantially reduce the wastewater volume. The use of drip pans to catch products, as in a dairy or ice cream manufacturing plant, instead of flushing this material to the sewer considerably reduces the organic load. A similar case exists in the plating industry where a drip pan placed between the plating bath and the rinse tanks will reduce the metal dragout.

5. *Substitution.* The substitution of chemical additives of a lower pollutional effect in processing operations, e.g., substitution of surfactants for soaps in the textile industry.

A summary of cost-effective pollution control is shown in Table 1.13.

Although it is theoretically possible to completely close up many industrial process systems through water reuse, an upper limit on reuse is imposed by product quality control. For example, a closed system in a paper mill will result in a continuous buildup of dissolved organic solids. This can increase the costs of slime control, cause more downtime on the paper machines, and under some conditions cause discoloration of the paper stock. Obviously, the maximum degree of reuse is reached before these problems occur.

Water requirements are also significant when reuse is considered. Water makeup for hydropulpers in a paper mill need not be treated for removal of suspended solids; on the other hand, solids must be removed from shower water on the paper machines to avoid clogging of the shower nozzles. Wash water for produce such as tomatoes does not have to be pure but usually requires disinfection to ensure freedom from microbial contamination.

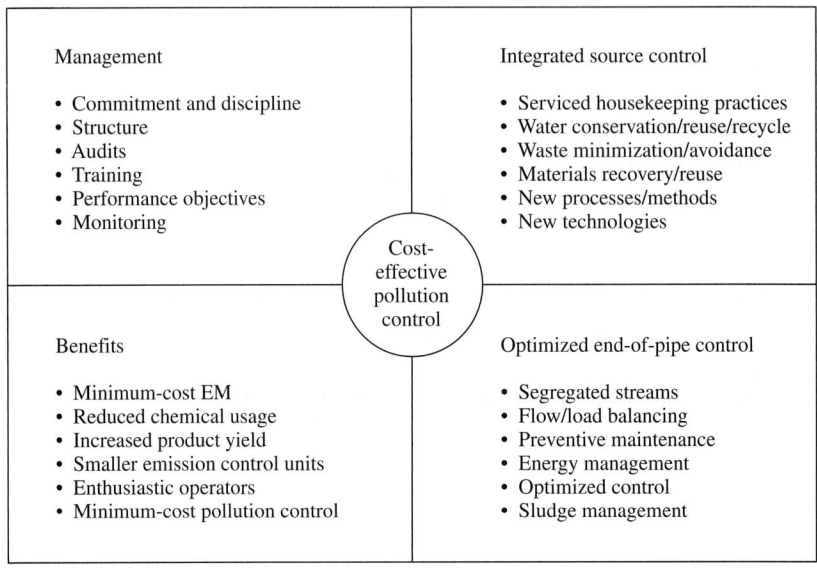

EM: environmental management

TABLE 1.13 Summary of Cost-Effective Pollution Control

By-product recovery frequently accompanies water reuse. The installation of save-alls in paper mills to recover fiber permits the reuse of treated water on cylinder showers. The treatment of plating-plant rinsewater by ion exchange yields a reusable chromic acid. Water conservation and materials recovery in nickel plating is shown in Fig. 1.21. There are many other similar instances in industry.

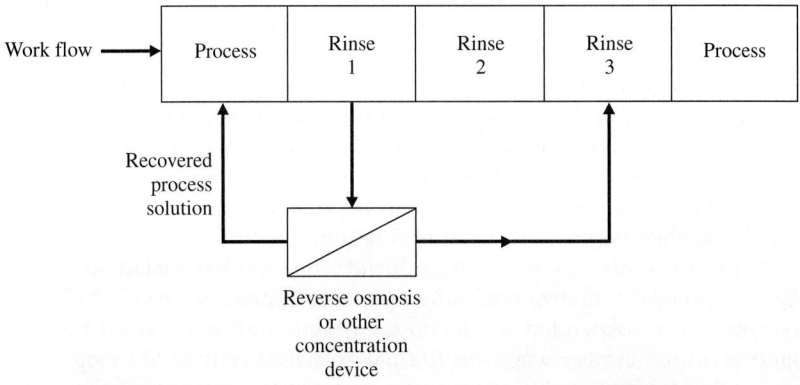

FIGURE 1.21 Water conservation and materials recovery (nickel plating).

- Improvements in process selectivity and/or conversion
- The ability to operate at lower temperatures and/or pressures
- Processes requiring fewer steps
- Products and/or catalysts with longer lives
- The use of feedstocks having fewer inherent by-products
- More efficient equipment design
- Innovative unit operations
- Innovative process integration
- New uses for otherwise valueless by-products
- The avoidance of heat degradation of reaction products
- Novel selective energy sources
- In-plant conversion of wastestream to clean fuels or feeds

TABLE 1.14 Waste Reduction Techniques in the Chemical/Petrochemical Industry

As evidenced by the Pollution Prevention Act of 1990, Pollution Prevention (P2) in industry is a key national environmental policy. As such the EPA has developed a number of partnership programs. The EPA Web site includes a pollution prevention site which incorporates a pollution prevention information clearinghouse (PPIC). Industrial sector notebooks are presented here for numerous industries such as agricultural chemicals, electronics and computer industry, inorganic chemicals, organic chemicals, petroleum refining, pharmaceutical industry, pulp and paper, and many others. Some are translated into Spanish. These notebooks include comprehensive environmental profiles, industrial process information, pollution prevention techni-ques, regulatory requirements, and case histories. Some waste reduction techniques practiced in the chemical and petrochemical industries are given in Table 1.14. A more detailed description of waste minimization and water reuse strategies, including case histories, can be found in Chap. 15.

ISO 14000 encourages industry to move beyond compliance to improve environmental performance by using pollution prevention strategies and environmental management systems. An important tool in this effort is life cycle assessment (LCA). This "cradle-to-grave" or, optimistically, "cradle-to-cradle" approach enables the estimation of cumulative environmental impacts resulting from all stages in the product life cycle (raw materials extraction, material transportation, processing, maintenance, final product disposal, and so on). A mass balance of materials and energy is made and output releases evaluated which represent environmental trade-offs in product and process selection. This approach includes four components:

1. *Goal.* Define and describe the product, process, or activity. Establish the context in which the assessment is to be made and identify the boundaries and environmental effects to be reviewed for the assessment.
2. *Inventory Analysis.* Identify and quantify energy, water, and materials usage and environmental releases (e.g., air emissions, solid waste disposal, wastewater discharges).
3. *Impact Assessment.* Assess the potential health, ecological, and human impacts of those elements identified in the inventory and analysis and consider opportunities for improvement.
4. *Interpretation.* Evaluate the results of the inventory analysis and impact phases to select the preferred options in relationship to the objectives of the study with a clear understanding of the uncertainty and the assumptions used to generate the results.

The EPA and United Nations Environment Programme (UNEP) have dedicated a portion of their Web sites to provide comprehensive and up-to-date information relevant to LCA.

1.9 Problems

1.1. The 4-h composite COD data from a brewery effluent over a 7-day period is as follows:

980			
2800	3200	6933	3325
1380	3175	1240	6000
1250	3850	580	3100
720	2870	710	2500
8650	2600	3410	1830
7200	2743	2910	3225
2800	3600	8300	2370
2570	4066	2950	1380
1780	1550	2230	2600

Develop a statistical plot of these data in order to define the 50 and the 90 percent values.

1.2. A tomato-processing plant is shown in Fig. P1.2. The industrial waste survey data are shown in the Table below. Develop a flow-and-material balance diagram for the plant sewer system. Show possible changes to reduce the flow and loading. The waste from the trimming Tables is presently flumed to the sewer.

Source and Characteristics of Industrial Wastewaters

Sampling Station	Process Unit	Flow (gal/min)	BOD (mg/L)	SS (mg/L)
1	Niagra washers	500	75	180
2	Rotary washers	—	90	340
3	Pasteurizer	300	30	20
4	Source cooker	10	3520	7575
5	Cook room	20	4410	8890
6	Source finisher	150	230	170
7	Trimming tables	100	450	540
8	Main outfall	1080	240	470

Note: gal/min = 3.78×10^{-3} m^3/min

1.3. A wastewater contains the following constituents:

40 mg/L phenol
350 mg/L glucose
3 mg/L S^{2-}

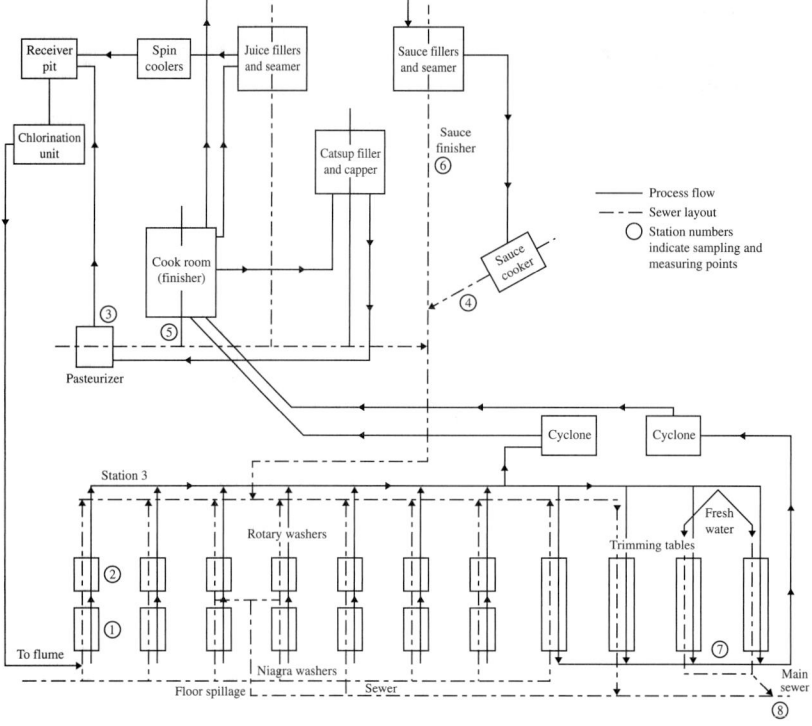

FIGURE P1.2 Flow diagram of a tomato-processing plant.

50 mg/L methyl alcohol, CH_3OH
100 mg/L isophorene, $C_9H_{14}O$ (nondegradable)

(a) Compute the THOD, the COD, the TOC, and the BOD_5 assuming the k_{10} for the mixed wastewater is 0.25/day.
(b) After treatment the soluble BOD_5 is 10 mg/L with a k_{10} of 0.1/day. Compute the residual COD and TOC.

References

1. Schroepfer, G. J.: *Advances in Water Pollution Control*, vol. 1, Pergamon Press, New York, 1964.
2. Stack, V. T.: *Proc. 8th Ind. Waste Conf.*, 1953, p. 492, Purdue University.
3. Busch, A. W.: *Proc. 15th Ind. Waste Conf.*, 1961, p. 67, Purdue University.
4. Ford, D. L.: *Proc. 23rd Ind. Waste Conf.*, 1968, p. 94, Purdue University.
5. Dorn, P. B.: *Toxicity Reduction—Evaluation and Control*, Technomic Publishing Co., 1992.

CHAPTER 2
Wastewater Treatment Processes

2.1 Introduction

Once the industrial wastewater has been adequately characterized, it is necessary that the appropriate options of unit processes be selected to obtain desired water quality goals. The selection and sequencing of these unit operations is essential for a cost-effective implementation of the waste management program. This chapter considers screening procedures and alternative technologies' applicabilities for both biological and physical chemical waste treatment.

The schematic diagram in Fig. 2.1 illustrates an integrated system capable of treating a variety of plant wastewaters. The scheme centers on the conventional series of primary and secondary treatment processes, but also includes tertiary treatment and individual treatment of certain streams.

Primary and secondary treatment processes handle most of the nontoxic wastewaters; other waters have to be pretreated before being added to this flow. These processes are basically the same in an industrial plant as in a publicly owned treatment works (POTW).

Primary treatment prepares the wastewaters for biological treatment. Large solids are removed by screening, and grit, if present, is allowed to settle out. Equalization, in a mixing basin, levels out the hour-to-hour variations in flows and/or concentrations. There should be a spill pond to retain slugs of concentrated wastes that could upset the downstream processes. Neutralization, where required, usually follows equalization because streams of different pH partly neutralize each other when mixed. Oils, greases, and suspended solids are removed by flotation, sedimentation, or filtration.

Secondary treatment is the biological degradation of soluble organic compounds—from input levels of 50 to 1000 mg/L BOD (biochemical

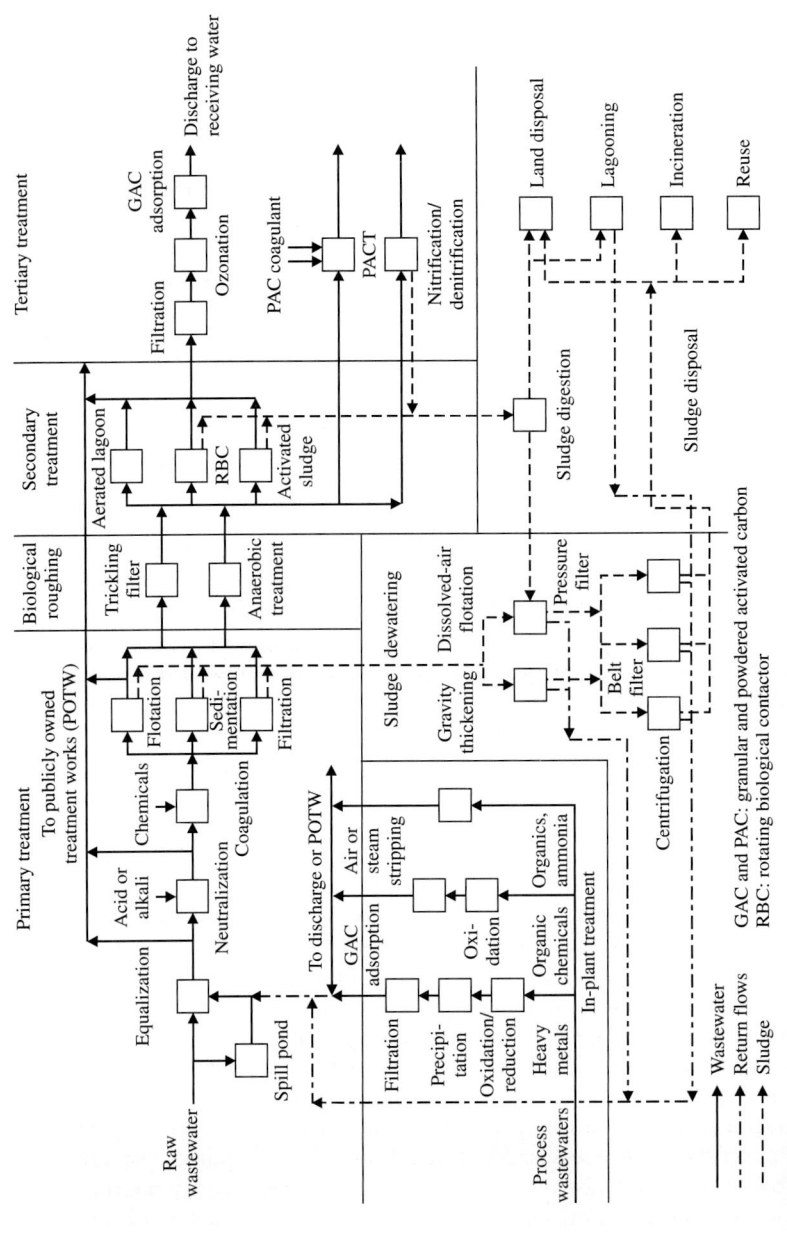

Figure 2.1 Alternative technologies for the treatment of industrial wastewaters.

oxygen demand) and even greater to effluent levels typically under 15 mg/L. This is usually done aerobically, in an open, aerated vessel or lagoon, but wastewaters may be pretreated anaerobically, in a pond or a closed vessel. After biotreatment, the microorganisms and other carried-over solids are allowed to settle. A fraction of this sludge is recycled in certain processes, but ultimately the excess sludge, along with the sedimented solids, has to be disposed of.

Many existing wastewater-treatment systems were built just for primary and secondary treatment, though a plant might also have systems for removing materials that would be toxic to microorganisms. Until recently, this was adequate, but now it is not, and so new facilities have to be designed and old facilities retrofitted to include additional capabilities to remove priority pollutants and residuals toxic to aquatic life and in some cases to remove nutrients.

Tertiary-treatment processes are added on after biological treatment in order to remove specific types of residuals. Filtration removes suspended or colloidal solids; adsorption by granular activated carbon (GAC) removes organics; and chemical oxidation also removes organics. Unfortunately, tertiary systems have to treat a large volume of wastewater, and so they are expensive. They can also be inefficient because the processes are not pollutant-specific. For example, dichlorophenol can be removed by ozonation or GAC adsorption, but those processes will remove most other organics as well, and this adds greatly to the treatment costs for removing the dichlorophenol.

In-plant treatment is necessary for streams rich in heavy metals, pesticides, and other substances that would pass through primary treatment and inhibit biological treatment. In-plant treatment also makes sense for low-volume streams rich in nondegradable materials, because it is easier and much less costly to remove a specific pollutant from a small, concentrated stream than from a large, dilute one. Processes used for in-plant treatment include precipitation, activated carbon adsorption, chemical oxidation, air or steam stripping, ion exchange, ultrafiltration, reverse osmosis, electrodialysis, and wet air oxidation.

Existing treatment systems can also be modified so as to broaden their capabilities and improve their performance; this is more widely practiced than the above options. One example is adding powdered activated carbon (PAC) to the biological-treatment process to adsorb organics that the microorganisms cannot degrade or slowly degrade; this is marketed as the PACT process. Another example is adding coagulants at the end of the biological-treatment basin so as to remove phosphorus and residual suspended solids.

All these processes have their place in the overall wastewater-treatment scheme. The selection of a wastewater-treatment process or a combination of processes depends upon:

1. The characteristics of the wastewaters. This should consider the form of the pollutant, that is, suspended, colloidal, or dissolved; the biodegradability; and the inhibition or toxicity of the organic and inorganic components.

52 Chapter Two

2. The required effluent quality. Consideration should also be given to possible future restrictions such as an effluent bioassay aquatic toxicity limitation.
3. The costs and availability of land for any given wastewater-treatment problem. One or more treatment combinations can produce the desired effluent. Only one of these alternatives, however, is the most cost-effective. A detailed cost analysis should therefore be made prior to final process design selection.

2.2 Process Selection

A preliminary analysis should be carried out to define the wastewater treatment problems as shown in Fig. 2.2.

For wastewaters containing nontoxic organics, process design criteria can be obtained from available data or from a laboratory or pilot plant program. Examples are pulp and paper mill wastewaters and food processing wastewaters. In the case of complex chemical wastewaters containing toxic and nontoxic organics and inorganics, a more defined screening procedure is necessary to select candidate processes for treatment. A protocol has been established as shown in Fig. 2.3. If the

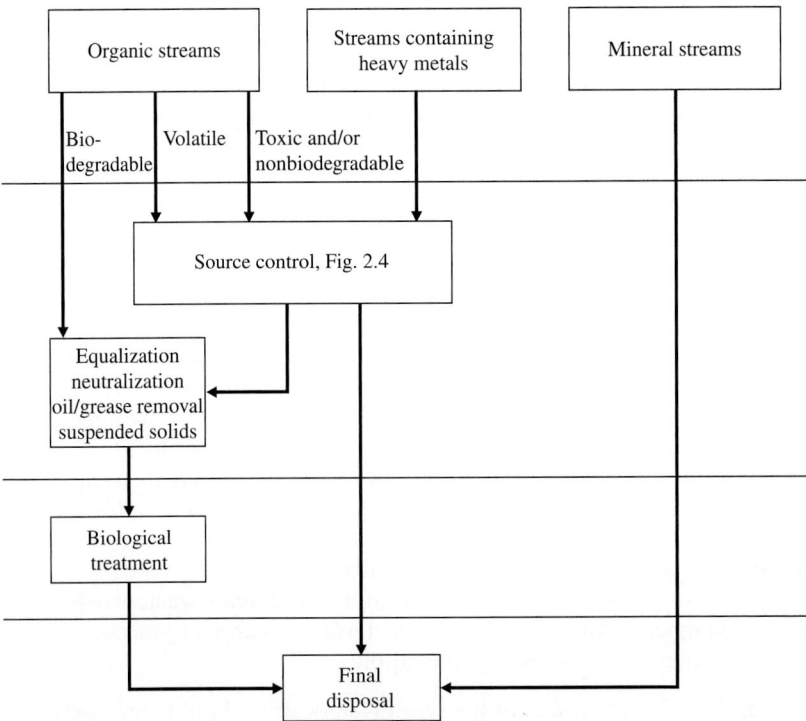

FIGURE 2.2 Conceptual approach of treatment/management program for a high (organic) strength and toxic industrial wastewater.

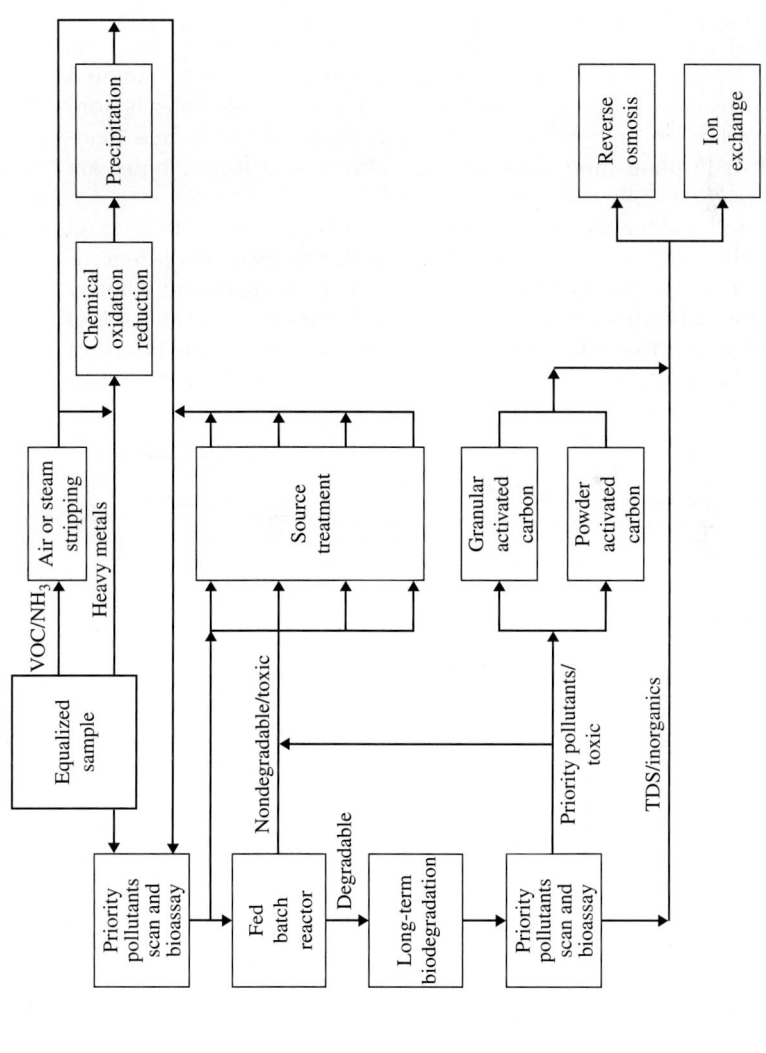

FIGURE 2.3 Screening laboratory procedures.

wastewater contains heavy metals, they are removed by precipitation. Volatile organics are removed by air stripping.

Following laboratory screening, pretreatment-as-required analyses are conducted on an equalized sample as shown in Fig. 2.3. It should be noted that all significant wastewater streams in the industrial complex should be evaluated. The next step is to determine whether the wastewater is biodegradable and whether it will be toxic to the biological process at some level of concentration. The fed batch reactor (FBR) procedure is employed for this purpose. Details of this procedure are discussed later. Acclimated sludge as discussed later should be used. If the wastewater is nonbiodegradable or toxic, it should be considered for source treatment or in-plant modification. Source treatment technologies are shown in Fig. 2.4.

If the wastewater is biodegradable, it is subjected to a long-term biodegradation, usually 48 h, in order to remove all degradable organics. It is then evaluated for aquatic toxicity and priority pollutants. If nitrification is required, a nitrification rate analysis is conducted. If the effluent is toxic or if priority pollutants have not been removed, it should be considered for source treatment or

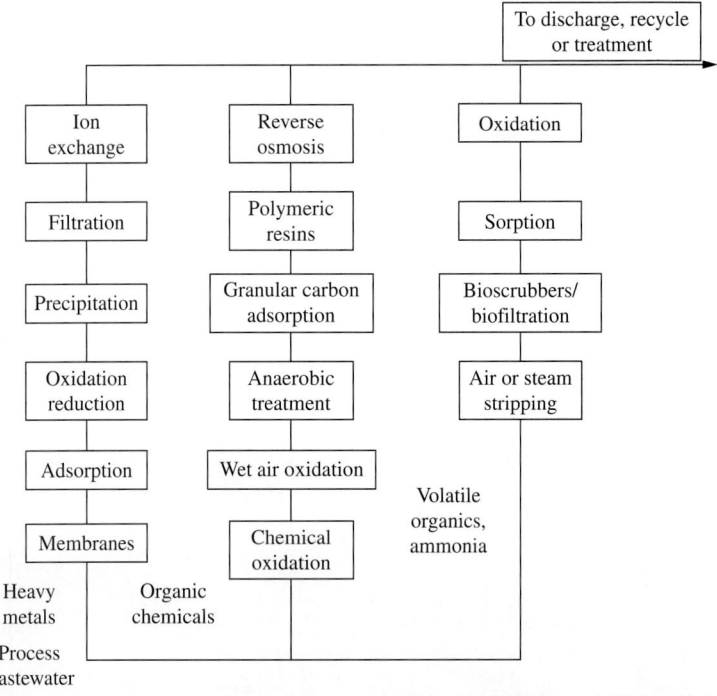

FIGURE 2.4 Application technologies for source treatment of toxic persistent wastewaters.

tertiary treatment using powdered or granular activated carbon. It should be noted that toxicity due to soluble microbial products (SMP) will require powdered activated carbon or tertiary treatment for its removal.

When biological treatment is considered, several options exist. A screening procedure has been developed to determine the most cost-effective alternative, as shown in Fig. 2.5.

The alternatives for biological treatment are summarized in Table 2.1 and discussed in detail in Chap. 7. A screening and identification matrix for physical-chemical treatment is shown in Table 2.2. Table 2.2 should provide a guide to select applicable technologies for specific problems. A summary of physical-chemical technology applications is shown in Table 2.3. The maximum attainable effluent quality based on the authors' experiences for conventional wastewater treatment processes is shown in Table 2.4.

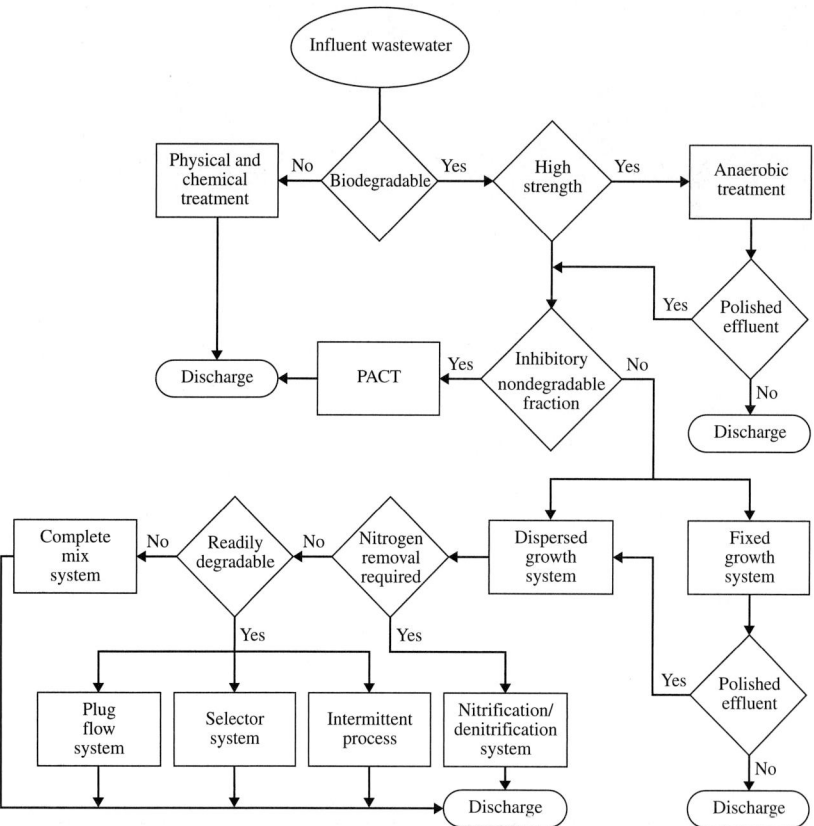

FIGURE 2.5 Simplified process selection flowsheet for biological treatment.

Treatment Method	Mode of Operation	Degree of Treatment	Land Requirements	Equipment	Remarks
Lagoons	Intermittent or continuous discharge; facultative or anaerobic	Intermediate	Earth dug; 10–60 days' retention or more		Odor control frequently required; acceptable linear needed; RCRA issues may apply
Aerated lagoons	Completely mixed or facultative continuous basins	High in summer; less in winter	Earth basin, 8–16 ft (2.44–4.88 m) deep, 8–16 acres/(million gal/d) (8.55–17.1 m^2/ [m^3/d])	Pier-mounted or floating surface aerators or subsurface diffusers	Solids separation in lagoon; periodic dewatering and sludge removal; acceptable linear needed; RCRA issues may apply
Activated sludge	Completely mixed or plug flow; sludge recycle	>90% removal of organics	Earth or concrete basin; 12–20 ft (3.66–6.10 m) deep; 75,000–350,000 ft^3/(million gal/d) (0.561–2.262 m^3/ [m^3/d])	Diffused or mechanical aerators; clarifier for sludge separation and recycle	Excess sludge dewatered and disposed of
Trickling filter	Continuous application; may employ effluent recycle	Intermediate or high, depending on loading	225–1400 ft^2/(million gal/d) (5.52–34.4 m^2/10^3 [m^3/d])	Plastic packing 20–40 ft deep (6.10–12.19 m)	Pretreatment before POTW or activated sludge plant

RBC	Multistage continuous	Intermediate or high		Plastic disks	Solids separation required
Anaerobic	Complete mix with recycle; upflow or downflow filter, fluidized bed; upflow sludge blanket	Intermediate		Gas collection required; pretreatment before POTW or activated sludge plant	
Spray irrigation	Intermittent application of waste	Complete; water percolation into groundwater and runoff to stream	40–300 gal/(min · acre) (6.24 × 10^{-7} – 4.68 × 10^{-6} m³/[s · m²])	Aluminum irrigation pipe and spray nozzles; movable for relocation	Solids separation required; salt content in waste limited

Notes:
ft = 0.305 m
acre/(million gal · d) = 1.07 m²/(m³ · d)
ft³/(million gal · d) = 7.48 × 10^{-3} m³/(thousand m³ · d)
ft²/(million gal · d) = 2.45 × 10^{-2} m²/(thousand m³ · d)
gal/(min · acre) = 1.56 × 10^{-8} m³/(s · m²)

TABLE 2.1 Biological Waste Treatment

		Stripping Parameter									
Process	Organic Compounds	Non-Condensable	Temperature	Pressure	pH	O&G, mg/L	SS, mg/L	TDS, mg/L	Fe, Mn	Sol	Comments
Air	<100 mg/L	A	DP	DP	DP	R	R	DP	R	L	$H_c > 0.005$ is recommended.
Steam	<100 mg/L to 10%	R	DP	DP	NI	R	R	DP	NI	M	Relative volatility referred to water >1.05 is recommended. Azeotrope formation is important.

		Oxidation Processes Parameter								
Process	Organic Compounds	Temperature °F	Pressure, Psig	pH	OD, g/l	O&G, mg/L	SS, mg/L	TDS, mg/L	Fe, Mn	MW	Comments
Wet air oxidation	A	350–650	300–3000	NI	20–200	NI	NI	DP	NI	NI	Not recommended for aromatic halogenated organics. Recommended for high COD/BOD ratio.
Supercritical water oxidation	A	750–1200	3675	NI	<10	NI	NI	L	NI	NI	
Chemical oxidation	O_3 <10,000 mg/L H_2O_2 A	DP DP	DP NI	DP DP	DP DP	R R	R R	DP DP	A A	NI H	Catalyst and additional source of energy, e.g., ultraviolet, are important factors.

Adsorption and Precipitation

	Parameter										
	Organic Compounds		Inorganic Ionic Species	Chemical Oxidants	Temperature	pH	O&G, mg/L	SS, mg/L	TDS, mg/L	Fe, Mn	MW	Sol	Comments
Process	Volatile	Semi-Volatile											
Activated carbon adsorption	<10,000 mg/L		NA	NA	DP	DP	<10	<50	<10	NI	H	L	$K > 5$ mg/g, high K_{ow}, inorganics <1000 mg/L are recommended. Heavy metals may poison the carbon.
Resin adsorption	A		NA	R	DP (L)	DP	<10	<10	DP	NI	DP	M	High K_{ow} and C_o <0.1 (resin capacity/3 BV) are recommended. K is a design parameter.
Chemical precipitation	NA		A	NI	DP	DP	R	NI	DP	A	NI	DP	Chelating and complexing agents interference may occur.

Membrane Processes and Ion Exchange

	Parameter								
Process	Organic Compounds		Inorganic Ionic Species	Chemical Oxidants	Temperature	Pressure, Psig	pH	O&G, mg/L	SS, mg/L	TDS, mg/L	Fe, Mn	MW, Amu	Comments
	Volatile	Semi-Volatile											
Reverse osmosis	R	A	A	R	DP	<1500	DP	R	R	<10,000	R	>150	Differential osmotic pressure <400 psi, LSI <0, SDI <5, and turbidity <1 NTU are recommended.

TABLE 2.2 Screening and Identification Matrix for Industrial Wastewaters Physical-Chemical Treatment

Membrane Processes and Ion Exchange

	Organic Compounds		Inorganic ionic Species	Parameter									
Process	Volatile	Semi-Volatile		Chemical Oxidants	Tempera-ture	Pressure, Psig	pH	O&G, mg/L	SS, mg/L	TDS mg/L	Fe, Mn	MW, Amu	Comments
Hyper-filtration	NA	A	NA	NI	DP	DP	DP	R	R	NI	NI	100–500	Molecular size, shape, and flexibility are important factors.
Ultrafiltra-tion	NA	A	NA	NI	DP	10–100	DP	R	R	NI	NI	500–1,000,000	Molecular size, shape, and flexibility are important factors.
Electro-dialysis/electro-dialysis reverse	R	R	A	R	DP	40–60	DP	R	R	<5000	<0.3	NI	Applied voltage is one of the design parameters. Ca < 900 mg/L is recommended.
Ion exchange resin	R	R	A	R	DP	NI	DP	R	<50 (<35)	<20,000	NI	NI	Selectivity quotient is one of the design parameters. Ion charge and volume are important factors.

A = applicable
BV = bed volume
DP = design parameter
H = high
H_C = Henry's constant
K = Freundlich isotherm coefficient
K_{ow} = octanol/water partition coefficient
M = moderate
MW = molecular weight
NA = not applicable
NTU = nephelometric turbidity units
O&G = oil and grease
R = must be removed
SDI = silt density index
Sol = solubility
SS = suspended solids
TDS = total dissolved solids

Note:
°C = $\frac{5}{9}$ (°F − 32)
kPa = 6.894 psi

TABLE 2.2 Screening and Identification Matrix for Industrial Wastewaters Physical-Chemical Treatment (*Continued*)

Treatment Method	Type of Waste	Mode of Operation	Degree of Treatment	Remarks
Ion exchange	Plating, nuclear	Continuous filtration with resin regeneration	Demineralized water recovery; product recovery	May require neutralization and solids removal from spent regenerant
Reduction and precipitation	Plating, heavy metals	Batch or continuous treatment	Complete removal of chromium and heavy metals	One day's capacity for batch treatment; 3-h retention for continuous treatment; sludge disposal or dewatering required
Coagulation	Paperboard, refinery, rubber, paint, textile	Batch or continuous treatment	Complete removal of suspended and colloidal matter	Flocculation and settling tank or sludge blanket unit; pH control required
Adsorption	Toxic or organics, refractory	Granular columns of powdered carbon	Complete removal of most organics	Powdered activated carbon (PAC) used with activated sludge process
Chemical oxidation	Toxic and refractory organics	Batch or continuous ozone or catalyzed hydrogen peroxide	Partial or complete oxidation	Partial oxidation to render organics more biodegradable

TABLE 2.3 Physical-Chemical Waste Treatment

Process	BOD	COD	SS	N	P	TDS
Sedimentation, % removal	10–30	—	59–90	—	—	—
Flotation, % removal*	10–50	—	70–95	—	—	—
Activated sludge, mg/L	<25	†	<20	†	†	—
Aerated lagoons, mg/L	<50	—	>50	—	—	—
Anaerobic ponds, mg/L	>100	—	<100	—	—	—
Deep-well disposal	Total disposal of waste					
Carbon adsorption, mg/L	<2	<10	<1	—	—	—
Denitrification and nitrification, mg/L	<10	—	—	<5	—	—
Chemical precipitation, mg/L	—	—	<10	—	<1	—
Ion exchange, mg/L	—	—	<1	§	§	§

*Higher removals are attained when coagulating chemicals are used.
†$COD_{inf} - [BOD_{ult}\ (removed)/0.9]$.
‡$N_{inf} - 0.12$ (excess biological sludge), lb.; $P_{inf} - 0.026$ (excess biological sludge), lb.
§Depends on resin used, molecular state, and efficiency desired.
Note:
lb = 0.45 kg.

TABLE 2.4 Maximum Quality Attainable from Waste Treatment Processes

CHAPTER 3
Pre- and Primary Treatment

3.1 Introduction

The objective of pre- and primary treatment is to render a wastewater suitable for discharge to a POTW (publicly owned treatment works) or subsequent biological or physical chemical treatment. Pretreatment of industrial wastewaters discharged to POTWs is a major technical issue specifically regulated by EPA. This issue and case histories are presented in Chap. 17. The concentrations of pollutants that make prebiological treatment desirable are summarized in Table 3.1. Pretreatment is often used to reduce or eliminate inhibition or toxicity to the biological process. It may also be used to render biorefractory organics more biodegradable through chemical oxidation or other methods.

Some potential toxicants include heavy metals, pesticides, priority pollutants, and so on. Those found to be most significant of one organic chemical manufacturing corporation are:

1. Surfactants
2. Cationic polymers
3. Free ammonia
4. Nitrite
5. Chlorine
6. Pesticides
7. Metals
8. Temperature

Some potential inhibitors may include:

- TDS (16,000 mg/L)
- Chlorides (8000 to 10,000 mg/L)

65

Pollutant or System Condition	Limiting Concentration	Kind of Pretreatment
Suspended solids	>125 mg/L	Sedimentation, flotation, lagooning
Oil or grease	>35	Skimming tank or separator
Toxic ions		Precipitation or ion exchange
Pb	≤0.1 mg/L	
Cu + Ni + CN	≤1 mg/L	
Cr^{+6} + Zn	≤3 mg/L	
Cr^{+3}	≤10 mg/L	
pH	6 to 9	Neutralization
Alkalinity	0.5 lb alkalinity as $CaCO_3$/lb BOD removed	Neutralization for excessive alkalinity
Acidity	Free mineral acidity	Neutralization
Organic load variation	>2:1	Equalization
Sulfides	>100 mg/L	Precipitation or stripping with recovery
Ammonia	>500 mg/L (as N)	Dilution, ion exchange, pH adjustment, and stripping
Temperature	>38°C in reactor	Cooling

TABLE 3.1 Concentrations of Pollutants That Make Prebiological Treatment Desirable

- H_2S-bacteria (100 mg/L, algae 7–10 mg/L)
- Heavy metals (1 mg/L [metal, its species, hardness, and pH])
- Ammonia 500 mg/L, f(pH), 0.02 mg/L (free ammonia) as aquatic criteria
- Weak organic acids—<6.5 or >8.5 [microorganism metabolism may increase pH, e.g., acetic acid and formic acid]

Microorganism metabolism may increase pH (e.g., acetic acid and formic acid)

- Strong bases—Microbial respiration decreases BOD thereby increasing CO_2 and forming H_2CO_3 (about 0.5 lb alkalinity as $CaCO_3$/lb BOD removed or about 0.63 lb alkalinity as $CaCO_3$/lb COD removed)

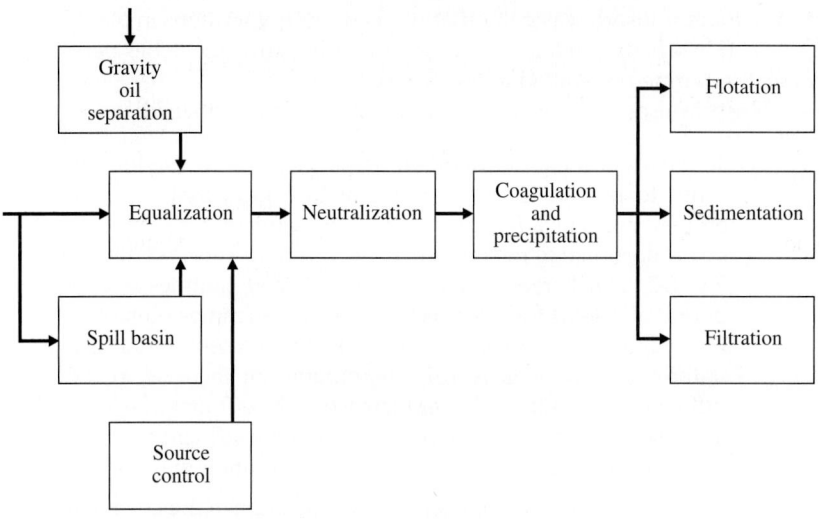

FIGURE 3.1 Pretreatment technologies.

- Contaminants which require specific pretreatment (i.e., pH, alkalinity, TSS, fats, oils and grease [FOG], metals, etc.)
- Nutrient N, P, and trace minerals availability—Microbes require ammonia and P plus trace metals for metabolism
- Variability

Common pretreatment technologies are shown in Fig. 3.1. This chapter will discuss several pre- and primary treatment processes often employed in industrial treatment. Other pretreatment technologies will be reviewed in subsequent chapters of this book.

3.2 Equalization

Variation in wastestream flows, constituents, and concentrations can significantly reduce the efficiency of unit processes and effluent quality. Most equations utilized for design assume steady-state conditions. Effluent permits are typically based on average as well as maximum discharge limits. Hence, variability reduction is critical to compliance with treatment and sustainability goals. Equalization therefore is a very important unit process typically preceding other pre- and primary treatment and subsequent treatment methods.

The objective of equalization is to minimize or control fluctuations in wastewater characteristics in order to provide optimum conditions for subsequent treatment processes. The size and type of the equalization basin varies with the quantity of waste and the variability of the wastewater stream. The basin should be of a sufficient size to adequately absorb waste fluctuations caused by variations in plant-production scheduling and to dampen the concentrated batches periodically dumped or spilled to the sewer.

The purposes of equalization for industrial treatment facilities are:

1. To provide adequate dampening of organic fluctuations in order to prevent shock loading of biological systems. The effluent concentration from a biological treatment plant will be proportional to the influent concentration. This is shown in Fig. 3.2, which represents 24-h composited samples over a period of 3 years for a petroleum refinery. As can be seen from Fig. 3.2, the effluent variability tracks the influent variability. If the wastewater is readily degradable an increase in the influent will result in a lesser increase in the effluent due to an increase in biometabolism. By contrast, if the influent contains bioinhibitors, an increased effluent concentration will result.

2. To provide adequate pH control or to minimize the chemical requirements necessary for neutralization.

3. To minimize flow surges to physical-chemical treatment systems and permit chemical feed rates compatible with feeding equipment.

4. To provide continuous feed to biological systems over periods when the manufacturing plant is not operating.

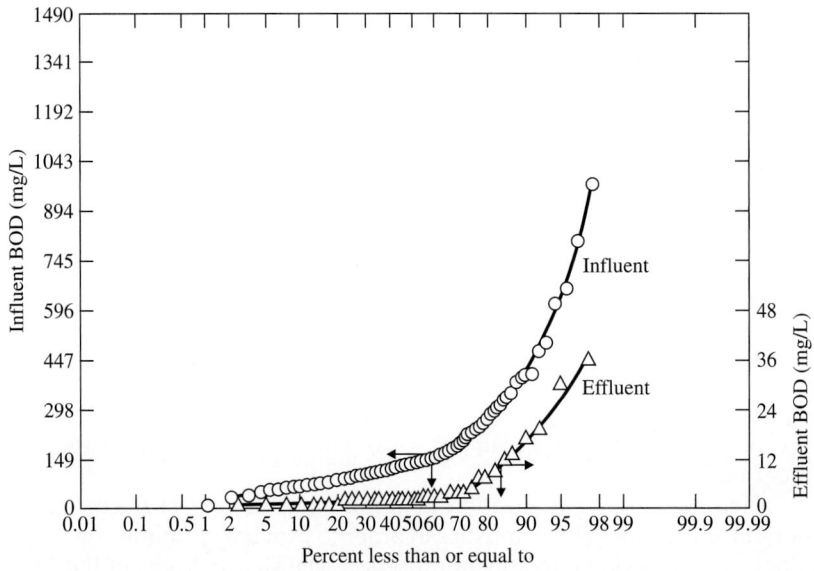

FIGURE 3.2 Variability in influent and effluent BOD.

5. To provide capacity for controlled discharge of wastes to municipal systems in order to distribute waste loads more evenly.
6. To prevent high concentrations of toxic materials from entering the biological treatment plant.

Complete mixing is usually provided to ensure adequate equalization and to prevent settleable solids from depositing in the basin. In addition, the oxidation of reduced compounds in the waste-stream or the reduction of BOD by air stripping may be achieved through mixing and aeration. Methods that have been used for mixing include:

1. Distribution of inlet flow and baffling
2. Turbine mixing
3. Diffused air aeration
4. Mechanical aeration
5. Submerged mixers

The most common method is to provide submerged mixers or, in the case of readily degradable wastewater, such as that from a brewery, to use surface aerators employing a power level of approximately 15 to 20 hp/million gal (0.003 to 0.0045 kW/m^3). Air requirements for diffused air aeration are approximately 0.5 ft^3 air/gal waste (3.74 m^3/m^3). Equalization basin types are shown in Fig. 3.3. It should be noted that in compliance with NESHAP regulations many industries must cover headworks, including the equalization basin, and satisfactorily treat the off-gas emissions to an acceptable level. In some cases, the secondary treatment facility will also be affected.

The equalization basin may be designed with a variable volume to provide a constant effluent flow or with a constant volume and an effluent flow that varies with the influent. The variable-volume basin is particularly applicable to the chemical treatment of wastes having a low daily volume. This type of basin may also be used for discharge of wastes to municipal sewers. It may be desirable to program the effluent pumping rate to discharge the maximum quantity of waste during periods of normally low flow to the municipal treatment facility, as shown in Fig. 3.4. Ideally, the organic loading to the treatment plant is maintained constant over a 24-h period.

Equalization basins may be designed to equalize flow, concentration, or both. For flow equalization, the cumulative flow is plotted versus time over the equalization period (e.g., 24 h). The maximum volume with respect to the constant-discharge line is the

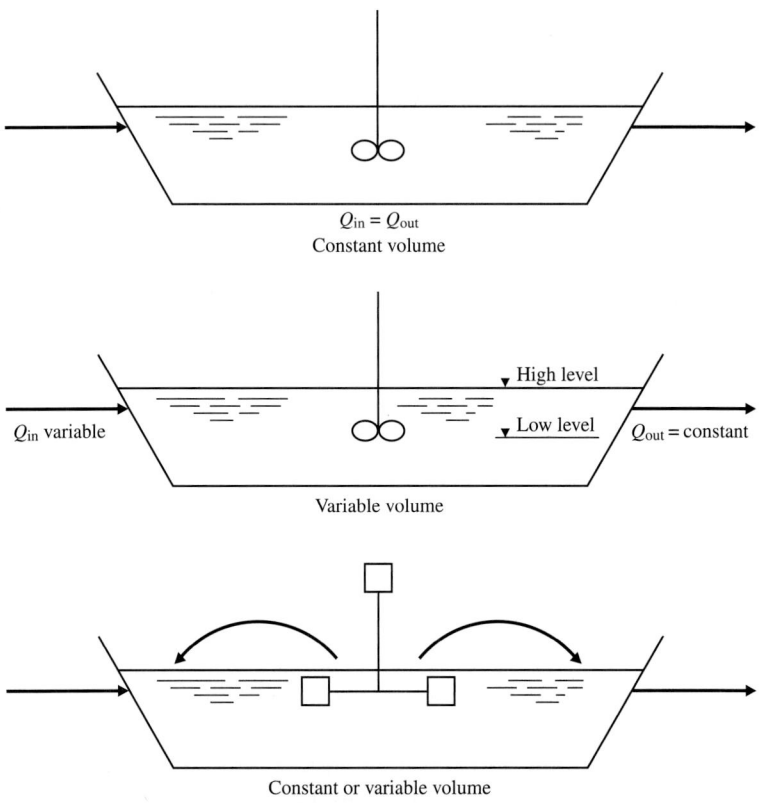

FIGURE 3.3 Constant-volume and variable-volume equalization basins.

equalization volume required. The required calculations are shown in Example 3.1.

Example 3.1. Given the data in Table 3.2, design an equalization basin for a constant outflow from the basin. The data are plotted as the summation of inflow versus time as shown in Fig. 3.5. The treatment rate is (193,300 gal/day)/(1440 min/day) = 134 gal/min (507 L/min). The required storage volume is 41,000 gal + 8000 gal residual or 49,000 gal (186 m³).

The equalization basin may be sized to restrict the discharge to a maximum concentration commensurate with the maximum permissible discharge from subsequent treatment units. For example, if the maximum effluent from an activated sludge unit is 50 mg/L BOD_5, the maximum effluent from the equalization basin may be computed and thereby provide a basis for sizing the unit.

For the case of near-constant wastewater flow and random input variations that have a normal statistical distribution of wastewater

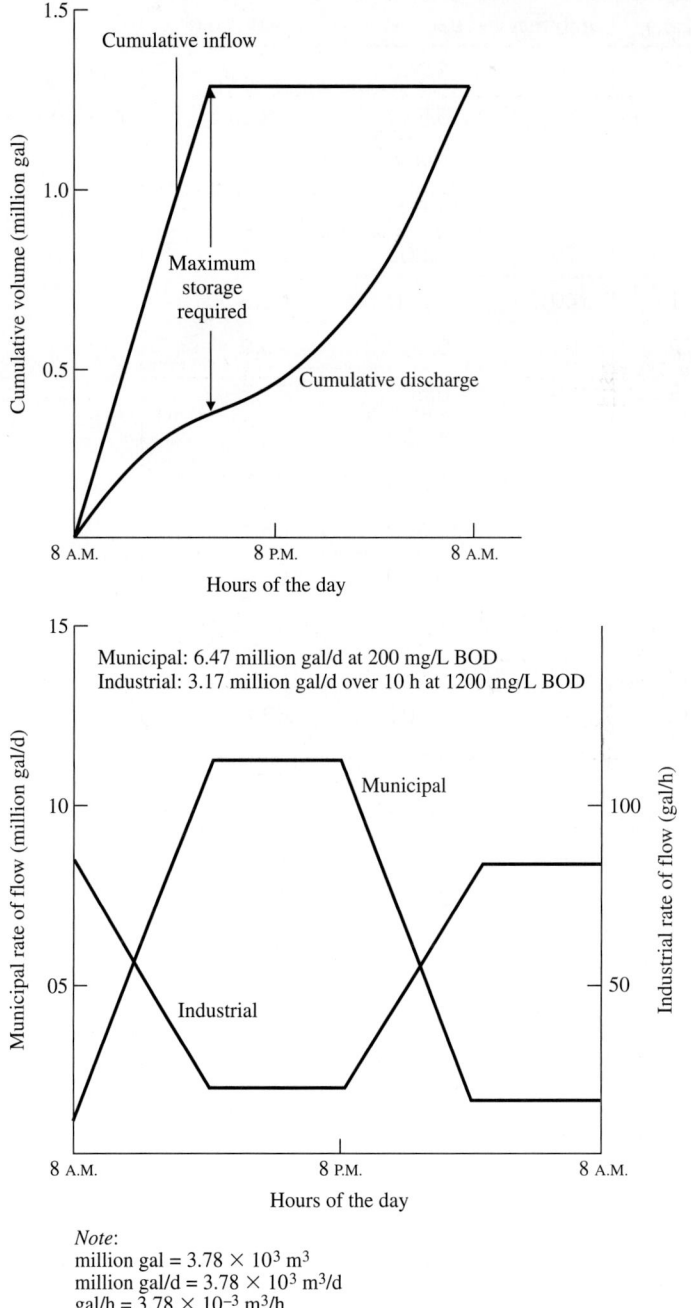

FIGURE 3.4 Controlled discharge of an industrial wastewater to a municipal plant.

72 Chapter Three

Time	gal/min	gal	Σ gal $\times 10^3$
8	50	3,000	3.0
9	92	5,520	8.5
10	230	13,800	22.3
11	310	18,600	40.9
12	270	16,200	57.1
1	140	8,400	65.5
2	90	5,400	70.9
3	110	6,600	77.5
4	80	4,800	82.3
5	150	9,000	91.3
6	230	13,800	105.1
7	305	18,300	123.4
8	380	22,800	146.2
9	200	12,000	158.2
10	80	4,800	163.0
11	60	3,600	166.6
12	70	4,200	170.8
1	55	3,300	174.1
2	40	2,400	176.5
3	70	4,200	180.7
4	75	4,500	185.2
5	45	2,700	187.9
6	55	3,300	191.2
7	35	2,100	193.3

Note: gal/min = 3.78×10^{-3} m^3/min
gal = 3.78×10^{-3} m^3

TABLE 3.2 Equalization Basin Constant Outflow Design Data

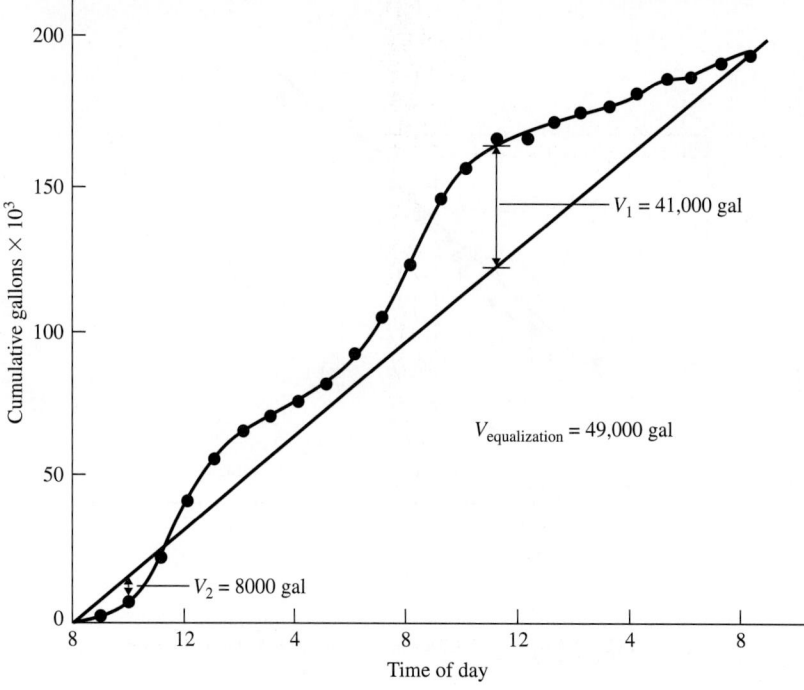

FIGURE 3.5 Constant outflow equalization basin design.

composite analyses, the required equalization retention time for a constant volume basin is[1]

$$t = \frac{\Delta t(S_i^2)}{2(S_e^2)} \quad (3.1)$$

where t = equalization detention time, h
   Δt = time interval over which samples were composited, h
   S_i^2 = variance of the influent wastewater concentration (the square of the standard deviation)
   S_e^2 = variance of the effluent concentration at a specified probability (e.g., 99 percent)

Example 3.2 illustrates this calculation.

Example 3.2. A waste with a total flow of 5 million gal/d (0.22 m³/s or 19,000 m³/d) was characterized as shown in Fig. 3.6. Extensive data were collected every 4 h for 17 d. The average BOD was 690 mg/L and the maximum value was 1185 mg/L.

Design calculations with activated sludge systems have indicated that the effluent from the equalization basin must not exceed 896 mg/L in order to meet the effluent quality criteria of an average BOD of 15 mg/L and a maximum concentration of 25 mg/L from the activated sludge system.

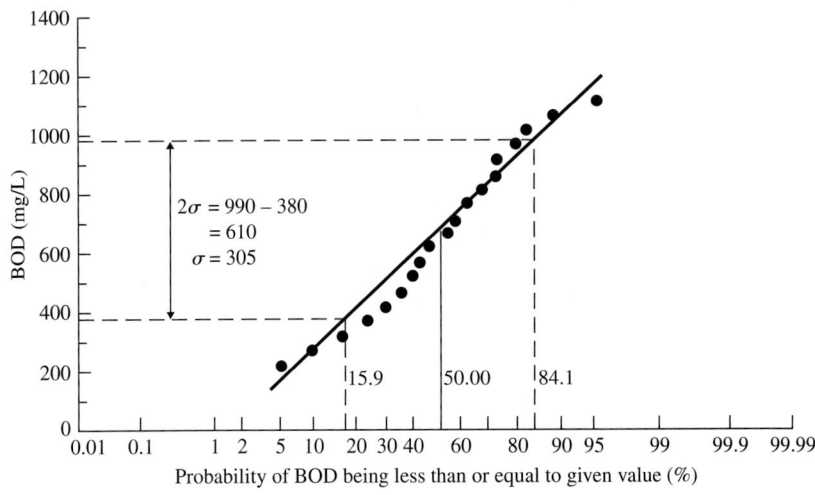

FIGURE 3.6   BOD probability analysis.

Design an equalization basin to meet the desired effluent requirements. Base calculations on a 95 percent probability that the equalized effluent will be equal to or less than 896 mg/L.

Solution
(a) Calculate the mean, standard deviation, and variance of the influent. These parameters may be calculated graphically from Fig. 3.6.

(b) From this plot obtain the 50 percent value:

$$50 \text{ percent value} \approx \overline{X} \approx 690 \text{ mg/L}$$

(c) Calculate the standard deviation, S_i, as half the difference in the values that occur at the 15.9 (50.0 minus 34.1) and 84.1 (50.0 plus 34.1) percentile levels from Fig. 3.6:

$$S_i = \frac{\text{value at 84.1\%} - \text{value at 15.9\%}}{2}$$

$$= \frac{990 - 380}{2}$$

$$= 305 \text{ mg/L}$$

(d) Calculate the variance as the square of the standard deviation:

$$S_i^2 = (305)^2$$

$$= 93{,}025 \text{ mg}^2/\text{L}^2$$

(e) Calculate the standard deviation of the effluent:

$$\overline{X} = 690 \text{ mg/L}$$

$$X_{max} = 896 \text{ mg/L}$$

The necessary condition for 95 percent of effluent BOD values to be less than 896 mg/L is

$$S_e = \frac{X_{max} - \overline{X}}{Z}$$

$$= \frac{896 - 690}{1.65}$$

$$= 125 \text{ mg/L}$$

where $Z = 1.65$ is from the normal probability tables for a 95 percent confidence level.

(f) Calculate the allowed effluent variance:

$$S_e^2 = (125)^2$$

$$= 15{,}625 \text{ mg}^2/L^2$$

(g) Calculate the required detention time:

$$t = \frac{\Delta t(S_i^2)}{2(S_e^2)}$$

$$= \frac{4(93{,}025)}{2(15{,}625)}$$

$$= 11.9 \text{ h}$$

$$\approx 0.5 \text{ d}$$

Where a completely mixed basin is to be used for treatment, such as in an activated sludge basin or an aerated lagoon, this volume can be considered as part of the equalization volume. For example, if the completely mixed aeration basin retention time is 8 h and the total required equalization retention time is 16 h, then the equalization basin needs to have a retention time of only 8 h.

Patterson and Menez[2] have developed a method to define equalization requirements when both the flow and the strength vary randomly. A material balance can be established for the equalization basin:

$$C_i QT + C_0 V = C_2 QT + C_2 V \tag{3.2}$$

where C_i = concentration entering the equalization basin over the sampling interval T
T = sampling interval, that is, 1 h
Q = average flow rate over the sampling interval
C_0 = concentration in the equalization basin at the start of the sampling interval
V = volume of the equalized basin
C_2 = concentration leaving the equalization basin at the end of the sampling interval

It is assumed that the effluent concentration is almost constant during one time interval. This is valid, assuming the time intervals are appropriately spaced.

Equation (3.2) can be rearranged to compute the effluent concentration after each time interval:

$$C_2 = \frac{C_i T + C_0 V/Q}{T + V/Q} \tag{3.2a}$$

The range of effluent concentrations can then be calculated for a range of equalization volumes V. A peaking factor PF is computed for the influent strength and flow. The effluent PF for design purposes is the ratio of the maximum concentration to the average concentration. An equalization basin design is shown in Example 3.3.

Example 3.3. A survey of the discharge from a chemical plant showed the following results:

Time Period	Mean Flow gal/min	Mean TOC, mg/L
8 to 10 A.M.	450	920
10 to 12 noon	620	1130
12 to 2 P.M.	840	1475
2 to 4 P.M.	800	1525
4 to 6 P.M.	340	910
6 to 8 P.M.	270	512
8 to 10 P.M.	570	1210
10 to 12 midnight	1100	1520
0 to 2 A.M.	1200	1745
2 to 4 A.M.	800	820
4 to 6 A.M.	510	410
6 to 8 A.M.	570	490

Develop a plot of peaking factor versus basin volume for a variable-volume equalization system and determine the volume to yield a peaking factor of 1.2 based on mass discharge.

Solution. Assuming the equalization basin is a completely mixed basin where no significant degradation or evaporation occurs, the differential equations that govern the constant effluent flow system for each time interval are:

$$\frac{dVi}{dt} = Q_{0i} - Q_{ei} = Q_{0i} - Q_{0avg} \tag{1}$$

$$\frac{d(V_i C_i)}{dt} = V_i \frac{dC_i}{dt} + C_i \frac{dV_i}{dt} = V_i \frac{dC_i}{dt} + C_i (Q_{0i} - Q_{0avg}) = Q_{0i} C_{0i} - Q_{ei} C_i \tag{2}$$

where V_i = volume in the basin at time t for time interval i, gal
Q_{0i} = influent flow rate at time interval i, gpm
Q_{ei} = effluent flow rate at time interval i, gpm
Q_{0avg} = daily average influent flow rate, gpm
C_{0i} = influent concentration at time interval i, mg/L
C_i = concentration in the basin and effluent concentration at time t for time interval i, mg/L

Assuming that Q_{0i}, Q_{ei}, and C_{0i} are constant during each time interval, separation of variables, integration, and rearrangement of Eqs. (1) and (2) lead to the following expressions for the volume $V_i(f)$ and the TOC concentration $C_i(f)$ in the basin at the end of each time interval, respectively.

$$V_i(f) = V_{(i-1)}(f) + (Q_{0i} - Q_{0avg})\Delta t_i \tag{3}$$

$$C_i(f) = C_{0i} - \frac{A}{(1 + B\Delta t_i)^D} \tag{4}$$

where

$$A = C_{0i} - C_{(i-1)}(f)$$

$$B = \frac{Q_{0i} - Q_{0avg}}{V_{(i-1)}(f)}$$

$$D = \frac{Q_{0i}}{Q_{0i} - Q_{0avg}}$$

$V_{(i-1)}(f)$ = volume in the basin at the end of time interval $i-1$, gal
Δt_i = time interval i, min
$C_{(i-1)}(f)$ = concentration in the basin and effluent concentration at the end of time interval $i-1$, mg/L

The TOC mass $M_{ei}(f)$ that left the equalization basin during each time interval can be calculated by

$$M_{ei}(f) = Q_{0avg} \int_0^{\Delta t_i} C_i dt = Q_{0avg}\left[C_{0i}\Delta t_i - \frac{A[(1+B\Delta t_i)^{1-D} - 1]}{B(1-D)}\right] \tag{5}$$

The procedure to developing the spreadsheet to calculate the masses leaving the basin at each time interval for a given volume of basin is:

1. The values for the first time interval (first row in the following table) are calculated as:
   - With the mean influent flow and concentration, the influent volume and mass of TOC are calculated.
   - The effluent flow rate, which is the same for all time intervals, is calculated by dividing the total influent volume by the total time.
   - For an initial basin volume (guess), the final volume is calculated by using Eq. (3).
   - For an initial concentration (guess), the final concentration is calculated by using Eq. (4).
   - The mass of TOC that left the basin is calculated by using Eq. (5).
2. The values for the other time intervals are calculated in a similar fashion but the initial volume and concentration for each time interval are the final volume and concentration of the previous time interval.

Time Period	Mean Flow, gpm	Mean TOC, mg/L	Influent			Effluent			Basin				
			Volume, gal	TOC, lb		Flow, gpm	TOC, lb		Volume		Concentration		
									Initial gal	Final gal		Initial mg/L	Final mg/L
8 to 10	450	920	54,000	414		672.5	673		278,200	251,500		1009	993
10 to 12	620	1130	74,400	701		672.5	681		251,500	245,200		993	1028
12 to 14	840	1475	100,800	1240		672.5	745		245,200	265,300		1028	1174
14 to 16	800	1525	96,000	1221		672.5	828		265,300	280,600		1174	1278
16 to 18	340	910	40,800	310		612.5	842		280,600	240,700		1278	1225
18 to 20	270	512	32,400	138		672.5	791		240,700	192,400		1225	1125
20 to 22	570	1210	68,400	690		672.5	767		192,400	180,100		1125	1151
22 to 24	1100	1520	132,000	1673		672.5	843		180,100	231,400		1151	1327
0 to 2	1200	1745	144,000	2096		672.5	960		231,400	294,700		1327	1504
2 to 4	800	820	96,000	657		672.5	946		294,700	310,000		1504	1318
4 to 6	510	410	61,200	209		672.5	829		310,000	290,500		1318	1150
6 to 8	570	490	68,400	280		672.5	725		290,500	278,200		1150	1009

Mean = 802 lb
Maximum = 960 lb
Peaking factor 1.20

Equalization Basin Design—Variable Volume

3. The volume guessed for the first time interval is modified until the maximum initial or final volume in the basin is the selected volume.
4. The concentration guessed for the first time interval is modified until it matches the final concentration for the last time interval.
5. The maximum and average masses leaving the basin are determined and the peaking factor is calculated as the ratio of the maximum to average values.

The results of the calculations for a basin volume of 310,000 gal are presented in the following table. Repetition of the calculations for different volumes allows the plotting of the peaking factor as a function of the basin volume as presented in Fig. 3.7.

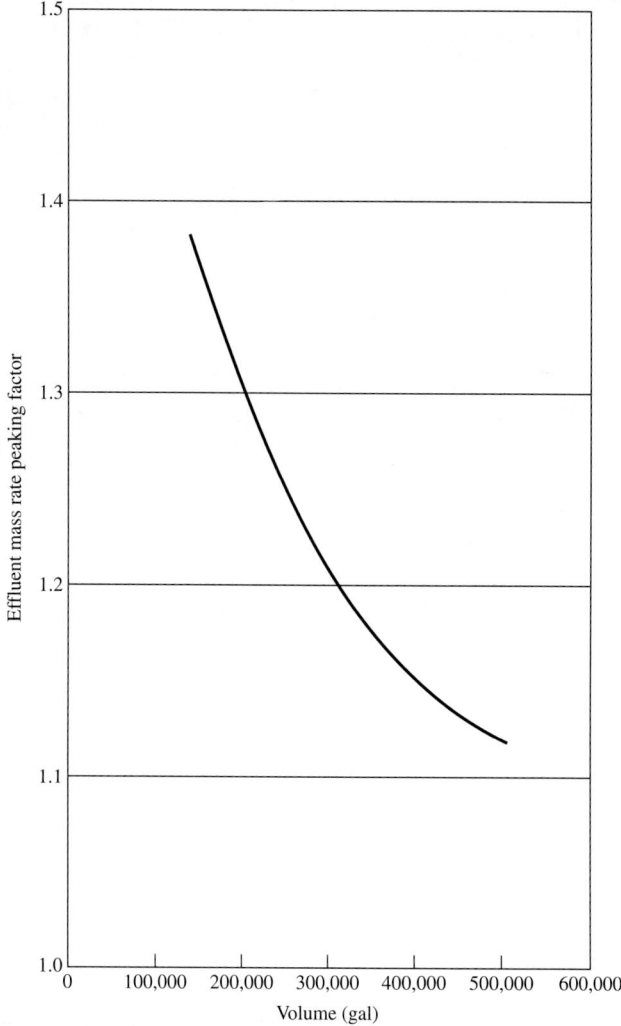

Figure 3.7 Variation in effluent mass rate peaking factor with basin volume for a variable-volume equalization basin.

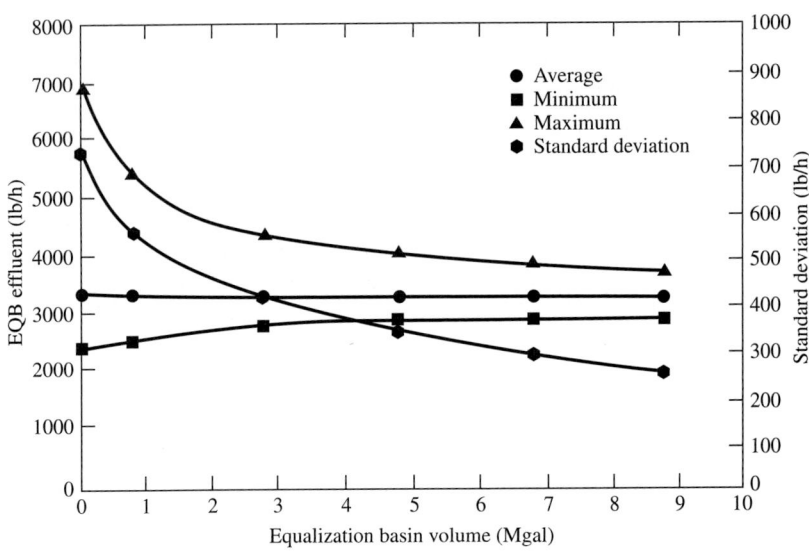

Figure 3.8a Effect of variable-volume equalization on EQB effluent BOD loading.

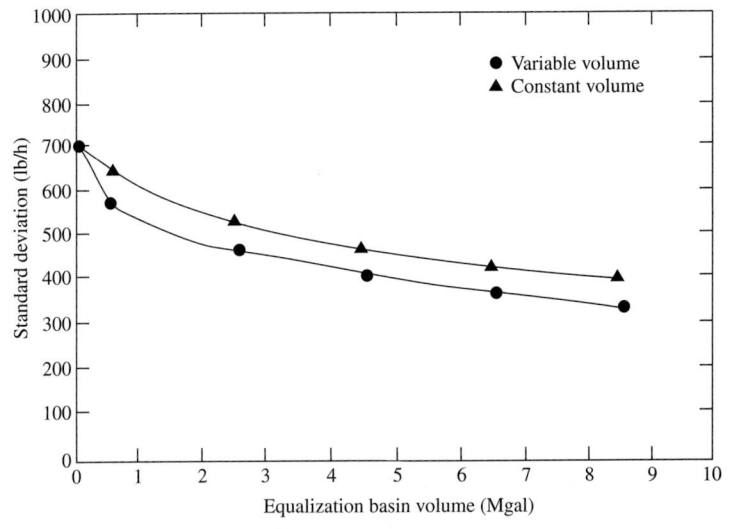

Figure 3.8b Comparison of variable-volume and constant-volume equalization on EQB effluent BOD loading.

In most cases with a variable flow, a variable-volume basin will be most effective, as shown in Fig. 3.8. Load balancing for a pharmaceutical wastewater is shown in Fig. 3.9.

When occasional dumps or spills are anticipated, for example, 1 percent of the time, a spill basin with an automatic bypass activated by a monitor should be used, as shown in Fig. 3.10. The spill basin

Pre- and Primary Treatment

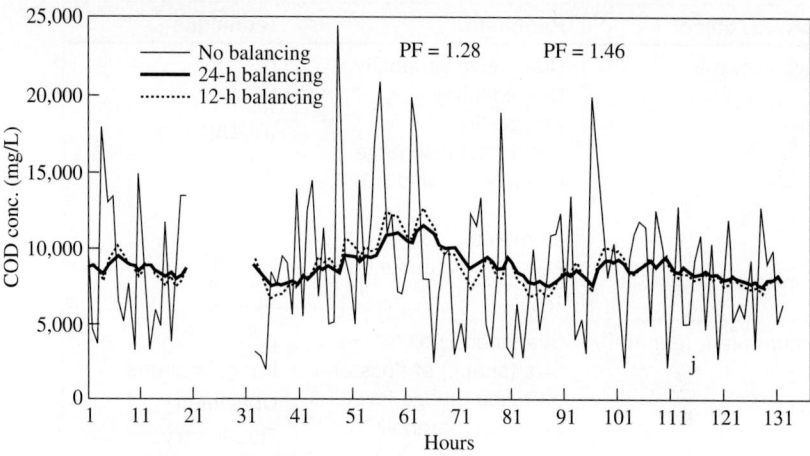

FIGURE 3.9 Load balancing analysis.

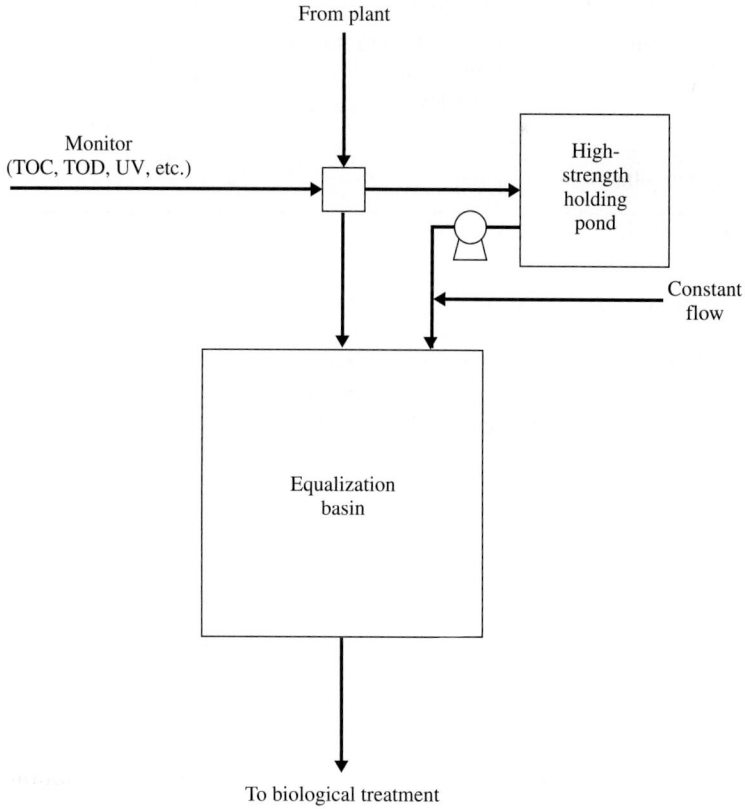

FIGURE 3.10 Use of a high-strength holding pond for spills.

Objectives/object	Parameter	Technique
Sewage network	Qualitative variability	UV
	Degradability,	"
	treatability	UV/UV*
	Accidental discharge	UV
	Phenols, sulphide, TOC	"
	ammonia	UV/UV*
	Benzenic compounds	Fluorescence
	TPH	IR
Treatment plant (general)	Suspended solids	UV-vis., NIR
	Size (shape) of flocs and	Image analysis
	particles Sludge level	Opacimetry
	Biomass metabolism	Fluorimetry
Treated effluent	COD, BOD, TOC, nitrate, phenol, sulphide	UV
External wastes	Variability, treatability, fingerprint	UV
Receiving medium	COD, TOC, nitrate, Water surface (oil)	UV
		NIR
	Turbidity	"

*Including a UV photodegradation step.

TABLE 3.3 Optical Monitoring Applications

may be required for surges in organics (e.g., TOC), TDS, temperature, or specific toxic compounds.

In Europe, monitoring and management of spills and other upset conditions in the process and treatment train have included optical methods as tabulated in Table 3.3. For some industries these have proven cost effective as compared to conventional methods. Typical design parameters for equalization basins are given in Table 3.4.

Detention time	12–24 h
Volume	Daily plant flow
Mixing requirement	0.02–0.04 hp/1000 gal
Maintain aerobic conditions	ORP > –100 mV
Depth	Approximately 15 ft
Freeboard	3 ft
Minimum operating level	5 ft

TABLE 3.4 Design Parameters for Equalization

3.3 Neutralization

Many industrial wastes contain acidic or alkaline materials that require neutralization prior to discharge to receiving waters or prior to chemical or biological treatment. For biological treatment, a pH in the biological system generally should be maintained between 6.5 and 8.5 to ensure optimum biological activity. The biological process itself provides a neutralization and a buffer capacity as a result of the production of CO_2, which reacts with caustic and acidic materials. The degree of preneutralization required depends, therefore, on the ratio of BOD removed and the causticity or acidity present in the waste. These requirements are discussed in Chap. 6. It should be noted that some organic acids, such as acetic or formic, may result in an increase in pH during biological treatment.

Types of Processes

Mixing Acidic and Alkaline Wastestreams
This process requires sufficient equalization capacity to effect the desired neutralization.

Acid Wastes Neutralization through Limestone Beds
These may be downflow or upflow systems. The maximum hydraulic rate for downflow systems is 1 gal/(min · ft²) (4.07×10^{-2} m³/[min · m²]) to ensure sufficient retention time. The acid concentration should be limited to 0.6 percent H_2SO_4 if H_2SO_4 is present to avoid coating of limestone with nonreactive $CaSO_4$ and excessive CO_2 evolution, which limits complete neutralization. High dilution or dolomitic limestone requires longer detention periods to effect neutralization. Hydraulic loading rates can be increased with upflow beds, because the products of reaction are swept out before precipitation. Since pH control is related to the bed depth, limestone beds are applicable only to wastewaters in which the influent acidity is relatively constant with time. A limestone bed system is shown in Fig. 3.11. The design of a limestone bed is shown in Example 3.4.

Example 3.4. A wastewater flow of 100 gpm (0.38 m³/min) with 0.1 N H_2SO_4 requires neutralization prior to secondary treatment. This flow is to be neutralized to a pH of 7.0 using a limestone bed. Figure 3.12 presents the results of a series of laboratory pilot tests using a 1-ft- (30.5-cm-) diameter limestone bed. These data are for upflow units, with the effluent being aerated to remove residual CO_2. Assume limestone is 60 percent reactive.

Design a neutralization system specifying:

(a) Most economical bed depth of limestone

(b) Weight of acid per day to be neutralized

(c) Limestone requirements on an annual basis

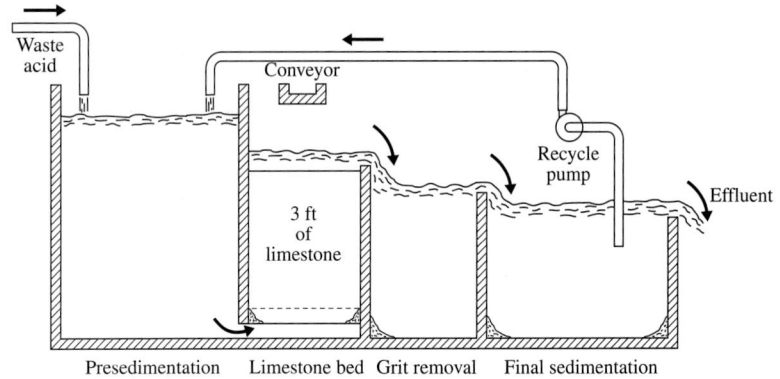

Figure 3.11 Simplified flow diagram of limestone neutralization. (*Adapted from Tully, 1958.*)

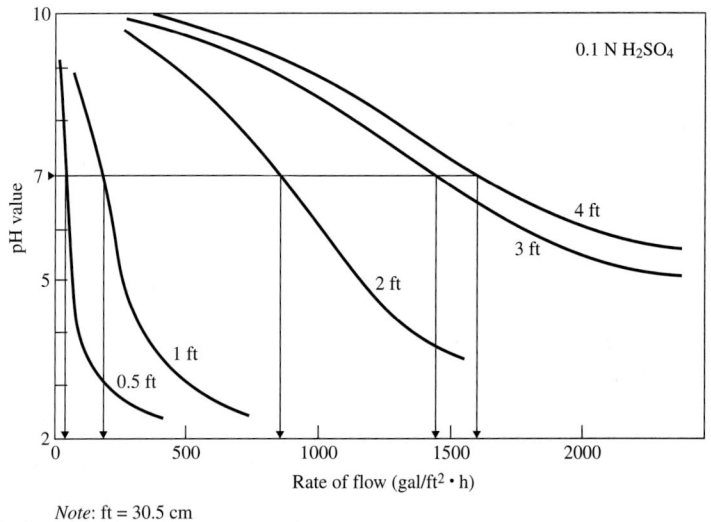

Note: ft = 30.5 cm

Figure 3.12 Allowable limestone bed loading rate vs. bed depth determination.

Solution

(a) *Most economical bed depth.* Hydraulic loadings to get pH 7.0 with various depths of limestone bed: From the figure the allowable hydraulic loadings are estimated to be:

Depth, ft	0.5	1.0	2.0	3.0	4.0
Hydraulic loading, gal/(ft² · h)	42	180	850	1440	1600

Note: ft = 30.5 cm
gal/(ft² · h) = 4.07 × 10⁻² m³/(m² · h)

Pre- and Primary Treatment

Flow rate per unit limestone volume: The required flow rate per unit volume of bed can be calculated by:

$$Q/V = \frac{\text{hydraulic loading}}{\text{bed depth}}$$

Depth, ft	0.5	1.0	2.0	3.0	4.0
Q/V, gal/(ft$^3 \cdot$ h)	84	180	425	480	400

Note: gal/(ft$^3 \cdot$ h) = 0.134 m^3/(m$^3 \cdot$ h)

By plotting the flow rate per unit limestone volume against the limestone bed depth, the most economical bed depth of limestone is found to be ≈ 3 ft (0.91 m). This is the depth that gives the maximum flow per unit volume; see Fig. 3.13.

(b) *Weight of acid per day to be neutralized.* The weight of acid can be calculated by

$$\frac{100 \text{ gal}}{\text{min}} \times \frac{4900 \text{ mg H}_2\text{SO}_4}{\text{L}} \times \frac{1440 \text{ min}}{\text{day}} \times \frac{8.34 \times 10^{-6} \text{ lb}}{(\text{mg/L}) \text{ gal}}$$

$$= 5890 \text{ lb/d} \quad (2670 \text{ kg/d})$$

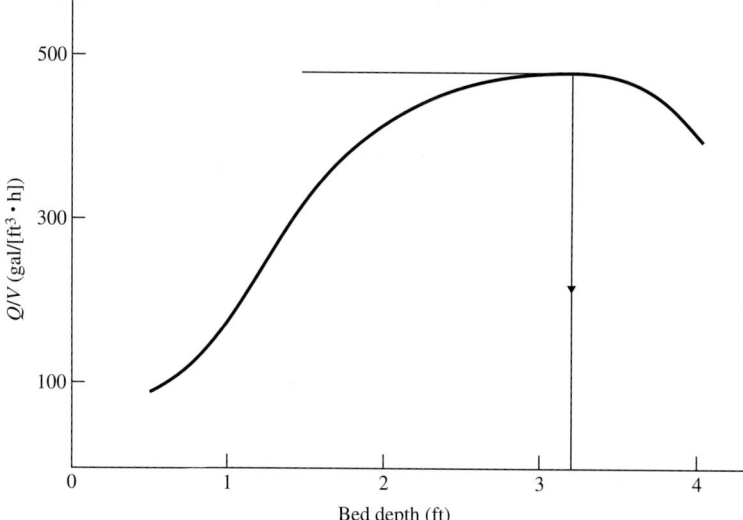

Note: ft = 30.5 cm
gal/(ft$^3 \cdot$ h) = 0.134 m^3/(m$^3 \cdot$ h)

FIGURE 3.13 Determination of optimal limestone bed depth.

(c) *Annual limestone requirements.* The limestone requirements can be calculated by:

$$\frac{5890 \text{ lb}}{\text{d}} \times \frac{50 \text{ g CaCO}_3}{49 \text{ g H}_2\text{SO}_4} \times \frac{365 \text{ d}}{\text{yr}} \times \frac{1 \text{ lb limestone}}{0.60 \text{ lb CaCO}_3}$$

$$= 3{,}660{,}000 \text{ lb/yr} \quad (1{,}660{,}000 \text{ kg/yr})$$

Mixing Acid Wastes with Lime Slurries

This neutralization depends on the type of lime used. The magnesium fraction of lime is most reactive in strongly acid solutions and is useful below pH 4.2. Neutralization with lime can be defined by a basicity factor obtained by titration of a 1-g sample with an excess of HCl, boiling 15 min, followed by back titration with 0.5 N NaOH to the phenolphthalein endpoint.

In lime slaking, the reaction is accelerated by heat and physical agitation. For high reactivity, the lime reaction is complete in 10 min. Storage of lime slurry for a few hours before neutralization may be beneficial. Dolomitic quicklime (only the CaO portion) hydrates except at elevated temperature. Slaked quicklime is used as an 8 to 15 percent lime slurry. Neutralization can also be accomplished by using NaOH, Na_2CO_3, NH_4OH, or $Mg(OH)_2$.

Basic (Alkaline) Wastes

Any strong acid can be used effectively to neutralize alkaline wastes, but cost considerations usually limit the choice to sulfuric or possibly hydrochloric acid. The reaction rates are practically instantaneous, as with strong bases.

Flue gases, which may contain 14 percent CO_2, can be used for neutralization. When bubbled through the waste, the CO_2 forms carbonic acid, which then reacts with the base. The reaction rate is somewhat slower, but is sufficient if the pH need not be adjusted to below 7 to 8. Another approach is to use a spray tower in which the stack gases are passed countercurrent to the waste liquid droplets.

All of the above processes usually work better with the stepwise addition of reagents, that is, a staged operation. Two stages, with possibly a third tank to even out any remaining fluctuations, are generally optimum.

There are a number of neutralizing agents available. Selection criteria should consider:

- Type and availability
- Reaction rate
- Sludge production and disposal
- Safety and ease of handling for addition and storage
- Total cost including chemical feed and feed and storage equipment

- Side reactions, including dissolved salts, scale formation, and heat produced
- The effect of overdosage

The primary neutralizing agents are:

Basic agents

- Lime in various forms—strong
- Caustic soda—strong
- Magnesium hydroxide—medium
- Sodium carbonate—weak
- Sodium bicarbonate—weak

Acidic agents

- Sulfuric acid—strong
- Carbon dioxide—weak

The characteristics of typical neutralizing chemicals are shown in Tables 3.5 and 3.6.

System

Batch treatment is used for waste flows to 100,000 gal/d (380 m^3/d). Continuous treatment employs automated pH control. Where air is used for mixing, the minimum air rate is 1 to 3 ft^3/(min · ft^2) (0.3 to 0.9 m^3/[min · m^2]) at 9 ft (2.7 m) liquid depth. If mechanical mixers are used, 0.2 to 0.4 hp/thousand gal (0.04 to 0.08 kW/m^3) is required.

Control of Process

The automatic control of pH for wastestreams is one of the most troublesome, for the following reasons:

1. The relation between pH and concentration or reagent flow is highly nonlinear for strong acid–strong base neutralization, particularly when close to neutral (pH 7.0). The nature of the titration curve as shown in Fig. 3.14 favors multistaging in order to ensure close control of the pH.
2. The influent pH can vary at a rate as fast as 1 pH unit per minute.
3. The wastestream flow rates can double in a few minutes.
4. A relatively small amount of reagent must be thoroughly mixed with a large liquid volume in a short time interval.
5. Changes in buffer capacity (i.e., alkalinity or acidity) will change neutralization requirements.

Property	Calcium Carbonate ($CaCO_3$)	Calcium Hydroxide [$Ca(OH)_2$]	Calcium Oxide (CaO)	Hydrochloric Acid (HCl)	Sodium Carbonate (Na_2CO_3)	Sodium Hydroxide (NaOH)	Sulfuric Acid (H_2SO_4)
Available form	Powder, crushed (various sizes)	Powder, granules	Lump, pebble, ground	Liquid	Powder	Solid flake, ground flake, liquid	Liquid
Shipping container	Bags, barrel, bulk	Bags (50 lb),* bulk	Bags (80 lb), barrels, bulk	Barrels, drums, bulk	Bags (100 lb), bulk	Drum (735, 100, 450 lb)	Carboys, drums (825 lb), bulk
Bulk weight, lb/ft^3	Powder 48 to 71; crushed 70 to 100	25 to 50	40 to 70	27.9%, 0.53 lb/gal†; 31.45%, 9.65 lb/gal	34 to 62	Varies	106, 114
Commercial strength	—	Normally 13% $Ca(OH)_2$	75 to 99%, normally 90% CaO	27.9, 31.45, 35.2%	99.2%	98%	60° Be, 77.7%; 66° Be, 93.2%
Water solubility, lb/gal	Nearly insoluble	Nearly insoluble	Nearly insoluble	Complete	0.58 @ 32°F, 1.04 @ 50°F, 1.79 @ 68°F, 3.33 @ 86°F	3.5 @ 32°F, 4.3 @ 50°F, 9.1 @ 68°F, 9.2 @ 86°F	Complete
Feeding form	Dry slurry used in fixed beds	Dry or slurry	Dry or slurry (must be slaked to $Ca(OH)_2$)	Liquid	Dry, liquid	Solution	Liquid

Feeding type	Volumetric pump	Volumetric metering pump	Dry-volumetric, wet slurry (centrifugal pump)	Metering pump	Volumetric feeder, metering pump	Metering pump	Metering pump
Accessory equipment	Slurry tank	Slurry tank	Slurry tank, slaker	Dilution tank	Dissolving tank	Solution tank	—
Suitable handling materials	Iron, steel	Iron, steel, plastic, rubber hose	Iron, steel, plastic, rubber hose	Hastelloy A, selected plastic and rubber types	Iron, steel	Iron, steel	—
Comments	—	—	Provide means for cleaning slurry transfer pipes	—	Can cake	Dissolving solid forms generate much heat	Provide for spill cleanup and neutralization

*lb × 0.4536 = kg
†lb/gal × 0.1198 = kg/L
‡0.555 (°F − 32) = °C

TABLE 3.5 Summary of Properties for Typical Neutralization Chemicals

Chemical	Formula	Equivalent Weight	To neutralize 1 mg/L Acidity or Alkalinity (Expressed as CaCO$_3$) Requires n mg/L	Neutralization Factor, Assuming 100% Purity for all Compounds
				Basicity
Calcium carbonate	CaCO$_3$	50	1.000	1.000/0.56 = 1.786
Calcium oxide	CaO	28	0.560	0.560/0.56 = 1.000
Calcium hydroxide	Ca(OH)$_2$	37	0.740	0.740/0.56 = 1.321
Magnesium oxide	MgO	20	0.403	0.403/0.56 = 0.720
Magnesium hydroxide	Mg(OH)$_2$	29	0.583	0.583/0.56 = 1.041
Dolomitic quicklime	(CaO)$_{0.6}$(MgO)$_{0.4}$	24.8	0.497	0.497/0.56 = 0.888
Dolomitic hydrated lime	[Ca(OH)$_2$]$_{0.6}$[Mg(OH)$_2$]$_{0.4}$	33.8	0.677	0.677/0.56 = 1.209
Sodium hydroxide	NaOH	40	0.799	0.799/0.56 = 1.427
Sodium carbonate	Na$_2$CO$_3$	53	1.059	1.059/0.56 = 1.891
Sodium bicarbonate	NaHCO$_3$	84	1.680	1.680/0.56 = 3.00
				Acidity
Sulfuric acid	H$_2$SO$_4$	49	0.980	0.980/0.56 = 1.750
Hydrochloric acid	HCl	36	0.720	0.720/0.56 = 1.285
Nitric acid	HNO$_3$	62	1.260	1.260/0.56 = 2.250
Carbonic acid	H$_2$CO$_3$	31	0.620	0.620/0.56 5 1.107

TABLE 3.6 Neutralization Factors for Common Alkaline and Acid Reagents

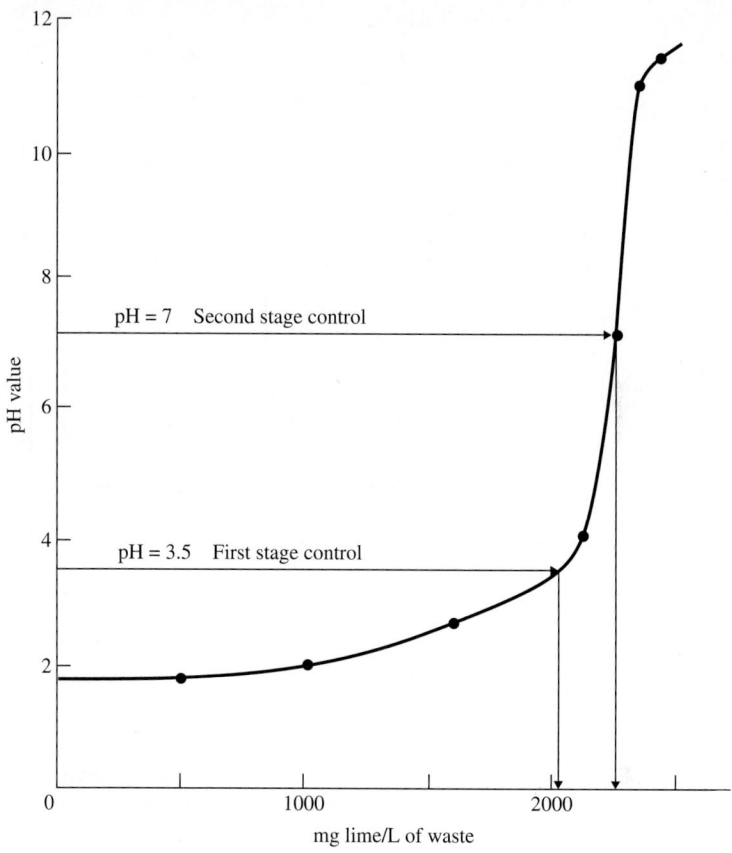

FIGURE 3.14 Lime-waste titration curve for strong acid.

Advantage is usually gained by the stepwise addition of chemicals (see Fig. 3.15). In reaction tank 1, the pH may be raised to 3 to 4. Reaction tank 2 raises the pH to 5 to 6 (or any other desired endpoint). If the wastestream is subject to slugs or spills, a third reaction tank may be desirable to effect complete neutralization. Okey et al.[5] have shown an advantage to combining NaOH and $NaHCO_3$. NaOH is employed in the first stage and $NaHCO_3$ in a second stage to provide a buffer and avoid pH swings due to a change in influent characteristics. Neutralization system design parameters are shown in Table 3.7. Acid waste neutralization is shown in Example 3.5.

Example 3.5. A wastewater (100 gal/min) (0.38 m³/min) is highly acidic and requires neutralization prior to secondary treatment. This flow is to be neutralized to a pH of 7.0 by lime. The titration curve for the wastewater is shown in Fig. 3.14, from which a two-stage control neutralization system will be used with a total lime consumption of 2250 mg/L. The first stage requires 2000 mg/L and the second stage 250 mg/L.

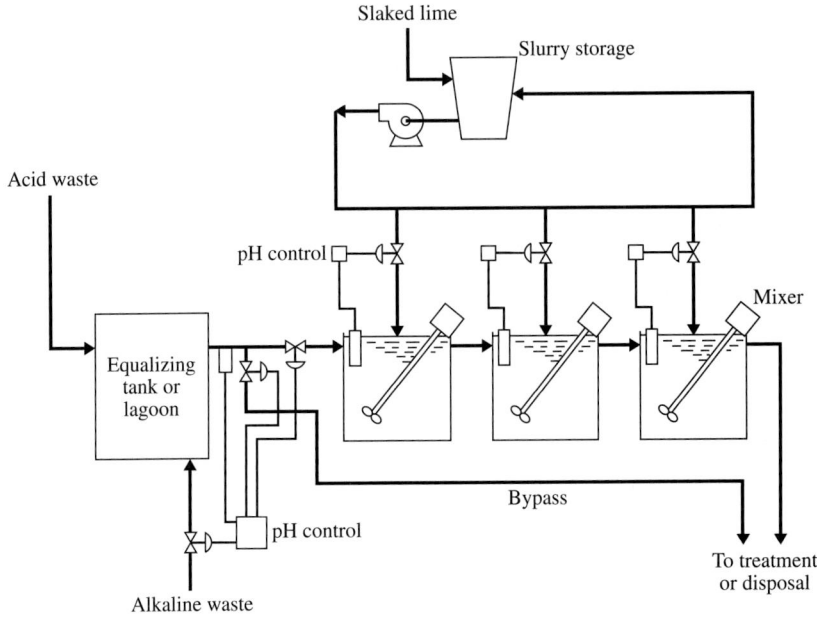

FIGURE 3.15 Multistage neutralization process.

Chemical storage tank	Liquid—use stored supply vessel Dry—dilute in a mix or day tank

Reaction tank:
  Size
 	
  Retention time
  Influent
  Effluent
	
	Cubic or cylindrical with liquid depth equal to diameter
	5 to 30 min (lime—30 min)
	Locate at tank top
	Locate at tank bottom

Agitator:
  Propeller type
  Axial-flow type
  Peripheral speeds
	
	Under 1000-gal tanks
	Over 1000-gal tanks
	12 ft/s for large tanks
	25 ft/s for tanks less than 1000 gal

pH sensor	Submersible preferred to flow-through type
Metering pump or control valve	Pump delivery range limited to 10 to 1; valves have greater ranges.

Note: The selection of neutralizing agent will depend on availability, chemical cost, and feeding methods.

TABLE 3.7 Neutralization System Design Parameters

The average lime dosage in the first stage is

(100 gal/min)(1440 min/d)(8.34 lb/million gal/mg/L)(2000 mg/L)

$\times 10^{-6}$ (million gal/gal) = 2400 lb/d   (1090 kg/d)

The average lime dosage in the second stage is

$100 \times 1440 \times 8.34 \times 250 \times 10^{-6} = 300$ lb/d   (140 kg/d)

The average lime dosage is 2700 lb/d (1.35 ton/d) (1230 kg/d). With this dosage and type of lime each basin should be designed with a detention time of 5 min.

Volume = 100 gal/min × 5 min = 500 gal (1.9 m³)

Use two tanks, 4.6 ft (1.40 m) diameter × 4.1 ft (1.25 m) deep, so that diameter and depth are approximately equal.

To maintain proper mixing in the reactor tanks, the power level required for 5 min detention time, $D/T = 0.33$, is 0.2 hp/thousand gal (40 W/m³) (Fig. 3.16). Use 0.1-hp (75-W) mixers in each reaction tank.

Either one or two standard wall baffles, 180° apart, and one-twelfth to one-twentieth of the width of the tank diameter, located 24 in (61 cm) from the periphery of the impeller, are recommended for this operation.

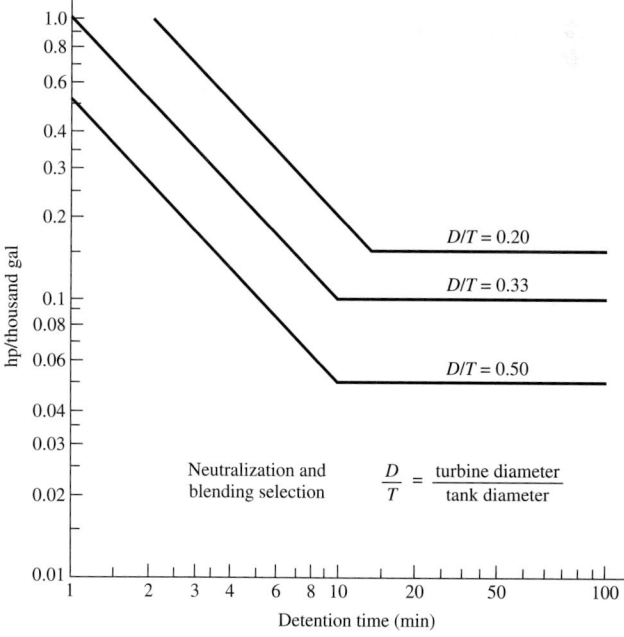

FIGURE 3.16 Neutralizing tank power requirements.

3.4 Sedimentation

Sedimentation is employed for the removal of suspended solids from wastewaters. The process can be considered in three basic classifications, depending on the nature of the solids present in the suspension: discrete, flocculent, and zone settling. In discrete settling, the particle maintains its individuality and does not change in size, shape, or density during the settling process. Flocculent settling occurs when the particles agglomerate during the settling period with a resulting change in size and settling rate. Zone settling involves a flocculated suspension which forms a lattice structure and settles as a mass, exhibiting a distinct interface during the settling process. Compaction of the settled sludge occurs in all sedimentation but will be considered separately under thickening.

Discrete Settling

A particle will settle when the impelling force of gravity exceeds the inertia and viscous forces. The terminal settling velocity of a particle is defined by the relationship

$$v = \sqrt{\frac{4g(\rho_s - \rho_l)D}{3C_d \rho_l}} \qquad (3.3)$$

where ρ_l = density of the fluid
ρ_s = density of the particle
v = terminal settling velocity of the particle
D = diameter of the particle
C_d = drag coefficient, which is related to the Reynolds number and particle shape
g = acceleration due to gravity

When the Reynolds number is small, i.e., less than 1.0 (small particles at low velocity), the viscous forces are predominant and

$$C_d = \frac{24}{N_{Re}} \qquad (3.4)$$

where

$$N_{Re} = \frac{vD\rho_l}{\mu} \qquad (3.5)$$

ρ_l and μ are the density and viscosity of the liquid, respectively. Substitution of Eq. (3.4) in Eq. (3.3) yields Stokes' law:

$$v = \frac{\rho_s - \rho_l}{18\mu} gD^2 \qquad (3.6)$$

As the Reynolds number increases, a transition zone occurs in which both inertia and viscous forces are effective. This occurs over a Reynolds number range of 1 to 1000, where

$$C_d = \frac{18.5}{N_{Re}^{0.6}} \qquad (3.7)$$

Above a Reynolds number of 1000, viscous forces are not significant and the coefficient of drag is constant at 0.4.

The settling velocities of discrete particles, as related to diameter and specific gravity, are shown in Fig. 3.17.

Hazen[6] and Camp[7] developed relationships applicable to the removal of discrete particles in an ideal settling tank, based on the premises that the particles entering the tanks are uniformly distributed over the influent cross section and that a particle is considered removed when it hits the bottom of the tank. The settling velocity of a particle that settles through a distance equal to the effective depth of the tank in the theoretical detention period can be considered as an overflow rate:

$$v_o = \frac{Q}{A} \qquad (3.8)$$

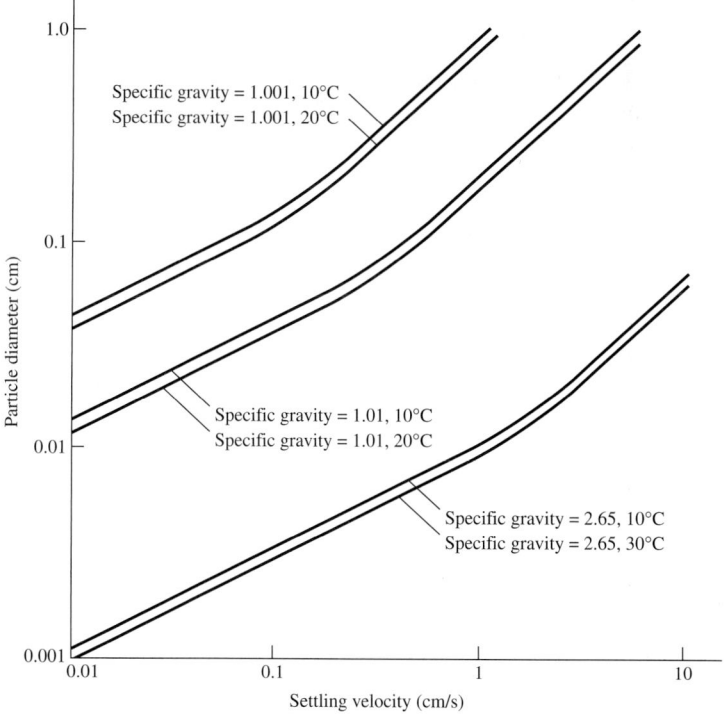

Figure 3.17 Settling properties of discrete particles.

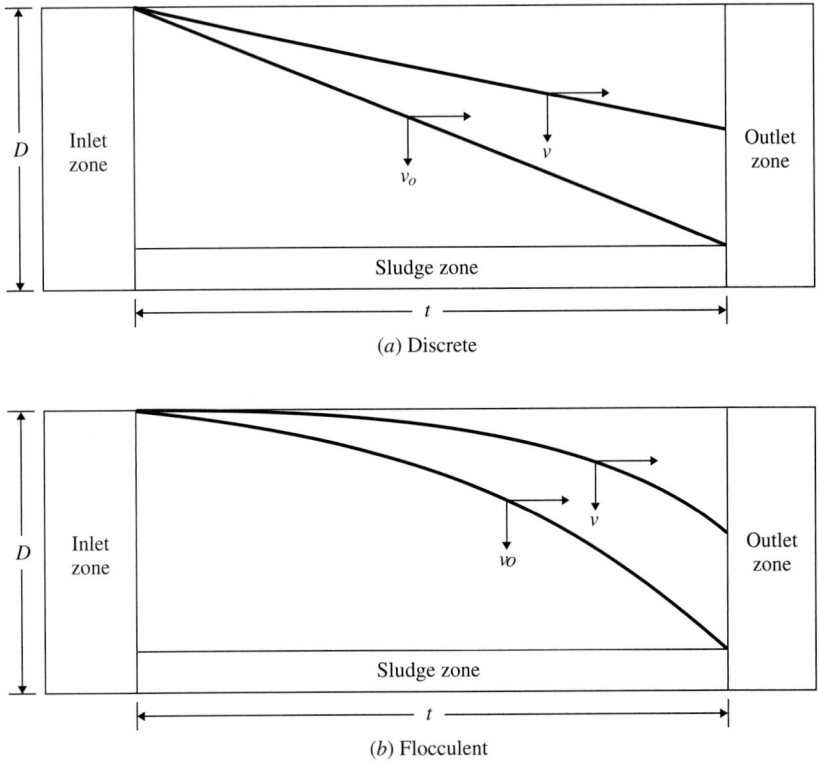

FIGURE 3.18 Ideal settling tank.

where Q = rate of flow through the tank and A = tank surface area. All particles with settling velocities greater than v_o will be completely removed, and particles with settling velocities less than v_o will be removed in the ratio v/v_o, as shown in Fig. 3.18. The removal of discrete particles is independent of tank depth and is a function of the overflow rate only.

When the suspension to be removed has a wide range of particle sizes, the total removal is defined by the relationship

$$\text{Total removal} = (1 - f_o) + \frac{1}{v_o} \int_0^{f_o} v \, df \qquad (3.9)$$

where f_o is the fraction of particles having a settling velocity equal to or less than v_o. Equation (3.9) must usually be solved by graphical integration to obtain the total removal.

The foregoing analysis is based on the performance of an ideal settling tank under quiescent conditions. In practice, however, short circuiting, turbulence, and bottom scour will affect the degree of solids removal. Dobbins[8] and Camp[7] have developed a relationship to compensate for the decrease in removals caused by turbulence (Fig. 3.19).

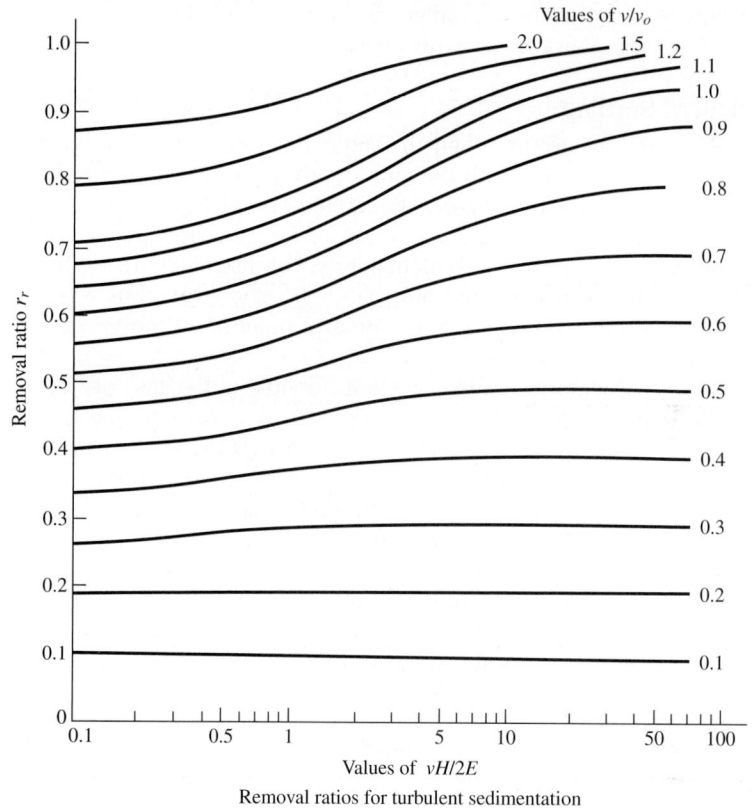

FIGURE 3.19 Effect of turbulence on the subsidence of particles. (*After Dobbins*[8])

Figure 3.19 shows the effect of turbulence in reducing the removal ratio of particles having a settling velocity v from the theoretical value of v/v_o, where v_o refers to the overflow rate. The ratio $vH/2E$ is a parameter of turbulent intensity, where H is the depth and E is a coefficient of turbulent transport. For narrow channels, Camp has shown the ratio to be equal to $122v/V$, where V is the average channel velocity.

Scour occurs when the flow-through velocity is sufficient to resuspend previously settled particles; it is defined by the relationship:

$$v_c = \sqrt{\frac{8\beta}{f} gD(S-1)} \qquad (3.10)$$

where v_c = velocity of scour
β = constant (0.04 for unigranular sand, 0.06 for nonuniform sticky material)
f = Weisbach-D'Arcy friction factor, 0.03 for concrete
S = particle specific gravity
g = acceleration due to gravity

Flocculent Settling

Flocculent settling occurs when the settling velocity of the particle increases as it settles through the tank depth, because of coalescence with other particles. This increases the settling rate, yielding a curvilinear settling path, as shown in Fig. 3.18b. Most of the suspended solids in industrial wastes are of a flocculent nature. For discrete particles, the efficiency of removal is related only to the overflow rate, but when flocculation occurs, both overflow rate and detention time become significant.

Since a mathematical analysis is not possible in the case of flocculent suspensions, a laboratory settling analysis is required to establish the necessary parameters. The laboratory settling study can be conducted in a column of the type shown in Fig. 3.20. A minimum diameter of 5 in (12.7 cm) is recommended to minimize wall effects. Taps are located at 2-ft (0.61-m) depth intervals.

The concentration of suspended solids must be uniform at the start of the test; sparging air into the bottom of the column for a few minutes will accomplish this. It is also essential that the temperature be maintained constant throughout the test period to eliminate settling interference by thermal currents. Suspended solids are determined on samples drawn off at selected time intervals up to 120 min. The data collected from the 2-ft (0.61 m), 4-ft (1.22 m), and 6-ft (1.83-m) depth taps are used to develop the settling rate–time relationships.

The results obtained are expressed in terms of percent removal of suspended solids at each tap and time interval. These removals are then plotted against their respective depths and times, as shown in Fig. 3.21. Smooth curves are drawn connecting points of equal removal. The curves thus drawn represent the limiting or maximum settling path for the indicated percent; that is, the specified percent suspended solids will have a net settling velocity equal to or greater than that shown, and would therefore be removed in an ideal settling tank of the same depth and detention time. The calculation of removal can be illustrated from the data of Fig. 3.21.

The overflow rate v_o is the effective depth, 6 ft (1.83 m), divided by the time required for a given percent to settle this distance. All particles having a settling velocity equal to or greater than v_o will be 100 percent removed. Particles with a lesser settling velocity v will be removed in the proportion v/v_o. For example, referring to Fig. 3.21a, at a detention period of 60 min and a 6-ft (1.83-m) settling depth ($v_o = 6$ ft/h [1.83 m/h]), 50 percent of the suspended solids are completely removed; that is, 50 percent of the particles have a settling velocity equal to or greater than 6 ft/h (1.83 m/h). Particles in each additional 10 percent range will be removed in the proportion v/v_o or in the proportion to the average

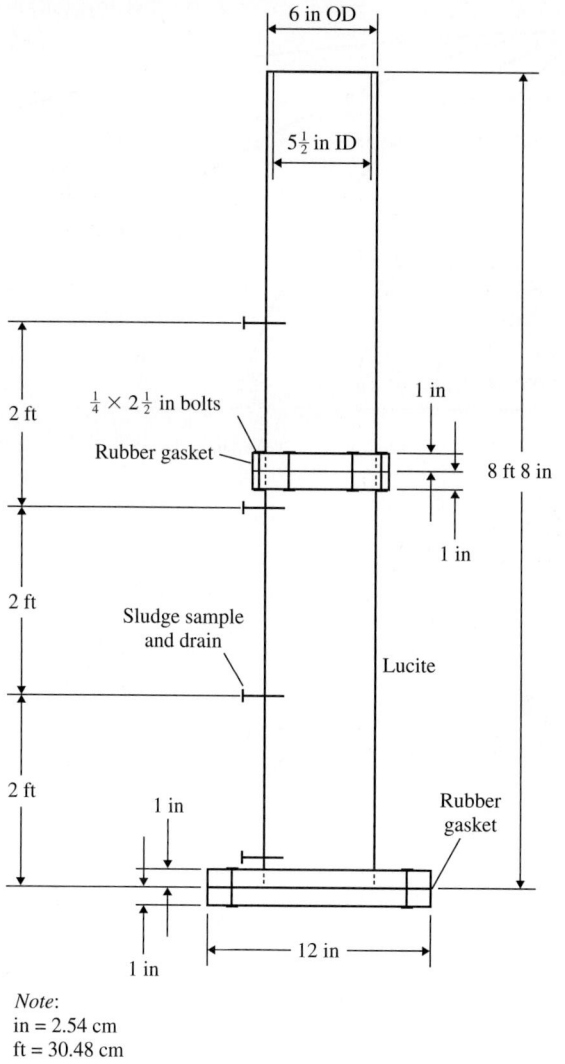

FIGURE 3.20 Laboratory settling column for the evaluation of flocculent settling.

depth settled to the total depth of 6 ft (1.83 m). The average depth to which the 50 to 60 percent range has settled in Fig. 3.21b is 3.8 ft (1.16 m). The percent removal of this fraction is therefore 3.8 ft/6.0 ft (1.16 m/1.83 m), or 63 percent of the 10 percent. Each subsequent percent range is computed in a similar manner, and the total removal developed as shown in Table 3.8.

The total removal of suspended solids of 62.4 percent can be accomplished at an overflow rate of 6 ft/h = 1080 gal/(d · ft²) (44 m³/[d · m²]) at a retention period of 60 min. In a similar manner various percent

100 Chapter Three

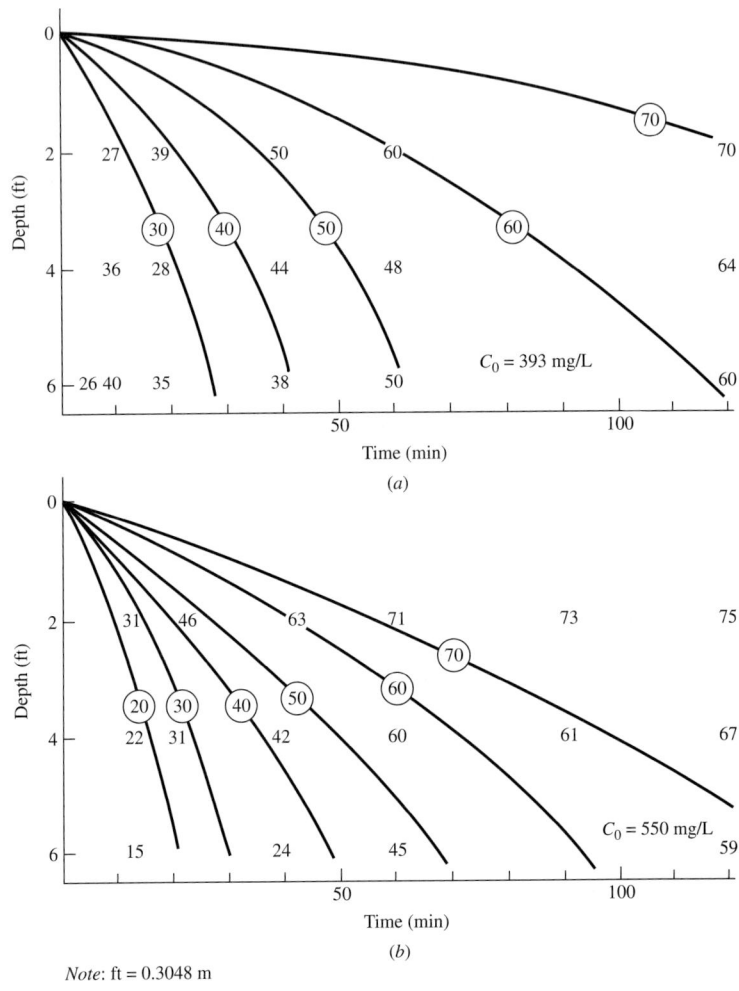

FIGURE 3.21 Flocculent settling relationships.

SS Range, %	d/d_o	SS Removal, %
0–50	1.0	50
50–60	0.64	6.4
60–70	0.25	2.5
70–100	0.05	1.5
	Total removal	62.4

TABLE 3.8 Percent Ranges for Total Removal of Suspended Solids

removals and their associated overflow rates and detention periods can be computed.

Since the degree of flocculation is influenced by the initial concentration of suspended solids, settling tests should be run over the anticipated range of suspended solids in the influent waste. In many wastes, a fraction of the measured suspended solids is not removed by settling, so the curves developed from the laboratory analysis will approach the removal as a limit.

Since the data obtained from the laboratory analysis represent ideal settling conditions, criteria for prototype design must account for the effects of turbulence, short circuiting, and inlet and outlet losses. The net effect of these factors is a decrease in the overflow rate and an increase in the detention time over that derived from the laboratory analysis. As a general rule, the overflow rate will be decreased by a factor of 1.25 to 1.75, and the detention period increased by a factor of 1.50 to 2.00. The development of these relationships is shown in Example 3.6.

Example 3.6. Laboratory data were obtained on the settling of a paper mill waste (Table 3.9). Design a settling tank to produce a maximum effluent suspended solids of 150 mg/L.

Time, min	Removal, %		
	2 ft	4 ft	6 ft
Initial solids, 393 mg/L			
5			26
10	27	36	40
20	39	28	35
40	50	44	38
60	60	48	50
120	70	64	60
Initial solids, 550 mg/L			
15	31	22	15
20	46	31	
40	63	42	24
60	71	60	45
90	73	61	
120	75	67	59

Note: ft = 30.48 cm

TABLE 3.9 Settling Data

Time, min	Velocity, ft/h	Removal of SS, %	Overflow Rate, gal/(d · ft²)
C_0 = 393 mg/L			
27.5	13.1	41.6	2360
42.0	8.6	49.5	1550
62.0	5.8	57.3	1050
115.0	3.1	65.8	560
C_0 = 550 mg/L			
20	18.0	36.7	3250
27	13.3	46.8	2400
47	7.7	56.5	1400
66	5.4	62.5	980
83	3.8	70.8	690

Note: ft/h = 0.305 m/h
gal/(d · ft²) = 4.07 × 10^{-2} m³/(d · m²)

TABLE 3.10 Overflow Rate and % ISS Removal as Function of Time

Solution. With the values given in Table 3.9, Fig. 3.21 can be drawn and Table 3.10 can be constructed by calculating the percent removal, the velocity, and the overflow rate of different times as indicated above.

By plotting the SS removal versus overflow rate and time, Figs. 3.22 and 3.23 can be drawn.

From Figs. 3.22 and 3.23:

For 393 mg/L:

$$\text{Percent removal} = \frac{393-150}{393} \times 100 = 62$$

Overflow rate = 770 gal/(d · ft²)(31.4 m³/[d · m²])

Detention time = 74 min

For 550 mg/L:

$$\text{Percent removal} = \frac{550-150}{550} \times 100 = 73$$

Overflow rate = 540 gal/(d · ft²)(22.0 m³/[d · m²])

Detention time = 104 min

Design:

$$\text{Overflow rate} = \frac{540}{1.5} = 360 \text{ gal/(d · ft²)}(14.7 \text{ m}^3/[\text{d · m}^2])$$

$$\text{Detention time } t = \frac{104 \text{ min}}{60 \text{ min/h}} \times 1.75 = 3 \text{ h}$$

For 1 million gal/d:

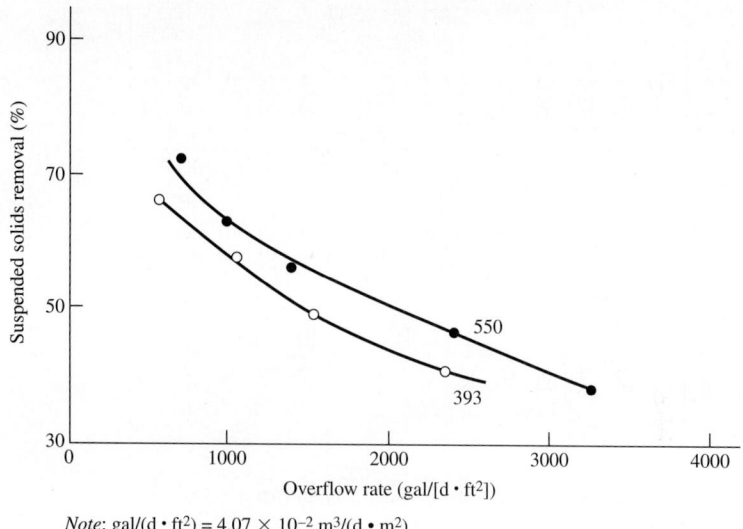

FIGURE 3.22 % TSS removal vs. overflow rate.

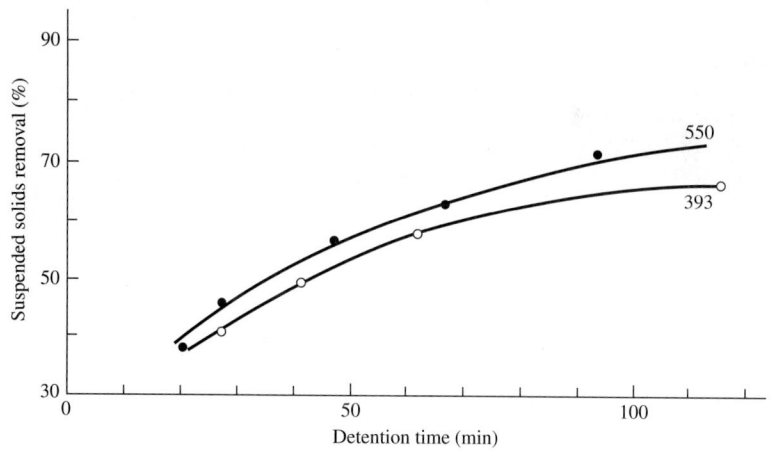

FIGURE 3.23 % TSS removal vs. detention time.

$$\text{Area } A = \frac{\text{flow}}{\text{overflow rate}} = \frac{10^6 \text{ gal/d}}{360 \text{ gal/(d} \cdot \text{ft}^2)} = 2780 \text{ ft}^2 \ (258 \text{ m}^2)$$

$$\text{Effective depth} = \frac{t \times \text{flow}}{A}$$

$$= \frac{3 \text{ h} \times 10^6 \text{ gal/d}}{2780 \text{ ft}^2} \times \frac{0.134 \text{ ft}^3/\text{gal}}{24 \text{ h/d}}$$

$$= 6 \text{ ft } (1.8 \text{ m})$$

The sedimentation performances for various pulp and paper-mill wastes are summarized in Table 3.11.

Type of Waste	Flow, million gal/day	Raw SS, ppm	Raw BOD, ppm	Temperature, °F	Removal %		Detention Time, h	OR, gal/(d·ft²)
					SS	BOD		
Paperboard	4.5	2,500	450	85	90	67	5.35	504
	0.75	136		85	90	50	1.15	940
	1.36	10,000	360	62	85	24	5.40	430
	2.5	1,185	395		96.1	19	5.3	525
	31	524	195	110	42	25	9.4	438
	30	850	250	95	80	25	0.5	1910
	3.3	2,000		90	85		2.6	1028
	0.25	50	100	100	80	25	4.5	39
	0.301	1,150	250	110	98	50		90
	35	4,000	200	100	90	10–15	1.5	374
Specialty	9.4	203	97	81	94	86	2.56	832
	2.2	6,215	120	120		90	1.5	157
	1.8	665	620	95	91	58	0.5	406
	50	120	85	100	80	16	18.2	477

Fine paper	6	200		65	95	90	3.9	695
	6.0	254	235	90	50	34	2.2	890
	9.9	500	364	70–100	90	35	2.4	1120
	3.5	300	250	65	95	48	6.0	372
	7.5–9.0	560	126	65	80	42	4.0	670
Miscellaneous	7	430	250	70	70	20	1.8	505
	14	1,000	330	73	65	60		911
	25	75	100		90	0.0	6.9	17
	17	100	425		95	50	5.9	846
	0.5	200	200	85	90		1.9	1590
	1.0	1,000	900	100		95	2.9	509

Note: million gal/d = 3.75×10^3 m^3/d.

°C = $\frac{5}{9}$ (°F − 32).

gal/(d · ft^2) = 4.07×10^{-2} m^3/(d · m^2).

Source: Adapted from Committee on Industrial Waste Practice of SED.

TABLE 3.11 Settling Characteristics of Pulp and Paper-Mill Wastes[9]

Tanning wastewaters[10] with an initial suspended solids of 1200 mg/L showed 69 percent reduction with a retention period of 2 h. BOD removals of 86.9 percent were obtained from the settling of cornstarch wastes.[11]

The effect of the initial concentration of suspended solids on the sedimentation efficiency of a pulpwood waste is shown in Fig. 3.24.

Zone Settling

Zone settling is characterized by activated sludge and flocculated chemical suspensions when the concentration of solids exceeds approximately 500 mg/L. The floc particles adhere together and the mass settles as a blanket, forming a distinct interface between the floc and the supernatant. The settling process is distinguished by four zones, as shown in Fig. 3.25. Initially, all the sludge is at a uniform concentration A, as shown in Fig. 3.25.

During the initial settling period the sludge settles at a uniform velocity. The settling rate is a function of the initial solids concentration A. As settling proceeds, the collapsed solids D on the bottom of the settling unit build up at a constant rate. C is a zone of transition through which the settling velocity decreases as the result of an increasing concentration of solids. The concentration of solids in the zone settling layer remains constant until the settling interface approaches the rising layers of collapsed solids, III, and a transition zone occurs. Through the transition zone C, the settling velocity will decrease because of the increasing density and viscosity of the suspension surrounding the particles. When the rising layer of settled solids reaches the interface, a compression zone occurs in stage IV.

In the separation of flocculent suspensions, both clarification of the liquid overflow and thickening of the sludge underflow are involved. The overflow rate for clarification requires that the average rise velocity of the liquid overflowing the tank be less than the zone settling velocity of the suspension. The tank surface area requirements for thickening the underflow to a desired concentration level are related to the solids loading to the unit and are usually expressed in terms of a mass loading (pounds of solids per square foot per day or kilograms per square meter per day) or a unit area (square feet per pound of solids per day or square meters per kilogram per day).

The mass loading concept for the thickening of industrial sludges is developed in Chap. 11.

Laboratory Evaluation of Zone Settling and Calculation of Solids Flux

The settling properties of flocculated sludges can be evaluated in a liter cylinder equipped with a slow-speed stirrer rotating at a speed of 4 to 5 revolutions per hour (r/h). The effect of the stirrer is to simulate the hydraulic motion and rake action in a clarifier and to break the stratification and arching action of the settling sludge. In some cases,

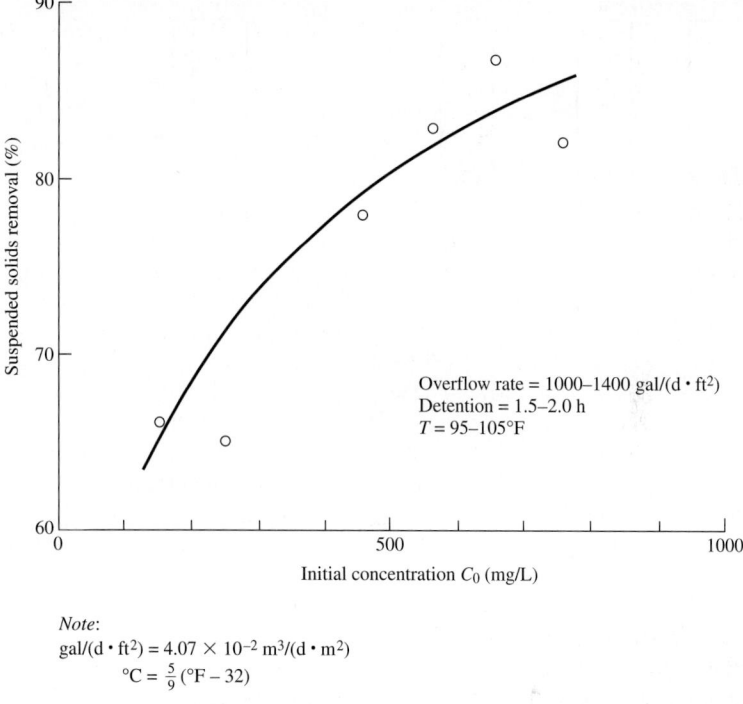

FIGURE 3.24 Effect of initial suspended solids concentration on percent removal from a pulp and paper-mill wastewater.

an initial flocculation of the sludge occurs when it is added to the graduated cylinder. The settling and compaction curve is developed by plotting the height of the sludge interface versus the time of settling.

Clarifiers

Clarifiers may be either rectangular or circular. In most rectangular clarifiers, scraper flights extending the width of the tank move the settled sludge toward the inlet end of the tank at a speed of about 1 ft/min (0.3 m/min). Some designs move the sludge toward the effluent end of the tank, corresponding to the direction of flow of the density current. A typical unit is shown in Fig. 3.26.

Circular clarifiers may employ either a center-feed well or a peripheral inlet. The tank can be designed for center sludge withdrawal or vacuum withdrawal over the entire tank bottom.

Circular clarifiers are of three general types. With the center-feed type, the water is fed into a center well and the effluent is pulled off at a weir along the outside. With a peripheral-feed tank, the effluent is pulled off at the tank center. With a rim-flow clarifier, the peripheral feed and effluent drainoff are also along the clarifier rim, but this type is usually used for larger clarifiers.

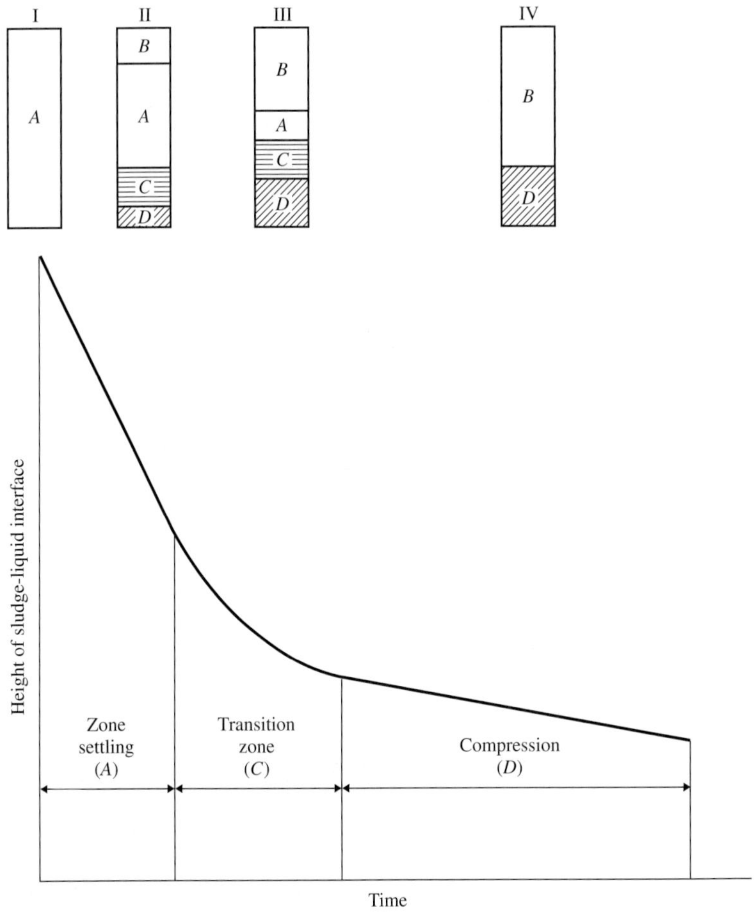

FIGURE 3.25 Settling properties of a flocculated sludge.

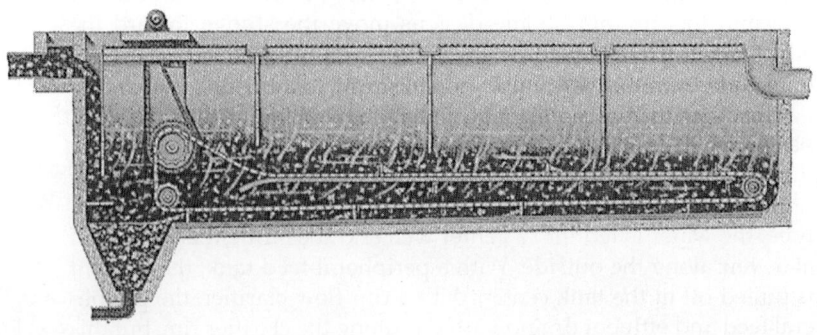

FIGURE 3.26 Rectangular clarifier.

The circular clarifier usually gives the optimal performance. Rectangular tanks may be desired where construction space is limited and multiple tanks are to be constructed. In addition, a series of rectangular tanks would be cheaper to construct because of the "shared wall" concept.

The reactor clarifier is another variation where the functions of chemical mixing, flocculation, and clarification are combined in the highly efficient solids contact unit. This combination achieves the highest overflow rate and the highest effluent quality of all clarifier designs.

The circular clarifier can be designed for center sludge withdrawal or vacuum withdrawal over the entire tank bottom. Center sludge withdrawal requires a minimum bottom slope of 1 in/ft (8.3 cm/m). The flow of sludge to the center well is largely hydraulically motivated by the collection mechanism, which serves to overcome inertia and avoid sludge adherence to the tank bottom. The vacuum drawoff is particularly adaptable to secondary clarification and thickening of activated sludge. A circular clarifier is shown in Fig. 3.27.

The mechanisms can be of the plow type or the rotary-hoe type. The plow-type mechanism employs staggered plows attached to two opposing arms that move at about 10 ft/min (3 m/min). The rotary-hoe mechanism consists of a series of short scrapers suspended from a rotating supporting bridge on endless chains that make contact with the tank bottom at the periphery and move to the center of the tank.

An inlet device is designed to distribute the flow across the width and depth of the settling tank. The outlet device is likewise designed to collect the effluent uniformly at the outlet end of the tank. Well-designed

FIGURE 3.27 Circular clarifier.

inlets and outlets will reduce the short-circuiting characteristics of the tank. Increased weir length can be provided by extending the effluent channels back into the basin or by providing multiple effluent channels. In circular basins, inboard or radial weirs will ensure low takeoff velocities. Relocation of weirs is sometimes necessary to minimize solids carryover induced by density currents and resulting in upwelling swells of sludge at the end of the settling tank. The installation of a plate below the effluent weir extending 18 in (45.7 cm) into the clarifier will deflect rising solids and permit them to resettle. Improved performance of secondary clarifiers has been achieved with this modification.

Tube settlers offer increased removal efficiency at higher loading rates and lower detention times. An immediate advantage is that modules of inclined tubes constructed of plastic can be installed in existing clarifiers to upgrade performance. However, it should be noted that in some cases biological sloughing has resulted from placement in secondary clarifiers following biological treatment.

There are two types of tube clarifiers, the slightly inclined and the steeply inclined units. The slightly inclined unit usually has the tubes inclined at a 5° angle. For the removal of discrete particles, an inclination of 5° has proven most efficient.

The steeply inclined unit is less efficient in removal of discrete particles, but can be operated continuously. When the tubes are inclined greater than 45°, the sludge is deposited and slides back out of the tube, forming a countercurrent flow. In practice most wastes are flocculent in nature, and removal efficiency is improved when the tubes are inclined to 60° to take advantage of the increased flocculation that occurs as the solids slide back out of the tube. Steeply inclined units are usually used where sedimentation units are being upgraded.

The hydraulic characteristics of a settling tank can be defined by a dispersion test in which dye or tracer is injected into the influent as a slug, and the concentration measured in the effluent as a function of time.

It is frequently possible to improve the performance of an existing settling basin by making modifications based on the results of a dispersion test. A comparison of a hydraulically overloaded center-feed tank converted to a peripheral-feed tank is shown in Fig. 3.28.

3.5 Oil Separation

In an oil separator, free oil is floated to the surface of a tank and then skimmed off. The same conditions holding for the subsidence of particles apply, except that the lighter-than-water oil globules rise through the liquid. The design of gravity separators as specified by the American Petroleum Institute[12] is based on the removal of all free oil globules larger than 0.015 cm.

The Reynolds number is less than 0.5, so Stokes' law applies. A design procedure considering short circuiting and turbulence has been developed.[13]

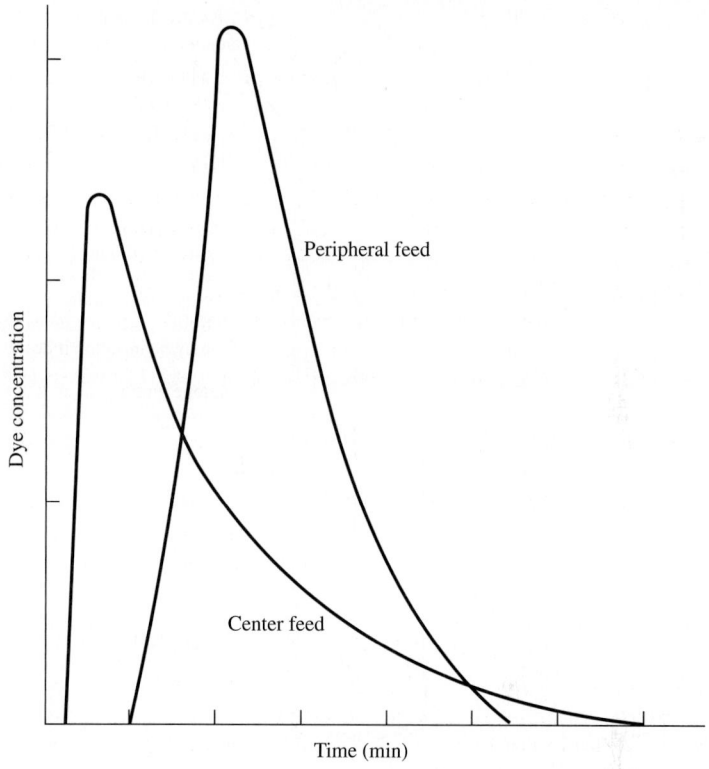

FIGURE 3.28 Dispersion characteristics of two settling tanks.

An API oil separator is shown in Fig. 3.29. The performance of an API separator is shown in Figs. 3.30 and 3.31 and typical efficiencies shown in Table 3.12.[14] Design criteria suggest that the flow rate should not exceed 2 ft/min, the length to width ratio should be at least 5 to avoid dead spots in the separator, and the minimum depth set at 4 ft. Variability in performance reported by Rebhun and Galil[15] is shown in Fig. 3.32. The influence of initial oil concentration on separator efficiency is shown in Fig. 3.33.[14]

Plate separators include parallel plate separators and corrugated plate separators (CPS). Plate separators are designed to separate oil droplets larger than 0.006 cm. It has been found by experience that a 0.006-cm separation will generally produce a 10-mg/L free oil non-emulsified effluent. This quality can usually be met when the influent oil content is less than 1 percent. One problem with the CPS is that reduced efficiency results from high oil loadings caused by oil-droplet shear and reentrainment of the oil droplet. This is largely overcome using a cross-flow corrugated plate separator in which the separated oil rises across the direction of flow rather than against the direction of flow (plates are angled at 45° and spaced at 10 mm).

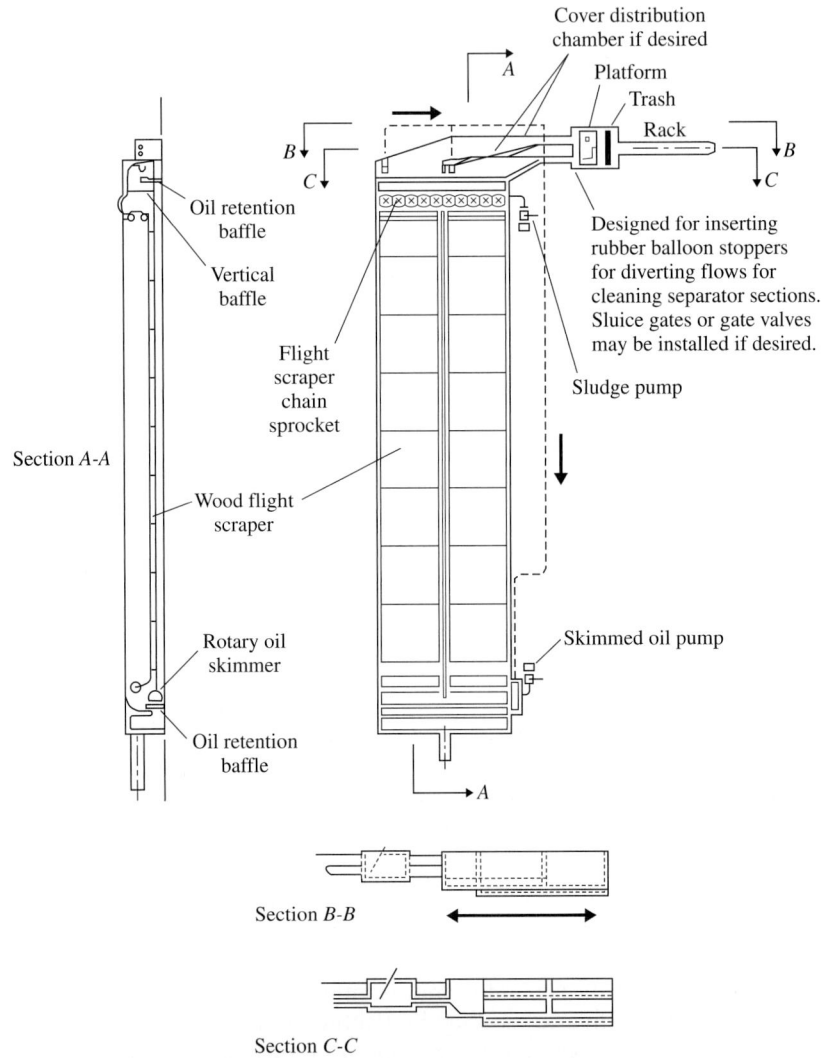

FIGURE 3.29 Example of general arrangement for API separator. (*Courtesy of the American Petroleum Institute*)

The hydraulic loading varies with temperature and the specific gravity of the oil. Nominal flow rates are specified for a temperature of 20°C and a specific gravity of 0.9 for the oil. A hydraulic loading of 0.5 m^3/(h · m^2) of actual plate area will usually result in separation of 0.006-cm droplets. A 50 percent safety factor is usually employed for design purposes. A plate separator is shown in Fig. 3.34.

Several types of filtration devices have proved effective in removing free and emulsified oils from refinery-petrochemical wastewaters. These vary from filters with sand media to those containing special

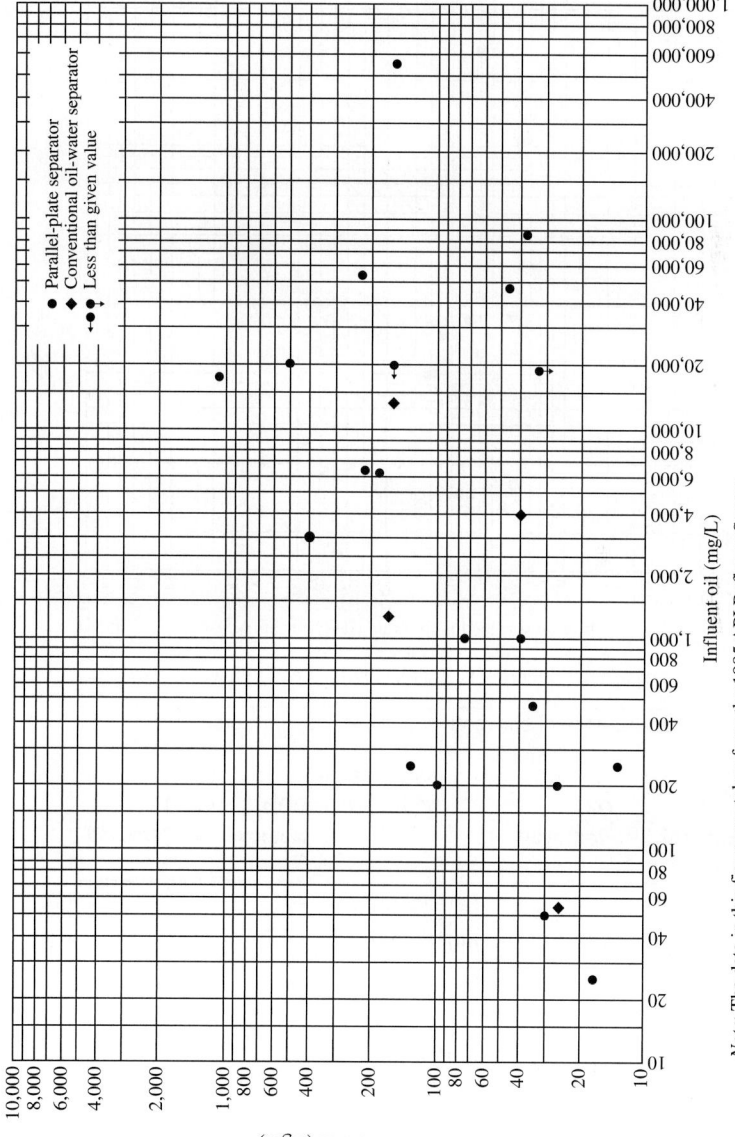

Figure 3.30 Correlation between influent and effluent oil levels for existing oil-water separators.

114 Chapter Three

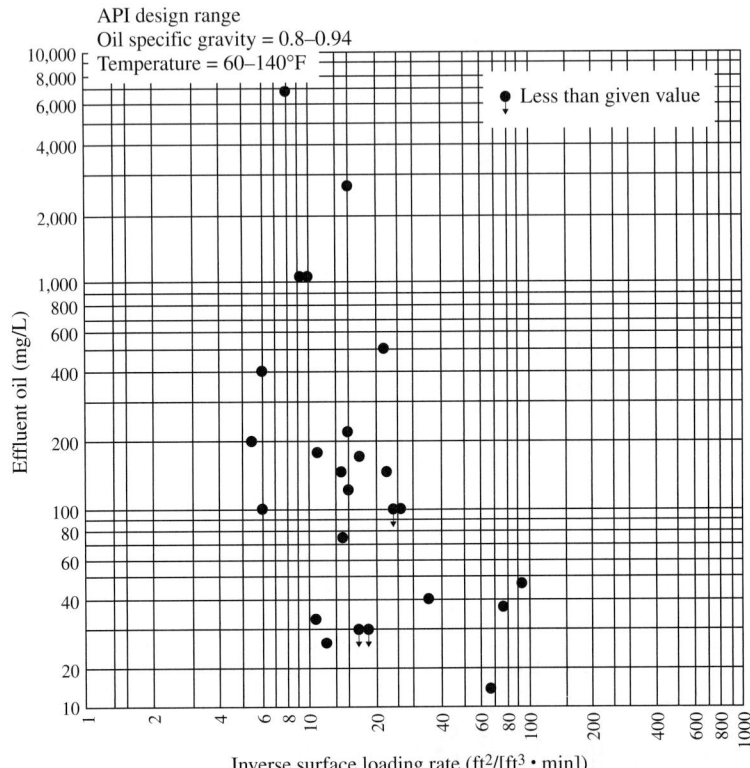

FIGURE 3.31 Performance of conventional oil-water separators.

Oil Content		Oil Removed, %	Type	COD Removed, %	SS Removed, %
Influent, mg/L	Effluent, mg/L				
300	40	87	Parallel plate	—	—
220	49	78	API	45	—
108	20	82	Circular	—	—
108	50	54	Circular	16	—
98	44	55	API	—	—
100	40	60	API	—	—
42	20	52	API	—	—
2000	746	63	API	22	33
1250	170	87	API	—	68
1400	270	81	API	—	35

TABLE 3.12 Typical Efficiencies of Oil Separation Units

Pre- and Primary Treatment 115

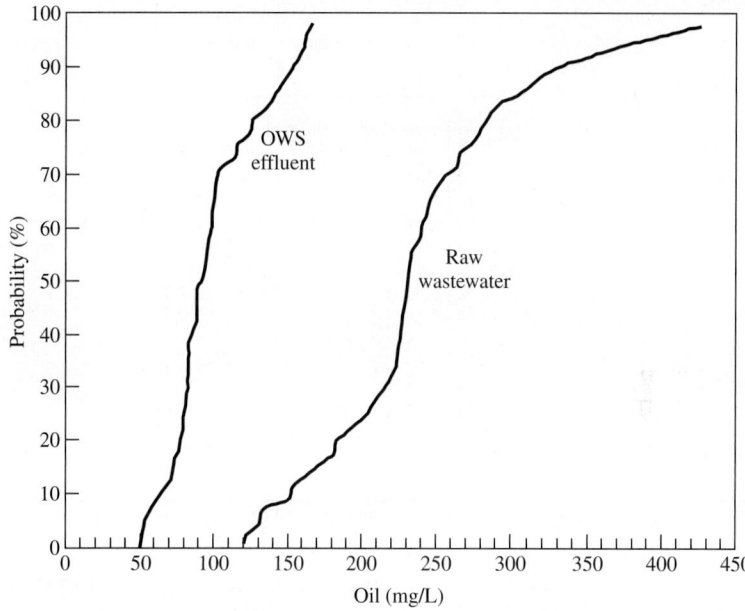

FIGURE 3.32 Oil removal by the oil-water separation (OWS) unit.

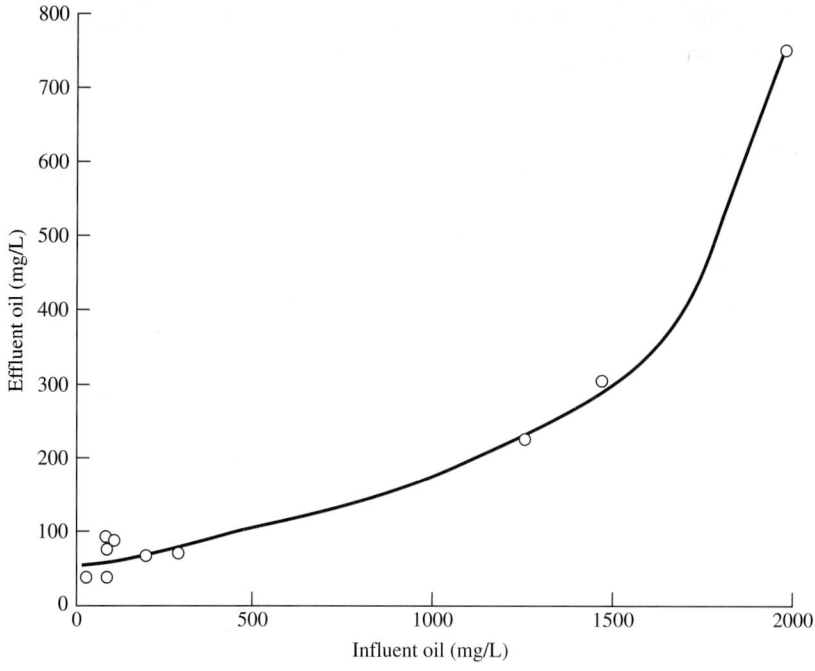

FIGURE 3.33 Effect of influent oil concentration separator efficiency.

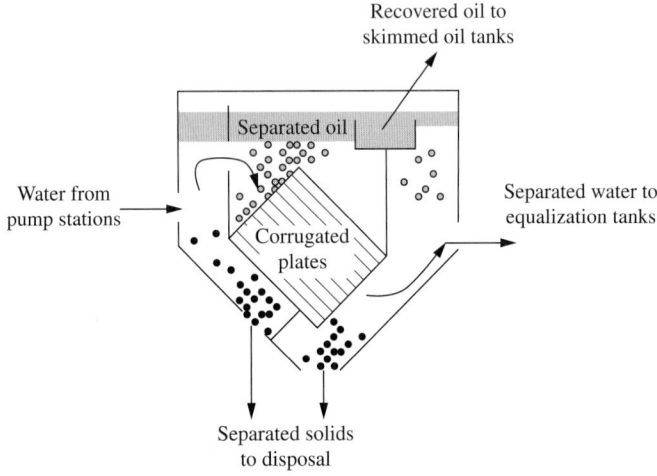

FIGURE 3.34 Corrugated plate interceptor oil/solids/water separator.

media which exhibit a specific affinity for oil. One type is an upflow unit using a graded silica medium as the filtering and coalescing section. Even the small particles and globules are separated and retained on the medium. The oil particles, which flow upward by gravity differential and fluid flow, rise through the coalescent medium and through the water phase from which the oil is separated and collected near the top of the separator. The bed is regenerated by introducing wash water at a rapid rate and evacuating the solids and remaining oil. This filtration and coalescing process is often enhanced by the use of polymer resin media. The primary application of these units are for selected in-plant streams which are dirt-free. Another application would be ballast water treatment following phase separation.

Emulsified oily materials require special treatment to break the emulsions so that the oily materials will be free and can be separated by gravity, coagulation, or air flotation. The breaking of emulsions is a complex art and may require laboratory or pilot-scale investigations prior to developing a final process design.

Emulsions can be broken by a variety of techniques. Quick-breaking detergents form unstable emulsions which break in 5 to 60 min to 95 to 98 percent completion. Emulsions can be broken by acidification, the addition of alum or iron salts, or the use of emulsion-breaking polymers. The disadvantage of alum or iron is the large quantities of sludge generated.

3.6 Oil Processing in a Typical Petroleum Refinery

As mentioned in this chapter, there are several primary and secondary oil removal options. It should be recognized that the process of oil recovery and removal systems result in the creation of listed hazardous wastes under the Resource Conservation and Recovery Act (RCRA). These include:

	RCRA Hazardous Waste Designation
• Oil separator bottom sludges	K051
• Slop oil emulsions	K049
• DAF float	K048
• Heat exchanger bundles	K050
• Leaded tank bottoms	K052
• Primary oil separator sludges and desalter muds	F037
• Miscellaneous sludge floats	F038

The handling of these listed hazardous residuals must be factored into the design, reprocessing, and disposal in accordance with RCRA regulations.

The two most typical methods of crude oil desalting, chemical and electrostatic separation, use hot water as the extraction agent. In chemical desalting, water and chemical surfactant (demulsifiers) are added to the crude, heated so that salts and other impurities dissolve into the water or attach to the water, and then held in a tank where they settle out. *Electrical desalting* is the application of high voltage electrostatic charges to concentrate suspended water globules in the bottom of the settling tank. Surfactants are added only when the crude has a large amount of suspended solids. Both methods of desalting are continuous. A third and less common process involves filtering heated crude using diatomaceous earth.

The feedstock crude oil is heated to between 150 and 350°F to reduce viscosity and surface tension for easier mixing and separation of the water. The temperature is limited by the vapor pressure of the crude oil feedstock. In both methods other chemicals may be added. Ammonia is often used to reduce corrosion. Caustic or acid may be added to adjust the pH of the water wash. Wastewater and contaminants are discharged from the bottom of the settling tank to the wastewater treatment facility. The desalted crude is continuously drawn from the top of the settling tanks and sent to the crude distillation (fractionating) tower. A diagram of the electrostatic desalting process is shown in Fig. 3.35.

As noted in Table 15.3, the desalter effluent water constitutes a significant fraction of the total wastewater flow from a petroleum refinery. As this flow contains high salt content (TDS), suspended muds, and unremoved oil (O&G), the must be routed through the appropriate oil removal process prior to the wastewater treatment plant. The amount of sludge (muds) associated with the raw crude must be removed to the maximum extent possible. A number of

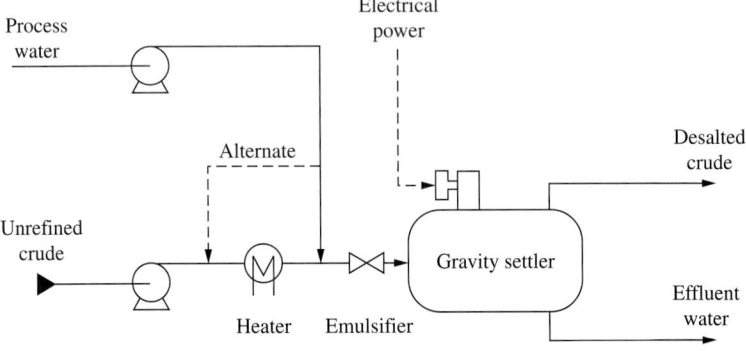

Figure 3.35 Electrostatic desalting.

techniques can be used such as low shear mixing devices to mix desalter wash water and crude oil, using lower pressure process water in the desalter to avoid turbulence, and replacing waste jets used in some refineries with mud rates which add less turbulence when removing settled solids. The desalting process combined with the hazardous waste designation of oil residuals, and vapor control and recovery (subject to national emission standards for hazardous air pollutants [NESHAPs]) which includes benzene and approximately 20 other chemicals which potentially can be emitted from petroleum refineries. (40CFR, Part 60, Subpart QQQ).

A simplified process flow diagram of the desalter, oil removal/recovery system, and wastewater treatment conveyance, using a CPI (Fig. 3.34) as the primary removal process, is shown in Fig. 3.36.

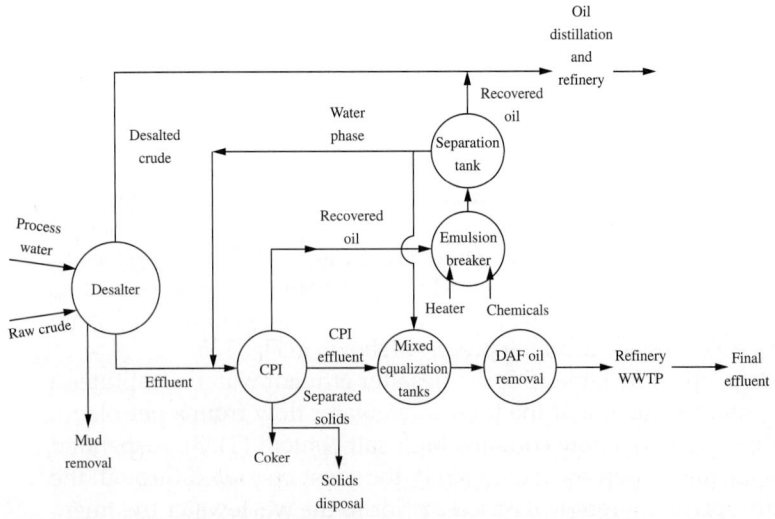

Figure 3.36 Process flow diagram desalter-oil recovery-wastewater treatment.

3.7 Sour Water Strippers

Stripping processes are used to remove selected constituents from liquid streams. The two most prevalent pollutants found in refinery wastewaters which are susceptible to stripping are hydrogen sulfide and ammonia resulting from the destruction of essentially all the organic nitrogen and sulfur compounds during desulfurization, denitrification, and hydrotreating. The use of steam within the processes is the primary source of conveyance, as the condensation occurs simultaneously with the condensation of hydrocarbon liquids and in the presence of a hydrocarbon vapor phase which contains H_2S and NH_3.[13] Phenols also may be present in these "sour water" condensates and can be stripped from solutions, although the efficiency of removal is less than for sulfide and ammonia. Other aromatics also can be stripped from solution at various levels of efficiency. As the regulatory agencies are imposing increasingly stringent quality standards for refinery wastewater in terms of immediate oxygen demand (sulfide causative) and ammonia, the necessity of in-plant control through stripping towers may be required whether it can be justified in terms of product recovery or not.

The design criteria for sour water strippers is well documented and is outlined in detail elsewhere.[13,16] There are various types of strippers but most involve a single tower equipped with trays or some type of packing. The feed water enters at the top of the tower and steam or stripping gas is introduced at the bottom. As H_2S is less soluble in water than NH_3, it is more readily stripped from the solution. High temperatures (230°F or more) are required to remove NH_3, where H_2S could be stripped at 100°F if NH_3 were fixed or not present. Therefore, acidification with a mineral acid or flue gas is often used to fix the NH_3 and allow more efficient H_2S removal. Average operating characteristics of some sour water strippers are cited in Table 3.13.[16] Although acidification enhances sulfide removal, it fixes the ammonia and prevents its removal. This has led to a two-stage stripping and recovery process developed by the Chevron Research Company. The process includes a degasser-surge tank combination which allows operational flexibility. After any floating hydrocarbon is skimmed, it is routed to the first column where the H_2S is stripped and sent to a sulfur recovery plant. The water-ammonia mixture then goes to a second fractionator for ammonia stripping. The overhead ammonia, approximately 98 percent pure as it leaves the condenser, is further purified passing through a scrubber system, and is then liquefied as high-purity ammonia. The cooled water bottoms from the system are then sufficiently free of H_2S and NH_3 to satisfy most quality criteria, containing less than 5 mg/L H_2S and 50 mg/L NH_3.

Type of Stripper	Flow Rate of Stripping Medium, SCF/gal	Removal		Temperature	
		H_2S, %	NH_3, %	Tower feed, °F	Tower Bottom, °F
Steam					
With acidifying*	8–32	96–100	69–95	150–240	230–270
Without acidifying†	4–6	97–100	0	200	230–250
Flue gas					
With steam‡	12.7	88–98	77–90	235	235
Without steam‡	11.9	99	8	135	140
Natural gas					
With acidifying	7.5	98	0	70–100	70–100

*Data from eight towers.
†Data from only one tower.
‡Data from two towers.

TABLE 3.13 Average Operating Characteristics of Sour Water Strippers

3.8 Flotation

Flotation is used for the removal of suspended solids and oil and grease from wastewaters and for the separation and concentration of sludges. The waste flow or a portion of clarified effluent is pressurized to 50 to 70 lb/in² (345 to 483 kPa or 3.4 to 4.8 atm) in the presence of sufficient air to approach saturation. When this pressurized air-liquid mixture is released to atmospheric pressure in the flotation unit, minute air bubbles are released from solution. The sludge flocs, suspended solids, or oil globules are floated by these minute air bubbles, which attach themselves to and become enmeshed in the floc particles. The air-solids mixture rises to the surface, where it is skimmed off. The clarified liquid is removed from the bottom of the flotation unit; at this time a portion of the effluent may be recycled back to the pressure chamber. When flocculent sludges are to be clarified, pressurized recycle will usually yield a superior effluent quality since the flocs are not subjected to shearing stresses through the pumps and pressurizing system.

Air Solubility and Release

The saturation of air in water is directly proportional to pressure and inversely proportional to temperature. Pray[17] and Frolich[18] found that

Temperature		Volume Solubility		Weight Solubility		Density	
°C	°F	mL/L	ft³/Thousand gal	mg/L	lb/Thousand gal	g/L	lb/ft³
0	32	28.8	3.86	37.2	0.311	1.293	0.0808
10	50	23.5	3.15	29.3	0.245	1.249	0.0779
20	68	20.1	2.70	24.3	0.203	1.206	0.0752
30	86	17.9	2.40	20.9	0.175	1.166	0.0727
40	104	16.4	2.20	18.5	0.155	1.130	0.0704
50	122	15.6	2.09	17.0	0.142	1.093	0.0682
60	140	15.0	2.01	15.9	0.133	1.061	0.0662
70	158	14.9	2.00	15.3	0.128	1.030	0.0643
80	176	15.0	2.01	15.0	0.125	1.000	0.0625
90	194	15.3	2.05	14.9	0.124	0.974	0.0607
100	212	15.9	2.13	15.0	0.125	0.949	0.0591

Values presented in absence of water vapor and at 14.7 lb/in² (1 atm).

TABLE 3.14 Air Characteristics and Solubilities

the oxygen and nitrogen solubilities in water follow Henry's law over a wide pressure range. Vrablick[19] has shown that although a linear relationship exists between pressure and solubility for most industrial wastes, the slope of the curve varies, depending on the nature of the waste constituents present. The solubility of air in water at atmospheric pressure is shown in Table 3.14.

The quantity of air that will theoretically be released from solution when the pressure is reduced to 1 atm can be computed from

$$s = s_a \frac{P}{P_a} - s_a \qquad (3.11)$$

where s = air released at atmospheric pressure per unit volume at 100 percent saturation, cm³/L
s_a = air saturation at atmospheric pressure, cm³/L
P = absolute pressure
P_a = atmospheric pressure

The actual quantity of air released will depend upon turbulent mixing conditions at the point of pressure reduction and on the degree of saturation obtained in the pressurizing system. Since the solubility in industrial wastes may be less than that in water, a correction may have to be applied to Eq. (3.11). Retention tanks will generally yield 86 to 90 percent saturation. Equation (3.11) can be modified to account for air saturation:

$$s = s_a \left(\frac{fP}{P_a} - 1 \right) \qquad (3.12)$$

where f is the fraction of saturation in the retention tank.

The performance of a flotation system depends on having sufficient air bubbles present to float substantially all of the suspended solids. An insufficient quantity of air will result in only partial flotation of the solids, and excessive air will yield no improvement. The performance of a flotation unit in terms of effluent quality and solids concentration in the float can be related to an air/solids (A/S) ratio, which is usually defined as mass of air released per mass of suspended solids in the influent waste:

$$\frac{A}{S} = \frac{s_a R}{S_a Q} \left(\frac{fP}{P_a} - 1 \right) \qquad (3.13)$$

where Q = wastewater flow
 R = pressurized recycle
 S_a = influent oil and/or suspended solids

The relationship between the air/solids ratio and effluent quality is shown in Fig. 3.37. It should be noted that the shape of the curve will vary with the nature of the solids in the feed.

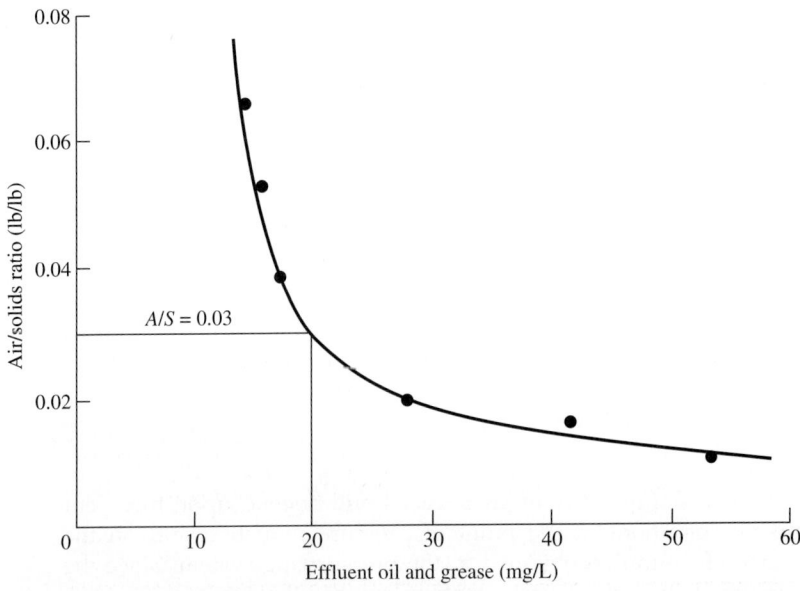

FIGURE 3.37 Effects on A/S on effluent quality.

Vrablick[19] has shown that the bubbles released after pressurization (20 to 50 lb/in² or 1.36 to 3.40 atm) range in size from 30 to 100 μm. The rise velocity closely follows Stokes' law. The rise velocity of a solids-air mixture has been observed to vary from 1 to 5 in/min (2.56 to 12.7 cm/min) and will increase with an increasing air/solids ratio. In the flotation of activated sludge of 0.91 percent solids and 40 lb/in² (276 kPa or 2.72 atm), Hurwitz and Katz[20] observed free rises of 0.3 (9), 1.2 (37), and 1.8 ft/min (55 cm/min) for recycle ratios of 100, 200, and 300 percent, respectively. The initial rise rate will vary with the character of the solids.

The primary variables for flotation design are pressure, recycle ratio, feed solids concentration, and retention period. The effluent suspended solids decrease and the concentration of solids in the float increase with increasing retention period. When the flotation process is used primarily for clarification, a detention period of 20 to 30 min is adequate for separation and concentration. Rise rates of 1.5 to 4.0 gal/(min · ft²) (0.061 to 0.163 m³/[min · m²]) are commonly employed. When the process is employed for thickening, longer retention periods are necessary to permit the sludge to compact.

The principal components of a flotation system are a pressurizing pump, air-injection facilities, a retention tank, a backpressure regulating device, and a flotation unit, as shown in Fig. 3.38. The

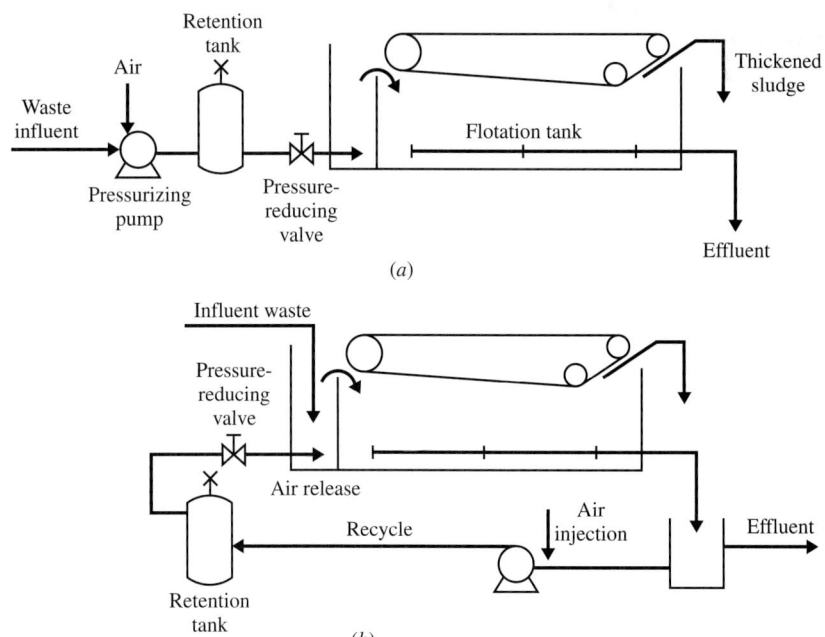

Figure 3.38 Schematic representation of flotation systems. (*a*) Flotation system without recirculation. (*b*) Flotation system with recirculation.

FIGURE 3.39 Clarifier flotation unit.

pressurizing pump creates an elevated pressure to increase the solubility of air. Air is usually added through an injector on the suction side of the pump or directly to the retention tank.

The air and liquid are mixed under pressure in a retention tank with a detention time of 1 to 3 min. A backpressure regulating device maintains a constant head on the pressurizing pump. Various types of values are used for this purpose. The flotation unit may be either circular or rectangular with a skimming device to remove the thickened, floated sludge, as shown in Fig. 3.39.

The induced-air flotation system, shown in Fig. 3.40, operates on the same principles as a pressurized-air dissolved air flotation (DAF) unit. The gas, however, is self-induced by a rotor-disperser mechanism. The rotor, the only moving part of the mechanism that is submerged in the liquid, forces the liquid through the disperser openings, thereby creating a negative pressure. It pulls the gas downward into the liquid, causing the desired gas-liquid contact. The liquid moves through a series of four cells before leaving the tank, and the float skimmings pass over the overflow weirs on each side of the unit. This type of system offers the advantages of significantly lower capital cost and smaller space requirements than the pressurized system, and current performance data indicate that these systems have the capacity to effectively remove free oil and suspended materials.

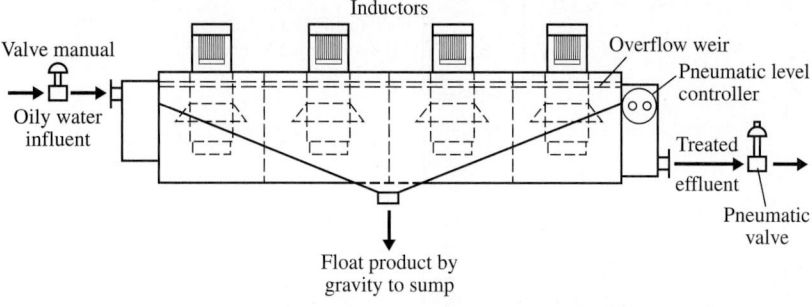

FIGURE 3.40 Induced-air flotation system (*Wemco Envirotech Company*).

The disadvantages include higher connected power requirements than the pressurized system, performance dependent on strict hydraulic control, less chemical addition and flocculation flexibility, and relatively high volumes of float skimmings as a function of liquid throughout (3 to 7 percent of the incoming flow for induced-air systems is common compared to less than 1 percent for pressurized-air systems).

It is possible to estimate the flotation characteristics of a waste by the use of a laboratory flotation cell, as shown in Fig. 3.41. The procedure is as follows:

1. Partially fill the calibrated cylinder with waste or flocculated sludge mixture and the pressure chamber with clarified effluent or water.

2. Apply compressed air to the pressure chamber to attain the desired pressure.

3. Shake the air-liquid mixture in the pressure chamber for 1 min and allow to stand for 3 min to attain saturation. Maintain the pressure on the chamber for this period.

4. Release a volume of pressurized effluent to the cylinder and mix with the waste or sludge. The volume to be released is computed from the desired recycle ratio. The velocity of release through the inlet nozzle should be of such a magnitude as not to shear the suspended solids in the feed mixture but to maintain adequate mixing.

5. Measure the rise of the sludge interface with time. Correction must be applied to scale up the height of rise in the test cylinder to the depth of the prototype unit.

6. After a detention time of 20 min, the clarified effluent and the floated sludge are drawn off through a valve in the bottom of the cylinder.

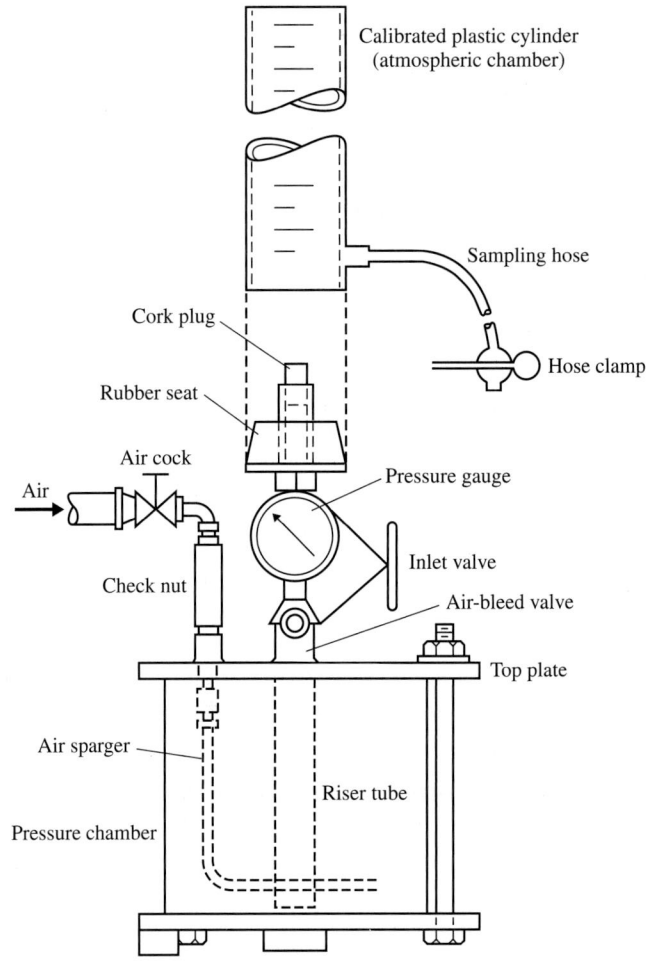

Figure 3.41 Laboratory flotation cell.

7. Relate the effluent suspended solids to the calculated air/solids ratio as shown in Fig. 3.37. When pressurized recycle is used, the air/solids ratio is computed:

$$\frac{A}{S} = \frac{1.3 s_a R(P-1)}{Q S_a}$$

where s_a = air saturation, cm³/L
  R = pressurized volume, L
  P = absolute pressure, atm
  Q = waste flow, L
  S_a = influent suspended solids, mg/L

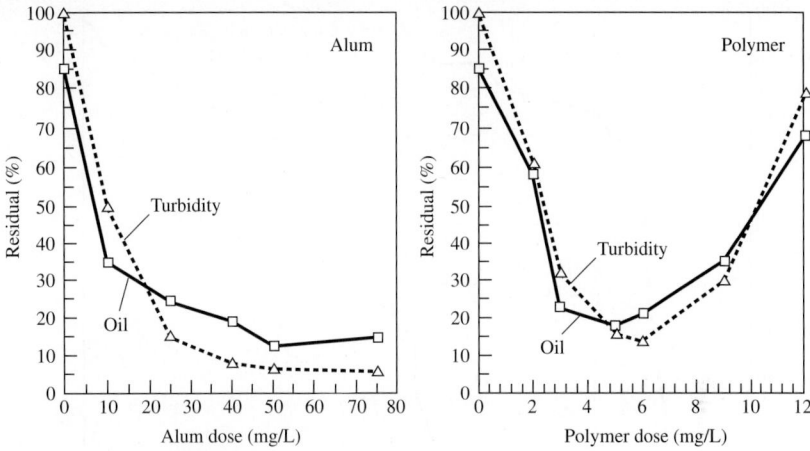

FIGURE 3.42   Treatment efficiency in chemical flocculation.[14]

High-degree clarification frequently requires that flocculating chemicals be added to the influent before it is mixed with the pressurized recycle. Alum or polyelectrolytes are used as flocculating agents. Hurwitz and Katz[20] have shown that the rise rate of chemical floc will vary from 0.65 to 2.0 ft/min (20 to 61 cm/min), depending on the floc size and characteristics. Treatment efficiency with chemical addition is shown in Fig. 3.42.

Flotation design for an oily wastewater is shown in Example 3.7. Flotation units have been used for the clarification of wastewaters, for the removal of oil, and for the concentration of waste sludges. Variability in DAF performance is shown in Figs. 3.43 and 3.44. Some reported data for petroleum refinery effluents are shown in Table 3.15 and for a variety of wastewaters in Table 3.16.

Example 3.7. A wastestream of 150 gal/min (0.57 m^3/min) and a temperature of 103°F (39.4°C) contains significant quantities of nonemulsified oil and nonsettleable suspended solids. The concentration of oil is 120 mg/L. Reduce the oil to less than 20 mg/L. Laboratory studies showed:

Alum dose = 50 mg/L

Pressure = 60 lb/in^2 gage (515 kPa absolute or 4.1 relative atm)

Sludge production = 0.64 mg/mg alum

Sludge = 3 percent by weight

Calculate:

(a) The recycle rate

(b) Surface area of the flotation unit

(c) Sludge quantities generated

128 Chapter Three

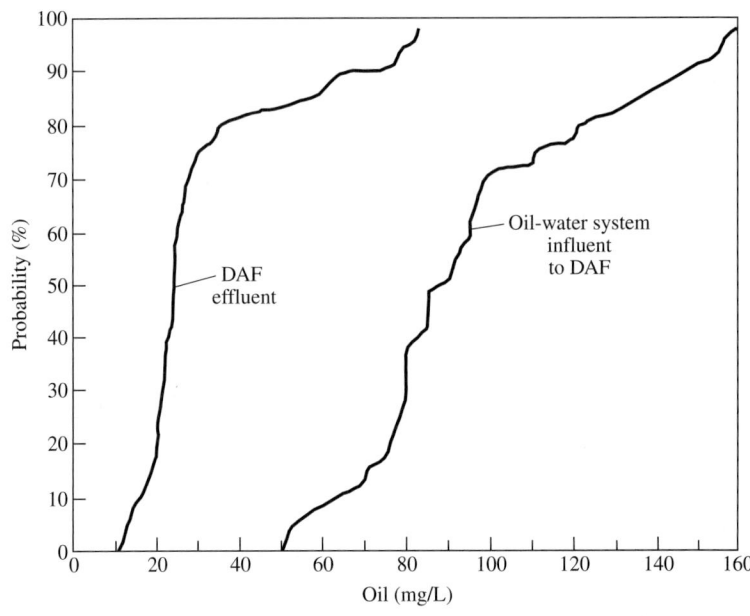

Figure 3.43 Oil removal by chemical flocculation and DAF. (*Adapted from Galil and Rebhum, 1993.*)

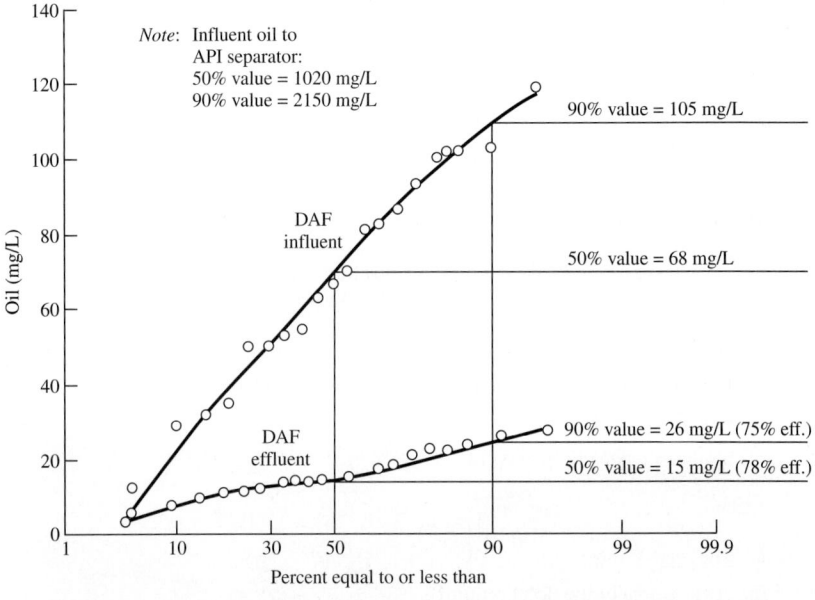

Figure 3.44 DAF performance variability.

Pre- and Primary Treatment

Influent Oil, mg/L	Effluent Oil, mg/L	% Removal	Chemicals*	Configuration
1930 (90%)	128 (90%)	93	Yes	Circular
580 (50%)	68 (50%)	88	Yes	Circular
105 (90%)	26 (90%)	78	Yes	Rectangular
68 (50%)	15 (50%)	75	Yes	Rectangular
170	52	70	No	Circular
125	30	71	Yes	Circular
100	10	90	Yes	Circular
133	15	89	Yes	Circular
94	13	86	Yes	Circular
838	60	91	Yes	Rectangular
153	25	83	Yes	Rectangular
75	13	82	Yes	Rectangular
61	15	75	Yes	Rectangular
360	45	87	Yes	Rectangular
315	54	83	Yes	Rectangular

*Alum most common, 100–300 mg/L. Polyelectrolyte, 1–5 mg/L, is occasionally added

TABLE 3.15 DAF Performance Data

Wastewater	Coagulant, mg/L	Oil Concentration, mg/L		
		Influent	Effluent	Removal, %
Refinery	0	125	35	72
	100 alum	100	10	90
	130 alum	580	68	88
	0	170	52	70
Oil tanker ballast water	100 alum + 1 mg/L polymer	133	15	89
Paint manufacture	150 alum + 1 mg/L polymer	1900	0	100
Aircraft maintenance	30 alum + 10 mg/L activated silica	250–700	20–50	>90
Meat packing		3830	270	93
		4360	170	96

TABLE 3.16 Air Flotation Treatment of Oily Wastewaters

The air/solids ratio for effluent oil and grease of 20 mg/L is found from Fig. 3.37:

$$\frac{A}{S} = 0.03 \text{ lb air released/lb solids applied}$$

At 103°F (39.4°C) the weight solubility of air is 18.6 mg/L from Table 3.14. The value of f is assumed to be 0.85.

Solution

(a) The recycle rate is

$$R = \frac{(A/S)QS_a}{s_a(fP/P_a - 1)}$$

$$= \frac{0.03 \times 150 \times 120}{18.6([0.85 \times 515/101.3] - 1)}$$

$$= 8.75 \text{ gal/min (33.1 L/min)}$$

(b) The hydraulic loading for oil removal is determined from Fig. 3.45, and is 2.6 gal/(min · ft²) (0.11 m³/[min · m²]) for an effluent of 20 mg/L. The required surface area is

$$A = \frac{Q+R}{\text{loading}}$$

$$= \frac{150 + 8.75}{2.6}$$

$$= 61 \text{ ft}^2 \text{ (5.7 m}^2\text{)}$$

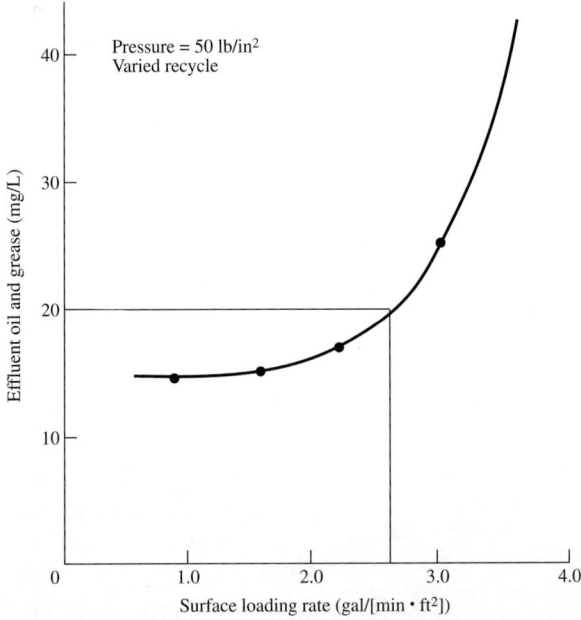

FIGURE 3.45 Hydraulic loading rate determination.

(c) Sludge quantities generated:

$$\text{Oil sludge} = (120 - 20) \text{ mg/L} \times 150 \text{ gal/min} \times 1440 \text{ min/d}$$
$$\times \text{ (million gal/10}^6 \text{ gal)} \left(8.34 \ \frac{\text{lb/million gal}}{\text{mg/L}}\right)$$
$$= 180 \text{ lb/d (82 kg/d)}$$

$$\text{Alum sludge} = 0.64 \text{ mg sludge/mg alum} \times 50 \text{ mg/L alum}$$
$$\times 150 \text{ gal/min}$$
$$\times (1440 \text{ min/d})(\text{million gal/10}^6 \text{ gal})(8.34)$$
$$= 58 \text{ lb/d (26 kg/d)}$$

$$\text{Total sludge} = 238 \text{ lb/d (108 kg/d)}$$
$$\text{Total sludge volume} = 238/0.03 \text{ lb/d (gal/8.34 lb)(day/1440 min)}$$
$$= 0.66 \text{ gal/min (2.5 L/min)}$$

3.9 Problems

3.1. An industry is required to equalize its wastewater and discharge it such that the load in BOD/day to a POTW is constant over 24 h. The POTW sewage flow is 6.47 million gal/d, with a BOD of 200 mg/L. The diurnal sewage variation in flow is shown in Fig. P3.1. The industrial waste flow is 3.17 million gal/d, with a BOD of 1200 mg/L, and is constant over 10 h (8 A.M. to 6 P.M.).

(a) Compute the volume of equalization (or holding) tank required.

(b) Plot the discharge curve of the industrial waste over 24 h.

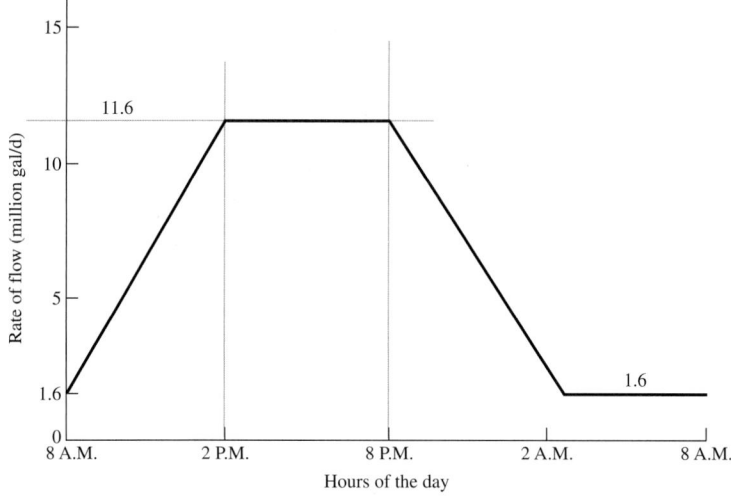

FIGURE P3.1

3.2. A survey of the discharge from a pharmaceutical plant showed the following data.

Time	Flow, gph	COD, mg/L
Day 1		
7:00 A.M.	1025	80
8:00 A.M.	600	55
9:00 A.M.	1200	48
10:00 A.M.	600	45
11:00 A.M.	720	95
12:00 P.M.	1080	66
1:00 P.M.	1200	41
2:00 P.M.	1620	39
3:00 P.M.	1200	29
4:00 P.M.	1320	138
5:00 P.M.	1020	146
6:00 P.M.	720	154
Day 2		
7:00 A.M.	960	47
8:00 A.M.	900	40
9:00 A.M.	1020	139
10:00 A.M.	900	1167
11:00 A.M.	1140	491
12:00 P.M.	1320	163
1:00 P.M.	900	90
2:00 P.M.	1320	143
3:00 P.M.	1200	88
4:00 P.M.	900	35
5:00 P.M.	1140	35
6:00 P.M.	960	47

Compute the volume of an equalization basin to yield a peaking factor of 1.2 and 1.4 based on flow through the system at constant volume and on a constant discharge rate (i.e., variable volume). For the variable-volume case the low level in the basin will be 20 percent of the daily flow.

3.3. An acidic industrial wastewater flow of 150 gal/min (0.57 m^3/min) with a peak factor of 1.2 is to be neutralized to pH 6.0 using lime. The titration curve is as follows:

pH	mg Lime/L Waste
1.8	0
1.9	500
2.05	1000
2.25	1500
3.5	2000
4.1	2100
5.0	2150
7.0	2200

Consider providing the necessary capacity for a maximum 2-week lime requirement for pH 7.0 with an hourly requirement 20 percent greater.

Determine:

(a) The lime feed rate for each stage in a two-stage system.

(b) The lime storage capacity (ft³) based on the controlling requirement of either the mean monthly requirement or the maximum two-week requirement. Assume the use of pebbled quicklime (CaO), which has a bulk density of 65 lb/ft³ (1060 kg/m³).

(c) The capacity of the lime slaker and mechanism for transporting the bulk lime based on the maximum estimated requirements.

(d) The mean and maximum slaking water requirements assuming a 10 percent by weight slurry.

(e) The size of the slurry control tank based on a minimum resident time of 5 min.

(f) The size of the caustic storage tank based on 24-h caustic feed (NaOH) at the maximum extended usage equivalent to a hydrated lime requirement of 1000 mg/L. Assume NaOH is available with a purity of 98.9 percent and a solubility of 2.5 lb/gal (300 kg/m³).

(g) The maximum caustic (NaOH) feed rate for backup of the lime system.

3.4. A laboratory settling analysis for a pulp and paper mill effluent gave the following results for $C_o = 430$ mg/L, $T = 29°C$:

(a) Design a settling tank to remove 70 percent of the suspended solids for 1 million gal/d flow (3785 m³/d). (Apply appropriate factors and neglect initial solids effects.)

(b) What removal will be attained if the flow is increased to 2 million gal/d (7570 m³/d)?

Time, min	Suspended Solids Removed at Indicated Depth, %		
	2 ft	4 ft	6 ft
10	47	27	16
20	50	34	43
30	62	48	47
45	71	52	46
60	76	65	48

Note: ft = 30.5 cm

3.5. A wastewater has a flow of 250 gal/min (0.95 m³/min) and a temperature of 105°F (40.5°C). The concentration of oil and grease is 150 mg/L and the suspended solids 100 mg/L. An effluent concentration of 20 mg/L is required. An alum dosage of 30 mg/L is required. The required *A/S* is 0.04 and the surface loading rate is 2 gal/(min · ft²) (0.081 m³/[min · m²]). The operating pressure is 65 lb/in² gage (4.4 relative atm).

Compute:

(a) The required recycle flow

(b) The surface area of the unit

(c) The volume of sludge produced if the skimmings are 3 percent by weight

References

1. Novotny, V., and A. J. England: *Water Res.*, vol. 8, p. 325, 1974.
2. Patterson, J. W., and J. P. Menez: *Am. Inst. Chem. Engrs. Env. Prog.*, vol. 3, p. 2, 1984.
3. Thomas, O., and D. Constant: "Trends in Optical Monitoring, Trends in Sustainable Production—From Wastewater Diagnosis to Toxicity Management and Ecological Protection," IWA Press, issue 1, vol. 49, London, 2004.
4. Tully, T. J.: *Sewage Ind. Wastes*, vol. 30, p. 1385, 1958.
5. Okey, R. W. et al.: *Proc. 32nd Purdue Industrial Waste Conf.*, Ann Arbor Science Pub., 1977.
6. Hazen, A.: *Trans. ASCE*, vol. 53, p. 45, 1904.
7. Camp, T. R.: *Trans. ASCE*, vol. 111, p. 909, 1946.
8. Dobbins, W. E.: "Advances in Sewage Treatment Design," Sanitary Engineering Division, Met. Section, Manhattan College, May 1961.
9. Committee on Industrial Waste Practice of SED: *J. Sanit. Engrg. Div. ASCE*, December 1964.
10. Sutherland, R.: *Ind. Eng. Chem.*, p. 630, May 1947.
11. Greenfield, R. E., and G. N. Cornell: *Ind. Eng. Chem.*, p. 583, May 1947.
12. American Petroleum Institute: *Manual on Disposal of Refinery Wastes*, vol. 1, New York, 1959.
13. Azad, H. S. (editor): *Industrial Wastewater Management Handbook*, McGraw-Hill, New York, 1976.

14. Ford, D. L., private communication.
15. Galil, N., and M. Rebhum: *Water Sci. Tech.*, vol. 27, no. 7–8, p. 79, 1993.
16. Jones, H. R.: *Pollution Control in the Petroleum Industry,* Noyes Data Corp., Princeton, New Jersey, 1973.
17. Pray, H. A.: *Ind. Eng. Chem.*, vol. 44, pt. 1, p. 146, 1952.
18. Frolich, R.: *Ind. Eng. Chem.*, vol. 23, p. 548, 1931.
19. Vrablick, E. R.: *Proc. 14th Ind. Waste Conf.,* 1959, Purdue University.
20. Hurwitz, E., and W. J. Katz: "Laboratory Experiments on Dewatering Sewage Sludges by Dissolved Air Flotation," unpublished report, Chicago, 1959.

CHAPTER 4
Coagulation, Precipitation, and Metals Removal

4.1 Introduction

Coagulation is a unit process used for removing colloids and other suspended particles from water and wastewater. It may be employed as source treatment for the removal of contaminants such as metals, within the treatment train, or with filtration as a polishing step. Coagulation destabilizes colloidal particles by charge neutralization and promoting collisions between neutralized particles, resulting in cohesion, floc growth, and eventual sedimentation and filtration. This chapter considers the principles of the coagulation process, coagulant properties, coagulation equipment, laboratory determinations for coagulant selection, and case studies. It also includes removal technologies for heavy metals of frequent concern.

4.2 Coagulation

Colloids are particles within the size range of 1 nm (10^{-7} cm) to 0.1 nm (10^{-8} cm). These particles do not settle out on standing and cannot be removed by conventional physical treatment processes. Colloids present in wastewater can be either hydrophobic or hydrophilic. The hydrophobic colloids (clays, etc.) possess no affinity for the liquid medium and lack stability in the presence of electrolytes. They are readily susceptible to coagulation. Hydrophilic colloids, such as proteins, exhibit a marked affinity for water. The absorbed water retards flocculation and frequently requires special treatment to achieve effective coagulation.[1]

Colloids possess electrical properties which create a repelling force and prevent agglomeration and settling. Stabilizing ions are strongly absorbed to an inner fixed layer which provides a particle charge that varies with the valence and number of adsorbed ions. Ions of an opposite charge form a diffuse outer layer which is held near the surface by electrostatic forces. The psi (ψ) potential is defined as the potential drop between the interface of the colloid and the body of solution. The zeta potential (ζ) is the potential drop between the slipping plane and the body of solution and is related to the particle charge and the thickness of the double layer. The thickness of the double layer is composed of a compact stern layer and a diffuse layer where the potential drops to 0 in bulk solution as shown in Fig. 4.1. A van der Waals attractive force is effective in close proximity to the colloidal particle.

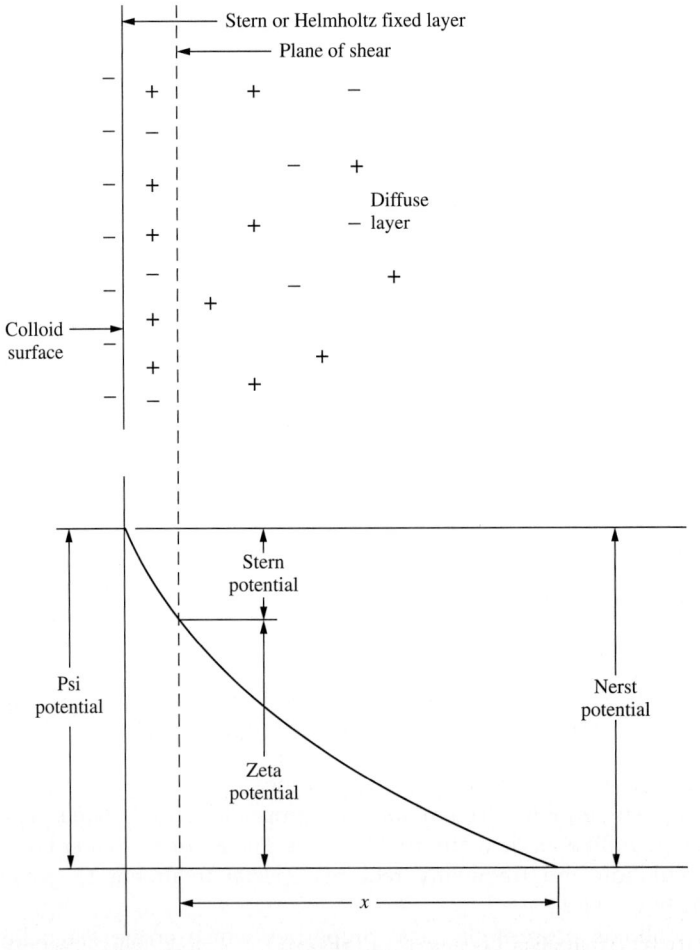

Figure 4.1 Electrochemical properties of a colloidal particle.

Coagulation, Precipitation, and Metals Removal

The stability of a colloid is due to the repulsive electrostatic forces, and in the case of hydrophilic colloids, also to solvation in which an envelope of water retards coagulation.

Zeta Potential

Since the stability of a colloid is primarily due to electrostatic forces, neutralization of this charge is necessary to induce flocculation and precipitation. Although it is not possible to measure the psi potential, the zeta potential can be determined, and hence the magnitude of the charge and resulting degree of stability can be determined as well. The zeta potential is defined as

$$\zeta = \frac{4\pi\eta v}{\varepsilon X} = \frac{4\pi\eta \mathrm{EM}}{\varepsilon} \qquad (4.1)$$

where v = particle velocity
ε = dielectric constant of the medium
η = viscosity of the medium
X = applied potential per unit length of cell
EM = electrophoretic mobility

For practical usage in the determination of the zeta potential, Eq. (4.1) can be re-expressed:

$$\zeta(\mathrm{mV}) = \frac{113{,}000}{\varepsilon} \, \eta(\mathrm{poise})\mathrm{EM}\left(\frac{\mu\mathrm{m/s}}{\mathrm{V/cm}}\right) \qquad (4.2)$$

where EM = electrophoretic mobility, $(\mu\mathrm{m/s})/(\mathrm{V/cm})$.
At 25°C, Eq. (4.2) reduces to

$$\zeta = 12.8 \, \mathrm{EM} \qquad (4.3)$$

The zeta potential is determined by measurement of the mobility of colloidal particles across a cell, as viewed through a microscope.[2,3] Several types of apparatus are commercially available for this purpose. A recently developed Lazer Zee meter does not track individual particles, but rather adjusts the image to produce a stationary cloud of particles using a rotating prism technique. This apparatus is shown in Fig. 4.2. The computations involved in determining the zeta potential are illustrated in Example 4.1.

Example 4.1. In a electrophoresis cell 10 cm in length, grid divisions are 160 μm at 6 × magnification. Compute the zeta potential at an impressed voltage of 35 V. The time of travel between grid divisions is 42 s and the temperature is 20°C.

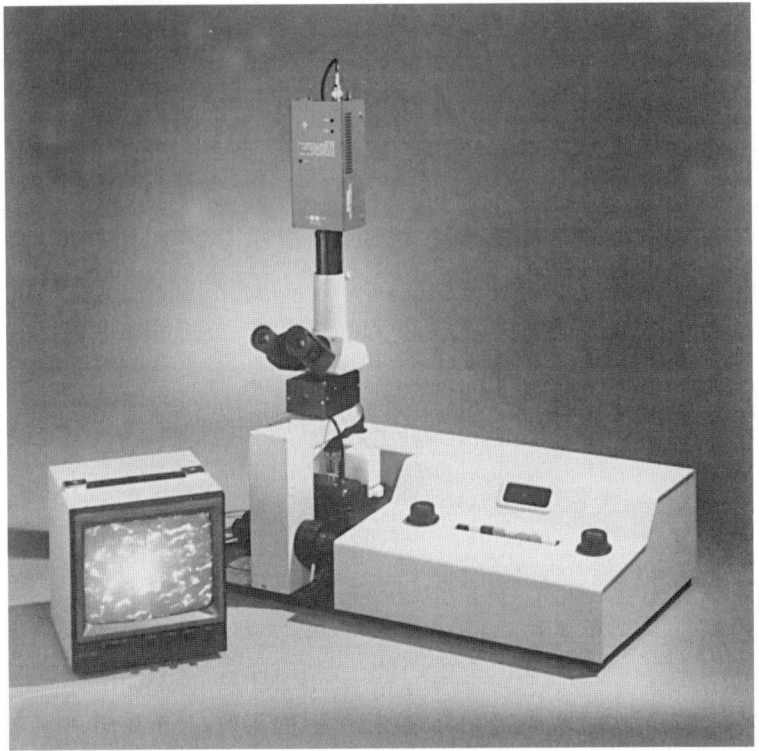

FIGURE 4.2 Lazer Zee meter for zeta potential measurement. (*Courtesy of Penkem Inc.*)

Solution
At 20°C:

$$\eta = 0.01 \text{ poise}$$

$$\varepsilon = 80.36$$

$$\text{EM} = \frac{v}{X} = \frac{160 \text{ μm}/42 \text{ s}}{35 \text{ V}/10 \text{ cm}} = 1.09 \left(\frac{\text{μm/s}}{\text{V/cm}}\right)$$

$$\zeta(\text{mV}) = 113{,}000 \times \frac{\eta(\text{poise})\text{EM}(\text{μm/s})/(\text{V/cm})}{\varepsilon}$$

$$= 113{,}000 \times \frac{0.01 \times 1.09}{80.36}$$

$$= 15.3 \text{ mV}$$

Since there will usually be a statistical variation in the mobility of individual particles, around 20 to 30 values should be averaged for any one determination. The magnitude of the zeta potential for water and waste colloids has been found to average from −16 to −22 mV with a range of −12 to −40 mV.[3]

The zeta potential is lowered by:

1. Change in the concentration of the potential determining ions.
2. Addition of ions of opposite charge.
3. Contraction of the diffuse part of the double layer by increase in the ion concentration in solution.

Since a vast majority of colloids in industrial wastes possess a negative charge, the zeta potential is lowered and coagulation is induced by the addition of high-valence cations. The precipitating power of effectiveness of cation valence in the precipitation of arsenious oxide is

$$Na^+ : Mg^{2+} : Al^{3+} = 1 : 63 : 570$$

Optimum coagulation will occur when the zeta potential is zero; this is defined as the isoelectric point. Effective coagulation will usually occur over a zeta potential range of ± 0.5 mV.

Mechanism of Coagulation

Coagulation results from two basic mechanisms: perikinetic (or electrokinetic) coagulation, in which the zeta potential is reduced by ions or colloids of opposite charge to a level below the van der Waals attractive forces, and orthokinetic coagulation, in which the micelles aggregate and form clumps that agglomerate the colloidal particles.

The addition of high-valence cations depresses the particle charge and the effective distance of the double layer, thereby reducing the zeta potential. As the coagulant dissolves, the cations serve to neutralize the negative charge on the colloids. This occurs before visible floc formation, and rapid mixing which "coats" the colloid is effective in this phase. Microflocs are then formed which retain a positive charge in the acid range because of the adsorption of H^+. These microflocs also serve to neutralize and coat the colloidal particle. Flocculation agglomerates the colloids with a hydrous oxide floc. In this phase, surface adsorption is also active. Colloids not initially adsorbed are removed by enmeshment in the floc.

Riddick[3] has outlined a desired sequence of operation for effective coagulation. If necessary, alkalinity should first be added. (Bicarbonate has the advantage of providing alkalinity without raising the pH.) Alum or ferric salts are added next; they coat the colloid with Al^{3+} or Fe^{3+} and positively charged microflocs. Coagulant aids, such as activated silica and/or polyelectrolyte for floc buildup and zeta potential control, are added last. After addition of alkali and coagulant, a rapid mixing of 1 to 3 min is recommended, followed by flocculation, with addition of coagulant aid, for 20 to 30 min. Destabilization can also be accomplished by the addition of cationic polymers, which can bring the system to the isoelectric point without a change in pH. Although polymers are 10 to 15 times as effective as alum as a coagulant, they are considerably more expensive. The mechanism of the coagulation process is shown in Fig. 4.3.

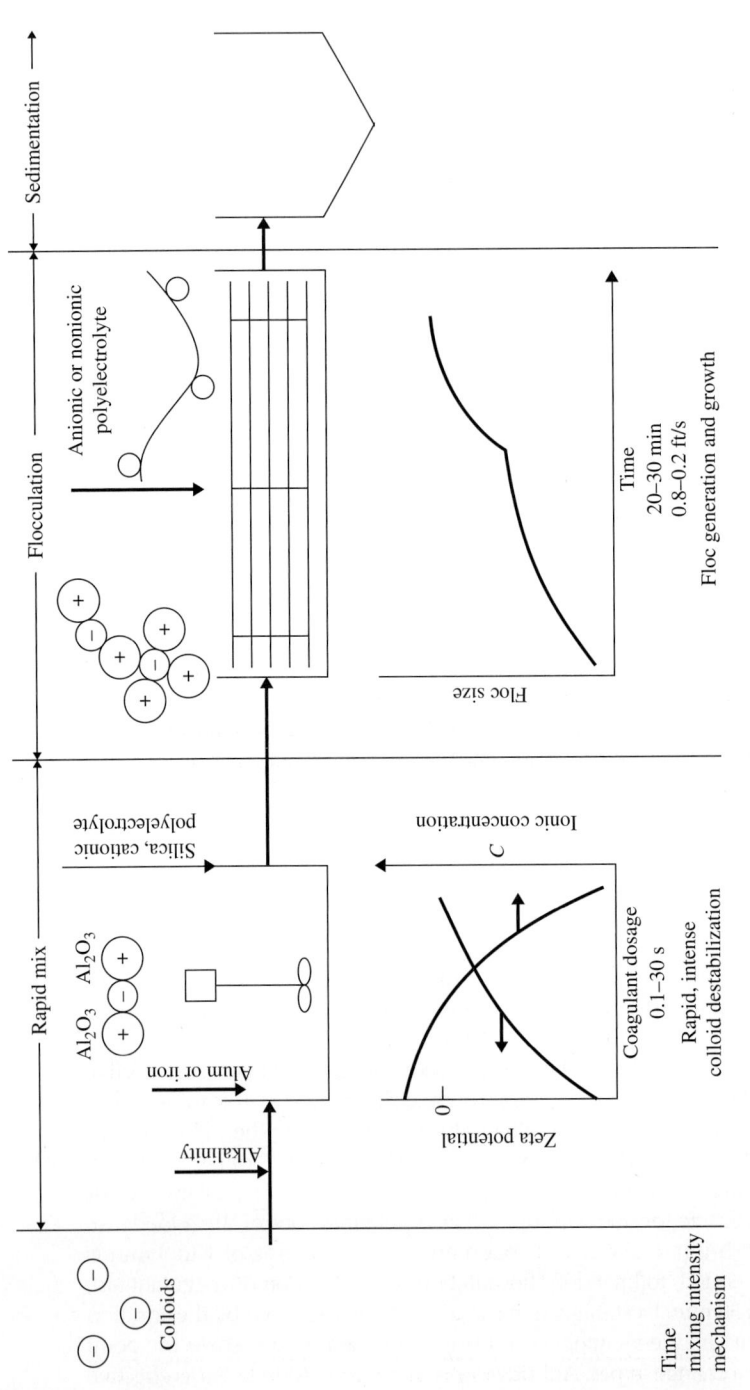

FIGURE 4.3 Mechanisms of coagulation.

Properties of Coagulants

The most popular coagulant in waste-treatment application is aluminum sulfate, or alum [$Al_2(SO_4)_3 \cdot 18H_2O$], which can be obtained in either solid or liquid form. When alum is added to water in the presence of alkalinity, the reaction is

$$Al_2(SO_4)_3 \cdot 18H_2O + 3Ca(OH)_2 \rightarrow 3CaSO_4 + 2Al(OH)_3 + 18H_2O$$

The aluminum hydroxide is actually of the chemical form $Al_2O_3 \cdot xH_2O$ and is amphoteric in that it can act as either an acid or a base. Under acidic conditions

$$[Al^{3+}][OH^-]^3 = 1.9 \times 10^{-33}$$

At pH 4.0, 51.3 mg/L of Al^{3+} is in solution. Under alkaline conditions, the hydrous aluminum oxide dissociates:

$$Al_2O_3 + 2OH^- \rightarrow 2AlO_2^- + H_2O$$

$$[AlO_2^-][H^+] = 4 \times 10^{-13}$$

At pH 9.0, 10.8 mg/L of aluminum is in solution.

The alum floc is least soluble at a pH of approximately 7.0. The floc charge is positive below pH 7.6 and negative above pH 8.2. Between these limits the floc charge is mixed. These relationships with respect to the zeta potential are shown in Fig. 4.4. High alum dosages used in the

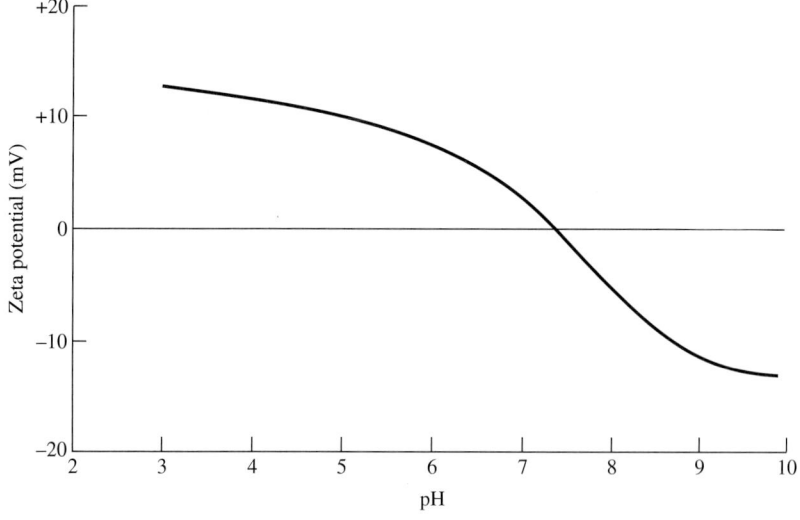

Figure 4.4 Zeta potential–pH plot for electrolytic aluminum hydroxide. (*After Riddick, 1964.*)

treatment of some industrial wastes may bring about postprecipitation of alum floc, depending on the pH of flocculation.

Ferric salts are also commonly used as coagulants but have the disadvantage of being corrosive and more difficult to handle. An insoluble hydrous ferric oxide is produced over a pH range of 3.0 to 13.0:

$$Fe^{3+} + 3OH^- \rightarrow Fe(OH)_3$$

$$[Fe^{3+}][OH^-]^3 = 10^{-36}$$

The floc charge is positive in the acid range and negative in the alkaline range, with mixed charges over the pH range 6.5 to 8.0.

The presence of anions will alter the range of effective flocculation. Sulfate ion will increase the acid range but decrease the alkaline range. Chloride ion increases the range slightly on both sides.

Lime is not a true coagulant but reacts with bicarbonate alkalinity to precipitate calcium carbonate and with *ortho*-phosphate to precipitate calcium hydroxyapatite. Magnesium hydroxide precipitates at high pH levels. Good clarification usually requires the presence of some gelatinous $Mg(OH)_2$, but this makes the sludge more difficult to dewater. Lime sludge can frequently be thickened, dewatered, and calcined to convert calcium carbonate to lime for reuse.

Coagulant Aids

The addition of some chemicals will enhance coagulation by promoting the growth of large, rapid-settling flocs. Activated silica is a short-chain polymer that serves to bind together particles of microfine aluminum hydrate and produces a tougher more durable floc. At high dosages, silica will inhibit floc formation because of its electronegative properties. The usual dosage is 5 to 10 mg/L and is usually used with alum.

Polyelectrolytes are high-molecular-weight polymers which contain adsorbable groups and form bridges between particles or charged flocs. Large flocs (0.3 to 1 mm) are thus created when small dosages of polyelectrolyte (1 to 5 mg/L) are added in conjunction with alum or ferric chloride. The polyelectrolyte is substantially unaffected by pH and can serve as a coagulant itself by reducing the effective charge on a colloid. There are three types of polyelectrolytes: a cationic, which adsorbs on a negative colloid or floc particle; an anionic, which replaces the anionic groups on a colloidal particle and permits hydrogen bonding between the colloid and the polymer; and a nonionic, which adsorbs and flocculates by hydrogen bonding between the solid surfaces and the polar groups in the polymer. Polymers do not add a significant amount of dissolved ions and result in a reduced sludge volume and, often, enhanced dewaterability. The general application of coagulants is shown in Table 4.1.

Chemical Process	Dosage Range, mg/L	pH	Comments
Lime	150–500	9.0–11.0	For colloid coagulation and P removal Wastewater with low alkalinity, and high and variable P Basic reactions: $Ca(OH)_2 + Ca(HCO_3)_2 \rightarrow 2CaCO_3 + 2H_2O$ $MgCO_3 + Ca(OH)_2 \rightarrow Mg(OH)_2 + CaCO_3$
Alum	75–250	4.5–7.5	For colloid coagulation and P removal Wastewater with high alkalinity and low and stable P Basic reactions: $Al_2(SO_4)_3 + 6H_2O \rightarrow 2Al(OH)_3 + 3H_2SO_4$
$FeCl_3$, $FeCl_2$	35–150	4.0–9.0	For colloid coagulation and P removal
$FeSO_4 \cdot 7H_2O$	70–200	4.0–6.0 8.0–10.0	Wastewater with high alkalinity and low and stable P Where leaching of iron in the effluent is allowable or can be controlled Where economical source of waste iron is available (steel mills, etc.) Basic reactions: $FeCl_3 + 3H_2O \rightarrow Fe(OH)_3 + 3HCl$
Cationic polymers	2–5	No change	For colloid coagulation or to aid coagulation with a metal Where the buildup of an inert chemical is to be avoided
Anionic and some nonionic polymers	0.25–1.0	No change	Use as a flocculation aid to speed flocculation and settling and to toughen floc for filtration
Weighting aids and clays	3–20	No change	Used for very dilute colloidal suspensions for weighting

TABLE 4.1 Chemical Coagulant Applications

Laboratory Control of Coagulation

Because of the complex reactions involved, laboratory experimentation is essential to establish the optimum pH and coagulant dosage for coagulation of a wastewater. Two procedures can be followed for this purpose: the jar test, in which pH and coagulant dosage are varied to attain the optimum operating conditions; and zeta potential control as proposed by Riddick,[3] in which coagulant is added to zero zeta potential. The procedures to determine the optimum coagulant dosage using these two tests are outlined below:

1. By zeta potential measurement[3]:
   (a) Place 1000 mL of sample in a beaker.
   (b) Add the coagulant in known increments. (The optimum pH should be established either by zeta potential or by a jar-test procedure.)
   (c) Rapid-mix the sample for 3 min after each addition of coagulant; follow with a slow mix.
   (d) Determine the zeta potential after each reagent addition and plot the results as shown in Fig. 4.5. To maintain constant volume, return the sample after each determination.
   (e) If a polyelectrolyte is to be used as a coagulant aid, it should be added last.

2. By jar-test procedure:
   (a) Using 200 mL of sample on a magnetic stirrer, add coagulant in small increments to natural or neutral pH. After each

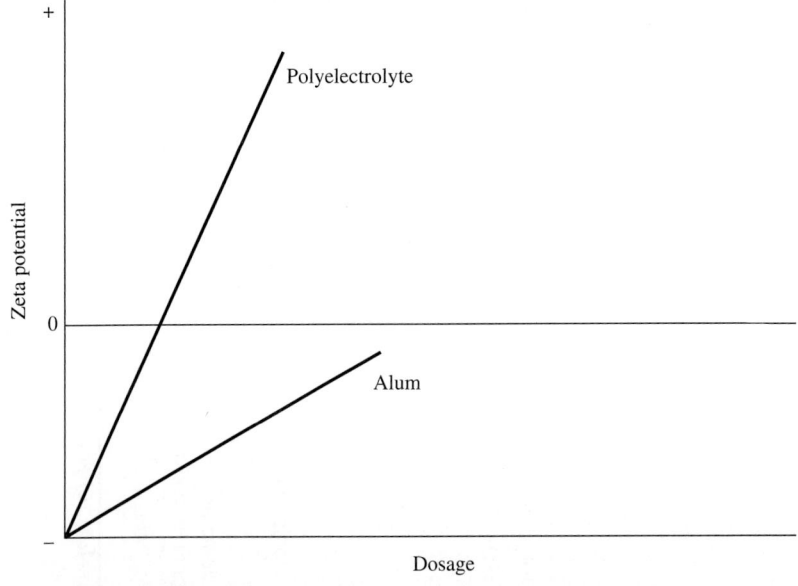

FIGURE 4.5 Typical zeta meter results.

Coagulation, Precipitation, and Metals Removal

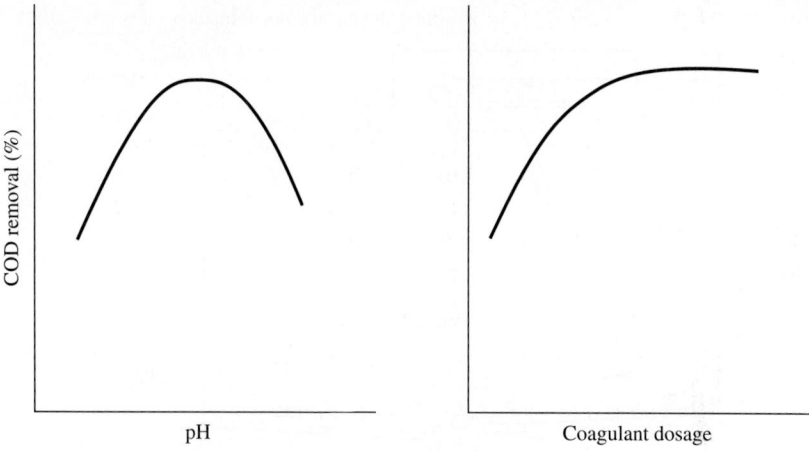

FIGURE 4.6 Characteristic plots of jar-test analysis.

addition, provide a 1-min rapid mix followed by a 3-min slow mix. Continue addition until a visible floc is formed.
(b) Using this dosage, place 1000 mL of sample in each of six beakers.
(c) Adjust the pH to 4.0, 5.0, 6.0, 7.0, 8.0, and 9.0 with standard alkali or acid. Add buffering agent as necessary to maintain pH after coagulant addition.
(d) Rapid-mix each sample for 3 min; follow this with 12-min flocculation at slow speed.
(e) Measure the effluent concentration of each settled sample.
(f) Plot the percent removal of characteristic versus pH and select the optimum pH (Fig. 4.6).
(g) Using this pH, repeat steps (b), (d), and (e), varying the coagulant dosage.
(h) Plot the percent removal versus the coagulant dosage and select the optimum dosage (Fig. 4.6).
(i) If a polyelectrolyte is used, repeat the procedure, adding polyelectrolyte toward the end of a rapid mix.

Coagulation Equipment

There are two basic types of equipment adaptable to the flocculation and coagulation of industrial wastes. The conventional system uses a rapid-mix tank, followed by a flocculation tank containing longitudinal paddles which provide slow mixing. The flocculated mixture is then settled in conventional settling tanks.

A sludge-blanket unit combines mixing, flocculation, and settling in a single unit. Although colloidal destabilization might be less effective than in the conventional system, there are distinct advantages in

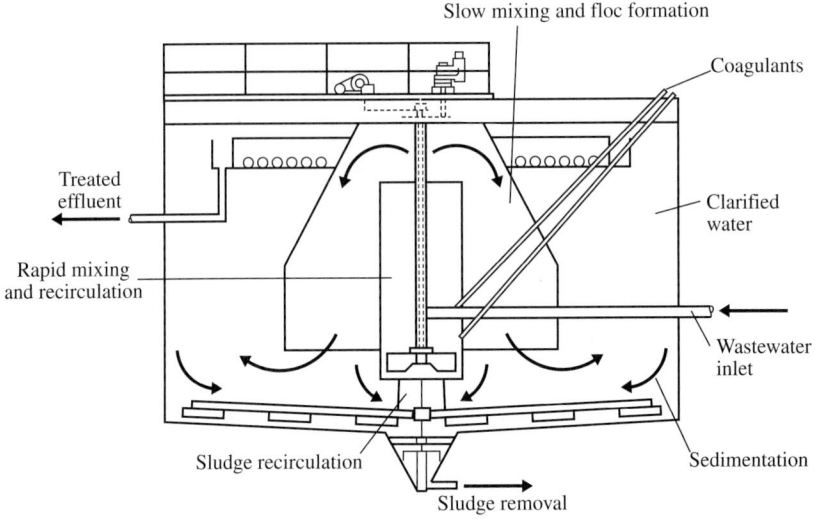

FIGURE 4.7 A reactor clarifier designed for both coagulation and settling.

recycling preformed floc. With lime and a few other coagulants, the time required to form a settleable floc is a function of the time necessary for calcium carbonate or other calcium precipitates to form nuclei on which other calcium materials can deposit and grow large enough to settle. It is possible to reduce both coagulant dosage and the time of floc formation by seeding the influent wastewater with previously formed nuclei or by recycling a portion of the precipitated sludge. Recycling preformed floc can frequently reduce chemical dosages, the blanket serves as a filter for improved effluent clarity, and denser sludges are frequently attainable. A sludge-blanket unit is shown in Fig. 4.7.

Coagulation of Industrial Wastes

Coagulation may be used for the clarification of industrial wastes containing colloidal and suspended solids. Paperboard wastes can be effectively coagulated with low dosages of alum. Silica or polyelectrolyte will aid in the formation of a rapid-settling floc. Typical data reported are summarized in Table 4.2.

Wastes containing emulsified oils can be clarified by coagulation.[7] An emulsion can consist of droplets of oil in water. The oil droplets are of approximately 10^{-5} cm and are stabilized by adsorbed ions. Emulsifying agents include soaps and anion-active agents. The emulsion can be broken by "salting out" with the addition of salts, such as $CaCl_2$. Flocculation will then effect charge neutralization and entrainment, resulting in clarification. An emulsion can also frequently be broken by lowering of the pH of the waste solution. An example of such a waste is that produced by ball-bearing manufacture, which contains cleaning

Waste	Influent BOD, ppm	Influent SS, ppm	Effluent BOD, ppm	Effluent SS, ppm	Effluent pH	Coagulants Alum, ppm	Coagulants Silica, ppm	Coagulants Other, ppm	Detentions, h	Sludge, % Solids	Remarks	Reference
Board		350–450		15–60		3	5		1.7	2–4		4
Board		140–420		10–40		1		10*	0.3	2	Flotation	4
Board		240–600		35–85					2.0	2–5	950 gal/ (d · ft^2)	5
Board†	127	593	68	44	6.7	10–12	10		1.3	1.76		5
Tissue	140	720	36	10–15		2	4					6
Tissue	208		33		6.6		4					6

*Gluc.
†15,000 gal/ton waste paper.
Note: gal/(d · ft^2) = 4.075×10^{-2} m^3/(d · m^2)
gal/ton = 4.17×10^{-3} m^3/t

TABLE 4.2 Chemical Treatment of Paper and Paperboard Wastes

soaps and detergents, water-soluble grinding oils, cutting oils, and phosphoric acid cleaners and solvents. Treatment of this waste has been effected by the use of 800 mg/L alum, 450 mg/L H_2SO_4, and 45 mg/L polyelectrolyte. The results obtained are summarized in Table 4.3a.

The presence of anionic surface agents in a waste will increase the coagulant dosage. The polar head of the surfactant molecule enters the double layer and stabilizes the negative colloids. Industrial laundry wastes have been treated with H_2SO_4 followed by lime and alum; this has resulted in a reduction of COD of 12,000 to 1800 mg/L and a reduction of suspended solids of 1620 to 105 mg/L. Chemical dosages of 1400 mg/L H_2SO_4, 1500 mg/L lime, and 300 mg/L alum were required, yielding 25 percent by volume of settled sludge.

(a) Ball Bearing Manufacture*		
	Analysis	
	Influent	Effluent
pH	10.3	7.1
Suspended solids, mg/L	544	40
Oil and grease, mg/L	302	28
Fe, mg/L	17.9	1.6
PO_4, mg/L	222	8.5
(b) Laundromat		
	Influent, mg/L	Effluent, mg/L
ABS	63	0.1
BOD	243	90
COD	512	171
PO_4	267	150
$CaCl_2$	480	
Cationic surfactant	88	
pH	7.1	7.7
(c) Latex-Base Paint Manufacture†		
	Influent, mg/L	Effluent, mg/L
COD	4340	178
BOD	1070	90
Total solids	2550	446

*800 mg/L alum, 450 mg/L H_2SO_4, 45 mg/L polyelctrolyte.
†345 mg/L alum, pH 3.5–4.0.

TABLE 4.3 Coagulation of Industrial Wastewaters

Laundromat wastes containing synthetic detergent have been coagulated with a cationic surfactant to neutralize the anionic detergent and the addition of a calcium salt to provide a calcium phosphate precipitate for flocculation. Typical results obtained are summarized in Table 4.3b. Operating the system above pH 8.5 will produce nearly complete phosphate removal.

BOD removals of 90 percent have been achieved on laundry wastes at pH 6.4 to 6.6 with coagulant dosages of 2 lb $Fe_2(SO_4)_3$/thousand gal waste[8] (0.24 kg/m^3).

Polymer waste from latex manufacture has been coagulated with 500 mg/L ferric chloride and 200 mg/L lime at pH 9.6. COD and BOD reductions of 75 and 94 percent, respectively, were achieved from initial values of 1000 and 120 mg/L. The resulting sludge was 1.2 percent solids by weight, containing 101 lb solids/thousand gal waste (12 kg/m^3) treated. Waste from the manufacture of latex base paints[8] has been coagulated with 345 mg/L alum at pH 3.0 to 4.0, yielding 20.5 lb sludge/thousand gal waste (2.5 kg/m^3) of 2.95 percent solids by weight. The treatment results are summarized in Table 4.3c. Wastes from paint-spray booths in automobile assembly plants have been clarified with 400 mg/L $FeSO_4$ at pH 7.0, yielding 8 percent sludge by weight.

Synthetic rubber wastes have been treated at pH 6.7 with 100 mg/L alum, yielding a sludge that was 2 percent of the original waste volume. COD was reduced from 570 to 100 mg/L, and BOD from 85 to 15 mg/L.

Vegetable-processing wastes have been coagulated[9] with lime, yielding BOD removals of 35 to 70 percent with lime dosages of approximately 0.5 lb lime/lb influent BOD (0.5 kg/kg). Results from the coagulation of textile wastewaters are shown in Table 4.4, and color removal from pulp and paper mill effluents in Table 4.5.

BOD removals of 75 to 80 percent have been obtained on wool-scouring wastes with 1 to 3 lb $CaCl_2$/lb BOD (1 to 3 kg/kg). Carbonation was used for pH control.[13]

Talinli[14] treated tannery wastewaters with 2000 mg/L lime at pH 11 and 2 mg/L nonionic polyelectrolyte, as shown in Table 4.6. The resulting sludge volume was 30 percent.

4.3 Heavy Metals Removal

There are a number of technologies available for the removal of heavy metals from a wastewater. These are summarized in Table 4.7. Chemical precipitation is most commonly employed for most of the metals. Typically, source reduction and stream segregation are practiced before these streams intermingle with others. Common precipitants include OH^-, CO_3^{2-}, and S^{2-}. Metals are precipitated as the hydroxide through the addition of lime or caustic to a pH of minimum solubility. However, several of these compounds are amphoteric and exhibit a point of minimum solubility. The pH of minimum solubility varies

Plant	Coagulant	Dosage, mg/L	pH	Color* Influent	Color* Removal, %	COD Influent, mg/L	COD Removal, %
1	$Fe_2(SO_4)_3$	250	7.5–11.0	0.25	90	584	33
	Alum	300	5–9		86		39
	Lime	1200			68		30
2	$Fe_2(SO_4)_3$	500	3–4, 9–11	0.74	89	840	49
	Alum	500	8.5–10		89		40
	Lime	2000			65		40
3	$Fe_2(SO_4)_3$	250	9.5–11	1.84	95	825	38
	Alum	250	6–9		95		31
	Lime	600			78		50
4	$Fe_2(SO_4)_3$	1000	9–11	4.60	87	1570	31
	Alum	750	5–6		89		44
	Lime	2500			87		44

*Color sum of absorbances at wavelengths of 450, 550, and 650 nm.
Source: Adapted from Olthof and Eckenfelder, 1975.

TABLE 4.4 Coagulation of Textile Wastewaters[10]

Plant	Coagulant	Dosage, mg/L	pH	Color		COD	
				Influent	Removal, %	Influent, mg/L	Removal, %
1	$Fe_2(SO_4)_3$	500	3.5–4.5	2250	92	776	60
	Alum	400	4.0–5.0		92		53
	Lime	1500	—		92		38
2	$Fe_2(SO_4)_3$	275	3.5–4.5	1470	91	480	53
	Alum	250	4.0–5.5		93		48
	Lime	1000	—		85		45
3	$Fe_2(SO_4)_3$	250	4.5–5.5	940	85	468	53
	Alum	250	5.0–6.5		91		44
	Lime	1000	—		85		40

Source: Adapted from Olthof and Eckenfelder, 1975.

TABLE 4.5 Color Removal from Pulp and Paper Mill Effluents[12]

Parameter	Influent	Effluent	% Removal
COD	7800	2900	63
BOD	3500	1450	58
SO_4	1800	1200	33
Chromium	100	3	97

Source: Adapted from Talinli, 1994.

TABLE 4.6 Coagulation of Tannery Wastewaters

Conventional precipitation
Hydroxide
Sulfide
Carbonate
Coprecipitation
Enhanced precipitation
Dimethyl thio carbamate
Diethyl thio carbamate
Trimercapto-s-triazine, trisodium salt
Other methods
Oxidation/reduction
Ion exchange
Adsorption
Biosorption
Recovery opportunities
Ion exchange
Membranes
Electrolytic techniques

TABLE 4.7 Heavy Metals Removal Technologies

with the metal in question as shown in Fig. 4.8. Metals can also be precipitated as the sulfide (Fig. 4.8) or in some cases as the carbonate (Fig. 4.9). Table 4.8 summarizes effective types of precipitation for selected metal ions. Removal of metals to acceptable levels may be needed for reuse application, to reduce toxic effects in biological treatment, and/or to comply with discharge standards.

Coagulation, Precipitation, and Metals Removal

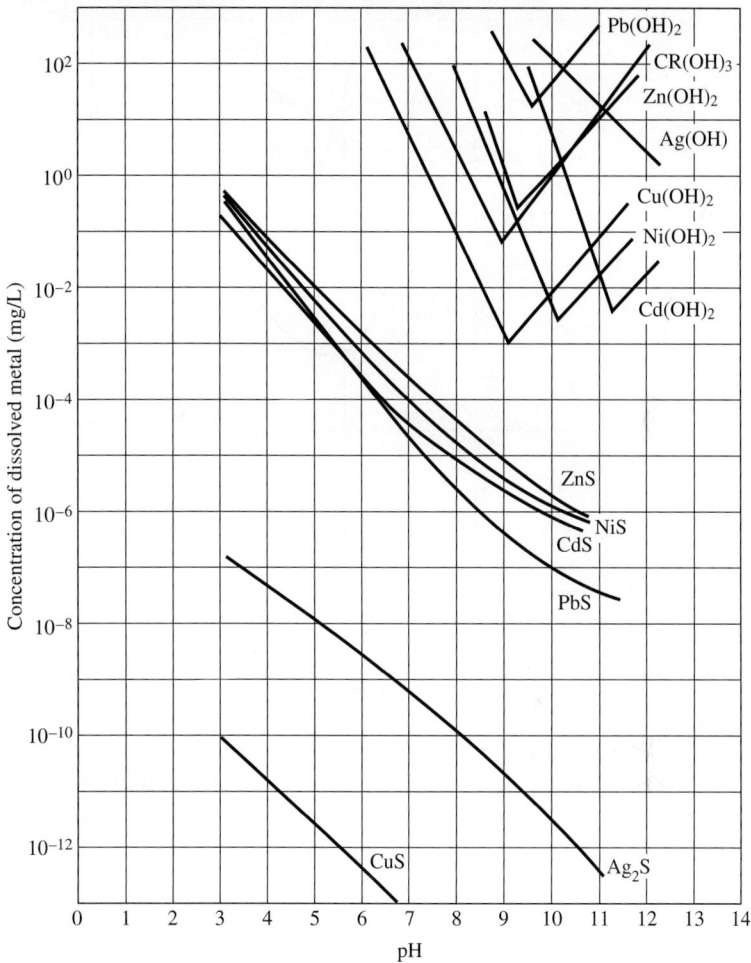

FIGURE 4.8 Heavy metals precipitation as the hydroxide and the sulfide.

In treating industrial wastewaters containing metals, it is frequently necessary to pretreat the wastewaters to remove substances that will interfere with the precipitation of the metals. Cyanide and ammonia form complexes with many metals that limit the removal that can be achieved by precipitation as shown in Fig. 4.10 for the case of ammonia. Cyanide can be removed by alkaline chlorination or other processes such as catalytic oxidation of carbon. Cyanide wastewaters containing nickel or silver are difficult to treat by alkaline chlorination because of the slow reaction rate of these metal complexes. Ferrocyanide [$Fe(CN)_6^{4-}$] is oxidized to ferricyanide [$Fe(CN)_6^{3-}$], which resists further oxidation. Ammonia can be removed

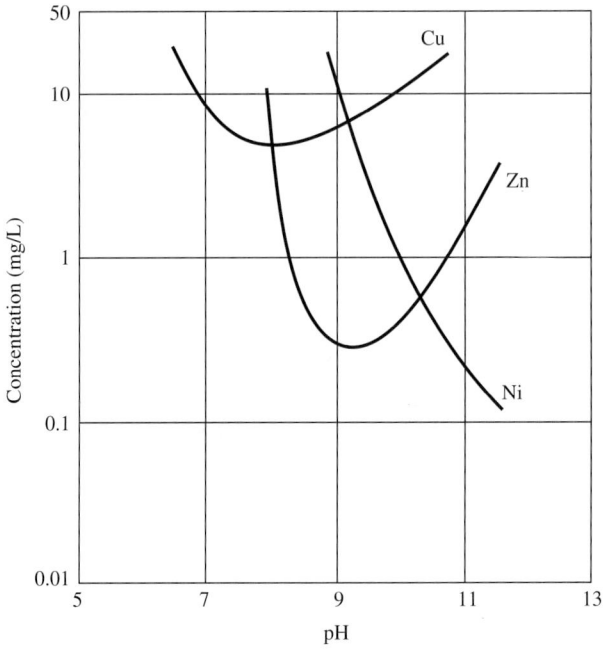

FIGURE 4.9 Residual soluble metal concentrations using Na_2CO_3.

by stripping, break-point chlorination, or other suitable methods prior to the removal of metals.

For many metals, such as arsenic and cadmium, coprecipitation with iron or aluminum is highly effective for removal to low residual levels. In this case the metal adsorbs to the alum or iron floc. In order to meet low effluent requirements, it may be necessary in some cases to provide filtration to remove floc carried over from the precipitation process. With precipitation and clarification alone, effluent metals concentrations may be as high as 1 to 2 mg/L. Filtration should reduce these concentrations to 0.5 mg/L or less. Carbamate salts can be employed for enhanced precipitation. Because of the chemical cost, this type of precipitation is usually employed as a polishing step following conventional precipitation. Typical results are shown in Table 4.9. Metals can be removed by adsorption on activated carbon, aluminum oxides, silica, clays, and synthetic material such as zeolites and resins (see Chap. 9). In the case of adsorption, higher pH favors the adsorption of cations while a lower pH favors the adsorption of anions. Complexing agents will interfere with cationic species. There will be competition from major background ions such as calcium or sodium. For chromium waste treatment, hexavalent chromium must first be reduced to the trivalent state, Cr^{3+}, and then precipitated with lime. This is referred to as the process of reduction and precipitation.

	Type of Precipitation		
Metal Ion	Hydroxide	Sulfide	Carbonate
Antimony			
Arsenic	X	X	
Beryllium		X	T
Cadmium	X		X
Chromium	X	X	
Copper	X	X	
Lead	X	X	X
Mercury		X	
Nickel	X	X	X
Selenium			
Silver	X	T	
Thallium	.	T	
Zinc	X	X	T
Iron	X	X	
Manganese	X	T	

Notes: X indicates process is applicable for removal of the metal ion. Bench or pilot scale data are available to affirm precipitation occurrence.
T indicates process may be applicable for removal of the metal ion. Bench or pilot scale data are not available for confirmation. However, the metal salt's solubility indicates precipitation may occur.

TABLE 4.8 Effective Types of Precipitation for Selected Metal Ions

Arsenic

Arsenic and arsenical compounds are present in wastewaters from the metallurgical industry, glassware and ceramic production, tannery operation, dyestuff manufacture, pesticide manufacture, some organic and inorganic chemicals manufacture, petroleum refining, and the rare-earth industry. Arsenic is removed from wastewater by chemical precipitation. Enhanced performance is achieved as arsenate (AsO_4^{3-}, As^{5+}) rather than arsenite (AsO_2^-, As^{3+}). Arsenite is therefore usually oxidized to arsenate prior to precipitation. Effluent arsenic levels of 0.05 mg/L are obtainable by precipitation of the arsenic as the sulfide by the addition of sodium or hydrogen sulfide at pH of 6 to 7. In order to meet reported effluent levels, polishing of the effluent by filtration is usually required.

Arsenic present in low concentrations can also be reduced by filtration through activated carbon. Effluent concentrations of 0.06 mg/L arsenic have been reported from an initial concentration of 0.2 mg/L. Arsenic is removed by coprecipitation with a ferric hydroxide floc that ties up the arsenic and removes it from solution. Effluent concentrations of less than 0.005 mg/L have been reported from this process.

Chapter Four

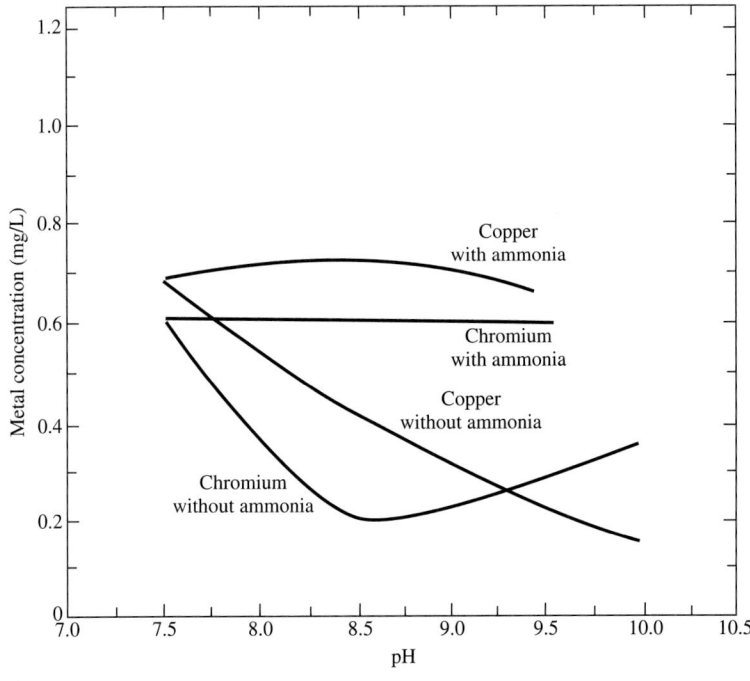

FIGURE 4.10 Comparative optimum pH values for metal removal with and without ammonia.

Metal	Input, mg/L	Ca(OH)$_2$, mg/L	X,* mg/L
Cadmium	0.4	0.2	0.04
Chromium	1.2	0.1	0.05
Copper	1.3	0.1	0.05
Nickel	3.5	0.9	0.67
Lead	7.4	0.4	0.35
Mercury	1.4	0.1	0.01
Zinc	13.5	0.2	0.09

*Enhancing chemicals:
TMT15—trimercapto triazine trisodium salt
Nalfloc Nalmet 8154 (diethyl thio carbamate)
IMP HM1 (dimethyl thio carbamate)

TABLE 4.9 Enhanced Removal of Soluble Metals by Precipitation

Barium

Barium is present in wastewaters from the paint and pigment industry; the metallurgical industry; glass, ceramic, and dye manufacturers; and the vulcanizing of rubber. Barium has also been reported in explosives manufacturing wastewater. Barium is removed from solution by precipitation as barium sulfate.

Barium sulfate is extremely insoluble, having a maximum theoretical solubility at 25°C of approximately 1.4 mg/L as barium at stoichiometric concentrations of barium and sulfate. The solubility level of barium can be reduced in the presence of excess sulfate. Coagulation of barium salts as the sulfate is capable of reducing barium to effluent levels of 0.03 to 0.3 mg/L. Barium can also be removed from solution by ion exchange and electrodialysis, although these processes are more expensive than chemical precipitation.

Cadmium

Cadmium is present in wastewaters from metallurgical alloying, ceramics, electroplating, photography, pigment works, textile printing, chemical industries, and lead mine drainage. Cadmium is removed from wastewaters by precipitation or ion exchange. In some cases, electrolytic and evaporative recovery processes can be employed, provided the wastewater is in a concentrated form. Cadmium forms an insoluble and highly stable hydroxide at an alkaline pH. Cadmium in solution is approximately 1 mg/L at pH 8 and 0.05 mg/L at pH 10 to 11. Coprecipitation with iron hydroxide at pH 6.5 will reduce cadmium to 0.008 mg/L; iron hydroxide at pH 8.5 reduces cadmium to 0.05 mg/L. Sulfide and lime precipitation with filtration will yield 0.002 to 0.03 mg/L at pH 8.5 to 10. Cadmium is not precipitated in the presence of complexing ions, such as cyanide. In these cases, it is necessary to pretreat the wastewater to destroy the complexing agent. In the case of cyanide, cyanide destruction is necessary prior to cadmium precipitation. A hydrogen peroxide oxidation precipitation system has been developed that simultaneously oxidizes cyanides and forms the oxide of cadmium, thereby yielding cadmium, whose recovery is feasible. Results for the hydroxide precipitation of cadmium are shown in Table 4.10.

Chromium

The reducing agents commonly used for chromium wastes are ferrous sulfate, sodium meta-bisulfite, or sulfur dioxide. Ferrous sulfate and sodium meta-bisulfite may be dry- or solution-fed; SO_2 is diffused into the system directly from gas cylinders. Since the reduction of chromium is most effective at acidic pH values, a reducing agent with acidic properties is desirable. When ferrous sulfate is used as the

Method	Treatment pH	Initial Cd, mg/L	Final Cd, mg/L
Hydroxide precipitation	8.0	—	1.0
	9.0	—	0.54
	10.0	—	0.10
	9.3–10.6	4.0	0.20
Hydroxide precipitation plus filtration	10.0	0.34	0.054
	10.0	0.34	0.033
Hydroxide precipitation plus filtration	11.0	—	0.00075
	11.0	—	0.00070
Hydroxide precipitation plus filtration	11.5	—	0.014
Hydroxide precipitation plus filtration	—	—	0.08
Coprecipitation with ferrous hydroxide	6.0	—	0.050
Coprecipitation with ferrous hydroxide	10.0	—	0.044
Coprecipitation with alum	6.4	0.7	0.39

TABLE 4.10 Hydroxide Precipitation Treatment for Cadmium

reducing agent, the Fe^{2+} is oxidized to Fe^{3+}; if meta-bisulfite or sulfur dioxide is used, the negative radical SO_3^{2-} is converted to SO_4^{2-}. The general reactions are

$$Cr^{6+} + Fe^{2+} \text{ or } SO_2 \text{ or } Na_2S_2O_5 + H^+ \rightarrow Cr^{3+} + Fe^{3+} \text{ or } SO_4^{2-}$$

$$Cr^3 + 3OH^- \rightarrow Cr(OH)_3 \downarrow$$

Ferrous ion reacts with hexavalent chromium in an oxidation-reduction reaction, reducing the chromium to a trivalent state and oxidizing the ferrous ion to the ferric state. This reaction occurs rapidly at pH levels below 3.0. The acidic properties of ferrous sulfate are low at high dilution; acid must therefore be added for pH adjustment. The use of ferrous sulfate as a reducing agent has the disadvantage that a contaminating sludge of $Fe(OH)_3$ is formed when an alkali is added. In order to obtain a complete reaction, an excess dosage of 2.5 times the theoretical addition of ferrous sulfate must be used.

Reduction of chromium can also be accomplished by using either sodium meta-bisulfite or SO_2. In either case reduction occurs by reaction with the H_2SO_3 produced in the reaction. The H_2SO_3 ionizes according to the mass action:

$$\frac{(H^+)(HSO_3^-)}{(H_2SO_3)} = 1.72 \times 10^{-2}$$

Above pH 4.0, only 1 percent of sulfite is present as H_2SO_3 and the reaction is very slow. During this reaction, acid is required to neutralize the NaOH formed. The reaction is highly dependent on both pH and temperature. At pH levels below 2.0 the reaction is practically instantaneous and close to theoretical requirements.

The theoretical quantities of chemicals required for 1 ppm Cr are:

2.81 ppm $Na_2S_2O_5$ (97.5 percent)
1.52 ppm H_2SO_4
2.38 ppm lime (90 percent)
1.85 ppm SO_2

At pH levels above 3, when a basic chrome sulfate is produced, the quantities of lime required for subsequent neutralization are reduced. At pH 8.0 to 9.9, $Cr(OH)_3$ is virtually insoluble. Experimental investigations have shown that the sludge produced will compact to 1 to 2 percent by weight.

Since dissolved oxygen is usually present in wastewaters, an excess of SO_2 must be added to account for the oxidation of the SO_3^{2-} to SO_4^{2-}:

$$H_2SO_3 + \frac{1}{2}O_2 \rightarrow H_2SO_4$$

An excess dosage of 35 ppm SO_2 will usually be sufficient for reaction with the dissolved oxygen present.

The acid requirements for the reduction of Cr^{6+} depend on the acidity of the original waste, the pH of the reduction reaction, and the type of reducing agent used (e.g., SO_2 produces an acid but meta-bisulfite does not). Since it is difficult, if not impossible, to predict these requirements, it is usually necessary to titrate a sample to the desired pH endpoint with standardized acid.

Many small plating plants have a total daily volume of waste of less than 30,000 gal/d (114 m³/d). The most economical system for such plants is a batch treatment in which two tanks are provided, each with a capacity of one day's flow. One tank is undergoing treatment while the other is filling. Accumulated sludge is either drawn off and hauled to disposal or dewatered on sand drying beds. A separable dry

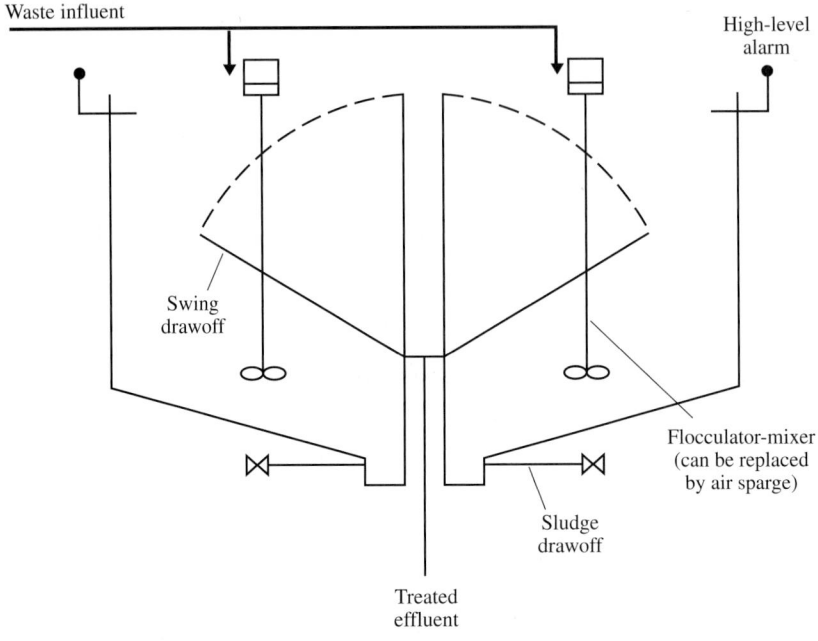

FIGURE 4.11 Batch treatment of chromium wastes.

cake can be obtained after 48 h on a sand bed. A typical batch treatment system is schematically shown in Fig. 4.11.

When the daily volume of waste exceeds 30,000 to 40,000 gal (114 to 151 m^3), batch treatment is usually not feasible because of the large tankage required. Continuous treatment requires a tank for acidification and reduction, then a mixing tank, for lime addition, and a settling tank. The retention time in the reduction tank is dependent on the pH employed but should be at least 4 times the theoretical time for complete reduction. Twenty minutes will usually be adequate for flocculation. Final settling should not be designed for an overflow rate in excess of 500 gal/(d · ft^2) (20 m^3/[d · m^2]).

In cases where the chrome content of the rinsewater varies markedly, equalization should be provided before the reduction tank to minimize fluctuations in the chemical feed system. The fluctuation in chrome content can be minimized by provision of a drain station before the rinse tanks.

Successful operation of a continuous chrome reduction process requires instrumentation and automatic control. Redox control and pH are provided for the reduction tank. The addition of lime should be modulated by a second pH control system. A continuous chrome reduction/precipitation process is shown in Fig. 4.12. Reported results for the precipitation of chromium are shown in Table 4.11.

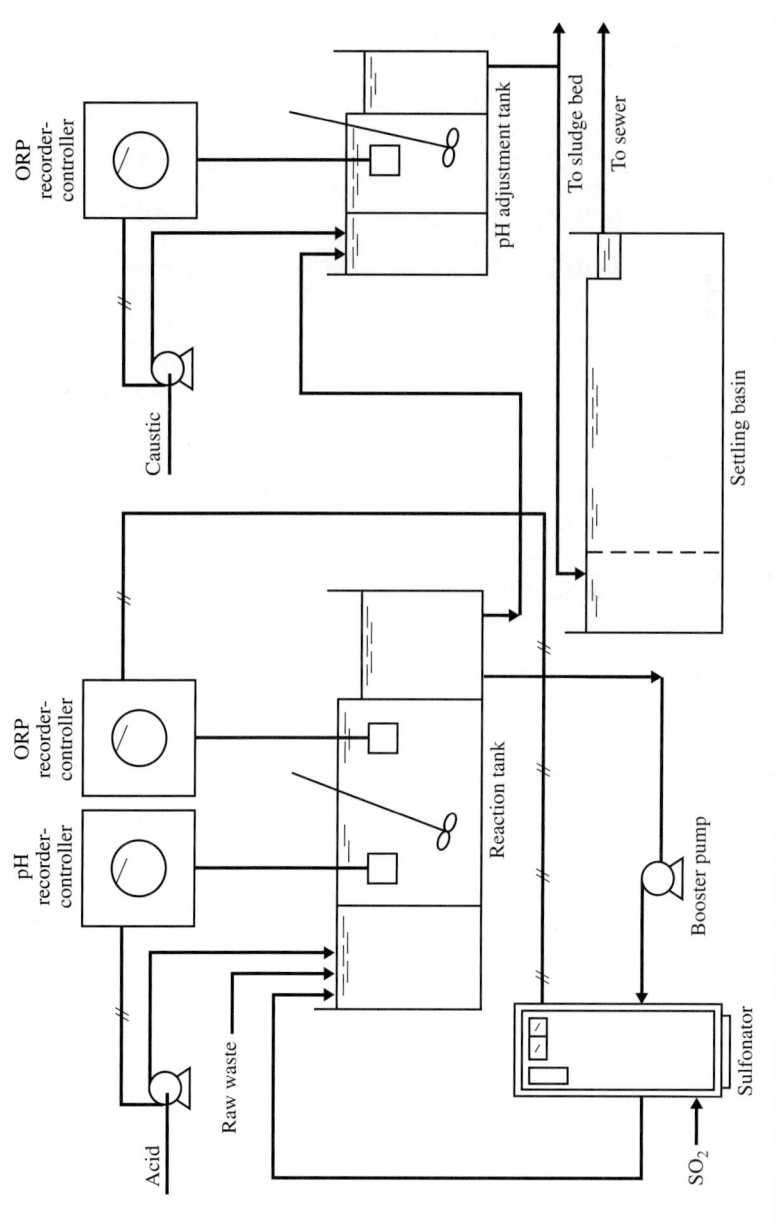

FIGURE 4.12 Continuous chrome waste treatment system. (*Courtesy of Fischer-Porter, Inc.*)

		Chromium, mg/L	
Method	pH	Initial	Final
Precipitation	7–8	140	1.0
	7.8–8.2	16.0	0.06–0.15
	8.5	47–52	0.3–1.5
	8.8	650	18
	8.5–10.5	26.0	0.44–0.86
	8.8–10.1	—	0.6–30
	12.2	650	0.3
Precipitation with sand filtration	8.5	7400	1.3–4.6
	8.5	7400	0.3–1.3
	9.8–10.0	49.4	0.17
	9.8–10.0	49.4	0.05

TABLE 4.11 Summary of Trivalent Chromium Treatment Results

	Chromium, mg/L		Resin Capacity*
Wastewater Source	Influent	Effluent	
Cooling tower blowdown	17.9	1.8	5–6
	10.0	1.0	2.5–4.5
	7.4–10.3	1.0	—
	9.0	0.2	2.5
Plating rinsewater	44.8	0.025	1.7–2.0
	41.6	0.01	5.2–6.3
Pigment manufacture	1210	<0.5	—

*lb chromate/ft^3 resin.
Source: Adapted from Patterson, 1985.

TABLE 4.12 Ion Exchange Performance in Hexavalent Chromium Removal

Coagulation, Precipitation, and Metals Removal

Removal of chromium by ion exchange is shown in Table 4.12. An example of metals removal is shown in Example 4.2.

Example 4.2. 30,000 gal/d (114 m^3/d) of a waste containing 49 mg/L Cr^{6+}, 11 mg/L Cu, and 12 mg/L Zn is to be treated daily by using SO$_2$. Compute the chemical requirements and the daily sludge production. (Assume the waste contains 5 mg/L O$_2$.)

Solution
(a) SO$_2$ requirements are as follows. For Cr^{6+}

$$1.85\left(\frac{\text{mg SO}_2}{\text{mg Cr}^{6+}}\right) \times 49(\text{mg Cr}^{6+}/\text{L}) \times 8.34\left(\frac{\text{lb/million gal}}{\text{mg/L}}\right)$$
$$\times 0.03 \text{ million gal/d} = 22.7 \text{ lb/d} \quad (10.3 \text{ kg/d})$$

and for O$_2$, where 1 part of O$_2$ requires 4 parts of SO$_2$:

$$4\left(\frac{\text{mg SO}_2}{\text{mg O}_2}\right) \times 5(\text{mg O}_2/\text{L}) \times 8.34 \times 0.03 = 5.0 \text{ lb/d} \quad (2.3 \text{ kg/d})$$

$$\text{Total} = 27.7 \text{ lb/d} \quad (12.6 \text{ kg/d})$$

(b) Lime requirements are as follows. For Cr^{3+}:

$$2.38\left(\frac{\text{mg lime}}{\text{mg Cr}^{3+}}\right) \times 49 \times 8.34 \times 0.03 = 29.2 \text{ lb/d} \quad (13.3 \text{ kg/d})$$

and for Cu and Zn (each part of Cu and Zn requiring 1.3 parts of 90 percent lime for precipitation):

$$1.3\left(\frac{\text{mg lime}}{\text{mg Cu or Zn}}\right) \times 23(\text{mg Cu and Zn/L}) \times 8.34 \times 0.03 = 7.5 \text{ lb/d} \quad (3.4 \text{ kg/d})$$

$$\text{Total} = 36.7 \text{ lb/d} \quad (16.7 \text{ kg/d})$$

(c) Sludge production is:

$$1.98\left(\frac{\text{mg Cr(OH)}_3}{\text{mg Cr}^{6+}}\right) \times 49 \times 8.34 \times 0.03 = 24.3 \text{ lb/d Cr(OH)}_3 \quad (11 \text{ kg/d})$$

$$1.53\left(\frac{\text{mg sludge}}{\text{mg Cu or Zn}}\right) \times 23 \times 8.34 \times 0.03$$
$$= 8.8 \text{ lb/d Cu(OH)}_2 \text{ and Zn(OH)}_2 \quad (4 \text{ kg/d})$$

$$\text{Total} = 33.1 \text{ lb/d} \quad (15 \text{ kg/d})$$

If the sludge concentrates to 1.5 percent by weight, the volume that will require disposal each day can be calculated as follows:

$$\frac{33.1 \text{ lb/d}}{0.015 \text{ lb solids/lb sludge} \times 8.34 \text{ lb/gal}} = 265 \text{ gal/d} \quad (1.0 \text{ m}^3/\text{d})$$

It should be noted that some of the copper and zinc will be soluble unless the final pH after lime addition exceeds pH 9.0.

Copper

The primary sources of copper in industrial wastewaters are metal-process pickling baths and plating baths. Copper may also be present in wastewaters from a variety of chemical manufacturing processes employing copper salts or a copper catalyst. Copper is removed from wastewaters by precipitation or recovery processes, which include ion exchange, evaporation, and electrodialysis. The value of recovered copper metal will frequently make recovery processes attractive. Ion exchange or activated carbon are feasible treatment methods for wastewaters containing copper at concentrations of less than 200 mg/L. Copper is precipitated as a relatively insoluble metal hydroxide at alkaline pH. In the presence of high sulfates, calcium sulfate will also be precipitated and will interfere with the recovery value of the copper sludge. This may dictate the use of a more expensive alkali such as NaOH to obtain a pure sludge. Cupric oxide has a minimum solubility between pH 9.0 and 10.3 with a reported solubility of 0.01 mg/L. Field practice has indicated that the maximum technically feasible treatment level for copper by chemical precipitation is 0.02 to 0.07 mg/L as soluble copper. Precipitation with sulfide at pH 8.5 will result in effluent copper concentrations of 0.01 to 0.02 mg/L. Low residual concentrations of copper are difficult to achieve in the presence of complexing agents such as cyanide and ammonia. Removal of the complexing agent by pretreatment is essential for high copper removal. Copper cyanide is effectively removed on activated carbon. A summary of copper results after hydroxide precipitation is shown in Table 4.13.

Fluorides

Fluorides are present in wastewaters from glass manufacturing, electroplating, steel and aluminum production, and pesticide and fertilizer manufacture. Fluoride is removed by precipitation with lime as calcium fluoride. Effluent concentrations in the order of 10 to 20 mg/L are readily obtainable. Lime precipitation at a pH above 12 has created problems with solids removal, poor settling, and cementation of filters. Enhanced removal of fluoride has been reported in the presence of magnesium. The increased removal is attributed to adsorption of the fluoride ion into the magnesium hydroxide floc, resulting in effluent fluoride concentrations of less than 1.0 mg/L. Alum coprecipitation

Source (Treatment)	Copper Concentration, mg/L	
	Initial	Final
Metal processing (lime)	204–385	0.5
Nonferrous metal processing (lime)	—	0.2–2.3
Metal processing (lime)	—	1.4–7.8
Electroplating (caustic, soda ash + hydrazine)	6.0–15.5	0.09–0.24 (solution) 0.30–0.45 (total)
Machine plating (lime + coagulant)	—	2.2
Metal finishing (lime)	—	0.19 average
Brass mill (lime)	10–20	1–2
Plating (CN oxidation, Cr reduction neutralization)	11.4	2.0
Wood preserving (lime)	0.25–1.1	0.1–0.35
Brass mill (hydrazine + caustic)	75–124	0.25–0.85
Silver plating (CN oxidation, lime + $FeCl_3^+$ + filtration)	30 (average)	0.16–0.3
Copper sulfate manufacture (lime)	433	0.14–1.25 (0.48 average)
Integrated circuit manufacture (lime)	0.23	0.05

Source: Adapted from Patterson, 1985.

TABLE 4.13 Summary of Copper Results After Hydroxide Precipitation Treatment

will result in effluent levels of 0.5 to 2.0 mg/L. Low concentrations of fluoride can be removed by ion exchange. Fluoride removal through ion exchange pretreated and regenerated with aluminum salts is attributable to aluminum hydroxide precipitated in a column bed. Fluoride is removed through contact beds of activated alumina, which may be employed as a polishing unit to follow lime precipitation. Fluoride concentrations of 30 mg/L from the lime precipitation process have been reduced to approximately 2 mg/L upon passage through an activated alumina contact bed. A summary of fluoride treatment processes and levels of treatment achieved is in Table 4.14.

Iron

Iron is present in a wide variety of industrial wastewaters including mining operations, ore milling, chemical industrial wastewater, dye manufacture, metal processing, textile mills, petroleum refining, and others. Iron exists in the ferric or ferrous form, depending on pH and dissolved oxygen concentration. At neutral pH and in the presence

Treatment Process	Fluoride Concentration, mg/L		Current Application
	Initial	Final	
Lime		10	Industrial
Lime	1000–3000	20	Industrial
Lime	500–1000	20–40	Industrial
Lime	200–700	6 (16-h settling)	Industrial
Lime	45	8	Industrial
Lime	4–20	5.9 (average)	Industrial
Lime	590	80	Industrial
Lime	57.8	29.1 (average) 14–16 (best)	Industrial
Lime	93,000	0.8–8.8	Industrial (pilot-scale)
Lime	—	10.6 (clarified)	Industrial
		10.4 (filtered)	
Lime, two-stage	1,460	9	Industrial
Lime + calcium chloride	—	12	Industrial
Lime + alum	—	1.5	Industrial
Lime + alum	2,020	2.4	Industrial (pilot-scale)
Calcium carbonate + lime, two-stage	11,100	6	Industrial

Source: Adapted from Patterson, 1985.

TABLE 4.14 Summary of Fluoride Treatment Processes and Levels of Treatment Achieved

of oxygen, soluble ferrous iron is oxidized to ferric iron, which readily hydrolyzes to form the insoluble ferric hydroxide precipitate. At high pH values ferric hydroxide will solubilize through the formation of the $Fe(OH)_4^-$ complex. Ferric and ferrous iron may also be solubilized in the presence of cyanide by the formation of ferro- and ferricyanide complexes. The primary removal process for iron is conversion of the ferrous to the ferric state and precipitation of ferric hydroxide at a pH of near 7, corresponding to minimum solubility. Conversion of ferrous to ferric iron occurs rapidly upon aeration at pH 7.5. In the presence of dissolved organic matter, the iron oxidation rate is reduced. Two-stage hydroxide precipitation or sulfate precipitation will reduce iron to 0.01 mg/L.

Lead

Lead is present in wastewaters from storage-battery manufacture. Lead is generally removed from wastewaters by precipitation as the carbonate, $PbCO_3$, or the hydroxide, $Pb(OH)_2$. Lead is effectively precipitated as the carbonate by the addition of soda ash, resulting in effluent-dissolved lead concentrations of 0.01 to 0.03 mg/L at a pH of 9.0 to 9.5. Precipitation with lime at pH 11.5 resulted in effluent concentrations of 0.019 to 0.2 mg/L. Precipitation as the sulfide to 0.01 mg/L can be accomplished with sodium sulfide at a pH of 7.5 to 8.5.

Manganese

Manganese and its salts are found in wastewaters from manufacture of steel alloy, dry-cell batteries, glass and ceramics, paint and varnish, and inks and dye. Among the many forms and compounds of manganese only the manganous salts and the highly oxidized permanganate anion are appreciably soluble. The latter is a strong oxidant that is reduced under normal circumstances to insoluble manganese dioxide. Treatment technology for the removal of manganese involves conversion of the soluble manganous ion to an insoluble precipitate. Removal is effected by oxidation of the manganous ion and separation of the resulting insoluble oxides and hydroxides. Manganous ion has a low reactivity with oxygen and simple aeration is not an effective technique below pH 9. It has been reported that even at high pH levels, organic matter in solution can combine with manganese and prevent its oxidation by simple aeration. A reaction pH above 9.4 is required to achieve significant manganese reduction by precipitation. The use of chemical oxidants to convert manganous ion to insoluble manganese dioxide in conjunction with coagulation and filtration has been employed. The presence of copper ion enhances air oxidation of manganese, and chlorine dioxide rapidly oxidizes manganese to the insoluble form. Permanganate has successfully been employed in the oxidation of manganese. Ozone has been employed in conjunction with lime for the oxidation and removal of manganese. The drawback in the application of ion exchange is the nonselective removal of other ions, which increases operating costs.

Mercury

The major consumptive user of mercury in the United States is the chlor-alkali industry. Mercury is also used in the electrical and electronics

Technology	Effluent, µg/L
Sulfide precipitation	10–20
Alum coprecipitation	1–10
Iron coprecipitation	0.5–5
Ion exchange	1–5
Carbon adsorption	
Influent	—
High	20
Moderate	2
Low	0.25

TABLE 4.15 Mercury Removal, Effluent Levels

industry, explosives manufacturing, the photographic industry, and the pesticide and preservative industry. Mercury is used as a catalyst in the chemical and petrochemical industry. Mercury is also found in most laboratory wastewaters. Power generation is a large source of mercury release into the environment through the combustion of fossil fuel. When scrubber devices are installed on thermal power plant stacks for sulfur dioxide removal, accumulation of mercury is possible if extensive recycle is practiced. Mercury can be removed from wastewaters by precipitation, ion exchange, and adsorption. Mercury ions can be reduced upon contact with other metals such as copper, zinc, or aluminum. In most cases mercury recovery can be achieved by distillation. For precipitation, mercury compounds must be oxidized to the mercuric ion. Table 4.15 shows effluent levels achievable by candidate technology.

Nickel

Wastewaters containing nickel originate from metal-processing industries, steel foundries, motor vehicle and aircraft industries, printing, and in some cases, the chemicals industry. In the presence of complexing agents such as cyanide, nickel may exist in a soluble complex form. The presence of nickel cyanide complexes interferes with both cyanide and nickel treatment. Nickel forms insoluble nickel hydroxide upon the addition of lime, resulting in a minimum solubility of 0.12 mg/L at pH 10 to 11. Nickel hydroxide precipitates have poor settling properties. Nickel can also be precipitated as the carbonate or the sulfate associated with recovery systems. In practice, lime addition (pH 11.5) may be expected to yield residual nickel concentrations in the order of 0.15 mg/L after sedimentation and filtration. Recovery of nickel can be accomplished by ion exchange or evaporative recovery, provided

| | Wastewater | | |
Parameter	A	B	C
Treatment pH	8.5	8.75	9.0
Initial nickel, mg/L	119.0	99.0	3.2
Lime treatment			
Clarifier effluent	12.0	16.0	0.47
Filter effluent	9.4	12.0	0.07
Lime plus sulfide			
Clarifier effluent	11.0	7.0	0.35
Filter effluent	3.5	4.2	0.20

Source: Adapted from Robinson and Sum, 1980.

TABLE 4.16 Comparison of Lime Versus Lime-Plus-Sulfide Precipitation of Nickel in Electroplating Wastewaters

the nickel concentrations in the wastewaters are at a sufficiently high level. Results of nickel precipitation are shown in Table 4.16.

Selenium

Selenium may be present in various types of paper, fly ash, and metallic sulfide ores. The selenious ion appears to be the most common form of selenium in wastewater, except for pigment and dye wastes, which contain selenide (yellow cadmium selenide). Selenium can be removed from wastewaters by precipitation as the sulfide at a pH of 6.6. Effluent levels of 0.05 mg/L are reported. Ferric hydroxide coprecipitation at pH 6.2 will reduce selenium to a range of 0.01 to 0.05 mg/L. Alumina adsorption results in effluent levels of 0.005 to 0.02 mg/L.

Silver

Soluble silver, usually in the form of silver nitrate, is found in wastewaters from the porcelain, photographic, electroplating, and ink manufacturing industries. Treatment technology for the removal of silver usually considers recovery because of the high value of the metal. Basic treatment methods include precipitation, ion exchange, reductive exchange, and electrolytic recovery. Silver is removed from wastewater by precipitation as silver chloride, which is an extremely insoluble precipitate resulting in the maximum silver concentration at 25°C of approximately 1.4 mg/L. An excess of chloride will reduce this value, but greater excess concentrations will increase the solubility of silver through the formation of soluble silver chloride complexes. Silver can be selectively precipitated as silver chloride from a mixed-metal wastestream without initial wastewater segregation or concurrent precipitation of other metals. If the treatment conditions are alkaline, resulting in precipitation of hydroxides of other metals along with the silver chloride, acid washing of the precipitated sludge will remove contaminated metal ions, leaving the insoluble silver chloride. Plating wastes contain silver in the form of silver cyanide, which interferes with the precipitation of silver as the chloride salt. Oxidation of the cyanide with chlorine releases chloride ions into solution, which in turn react to form silver chloride directly. Sulfide will precipitate silver from photographic solutions as the extremely insoluble silver sulfide. Ion exchange has been employed for the removal of soluble silver from wastewaters. Activated carbon will remove low concentrations of silver. The mechanism reported is one of reductive recovery by formation of elemental silver at the carbon surface. Reported results indicate that the carbon is capable of retaining silver to 9 percent of its weight at a pH of 2.1 and 12 percent of its weight at a pH of 5.4. Alum or iron coprecipitation will reduce silver to 0.025 mg/L and hydroxide precipitation at pH 11 to 0.02 mg/L.

Zinc

Zinc is present in wastewater streams from steelworks, rayon yarn and fiber manufacture, ground wood-pulp production, and recirculating cooling water systems employing cathodic treatment. Zinc is also present in wastewaters from the plating and metal-processing industry. Zinc can be removed by precipitation as zinc hydroxide with either lime or caustic. The disadvantage of lime addition is the concurrent precipitation of calcium sulfate in the presence of high sulfate levels in the wastewater. An effluent soluble zinc of less than 0.1 mg/L has been achieved at pH 11.0. Zinc is an amphoteric metal with increasing solubility at both higher and lower pH values. A summary of hydroxide precipitation results is shown in Table 4.17. Results for reverse osmosis treatment of zinc wastewaters are shown in Table 4.18. Electrolytic treatment of zinc cyanide wastewaters results are shown in Table 4.19.

4.4 Summary

Effluent concentrations achievable by metals removal processes are summarized in Table 4.20. A summary of BAT equivalent metals removal performance is presented in Table 4.21. A detailed discussion of metals removal has been presented by Patterson.[15]

Industrial Source	Zinc Concentration, mg/L		Comments*
	Initial	Final	
Zinc plating	—	0.2–0.5	pH 8.7–9.3
General plating	18.4	2.0	pH 9.0
	—	0.6	Sand filtration
	55–120	1.0	pH 7.5
	46	2.9	pH 8.5
		1.9	pH 9.2
		2.8	pH 9.8
		2.9	pH 10.5
Vulcanized fiber	100–300	1.0	pH 8.5–9.5
Tableware plant	16.1	0.02–0.23	Sand filtration
Viscose rayon	26–120	0.86–1.5	—
	70	3–5	pH 5
	20	1.0	—
Metal fabrication	—	0.5–1.2	Sedimentation
		0.1–0.5	Sand filtration
Radiator manufacture		0.33–2.37	Sedimentation
		0.03–0.38	Sand filtration
Blast furnace gas scrubber water	50	0.2	pH 8.8
Zinc smelter	744	50	
	1500	2.6	
Ferroalloy waste	11.2–34	0.29–2.5	
	3–89	4.2–7.9	
Ferrous foundry	72	1.26	Sedimentation
		0.41	Sand filtration
Deep coal mine—acid water	33–7.2	0.01–10	

*All treatment involved precipitation plus sedimentation. Special or additional aspects of treatment are indicated under Comments.
Source: Adapted from Patterson, 1985.

TABLE 4.17 Summary of Hydroxide Precipitation Treatment Results for Zinc Wastewaters

Industrial Source	Zinc Concentration, μg/L		
	Feed	Permeate	% Removal
Zinc cyanide plating rinse	1,700	30	98
Steam electric power plant	300	53	82
	780	3	99
Textile mill	7,200	140	98
	5,400	6,600	−20
	460	250	46
	520	360	31
	7,200	360	95
	1,400	30	98
	4,100	180	96
	1,200	22	98
	24,000	430	98
	9,700	37	>99
Cooling tower blowdown	10,000	300	97

Source: Adapted from Cawley, 1980.

TABLE 4.18 Pilot-Scale Results for Reverse Osmosis Treatment of Zinc Wastewaters

Waste	Parameter	Concentration, mg/L	
		Initial	Final
A	Zinc	352	0.7
	Cyanide	258	12.0
B	Zinc	117	0.3
	Copper	842	0.5
	Cyanide	1230	<0.1

TABLE 4.19 Electrolytic Treatment of Zinc Cyanide Wastes

Coagulation, Precipitation, and Metals Removal

Metal	Achievable Effluent Concentration, mg/L	Technology
Arsenic	0.05	Sulfide ppt with filtration
	0.06	Carbon adsorption
	0.005	Ferric hydroxide co-ppt
Barium	0.5	Sulfate ppt
Cadmium	0.05	Hydroxide ppt at pH 10–11
	0.05	Co-ppt with ferric hydroxide
	0.008	Sulfide precipitation
Copper	0.02–0.07	Hydroxide ppt
	0.01–0.02	Sulfide ppt
Mercury	0.01–0.02	Sulfide ppt
	0.001–0.01	Alum co-ppt
	0.0005–0.005	Ferric hydroxide co-ppt
	0.001–0.005	Ion exchange
Nickel	0.12	Hydroxide ppt at pH 10
Selenium	0.05	Sulfide ppt
Zinc	0.1	Hydroxide ppt at pH 11

TABLE 4.20 Effluent Levels Achievable in Heavy Metals Removal[14]

Constituent	BAT-Equivalent Concentration, µg/L (30-d Average)	Treatment Technology
Arsenic	200	1. Arsenite (As^{+3}) oxidation to arsenate (As^{+5}) 2. Lime precipitation, or iron or alum co-precipitation (3–6 µg/L) 3. Gravity clarification
Barium	1000	1. Sulfite precipitation 2. Coagulation; ppt as BaSO$_4$; 30–300 µg/L 3. Gravity clarification
Cadmium	100	1. High pH precipitation; 50 µg/L; pH 10–11 co-ppt Fe(OH)$_3$; 8 µg/L at pH 6.5 2. Gravity clarification for lime or filtration for caustic; sulfide ppt: 5–10 µg/L

TABLE 4.21 Summary of BAT-Equivalent Treatment Performance

Constituent	BAT-Equivalent Concentration, µg/L (30-d Average)	Treatment Technology
Chromium, hexavalent	50	1. Acidic reduction for trivalent chromium or iron exchange at pH below 6.0; pH 2–3
Chromium, total	500	1. Precipitation (OH ppt) 2. Gravity clarification for lime or filtration for caustic; lime produces better settling
Copper	400	1. Precipitation; (OH ppt); pH 8.5; sulfide ppt 10 µg/L 2. Gravity clarification
Flouride	10,000	1. High pH lime precipitation 2. Gravity clarification
Iron	1500	1. Oxidation at neutral pH of ferrous to ferric iron 2. Precipitation 3. Gravity clarification or filtration
Lead	150	1. High pH precipitation (OH ppt); pH 11.5; carbonate pH 9–9.5, 10–30 µg/L; sulfide 10 µg/L 2. Gravity clarification for lime or filtration for caustic
Mercury	3	1. Ion exchange or coagulation plus filtration; Hg must be oxidized to ionic from; sulfide 10–20 µg/L; co-ppt and filtration
Nickel	750	1. High pH precipitation; (OH) ppt; pH 9–12; lime and sulfide, 40 µg/L 2. Gravity clarification and/or filtration
Silver	100	1. Ion exchange or ferric chloride co-precipitation plus filtration
Zinc	500	1. Precipitation at optimized pH; $Zn(OH)_2$ with lime or caustic; pH 9–9.5 and 11 2. Gravity clarification and/or filtration

TABLE 4.21 Summary of BAT-Equivalent Treatment Performace (*Continued*)

4.5 Problem

4.1. A metal-finishing plant has a wastewater flow of 72,000 gal/d (273 m^3/d) with the following characteristics:

Cr^{6+} 75 mg/L
Cu 10 mg/L
Ni 8 mg/L

Design a reduction and precipitation plant for

1. Continuous flow
2. Batch flow
   (a) Develop the ORP control points using SO_2 as the reducing agent.
   (b) Compute the quantity of sludge and drying bed area assuming the sludge concentrates to 2 percent is applied to the bed to a depth of 18 in (0.46 m), and is removed every 5 days at 12 percent.
   (c) Compute the residual soluble metal if a terminal pH of 8.5 is used for the precipitation.

References

1. Mysels, K. J.: *Introduction to Colloid Chemistry*, Interscience Publishers, New York, 1959.
2. Black, A. P., and H. L. Smith: *J. Am. Water Works Assoc.*, vol. 54, p. 371, 1962.
3. Riddick, T. M.: *Tappi*, vol. 47, pt. 1, p. 171A, 1964.
4. Palladino, A. J.: *Proc. 10th Ind. Waste Conf.*, May 1955, Purdue University.
5. Knack, M. F.: *Proc. 4th Ind. Waste Conf.*, 1949, Purdue University.
6. Leonard, A. G., and R. G. Keating: *Proc. 13th Ind. Waste Conf.*, 1946, Purdue University.
7. Bloodgood, D., and W. J. Kellenher: *Proc. 7th Ind. Waste Conf.*, 1952, Purdue University.
8. Eckenfelder, W. W., and D. J. O'Connor: *Proc. 10th Ind. Waste Conf.*, May 1955, p. 17, Purdue University.
9. Webster, R. A.: *Sewage Ind. Wastes*, vol. 25, pt. 12, p. 1432, December 1953.
10. Olthof, M. G., and W. W. Eckenfelder: *Textile Chemist and Colorist*, vol. 8, pt. 7, p. 18, 1976.
11. Olthof, M. G., and W. W. Eckenfelder: *Water Res.*, vol. 9, p. 853, 1975.
12. Southgate, B. A.: *Treatment and Disposal of Industrial Waste Waters*, His Majesty's Stationery Office, London, 1948, p. 186.
13. McCarthy, Joseph A.: *Sewage Works J.*, vol. 21, pt. 1, p. 75, January 1949.
14. Talinli, I.: *Wat. Sci. Tech.*, 29, 9, p. 175, 1994.
15. Patterson, J. W.: *Industrial Wastewater Treatment Technology*, Butterworth Publishers, Boston, 1985.
16. Robinson, A., and J. Sum: U.S. EPA 600/2-80-139, June 1980.
17. Cawley, W. (ed.): *Treatability Manual*, vol. 3rd, U.S. EPA 600/8-80-042-C, July 1980.

CHAPTER 5
Aeration and Mass Transfer

5.1 Introduction

Aeration is used for transferring oxygen to biological-treatment processes, for stripping solvents from wastewaters, and for removing volatile gases such as H_2S and NH_3. An adequate oxygen supply is critical to aerobic biological treatment. It is also energy intensive and one of the more expensive components of the treatment system. Due to turbulence provided by aeration equipment, mixing is a significant factor in design and operation. This chapter describes the mechanism of oxygen transfer, aeration equipment commonly used in industrial wastewater treatment, and air stripping of volatile organics.

5.2 Mechanism of Oxygen Transfer

Aeration is a gas-liquid mass-transfer process in which interphase diffusion occurs when a driving force is created by a departure from equilibrium. In the gas phase, the driving force is a partial pressure gradient; in the liquid phase, it is a concentration gradient.

The rate of molecular diffusion of a dissolved gas in a liquid is dependent on the characteristics of the gas and the liquid, the temperature, the concentration gradient, and the cross-sectional area across which diffusion occurs. The diffusional process is defined by Fick's law:

$$N = -D_L A \frac{dc}{dy} \qquad (5.1)$$

where N = mass transfer per unit time
  A = cross-sectional area through which diffusion occurs
  dc/dy = concentration gradient perpendicular to cross-sectional area
  D_L = diffusion coefficient through the liquid film

If it is assumed that equilibrium conditions exist at the interface, the mass-transfer process can be reexpressed as:

$$N = \left(-D_g A \frac{dp}{dy}\right)_1 = \left(-D_L A \frac{dc}{dy}\right)_2 = \left(-D_e A \frac{dc}{dy}\right)_3 \qquad (5.2)$$

where D_g = diffusion coefficient through the gas film
D_e = eddy diffusion coefficient of the gas in the body of the liquid
D_L = diffusion coefficient through the liquid film

Since the systems dealt with in waste treatment involve high degrees of turbulence, the eddy diffusivity will be several orders of magnitude greater than the coefficients of molecular diffusivity, and this need not be considered as a rate-controlling step. An exception may be large aerated lagoons or aeration in flowing rivers.

Lewis and Whitman[1] developed the two-film concept which considers stagnant films at the gas and liquid interfaces through which mass transfer must occur. Equation (5.2) can be reexpressed in terms of a liquid and gas film:

$$N = K_L A(C_s - C_L) = K_g A(P_g - P) \qquad (5.3)$$

where N = mass of oxygen transferred per unit area
A = interfacial surface area
C_s = oxygen saturation concentration
C_L = concentration of oxygen in the liquid
D_L = diffusion coefficient through liquid film (defined after Eq. 5.2)
D_g = diffusion coefficient through the gas film (defined after Eq. 5.2)
K_L = liquid-film coefficient, defined as D_L/Y_L
K_g = gas-film coefficient, defined as D_g/Y_g
P_g = gas concentration outside gas film
P = interfacial gaseous concentration
Y_L = liquid-film thickness
Y_g = gas-film thickness

For sparingly soluble gases, such as oxygen and CO_2, the liquid-film resistance controls the rate of mass transfer; for highly soluble gases, such as ammonia, the gas-film resistance controls the transfer rate. Most of the mass-transfer applications in waste treatment are liquid-film controlled. Increasing the fluid turbulence will decrease the film thickness and hence increase K_L. Danckwertz[2] has defined the liquid-film coefficient as the square root of the product of the diffusivity and the rate of surface renewal:

$$K_L = \sqrt{D_L r} \qquad (5.4)$$

The rate of surface renewal r can be considered as the frequency with which fluid with a solute concentration C_L is replacing fluid from the interface with a concentration C_s. High degrees of fluid turbulence will increase r.

Dobbins[3] has proposed a relationship that describes the aforementioned transfer mechanisms:

$$K_L = (D_L r)^{1/2} \coth\left(\frac{rY_L^2}{D_L}\right)^{1/2} \quad (5.5)$$

When the surface-renewal rate is zero, K_L is equal to D_L/Y_L and the transfer is controlled by molecular diffusion through the surface film. As r increases, K_L becomes equal to $\sqrt{D_L r}$ and the transfer is a function of the rate of surface renewal.

For liquid-film-controlled processes, Eq. (5.3) can be reexpressed in concentration units:

$$\frac{1}{V}N = \frac{dc}{dt} = K_L \frac{A}{V}(C_s - C_L) \quad (5.6)$$

in which V is the liquid volume and

$$K_L \frac{A}{V} = K_L a$$

$K_L a$ is an overall film coefficient and is usually used to compute the transfer rate.

The most important application of aeration in waste treatment is the transfer of oxygen to biological-treatment processes and the natural reaeration of streams and other watercourses.

The equilibrium concentration of oxygen in contact with water, C_s, is defined by Henry's law:

$$p = HC_s \quad (5.7)$$

where p = partial pressure of oxygen in the gas phase and H = Henry's constant, which is proportional to temperature and is influenced by the presence of dissolved solids.

The solubility of oxygen in water at various temperatures is summarized in Table 5.1. As temperature and dissolved solids concentrations increase, Henry's constant also increases, thereby reducing C_s. It is therefore usually necessary in industrial wastes to measure solubility experimentally.

In aeration tanks, where air is released at an increased liquid depth, the solubility of oxygen is influenced both by the increasing

Temperature		Elevation, ft							TDS (sea level), ppm			
°F	°C	0	1000	2000	3000	4000	5000	6000	400	800	1500	2500
32.0	0	14.6	14.1	13.6	13.1	12.6	12.1	11.7	—	—	—	—
35.6	2	13.8	13.3	12.8	12.4	11.9	11.5	11.1	13.74	13.68	13.58	13.42
39.2	4	13.1	12.6	12.2	11.8	11.4	10.9	10.5	13.04	12.98	12.89	12.75
42.8	6	12.5	12.0	11.6	11.2	10.8	10.4	10.0	12.44	12.38	12.29	12.15
46.4	8	11.9	11.4	11.0	10.6	10.2	9.9	9.5	11.85	11.80	11.70	11.58
50.0	10	11.3	10.9	10.5	10.1	9.8	9.4	9.1	11.25	11.20	11.12	11.00
53.6	12	10.8	10.4	10.1	9.7	9.4	9.0	8.6	10.76	10.71	10.64	10.52
57.2	14	10.4	10.0	9.6	9.3	8.9	8.6	8.3	10.36	10.32	10.25	10.15
60.8	16	10.0	9.6	9.2	8.9	8.6	8.3	8.0	9.96	9.92	9.85	9.75
64.4	18	9.5	9.2	8.9	8.5	8.2	7.9	7.6	9.46	9.43	9.36	9.27
68.0	20	9.2	8.8	8.5	8.2	7.9	7.6	7.3	9.16	9.13	9.06	8.97
7.16	22	8.8	8.5	8.2	7.9	7.6	7.3	7.1	8.77	8.73	8.68	8.60
75.2	24	8.5	8.2	7.9	7.6	7.3	7.1	6.8	8.47	8.43	8.38	8.30
78.8	26	8.2	7.9	7.6	7.3	7.1	6.8	6.6	8.17	8.13	8.08	8.00
82.4	28	7.9	7.6	7.4	7.1	6.8	6.6	6.3	7.87	7.83	7.78	7.70

86.0	30	7.6	7.4	7.1	6.9	6.6	6.4	6.1	7.57	7.53	7.48	7.40
89.6	32	7.4	7.1	6.9	6.6	6.4	6.2	5.9	7.4	—	—	—
93.2	34	7.2	6.9	6.7	6.4	6.2	6.0	5.8	7.2	—	—	—
96.8	36	7.0	6.7	6.5	6.3	6.0	5.8	5.6	7.0	—	—	—
100.4	38	6.8	6.6	6.3	6.1	5.9	5.6	5.4	6.8	—	—	—
104.0	40	6.6	6.4	6.1	5.9	5.7	5.5	5.3	6.6	—	—	—

Note: ft = 0.3048 m.

TABLE 5.1 Solubility of Oxygen (mg/L) at Various Temperatures, Elevations, and Total Dissolved Solids Levels

partial pressure of the air entering the aeration tank and by the decreasing partial pressure in the air bubble as oxygen is absorbed. For these cases, a mean saturation value corresponding to the aeration tank middepth is used:

$$C_{s,m} = C_s \times \frac{1}{2}\left(\frac{P_b}{P_a} + \frac{O_t}{20.9}\right) \tag{5.8}$$

where P_a = atmospheric pressure
P_b = absolute pressure at the depth of air release
O_t = percent concentration of oxygen in the air leaving the aeration tank

Mueller et al.[4] have shown that oxygen saturation is a function not only of submergence but also of diffuser type. Coarse bubble units provide lower clean water saturation values than the fine bubble or jet diffusers. Saturation appears to be related to both bubble size and mixing pattern. It should further be noted that field data seems to suggest that the 0.25 depth level may be more accurate for fine bubble diffusers (Schmit et al.[5]). Therefore, where possible, oxygen saturation should be determined from field data.

To account for wastewater constituents, a factor is used:

$$\beta = \frac{C_s \text{ wastewater}}{C_s \text{ tap water}}$$

The American Society for Civil Engineers (ASCE) committee on oxygen transfer[19] has recommended using a correction for total dissolved solids (TDS) as shown in Table 5.1 to determine β.

The oxygen-transfer coefficient $K_L a$ is affected by the physical and chemical variables characteristic of the aeration system:

1. *Temperature.* The liquid-film coefficient will increase with increasing temperature. When air bubbles are involved, changes in liquid temperature will also affect the size of bubbles generated in the system. The effect of temperature on the coefficient is

$$K_L(T) = K_L(20°C)\theta^{T-20} \tag{5.9}$$

For diffused aeration units, θ is usually taken as 1.02. Correlations of Imhoff and Albrect[6] showed θ to be higher for low-turbulence diffused systems and lower for high-turbulence surface aeration systems. Landberg et al.[7] suggested θ of 1.012 for surface aeration systems. The effect of temperature on $K_L a$ is shown in Fig. 5.1.

Aeration and Mass Transfer

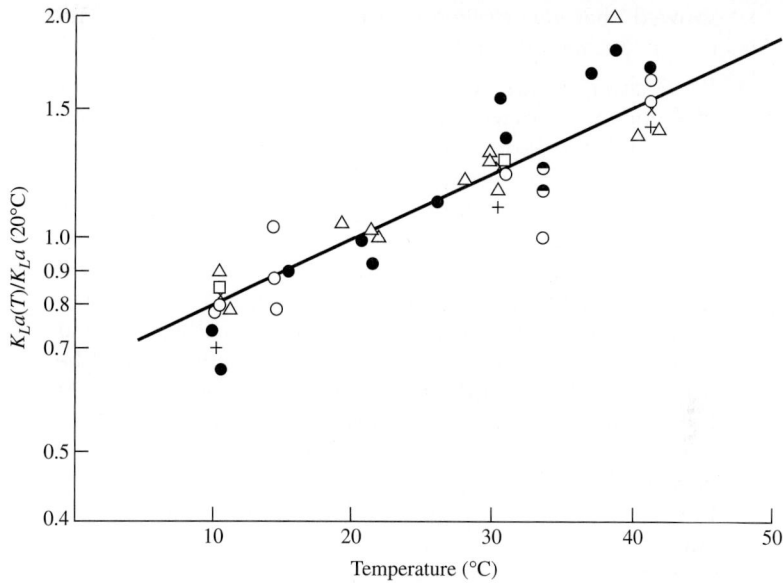

Symbol	Substance	Unit
○	Water	Spinnerette 20 holes, 0.035 mm diameter
+	Water	Spinnerette 10 holes, 0.05 mm diameter
●	Water	Aloxite stone
△	1% KCl	Aloxite stone
□	1% KCl	Spinnerette 10 holes, 0.05 mm diameter
×	50 ppm heptanoic acid	Spinnerette 20 holes, 0.035 mm diameter
	Water	Data of Carpani and Roxburgh[9]
△	Water	Data of Gameson and Robertson[10]

FIGURE 5.1 Relationship between the overall transfer coefficient $K_L a$ and temperature.

2. *Turbulent mixing.* Increasing the degree of turbulent mixing will increase the overall transfer coefficient.
3. *Liquid depth.* The effect of liquid depth H on $K_L a$ will depend in large measure on the method of aeration. For most types of bubble-diffusion systems, $K_L a$ will vary with depth according to the relationship

$$\frac{K_L a(H_1)}{K_L a(H_2)} = \left(\frac{H_1}{H_2}\right)^n \qquad (5.10)$$

The exponent n has a value near 0.7 for most systems. Wagner and Popel[8] evaluated several diffused aeration systems and showed that the oxygen-transfer efficiency increased 1.5 percent per foot of depth.

4. *Waste characteristics.* The presence of surface-active agents and other organics will have a profound effect on both K_L and A/V. Molecules of surface-active materials will orient themselves on the interfacial surface and create a barrier to diffusion. The excess surface concentration is related to the change in surface tension, as defined by the Gibbs equation, such that small concentrations of surface-active material will depress K_L while large concentrations will exert no further effect. The absolute effect of surfactants on K_L will also depend on the nature of the aeration surface. Less effect would be exerted at a highly turbulent liquid surface, since the short life of any interface would restrict the formation of an adsorbed film. Conversely, a greater effect would be exerted at a bubble surface because of the relatively long life of the bubble as it rises through an aeration tank. A decrease in surface tension will decrease the size of bubbles generated from an air-diffusion system. This in turn will increase A/V. In some cases, the increase in A/V will exceed the decrease in K_L, and the transfer rate will increase over that in water. The effect of waste characteristics on $K_L a$ is defined by a coefficient α in which

$$\alpha = \frac{K_L a \text{ waste}}{K_L a \text{ water}}$$

These relationships are shown in Fig. 5.2.

Turbulence has a significant effect on α. At high turbulence levels, oxygen transfer is dependent on surface renewal and not significantly affected by diffusion through interfacial resistances. Under these conditions α may be greater than 1.0 because of the increased A/V ratio. Under low-turbulence conditions, the bulk oxygen transfer resistance is reduced but surface renewal does not yet occur; thus surfactant interfacial resistance causes a significant reduction in the oxygen transfer rate, as shown in Fig. 5.3. Increasing the TDS will increase $K_L a$ due to the generation of finer bubbles, as shown in Fig. 5.4.

In order to compare the transfer rate in water to a waste with a particular aeration device, a coefficient α has been defined as $K_L a$ (waste)/$K_L a$ (water). The coefficient α can be expected to increase or decrease and approach unity during the course of biooxidation, since the substances affecting the transfer rate are being removed in the biological process, as shown in Fig. 5.5.

In view of the effects previously mentioned, the type of aeration will have a profound effect on α. For the fine-bubble diffuser systems, values are generally lower than for coarse-bubble or surface aeration

Aeration and Mass Transfer

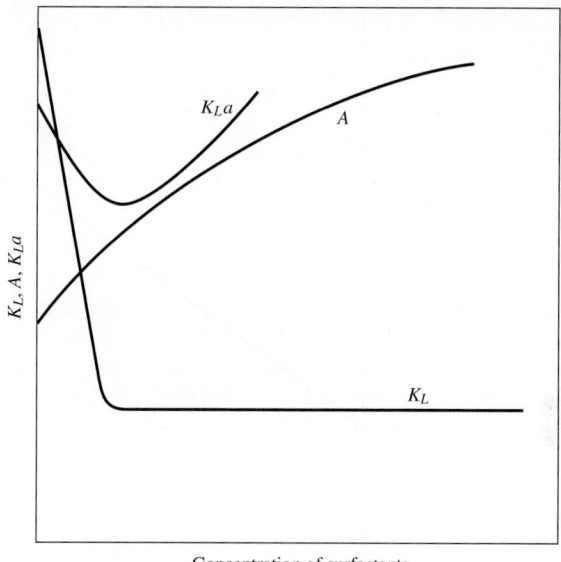

FIGURE 5.2 Effect of concentration of surface active agent on oxygen transfer.

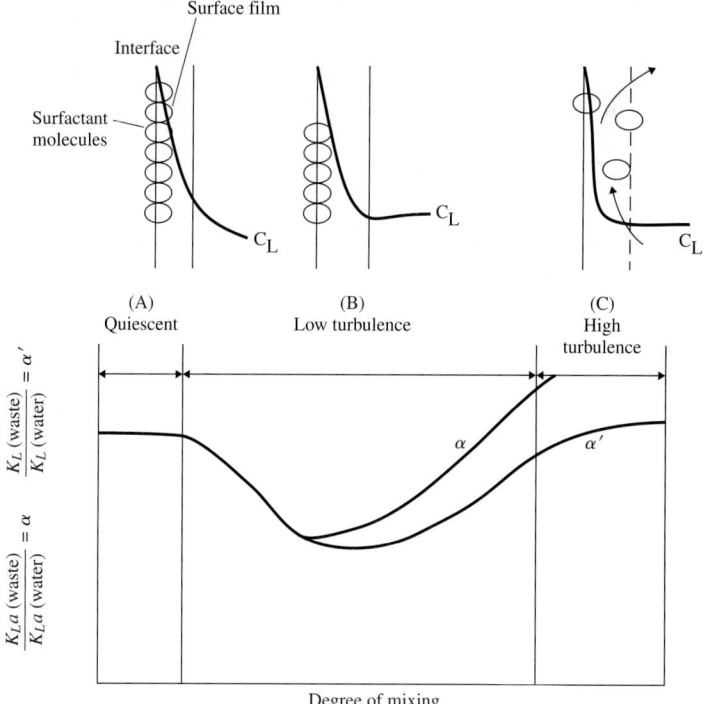

FIGURE 5.3 Effect of turbulence on oxygen transfer.

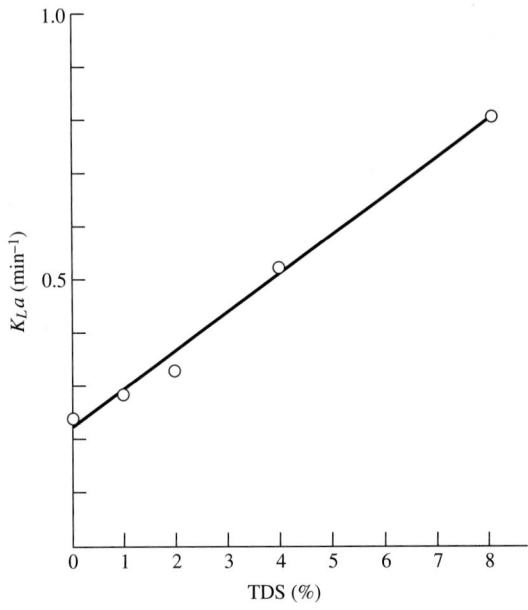

Figure 5.4 Effect of TDS on the oxygen transfer rate.

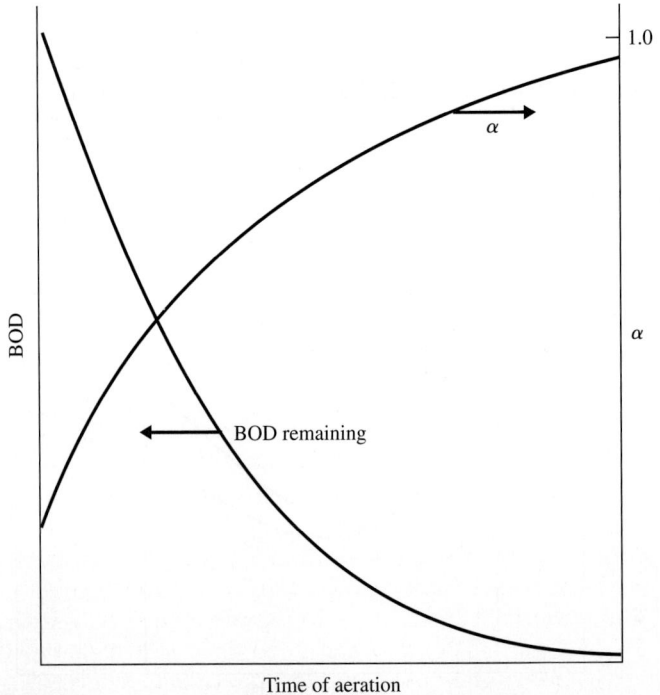

Figure 5.5 Change in transfer coefficient α in a biological oxidation process.

Aeration Device	Alpha Factor	Wastewater
Fine bubble diffuser	0.4–0.6	Tap water containing detergent
Brush	0.8	Domestic wastewater
Coarse bubble diffuser, sparger	0.7–0.8	Domestic wastewater
Coarse bubble diffuser, wide band	0.65–0.75	Tap water with detergent
Coarse bubble diffuser, sparger	0.55	Activated sludge contact tank
Static aerator	0.60–0.95	Activated sludge treating high-strength industrial waste
Static aerator	1.0–1.1	Tap water with detergent
Surface aerators	0.6–1.2	Alpha factor tends to increase with increasing power (tap water containing detergent and small amounts of activated sludge)
Turbine aerators	0.6–1.2	Alpha factor tends to increase with increasing power; 25, 50, 190 gal tanks (tap water containing detergent)

TABLE 5.2 Values of α for Different Aeration Devices

systems. Variation in mixed liquor suspended solids (MLSS) from 2000 to 7000 mg/L with dome diffusers had no significant effect on $K_L a$. Reported values of α for different aeration devices are summarized in Table 5.2. In plug flow basins the α value will increase as purification occurs through the tank length. This is shown in Fig. 5.6 for two types of aeration devices.[15]

In diffused aeration systems, air bubbles are formed at an orifice from which they break off and rise through the liquid, finally bursting at the liquid surface. The velocity and shape of the air bubbles is related to a modified Reynolds number. At N_{Re} of 300 to 4000, the bubbles assume an ellipsoidal shape and rise with a rectilinear rocking motion. At N_{Re} greater than 4000, the bubbles form spherical caps. The rising velocity of the bubble is increased at high airflows because of the proximity of other bubbles and the resulting disturbances of the bubble wakes.

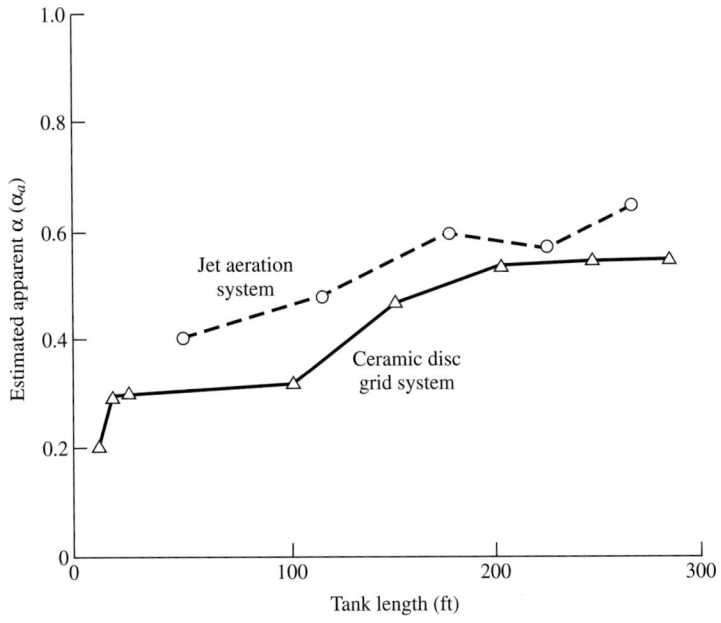

Figure 5.6 Estimated change in apparent α with tank length, Whittier Narrows, California.

A general correlation for oxygen transfer from air bubbles rising through a still water column has been developed by Eckenfelder,[11] as shown in Fig. 5.7. The general relationship obtained is

$$\frac{K_L d_B}{D_L} H^{1/3} = C\left(\frac{d_B v_B \rho}{\mu}\right) \tag{5.11}$$

where C = constant
  d_B = bubble diameter
  v_B = bubble velocity
  ρ = liquid density
  μ = liquid viscosity

Equation (5.11) can be expressed in terms of the overall coefficient $K_L a$ if A/V for air bubbles in an aeration tank is considered as

$$\frac{A}{V} = \frac{6 G_s H}{d_B v_B V} \tag{5.12}$$

where G_s is the airflow. Equation (5.11) neglects the aeration tank liquid surface as small compared to the interfacial bubble surface.

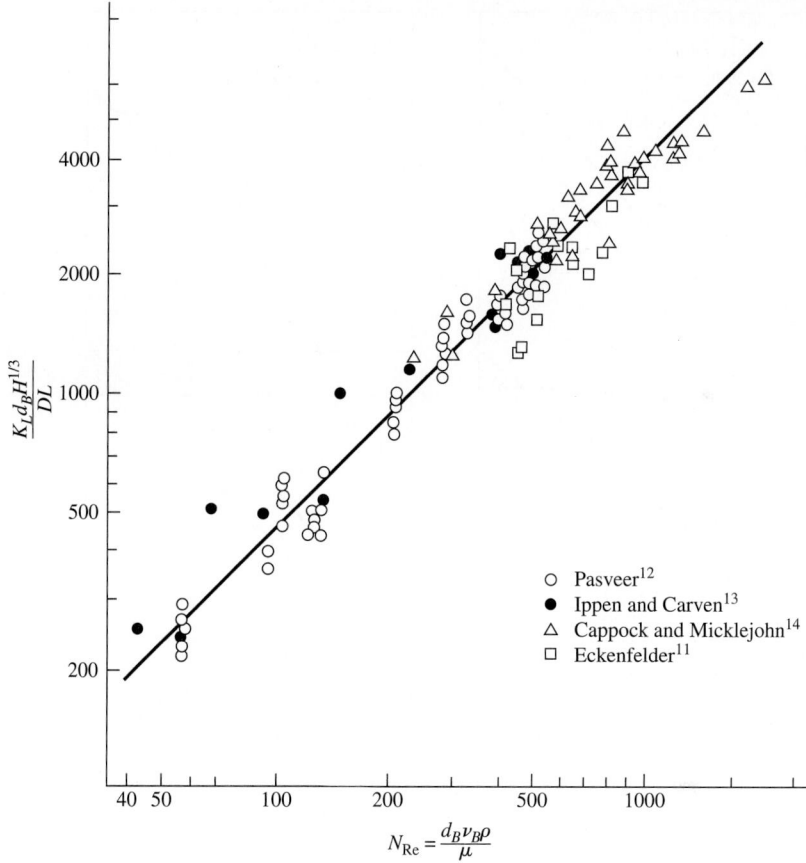

FIGURE 5.7 Correlation of bubble-aeration data.

Over the range of airflows normally encountered in aeration practice,

$$d_B \approx G_s^n \tag{5.13}$$

Equations (5.11) to (5.13) can be combined to yield a general relationship for oxygen transfer from air-diffusion systems:

$$K_L a = \frac{C' H^{2/3} G_s^{(1-n)}}{V} \tag{5.14}$$

The relationship between $K_L a$ and gas flow for several diffusion devices is shown in Fig. 5.8.

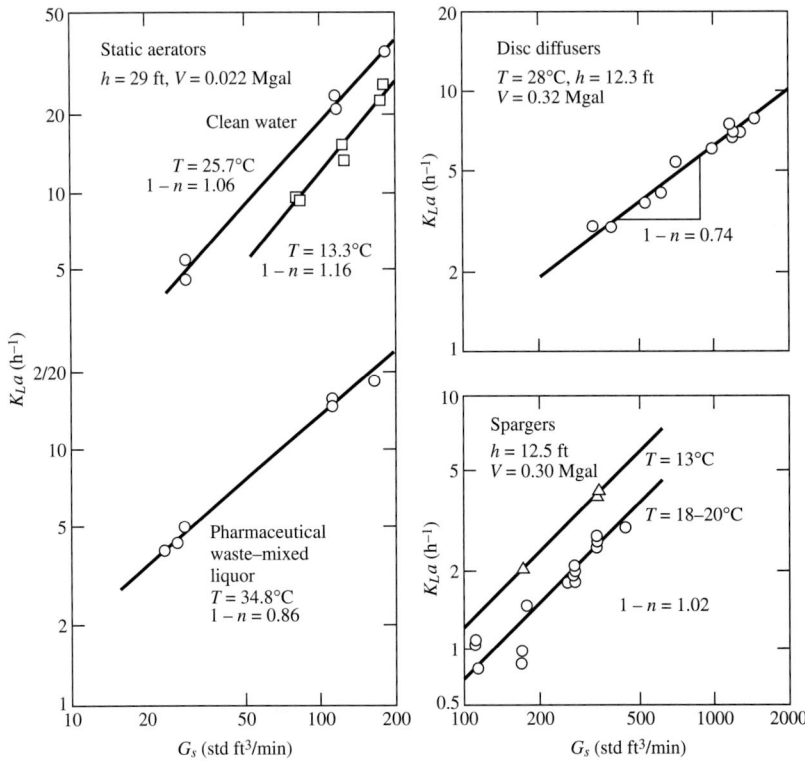

FIGURE 5.8 Effect of gas flow on $K_L a$ for various types of diffuser systems under process conditions. (*Adapted from Muller, 1996.*)

Equation (5.14) can be reexpressed in terms of the mass of oxygen transferred per diffusion unit:

$$N = C' H^{2/3} G_s^{(1-n)} (C_s - C_L) \qquad (5.15)$$

where N is the mass of O_2 per hour transferred per diffuser unit. The oxygen transfer efficiency of a unit is defined as

$$\text{Percent transfer} = \frac{\text{wt } O_2 \text{ adsorbed/unit time}}{\text{wt } O_2 \text{ supplied/unit time}} \times 100$$

$$= \frac{K_L a (C_s - C_L) \times V}{G_s \text{ (std ft}^3/\text{min)} \times 0.232 \text{ lb } O_2/\text{lb air}} \times 100$$

$$\times 0.075 \text{ lb air/std ft}^3 \qquad (5.16)$$

Example 5.1. The following data were obtained on the oxygen transfer capacity of an air-diffusion unit in clean water.

Airflow rate   25 std ft³/min · thousand ft³)(25 L/[min · m³])
Volume   1000 ft³ (28.3 m³)

Temperature   54°F (12°C)
Liquid depth   15 ft (4.6 m)
Average bubble diameter   0.3 cm
Average bubble velocity   32 cm/s

Time, min	C_L, mg/L
3	0.6
6	1.6
9	3.1
12	4.3
15	5.4
18	6.0
21	7.0

(a) Compute $K_L a$ and K_L.

(b) Compute the mass of O_2 per hour transferred per unit volume at 20°C and zero dissolved oxygen, and the oxygen transfer efficiency.

(c) How much oxygen will be transferred to a waste with an α of 0.82, a temperature of 32°C, and an operating dissolved oxygen of 1.5 mg/L?

Solution

(a) At a temperature of 54°F (12°C) saturation is 10.8 mg/L. The mean saturation in the aeration tank, assuming 10 percent oxygen adsorption, is

$$C_{s,m} = C_s \left(\frac{P_b}{29.4} + \frac{O_t}{42} \right)$$

where

$$P_b = \frac{15 \text{ ft}}{2.3 \text{ ft/(lb} \cdot \text{in}^2)} + 14.7 = 21.2 \text{ lb/in}^2 \text{ (1.44 atm)}$$

$$O_t = \frac{21(1-0.1)}{21(1-0.1)+79} \times 100 = 19.3\% \text{ O}_2$$

$$C_{s,m} = 10.8 \left(\frac{21.2}{29.4} + \frac{19.3}{42} \right) = 10.8 \times 1.18$$

$$= 12.7 \text{ mg/L}$$

Time, min	$C_{s,m} - C_L$
3	12.1
6	11.1
9	9.6
12	8.4
15	7.3
18	6.7
21	5.7

From Eq. (5.22),

$$\log(C_{s,m} - C_L) = \log(C_{s,m} - C_O) - \frac{K_L a}{2.3} t$$

which represents a straight line on a semilog plot of log $(C_{s,m} - C_L)$ versus time. $K_L a$ can be computed from the slope of the line in Fig. 5.9.

$$K_L a = 2.3 \, \frac{\log(14/9)}{10} \times 60 = 2.63 / h$$

The interfacial area/volume ratio, A/V, is

$$\frac{A}{V} = \frac{6 G_s H}{d_B v_B V} = \frac{6 \times 25 \times 15 \times 60}{(0.3/30.5) \times (32/30.5) \times 3600 \times 1000}$$

$$= 3.65 \text{ ft}^2/\text{ft}^3 \ (12 \text{ m}^2/\text{m}^3)$$

where d_B = average bubble diameter
H = aeration liquid depth
v_B = average bubble velocity
V = volume

$$K_L = \frac{K_L a}{A/V}$$

$$= \frac{2.63}{3.65} = 0.72 \text{ ft/h } (21.96 \text{ cm/h})$$

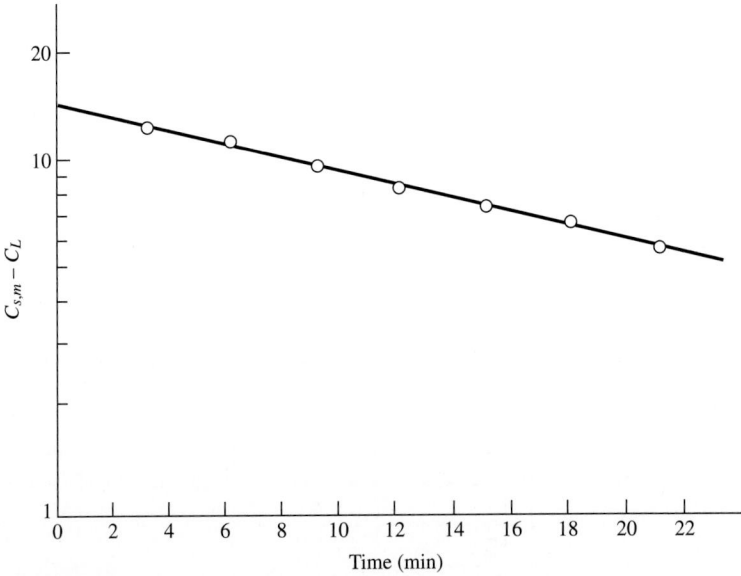

Figure 5.9 Determination of $K_L a$.

(b)

$$K_L a_T = K_L a_{20} \times 1.02^{T-20}$$

$$K_L a_{20} = 1.02^8 \times 2.63 = 1.17 \times 2.63 = 3.07/h$$

$$C_{s,m(20°)} = 9.1 \times 1.18 = 10.7 \text{ mg/L}$$

$$N_O = K_L a V C_{s,m}$$

$$= 3.07/h \times 1000 \text{ ft}^3 \times 10.7 \text{ mg/l} \times 7.48 \text{ gal/ft}^3$$

$$\times 8.34 \times 10^{-6}$$

$$= 2.05 \text{ lb } O_2/h \quad (0.93 \text{ kg } O_2/h)$$

$$\% \ O_2 \text{ transfer efficiency} = \frac{2.05 \text{ lb } O_2/h \text{ transferred}}{25 \times 60 \times 0.0746 \times 0.232 \text{ lb } O_2/h \text{ supplied}} \times 100$$

$$= 8.0$$

(c) The weight of oxygen transferred is determined from Eq. (5.17):

$$N = N_O \left(\frac{\beta C_{s,m} - C_L}{C_{s,m(20°)}} \right) \alpha \times 1.02^{T-20}$$

$$= 2.05 \left(\frac{0.99 \times 7.4 \times 1.18 - 1.5}{10.7} \right) \times 0.82 \times 1.02^{12}$$

$$= 1.42 \text{ lb/h} (0.64 \text{ kg/h})$$

5.3 Aeration Equipment

The aeration equipment commonly used in the industrial waste field consists of air-diffusion units, turbine aeration systems in which air is released below the rotating blades of an impeller, and surface aeration units in which oxygen transfer is accomplished by high surface turbulence and liquid sprays. Generic types of aeration equipment are shown in Fig. 5.10.

A manufacturer will generally designate the oxygen transfer capability of his equipment in terms of the pounds of O_2 transferred per horsepower-hour (kg O_2/[kW · h]) or the oxygen-transfer efficiency or pounds of O_2 transferred per hour (kg O_2/h) per diffusion unit. This is called the *standard oxygen rating* (SOR) in tap water, at 20°C, and zero dissolved oxygen at sea level.

The actual oxygen transferred to the wastewater (AOR) is computed from

$$N = N_O \left(\frac{\beta C_s - C_L}{9.2} \right) \alpha \times 1.02^{T-20} \quad (5.17)$$

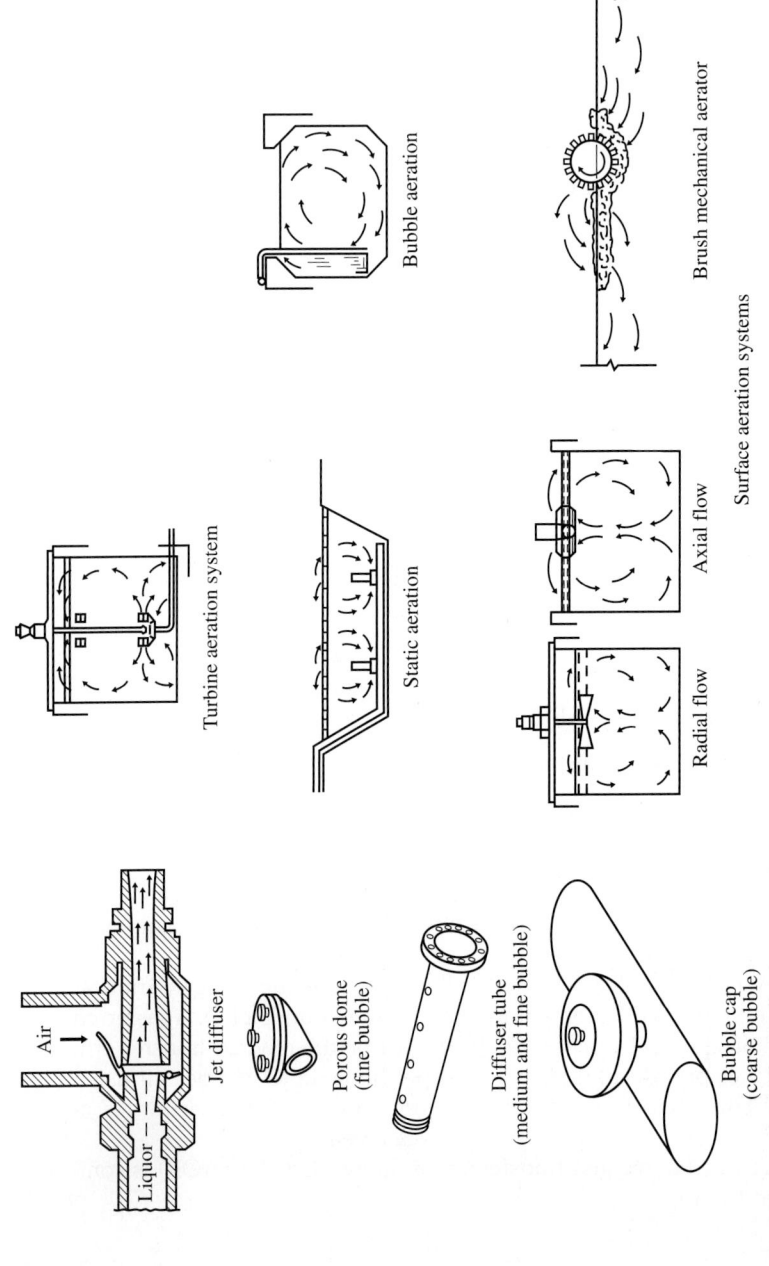

FIGURE 5.10 Aeration equipment.

where N = AOR, lb O_2/hp · h   [kg O_2/kW · h]
       N_O = SOR, lb O_2/hp · h [kg O_2/kW · h]

Diffused Aeration Equipment

Basically, there are two types of diffused aeration equipment: units producing a small bubble from a porous medium or membrane and units using a large orifice or a hydraulic shear device to produce large air bubbles.

Porous media are either tubes or plates constructed of carborundum or other finely porous media or membranes. Tubes are placed at the sidewall of the aeration tank, perpendicular to the wall, and generate a rolling motion to maintain mixing, or across the bottom of the basin. Maximum spacing is required to maintain solids in suspension; minimum spacing is required to avoid bubble coalescence. A diffused aeration system is shown in Fig. 5.11.

In order for adequate mixing to be maintained for sidewall mounts, the maximum width of the aeration tank is approximately twice its depth. This width can be doubled by placing a line of diffusion units along the centerline of the aeration tank. Figure 5.12 shows the performance of air-diffusion units in water in an aeration tank 15 ft (4.6 m) deep and 24 ft (7.3 m) wide. Fine bubble diffusers tend to clog with time, resulting in a reduced oxygen transfer efficiency as shown in Fig. 5.13. For bottom tank coverage, the tank must be drained and the diffusers cleaned.

Figure 5.11 Diffused aeration system.

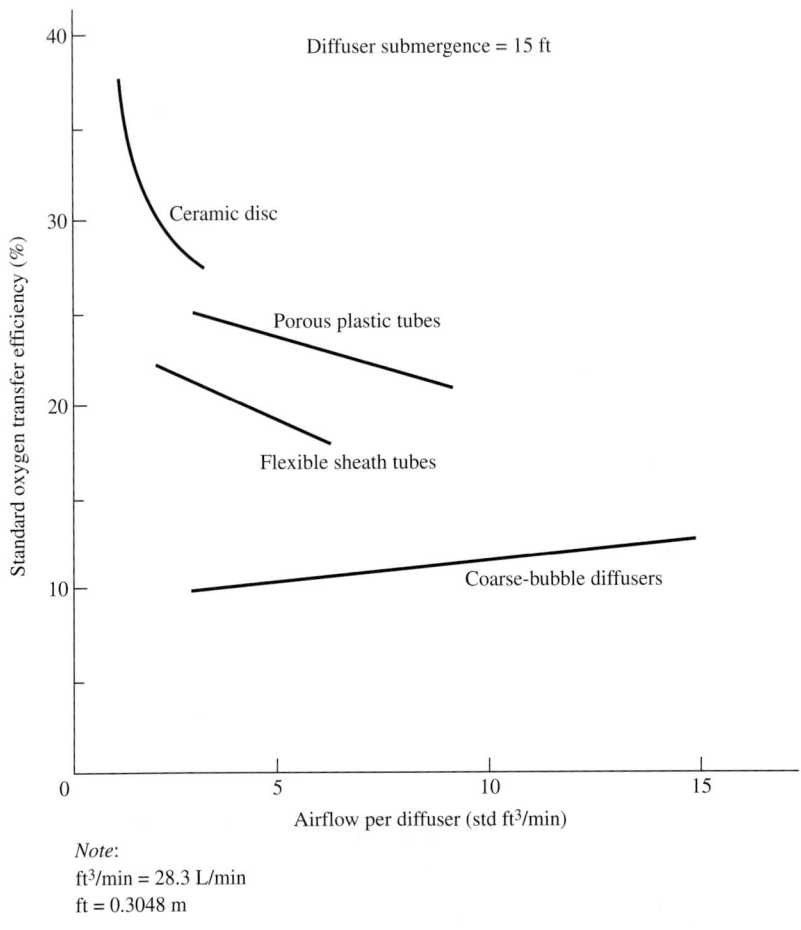

FIGURE 5.12 Effect of airflow rate per diffuser on oxygen transfer efficiency.

The effect of diffuser placement on clean water oxygen transfer efficiency is shown in Table 5.3. Fine-bubble and membrane diffusers will tend to clog on the liquid side because of the precipitation of metal hydroxides and carbonates or formation of a biofilm layer. The fouling factor F is site specific and can vary from 0.2 to 0.9. F appears to increase with increasing solids retention time (SRT). The average F value for municipal wastewater treatment facilities using medium- to fine-bubble membrane diffusers is 0.6. This value will apply for most industrial facilities unless other information is available. In cases where fouling is a factor, α in Eq. (5.17) should be corrected to $F\alpha$.

Large-bubble air-diffusion units will not yield the oxygen-transfer efficiency of fine-bubble diffusers since the interfacial area for transfer is considerably less. These units have the advantage, however, of not

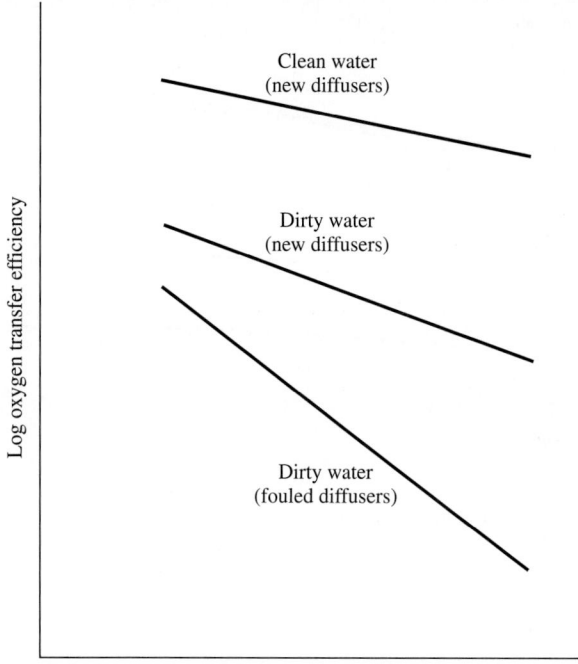

FIGURE 5.13 Change in oxygen transfer efficiency with fine-pore diffuser fouling—a hypothetical case.

Placement	Airflow (std ft³/min/Diffuser)	Standard OTE (%) at Water Depth		
		3 m	4.5 m	6 m
Floor cover (grid)	1–4	14–18	21–27	29–35
Quarter points	2–6	13–15	18–22	24–29
Mid-width	2–6	9–11	15–18	23–17
Single-spiral roll	2–6	7–11	14–18	21–28

Note: cfm × 0.47 = l/s
Source: Adapted from WPCF Aeration Manual of Practice, 1988.

TABLE 5.3 Effects of Diffuser Placement on Clean Water Oxygen Transfer Efficiencies (OTEs) of Flexible Sheath Tubes

requiring air filters and of generally requiring less maintenance. These units generally operate over a wider range of airflow per unit. Performance data for coarse-bubble aeration units are shown in Fig. 5.14.

Static aerators consist of vertical cylindrical tubes placed at specified intervals in an aeration basin and containing fixed internal elements.

200   Chapter Five

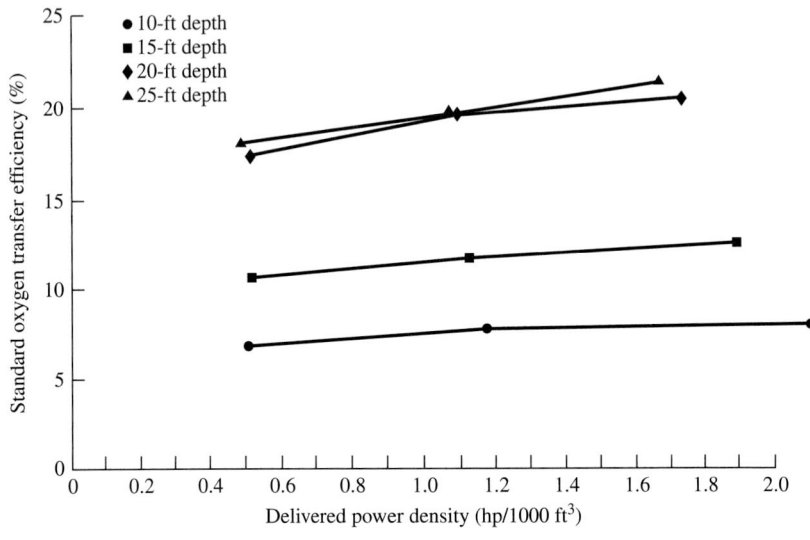

Note:
10- and 20-ft depths: 24-D-24 fixed-orifice diffusers applied in a total floor coverage configuration.
15- and 25-ft depths: 30-D-24 fixed-orifice diffusers applied in a wideband configuration along one centerline of the tank.

FIGURE 5.14 Oxygen transfer efficiency of coarse-bubble diffusers (*courtesy of Sanitaire Corp.*). Results are for clean water.

In aeration of sewage or industrial wastes, the coefficient α will usually be less than 1.0. There is some evidence that the large-bubble diffusers are less affected by the presence of surface-active materials than are the fine-bubble diffusers.

A diffused aeration design is shown in Example 5.2.

Example 5.2. An aeration system is to operate under the following conditions:

Aeration Design

1000 lb O_2/h
$T = 30°C$
$C_L = 2$ mg/L
$\beta = 0.95$
$\alpha = 0.85$
$D = 20$ ft

Use coarse bubble diffusers with a 12 percent oxygen transfer efficiency (OTE) at 15 std ft³/min per tube and 15-ft liquid depth. At 20 ft:

$$\text{OTE} = 12\left(\frac{20}{15}\right)^{0.7} = 14.7\%$$

O_2 saturation at 30°C is 7.6 mg/L. Assume 15% OTE. Then

$$O_t = \frac{21(1-0.15) \times 100}{79 + 21(1-0.15)}$$

$$= 0.18$$

$$C_{s,m} = \frac{7.6}{2}\left[\frac{14.7 + 0.433 \times 20}{14.7} + \frac{18}{21}\right]$$

$$= 9.3 \text{ mg/L}$$

$$N = 15 \text{ std ft}^3/\text{min/unit} \times 60 \times 0.232 \times 0.0746 \times 0.147$$

$$= 2.29 \text{ lb } O_2/\text{h/unit}$$

$$N = N_O\left[\frac{\beta C_{SW} - C_L}{C_s}\right]\alpha \times 1.02^{T-20}$$

$$= 2.29\left[\frac{0.95 \times 9.3 - 2.0}{9.3}\right]0.85 \times 1.02^{10}$$

$$= 1.74 \text{ lb } O_2/\text{h/unit}$$

$$\frac{1000 \text{ lb } O_2/\text{h}}{1.74} = 575 \text{ diffusers}$$

Airflow $= 575 \times 15 = 8620$ std ft^3/min

$$\text{hp} = \frac{(\text{std ft}^3/\text{min})(\text{lb/in}^2)(144)}{0.7 \times 33,000} \quad (195 \text{ kW})$$

$$= \frac{8620 \times 10 \times 144}{0.7 \times 33,000} = 537$$

Turbine Aeration Equipment

Turbine aeration units disperse compressed air by the shearing and pumping action of a rotating impeller. Since the degree of mixing is independently controlled by the power input to the turbine, there are no restrictive limitations to tank geometry. Figure 5.15 shows a typical turbine aeration unit.

Air is usually fed to the turbine through a sparge ring located just beneath the impeller blades. The ratio of the turbine diameter to the equivalent tank diameter varies from 0.1 to 0.2. Most aeration applications have employed impeller tip speeds of 10 to 18 ft/s (3.1 to 5.5 m/s).

Quirk[16] showed that the oxygen transfer for a turbine aerator can be estimated.

$$O_2 \text{ transfer efficiency} = CP_a^n \quad (5.18)$$

where $P_d = \text{hp}_R/\text{hp}_c$ in which hp_R and hp_c are the horsepower (kW) of the turbine and compressor, respectively. The optimum oxygenation

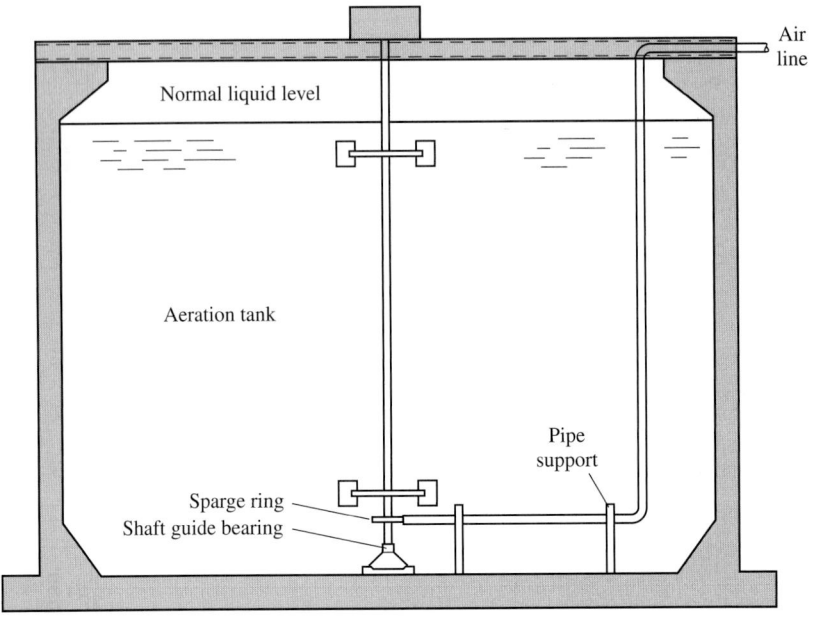

FIGURE 5.15 Typical turbine aerator installation.

efficiency occurs when the power split between turbine and blower is near a 1:1 ratio.[15]

The actual power drawn by the impeller will decrease as air is introduced under the impeller because of the decreased density of the aerating mixture. The transfer efficiency of turbine aeration units will vary from 1.6 to 2.9 lb O_2/hp · h (0.97 to 1.76 kg O_2/[kW · h]), depending on the turbine-blower power split.

To eliminate swirling and vortexing, baffles are normally required. In a round tank, four baffles are placed equally around the circumference of the wall. In a square tank, two baffles on opposite walls are used, while in a rectangular tank with a length/width ratio greater than 1.5, no baffles are required.

Surface-Aeration Equipment

Surface-aeration units are of two types: those employing a draft tube and those with only a surface impeller. In both types, oxygen transfer occurs through a vortexing action and from the surface area exposure of large volumes of liquid sprayed over the surface of the aeration tank. Examples of these units are shown in Figs. 5.16 and 5.17. The transfer rate is influenced by the diameter of the impeller and its speed of rotation, and by the submergence level of the rotating element. Under optimum submergence conditions, the transfer rate per unit horsepower (W) remains relatively constant over a wide range of impeller diameters.

Aeration and Mass Transfer 203

FIGURE 5.16 Low-speed surface aerator.

FIGURE 5.17 High-speed surface aerator. (*Courtesy of Aqua Aerobic Systems Inc.*)

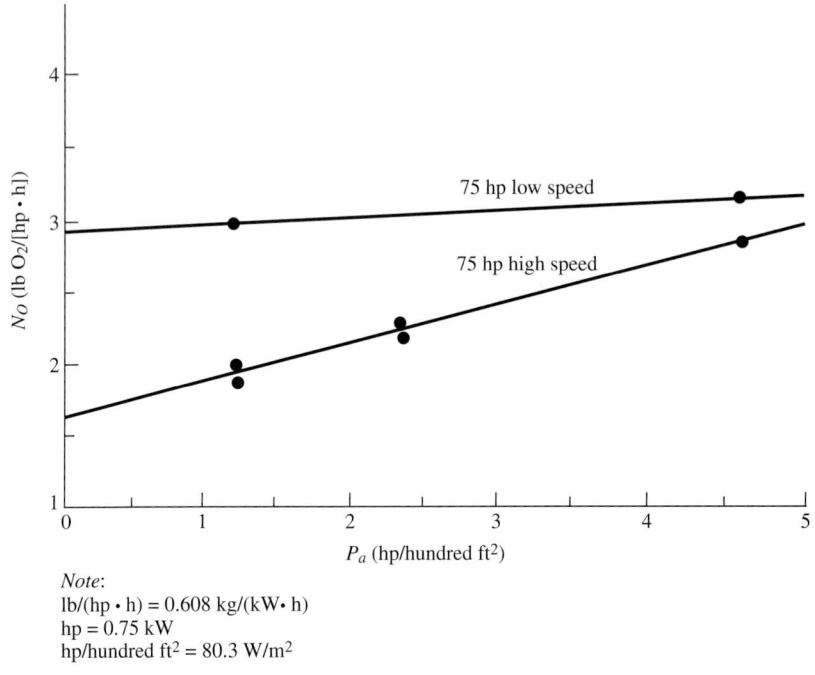

Figure 5.18 Comparative effect of surface area on high- and low-speed surface aerators. (*Adapted from Kormanik et al., 1973.*)

The quantity of oxygen transferred at the liquid surface is a function of the power level, and the overall oxygen transfer rate generally increases with increasing power level. Kormanik et al.[17] have shown a correlation between the oxygen-transfer rate and horsepower per unit surface area, as shown in Fig. 5.18. To maintain uniform dissolved oxygen concentrations requires power levels of 6 to 10 hp/ million gal (1.2 to 2.0 W/m³). The maximum area of influence for high-speed-and low-speed surface aerators is shown in Table 5.4. To maintain biological solids in suspension, a minimum bottom velocity of 0.4 ft/s (12 cm/s) is required for 5000 mg/L MLSS.

To prevent bottom scour, minimum depths of 6 to 8 ft (1.8 to 2.4 m) are recommended, and, to maintain solids in suspension, maximum depths of 12 ft (3.7 m) are recommended for high-speed aerators and 16 ft (4.9 m) for low-speed aerators. The maximum allowable F/M and power level for various aeration equipment are presented in Fig. 5.19.

The brush aerator, which has been very popular in Europe, uses a high-speed rotating brush that sprays liquid across the tank surface. A circular liquid motion is induced in the aeration tank. The performance of the brush aerator is related to the speed of rotation and submergence of the brush. The efficiencies of various aerators are shown in Table 5.5. Mixing requirements are shown in Table 5.6.

| | Radius of Influence, ft | |
Horsepower	High Speed	Low Speed
5	40	50
10	55	70
20	80	95
30	100	120
50	130	155
75	155	190
100	—	220
150	—	260

Source: Adapted from Arthur, 1986.

TABLE 5.4 Maximum Area of Influence for High-Speed-and Low-Speed Surface Aerators

Measurement of Oxygen Transfer Efficiency

Although several procedures have been followed to estimate the transfer efficiency of aeration devices, the non–steady-state aeration procedure has been generally adopted as a standard. This test involves the chemical removal of dissolved oxygen by the addition of sodium sulfite with cobalt added as a catalyst. The cobalt concentration should be 0.05 mg/L to avoid a "cobalt effect" on the magnitude of $K_L a$. The increase in oxygen concentration is measured during

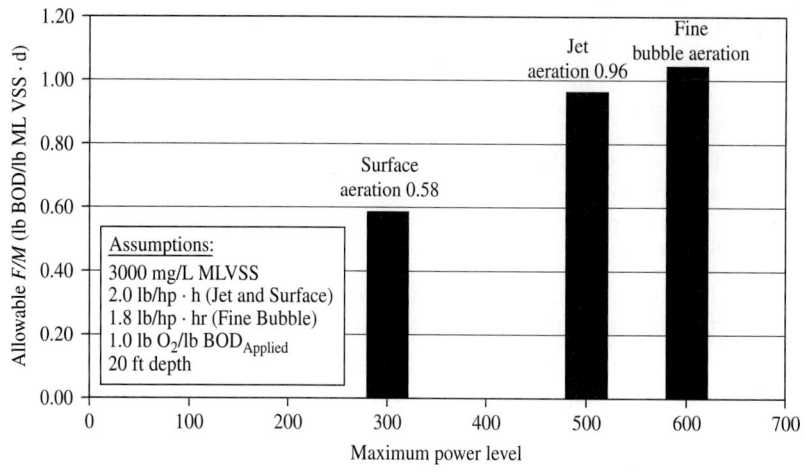

FIGURE 5.19 Maximum allowable power level and operating F/M for various aeration equipment.

Type of Aerator	Water Depth, ft	OTE, %	lb O_2/hp · h	Reference*
Fine bubble				
Tubes—spiral roll	15	15–20	6.0–8.0	20
Domes—full floor coverage	15	27–31	10.8–12.4	20
Coarse bubble				
Tubes—spiral roll	15	10–13	4.0–5.2	20
Spargers—spiral roll	14.5	8.6	3.4	21
Jet aerators	15	15–24	4.4–4.8	20
Static aerators	15	10–11	4.0–4.4	20
	30	25–30	6.0–7.5	
Turbine	15	10–25	—†	16
Surface aerator				
Low speed	12	—	5.9–7.5	17
High speed	12	—	3.3–5.0	17

*Wire horsepower must correct for overall blower efficiency.
†Horsepower depends on power split.
Note: ft = 0.3048 m
lb/(hp · h) = 0.608 kg/(kW · h)

TABLE 5.5 Summary of Aerator Efficiencies

aeration under specified conditions, and the overall transfer coefficient is calculated from Eq. (5.6). Dissolved oxygen can be measured by the Winkler test or by use of a dissolved oxygen probe. The suggested procedure for non–steady-state aeration is as follows:

1. Remove the dissolved oxygen in the aeration unit by adding sodium sulfite and cobalt chloride. A concentration of 0.05 mg/L of cobalt should be added, and 8 mg/L Na_2SO_3 per mg/L dissolved oxygen.

2. Thoroughly mix the tank contents. In a diffused aeration unit, aeration for 1 to 2 min will usually be sufficient.

3. Start the aeration unit at the desired operating rate. Sample for dissolved oxygen at selected time intervals. (At least 5 points should be obtained before 90 percent of saturation is reached.)

4. In a large aeration tank, multiple sampling points should be selected (both longitudinally and vertically) to compensate for concentration gradients.

Aeration Device*	hp/1000 ft³, Activated Sludge	hp/1000 ft³, Aerobic Digestion
High speed	1.30	2.00
Low speed	1.00	1.50
Brush ditch	0.60	1.00
Fine bubble (floor cover)	0.40	0.80
Fine bubble (gut roll)	1.00	1.50
Fine bubble (spiral roll)	0.75	1.25
Fine bubble (moving bridge)	0.60	1.00
Jet	1.00	1.50
Coarse bubble (floor cover)	0.60	1.00
Coarse bubble (gut roll)	1.00	1.50
Coarse bubble (spiral roll)	0.75	1.25
Submerged turbine	0.75	1.25

*Operation conditions:
hp, delivered or water horsepower
MLSS, 1500–3000 mg/L (activated sludge)
MLSS, 10,000–20,000 mg/L (aerobic digestion)
Source: Adapted from Arthur, 1986.

TABLE 5.6 Activated Sludge Process Mixing Horsepower Guide

5. If a dissolved oxygen probe is used, it can be left in the aeration tank; the values from the probe can be recorded at appropriate time intervals.

6. Record the temperature and measure oxygen saturation. If water is being aerated, it is usually satisfactory to select the saturation value from Table 5.1. For diffused aeration units, saturation should be corrected to the tank middepth according to Eq. (5.8).

7. Compute the oxygen-transfer rate in accordance with Eq. (5.6) (see Example 5.1).

Equation (5.6) can be integrated to yield

$$C_s - C_L = (C_s - C_0)e^{-K_L at} \qquad (5.19)$$

where C_0 = dissolved oxygen at time zero. To evaluate the constants in Eq. (5.19), a nonlinear least-squares program is recommended by ASCE,[19] one that provides the best set for $K_L a$, C_0, and C_s.

It is occasionally desirable to measure the oxygen-transfer rate in the presence of activated sludge. Either a steady-state or non–steady-state procedure can be followed for this purpose.

In the aeration of activated sludge, Eq. (5.6) must be modified to account for the effect of the oxygen-utilization rate of the sludge-liquid mixture:

$$\frac{dc}{dt} = K_L a(C_s - C_L) - r_r \tag{5.20}$$

where r_r is the oxygen-utilization rate in milligrams per liter per hour.

Under steady-state operation $dc/dt \rightarrow 0$, $K_L a$ can be computed from the relationship

$$K_L a = \frac{r_r}{C_s - C_L} \tag{5.21}$$

Aeration is continued until a constant oxygen uptake rate is obtained in order to ensure a steady state.

In the nonsteady procedure, aeration is stopped and the dissolved oxygen is allowed to approach zero by microbial respiration. Aeration is then begun and the buildup in dissolved oxygen recorded as in the non–steady-state water procedure; Fig. 5.20 shows how. $K_L a$ can be computed from the slope of a plot of dc/dt versus C_L, in accordance with a rearrangement of Eq. (5.20):

$$\frac{dc}{dt} = K_L a(C_s - r_r) - K_L a C_L \tag{5.22}$$

since $K_L a(C_s - r_r)$ will be constant for any specific operating condition. This is shown in Example 5.3.

Example 5.3. The following data were obtained for non–steady-state aeration in an activated sludge basin:

Time, min	C_L, mg/L
0.0	0.52
0.5	0.70
1.0	0.93
2.0	1.23
3.0	1.55
4.0	1.80
5.0	2.00
10.0	2.20

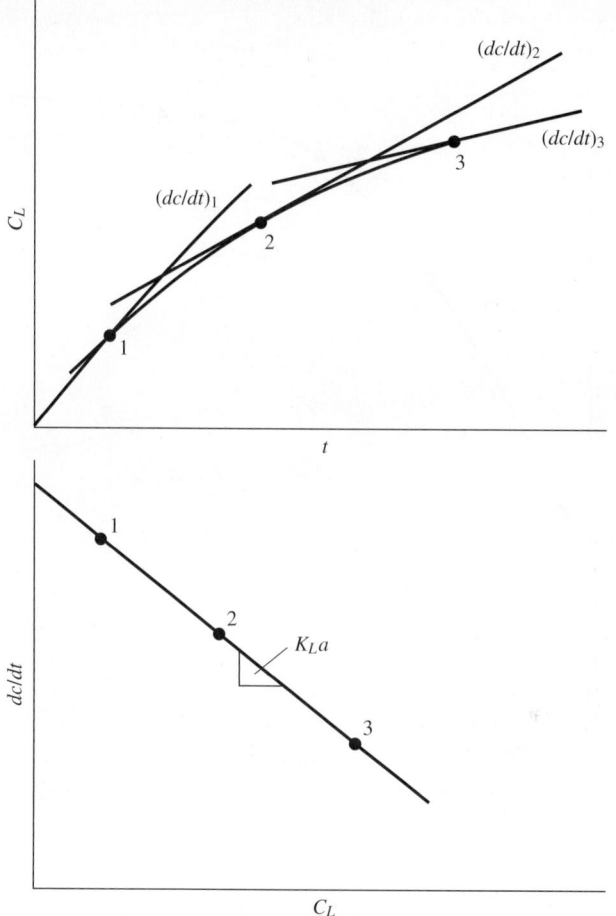

FIGURE 5.20 Non–steady-state evaluation of $K_L a$.

Solution
Determine the $K_L a$.
The non–steady-state equation is:

$$\frac{dC_L}{dt} = K_L a (C_s - C_L) - R_r$$

Rearrangement yields:

$$\frac{dC_L}{dt} = (K_L a C_s - R_r) - K_L a\, C_L$$

In a plot of dC_L/dt against C_L, the negative slope of the resulting straight line is $K_L a$. The dC_L/dt values are calculated and plotted in Table I and Figs. I and II.

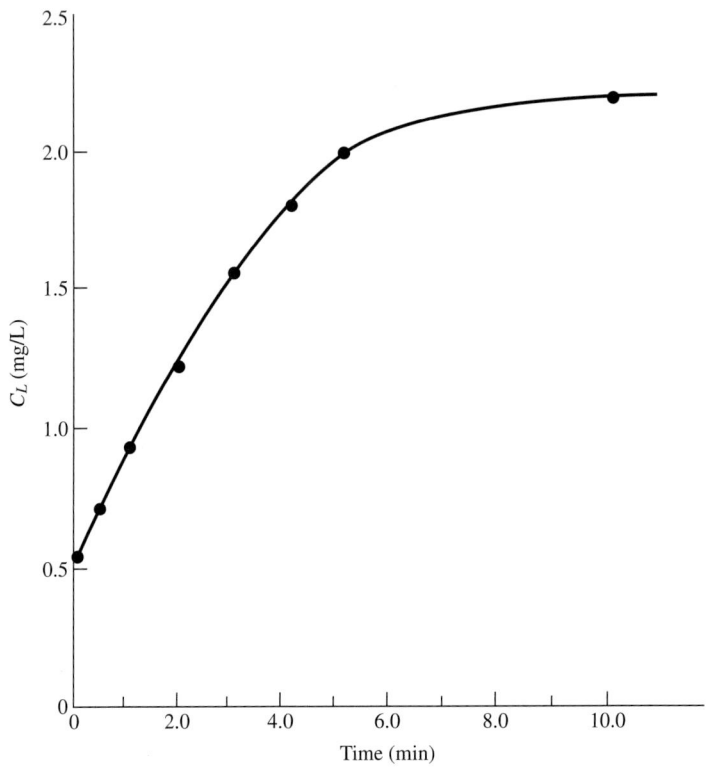

FIGURE I For Example 5.3

TABLE I		
t, min	C_L, mg/L	dC_L/dt
0	0.51	0.43
1	0.91	0.37
2	1.25	0.32
3	1.55	0.27
4	1.80	0.22
5	1.99	0.17

Other Measuring Techniques

The off-gas analysis utilizes a hood over a portion of the aeration tank to collect the off-gases for O_2 analysis with a non-membrane-type probe. This technique is applicable for diffused aeration systems with transfer efficiencies above 5 percent.[20]

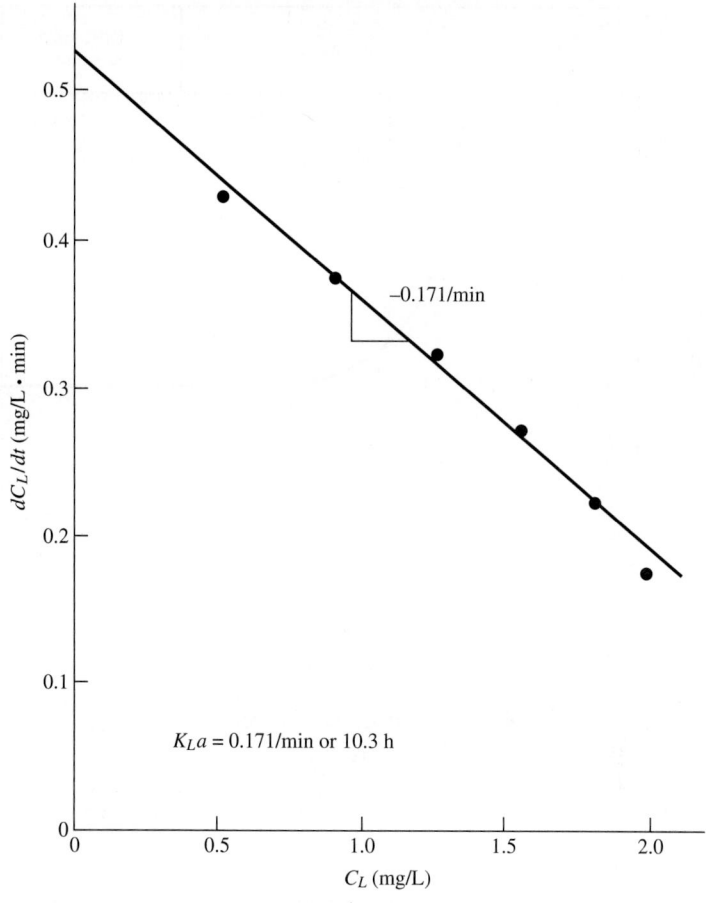

FIGURE II For Example 5.3.

A radioactive tracer technique has been developed by Neal and Tsivoglou.[21] Their technique uses the stripping of krypton from the aeration tank to measure $K_L a$ under the exact aeration tank mixing and waste conditions. The oxygen transfer coefficient is then calculated from the known ratio of $(K_L a)_{O_2} / (K_L a)_{krypton}$.

5.4 Air Stripping of Volatile Organic Compounds

The physical process of transferring volatile organic compounds (VOCs) from water into air is called *desorption* or *air stripping*. This can be accomplished by injection of water into air via spray systems, spray towers, or packed towers, or by injection of air into water through diffused or mechanical aeration systems. The most common systems in use today are packed towers or aeration systems. Aeration

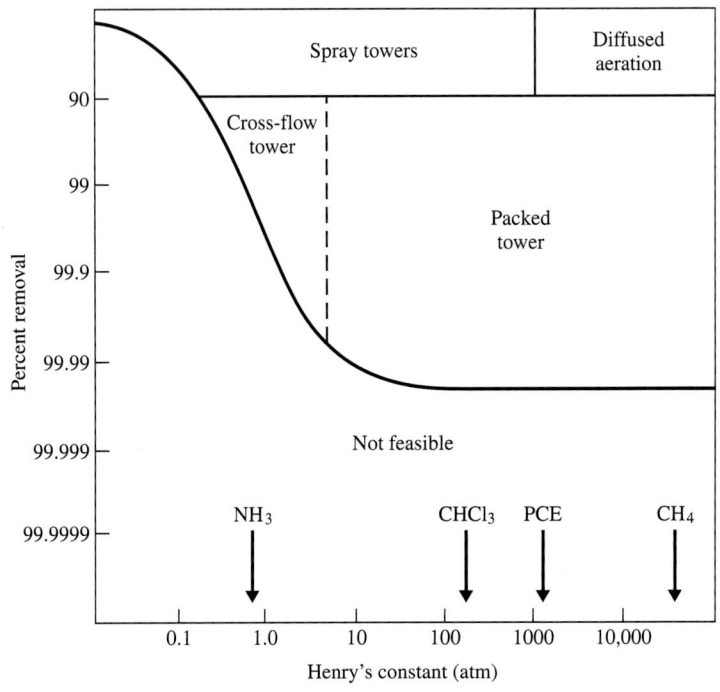

FIGURE 5.21 Stripping efficiency of various technologies.

systems are usually used in conjunction with biological wastewater-treatment processes. Stripping efficiencies for various techniques are shown in Fig. 5.21. A typical packed tower is shown in Fig. 5.22.

Packed tower media are open-structured, chemically inert materials, usually plastic, which are selected to give high surface areas for good contacting while offering a low pressure drop through the tower. Some of the factors that affect removal of VOCs are the contact area, the solubility of the contaminant, the diffusivity of the contaminant in the air and water, and the temperature. All of these factors except diffusivity and temperature are influenced by the airflow and water-flow rates and the type of packing media. General relationships are shown in Fig. 5.23. The efficiency of transfer of contaminant from water to air depends on the mass-transfer coefficient and Henry's law constant (see Eq. [5.7]). The mass-transfer coefficient defines the transfer of contaminant from water to air per unit volume of packing per unit time. The ability of a contaminant to be air stripped can be estimated from its Henry's law constant. A high Henry's constant indicates that the contaminant has low solubility in water and can therefore be removed by air stripping. In general, Henry's constant increases with increasing temperature and decreases with increasing solubility.

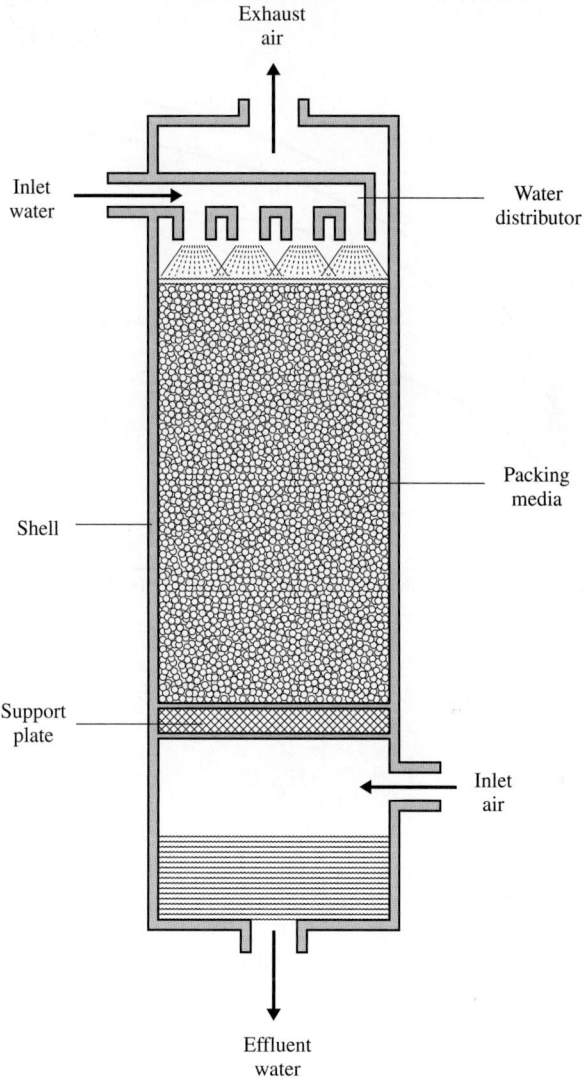

FIGURE 5.22 Typical packed tower.

Packed Towers

Packed-tower air strippers can be designed for a wide variety of flow ranges, temperatures, and organics. One of the first steps in applying air stripping is to estimate the maximum possible removal for a given contaminant, based on its Henry's law constant, temperature, and volumetric air/water ratio. By assuming equilibrium

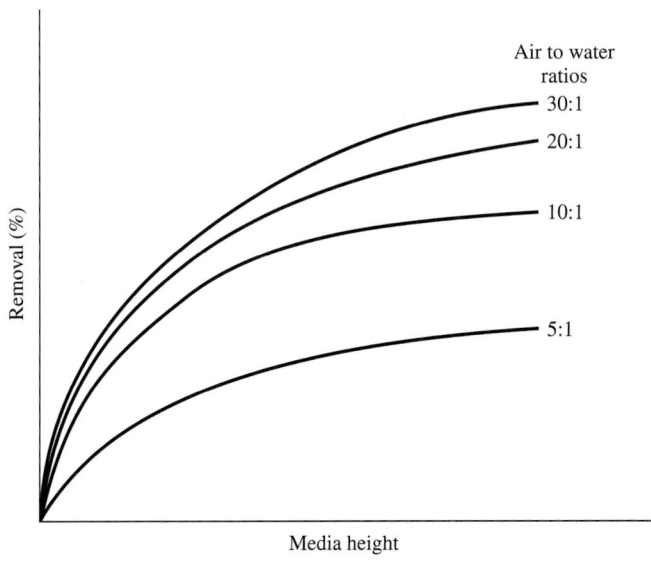

FIGURE 5.23 Illustrative relationships of air stripping removal efficiencies to media height and air to water ratios.

between a volume of gas and a volume of water, the following equation can be used:

$$\frac{C_2}{C_1} = \left(1 + \frac{H_M A_w}{RT}\right)^{-1} \quad (5.23)$$

where C_2 = final concentration of organic, µg/L
C_1 = initial concentration of organic, µg/L
H_M = Henry's law constant, atm · m³/mol
A_w = volumetric air/water ratio
R = universal gas constant, 8.206×10^{-5} atm · m³/(mol · K)
T = temperature, K

While Henry's constant is not accurately defined for many organics, Table 5.7 lists estimated values for some organics that are found in wastewaters. In the scientific literature Henry's constant is often expressed as H_M (atm · m³/mol). Another form of Henry's constant is calculated as $H_c = H_M/RT$ (expressed as m³ water/m³ air). At 20°C, $H_c = 41.6 H_M$, with the constant varying from 44.6 to 40.2 for a temperature range between 0 and 30°C.

The effect of temperature on the Henry's law constant is not well defined. Ashworth et al.[23] developed the following empirical relationship:

$$H_M = \exp\left(A - \frac{B}{T}\right) \quad (5.24)$$

Compound	Formula	Henry's Constant H_M, atm · m³/mol	H_c, (µg/L)/(µg/L)
Easy to strip			
Vinyl chloride	CH_2CHCl	6.38	265
Trichloroethylene	$CCHCl_3$	0.010	0.43
1,1,1-Trichloroethane	CCH_3Cl_3	0.007	0.29
Toluene	$C_6H_5CH_3$	0.006	0.25
Benzene	C_6H_6	0.004	0.17
Chloroform	$CHCl_3$	0.003	0.12
Difficult to strip			
1,1,2-Trichloroethane	CCH_3Cl_3	7.7×10^{-4}	0.032
Bromoform	$CHBr_3$	6.3×10^{-4}	0.026
Nonstrippable			
Pentachlorophenol	$C_6(OH)Cl_5$	2.1×10^{-6}	0.000087
Dieldrin	—	1.7×10^{-8}	0.0000007

TABLE 5.7 Henry's Constants for Selected Compounds at 20°C

where A = empirical constant = 5.534 for benzene, 7.845 for trichloroethylene, 8.483 for methylene chloride
B = empirical constant = 3194 for benzene, 3702 for trichloroethylene, 4268 for methylene chloride
T = temperature, K

In general, the ability of a contaminant to be air stripped increases with increasing temperature and decreases with increasing solubility. Compounds with $H_M > 10^{-3}$ atm · m³/mol are easy to strip, compounds with H_M between 10^{-4} and 10^{-3} atm · m³/mol are difficult to strip, and compounds with H_M less than 10^{-4} atm · m³/mol are nonstrippable.

The effect of feed temperature on the removal of soluble compounds is shown in Table 5.8.

Compound	Percent Removal at		
	12°C	35°C	73°C
2-Propanol	10	23	70
Acetone	35	80	95
Tetrahydrofuran	50	92	>99

TABLE 5.8 Influence of Feed Temperature on Removal of Water-Soluble Compounds from Groundwater

The height of the stripping tower can be computed from the relationship

$$H = \frac{L_v}{K_L a} \frac{S}{S-1} \ln\left[\frac{\frac{C_1}{C_2}(S-1)+1}{S}\right] \qquad (5.25)$$

where L_v = volumetric liquid loading rate, ft/s
H = packing height, ft
$K_L a$ = mass-transfer coefficient
S = stripping factor = $H_c a_w$
H_c = Henry's law constant

In order to calculate the amount of packing needed to obtain the removal found from Eq. (5.25), the mass transfer coefficient $K_L a$ must be known for the VOC of interest, the type of packing media used, and the design conditions. $K_L a$ can be determined through pilot testing or calculated as follows:

$$\frac{1}{K_L a} = \left(\frac{1}{k_L a}\right) + \left(\frac{1}{H_c k_g a}\right) \qquad (5.26)$$

Onda et al.[24] developed relationships to define k_L and k_g as summarized below.

$$k_L \left(\frac{\rho_L}{\mu_L g}\right)^{1/3} = 0.0051 \left(\frac{L}{a_w \mu_L}\right)^{2/3} \left(\frac{\mu_L}{\rho_L D_L}\right)^{-0.5} (a_t d_p)^{0.4} \qquad (5.27)$$

$$\frac{k_G}{a_t D_g} = 5.23 \left(\frac{G}{a_t \mu_g}\right)^{0.7} \left(\frac{\mu_g}{\rho_g D_g}\right)^{1/3} (a_t d_p)^{-2} \qquad (5.28)$$

$$\frac{a_w}{a_t} = 1 - \exp\left[-1.45\left(\frac{\sigma_c}{\sigma_L}\right)^{0.75} (\text{Re}_L)^{0.1} (\text{Fr}_L)^{-0.05} (\text{We}_L)^{-0.2}\right] \qquad (5.29)$$

where k_L = liquid-phase mass-transfer coefficient
μ_L = viscosity of liquid, Pa · s
ρ_L = density of liquid, kg/m³
g = gravitational acceleration, 9.81 m/s²
L = liquid loading rate, kg/(m² · s)
a_w = wetted specific surface area, m²/m³
D_L = liquid-phase diffusivity, m²/s
a_t = total specific surface area, m²/m³
d_p = nominal packing size, m

Aeration and Mass Transfer

k_g = gas-phase mass-transfer coefficient
D_g = gas-phase diffusivity, m²/s
G = gas loading rate, kg/(m² · s)
μ_g = viscosity of gas, Pa · s
ρ_g = density of gas, kg/m³
σ_c = surface tension of packing material, N/m
σ_L = surface tension of liquid, N/m
Re_L = liquid-phase Reynolds's number = $L/a_t\mu_L$
Fr_L = liquid-phase Froude number = $L^2 a_t/(\rho_L)^2 g$
We_L = liquid-phase Weber number = $L^2/\rho_L \sigma_L a_t$

The $K_L a$ can be adjusted for different temperatures as follows:

$$K_L a_{T_2} = K_L a_{T_1} \times 1.024^{T_2 - T_1}$$

Once the packing volume has been calculated, the tower diameter can be found from the desired pressure drop calculated in Fig. 5.24, which is a generalized correlation for predicting pressure drop in packed towers.[25] The following Eqs. (5.30) and (5.31) are the *x*- and *y*-axis values of Fig. 5.24, respectively. The packing factor is unique to the type of packing and can usually be obtained from the manufacturer.

$$\frac{L'}{G'} \left(\frac{\rho_G}{\rho_L - \rho_G} \right)^{1/2} \qquad (5.30)$$

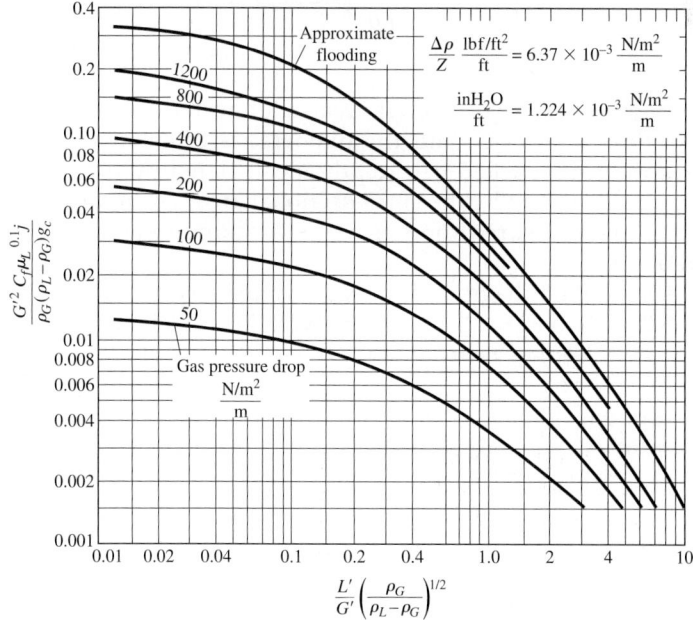

Figure 5.24 Generalized flooding and pressure drop curves for packed towers.

where L = liquid loading rate, lb/(ft²· s)
G = gas loading rate, lb/(ft² · s)
ρ_g = density of gas, lb/ft³
ρ_L = density of liquid, lb/ft³

$$\frac{G^2 F \mu_L{}^{0.1} J}{\rho_g (\rho_L - \rho_g) g_c} \tag{5.31}$$

where F = packing factor (97 for 16-mm [5/8-in] plastic Pall rings)
μ = liquid viscosity, cP [kg/m · s)]
$g_c = 4.18 \times 10^8$ for English units (1 for SI units)
$J = 1.502$ for English units (1 for SI units)

To estimate the tower size, an allowable pressure drop value is selected. Usually a pressure drop in the range of 0.25 to 0.50 in of water per foot of packed bed gives an average-size tower and a flexible operating range. Once the pressure drop is chosen, the x-axis value for Fig. 5.24 is calculated from Eq. (5.30). A vertical line is drawn from the x-axis to the intersection of the pressure drop curve corresponding to the selected value. At this intersection, the y-axis value is read. Equation (5.31) is rearranged and solved for G:

$$G = \left(\frac{(y\text{-axis value}) \rho_g (\rho_L - \rho_g) g_c}{F \mu_L{}^{0.1} J} \right)^{0.5} \tag{5.32}$$

The tower cross-sectional area is found by dividing the air mass flow rate by G, and the diameter is calculated from the area. The total packed-bed depth is calculated by dividing the volume of packing by the cross-sectional area. The total pressure drop is the pressure drop per foot of packing times the packed bed depth.

A schematic diagram of an air stripping system is shown in Fig. 5.25. The off gas containing the volatile organics must usually be treated. Applicable technologies are shown in Fig. 5.26.

Example 5.4. Air stripper removal efficiency can be calculated from the equation

$$\frac{C_1}{C_2} = \frac{(S) \exp \left[\left(\frac{S-1}{S} \right) \left(\frac{H K_L a}{L_v} \right) \right] - 1}{S - 1}$$

where C_1 = influent concentration
C_2 = effluent concentration

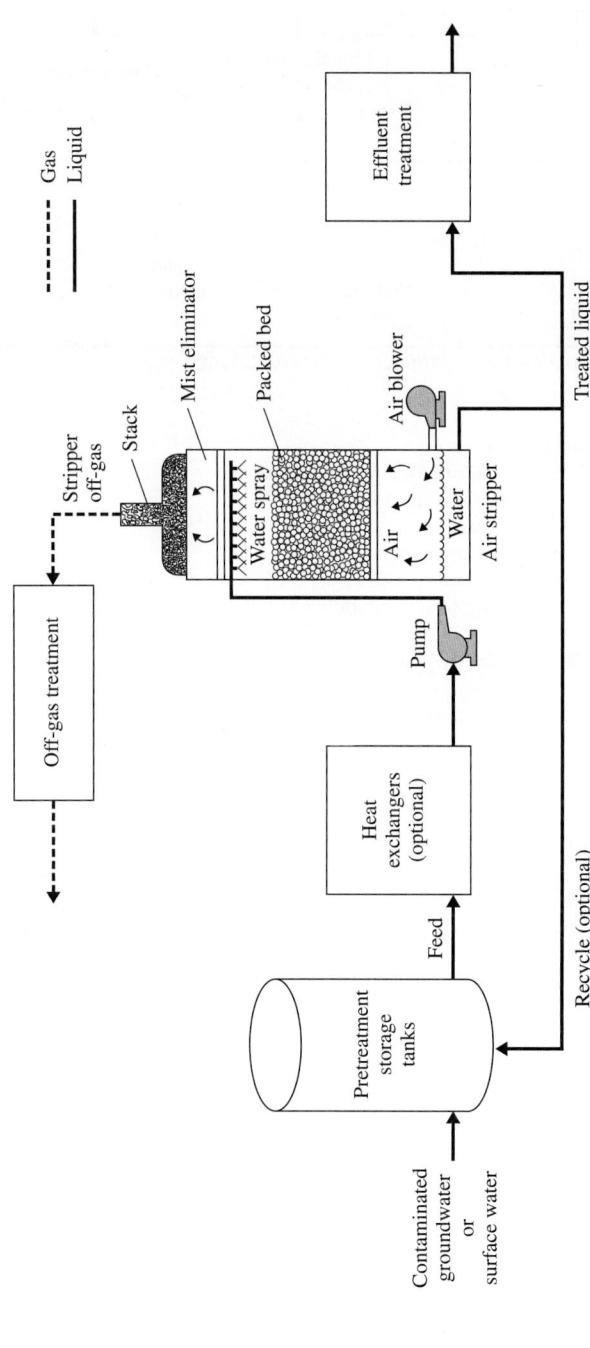

FIGURE 5.25 Schematic diagram of air stripping system.

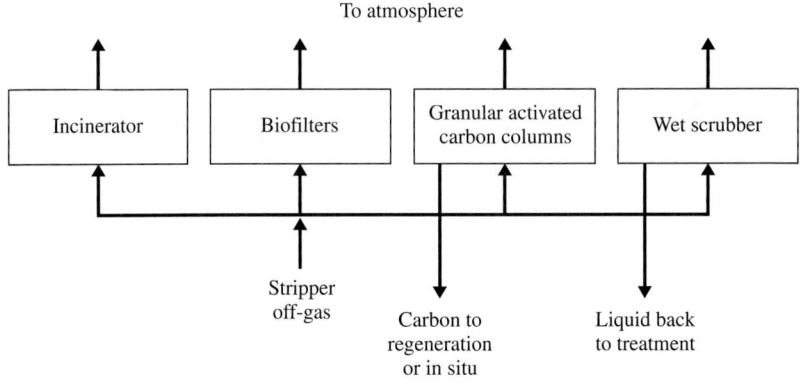

FIGURE 5.26 Process options for air stripper off-gas treatment.

H = packing height
L_v = volumetric liquid loading rate = Q_L/A
A = tower cross-sectional area = πr^2
Q_L = water-flow rate
S = stripping factor = $H_c a_w$
H_c = Henry's constant
A_w = volumetric air to water ratio = QG/QL
Q_G = airflow rate
$K_L a$ = mass-transfer coefficient

If the following conditions are assumed:

H = 40 ft
L_v = 10.03/153.9 = 0.06517 ft/s
$A = \sigma 7^2$ = 153.9 ft^2
Q_L = 4500 gal/min = 10.03 ft^3/s
S = (0.2315)(19.94) = 4.616
H_c = 0.2315 for TCE at 10°C
A_w = 12,000/(10.03)(60) = 19.94
Q_G = 12,000 ft^3/min
$K_L a$ = 0.0125 s^{-1}

then

$$\frac{C_1}{C_2} = \frac{(4.616)\exp\left[\left(\frac{4.616-1}{4.616}\right)\left(\frac{(40)(0.0125)}{0.06517}\right)\right]-1}{4.616-1} = 520$$

$$\frac{C_2}{C_1} = 0.00192$$

% Removal = $(1 - C_2/C_1) \times 100\%$
= 99.81

Steam stripping may be applied for the removal of high concentrations of volatile organics. The principal index used to

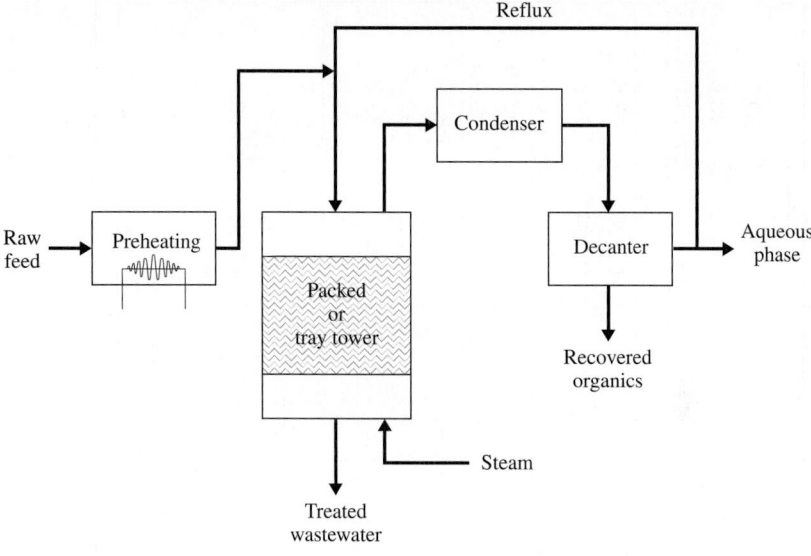

FIGURE 5.27 Typical steam stripping process.

estimate the steam stripping capability is the boiling point of the organic compound. A compound should exhibit a relatively low boiling point (150°C) and an acceptable Henry's law constant for effective steam stripping. A contaminated condensate is produced that may be recovered or treated further. A steam stripping system is shown in Fig. 5.27 and performance characteristics are given in Table 5.9.

5.5 Problems

5.1 An air diffuser yielded a $K_L a$ of 6.5/h at 10°C. The airflow is 50 cm³/min, the bubble diameter 0.15 cm, and the bubble velocity 28 cm/s. The depth of the aeration column is 250 cm with a volume of 4000 cm³.

(a) Compute K_L.
(b) Compute $K_L a$ at 25°C.

5.2 Design an aeration system using coarse bubble diffusers with a 12 percent OTE at 15-ft liquid depth under standard conditions.

1000 lb O_2/h
$T = 30°C$
$C_L = 2$ mg/L
$\beta = 0.95$
$\alpha = 0.85$
$D = 20$ ft

Plants Using Steam Stripping	Stripped Compound	Henry's Constant, atm	Vapour Pressure, mm Hg @ 25°C	Concentration, ppm		Percent Removal
				Influents	Effluent	
Pesticide industry						
Plant 1	Methylene chloride	177	425	<159	<0.01	99.9
Plant 2	Chloroform	188	180	70.0	<5.0	>92.6
Plant 3	Toluene	370	29	721	43.4	94.0
Organic chemicals industry						
Plant 4	Benzene	306	74	<15.4	<0.230	98.5
Plant 5	Methylene chloride	177	425	<3.02	<0.0141	99.5
	Toluene	370	29	178	<52.8	>70.3
Plant 6a	Methylene chloride	177	425	1430	<0.0153	>99.99
	Carbon tetrachloride	1280	113	<665	<0.0549	>99.99
	Chloroform	188	180	<8.81	1.15	<86.9
Plant 6b	Methylene chloride	177	425	4.73	<0.0021	>99.95
	Chloroform	188	180	<18.6	<1.9	89.8
	1,2-Dichloroethane	62	82	<36.2	<4.36	88.0
	Carbon tetrachloride	1280	113	<9.7	<0.030	99.7
	Benzene	306	74	24.1	<0.042	>99.8
	Toluene	370	29	22.3	<0.091	>99.6
Plant 7	Methylene chloride	177	425	34	<0.01	>99.97
	Chloroform	188	180	4509	<0.01	99.99
	1,2-Dichloroethane	62	82	9030	<0.01	>99.99

TABLE 5.9 Full-scale Industrial Steam Stripper Performance Summary

5.3 The following data were collected at 29°C for non–steady-state oxygen transfer in an activated sludge basin. Compute K_L and $K_L a$ at 20°C.

Time, min	C_L, mg/L
0	0.75
3	1.60
6	2.30
9	2.80
12	3.20
15	3.50
18	3.72

5.4 A pilot study was conducted to evaluate the removal of benzene by air stripping. The pilot packed-tower air stripper had a 10-in diameter and 10 ft of packing height. The tower was operated at 50°C with a flow rate of 10 gal/min and an air to water ratio of 50:1. The benzene concentration was reduced from 1400 to 200 µg/L. What is the required tower size to reduce the benzene concentration from 1600 to 10 µg/L at 50 gal/min? Assume the same packing type, liquid loading rate, air to water ratio, and temperature as the pilot study.

References

1. Lewis, W. K., and W. G. Whitman: *Ind. Eng. Chem.*, vol. 16, p. 1215, 1924.
2. Danckwertz, P. V.: *Ind. Eng. Chem.*, vol. 43, p. 6, 1951.
3. Dobbins, W. E.: *Advances in Water Pollution Research*, vol. 2, Pergamon Press, 1964, p. 61.
4. Mueller, J. A., et al.: *Proc. 37th Ind. Waste Conf.*, May 1982, Purdue University.
5. Schmit, F., and D. Redmon: *J. Water Pollution Control Fed.*, November 1975.
6. Imhoff, K., and D. Albrecht: *Proc. 6th International Conf. on Water Pollution Research*, Jerusalem, 1972.
7. Landberg, G., et al.: *Water Research*, vol. 3, p. 445, 1969.
8. Wagner, M. R., and H. J. Popel: *Proc. Wat. Env. Fed.*, Dallas, 1996.
9. Carpani, R. E., and J. M. Roxburgh: *Can. J. Chem. Engrg*, vol. 36, p. 73, April 1958.
10. Gameson, A. H., and H. B. Robertson: *J. Appl. Chem. Engrg.*, vol. 5, p. 503, 1955.
11. Eckenfelder, W. W.: *J. Sanit. Engrg. Div. ASCE*, vol. 85, pp. 88–99, 1959.
12. Pasveer, A.: *Sewage Ind. Wastes*, vol. 27, pt. 10, p. 1130, 1955.
13. Ippen, H. T., and C. E. Carver: MIT Hydrodynamics Lab. Tech. Rep. 14, 1955.
14. Cappock, P. D., and G. T. Micklejohn: *Trans. Inst. Chem. Engrs. London*, vol. 29, p. 75, 1951.
15. *WPCF Aeration Manual of Practice*, FD-13, 1988.
16. Quirk, T. P.: personal communication.
17. Kormanik, R., et al.: *Proc. 28th Ind. Waste Conf.*, 1973, Purdue University.
18. Arthur, R. M.: *Treatment Efficiency and Energy Use III, Activated Sludge Process Control*, Butterworths, Boston, 1986.
19. "ASCE Standard Measurement of Oxygen Transfer in Clean Water," ASCE, New York, 1992.

20. Redman, D., and W. C. Boyle: Report to Oxygen Transfer Subcommittee of ASCE, 1981.
21. Neal, L. A., and E. C. Tsivoglou: *J. Water Pollut. Control Fed.*, vol. 46, p. 247, 1974.
22. Muller, J.: Manhattan College Summer Institute, 1996.
23. Ashworth, R. A., G. B. Howe, M. E. Mullins, and T. N. Rogers: "Air-Water Partitioning Coefficients of Organics in Dilute Aqueous Solutions," *Jour. Haz. Mat.*, vol. 18, pp. 25–36, 1988.
24. Onda, K., H. Takeuchi, and Y. Okumoto: "Mass Transfer Coefficients between Gas and Liquid Phases in Packed Columns," *J. Chem. Engrg. Japan*, vol. 1, no. 1, 1968.
25. Eckert, J. S.: *Chem. Eng. Progress*, vol. 66, March 1970.

CHAPTER 6
Principles of Aerobic Biological Oxidation

6.1 Introduction

Almost all industrial wastewaters contain biodegradable organics. Typically, the most cost-effective stabilization of these wastewaters is by biological treatment. The principles to convert these organics to stabilized innocuous and/or reusable end products are described in this chapter. The processes are natural, common in streams, lakes, groundwaters, and other environmental media. It is necessary for environmental engineers to understand, apply, optimize, and implement these principles in the design and operation of wastewater treatment facilities. The background material and concepts needed are described in this and subsequent chapters.

The objective of wastewater treatment is to remove or reduce the concentration of organics, solids, and in some cases, inorganics to acceptable levels. With industrial wastewaters some of the contaminants may be inhibitory or toxic to the microbes; hence source or pretreatment may be required. This chapter presents pertinent biological principles including mechanisms of organics removal, bio-oxidation mechanisms, effect of environmental factors on biological reactions, bioinhibition, nitrification/denitrification, phosphorus removal, and laboratory and pilot plant procedures for process design criteria development.

6.2 Organics Removal Mechanisms

Organics are removed in a biological-treatment process by one or more mechanisms, namely sorption, stripping, or biodegradation. Table 6.1 identifies several organics and the mechanisms responsible for their removal.

	Percent Treatment Achieved		
Compounds	Stripping	Adsorption	Biodegradation
Nitrogen compounds			
Acrylonitrile			99.9
Phenols			
Phenol			99.9
2,4-DNP			99.3
2,4-DCP			95.2
PCP		0.58	97.3
Aromatics			
1,2-DCB	21.7		78.2
1,3-DCB	—	—	—
Nitrobenzene			97.8
Benzene	2.0		97.9
Toluene	5.1	0.02	94.9
Ethylbenzene	5.2	0.19	94.6
Malogenated hydrocarbons			
Methylene chloride	8.0		91.7
1,2-DCE	99.5	0.50	
1,1,1-TCE	100.0		
1,1,2,2-TCE	93.5		
1,2-DCP	99.9		
TCE	65.1	0.83	33.8
Chloroform	19.0	1.19	78.7
Carbon tetrachloride	33.0	1.38	64.9
Oxygenated compounds			
Acrolein			99.9
Polynuclear aromatics			
Phananthrene			98.2
Naphthalene			98.6
Phthalates			
Bis (2-ethylhexyl) phthalates			76.9
Other			
Ethyl acetate			98.8

TABLE 6.1 Specific Removal Efficiencies of Priority Pollutants

Sorption

Limited sorption of nondegradable organics on biological solids occurs for a variety of organics, and this phenomenon is not a primary mechanism of organic removal in the majority of cases. An exception is Lindane, as reported by Weber and Jones,[1] who showed that while no biodegradation occurred, there was significant sorption. It is probable that other pesticides will respond in a similar manner in biological wastewater-treatment processes.

This removal mechanism is termed *partitioning* and has been related to the octanol-water partition coefficient of the organic.

$$K_{SW} = kK_{OW}{}^n \qquad (6.1)$$

where K_{SW} = biosolids accumulation factor, ratio of organic sorbed and in solution, (mg/mg)(mg/L)
K_{OW} = octanol-water partition coefficient, $(mg/L)_o/(mg/L)_w$
k, n = factors that have been reported to vary from 1.38×10^{-5} to 4.3×10^{-7} (k) and from 0.58 to 1.0 (n)[2-5]

In most industrial wastewaters, partitioning provides negligible SCOD removal but may be a method of bioaccumulation of certain lipid-soluble organic compounds.

The removal by adsorption can be determined through the relationship

$$\frac{C_e}{C_i} = \frac{1}{\left[1 + \dfrac{k_{SW} \cdot Xt}{\theta_c}\right]} \qquad (6.1a)$$

where C_e = effluent concentration, mg/L
C_i = influent concentration, mg/L
X = mixed liquor suspended solids, mg/L
t = hydraulic detention time, d
θ_c = sludge age, d

Adsorption is not a significant factor when log K_{OW} is less than 4.

Example 6.1. Determine the adsorption of tetrachloroethane and Lindane in the activated sludge process for the following conditions.

Tetrachloroethane K_{OW} = 363
Lindane K_{OW} = 12,600
In Eq. (6.1), $k = 3.45 \times 10^{-7}$ 1/mg
X = 3500 mg/L
t = 0.23 d
θ_c = 6 d

Solution For tetrachlorethane:

$$\frac{C_e}{C_i} = \frac{1}{\left[1 + 3.45 \times 10^{-7} \frac{1}{\text{mg}}(363)(3500)\frac{\text{mg}}{\text{L}} \times \frac{0.23 \text{ d}}{6 \text{ d}}\right]}$$

$$= 0.984 \quad \text{or} \quad 1.6\% \text{ adsorbed}$$

For Lindane:

$$\frac{C_e}{C_i} = \frac{1}{\left[1 + 3.45 \times 10^{-7} (12,600)(3500) \times \frac{0.23}{6}\right]}$$

$$= 0.633 \quad \text{or} \quad 37\% \text{ adsorbed}$$

While sorption on biomass does not seem to be a significant removal mechanism for toxic organics, sorption on suspended solids in primary treatment may be significant. The importance of this phenomenon is the fate of the organics during subsequent sludge-handling operations. In some cases, toxicity to anaerobic digestion may result or land-disposal alternatives may be restricted.

While sorption of organics on biomass is usually not significant, this is not true of heavy metals. Metals will complex with the cell wall and bioaccumulate. While low concentrations of metals in the wastewater are generally not inhibitory to the organic removal efficiency, their accumulation on the sludges can markedly affect subsequent sludge treatment and disposal operations.

Stripping

Volatile organic carbon compounds (VOCs) will air strip in biological-treatment processes, that is, trickling filters, activated sludge, aerated lagoons. Depending on the VOC in question, both air stripping and biodegradation may occur, as described by Kincannon and Stover.[6] Stripping of VOCs in biological treatment processes is currently receiving considerable attention in the United States, as legislation is severely limiting the permissible atmospheric emissions of VOCs.

The volatilization of high vapor pressure compounds has been more of an issue with activated sludge systems, primarily because of air quality considerations. There are several factors which affect the magnitude of volatile organic compound (VOC) emissions from an aerated biological reactor such as the compound vapor pressure and Henry's constant, the relative concentration of the VOCs (Raoult's Law and partial pressures), the air flow rate or power level, and the degree of biological activity. When considering the removal of a constituent in an activated sludge system, the compound can be removed in only one of three ways, namely, biochemical degradation, air stripping, or physical sorption. This can be expressed mathematically as follows:

$$\frac{dS}{dt} \cdot V = QS_o - \left[QS_e + \frac{dS}{dt_{bio}} \cdot V + \frac{dS}{dt_{str}} \cdot V + \frac{dS}{dt_{sorp}} \cdot V \right] \quad (6.2)$$

where S_o = initial substrate concentration
S = substrate concentration
S_e = effluent substrate concentration
Q = flow
V = volume

The fraction stripped, which can be theoretically calculated using several models, also can be predicted using bench-scale studies using a small reactor such as the one depicted in Fig. 6.1. This is a closed system, with the stripped compounds being collected in an off-gas apparatus, trapped on activated carbon, and analyzed for content. Parallel units can be sterilized and using the same air flow rate, used as an abiotic base case to compare with the biologically active system. As shown in Table 6.2, only a small fraction of the VOCs benzene, ethylbenzene, xylene, and toluene are stripped in a biologically active reactor. With respect to chlorinated VOCs biodegradation rates generally decrease with an increasing number of chlorine atoms, while the stripping rate increases. An equation has been developed to predict the evaporation rates of various nonpolar organics from water. The equation for dissolved organic compounds below the saturation concentration is[7]

$$\text{Half life} = \Lambda = 12.48 \; L \; P_W C_i / EP_i M_i \quad (6.3)$$

where Λ = half life, min
L = water depth, m
P_W = partial pressure of water @ 20°C = 17.54 mm Hg
C_i = concentration of compound i in water
M_i = molecular weight of compound i
P_i = vapor pressure of compound i
E = water evaporation rate, $g/m^2/day$

Laboratory studies estimating the evaporation of chlorinated solvents dissolved in water indicated rapid evaporation (less than 20 minutes for chlorinated solvents greater than 1 ppm and less than 2 hours at concentrations in the 0.1 ppm (100 ppb) range[7].

Sorbability

Organic compounds can be adsorbed (or possibly absorbed) on biological floc and this will depend on the ability of a compound to effectively partition to the nonsoluble fraction. This is

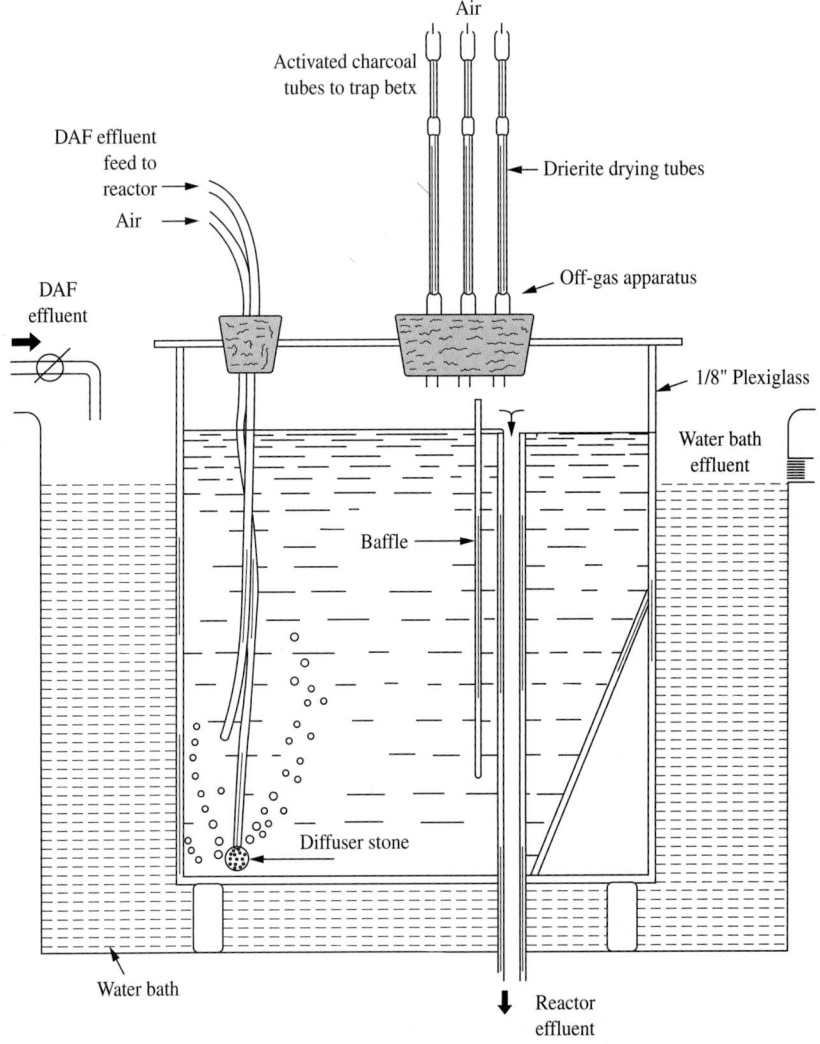

Figure 6.1 Schematic diagram of a longitudinal section of a reactor with off-gas apparatus attached.

normally not significant for water-soluble organic compounds, but needs to be recognized as a candidate removal mechanism. As shown in Table 6.2, the sorbed fraction is small for the BTEX compounds.

The general relationship between the volatilization, biodegradation, and sorption of certain organic compounds is shown in Fig. 6.2.

Principles of Aerobic Biological Oxidation

Case History	Pathway	Benzene	Toluene	Ethylbenzene	Xylene
No. 1	% Biodegraded	97.8	94.90	94.6	–
	% Air stripped	2.0	5.10	5.2	–
	% Sorbed	–	0.02	0.2	–
No. 2	% Biodegraded	83.6	83.8	80.0	78
	% Air stripped	15.9	15.6	19.5	21
	% Sorbed	0	0	0	0
No. 3	% Biodegraded	–	88.4	–	–
	% Air stripped	–	11.6	–	–
	% Sorbed	–	–	–	–

TABLE 6.2 Fate of Benzene, Ethylbenzene, Toluene, and Xylene in Activated Sludge Reactors

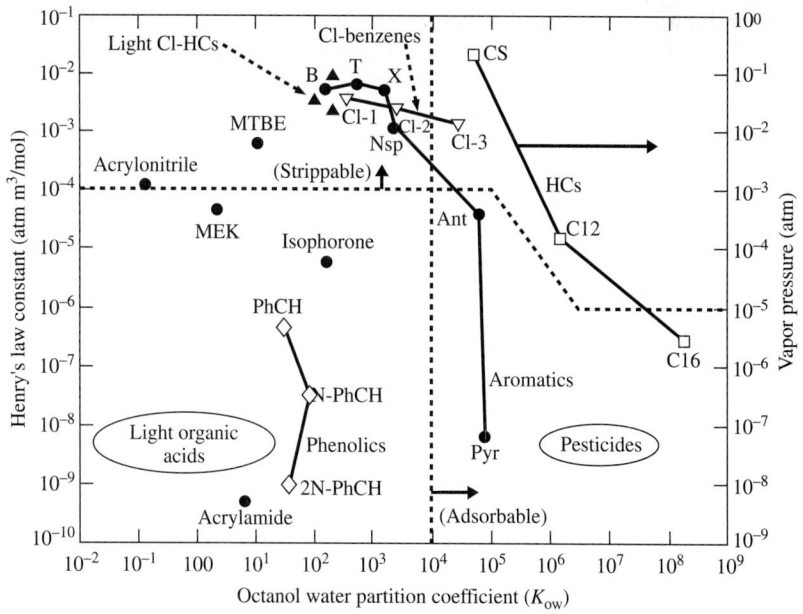

FIGURE 6.2 Categorization of organic compounds—strippability versus sorbability. (*Adapted from Grady et al., 1996.*)[8]

Biodegradation

When organic matter is removed from solution by aerobic microorganisms, two basic phenomena occur: Oxygen is consumed by the organisms for energy, and new cell mass is synthesized. The organisms also undergo progressive autooxidation in their cellular mass. These reactions can be illustrated by the following generic equations:

$$\text{Biodegradable organics} + a'O_2 + N + P \xrightarrow[K]{\text{cells}} a \text{ new cells} + CO_2$$
$$+ H_2O + \text{nonbiodegradable soluble residue (SMP)} \quad (6.4)$$

$$\text{Cells} + b'O_2 \xrightarrow[b]{} CO_2 + H_2O + N + P$$
$$+ \text{nonbiodegradable cellular residue} + \text{SMP} \quad (6.5)$$

Of primary concern to the engineer in the design and operation of industrial waste treatment facilities are the rate at which these reactions occur, the amounts of oxygen and nutrient they require, and the quantity of biological sludge they produce.

In Eq. (6.4), K is a rate coefficient and is a function of the biodegradability of the organic or the mixture of organics in the wastewater. The coefficient a' is the fraction of the organics removed that is oxidized to end products for energy and the coefficient a is the fraction of organics removed that is synthesized to cell mass. The coefficient b is the fraction per day of degradable biomass oxidized, and b' is the oxygen required for this oxidation.

A small portion of the organics removed in Eq. (6.4) remains as nondegradable by-products which appear in the effluent as TOC or COD but not BOD and are defined as soluble microbial products (SMP). A portion of the cell mass generated in Eq. (6.4) remains as a nondegradable residue.

For the design or operation of a biological treatment facility, a primary objective is to balance Eqs. (6.4) and (6.5) for the wastewater in question.

Disregarding the SMP, all of the organics removed are either oxidized to end products of CO_2 and H_2O or synthesized to biomass. Therefore,

$$a_{COD} + a'_{COD} \sim 1$$

Since the biomass is usually expressed as volatile suspended solids (VSS) and that 1.4 lb O_2 is required to oxidize 1 lb of cells as VSS, then:

$$1.4a_{VSS} + a'_{COD} \sim 1$$

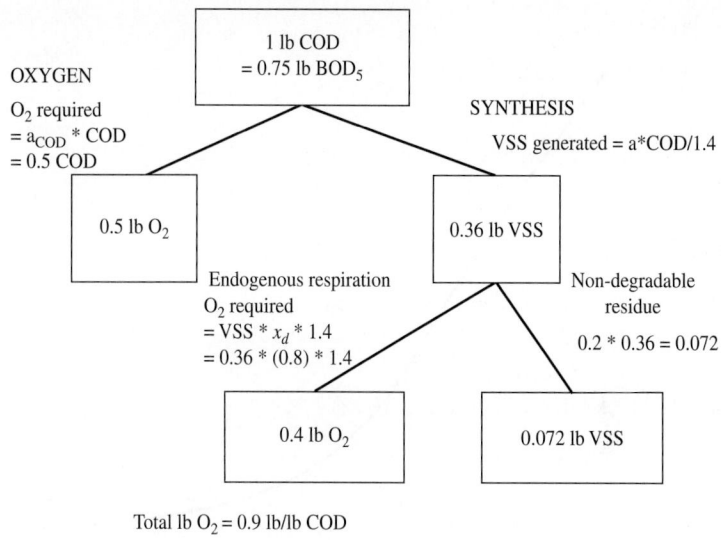

FIGURE 6.3 Conceptual material balance calculations for oxygen and VSS.

A schematic illustrating a conceptual material balance of the removal of 1 lb of COD for domestic sewage in terms of oxygen and VSS is presented in Fig. 6.3.

6.3 Mechanisms of Organic Removal by Biooxidation

The major organic-removal mechanism for most wastewaters is biooxidation in accordance with Eqs. (6.4) and (6.5).

It should be noted that, in treating industrial wastewaters, the active microbial population must be acclimated to the wastewater in question. For the more complex wastewaters, acclimation may take up to 6 weeks, as shown in Fig. 6.4 for benzidine.[9] When acclimating sludge the feed concentration of the organic in question must be less than the inhibition level, if one exists. Acclimation time required for several organics is shown in Fig. 6.5.

BOD removal from a wastewater by a biological sludge may be considered as occurring in two phases. An initial high removal of suspended, colloidal, and soluble BOD is followed by a slow progressive removal of remaining soluble BOD. Initial BOD removal is accomplished by one or more mechanisms, depending on the physical and chemical characteristics of the organic matter. These mechanisms are:

1. Removal of suspended matter by enmeshment in the biological floc. This removal is rapid and depends upon adequate mixing of the wastewater with the sludge.

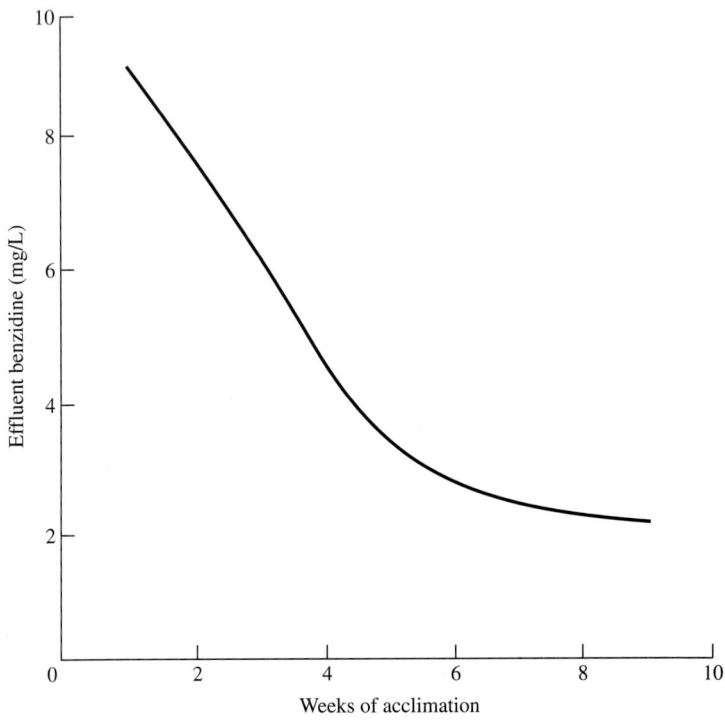

FIGURE 6.4 Acclimation for the degradation of benzidine.

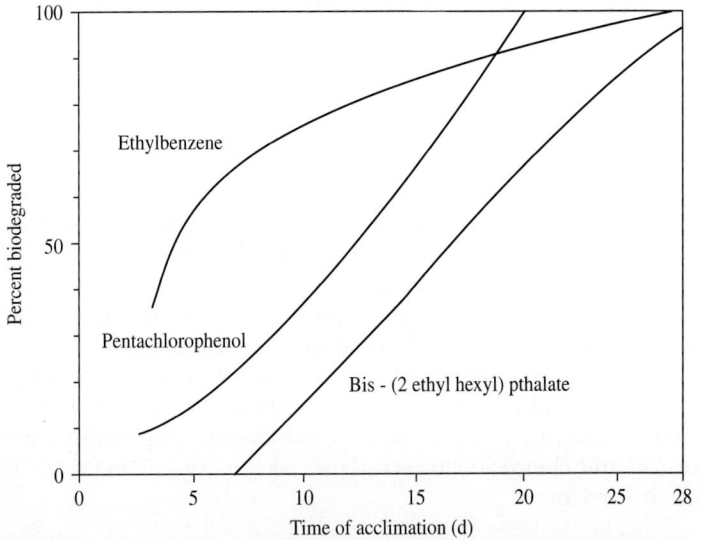

FIGURE 6.5 Acclimation of activated sludge to specific organics. (*Adapted from Tabak, H., et al., Biodegradability Studies with Organic Priority Pollutant Compounds, J. Water Pollution Control Federation, 1981.*)

2. Removal of colloidal material by physicochemical adsorption on the biological floc.
3. A biosorption of soluble organic matter by the microorganisms. There is some question as to whether this removal is the result of enzymatic complexing or is a surface phenomenon and whether the organic matter is held to the bacterial surface or is within the cell as a storage product or both. The amount of immediate removal of soluble BOD is directly proportional to the concentration of sludge present, the sludge age, and the chemical characteristics of the soluble organic matter.

The biosorption phenomenon is related to the microbial floc load in a contact time of 10 to 15 min:

$$\text{Floc load} = \frac{\text{mg BOD applied}}{\text{g VSS biological}} \quad (6.6)$$

where VSS = volatile suspended solids. The relationship between floc load and organic removal by biosorption is shown in Fig. 6.6.

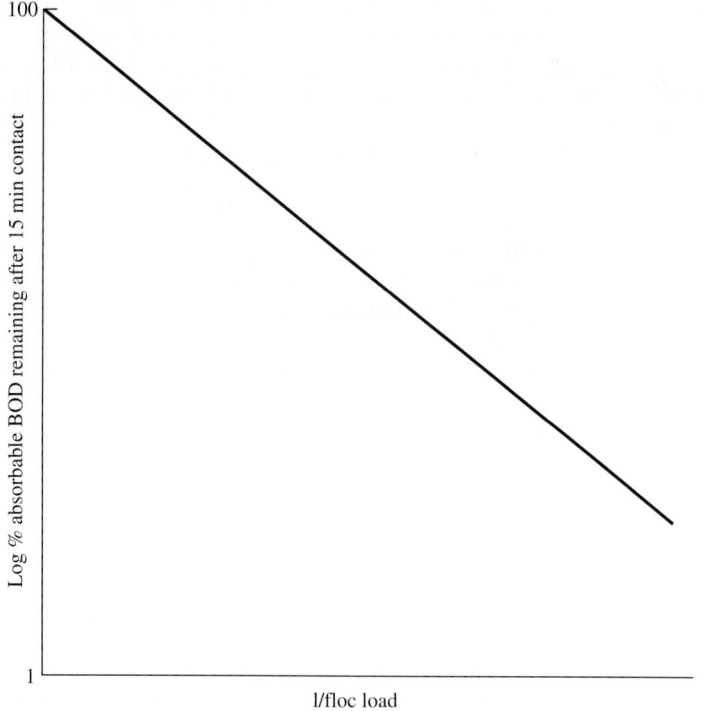

Figure 6.6 Biosorption relationship for soluble degradable wastewaters.

The type of sludge generated markedly affects its sorptive properties. In general, biomass generated from a batch or plug flow configuration will have better sorptive properties than that generated from a complex mix configuration.

The three mechanisms begin immediately on contact of biomass with wastewater. The colloidal and suspended material must undergo sequential breakdown to smaller molecules in order that it may be made available to the cell for oxidation and synthesis. The time required for this breakdown in an acclimated system is related primarily to the characteristics of the organic matter and to the concentration of active sludge. In a complex waste mixture at high concentrations of BOD, the rate of synthesis is independent of concentration as long as all components remain and, as a result, there is a constant and maximum rate of cellular growth. With continuing aeration, the more readily removable components are depleted and the rate of growth will decrease with decreasing concentration of BOD remaining in solution. This is shown in Fig. 6.19 on p. 257.

This causes a decrease in cellular mass and cellular carbon accompanied by a corresponding decrease in cellular nitrogen, as shown in Fig. 6.7. This phenomenon has been demonstrated by Gaudy,[10] by Englebrecht and McKinney,[11] and by McWhorter and Heukelekian.[12] Gaudy, in the treatment of pulp mill wastes, showed a peak in cellular carbohydrate after a 3-h aeration with a corresponding cellular protein peak after a 6-h aeration. The decrease in cellular mass after substantial exhaustion of substrate, shown by Engelbrecht and McKinney,[11] can be attributed to conversion of stored carbohydrate to cellular protoplasm.

In Fig. 6.7 declining removal occurs following biosorption. Stored carbohydrate is used by the cell over the time interval AC; this results in an increase in cellular nitrogen. The cellular nitrogen is at its peak at point C, when the stored carbohydrate is depleted. The cellular mass is shown to increase over the time interval CD (declining growth and removal). Depending on the concentration of BOD remaining at point C and the rate of removal, the sludge mass may tend to remain constant or even increase while the cellular nitrogen would remain substantially constant. Beyond point D, cell death and decay, the endogenous or autooxidation phase results in a decrease in both cell weight and cell nitrogen.

The oxygen uptake rate per unit of cell mass k_r will remain constant at a maximum rate during the log-growth phase, since substrate is not limiting the rate of synthesis.

The oxygen utilization will continue at a maximum rate until depletion of the sorbed BOD; after this it will decrease as the rate of BOD removal decreases. In wastes containing suspended and colloidal matter, the oxygen-uptake rate will also reflect the rate of solubilization and subsequent synthesis of colloidal and suspended BOD.

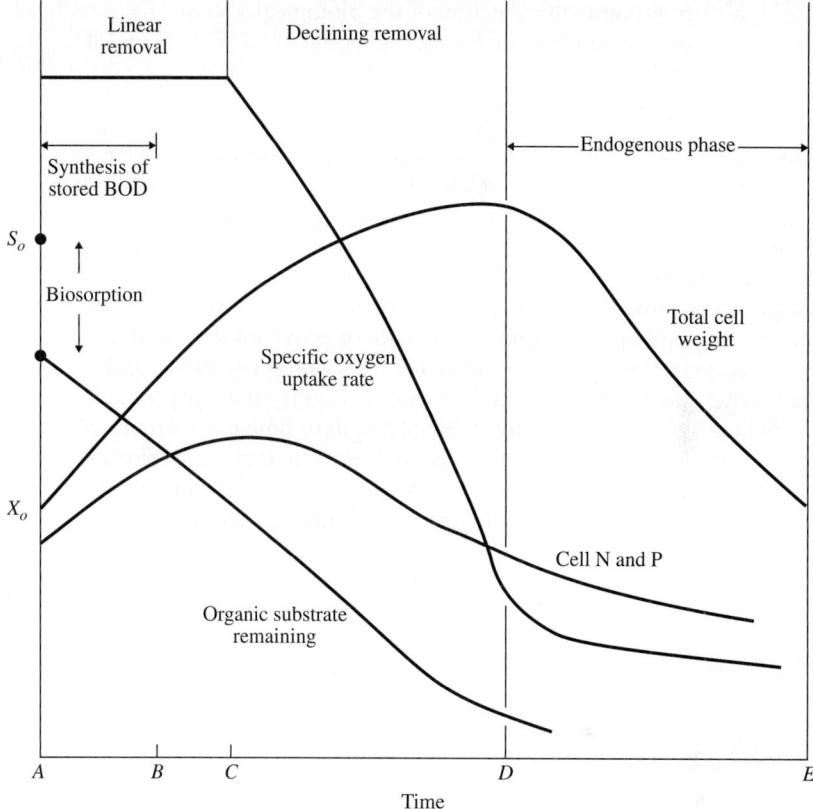

Figure 6.7 Reactions occurring during biooxidation.

Sludge Yield and Oxygen Utilization

As shown in Eq. (6.5), endogenous respiration results in a degradation of cell mass. A portion of the volatile cell mass, however, is nondegradable, that is, does not degrade in the time frame of the biological process. Quirk and Eckenfelder[13] have shown that a portion of the volatile biomass is nondegradable. As aeration proceeds the degradable portion of the biomass is oxidized, resulting in a decrease in the degradable fraction. Through kinetic and mass balances, the degradable fraction can be related to the endogenous rate coefficient and the sludge age:

$$X_d = \frac{X'_d}{1 + bX'_n \theta_c} \qquad (6.7)$$

where X_d = degradable fraction of the biological VSS
X'_d = degradable fraction of the biological VSS at generation, that is, Eq. (6.4) average of 0.8

X'_n = nondegradable fraction of the biological VSS at generation, that is, Eq. (6.4) average of $0.2(X'_d + X'_n) = 1.0$
b = endogenous rate coefficient, d^{-1}
θ_c = sludge age, d

This relationship is shown for a food processing wastewater in Fig. 6.8.

The degradable fraction is related to the viable or active mass. The degradable mass can be corelated to oxygen-uptake rates, ATP, dehydrogenase enzyme content, or plate count measurement. It is significant to note that, while volatile suspended solids are traditionally used as a measure of biomass, only the active mass is responsive in the process. In plant operation, some measure of active mass is necessary for process control in order to detect toxic shocks, shock loads, and so on. Oxygen-uptake rate is most commonly used for this purpose.

Sludge age is defined as the average length of time the microorganisms are under aeration. In a flow-through system, that is, no biomass recycle, sludge age is the reciprocal of the dilution rate Q/V. In order for growth to occur and to effect BOD removal, the growth rate becomes

$$\theta_c = \frac{V}{Q} \quad (6.8)$$

where θ_c is the sludge age.

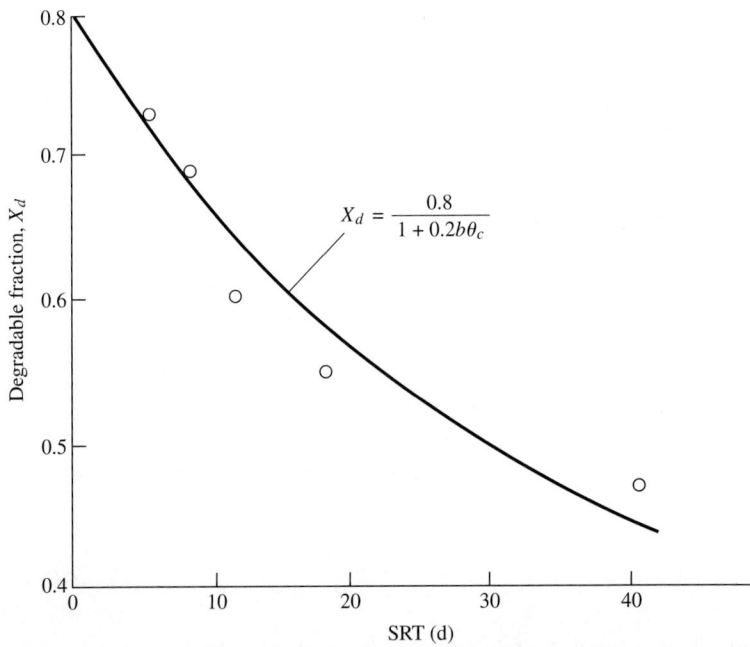

FIGURE 6.8 Relationship between degradable fraction and solids retention time (SRT) for a food processing wastewater.

In a recycle system such as an activated sludge plant, the sludge age is defined:

$$\theta_c = \frac{X_v t}{\Delta X_v} \tag{6.9}$$

where X_v = volatile suspended solids concentration, mg/L
$t = V/Q$, the hydraulic detention time, d
ΔX_v = the volatile suspended solids wasted per day in milligrams per liter, based on influent flow

Process performance can also be related to the organic loading of the process as defined by the food/microorganism ratio F/M:

$$\frac{F}{M} = \frac{S_0}{f_b X_v t} \tag{6.10}$$

in which X_v = the volatile suspended solids under aeration and f_b = the fraction which is biological.

The F/M is related to the sludge age θ_c by the relationship

$$\frac{1}{\theta_c} = a\frac{F}{M} - bX_d \tag{6.11}$$

when the effluent BOD is negligible, where a = the yield coefficient or the fraction of organic removed and synthesized to biomass.

Several investigations have shown that a constant mass of biological cells is synthesized from a given weight of organic matter removed (expressed as total oxygen demand, COD). McKinney[11] has indicated that one-third of the ultimate oxygen demand (COD) of a substrate is used for energy and that two-thirds results in synthesis. Using a factor of 0.7 g VSS/g O_2 for conversion of oxygen to cellular volatile solids, 0.47 g VSS (volatile suspended solids) will be synthesized for each gram of COD removed. Variations in this value have been attributed to endogenous respiration effects. Sawyer[14] and Gellman and Heukelekian[15] have shown that for sewage and several industrial wastes, 0.5 g VSS is synthesized per gram of BOD_5 removed. Busch and Myrick[16] showed total synthesis from glucose to be 0.44 g cells/g COD. Using NO_3^- as a nitrogen source, McWhorter and Heukelekian[12] found an average synthesis of 0.315 g VSS/g COD from glucose. As shown by Pipes,[17] the reduction of nitrate to amino nitrogen requires energy such that some of the COD is consumed for this reduction, resulting in a lower yield than when ammonia is used as a nitrogen source.

When the nutrient nitrogen level is lowered below the optimum value, cell yield tends to increase as more substrate is shunted to the buildup of insoluble cell polymers.

Sludge generation from the biological oxidation of soluble substrates ($f_b = 1.0$) has been summarized by Eckenfelder[18]:

$$\Delta X_v = aS_r - bX_d X_v t \tag{6.12}$$

where ΔX_v = biomass yield, mg/L
S_r = organic removal, mg/L
X_d = biodegradable fraction of biomass

The biomass generated in Eq. (6.12) is about 80 percent biodegradable. As the SRT is increased, the degradable portion of the biomass will be endogenously oxidized and the biodegradable fraction of the remaining volatile biomass (designated as X_d) will decrease. The degradable fraction of the volatile biomass can be calculated for a wastewater containing only soluble organic substrates by Eq. (6.13). This relationship is shown in Fig. 6.8 for treatment of a food processing wastewater.

$$X_d = \frac{0.8}{1 + 0.2b\theta_c} \tag{6.13}$$

Equations (6.11), (6.12), and (6.13) can be combined to determine X_d for a soluble organic substrate using the kinetic coefficients.

$$X_d = \frac{aS_r bX_v - [(aS_r + bX_v)^2 - (4bX_v)(0.8\,aS_r)]^{0.5}}{2bX_v} \tag{6.14}$$

If the influent contains VSS, such as in a pulp and paper mill wastewater, Eq. (6.12) is modified to include this contribution.

$$\Delta X_v = a[S_r + f_d f_x X_i] \cdot bX_d f_b X_v t + (1 - f_d)f_x X_i + (1 - f_x)X_i \tag{6.15}$$

where X_i = influent VSS, mg/L
f_x = fraction of influent VSS that is degradable
f_d = fraction of degradable influent VSS degraded
f_b = fraction of mixed liquor VSS that is biomass

The degradation rate of the influent degradable VSS is a function of the SRT and their specific degradation rate. Most will usually degrade within a 10-day sludge age. Figures 6.9 and 6.10 illustrate the oxidation of primary sludge degradable VSS and nondegradable VSS as a function of sludge age. The inert fraction will increase with increasing sludge age. The fraction of VSS remaining can be determined from Eq. (6.16).

$$(1 - f_d) = e^{-K'_p \theta_c} \tag{6.16}$$

where K_p = degradation rate coefficient of influent VSS, day^{-1}

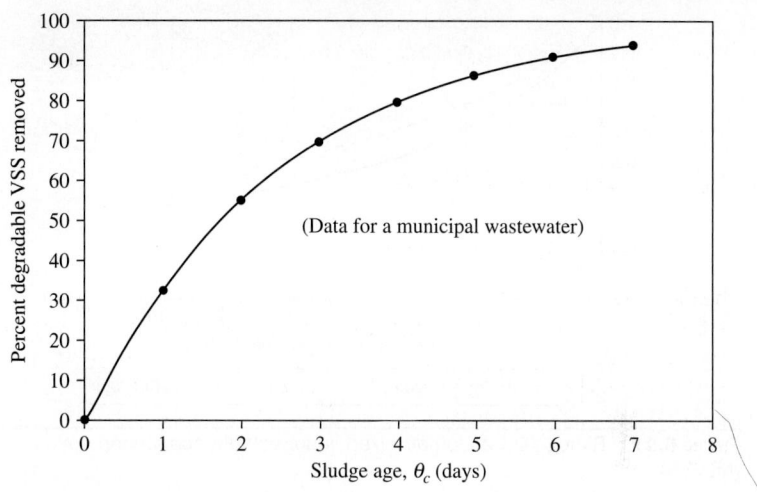

FIGURE 6.9 Oxidation of degradable VSS.

If it is assumed that 1 mg/L VSS solubilizes to generate 1 mg/L COD, then the fraction of biomass (f_b) in the overall mixed liquor can be determined as follows.

$$f_b = \frac{a[S_r + f_d f_x X_i] - b X_d f_b X_v t}{a[S_r + f_d f_x X_i] - b X_d f_b X_v t + (1 - f_d) f_x X_i + (1 - f_x) X_i} \quad (6.17)$$

Figure 6.11 illustrates the reduced biological fraction of ML VSS with increasing influent nonbiodegradable VSS.

Most pulp and fiber in pulp and paper mill wastewaters is essentially nondegradable, and hence $(1 - f_d)$ is approximately unity. In food processing wastewaters, however, $(1 - f_d)$ may be less than 0.2. If the influent contains high levels of VSS, the value $(1 - f_d)$ must be experimentally determined in order to accurately predict the volatile

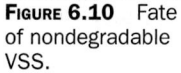

FIGURE 6.10 Fate of nondegradable VSS.

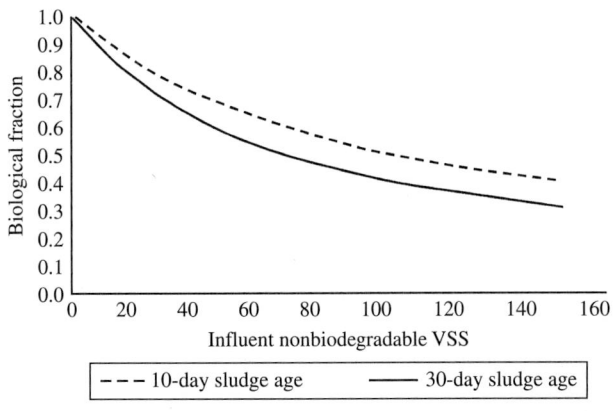

FIGURE 6.11 Biological fraction of mixed liquor volatile suspended solids (MLVSS).

sludge production rate and true biomass yield. When the wastewater contains influent VSS, Eq. (6.14) for X_d must be modified as follows:

$$X_d = \frac{J - [J^2 - (4bf_b X_v)(0.8\ aS_r)]^{0.5}}{2bf_b X_v} \quad (6.18)$$

where $J = aS_r + bf_b X_v - f_d f_x X_i$.

The impact of influent nonvolatile suspended solids on mixed liquor characteristics and sludge production rate can also be significant. The quantity of inert material (measured as nonvolatile suspended solids) generated is related to the SRT, the hydraulic retention time, the fraction of influent nonvolatile suspended solids that is nondegradable/nonsolubilized and the formation of nonbiomass particulates in the activated sludge process. This relationship is experessed as follows.

$$\Delta NVSS = a^* S_r f_{ibnd} + f_{oi} X_{oi} + \text{Nonbiomass particulate formation} \quad (6.19)$$

where $\Delta NVSS$ = inert suspended solids produced, mg/L
X_{oi} = influent inert solids, mg/L
f_{ibnd} = fraction of inert biomass
f_{oi} = fraction of influent inert solids not degraded or solubilized
a^* = biomass produced per unit of substrate removed, mg/TSS/mg BOD (or COD)

Inert material may also be produced in the activated sludge system through precipitation reactions of the wastewater. This later accumulation term is indicated but not characterized in Eq. (6.19) since it is difficult to quantify unless the influent suspended solids concentration is negligible and the inert accumulation is significant.

Total sludge production can be calculated by summing Eqs. (6.15) and (6.19). The impact of the total suspended solids production on the operating aeration basin MLSS concentration is a direct function of SRT. As SRT increases, the aeration basin MLSS will increase. Consideration must be given to secondary clarifier solids loading rate design to accommodate the inert and volatile suspended solids generation while maintaining the required SRT for substrate removal. A determination of yield coefficient, a, and endogenous respiration coefficient, b, can be made from a plot of Eq. (6.12) as shown in Fig. 6.12. Figure 6.13 illustrates sludge production coefficient determination for a soluble pharmaceutical wastewater. Example 6.2 illustrates the calculation for sludge yields. A summary of sludge production nomenclature is provided in Table 6.3.

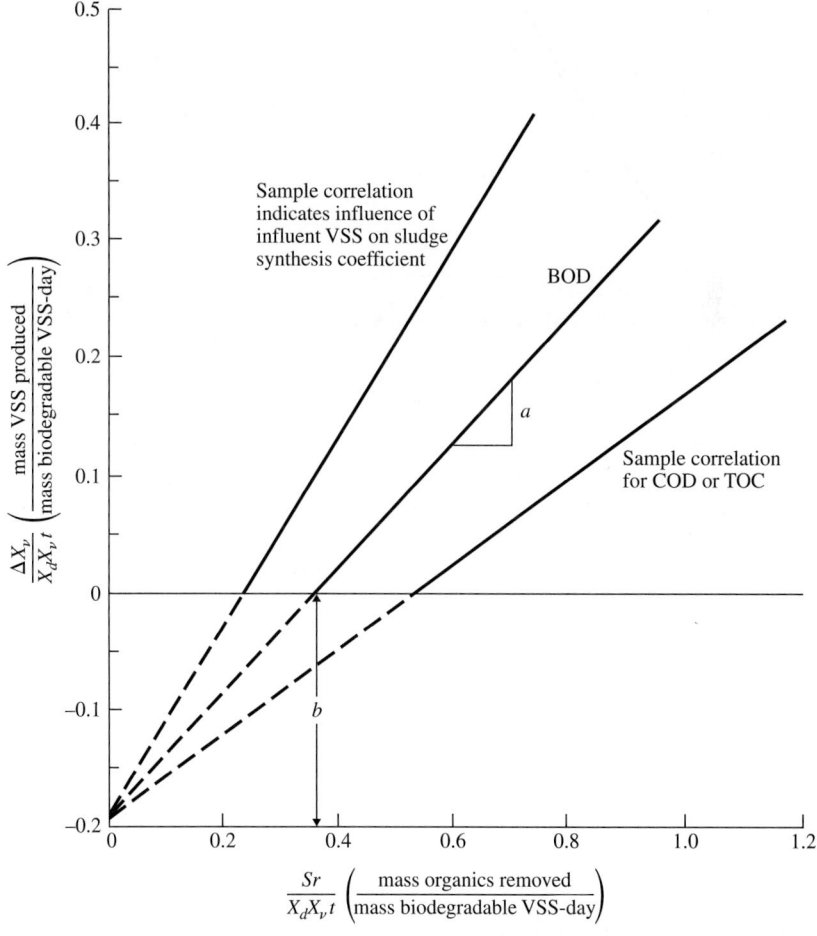

FIGURE 6.12 Determination of sludge production coefficients from Eq. (6.12).

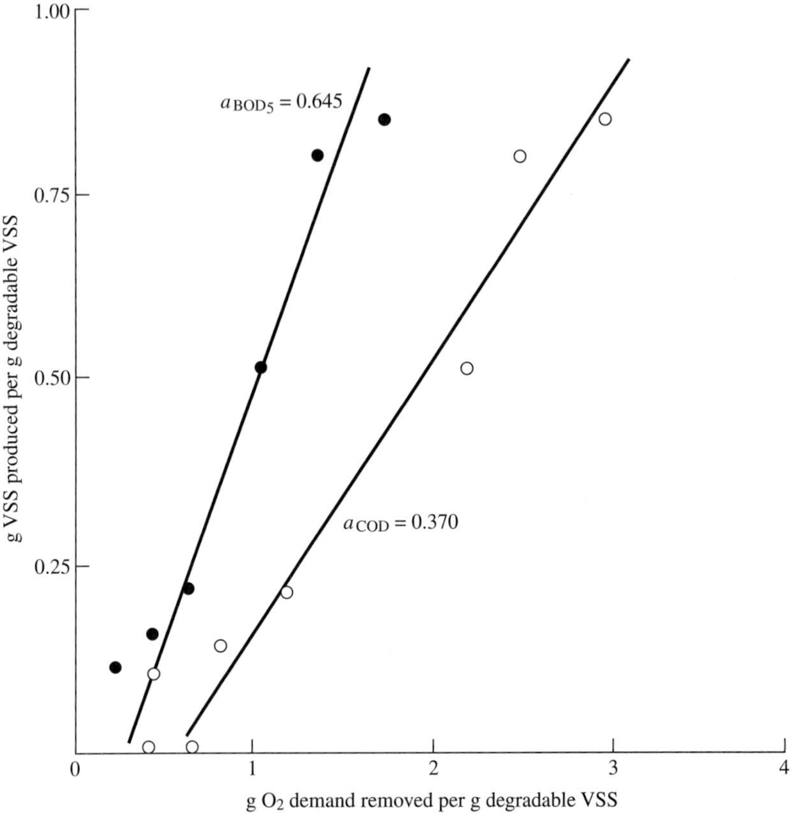

244 Chapter Six

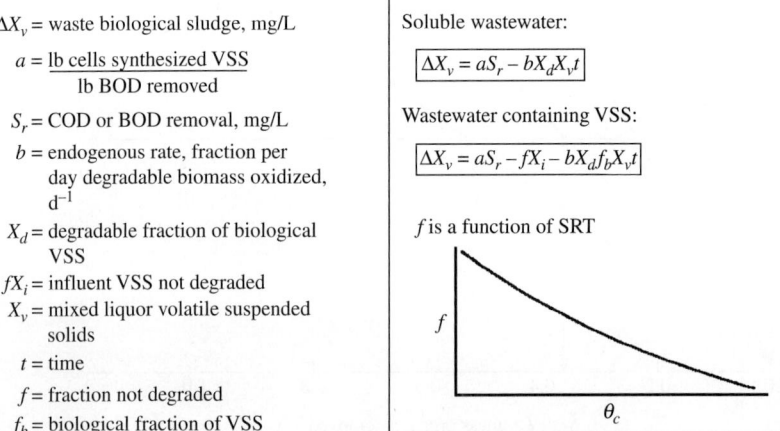

FIGURE 6.13 Cell synthesis relationship for a soluble pharmaceutical wastewater.

ΔX_v = waste biological sludge, mg/L

$a = \dfrac{\text{lb cells synthesized VSS}}{\text{lb BOD removed}}$

S_r = COD or BOD removal, mg/L

b = endogenous rate, fraction per day degradable biomass oxidized, d^{-1}

X_d = degradable fraction of biological VSS

fX_i = influent VSS not degraded

X_v = mixed liquor volatile suspended solids

t = time

f = fraction not degraded

f_b = biological fraction of VSS

Soluble wastewater:

$$\Delta X_v = aS_r - bX_d X_v t$$

Wastewater containing VSS:

$$\Delta X_v = aS_r - fX_i - bX_d f_b X_v t$$

f is a function of SRT

TABLE 6.3 Sludge Production Nomenclature

Example 6.2. Determine the operating F/M, MLVSS, and f_b for an activated sludge process under the following operating conditions:

$a = 0.45$
$b = 0.1$ at 20°C
$\theta_c = 10$ days
$X_i = 200$ mg/L
$f_x = 0$
$S_o = 1,000$ mg/L
$S_e = 20$ mg/L
$t = 0.9$ days

Solution The F/M is calculated as follows:

$$\frac{1}{\theta_c} = a(F/M) - bX_d$$

$$X_d = \frac{0.8}{1 + 0.2\,b\theta_c}$$

$$X_d = \frac{0.8}{1 + 0.2(0.1 \cdot 10)} = 0.67$$

$$\frac{1}{10} = 0.45(F/M) - (0.1 \cdot 0.67)$$

$$F/M = 0.37 / \text{day}$$

The MLVSS (X_v) is composed of biomass VSS and influent VSS that are not biodegradable and is calculated as:

$F/M = S_O / (f_b X_v t)$

$f_b X_v = 1,000 / (0.37 \cdot 0.9)$

$f_b X_v = 3,000$ mg/L

$$X_v = [(aS_r - bX_d f_b X_v t) + (1 - f_x)X_i] \frac{\theta_c}{t}$$

$$= [0.45\,(1,000 - 20) - (0.1 \cdot 0.67 \cdot 0.9 \cdot 3,000) + (1.0 - 0)200] \frac{10}{0.9}$$

$X_v = 5,112$ mg/L

The biomass fraction of X_v is:

$$f_b = \frac{3,000}{5,112} = 0.59$$

The waste sludge production rate per pass through the aeration basin is:

$$\Delta X_v = aS_r - bX_d f_b X_v t + \Delta X_i$$
$$= (0.45 \cdot 980) - (0.1 \cdot 0.67 \cdot 3{,}000 \cdot 0.9) + (1.0 - 0)200$$
$$= 441 - 181 + 200$$
$$\Delta X_v = 460 \text{ mg/L-pass}$$

Check that $\theta_c = 10$ days

$$\theta_c = \frac{5{,}112 \cdot 0.9}{460} = 10 \text{ days}$$

Oxygen Requirements

The biological oxygen requirements can be computed by Eq. (6.20):

$$O_2 = a'S_r + (1.4b)X_d f_b X_v t \qquad (6.20)$$

where O_2 = oxygen requirement, mg/L
 $f_b X_v$ = biomass under aeration, mg VSS/L
 X_d = total degradable fraction of biological VSS
 $a' = \dfrac{\text{lb} O_2 \text{ consumed}}{\text{lb BOD removed}}$
 S_r = BOD removed mg/L
 b = endogenous rate, fraction per day degradable biomass oxidized, d^{-1}

The "lumped" coefficient, $1.4b$, is frequently referred to as b'. Equation (6.20) is used to determine a' and b' and the system oxygen requirements. This relationship is shown in Fig. 6.14 for a food processing wastewater where $f_b \approx 1$. Table 6.4 presents a survey of oxygen utilization coefficients for a variety of wastewaters.

The oxygen requirements for a soluble substrate can be calculated as:

$$r_r = \frac{a's_r}{t} + b'X_d f_b X_v \qquad (6.21)$$

where r_r = the oxygen uptake rate, mg/L-d.

During the log-growth phase, the specific oxygen uptake rate (SOUR) $k_r = r_r/X_v$ is constant, and hence r_r will increase with increasing synthesis of new cells. As the substrate concentration decreases, the oxygen uptake rate decreases. When the available

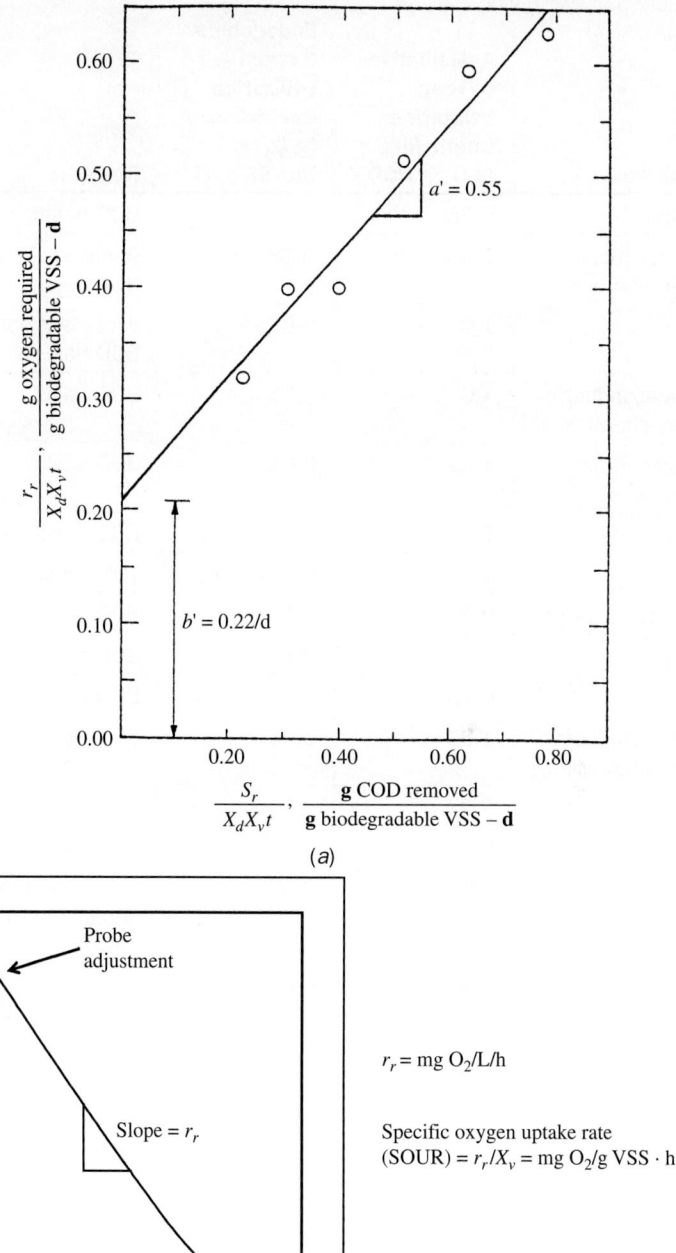

FIGURE 6.14 (a) Development of oxygen utilization coefficients for a food processing wastewater. (b) Oxygen update rate methodology.

Type of Waste	Assimilative Oxygen Utilization Coefficient a' (g O_2/g BOD_r)	Endogenous Oxygen Utilization Coefficient b' (g O_2/g MLVSS-day)	Remarks
Municipal	0.66	0.10	BOD basis
Pharmaceutical (fermentation)	1.30	0.18	Bench-scale data BOD basis
	1.0	0.01	Pilot plant data BOD basis
Municipal/industrial (organic chemicals)	1.0	0.05	BOD basis
Specialty organic chemicals	1.22	0.04	BOD basis
Pesticide mfg.	1.18	0.08	BOD basis
	1.43	0.08	TOC basis
Municipal/industrial	0.60	0.15	BOD basis
Municipal	0.90	0.15	BOD basis
Municipal	0.73	0.07	BOD basis
Municipal/industrial (organic chemicals)	1.00	0.06	BOD basis
Vegetable tannery	2.27	0.015	BOD basis
	0.54	0.015	COD basis
Plastics mfg.	1.02	0.095	Unadjusted for nitrification BOD basis
	0.95	0.05	Adjusted for nitrification BOD basis
	1.12	0.06	Selected for design BOD basis
Plastics mfg. (various combinations of production unit wastes)	1.25	0.05	BOD basis
	0.95	0.05	BOD basis
	0.71	0.05	BOD basis
	1.28	0.03	BOD basis
	1.20	0.03	BOD basis
	0.41	0.03	COD basis
Pharmaceutical	1.60	0.03	BOD basis

TABLE 6.4 Oxygen Utilization of Coefficients for a Variety of Wastes

Type of Waste	Assimilative Oxygen Utilization Coefficient a' (g O_2/g BOD_r)	Endogenous Oxygen Utilization Coefficient b' (g O_2/g MLVSS-day)	Remarks
Organic chemicals	1.0	0.14	Not determined experimentally BOD basis
Refinery	0.81	0.09	BOD basis
	0.60	0.09	COD basis
Coal liquefaction	0.52	0.04	BOD basis
	0.37	0.04	COD basis
	0.60	0.04	Selected for design BOD basis
Kraft pulping	1.28	0.04	BOD basis
	0.60	0.04	COD basis
Synthetic fibers mfg.	1.03	0.08	BOD basis time period I
	0.73	0.04	BOD basis time period II
	0.52	0.08	COD basis time period I
	0.47	0.025	COD basis time period II
	1.03	0.07	Selected for design BOD basis
Whey processing	0.70	0.08	Not determined experimentally BOD basis

TABLE 6.4 *(Continued)*

substrate is exhausted, the oxygen uptake rate decreases to the endogenous rate, which is approximately $bX_d f_b X_v$.

Figure 6.15 illustrates methodology used to determine the SOUR. It is important to note that the in-situ SOUR measurement will be greater than the ex-situ SOUR since depletion of substrate will reduce microbial activity and hence lower the oxygen uptake rate. This is shown by Fig. 6.16. At higher oxygen uptake rates the percent difference increases exponentially.

The immediate maximum specific oxygen uptake rate ($SOUR_{im}$) has been found to be a better measure of microbial response to

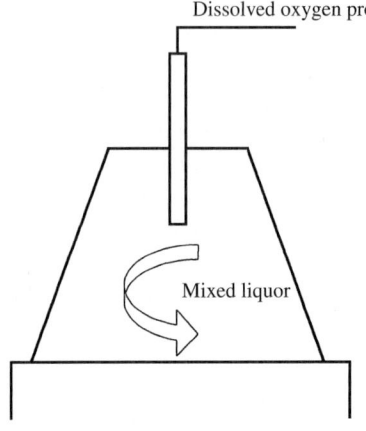

Figure 6.15 Oxygen uptake rate test methodology.

transient loadings and toxic/inhibitory inputs.[19] The test is essentially the same as that for SOUR, except that the substrate limitation is removed by the addition of a nontoxic, nonlimiting amount of substrate and the oxygen uptake is immediately recorded. Upon removal of substrate restriction, the biomass immediately increases its oxygen uptake rate to a maximum level dependent on culture history and physiological characteristics. This sudden increase is termed the immediate maximum oxygen uptake rate and can be correlated to the health and quality of the biomass. A strong

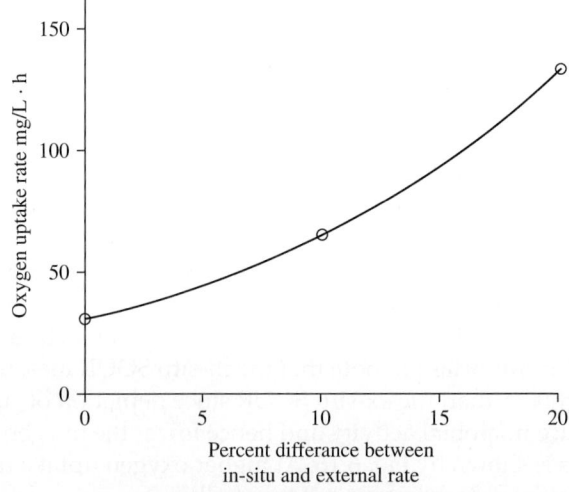

Figure 6.16 Relationship between in-situ and ex-situ SOUR.

Principles of Aerobic Biological Oxidation

relationship was found between immediate SOUR and RNA—precursor for protein synthesis and COD variation over time for a batch system treating organic chemical wastewater. No significant correlations were found using the conventional ex-situ SOUR method. Consequently $SOUR_{im}$ may better estimate biomass response to varying inputs and hence better suited for process control.

The oxygen utilization coefficients, a' and b', and the sludge yield coefficient, a, can be determined from wastewater treatment plant operating data using the above relationships. The calculations are illustrated in Examples 6.3 and 6.4.

Example 6.3. Operating data from a two-stage high purity oxygen activated sludge plant treating an organic chemicals wastewater were analyzed. The first stage of the oxygen process served as a reaeration basin for return activated sludge and the specific oxygen uptake rate (SOUR) was assumed to be the endogenous demand (b'). The organic removal (expressed as TOD) occurred in the second stage of the process and exerted an oxygen demand for synthesis (a').

The first-stage operating characteristics were:
OUR = 27 mg/L · h
VSS = 9,000 mg/L
$SOUR_e$ = 3 mg O_2/g VSS · h

The second-stage operating characteristics were:
OUR = 104 mg/L · h
VSS = 6,000 mg/L
$SOUR_r$ = 17.3 mg O_2/g VSS · h
TOD removed = 12,000 lb/d
Volume = 0.26 mg

Solution The coefficients a', a, and b are calculated as follows:

$$a' = \frac{V(SOUR_r - SOUR_e)X_v \times 24 \times 8.34}{S_r}$$

$$a' = \frac{0.26(17.3 - 3)\, 6.0 \times 24 \times 8.34}{12,000}$$

$$a' = 0.37 \text{ lb } O_2/\text{lb TOD removed}$$

Knowing $1.4\, a_{VSS} + a'_{COD} \sim 1$, the sludge yield coefficient is

$$a = \frac{1.0 - 0.37}{1.4} = 0.45 \text{ mg VSS/mg TOD removed}$$

and the endogenous decay coefficient is

$$b = \frac{(3 \cdot 24/1,000)}{1.4} = 0.05 \text{ mg VSS g/mg deg VSS-d}$$

Example 6.4. The average operating and performance conditions for an activated sludge process are illustrated below.

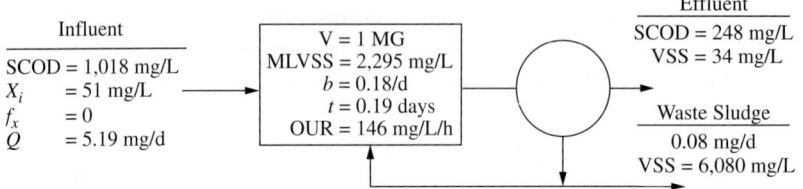

Solution Calculation of SRT and X_d:

$$\theta_c = \frac{2{,}295 \cdot 1.0}{(34 \cdot 5.19) + (0.08 \cdot 6{,}080)}$$

$$\theta_c = \frac{2{,}295}{5{,}528} = 3.5 \text{ days}$$

$$X_d = \frac{0.8}{1 + (0.18 \cdot 0.2 \cdot 3.5)} = 0.71$$

Calculation of nonbiological MLVSS:

$$(1 - f_b)X_v = \frac{(1 - f_x)X_i \theta_c}{t} = \frac{(1 - 0)(51 \cdot 3.5)}{0.19} = 939 \text{ mg/L}$$

$$f_b = \frac{1{,}356}{2{,}295} = 0.59$$

Calculation of biomass yield coefficient, a:

$$\Delta f_b X_v = aS_r - bX_d f_b X_v$$

$$0.59 \cdot 5{,}528 = [a(1{,}018 - 248) \cdot 5.19 \cdot 8.34] - [0.18 \cdot 0.71 \cdot 0.59 \cdot 2{,}295 \cdot 8.34]$$

$$3{,}261 = a(33{,}329) - 1{,}443$$

$$a = 0.14 \text{ mg VSS/mg } \Delta\text{COD}$$

Calculation of oxygen consumption and a':

$$O_2/d = 146 \cdot 1 \cdot 8.34 \cdot 24$$

$$= 29{,}223 \text{ lb/d}$$

$$29{,}223 = a'(33{,}329) + (1.4 \cdot 0.18 \cdot 0.71 \cdot 0.59 \cdot 2{,}295 \cdot 8.34)$$

$$a' = 0.81 \text{ mgO}_2/\text{mg } \Delta\text{COD}$$

Check oxygen equivalents of a and a':

$$(1.4 \cdot 0.14) + 0.81 = 1.0$$

Nutrient Requirements

Several mineral elements are essential for the metabolism of organic matter by microorganisms. All but nitrogen and phosphorus are usually present in sufficient quantity in the carrier water. An exception is process wastewater generated from deionized water or high-strength industrial wastewaters. Iron and other trace nutrients may be deficient in this case. Trace nutrient requirements are shown in Table 6.5.

Sewage provides a balanced microbial diet, but many industrial wastes (cannery, pulp and paper, etc.) do not contain sufficient nitrogen and phosphorus and require their addition as a supplement.

The quantity of nitrogen required for effective BOD removal and microbial synthesis has been the subject of much research. Early work by Helmets et al.[19] indicates a nitrogen requirement of 4.3 lb N/100 lb BOD_{rem} (4.3 kg N/100 kg BOD_{rem}) and a phosphorus requirement of 0.6 lb P/100 lb BOD_{rem} (0.6 kg P/100 kg BOD_{rem}). These represent average values derived from the treatment of several nitrogen-supplemented industrial wastes. When insufficient nitrogen is present, the amount of cellular material synthesized per unit of organic matter removed increases as an accumulation of polysaccharide. At some point, nitrogen-limiting conditions restrict the rate of BOD removal. The rule-of-thumb number is BOD:N:P of 100:5:1. Nutrient-limiting conditions will also stimulate filamentous growth, as discussed on p. 305.

	mg/mg BOD
Mn	10×10^{-5}
Cu	14.6×10^{-5}
Zn	16×10^{-5}
Mo	43×10^{-5}
Se	14×10^{-10}
Mg	30×10^{-4}
Co	13×10^{-5}
Ca	62×10^{-4}
Na	5×10^{-5}
K	45×10^{-4}
Fe	12×10^{-3}
CO_3	27×10^{-4}

TABLE 6.5 Trace Nutrient Requirements for Biological Oxidation

The nitrogen content of sludge as generated in the process has been shown to average 12.3 percent on the basis of the VSS. The nitrogen content of the sludge will decline during the endogenous phase. The nitrogen content of the nondegradable cellular mass has been shown to average 7 percent. This is shown in Fig. 6.17. The decrease in nitrogen content of an activated sludge with a soluble substrate as a function of sludge age is shown in Fig. 6.18. The phosphorus content of sludge at generation has been found to average 2.6 percent with the nondegradable cellular mass having a phosphorus content of 1 percent. The nitrogen and phosphorus requirements can be calculated by considering the nitrogen and phosphorus content of the biological sludge wasted from the process:

$$N = 0.123 \frac{X_d}{0.8} \Delta X_{v_b} + 0.07 \frac{0.8 - X_d}{0.8} \Delta X_{v_b} \qquad (6.22)$$

$$P = 0.026 \frac{X_d}{0.8} \Delta X_{v_b} + 0.01 \frac{0.8 - X_d}{0.8} \Delta X_{v_b} \qquad (6.23)$$

Not all organic nitrogen compounds are available for synthesis. Ammonia is the most readily available form, and other nitrogen compounds must be converted to ammonia. Nitrite, nitrate, and about 75 percent of organic nitrogen compounds are also available.

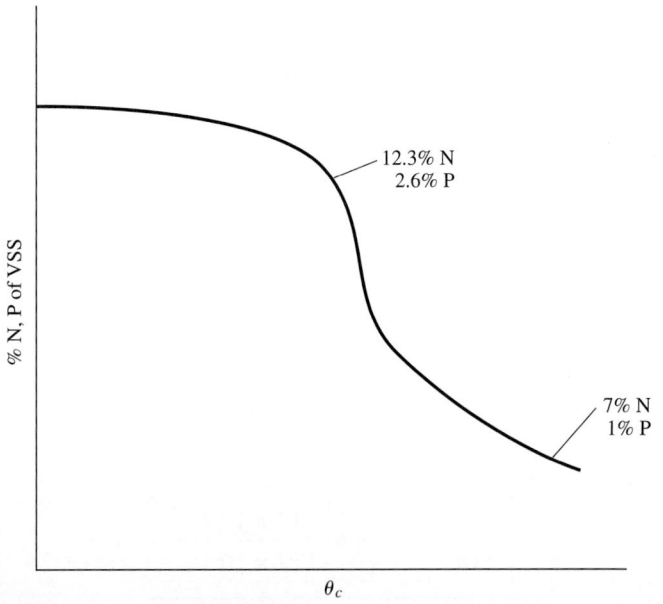

FIGURE 6.17 Nutrient requirements.

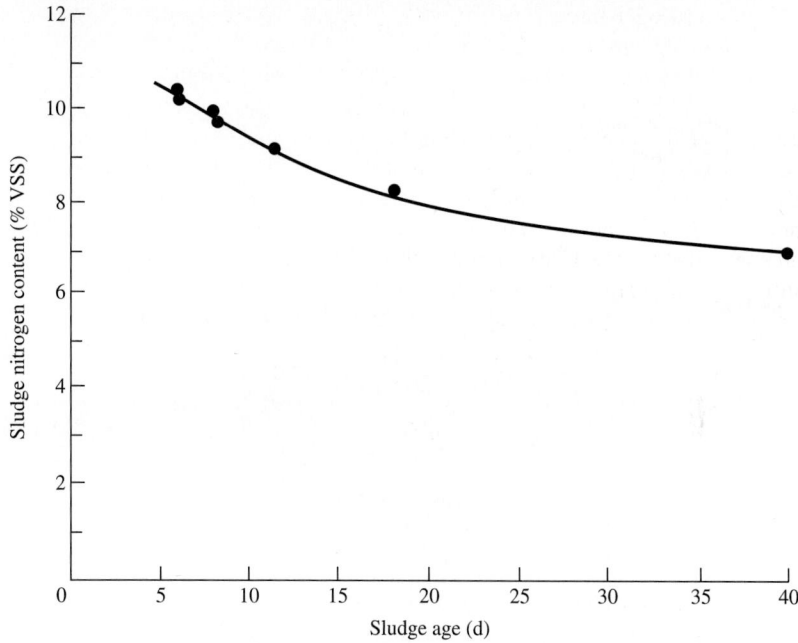

FIGURE 6.18 Nitrogen content of activated sludge as related to sludge age.

Phosphorus may be fed as phosphoric acid in larger plants and ammonia as anhydrous or aqueous ammonia. In small plants, nutrients may be fed as diammonium phosphate. In many cases, in aerated lagoons treating pulp and paper mill wastewaters, nitrogen and phosphorus have not been added, but rather the retention time has been increased. Calculated rate coefficients with and without the addition of nutrients is shown in Table 6.6. Nutrient requirements are computed in Example 6.5.

Example 6.5. An activated sludge plant treating an industrial wastewater operated under the following conditions:

Flow = 1.6 million gal/d
$S_o = 560$ mg/L (BOD basis)

	K, d^{-1}	
Waste	Without Nutrients	With Nutrients
Kraft paper	0.35	1.33
Board mill	0.70	3.20
Hardboard	0.34	1.66

TABLE 6.6

$S_e = 20$ mg/L
$X_v = 3000$ mg/L
$a = 0.55$
$b = 0.1/d$
$NH_3\text{-}N = 5$ mg/L
$P = 3$ mg/L
$F/M = 0.4/d^{-1}$
$\theta_c = 7$ d

Compute the N and P which must be added to the process.

Solution

$$t = \frac{S_o}{X_v F/M} = \frac{560}{3000 \cdot 0.4} = 0.47 \text{ d}$$

$$X_d = \frac{0.8}{1+(0.2 \cdot 0.1 \cdot 7)} = 0.7$$

$$\Delta X_v = aS_r - bX_d X_v t$$

$$\Delta X_v = 0.55(540) - 0.1 \cdot 0.7 \cdot 3000 \cdot 0.47$$

$$= 198 \text{ mg/L or } 2642 \text{ lb/d} \quad (1200 \text{ kg/d})$$

$$N = 0.123 \cdot \frac{0.7}{0.8} \cdot 2642 + 0.07 \cdot \frac{0.8-0.7}{0.8} \cdot 2642$$

$$= 284 + 23 = 307 \text{ lb/d} \quad (140 \text{ kg/d})$$

$N_{INFLUENT} = 5 \cdot 1.6 \cdot 8.34 = 67$ lb/d (30 kg/d)

$N_{ADDED} = 307 - 67 = 240$ lb/d

$$P = 0.026 \cdot \frac{0.7}{0.8} \cdot 2642 + 0.01 \cdot \frac{0.8-0.7}{0.8} \cdot 2642$$

$$= 60 + 3.3$$

$$= 63.3 \text{ lb/d} \quad (29 \text{ kg/d})$$

$P_{INFLUENT} = 3 \cdot 1.6 \cdot 8.34 = 40$ lb/d (18 kg/d)

$P_{ADDED} = 63.3 - 40 = 23.3$ lb/d (11 kg/d)

Mathematical Relationships of Organic Removal

Several mathematical models have been offered to explain the mechanism of BOD removal by biological oxidation processes. All these models have shown that, at high BOD levels, the rate of BOD removal per unit mass of cells will remain constant to a limiting BOD concentration, below which the rate will become concentration-dependent and will decrease as the concentration. Wuhrmann[20] and Tischler and Eckenfelder[21] have shown that single substances are removed by a zero-order reaction to very low substrate levels. Some reactions are shown in Fig. 6.19. In a mixture of substances being removed

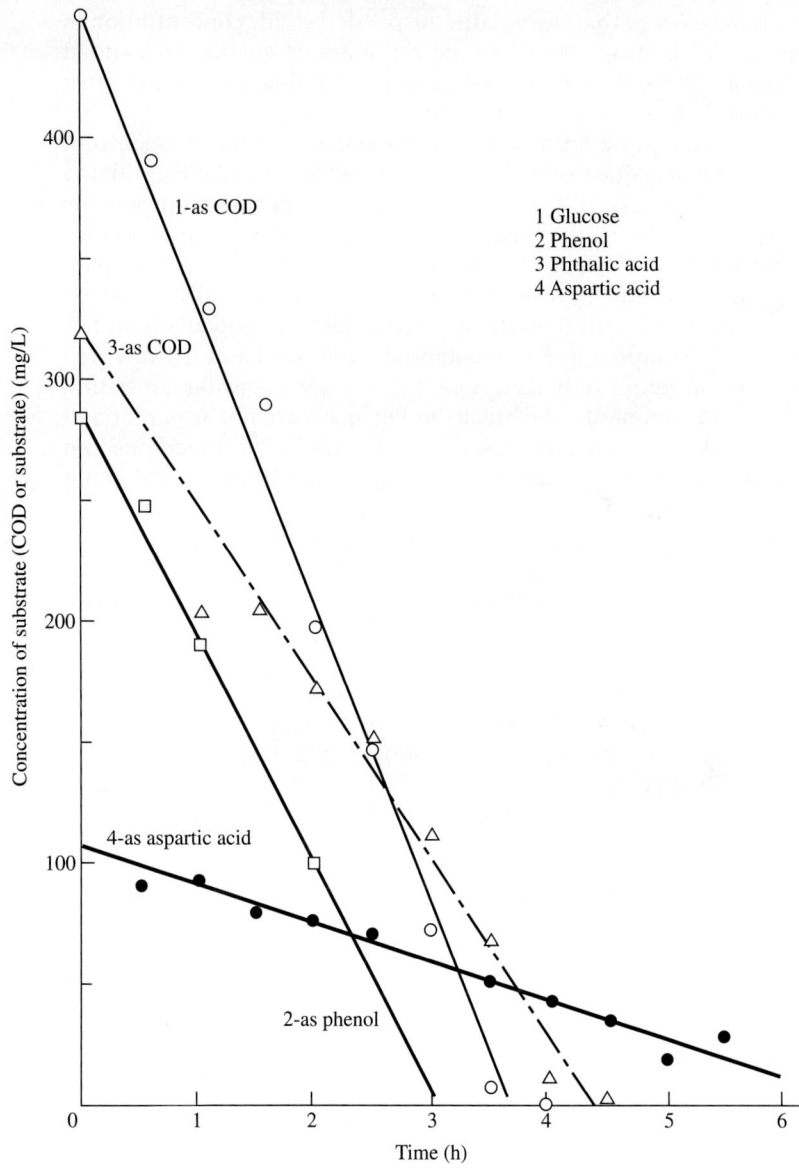

FIGURE 6.19 Zero-order removal rates for specific substrates.

at different rates, a constant maximum removal rate will prevail until one of the substances is completely removed. As other substances are progressively removed, the overall rate will decrease. As Gaudy, Komolrit, and Bhatla[22] have shown, sequential substrate removal will also yield a decreasing overall rate.

It is assumed that the volatile suspended solids concentration in the reactor is proportional to the cell mass. If volatile suspended solids are present in the influent wastewater, this assumption must be modified as discussed on p. 240.

In the case of a multicomponent wastewater where concurrent zero-order reactions occur, the overall removal can be formulated as shown in Fig. 6.20. The zero-order removal for three components is shown in Fig. 6.20a. When considering the total removal of all components as BOD, COD, or TOC, the overall rate will remain constant until time t_1, when component A is substantially removed. The overall rate will then decrease to reflect components B and C. At time t_2, component B is substantially removed and the rate will decrease to reflect only component C. For a wastewater consisting of many components, the breaks in Fig. 6.20b are not apparent and a curve results. In most cases, this curve can be linearized and can be fitted to another order reaction equation[21,23,24] of the following form:

$$\frac{dS}{dt} = -K_n X_a \left(\frac{S}{S_o}\right)^n \tag{6.24}$$

where S = COD concentration at time t, mg/L
S_o = COD concentration at time zero, mg/L
X_a = active biomass concentration, mg/L
t = time, d
K_n = curve fitting coefficient, d^{-1}
n = function power order

The active mass can be defined as

$$X_a = X_v \cdot \frac{X_d}{0.8} \cdot f_b \tag{6.25}$$

If the wastewater contains no influent VSS,

$$f_b = 1.0$$

If the influent VSS are nondegradable,

$$f_b = 1 - \frac{X_i \theta_c}{X_v t} \tag{6.26}$$

If the influent VSS are degradable, Eq. (6.17) applies.

Regardless of the actual degradation rates of the individual substrate components, the integrated form of Eq. (6.24) represents the

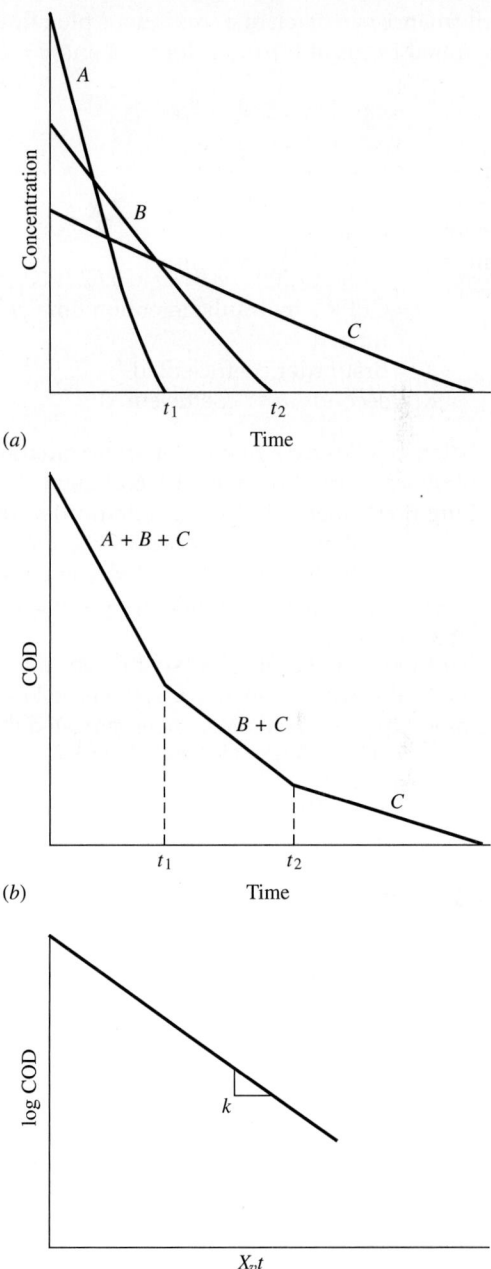

FIGURE 6.20
Schematic representation of multicomponent substrate removal.

performance of a batch or a continuous plug flow reactor (CPFR). The integrated forms of Eq. (6.24) for $n = 1$ and $n = 2$ are:

$$S_e = S_o e^{-K_1 f_b X_v t / S_o} \quad (6.27)$$

$$S_e = \frac{S_o^2}{S_o + K_2 f_b X_v t} \quad (6.28)$$

where S_e = effluent COD or BOD in a CPFR or a batch reactor, mg/L
 t = CPFR's hydraulic retention time or batch test reaction time, d
 K_1 = first-order coefficient, d^{-1}
 K_2 = second-order coefficient, d^{-1}

Batch kinetics for a first-order approximation and a second-order approximation are shown in Figs. 6.21 and 6.22.

Plug flow kinetics following a first-order function for a pulp and paper mill wastewater are shown in Fig. 6.23. Initial biosorption in the plug flow reactor should be noted. Generalized organic removals by activated sludge for various types of wastewaters are illustrated by Fig. 6.24.

If the individual components of the substrate are degraded at a zero-order rate, the overall rate for a CSTR will still follow a kinetic equation similar to Eq. (6.24), assuming a nonsegregated flow regime.[25,26]

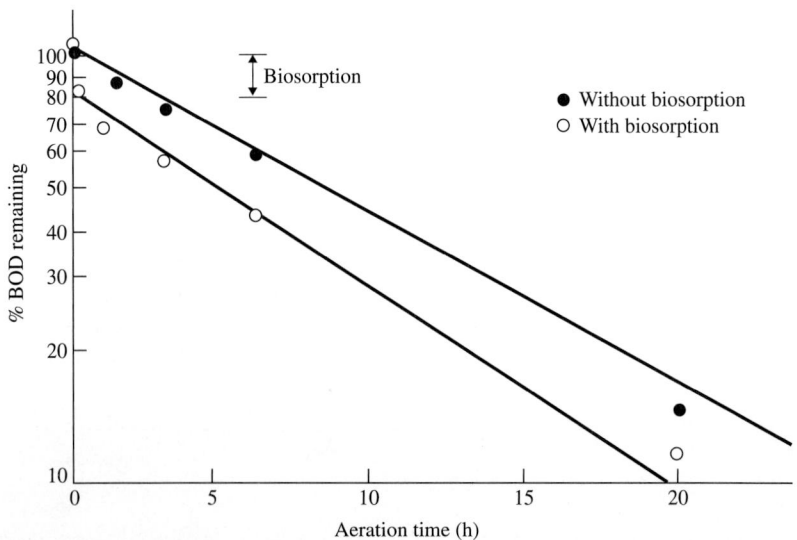

Figure 6.21 Batch-activated sludge with and without biosorption for a pulp and paper mill wastewater.

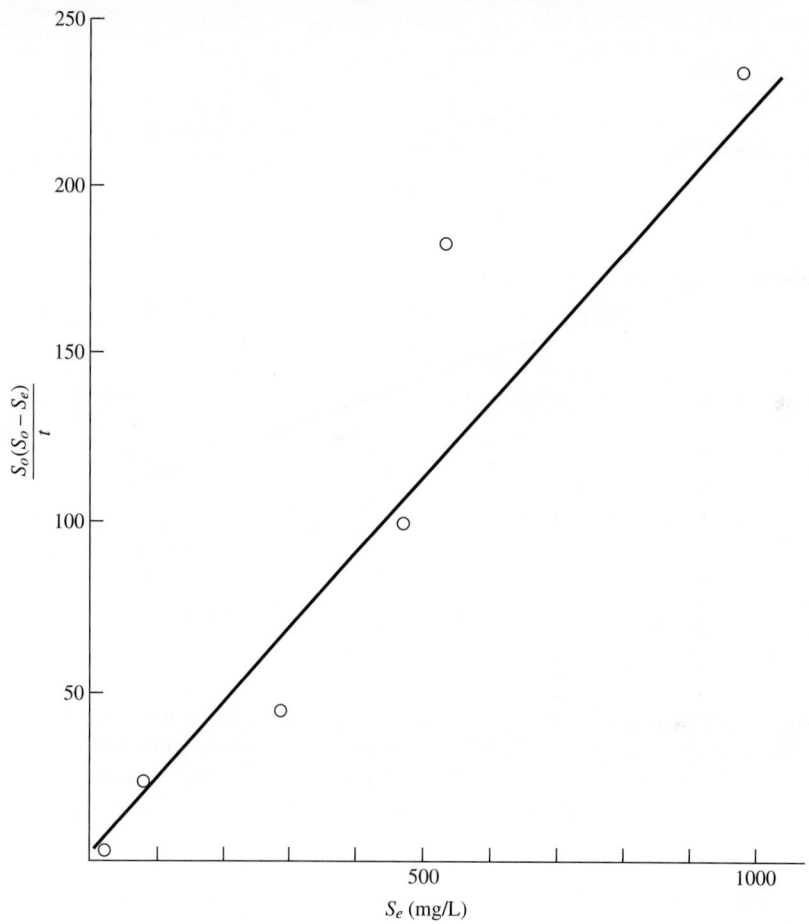

FIGURE 6.22 Batch oxidation of a chemical industry wastewater following second-order kinetics.

In a complete mix basin, the overall removal rate will decrease as the concentration of organics remaining in solution decreases, since the more readily degradable organics will be removed first. The kinetics of the oxidation are therefore limited by the effluent concentration and can be expressed as

$$\frac{S_o - S_e}{f_b X_v t} = K \frac{S_e}{S_o} \tag{6.29}$$

This relationship is shown in Fig. 6.25 for a soybean wastewater.

The relative biodegradability of various organic compounds is shown in Table 6.7. The rate coefficient K for various industrial

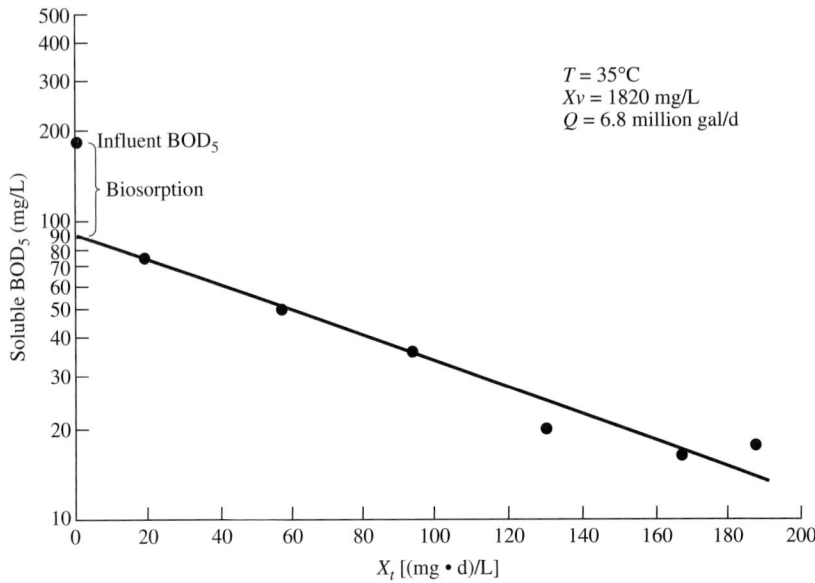

Figure 6.23 Plug flow BOD removal kinetics for a bleached kraft pulp and paper wastewater.

wastewaters in CMAS systems is shown in Table 6.8. Assuming no inhibition or toxicity, the plug flow reactor is more efficient in terms of organic removal as shown in Fig. 6.26. However, it does not offer the equalization capacity of a CSTR. Consequently, CSTR is preferred when potential inhibition or toxicity is of concern.

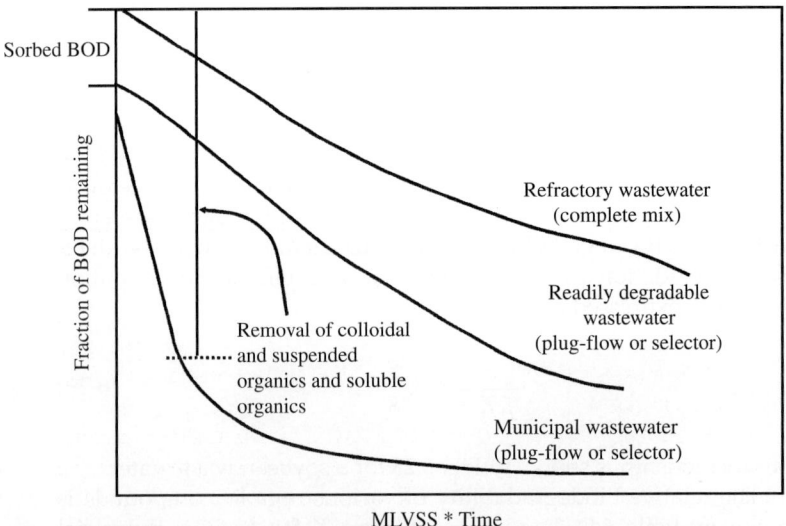

Figure 6.24 Organic removal characteristics by activated sludge.

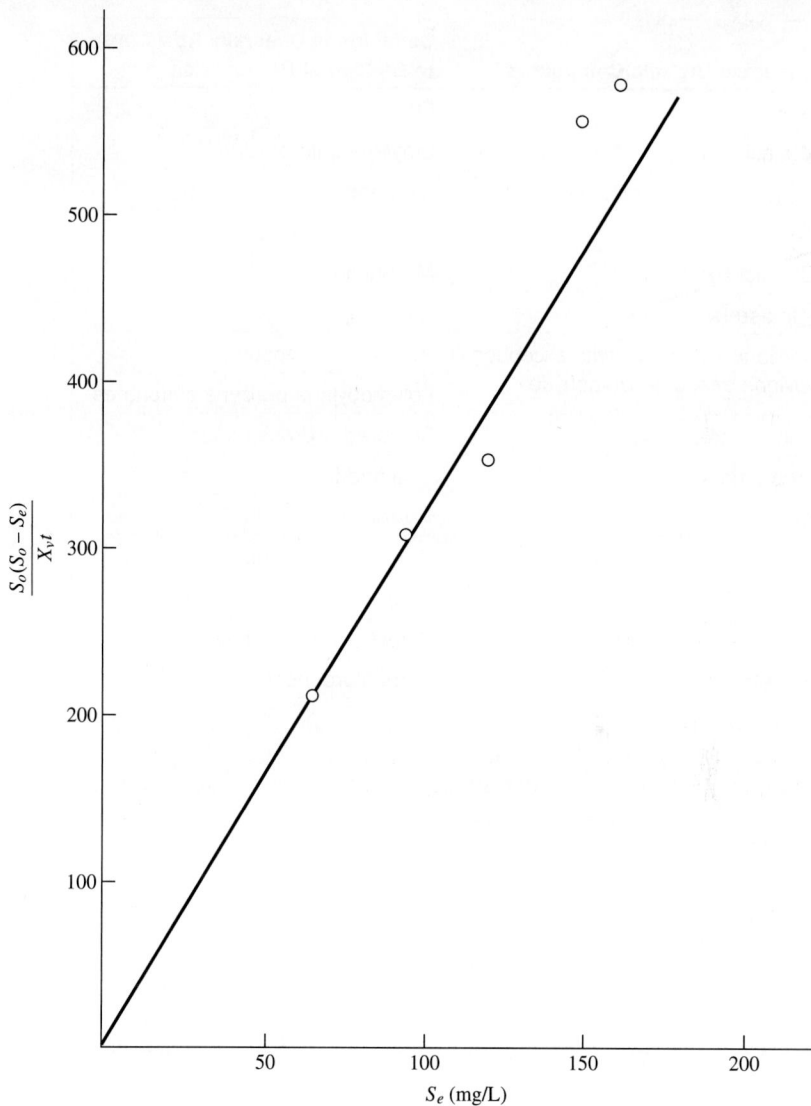

FIGURE 6.25 Complete mix kinetics for a soybean wastewater.

In all kinetic calculations the rate coefficient K should be based on active mass in the reactor.

There are two operating conditions which influence the organic removal rate, as defined by a rate coefficient K. These are the active fraction of the biomass in the MLVSS and the aerobic portion of the biofloc. The active fraction of the MLVSS is related to the sludge age

Biodegradable Organic Compounds*	Compounds Generally Resistant to Biological Degradation
Acrylic acid	Ethers
Aliphatic acids	Ethylene chlorohydrin
Aliphatic alcohols (normal, iso, secondary)	Isoprene
	Methyl vinyl ketone
Aliphatic aldehydes	Morpholine
Aliphatic esters	Oil
Alkyl benzene sulfonates with exception of propylene-based benzaldehyde	Polymeric compounds
	Polypropylene benzene sulfonates
Aromatic amines	Selected hydrocarbons
Dichlorophenols	Aliphatics
Ethanolamines	Aromatics
Glycols	Alkyl-aryl groups
Ketones	Tertiary aliphatic alcohols
Methacrylic acid	Tertiary aliphatic sulfonates
Methyl methacrylate	Trichlorophenols
Monochlorophenols	
Nitriles	
Phenols	
Primary aliphatic amines	
Styrene	
Vinyl acetate	

*Some compounds can be degraded biologically only after extended periods of seed acclimation.

Table 6.7 Relative Biodegradability of Certain Organic Compounds

or the F/M. Increasing the F/M or decreasing the θ_c will increase the active fraction of the biomass as shown in Fig. 6.27. The second condition defines that portion of the floc which is aerobic and is related to the turbulence level or mixing intensity and bulk dissolved oxygen level in the aeration basin.

The fraction of active biomass in the mixed liquor is defined here as the biodegradable fraction X_d divided by 0.80. This is illustrated in Examples 6.6 and 6.7.

Principles of Aerobic Biological Oxidation

Wastewater Source	K, d⁻¹	Temperature, °C
Vegetable tannery	1.2	20
Cellulose acetate	2.6	20
Peptone	4.0	22
Organic phosphates	5.0	21
Vinyl acetate monomer	5.3	20
Organic intermediates	5.8	8
	20.6	26
Viscose rayon and nylon	6.7	11
	8.2	19
Domestic sewage (solubles)	8.0	20
Polyester fiber	14.0	21
Formaldehyde, propanol, methanol	19.0	20
High-nitrogen organics	22.2	22
Potato processing	36.0	20

TABLE 6.8 Reaction Rate Coefficient for Selected Wastewaters

Example 6.6. A complete mix activated sludge plant with an influent BOD of 800 mg/L and a K of 6 d⁻¹ is operating at an SRT of 10 d. What is the effluent quality and what will the effluent quality be if the SRT is increased to 30 d? a is 0.5 and b is 0.1 d⁻¹.

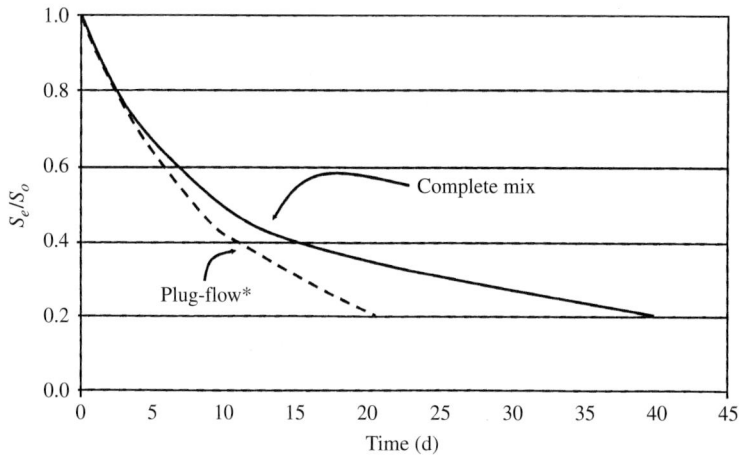

*Assuming no toxicity/inhibition

FIGURE 6.26 Comparison between plug-flow and complete mix performance.

266 Chapter Six

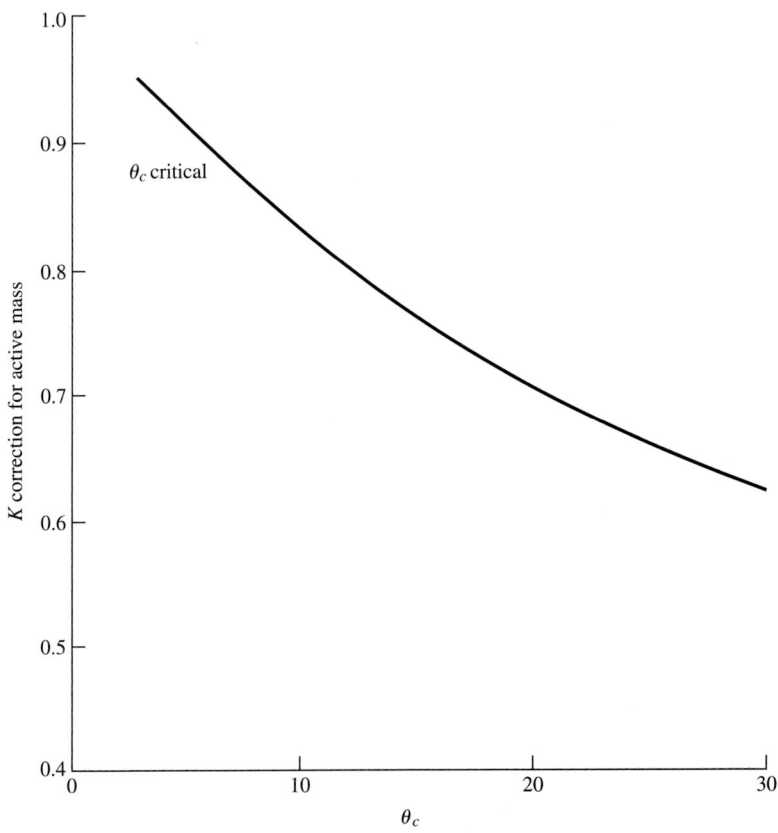

FIGURE 6.27 Relationship between K and θ_c, considering biological active mass.

Solution

$$\frac{1}{\theta_c} = aK\frac{S_e}{S_o} - bX_d$$

At $\theta_c = 10$ d,

$$\frac{1}{10} = 0.5 \cdot 6 \cdot \frac{S_e}{S_o} - 0.1 \cdot \left(\frac{0.8}{1+0.2\cdot 0.1\cdot 10}\right)$$

$$\frac{S_e}{S_o} = 0.056$$

and

$$S_e = 0.056 \cdot 800 = 45 \text{ mg/L}$$

At $\theta_c = 30$ d, the K can be adjusted by Fig. 6.27:

$$K = 6 \cdot \frac{0.625}{0.83} = 4.5 \text{ d}^{-1}$$

$$\frac{1}{30} = 0.5 \cdot 4.5 \frac{S_e}{S_o} \cdot 0.1 \cdot \left(\frac{0.8}{1 + 0.2 \cdot 0.1 \cdot 30}\right)$$

$$\frac{S_e}{S_o} = 0.0368$$

and

$$S_e = 0.0368 \cdot 800 = 30 \text{ mg/L}$$

If X_v is 2500 mg/L and $t = 0.9$ d at $\theta_c = 10$ d, the $X_v t$ at a 30-d SRT is

$$X_v t = \frac{\theta_c a S_r}{1 + \theta_c b X_d}$$

$$= \frac{30 \cdot 0.5 \cdot 770}{1 + 30 \cdot 0.1 \cdot 0.5} = 4620 \text{ (mg} \cdot \text{d)/L}$$

$$X_v = 5133 \text{ mg/L}$$

Example 6.7. Determine the hydraulic retention time and the sludge age for the following condition:

$S_o = 700$ mg/L
$S_e = 30$ mg/L
$K = 10/\text{d}$
$X_v = 3000$ mg/L
$a = 0.4$

What will be the required hydraulic retention time to produce the same effluent if the influent nondegradable VSS is 50 mg/L?

Solution Solving Eq. (6.29) for the hydraulic retention time t produces, $S_e = 30$ mg/L,

$$t = \frac{S_o S_r}{K X_v S_e}$$

$$= \frac{700 \cdot 670}{10 \cdot 3000 \cdot 30} = 0.52 \text{ d}$$

and

$$X_v t = 1560 \text{ (mg} \cdot \text{d)/L}$$

The SRT for these conditions is

$$\theta_c = \frac{X_v t}{a S_r - b X_d X_v t}$$

Assume $X_d = 0.7$. Then

$$\theta_c = \frac{1560}{0.4 \cdot 670 - 0.1 \cdot 0.7 \cdot 1560}$$

$$= \frac{1560}{159} = 9.8 \text{ d, say } 10 \text{ d}$$

Check assumption for $X_d = 0.7$ of SRT = 10 d:

$$X_d = \frac{0.8}{1 + 0.2 \cdot 0.1 \cdot 10}$$

$$X_d = 0.67$$

To determine the required hydraulic retention time to produce the same effluent $S_e = 30$ mg/L if the influent nondegradable VSS = 50 mg/L, assume $t = 0.69$ d (by trial and error) and calculate the accumulation of the influent VSS in the mixed liquor (MLVSS$_i$):

$$\text{MLVSS}_i = \frac{50 \cdot 10}{0.69} = 725 \text{ mg/L}$$

The residual biomass VSS in the mixed liquor is

$$X_{vb} = 3000 - 725 = 2275 \text{ mg/L}$$

Calculate the required t:

$$t = \frac{X_{vb} t}{X_{vb}} = \frac{1560}{2275} = 0.69 \text{ d}$$

$$f_b = 2275/3000 = 0.76$$

Check sludge age:

$$\theta_c = \frac{X_v t}{(aS_r - bX_{vb} X_d t) + X_i}$$

$$= \frac{3000 \cdot 0.69}{(268 - 0.1 \cdot 2275 \cdot 0.7 \cdot 0.69) + 50}$$

$$= \frac{2070}{208} = 9.9 \text{ d}$$

The greater hydraulic retention time is required to produce the same effluent quality since the nondegradable influent VSS accumulated in the constant MLVSS concentration of 3000 mg/L. This is shown in the figure on the next page.

The size and the aerobic fraction of the biological floc are related to the operating power level in the aeration basin and the mixed liquor bulk dissolved oxygen concentration. These in turn influence both the reaction rate coefficient and the endogenous decay rate.

The effect on the reaction rate coefficient has been demonstrated in treatment of wastewater from a bleached kraft pulp and paper mill. The biodegradation rate [K in Eq. (6.29)] under operation at an F/M of 0.3 d^{-1} and a conventional aeration basin power level of 200 hp/million gal averaged 4.5 d^{-1}. Operation at an F/M of 0.88 d^{-1} and a power level of 500 hp/million gal averaged a K of 12.5 d^{-1}. BOD removal of 92 percent was achieved under these conditions.

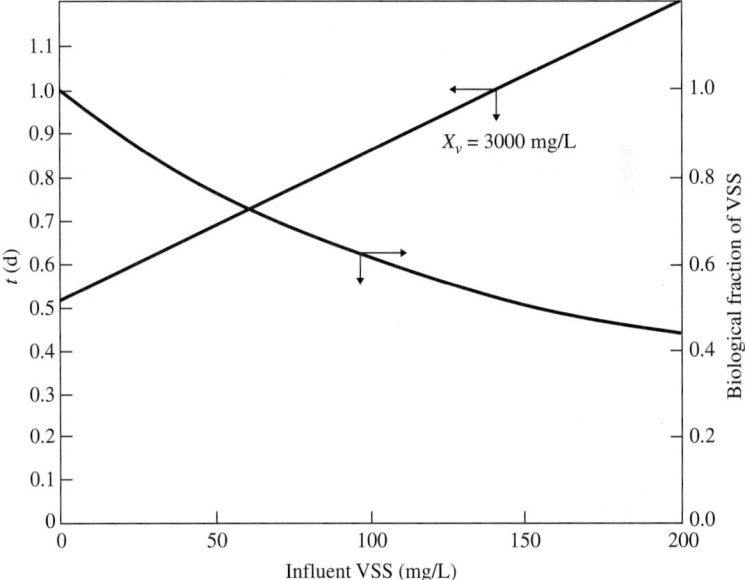

Two factors will influence the endogenous decay rate and the observed sludge yield. These are the degradable fraction X_d, which is a function of SRT or F/M, and the aerobic fraction of the MLVSS. Increasing the mixing intensity in the aeration basin will increase the aerobic fraction of the mixed liquor solids and the rate of endogenous respiration. This condition has been observed in the high-purity-oxygen activated sludge process in which high mixed liquor dissolved oxygen levels increased the aerobic fraction of the floc and decreased the observed sludge yield. Rickard and Gaudy[27] showed a decrease in observed sludge yield with increasing agitation at a constant F/M.

It has been observed that, at high energy levels in the aeration basin, filamentous bulking is suppressed. This was noted by Rickard and Gaudy,[27] who showed that, at a velocity gradient of 310 s^{-1}, a majority of the cells were of the filamentous type, while, at a velocity gradient of 1010/s^{-1}, filaments were reportedly absent. Zahradka[28] observed that a high power input caused the growth of sludge flocs of small and uniform size and suppressed the growth of filamentous organisms. This can be explained by the fact that, at high power levels, oxygen and BOD diffusion is not limiting with small flocs, and therefore the floc formers will dominate over the filaments. It is therefore important to consider the aeration power level and its effect on the reaction coefficient K and b in a process design or a pilot plant study.

In cases where multiple wastewaters are mixed for biological treatment, a mean reaction rate K is determined. In this case the average rate coefficient K_c can be computed:

$$\frac{1}{K_c} = \frac{\frac{1}{K_1}(Q_1 S_{o_1}) + \frac{1}{K_2}(Q_2 S_{o_2})}{Q_1 S_{o_1} + Q_2 S_{o_2}} \tag{6.30}$$

This is shown in Example 6.8. The multiple zero-order concept may also predict the performance of a combined system, where two or more different wastewater streams are treated in one unit.

Example 6.8. Treatment of mixed industrial wastewaters. Three industrial wastewaters are to be blended and treated in an activated sludge plant. Compute the sludge age and the hydraulic retention time required to produce an effluent of 20 mg/L BOD.

The characteristics of the three wastewaters are

1. $Q = 2$ million gal/d
   $S_o = 600$ mg/L
   $K = 5$ d^{-1}

2. $Q = 1$ million gal/d
   $S_o = 1200$ mg/L
   $K = 10$ d^{-1}

3. $Q = 5$ million gal/d
   $S_o = 300$ mg/L
   $K = 2$ d^{-1}

Solution The average influent BOD is

$$S_{o_{ave}} = \frac{2 \cdot 600 + 1 \cdot 1200 + 5 \cdot 300}{8} = 487 \text{ mg/L}$$

The average K is

$$\frac{1}{K_{ave}} = \frac{Q_1 S_{o_1}/K_1 + Q_2 S_{o_2}/K_2 + Q_3 S_{o_3}/K_3}{Q_1 S_{o_1} + Q_2 S_{o_2} + Q_3 S_{o_3}}$$

$$= \frac{0 \cdot 2 \cdot 600 + 0.1 \cdot 1 \cdot 1200 + 0.5 \cdot 5 \cdot 300}{2 \cdot 600 + 1 \cdot 1200 + 5 \cdot 300}$$

$$= 0.28$$

$$K_{ave} = 3.57 \text{ d}^{-1}$$

The mean yield coefficient a is 0.5, b is 0.1, and X_v is 3000 mg/L. For a K_{ave} of 3.57,

$$\frac{1}{\theta_c} = aK\frac{S_e}{S_0} - bX_d$$

$$= 0.5 \cdot 3.57 \cdot \frac{20}{487} - 0.1 \cdot 0.46$$

$$= 0.027$$

$$\theta_c = 37 \text{ d}$$

$$\theta_c = \frac{X_v t}{aS_r - bX_d X_v t}$$

or

$$X_v t = \frac{\theta_c a S_r}{1 + \theta_c b X_d}$$

$$= \frac{37 \cdot 0.5 \cdot 467}{1 + 37 \cdot 0.1 \cdot 0.46}$$

$$= 3200$$

$$t = \frac{3200}{3000} = 1.07 \text{ d}$$

In a two-stage activated sludge system without an intermediate clarifier, the effluent will be the same as a single-stage system as long as multiple zero-order kinetics apply and the second-stage kinetic coefficient is proportional to the squared ratio of the pretreated to raw substrate concentrations. The multistage operation kinetics approach plug flow and hence produces a higher quality effluent than a CSTR for a given HRT. This is shown by Fig. 6.28. Figure 6.29 demonstrates an improved effluent quality for multiple versus single-stage activated sludge operation. Table 6.9 presents multistage operational data for a Bleached Kraft pulp and paper mill wastewater.

In a two-stage system with an intermediate clarifier between the two stages, leading to the development of a biomass specifically acclimated to the substrate remaining in the first-stage effluent, the rate coefficient in the second stage will be lower than in the first stage since the more degradable compounds were removed in the first stage. The second stage K can be estimated from the relationship

$$K_2 = K_1 \left[\frac{S_1}{S_0}\right] \quad (6.31)$$

where K_2 = rate coefficient for the second stage
K_1 = rate coefficient for the first stage
S_0 = influent BOD or COD to stage 1
S_1 = influent BOD or COD to stage 2

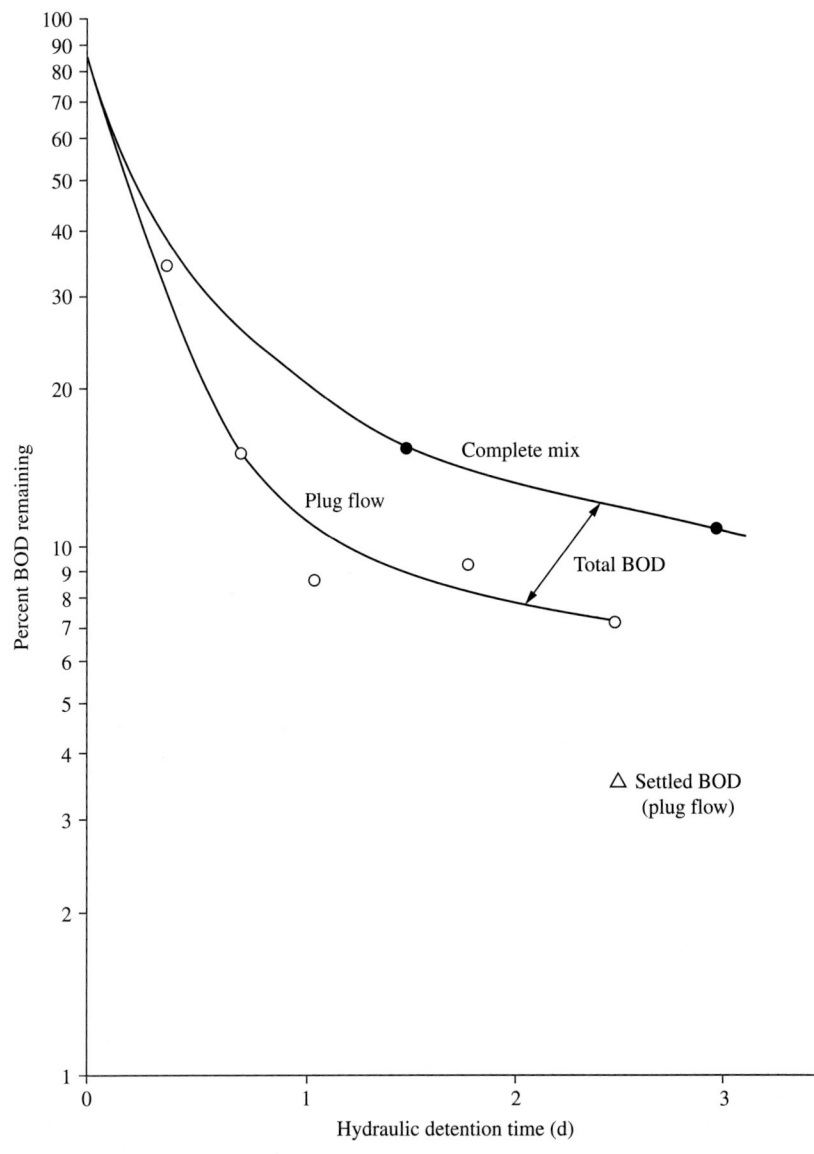

FIGURE 6.28 Pilot plant results for single- and multistage operation.

The K rate coefficient determination for a two-stage CMAS system compared to a single-stage CMAS for a synthetic fibers wastewater is shown in Fig. 6.30. Note that higher removal rates are obtained with the two-stage system.

The application of Eq. (6.31) can be illustrated by using data from a two-stage activated sludge plant treating a pharmaceutical

Principles of Aerobic Biological Oxidation

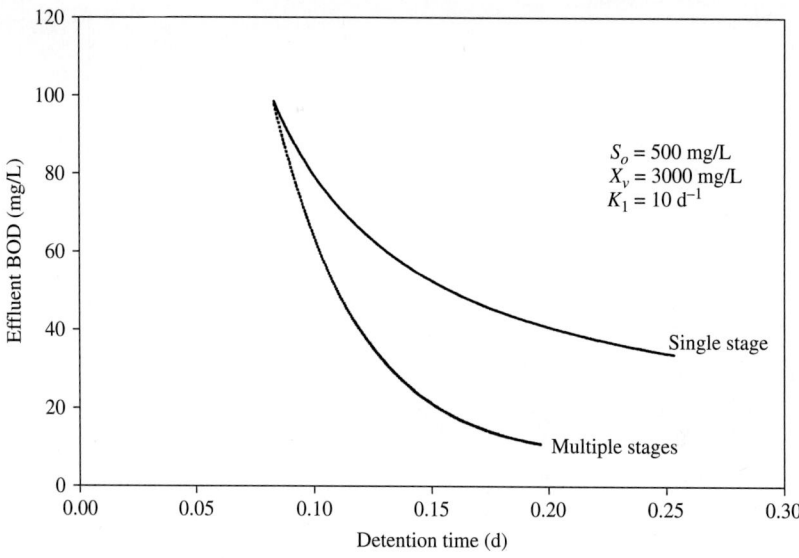

FIGURE 6.29 Multiple stages vs. single-style BOD removal comparison.

Parameter	Average	Maximum	Minimum
Flow, MGD	21.7	23.5	19.4
Influent TBOD, mg/L	241	260	210
Influent SBOD, mg/L	207	225	180
Influent TSS, mg/L	211	257	164
Detention time, d	0.132	0.147	0.121
MLSS, mg/L	2077	2219	1810
F/M, lb BOD/lb MLSS · d	0.88	0.97	0.75
Temperature, °F	72	104	57
SVI, mL/g	95	98	93
SRT, d	0.76	0.91	0.62
SLR, lb/d-sq ft	21.6	23.6	19.3
SOR, gal/d-sq ft	941	1021	840
Effluent TSS, mg/L	78	99	68
Effluent TBOD, mg/L	41	48	34
Effluent SBOD, mg/L	19	25	13

TABLE 6.9 Multistage Activated Sludge Treatment of Bleached Kraft Pulp and Paper Mill Wastewater

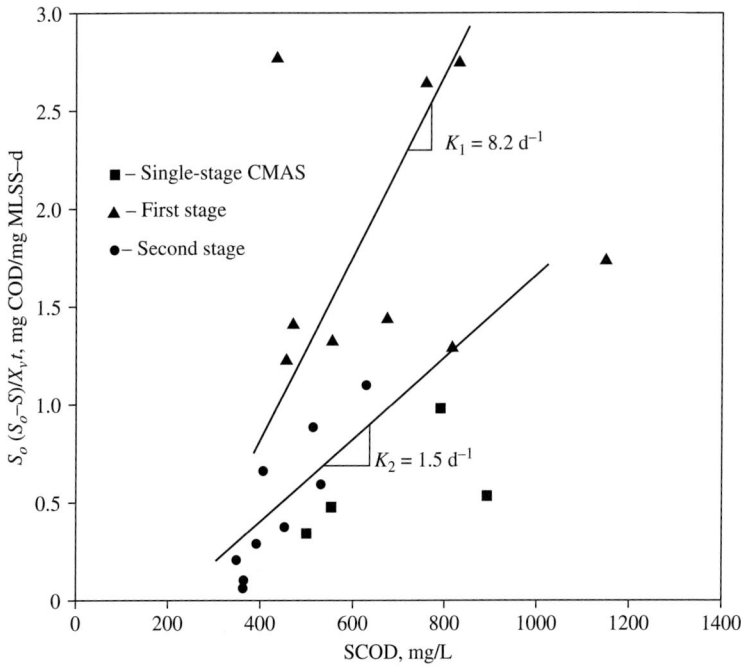

FIGURE 6.30 Effect of two-stage operation on reaction rate coefficient.

wastewater. The first stage has an influent BOD of 5825 mg/L, an effluent BOD of 540 mg/L, and a K of 3.9 d^{-1}. The influent BOD to the second stage is 540 mg/L and the K is 0.4 d^{-1}. The calculated K in stage 2, from Eq. (6.31), is

$$K_2 = K_1 \left(\frac{S_1}{S_o} \right)$$

$$= 3.9 \left(\frac{540}{5825} \right)$$

$$= 0.36 \text{ d}^{-1}$$

This is in good agreement with the measured value of 0.4 d^{-1}.

Kinetic parameters for a pulp and paper mill wastewater are shown in Table 6.10. Parameters for petroleum refinery wastewaters are shown in Table 6.11.

It is apparent that as the organic composition of the wastewater changes, the rate coefficient K in Eq. (6.29) will also change. This is not a problem for wastewaters such as those from a dairy or food processing plant since their composition remains substantially unchanged, and hence, K will remain nearly constant. Wastewaters

Type of Mill	K, d^{-1}	Temperature, °C
Oxygen bleached kraft	13.5	35
Virgin pulp and wastepaper	13.6	23
Unbleached kraft	4.5	38
Sulfite	5.0	18
Bleached sulfite	6.2	—
Bleached kraft	5.2	—
Bleached kraft	4.4	34

TABLE 6.10 Reaction Rate Coefficient for Pulp and Paper Mills

generated from plants with multi-products and campaign production, however, will experience a constantly changing wastewater composition and, hence, a highly variable K value. Daily data from a plant treating municipal sewage, a pulp and paper mill wastewater, and two organic wastewaters are shown in Fig. 6.31. The probability of occurrence of some compounds found in a petrochemical plant influent is illustrated by Table 6.12.

The rate coefficient combines the effects of all removal mechanisms: biosorption, biodegradation, and volatilization, unless steps are taken to separate the effects of an individual removal mechanism. Unusually high "apparent" reaction rate coefficients may be observed when volatile organics constitute a large portion of the wastewater. Volatilization of substrate should be considered when calculated K values exceed about 30/d^{-1} at 20 to 25°C.

Industrial wastewater discharge permits typically contain two limiting conditions: a monthly average limit and a daily or weekly maximum limit. The treatment process must be designed and operated to reliably satisfy both of these discharge conditions.

A suggested design approach is based on a statistical distribution of the removal rate coefficient and the performance of the upstream equalization basin. For average discharge conditions, the mean K value is based on the average discharge limit and average influent load. These values are substituted into Eq. (6.29) to yield

$$\frac{\bar{S}_o - \bar{S}_e}{f_b X_v t} = K_{50\%} \frac{\bar{S}_e}{\bar{S}_o} \tag{6.32}$$

where $\bar{S}_o$ = average influent BOD, mg/L
 $\bar{S}_e$ = soluble BOD at average permit limit, mg/L
 $K_{50\%}$ = the 50 percentile value of K, d^{-1}

Influent		Organic Removal Rate K*		Sludge Growth Coefficients				Oxygen Requirement Coefficients[†]		Residual COD, mg/L
				BOD Basis		COD Basis				
BOD, mg/L	COD, mg/L	BOD, d^{-1}	COD, d^{-1}	a	bX_d	a	bX_d	a'	$b'X_d$	
244	509	4.15	2.74	—	—	—	—	0.57	0.1	106
575	981	—	7.97	—	—	0.5	0.06	0.60	0.11	53
396	782	—	5.86	—	—	0.5	0.06	0.34	0.06	100
153	428	—	2.92	0.5	0.08	0.44	0.1	0.35	0.08	22
170	600	—	5.0	—	—	0.26	0.03	0.46	0.05	100[‡]
248	563	4.11	7.79	—	—	0.2	0.08	0.40	0.01	76
345	806	—	7.24	—	—	0.43	0.10	0.52	0.14	82
196	310	4.70	—	0.6	0.05	—	—	0.46	0.14	50
138	275	—	—	0.58	—	0.25	—	0.60	0.09	42

*At 24°C.
[†]COD basis.
[‡]TOD.

TABLE 6.11 Biological Treatment Coefficients for Petroleum Refinery Wastewaters

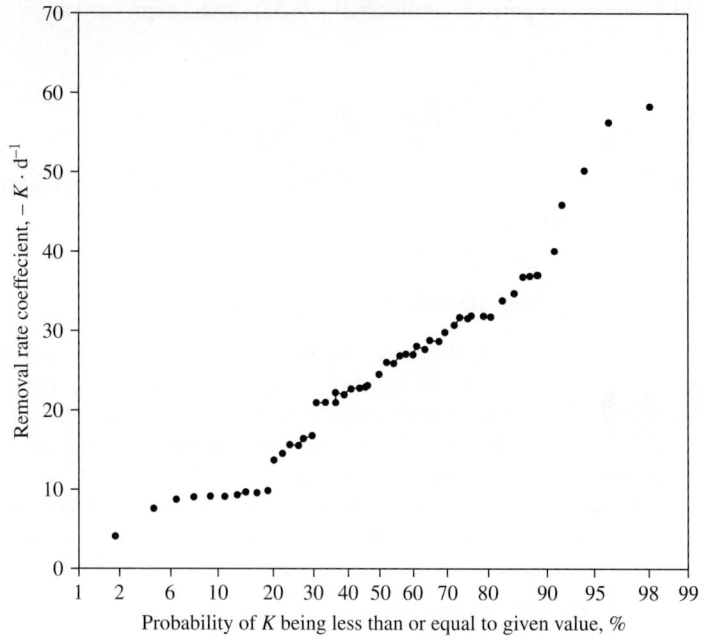

Figure 6.31 Variability in K as related to wastewater composition.

Compound	Occurrence Frequency, % of Daily Composites	Concentration When Present, mg/L	
		Range	Average
Acetone	100	2–430	52
Ethanol	100	4–280	38
Methanol	100	1–180	44
Isopropanol	95	3–620	60
2-Ethyl hexanol	85	2–140	21
Isopropyl acetate	80	1–47	11
Acrylonitrile	65	1–110	15
Acetaldehyde	48	1–98	21
Ethylene glycol	43	1–480	43
Diethylene glycol	27	1–500	64

Table 6.12 Variability of Several Compounds Found in a Petrochemical Plant Influent Wastewater

For the maximum permit condition, Eq. (6.32) can be expressed as

$$\frac{S_{o_m} - S_{e_m}}{f_b X_v t} = K_{5\%} \frac{S_{e_m}}{S_{o_m}} \qquad (6.33)$$

where S_{o_m} = maximum effluent BOD from the equalization basin, mg/L
S_{e_m} = soluble BOD at maximum permit limit, mg/L
$K_{5\%}$ = 5 percentile value of K, d^{-1}

The values of $X_v t$ are calculated for Eqs. (6.32) and (6.33), and the larger of the two values is used for design. However, if the $X_v t$ value computed for the maximum permit condition exceeds twice the value computed for the average condition, changes in the equalization capacity or plant production schedules should be considered in order to reduce the difference. Alternatively, a less conservative value of K ($> K_{5\%}$) could be used.

Specific Organic Compounds

The kinetic removal mechanism for specific organics in an aerobic biological process has been defined by Monod:

$$\mu = \frac{\mu_m S}{K_S + S} \quad \text{and} \quad q = \frac{q_m S}{K_S + S} \qquad (6.34)$$

where μ = specific growth rate, d^{-1}
μ_m = maximum specific growth rate, d^{-1}
S = substrate concentration, mg/L
K_S = substrate concentration when the rate is one-half the maximum rate, mg/L
q = specific substrate removal rate, d^{-1}
q_m = maximum substrate removal rate, d^{-1}

The relative biodegradability of specific organics in terms of the Monod relationship is shown in Fig. 6.32.

In complete mix activated sludge (CMAS) with sludge recycle, the Monod equation can be expressed as:

$$S_o - S = \frac{q_m S}{K_S + S} \cdot X_{vb} t \qquad (6.35)$$

where S_o = influent substrate concentration, mg/L
X_{vb} = biological volatile suspended solids under aeration, mg/L
t = liquid retention time, d

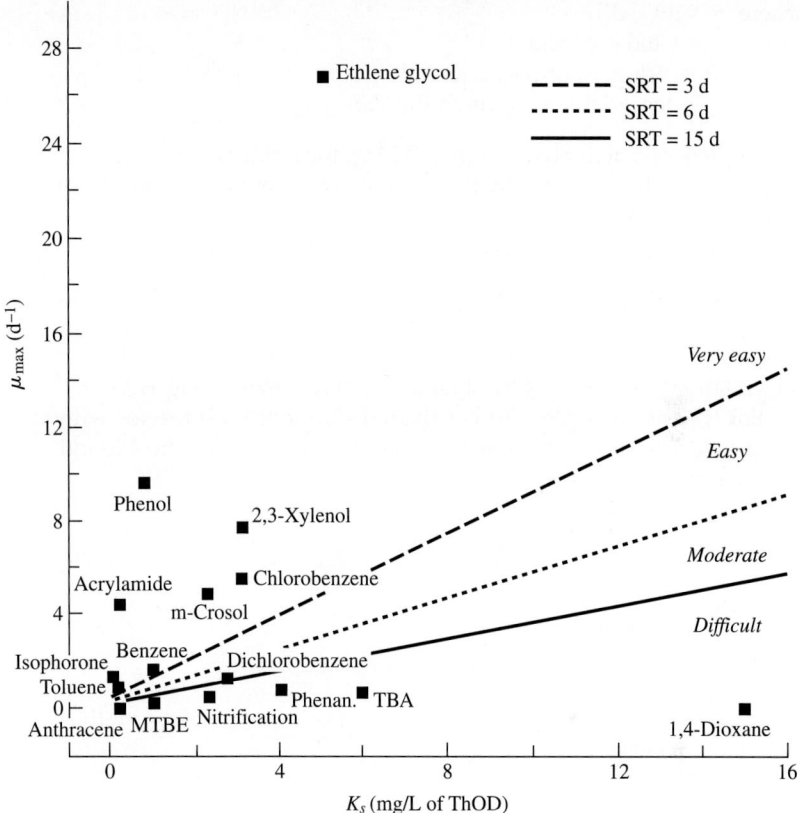

FIGURE 6.32 Relative biodegradability of specific organics at different SRT at 25°C (effluent concentration = 0.5 mg/L as COD and b = 0.11 per day). (*Adapted from Grady et al., 1996.*)

Solving for S yields:

$$S = \frac{-B + (B^2 + 4S_oK_S)^{1/2}}{2} \quad (6.36)$$

in which

$$B = q_mX_{vb}t + K_S - S_o$$

The SRT in the activated sludge process for a soluble substrate can be defined as follows:

$$\theta_c = \frac{X_{vb}t}{a(S_o - S) - bX_dX_{vb}t} \quad (6.37)$$

where θ_c = SRT, d
 a = yield coefficient, d^{-1}
 b = endogenous coefficient, d^{-1}
 X_d = degradable fraction of the VSS

In complete mix activated sludge (CMAS), the effluent substrate concentration is directly related to the SRT (θ_c). Combining Eqs. (6.34) and (6.37) yields:

$$S = \frac{K_S(1+bX_d\theta_c)}{\theta_c(q_m a - bX_d) - 1} \quad (6.38)$$

This relationship for dichlorophenol (DCP) is shown in Fig. 6.33.

For the case of a plug flow activated sludge (PFAS) reactor with sludge recycle, the performance equation derived from the Monod relationship is:

$$\frac{1}{\theta_c} = \frac{\mu_m(S_o - S)}{(S_o - S) + CK_S} - bX_d \quad (6.39)$$

where $C = (1 + \alpha)\ln[X(\alpha S + S_o)/(1 + \alpha)S]$
 $\alpha = R/Q$
 S_o = influent substrate concentration prior to mixing with the recycle

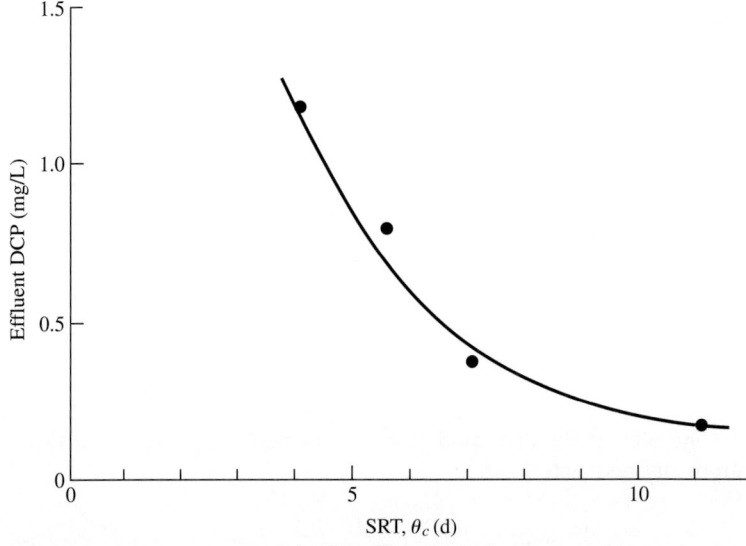

FIGURE 6.33 Effect of SRT on DCP removal.

Effluent levels of specific priority organics, for example, phenol, can be computed from Eq. (6.38) or (6.39), depending on the configuration of the reactor, complete mix, or plug flow.

The modified fed batch reactor (FBR) test described by Philbrook and Grady[29] is applicable to the determination of the kinetic coefficients q_m and K_S under field operating conditions. In the test, plant or pilot plant sludge at the desired SRT is placed in a 2-L reactor, and plant wastewater is added at a constant rate. In order to determine q_m, the addition rate must exceed the degradation rate. Since, in many wastewaters, the priority pollutant levels are low, the wastewater may have to be spiked to ensure a sufficient concentration of pollutant to meet the conditions of the test. It is important, however, that the concentration levels achieved in the test are below the inhibition threshold. This can be found by the shape of the concentration-time curve. The degradation rate q_m is computed as the difference in the slopes of the substrate addition rate and the residual substrate accumulation. A schematic illustrating the FBR and its methodology is detailed more, later in this chapter.

A second FBR test is then conducted with the addition rate of the priority pollutant equal to one-half the maximum rate determined in the first test. The steady-state concentration observed in the reactor will be K_S. FBR test data for phenol are shown in Fig. 6.34. Hoover[30] found a high variability in q_m with sludges operating under the same loading conditions with time. Based on these observations, a routine test program should be established at a treatment plant and values for q_m and K_S interpreted on a statistical basis.

In the pulp and paper industry attention is being focused on absorbable organic halides (AOX) and chlorinated phenolics in bleach plant effluents. Chlorophenols are amenable to aerobic mineralization, while methoxalated chlorophenols are recalcitrant to aerobic oxidation. Anaerobic pretreatment, however, results in readily degradable aerobic forms. Table 6.13 summarizes the observed removal of pollutants from pulp and paper mill wastewaters using activated sludge (AS), facultative stabilization basins (FSB), and aerated stabilization basins (ASB). Substituting oxygen bleaching for conventional chlorine bleaching has a marked effect on the removal of both AOX and chlorinated phenolics as shown in Table 6.14.

Example 6.9. Compute the SRT required in a complete mix activated sludge plant to reduce phenol from $S_o = 10$ mg/L to 15 µg/L

where $q_m = 1.8$ g/(g VSS · d) at 20°C
$\theta = 1.1$ (temperature coefficient)
$K_S = 100$ µg/L
$a = 0.6$
$bX_d = 0.05$ d^{-1} at 20°C = 0.033 at 10°C

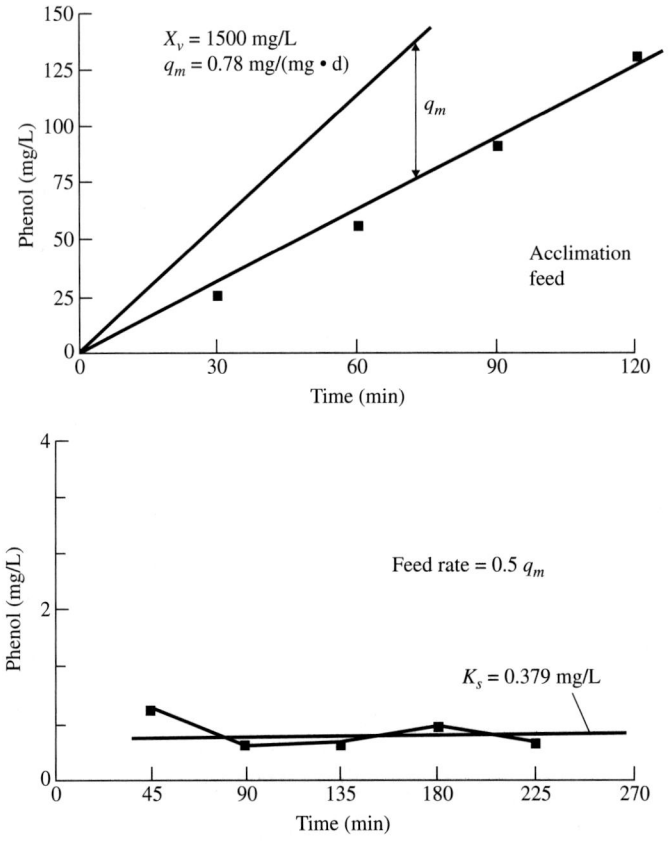

Figure 6.34 FBR test for the determination of q_m and K_S.

Solution Equation (6.38) can be rearranged to yield

$$\theta_c = \frac{K_S + S}{aq_m S - bX_d(K_S + S)}$$

$$= \frac{0.1 + 0.015}{0.6 \cdot 1.8 \cdot 0.015 - 0.05(0.115)}$$

$$= 11.0 \text{ d}$$

What is the required SRT if the temperature is reduced from 20 to 10°C?

$$q_{m(10°)} = q_{m(20°)} \cdot 1.1^{-10}$$

$$= 1.8/2.6$$

$$= 0.69 \text{ d}^{-1}$$

$$\theta_c = \frac{0.1 + 0.015}{0.6 \cdot 0.69 \cdot 0.015 - 0.033(0.115)}$$

$$= 47.6 \text{ d}$$

	50th and (90th) Percentile Values		
Parameter	AS	FSB	ASB
Conventional			
COD removal, %	54 (65)	55 (78)	57 (68)
BOD removal, %	96 (98)	96 (98)	96 (98)
NH_4-N_{effl}, mg/L	1.5 (10.1)	0.25 (5.8)	0.25 (4.5)
$(NO_2 + NO_3)$-N_{effl}, mg/L	1.4 (8.0)	4.3 (11.1)	8.0 (13.2)
VSS_{effl}, mg/L	32 (110)	62 (200)	75 (200)
AOX			
Total AOX removal, %	22 (1.7)	43 (1.3)	40 (1.3)
Filt-AOX removal, %	28 (1.5)	48 (1.2)	45 (1.2)
(Nonfilt-AOX/total AOX)$_{effl}$, %	8 (1.9)	8 (1.9)	8 (1.9)
(Nonfilt-AOX/CSS)$_{effl}$, mg/g	45 (2.8)	28 (1.9)	20 (2.6)

TABLE 6.13 Removal Performance Summary for Conventional Pollutants and AOX

	Reduction (%) for Bleaching Process	
Parameter	Conventional	Oxygen
AOX	22*	40*
Chlorinated Phenols		
Phenol	39	45
Guajacols	41	79
Catacols	50	63

*Based on influent AOX concentrations of 136 and 57 mg/L.
Source: Adapted from Nevalainen et al., 1991.[31]

TABLE 6.14 Comparison of Conventional and Oxygen Bleaching on Activated Sludge Performance (Kraft Hardwood)

In practice in a majority of cases, a high degree of back-mixing takes place so that conditions in a plug flow tank may approach that of complete mixing. However, in order to take advantage of the kinetics of priority pollutant removal, multiple basins in series will offer distinct advantages over a single complete mix basin.

Equation (6.38) can be applied to predict the performance of multiple reactors in series as shown by the following example.

Example 6.10. Design a single-stage and a three-stage activated sludge reactor to reduce phenol. Compute the effluent phenol level in each case.

$S_o = 10$ mg/L
$bX_d = 0.05$
$\theta_c = 10$ d
$q_m = 0.6$
$K_S = 0.2$
$a = 0.6$

Solution For the CMAS (one stage), $X_v t = 39.44$ mg/(L · d) is computed from Eqs. (6.35) and (6.36).

$$B = 0.6(39.44) + 0.2 - 10$$
$$= 13.86$$

$$S = \frac{-13.86 + [(13.86)^2 + 4 \cdot 10 \cdot 0.2]^{1/2}}{2}$$
$$= 0.14 \text{ mg/L}$$

Hence, S for a single complete mix basin is 0.14 mg/L. We can now compute the performance of three basins in series. Each basin is considered as a complete mix basin in which the retention time is based on $(Q + R)$. The effluent from the third basin is assumed negligible. The concentration entering the first basin with a 50 percent recycle is

$$S_o = \frac{10}{1.5}$$
$$= 6.67 \text{ mg/L}$$

For the same total basin volume as in the complete mix case, $X_v t$ is reduced by a factor of 1.5:

$$X_v t = 39.44/3 \cdot 1.5$$
$$= 8.76 \text{ mg/(L·d)}$$

$$B = 0.6 \cdot 8.76 + 0.2 - 6.67$$
$$= -1.21$$

$$S = \frac{+1.21 + \left[1.21^2 + 4 \cdot 0.2 \cdot 6.67\right]^{1/2}}{2}$$
$$= 1.9 \text{ mg/L}$$

In like manner, the effluent (S) from basin 2 is computed as 0.105 mg/L and from basin 3 as 0.005 mg/L.

6.4 Effect of Temperature

Variations in temperature affect all biological processes. There are three temperature regimes: the mesophilic over a temperature range of 4 to 39°C, the thermophilic which peaks at a temperature of 55°C, and the psychrophilic which operates at temperatures below 4°C. For economic and geographical reasons, most aerobic biological treatment

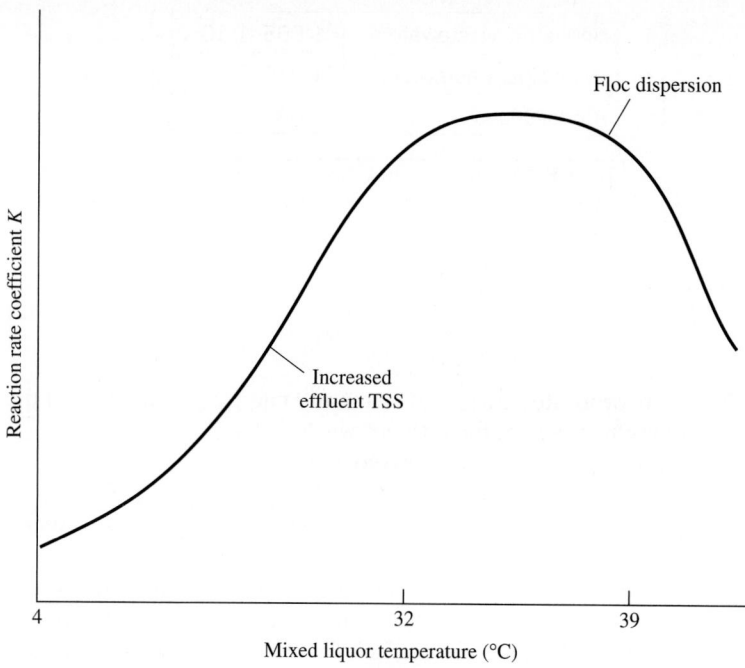

Figure 6.35 Effect of temperature on biological oxidation rate constant K.

processes operate in the mesophilic range, which is shown in Fig. 6.35. In the mesophilic range, the rate of the biological reaction will increase with temperature to a maximum value at approximately 31°C for most aerobic waste systems. A temperature above 39°C will result in a decreased rate for mesophilic organisms.

The liquid temperature therefore can have a significant effect on the organic removal velocity and subsequent organic removal efficiency. The most accepted approach is the applied use of the van't Hoff-Arrhenius equation:

$$\frac{d\ln k}{dT} = \frac{E}{RT^2} \tag{6.40}$$

where k = organic removal velocity
E = energy of activation
R = gas constant
T = temperature, °C

or

$$\ln \frac{k_1}{k_2} = \frac{E(T_2 - T_1)}{RT_1 T_2} \tag{6.41}$$

TABLE 6.15 Temperature Coefficient θ

Industrial wastewaters	1.065–1.10
Municipal wastewaters	1.015
Endogenous	1.04

or

$$\frac{k_1}{k_2} = \theta^{T_2 - T_1} \tag{6.42}$$

where θ incorporates the gas constant and energy of activation of the molecular composite of the influent wastewater.

Over a temperature range of 4 to 31°C

$$K_t = k_{20}\,\theta^{T-20} \tag{6.43}$$

The θ value ranges from 1.01 for simple organic compounds and municipal effluents to 1.1 and over for complex and stable organic wastes. Table 6.15 presents typical θ values for industrial versus municipal wastewater and endogenous respiration of the biomass. The effect of temperature on the latter can be represented by

$$b_T = b_{20°C} \times 1.04^{T-20} \tag{6.44}$$

The effect of θ temperature and reaction rate for an agricultural chemicals wastewater and a bleached sulfite mill wastewater is shown in Figs. 6.36 and 6.37, respectively. The effect of temperature on municipal wastewater is not as significant compared to most in industrial wastewaters since most BOD is present as colloidal or suspended organics, which are biocoagulated with very little temperature effect.

As the θ relationship is exponential in related k values, the effects of temperature on activated sludge systems treating complex organic wastes which exhibit high θ values is obvious. For example:

Temperature differential (summer & winter) = 16°C

Wastewater A $\theta = 1.01$

Therefore, $k_1/k_2 = \dfrac{1}{1.01^{16}} = 88\% \dfrac{\text{winter}}{\text{summer}}$ efficiency

Wastewater B $\theta = 1.10$

Therefore, $k_1/k_2 = \dfrac{1}{1.10^{16}} = 22\% \dfrac{\text{winter}}{\text{summer}}$ efficiency

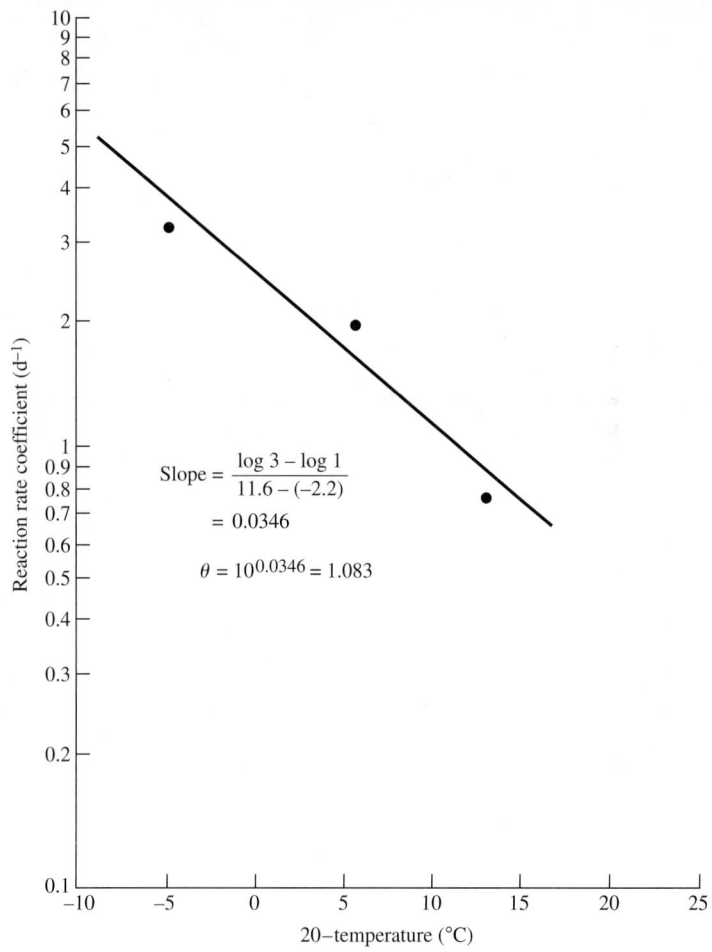

Figure 6.36 Effect of temperature on the reaction rate coefficient K for an agricultural chemicals wastewater.

As noted, elevated temperatures in an activated sludge system can have correspondingly negative effects on efficiency, probably attributable to denurturing the enzymatic reactions in the biochemical transformations. This is illustrated in Figs. 6.38 and 6.39.

Decreasing aeration basin temperature will also cause an increase in effluent suspended solids. The solids are of a dispersed nature and are nonsettleable. For example, the Union Carbide plant at South Charleston, West Virginia, had an effluent suspended solids of 42 mg/L during the summer and 104 mg/L during the winter. Removal of these suspended solids requires the addition of coagulating chemicals.

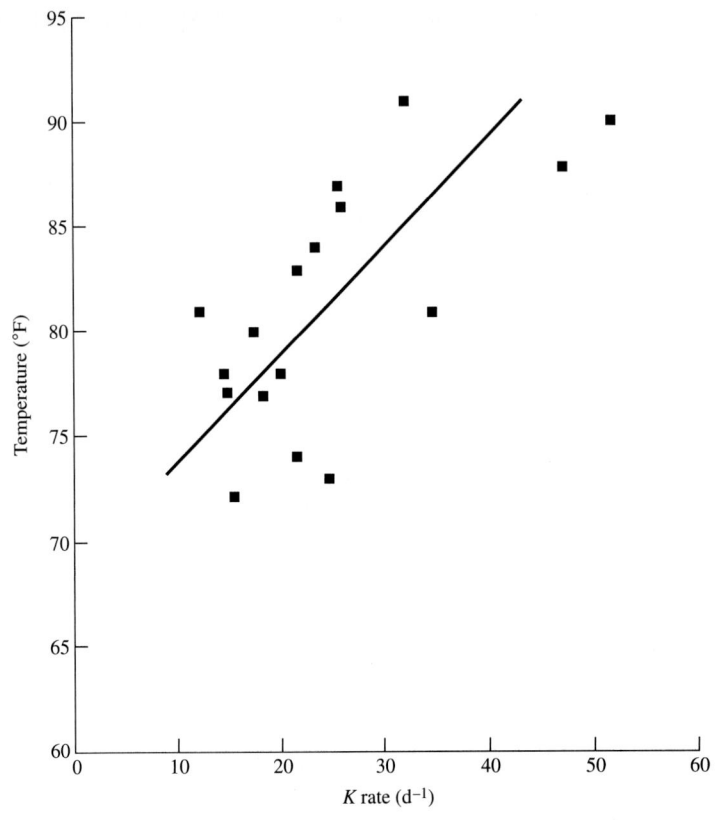

FIGURE 6.37 Effect of temperature on the reaction rate for a bleached sulfite mill wastewater.

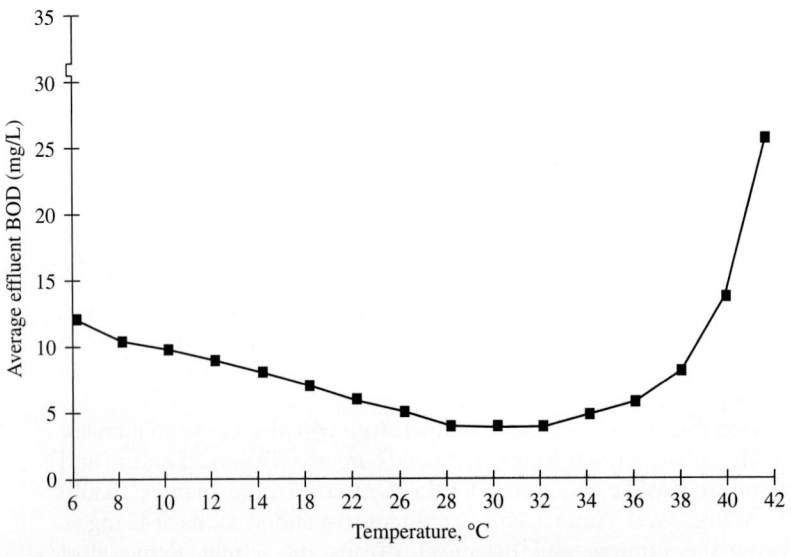

FIGURE 6.38 Effluent BOD as a function of liquid temperature.

Principles of Aerobic Biological Oxidation

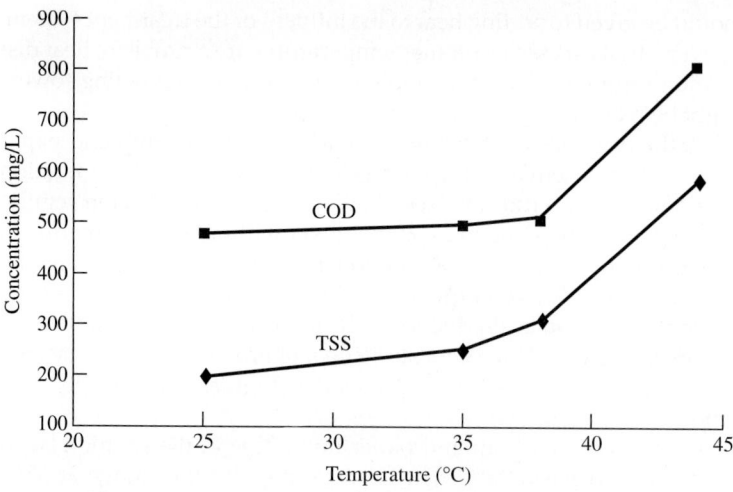

FIGURE 6.39 Effect of mixed liquor temperature on effluent quality of a Synfuels waste.

Figure 6.40 illustrates an increase in effluent TSS for an activated sludge system during cold weather operation in late January and early February.

At temperatures above 96°F (35.5°C) there is a deterioration in the biological floc. Protozoa have been observed to disappear at 104°F (40°C) and a dispersed floc with filaments to dominate at 110°F (43.3°C).

If the nature of the wastewater and severity of winter temperatures result in a significant diminution of efficiency, then consideration

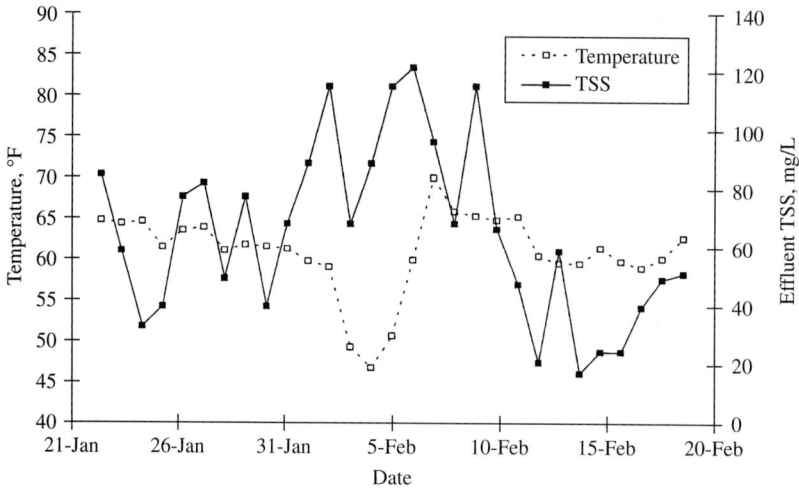

FIGURE 6.40 Effect of temperature on effluent TSS for an activated sludge system.

should be given to adding heat to the influent of the treatment system. Conversely, if excessive summer temperatures are a problem, heat dissipation through selected aeration/cooler systems or cooling towers might be required.

In the past, hot wastewaters such as those in the pulp and paper industry were pretreated through a cooling tower so that the aeration basin temperature did not exceed 35°C. Recent air pollution regulations preclude stripping in a cooling tower without off-gas treatment. As a result, in many cases, the aeration basins must be covered and the off-gas treated. The consequence of this is a substantial temperature rise in the aeration basin due to the heat release from the exothermal biological reaction. In a recent study on a pharmaceutical wastewater, the average heat release was calculated as high as 5000 Btu/lb COD removed.

In one case of a pulp and paper mill effluent, the aeration basin temperature reached 43°C. The characteristics of the sludge at 35°C and 43°C are shown in Fig. 6.41. One can see in Fig. 6.41 the absence of protozoans and the presence of filaments and dispersed floc at 43°C. The relationship between zone settling velocity and temperature is shown in Fig. 6.42. The net effect is a flux limitation on the final clarifier, resulting in decreased plant performance. The effect of temperature on solids flux rate is illustrated by Fig. 6.43.

As demonstrated in the figure, high solids flux rates can be experienced at higher temperatures allowing for higher solids loadings, provided temperatures are not excessive.

In a second case of an agricultural chemicals wastewater, floc dispersion resulted at an aeration basin temperature of 36°C. In order to maintain plant performance, massive polymer doses were required. The problem was solved by installing heat exchangers on the influent wastewater so that the influent temperature did not exceed 30°C.

Example 6.11. Estimate the maximum influent temperature for a wastewater with the following characteristics.

Influent degradable COD, S_o = 2580 mg/L
Effluent degradable COD, S_e = 70 mg/L
Reaction rate coefficient K = 5.0 d^{-1}
MLVSS, X_v = 3000 mg/L
Wastewater flow Q = 1 million gal/d

Solution The required aeration detention time can be calculated:

$$t = \frac{S_o(S_o - S_e)}{KX_v S_e}$$

$$= \frac{2580\,(2580 - 70)}{5 \cdot 3000 \cdot 70}$$

$$= 6.1 \text{ d}$$

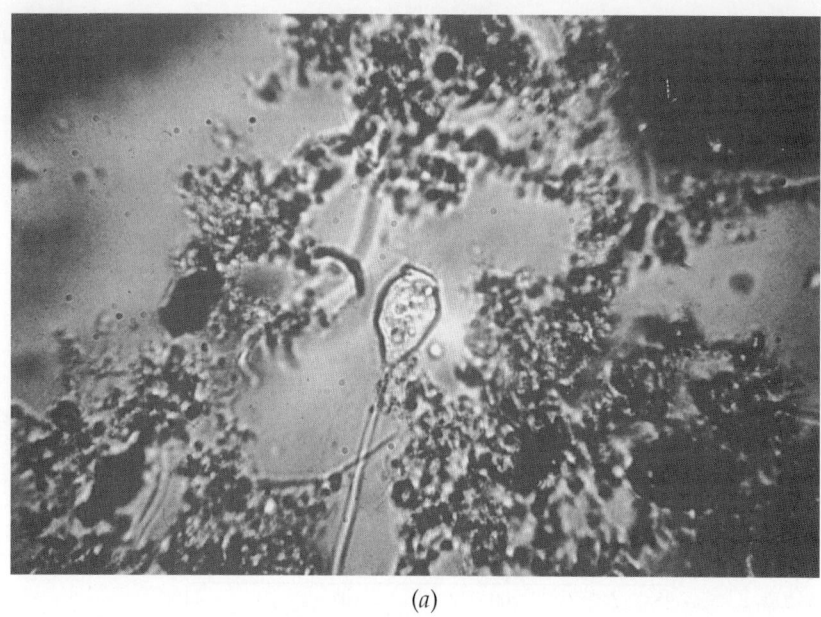

(a)

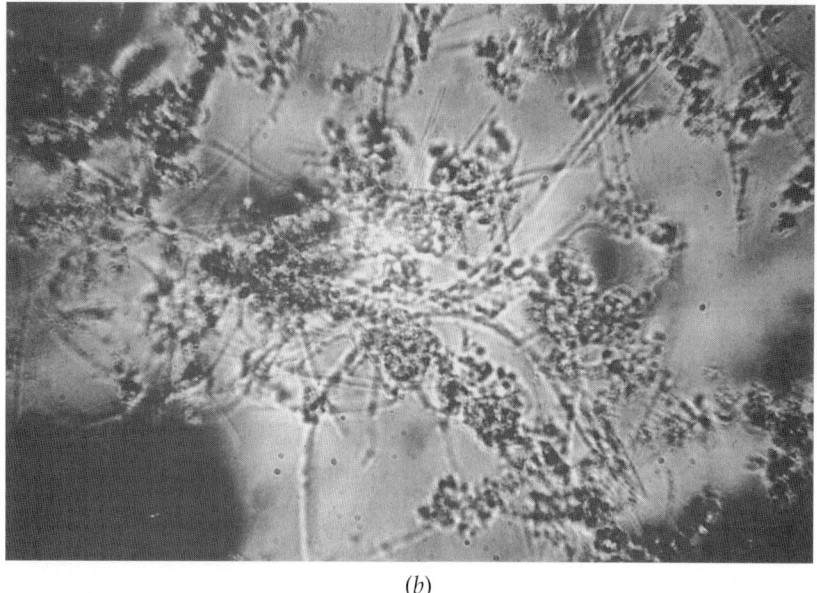

(b)

Figure 6.41 Sludge characteristics at (a) 35°C and (b) 43°C.

292 Chapter Six

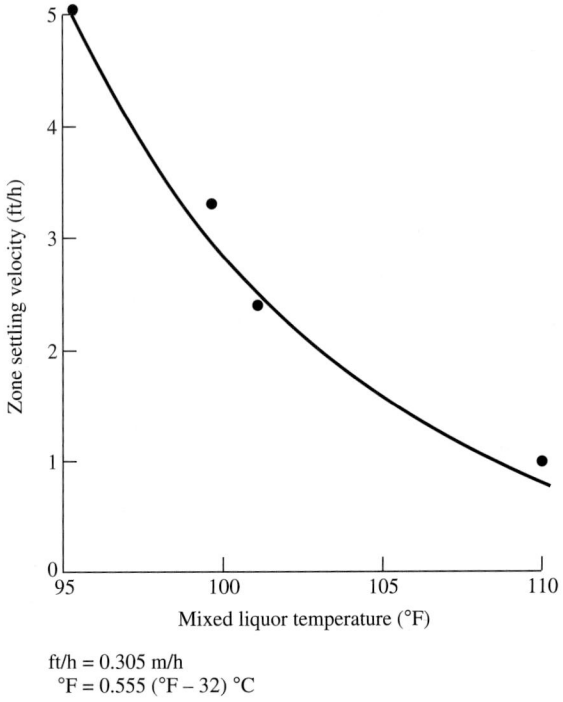

ft/h = 0.305 m/h
°F = 0.555 (°F − 32) °C

FIGURE 6.42 Effect of mixed liquor temperature on the zone settling velocity of activated sludge—pulp and paper wastewater.

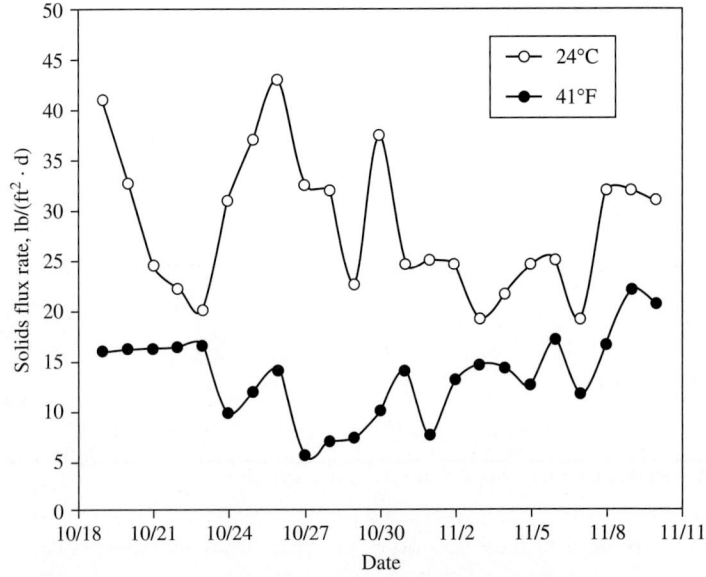

FIGURE 6.43 Effect of temperature on solids flux rate.

The F/M is:

$$F/M = \frac{S_o}{X_v t}$$

$$= \frac{2580}{3000 \cdot 6.1}$$

$$= 0.14 \text{ d}^{-1}$$

The aeration basin volume is 6.1 million gal. The heat loss from a concrete basin can be calculated assuming a total basin surface area of 56,650 ft².

If we assume the basin temperature at 96°F and an average air temperature of 85°F during the summer months, the heat loss can be estimated as:

$$q \text{ (Btu/h)} = UA\Delta T$$

where U = overall heat transfer coefficient, assumed as 0.35 Btu/ft² · °F · h
A = exposed surface area
ΔT = difference between the basin and air temperature
$= 0.35 \cdot 56{,}650 \cdot 11$
$= 218{,}102$ Btu/h $= 5.2 \times 10^6$ Btu/d (5.5 KJ/d)

The heat generated can be calculated by assuming that 5000 Btu is generated per pound of COD removed. COD removed is:

$$(2580 \cdot 70) \cdot 1 \cdot 8.34 = 20{,}933 \text{ lb/d } (9500 \text{ kg/d})$$

The Btu generated is:

$$20{,}933 \cdot 5000 \text{ Btu/16 COD}_{rem} = 104.6 \times 10^6 \text{ Btu/d}$$

The net Btu increase is:

$$104.6 - 5.2 = 99.4 \times 10^6 \text{ Btu/d}$$

The temperature increase can be calculated as:

$$\frac{99.4 \times 10^6 \text{ Btu/d}}{1 \text{ Btu/lb} \cdot °F \times 8.34 \times 10^6} = 11.9°F \quad (6.6°C)$$

Under these conditions, the maximum allowable influent temperature is:

$$96 - 11.9 = 84.1°F \quad (29°C)$$

Effect of pH

A relatively narrow effective pH range will exist for most biooxidation systems. For most processes, this covers a range of pH 5 to 9 with optimum rates occurring over the range pH 6.05 to 8.5. It is significant to note that this relates to the pH of the mixed liquor in contact with the biological growths and not the pH of the waste entering the system. The influent waste is diluted by the aeration tank contents and is neutralized by reaction with the CO_2 produced by microbial respiration. In the case of both caustic and acidic wastes, the end product is bicarbonate (HCO_3^-), which effectively buffers the aeration system at near pH 8.0. Application of the complete mixing concept is essential in order to take advantage of these reactions. For example, the oxidation of sulfonates will result in formation of sulfuric acid.

The amount of caustic that can be present in a wastestream is related to the BOD removal, which in turn will determine the CO_2 produced for reaction with the caustic. This can be represented as $OH + CO_2 \rightarrow HCO_3$ at pH ~ 8. If we consider 0.9 lb or kg of CO_2 produced per lb or kg of COD removed (assuming conventional loadings) and 70 percent of this is reactive with caustic alkalinity present, then 0.63 lb or kg of caustic alkalinity (as $CaCO_3$) will be neutralized per pound or kilogram of COD removed.

Since the oxidation of organic acids results in the production of CO_2, the permissible concentration of organic acids in the wastestream is related to the reaction rate of acid degradation to CO_2. For example, $HAC \rightarrow CO_2 + H_2O$ at pH ~ 6.5.

As long as the buffering capacity of the process is maintained, the pH of the aeration tank contents should remain near pH 8.0, even under conditions of fluctuating caustic or acidic loads.

Nitrification may result in a significant reduction in pH if insufficient alkalinity is not available for reaction. Approximately 2.15 mg alkalinity/mg N oxidized is needed according to the reaction:

$$NH_4^+ \rightarrow NO_3 + 2H^+$$

Toxicity

Toxicity in biological oxidation systems may have any of several causes:

1. An organic substance, such as phenol, which is toxic in high concentrations, but biodegradable in low concentrations

2. Substances such as heavy metals that have a toxic threshold depending on the operating conditions

3. Inorganic salts and ammonia, which exhibit a retardation at high concentrations

The toxic effects of organics can be minimized by employing the complete mixing system, in which the influent is diluted by the aeration tank contents and the microorganisms are in contact only with the effluent concentration. In this way wastes with concentrations many times the toxic threshold can be successfully treated. Toxicity may also be decreased by increasing solids retention time and reducing influent COD as shown in Figs. 6.44 and 6.45, respectively. Toxicity was reduced to acceptable levels by increasing the SRT from 4.5 to

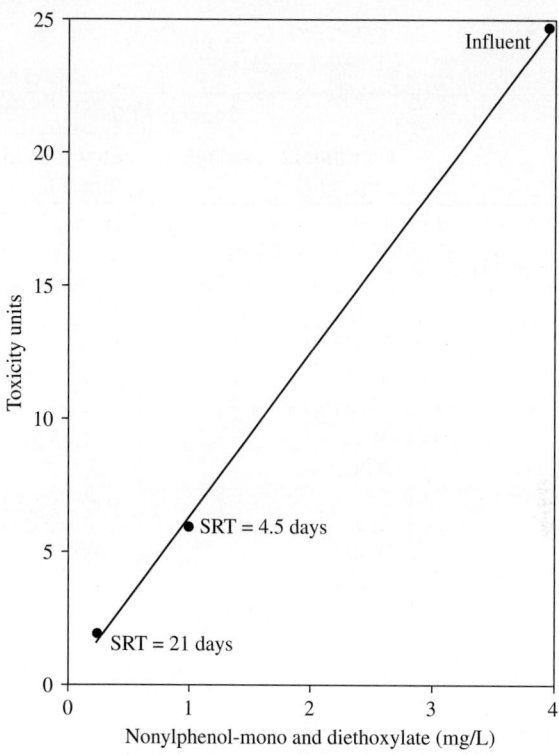

FIGURE 6.44 Effects of sludge age on toxicity reduction for nonylphenolics.

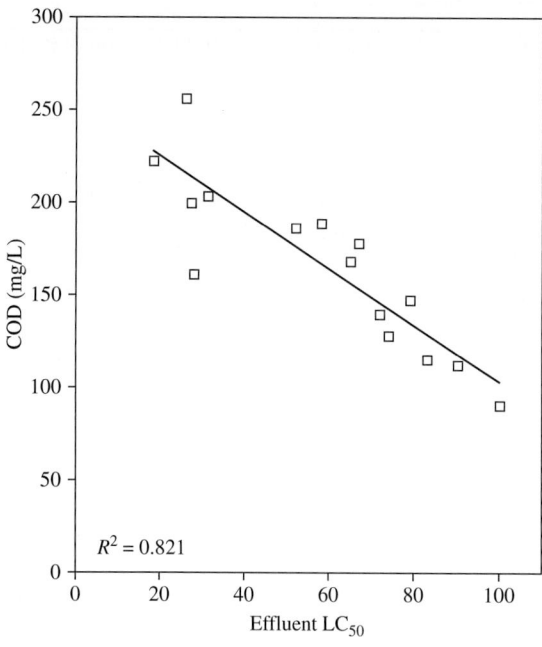

FIGURE 6.45 Toxicity/COD relationship for petroleum refining chemical plant effluent.

Metal	Concentration*	
	Continuous Loading (mg/L)	Shock Loading (mg/L)
Cadmium	1	10
Chromium (hexavalent)	2	2
Copper	1	1.5
Iron	35	100
Lead	1	–
Maganese	1	–
Mercury	0.002	0.5
Nickel	1	2.5
Silver	0.03	0.25
Zinc	1 to 5	10
Cobalt	>1	–
Cyanide	1	1–5
Arsenic	0.7	–

*The specific level can be variable, depending on biological acclimation, pH, sludge age, and degree of metal complexation.

TABLE 6.16 Threshold Concentrations for Heavy Metal Inhibition of Biological Treatment Processes

21 days in the case of a nonylphenolic wastewater. In the case of a petroleum wastewater LC_{50} toxicity was reduced as the influent COD was reduced owing to a decrease in SMP production. In some cases, powered activated carbon (PAC) may be needed to reduce inhibition to acceptable levels.

Heavy metals exhibit a toxicity in low concentrations to biological sludges. Acclimation of the sludge to the metal, however, will increase the toxic threshold considerably. Threshold concentrations that inhibit biological treatment processes are given in Table 6.16.

While an acclimated biological process is tolerant of the presence of heavy metals, the metal will concentrate in the sludge by complexing with the cell wall. Metal concentrations in the sludge as high as 4 percent have been reported. Data on heavy metal removal from petroleum refinery wastewater are shown in Table 6.17. Data on removal of metals at low concentration in a municipal plant are shown in Fig. 6.46. This in turn creates problems relative to ultimate sludge disposal. The accumulation of copper on activated sludge with increasing sludge age is shown in Fig. 6.47.

	Activated Sludge Plant	
Heavy Metal	Influent (mg/L)	Effluent (mg/L)
Cr	2.2	0.9
Cu	0.5	0.1
Zn	0.7	0.4

TABLE 6.17 Heavy Metals Removal in the Activated Sludge Process Treatment of Petroleum Refinery Wastewater

High concentrations of inorganic salts are not toxic in the conventional sense, but rather exhibit progressive inhibition and a decrease in rate kinetics. Figure 6.48 demonstrates the effect of increasing TDS concentration on effluent soluble BOD for a chemical plant effluent. The reduction of the BOD removal rate constant with increasing mixed liquor TDS is shown by Fig. 6.49. Biological sludges, however, can be acclimated to high concentrations of salt. Processes are successfully operating with as high as 6 percent salt by weight. Frequently, high salt content will increase the effluent suspended solids. The effect of high salt concentration on suspended solids is shown in Table 6.18, which indicates that most of the biomass was nonflocculated. Monovalent ions such as Na^+ and K^+ will disperse the biological flocs while divalent ions such as Ca^{2+} and Mg^{2+} will tend to aid flocculation.[33]

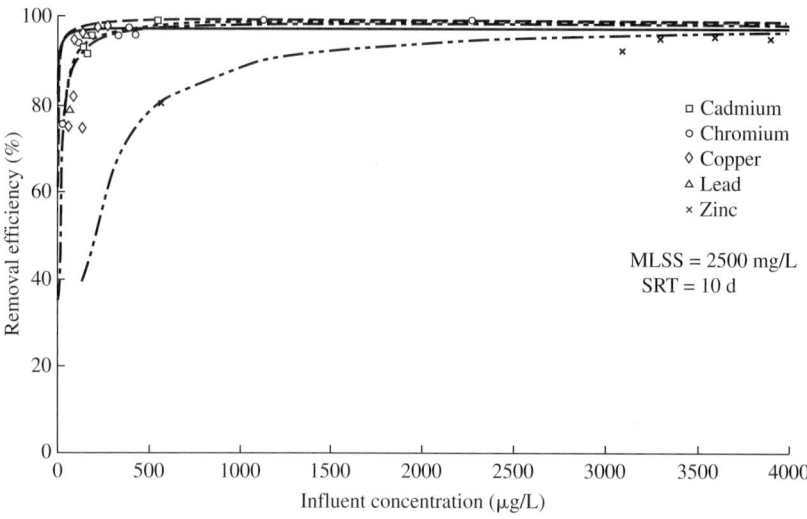

FIGURE 6.46 Heavy metals removal from municipal sewage in the activated sludge process.

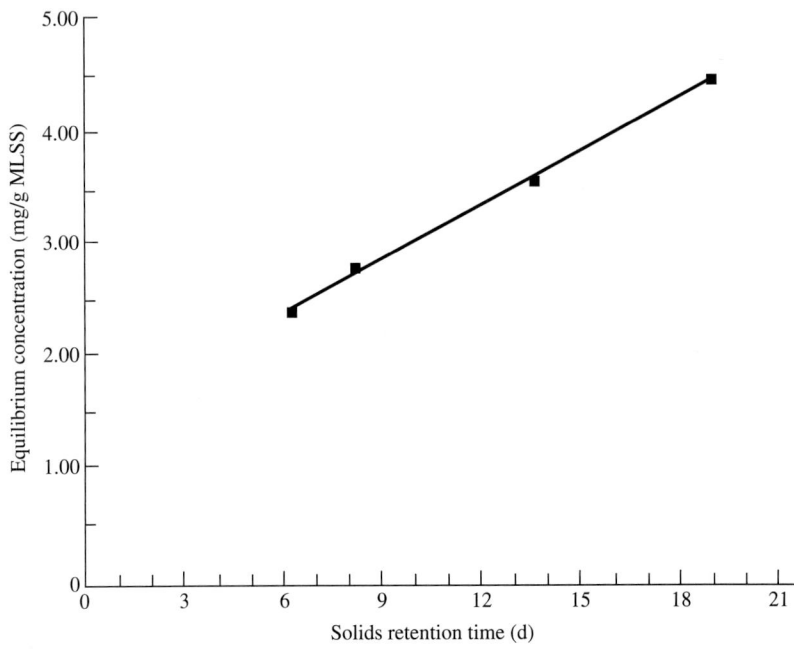

FIGURE 6.47 Copper accumulation in the activated sludge process as a function of solids retention time.

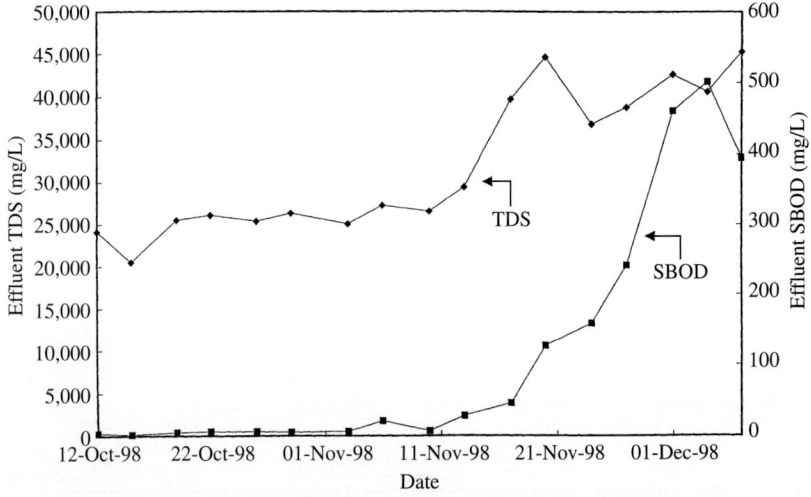

FIGURE 6.48 Effect of TDS on effluent BOD.

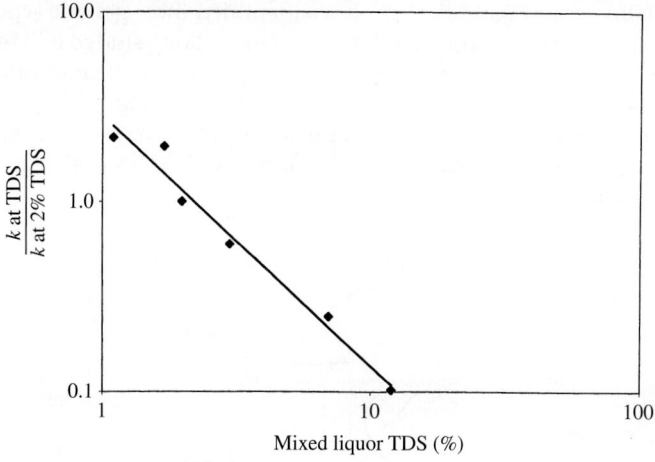

FIGURE 6.49 Effect of mixed liquor TDS on BOD removal rate constant.

Influent		Effluent			
COD (mg/L)	COD$_s$ (mg/L)	COD$_s$ (mg/L)	VSS (mg/L*)	VSS (mg/L†)	TDS (mg/L)
6437	1182	181	50	597	44,000

*1.5-micron filter
†0.45-micron filter

TABLE 6.18 Effect of High Salt Content on Process Performance

6.5 Sludge Quality Considerations

One of the factors essential to the performance of the activated sludge process is effective flocculation of the sludge, with subsequent rapid settling and compaction. McKinney[11] related flocculation to the food/microorganism ratio and showed that certain organisms normally present in activated sludge deflocculate rapidly under starvation conditions. More recently, it has been shown that flocculation results from the production of a sticky polysaccharide slime layer to which organisms adhere. Flagellates are also entrapped in this slimy material. Filamentous organisms are present in most activated sludges (exceptions are found in the chemical and petrochemical industry).

Palm, Jenkins, and Parker[34] have identified three generic types of activated sludge, as shown in Fig. 6.50. Nonbulking sludge will result from plug flow or selector plant configuration or from complex organic wastewaters. Bulking sludges result from degradable wastewaters treated in a complete mix process or from oxygen or nutrient deficiencies. Pin floc usually results from low F/M (long sludge age) operation.

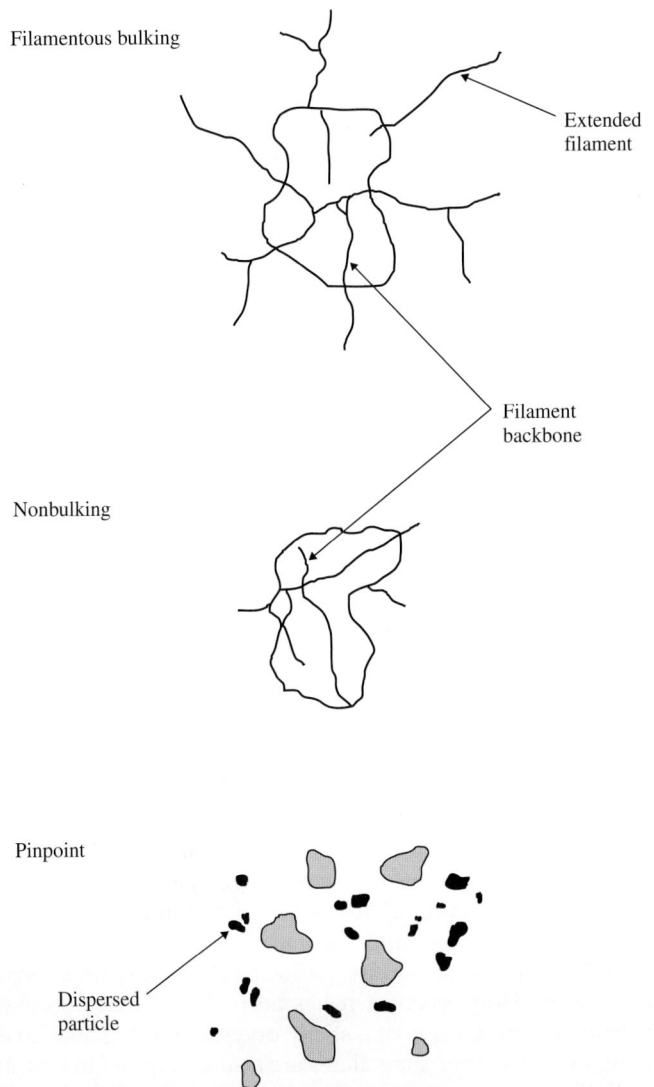

FIGURE 6.50 Activated sludge characteristics.

Suggested Causative Conditions	Indicative Filament Types
Low dissolved oxygen (DO)	Type 1701, S. natans, H. hydrossis
Low food/mass ratio (F/M)	M. parvicella, H. hydrossis, Nocardia spp., types 021N, 0041, 0675, 0092, 0581, 0961, and 0803
Septic wastewater sulfides	Thiothrix spp, Beggiatoa spp., and type 021N
Nutrient deficiencies	Thiothrix spp., S. natans, type 021N, and possibly H. hydrossis and types 0041 and 0675
Low pH	Fungi

TABLE 6.19 Dominant Filament Types as Indicators of Activated Sludge Operational Problems

A number of filamentous organisms have been identified in activated sludge treating municipal and industrial wastewaters. Depending on process operating conditions, one or more of these organisms may dominate in the process, as shown in Table 6.19. Identification and control of filaments have been reported. Filament types found in various industrial wastewaters are shown in Table 6.20.[35] The occurrence and rank of filaments causing bulking in activated sludge treating a pulp and paper wastewater are tabulated in Table 6.21. Proper process design and operation should not permit the filaments to overgrow the floc formers.

Filamentous overgrowth is affected by:

1. *Wastewater composition.* Wastewaters containing glucoselike saccharides (glucose, saccharin, lactose, maltose, and so on) promote filamentous growth while laundry, textile, and complex chemical wastewaters inhibit filamentous growth in a completely mixed system. In general, the more readily degradable the substrate, the more prone the system is to filamentous bulking. This can broadly be related to the reaction rate coefficient K as shown in Table 6.22.

2. *Dissolved oxygen concentration.* Oxygen must diffuse into the floc in order to be available to the organisms within the floc. The depth of oxygen penetration within the floc depends on the bulk concentration of oxygen in the surrounding liquid and on the oxygen-utilization rate of the floc. The oxygen-utilization rate is proportional to the organic loading (F/M). Hence, as the organic loading increases, the dissolved oxygen

	S. natans	Type 1701	H. hydrossis	Type 021N	Thiothrix I and II	Type 1851	Type 0581	Type 0041	Type 0803	Type 0675	Type 0211	Type 0092	Type 0914	M. parvicella	N. limicola	Type 0411	Nocardia
Food processing and brewing	•	•	•	•	•												•
Textile					•	•		•		•		•					
Slaughterhouse and meat processing	•	•	•	•	•								•				•
Petrochemical	•	•	•		•												
Organic chemicals	•			•	•	•						•	•				•
Pulp and paper mill		•	•	•	•			•	•	•	•	•	•		•	•	

Source: Adapted from Richard, 1997.

TABLE 6.20 Filament Types Found in Industrial Wastewaters[35]

Rank	Filament Type	Number of Plants	Percent
Occurrence in 29 industry plants during 1982–1990			
1	Type 0675	16	55
2	Type 1701	8	28
2	Type 1851	8	28
3	Thiothrix II	7	24
3	Type 0041	7	24
4	Nostocoida limicola II	5	17
Occurrence in 80 industry plants during 1996			
1	Thiothrix II	44	55
2	Thiothrix I	36	45
3	Nostocoida limicola II	20	25
4	Type 0914	19	24
5	Haliscomenobacter hydrossis	18	23
6	Nostocoida limicola III	10	13

Source: Adapted from Richard, 1997.

TABLE 6.21 Occurrence and Rank of Filament Types as a Cause of Bulking in Pulp and Paper Activated Sludge Plants

necessary to maintain a fully aerobic floc also increases. The thin filaments (1 to 4 μm) can readily obtain oxygen at concentrations less than 0.1 mg/L. The relationship between dissolved oxygen concentration to maintain a fully aerobic floc and F/M developed from the data of Palm, Jenkins, and Parker[34] is shown in Fig. 6.51. Performance data from a pulp and paper mill is shown in Fig. 6.52.

With degradable substrates at low concentrations, the filaments tend to grow. This explains why complete mix systems with low mixed liquor substrate concentrations favor filamentous growth.

Wastewater Characteristics	$K_{20°C}$, d^{-1}
Readily degradable (food process, brewery)	16–30
Moderately degradable (petroleum, pulp and paper)	8–15
Poorly degradable (chemical, textile)	2–6

TABLE 6.22 Composite Reaction Rate Coefficients for Industrial Wastewaters

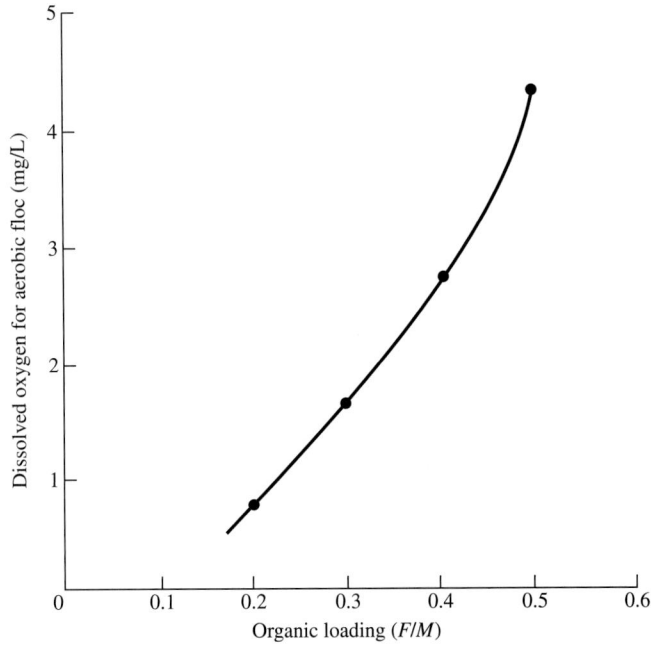

Figure 6.51 Relationship between dissolved oxygen and F/M for an aerobic floc. (*Adapted from Palm, Jenkins, and Parker, 1980.*)

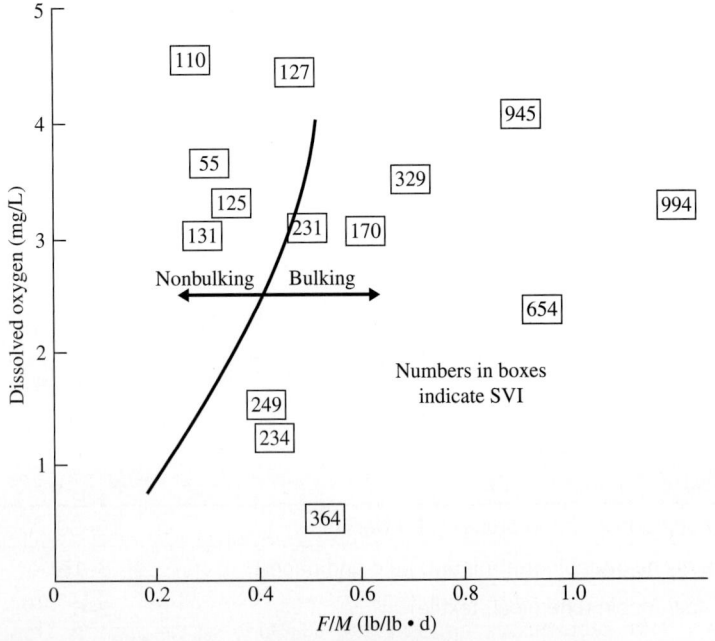

Figure 6.52 Relationship between F/M and dissolved oxygen relative to sludge bulking for a pulp and paper mill wastewater.

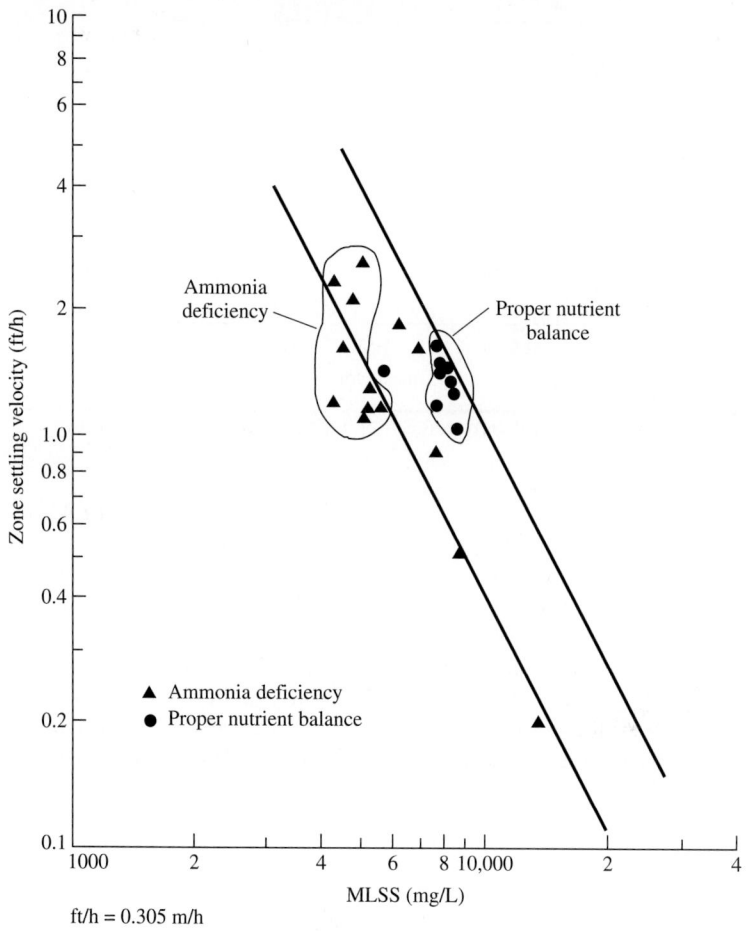

Figure 6.53 Effect of ammonia deficiency on zone settling velocity.

One of the more common causes of filamentous bulking in industrial wastewaters is inadequate nitrogen or phosphorus. There are numerous examples, particularly in the pulp and paper industry, in which severe filamentous bulking resulted from inadequate nitrogen. This is shown in Fig. 6.53. Restoration of adequate nitrogen restored a flocculated sludge within three sludge ages. Studies on a Wisconsin pulp mill indicated a minimum concentration of NH_3^-N in the effluent of 1.5 mg/L to favor zoogleal growth. Other studies indicate that higher ammonia concentrations may be required in some cases. Minimum soluble phosphorus concentrations in the effluent of 0.5 mg/L have been reported as required for optimal zoogleal growth. This is illustrated by Fig. 6.54.

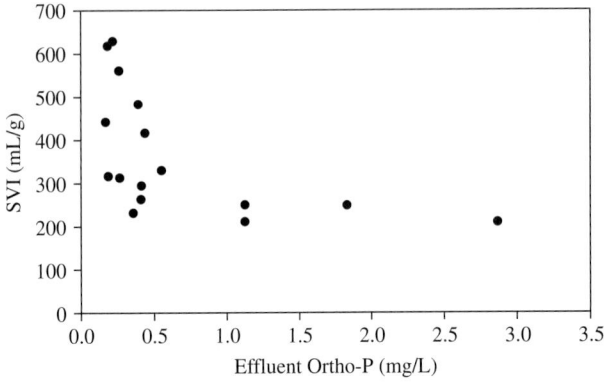

FIGURE 6.54 Correlation between SVI and effluent Ortho-P-concentration.

Therefore deficiency in substrates, such as the macro- or micronutrient concentration, residual soluble BOD, and/or dissolved oxygen concentration in the biological floc, can promote filamentous growth and sludge bulking. To illustrate these effects, consider the transfer of dissolved oxygen to the hypothetical biological floc particles, as illustrated in Fig. 6.55. Oxygen must diffuse from the bulk liquid through the floc in order to be available to the organisms within the interior of the floc particle. As it diffuses, it is consumed by the organisms within the floc. If there is an adequate residual of

FIGURE 6.55 A mechanism of sludge bulking.

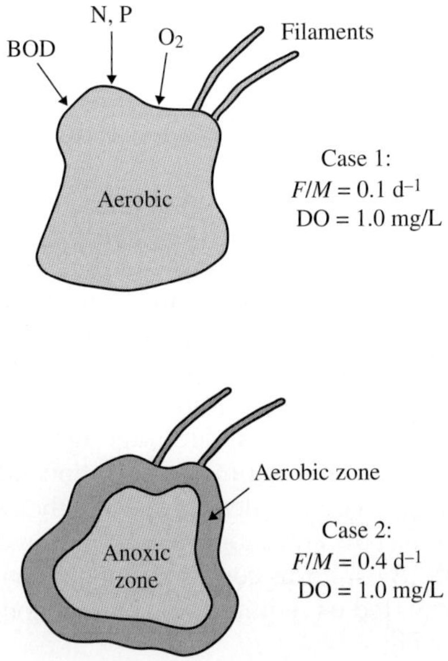

Case 1:
$F/M = 0.1$ d^{-1}
DO = 1.0 mg/L

Case 2:
$F/M = 0.4$ d^{-1}
DO = 1.0 mg/L

dissolved oxygen (and nutrients and organics), the rate of growth of the floc-forming organisms will exceed that of the filaments, and a flocculant well-settling sludge will result. If there is a deficiency in any of these substrates, however, the filaments, having a high surface area to volume ratio, will have a "feeding" advantage over the floc formers and will proliferate because of their higher growth rate under the adverse conditions.

Considering Case 1 in Fig. 6.55 at an F/M of 0.1 d^{-1}, the oxygen utilization rate is low, and even with a bulk liquid dissolved oxygen concentration of 1.0 mg/L, oxygen will fully penetrate the floc. Under these conditions, the floc formers will outgrow the filaments. In Case 2, the F/M is increased to 0.4 d^{-1}, causing a corresponding increase in oxygen uptake rate. If the bulk mixed liquor dissolved oxygen is maintained at 1.0 mg/L, the available oxygen will be rapidly consumed at the periphery of the floc, thus depriving a large interior portion of the floc particle of oxygen. Since the filaments have a competitive growth advantage at low dissolved oxygen levels, they will be favored and will outgrow the floc formers.

In summary, the following principles apply:

- Most filaments can only degrade readily degradable substrates.
- All things being equal (i.e., adequate oxygen, nutrients, and substrate), the floc formers will outgrow the filaments. This may be a slow process, however, if filaments are already present.
- Most floc formers have the ability to biosorb readily degradable substrates while most filaments do not.

Filamentous Bulking Control

The bulk mixed liquor SBOD concentration must be sufficient to provide a driving force to penetrate the biological floc. In a complete mix basin, the concentration of soluble BOD in the mixed liquor is essentially equal to the effluent concentration and is, therefore, low (< 10 mg/L) for readily degradable wastewaters. As a result, substrate penetration of the floc is not achieved, and filaments dominate the interior floc population. In order to shift the population in favor of the floc formers, sufficient driving force must be developed to penetrate the floc and favor their growth. This can be achieved by a batch or plug flow operating configuration in which a high substrate gradient (driving force) exists. Maximum growth of floc formers occurs at the influent end of the plug flow basin or in the initial period of each feed cycle of a batch-activated sludge process. Complete-mix reactors may be modified by the addition of a selector as described in the following section. A schematic illustrating these various flow regimes is shown in Fig. 6.56.

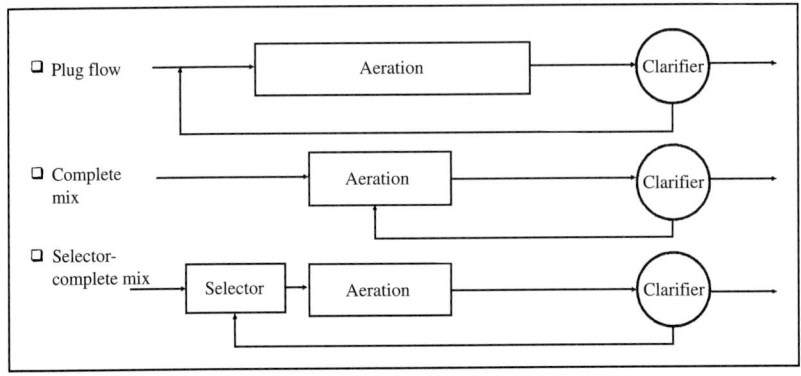

FIGURE 6.56 Types of activated sludge processes.

Biological Selectors

A biological selector may be used for filament control instead of a plug flow or batch treatment process. In the selector, a significant portion of the soluble substrate removal occurs by biosorption. Under these conditions, the substrate gradient is high and promotes the growth of floc formers over the filaments since they have a high "sorption" capacity, whereas the filamentous organisms do not. When the wastewater is discharged from the selector to the downstream CMAS, the soluble substrate concentration is relatively low and is available for utilization by both the floc formers and the filaments. The filaments do not predominate in the mixed liquor, however, since the principal mass of substrate removed in the selector has been initially directed to storage and subsequent growth of floc-forming biomass.

Results from parallel treatment studies of a grapefruit processing wastewater are shown in Table 6.23. Reactors 1 and 2 used an aerobic selector followed by a completely mixed aeration basin, whereas reactor 3 used a plug flow regime aeration basin. In all three cases, the SVI was below 100 mL/g, with effluent BOD concentrations ranging from 4 to 18 mg/L. The plug flow reactor, however, produced a sludge that was more readily dewatered and had superior thickening properties. A parallel complete mix activated sludge process (without selector) that was operated at the same organic loading rate as these systems suffered severe bulking problems and was shut down because it was inoperable.

Operational data representing the enhanced floc settleability in terms of SVI and effluent turbidity following the addition of an aerobic selector prior to a CMAS system is shown in Fig. 6.57.

Design of Aerobic Selectors

Several methods have been proposed for process design of the aerobic biological selector. Each of these is based on the selector F/M or the floc loading relationship for the wastewater-sludge mixture, as defined

	Aerobic Selector* with CMAS		Plug Flow Activated Sludge
	Reactor 1	Reactor 2	
Operating characteristics			
Influent BOD, mg/L	2543	3309	3309
Effluent BOD, mg/L	18	4	6
Influent COD, mg/L	4768	4460	4460
Effluent COD, mg/L	221	139	135
MLVSS, mg/L	3431	5975	5333
SRT, d	7.2	13.2	13.5
Temperature, °C	22	22	22
SVI, mL/g	71	67	69
F/M, d^{-1}	0.32	0.24	0.20
Sludge characteristics			
	Limiting Flux to Achieve a 1.5% Underflow		
	Solids Flux without Polymer, lb/ft² · d		Solids Flux with Polymer, lb/ft² · d
Selector with CMAS	5.5		47
Plug flow	24.5		62
Specific resistance			
	Polymer Dosage, lb/ton		Specific Resistance, s²/g
Selector with CMAS	3.2		190 × 10⁶
Plug flow	2.4		104 × 10⁶

*Floc load = 120 mg COD/g VSS.

TABLE 6.23 Treatment of a Grapefruit Processing Wastewater

by Eq. (6.6). The design objective is to provide sufficient biomass-wastewater contact time to remove a significant portion of the influent degradable substrate. If 60 to 75 percent of the influent degradable substrate is sorbed in the selector, then the subsequent metabolism and growth of the floc formers is usually adequate to establish a well-settling sludge. If less degradable substrate is removed because of an excessive floc load, then higher concentrations "leak" into the activated sludge reactor and support filamentous growth.

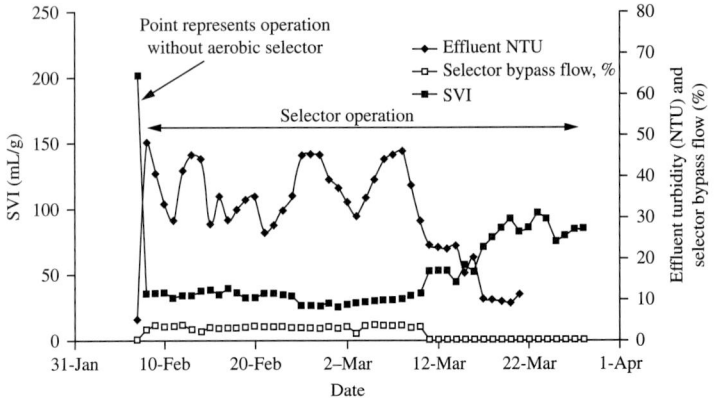

Figure 6.57 Selection operation for a deinking mill wastewater.

The results of multiple batch floc load tests on a readily degradable pulp and paper mill wastewater are shown in Fig. 6.58. These data are correlated according to the relationship

$$\frac{S}{S_o} = e^{-KFL^{-1}}$$

as shown in Fig. 6.59. These data were used to select a floc loading of 100 to 150 mg COD/g VSS for operation of the aerobic selector of a bench-scale selector CMAS system.

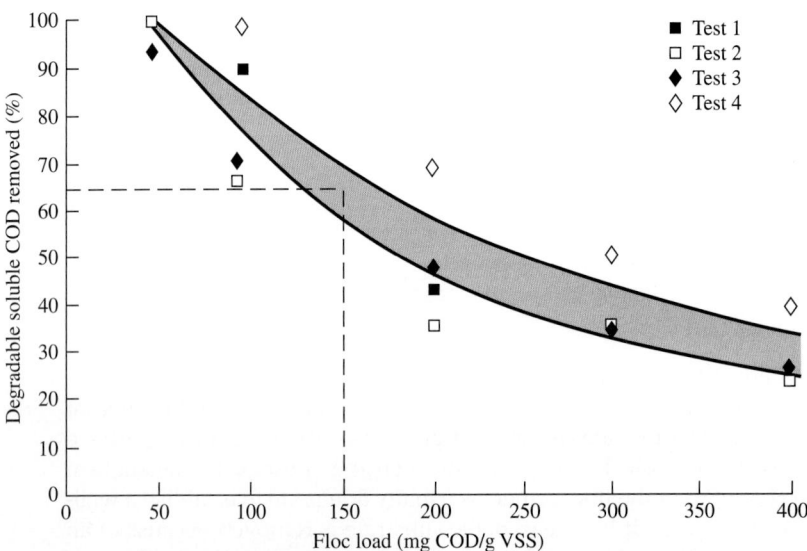

Figure 6.58 Floc load test results for a pulp and paper mill wastewater.

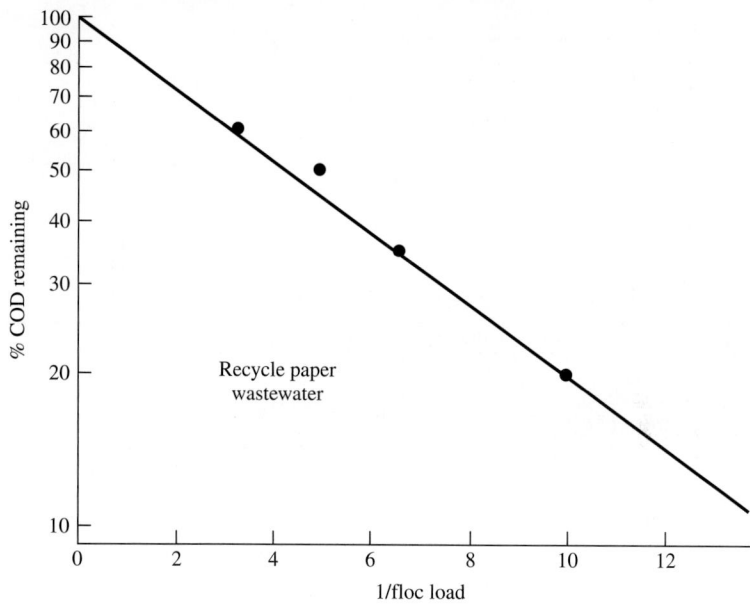

FIGURE 6.59 COD removal—floc load relationship.

Equation (6.6) can be rearranged to include mixed liquor recycle

$$\text{FL} = \frac{S_o}{rX_R + r_R X_v} \tag{6.45}$$

where S_o = the degradable COD, mg/L
 X_R = recycle VSS, mg/L
 X_v = mixed liquor VSS, mg/L
 r = sludge recycle ratio, R/Q
 r_R = internal recycle ratio, $Q_{R/Q}$

The aerobic selector design is shown in Example 6.12.

Example 6.12. Aerobic selector design for industrial wastewaters. Biosorption correlations are shown in Fig. 6.58. The design basis is 65 percent removal of sorbable COD in the selector. For the recycle paper case, design an aerobic selector for an influent COD of 2000 mg/L and an MLVSS of 4000 mg/L. The floc load is 150 mg COD/g VSS. The recycle sludge concentration is 8000 mg/L VSS.

Solution

$$r = \frac{X_v}{X_r - X_v} = \frac{4000}{8000 - 4000} = 1.0$$

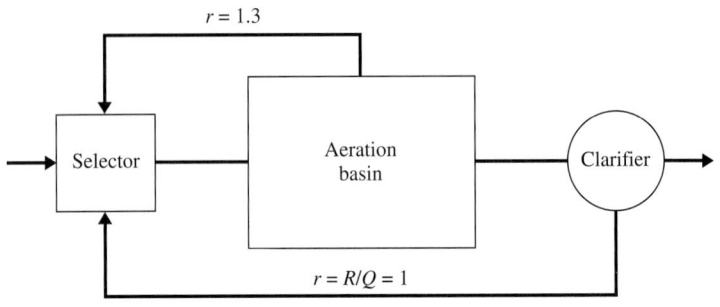

FIGURE 6.60 Selector flow sheet.

Rearranging Eq. (6.45) gives

$$r_R = \frac{S_o - FL_r X_R}{FLX_v}$$

$$= \frac{2000 - 0.15 \cdot 1 \cdot 8000}{0.15 \cdot 4000}$$

$$= 1.3$$

If the return sludge recycle rate is 100 percent, the internal recycle will be 130 percent. This is shown in Fig. 6.60. The selector detention time based on Q will be 0.825 h.

The sorption phenomenon provides the basis for selector design such that sufficient organics are sorbed for subsequent utilization and growth by the floc-forming organisms in preference to the growth of filamentous organisms.[36] After sorption is complete, a minimum aeration period must be provided to oxidize the sorbed organics. This is illustrated by plug flow data for a grapefruit processing wastewater, as shown in Fig. 6.61. When these date are replotted, the minimum required downstream aeration time can be determined as shown in Fig. 6.62. Metabolism of the sorbed substrate was completed at a substrate removal velocity of approximately 0.8 d^{-1}. The required minimum aeration time can then be calculated as

$$0.8 = \frac{S_r}{X_v t}$$

$$t = \frac{3600 \text{ mg/L}}{3618 \text{ mg/L} \cdot 0.8/d}$$

$$= 1.24 \text{ d}$$

Many filamentous organisms are aerobic and can be destroyed by prolonged periods of anaerobiosis. Most of the bacteria, on the other hand, are facultative and can exist for extended periods without oxygen. Although available data are somewhat contradictory, it

Principles of Aerobic Biological Oxidation

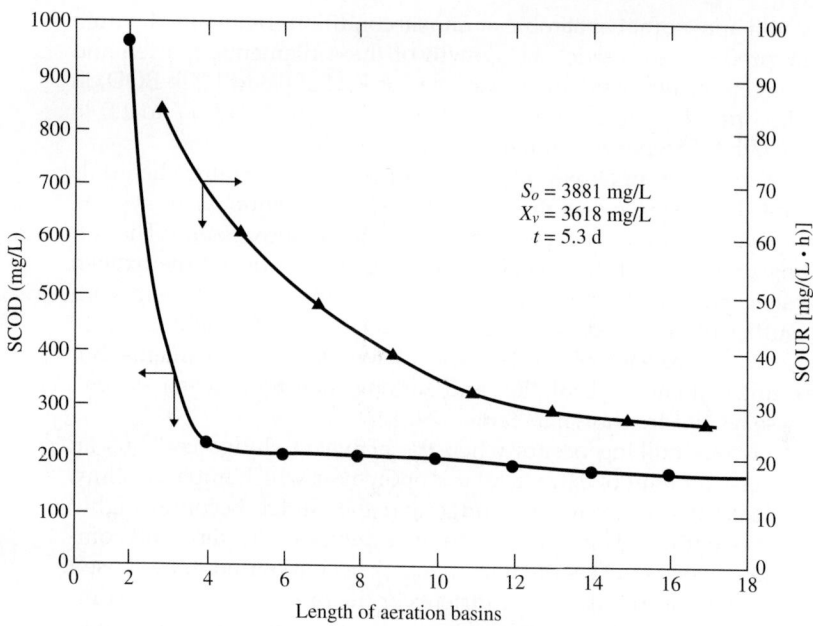

FIGURE 6.61 Change in SCOD and SOUR in a plug flow aeration basin.

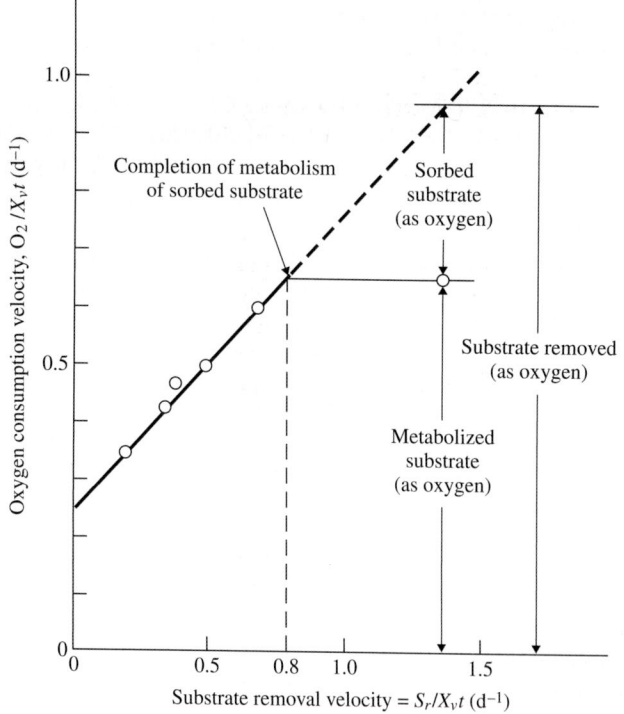

FIGURE 6.62 Stabilization time required for sorbed substrate.

would appear that anaerobic or anoxic conditions maintained within the process will restrict the growth of these filaments. Marten and Daigger[37] recommend an anoxic selector F/M of 0.8 to 1.2 lb BOD/lb MLSS per day at temperatures greater than 18°C and 0.7 to 1.0 lb BOD/lb MLSS per day at temperatures less than 18°C.

The use of an anoxic zone was studied at an organic chemicals plant. Three systems were run in parallel: a conventional air system, an oxygen system, and an air system with an anoxic zone. The settling properties of the respective sludges indicated that the oxygen system yielded the best settling sludge, while the air system with insufficient dissolved oxygen demonstrated severe filamentous bulking. The anoxic-aerobic system was almost devoid of filaments, but exhibited poorer flocculation and settling than the oxygen system. These results are shown in Table 6.24.

Viscous bulking occurs when the activated sludge contains an excessive amount of extracellular biopolymers which impart a slimy, jellylike consistency to the sludge and the sludge becomes highly water retentive.[38] This hydrous sludge exhibits low settling and compaction velocities. This phenomenon has been reported to be caused by a lack of nutrients/micronutrients or the presence of a toxic compound. Low dissolved oxygen has also been shown to contribute to viscous bulking.

Chlorine or hydrogen peroxide can be added to the return sludge or the aeration basin to reduce bulking.[39] Hydrogen peroxide is selective for some filament types. Hydrogen peroxide dosages are in the order of 20 to 50 mg/L. Chlorine dosages may vary from 9 to 10 lb Cl_2/(d · 1000 lb MLSS) [9 to 10 kg Cl_2/(d · 1000 kg MLSS)] in cases of severe bulking and 1 to 2 lb Cl_2/(d · 1000 lb MLSS) [1 to 2 kg Cl_2/(d · 1000 kg MLSS)] in cases of moderate bulking. A maximum of 4.5 lb Cl_2 per 1000 lb MLSS per day may be added to suppress filamentous bulking during nitrification. Because of the growth rate of the filaments, if the hydraulic retention time exceeds 8 h, chlorination must be applied directly to the aeration basin. Since the filaments exhibit a high negative zeta potential, they can be flocculated by the addition of cationic polyelectrolytes. This treatment is expensive and the filaments are not destroyed in the process.

Nocardia foams cause significant problems in plant operation. *Nocardia* actinomycetes are slower-growing organisms compared to floc-forming organisms. The growth of *Nocardia* can be controlled by reducing the SRT to less than 3 d to create washout conditions. These short SRTs are usually not feasible if nitrification is to be employed. Aerobic selectors or anoxic zones have been successful at controlling *Nocardia* foaming when operated at low SRTs. These methods are less effective at controlling *M. parvicella*. Successful control of biological foaming problems by applying powdered hypochlorite or a hypochlorite spray directly on the foam has been reported.

System	ZSV, ft/hr	SVI, mL/g	MLSS, mg/L	F/M,* g/g · d	Temperature, °C	Effluent TSS, mg/L
Low DO	0.6	222	3500	0.44	33	120
Anoxic	2.0	116	3850	0.44	34	50
	2.0	129	3600	0.35	40	140
	1.4	228	2800	0.44	45	130
High DO	3.6	100	3450	0.31	35	40
	3.5	96	3350	0.61	39	160
	5.5	90	3200	0.55	45	200
Low loading	3.2	133	2700	0.28	24	20
	4.2	133	2200	0.25	35	275

*BOD basis.

TABLE 6.24 Sludge Characteristics Under Various Operating Conditions

6.6 Soluble Microbial Product Formation

Soluble microbial products (SMP) are generated in the activated sludge process through the biodegradation of organics and through endogenous degradation of the biomass. The SMP are oxidation by-products that are nondegradable. Pitter and Chudoba[40] have indicated that, depending on cultivation conditions, the nonbiodegradable waste products can amount to 2 to 10 percent of the COD removed. The COD, BOD, and SMP_{nd} relationships for several industrial wastewaters are summarized in Table 1.8. Data for biodegradation of a peptone-glucose mixture and a synthetic fiber wastewater are shown in Fig. 6.63. They indicate that approximately 0.20 mg of nondegradable TOC (TOC_{nd}) was produced per milligram of influent TOC for the synthetic fiber wastewater. The peptone-glucose–containing wastewater produced approximately 0.12 mg TOC_{nd} per mg influent TOC. In both cases, the ratios were constant over the range of influent loading conditions. This indicates that there was a constant metabolic by-product or a portion of the original substrates that was nondegradable.

Many of the metabolic by-products are of high molecular weight. The molecular weight distribution (expressed as the TOC and COD fractions) of the influent and biologically treated effluent from a plastic additives wastewater and a biologically treated effluent from a glucose-based

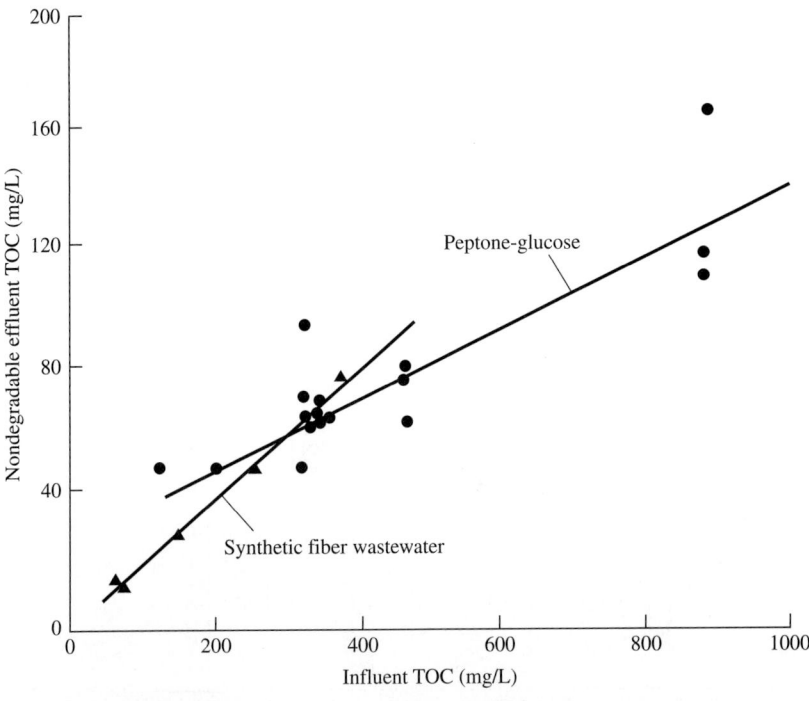

Figure 6.63 Nondegradable TOC as related to influent TOC.

| | Plastic Additives Wastewater | | Glucose[39] |
Molecular Weight	Influent TOC, %	Bioeffluent TOC, %	Wastewater COD, %
>10,000	—	11.5	45
500–10,000	—	14.5	16
<500	100	74.0	39

TABLE 6.25 Molecular Weight Distribution of a Biological Effluent

synthetic wastewater are presented in Table 6.25. Pitter and Chudoba[40] have indicated that approximately 75 percent of the SMP_{nd} from treatment of a wastewater containing only phenol had a molecular weight above 1000. It has been further determined that some of these high-molecular-weight fractions are toxic to some aquatic organisms. Results showing the aquatic toxicity of several wastewaters from the plastics industry before and after biological oxidation are shown in Fig. 6.64.

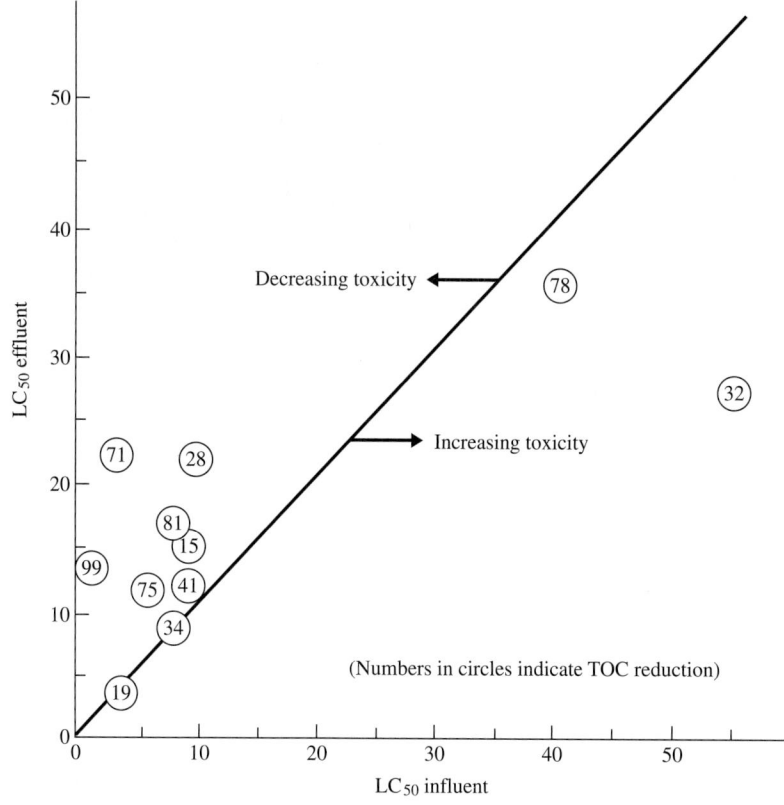

FIGURE 6.64 Effect of SMP on effluent toxicity for wastewaters from the plastics and dyestuffs industry.

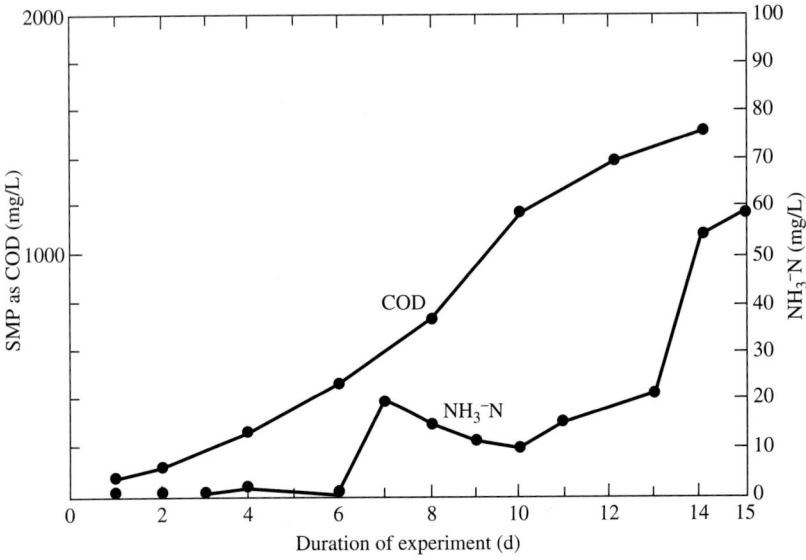

Figure 6.65 Relationship between SMP as COD on ammonia buildup.[33]

They indicate that biological treatment reduced the TOC and toxicity in most cases. In two wastewaters, however, the TOC was reduced by 32 and 78 percent, but the treated effluent was more toxic than the influent wastewater, making the oxidation by-products suspect toxicants. There is reason to believe that the high-molecular-weight SMP_{nd} strongly adsorb on activated carbon. This characteristic makes granular-activated carbon (GAC) or powdered-activated carbon (PAC) an excellent candidate process for toxicity reduction when toxicity is caused by SMP_{nd}. Inhibition to nitrification by SMP_{nd} is shown in Fig. 6.65 from the data of Chudoba.[41]

6.7 Bioinhibition of the Activated Sludge Process

Many organics will exhibit a threshold concentration at which they inhibit the heterotropic and/or nitrifying organisms in the activated sludge process. Inhibition has been defined by the Haldane equation (or its modifications) using Monod kinetics:

$$\mu = \frac{\mu_m S}{S + K_S + S^2/K_I} - b \qquad (6.46)$$

in which K_I is the Haldane inhibition coefficient. Inhibition to biotreatment of pulp mill wastewater by the addition of resin as characterized by K rate coefficient is shown in Fig. 6.66.

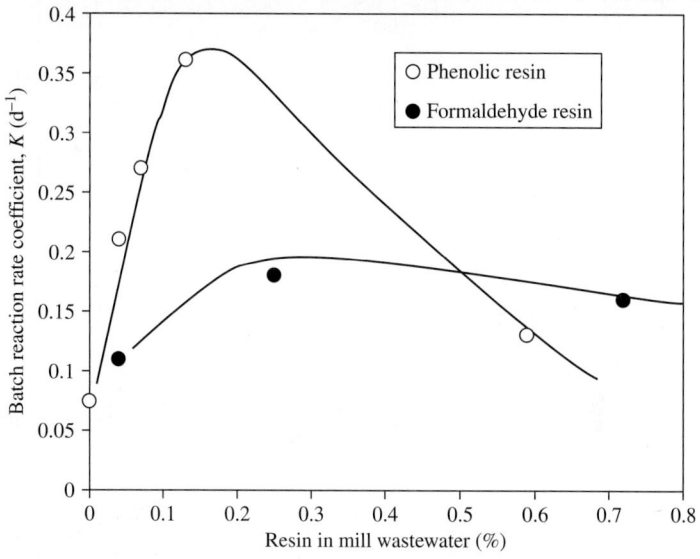

FIGURE 6.66 Inhibition from resin addition to a pulp mill wastestream.

An example of bioinhibition by a plastics additives wastewater is shown in Fig. 6.67. The extent of bioinhibition is expressed as the ratio of the SOUR at the selected influent loading concentration (K_r) to the SOUR at the no-observed-effect loading (K_o). As the concentration of influent COD (and inhibitory agent) increases, the SOUR decreases, resulting in higher effluent SBOD concentrations. In this case, the inhibition was removed by pretreating the wastewater with hydrogen peroxide (H_2O_2), thereby effecting detoxification and enhanced biodegradability. These effects are shown in Fig. 6.68. Addition of powdered activated carbon to the mixed liquor to adsorb the toxicant was also shown to reduce the inhibition.

Several acidic, aromatic, and lipid-soluble organic compounds have been demonstrated to "uncouple" oxidative phosphorylation. The result of this uncoupling effect is uncontrolled respiration and oxidation of primary substrates and intracellular metabolites. At low concentrations, uncoupling is evidenced by highly elevated oxygen utilization rates but no effect on cell growth or substrate removal. At higher concentrations, inhibition and toxicity are demonstrated by dramatic reductions in both the oxygen utilization rate and cell growth.

Volskay and Grady[42] and Watkin[43] have shown that inhibition can be competitive (the inhibitor affects the base substrate utilization), noncompetitive (the inhibitor rate is influenced), or mixed, in which both rates are influenced. The effects of substrate and inhibitor concentrations on the respiration rate of a microbial culture expressed as a fraction of the rate in the absence of the inhibitor have been shown by Volskay and Grady.[42]

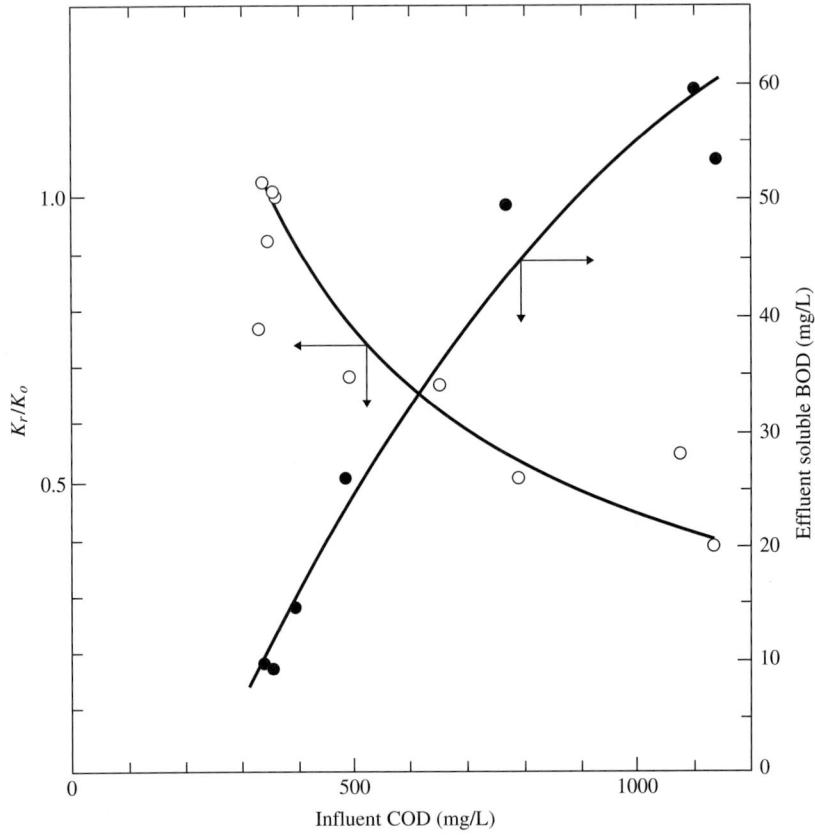

Figure 6.67 Activated sludge inhibition from a plastics additives wastewater.

While the relationships described above define the mechanism of inhibition, they are of limited use in evaluating industrial wastewaters. In most cases, the inhibitor itself is not defined, variable sludge and substrate composition will influence inhibition, and interactions will frequently exist between inhibitors. The inhibition constant K_I is highly dependent on the specific enzyme system involved, which, in turn, is dependent on the history and population dynamics of the sludge. In some cases, the inhibition constant may be dependent on the particular metabolic pathways that are present in any given microbial population. For example, Watkin and Eckenfelder[44] showed a variation in K_I of 6.5 to 40.4 for different sludges and operating conditions treating 2,4-dichlorophenol and glucose. Volskay and Grady[44] showed a variation of 2.6 to 25 mg/L in the concentration of pentachlorophenol, which would cause 50 percent inhibition of oxygen utilization rates.

It is apparent, therefore, that each wastewater must be independently evaluated for its bioinhibition effects. Several protocols have

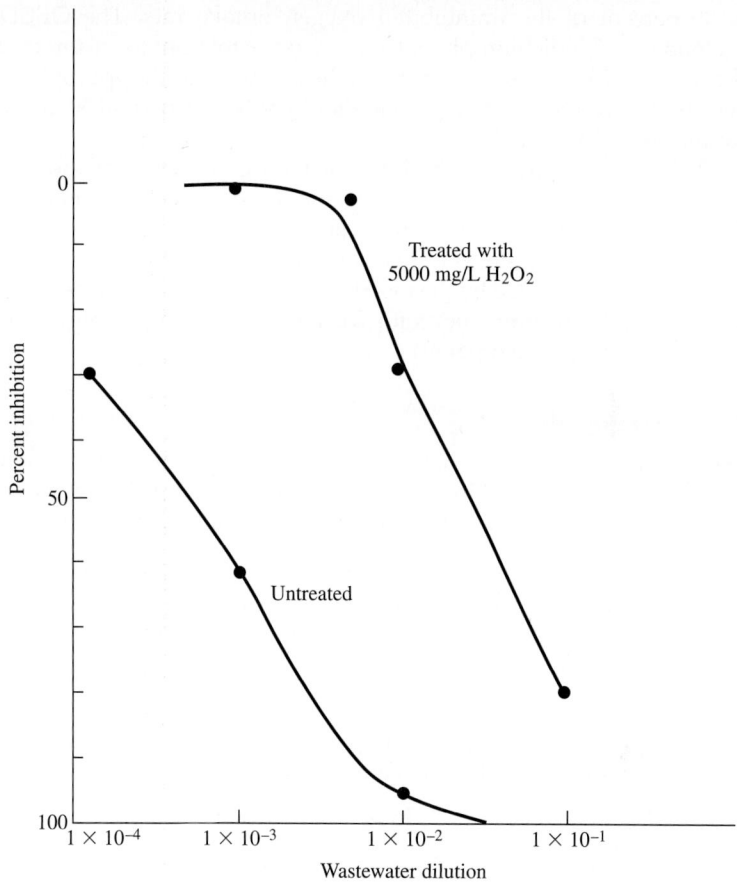

Figure 6.68 Detoxification of a plastics additives wastewater.

been developed for this purpose. These are the fed-batch reactor (FBR) of Philbrook and Grady[29] and Watkin and Eckenfelder,[44] the OECD method 209 of Volskay and Grady,[42] and the glucose inhibition test of Larson and Schaeffer.[45] Depending on the particular wastewater, one or more of these test protocols will be applicable. Each of these methods is discussed below.

OECD Method 209

The OECD method 209 involves measurement of activated sludge oxygen uptake rate from a synthetic substrate to which the test compound has been added at various concentrations. The oxygen uptake rate is measured immediately after addition of the test compound and after 30 min of aeration. The EC_{50} value is determined as the concentration of test compound at which the oxygen uptake rate (at 30 min) is 50 percent of the uninhibited oxygen uptake rate. The OECD method uses 3,5-dichlorophenol as a reference toxicant to ensure that the test is working properly and that the biomass has the appropriate sensitivity. The reference EC_{50} value should be between 5 and 30 mg/L for the test to be valid.

Volskay and Grady[44] employed a modified OECD method to determine the toxicity of selected organic compounds. Since many of the compounds were volatile, the test was modified by using more dilute cell and substrate concentrations and by conducting the test in vessels that were sealed by the insertion of a polyfluoroethylene plug. This protocol is recommended for wastewaters containing high concentrations of volatile organics.

Fed Batch Reactor

Fed batch reactors (FBRs) have been used to determine nitrification kinetics and removal kinetics of specific pollutants in activated sludge. The essential characteristics of the FBR procedure are that:

1. Substrate is continuously introduced at a sufficiently high concentration and low flow rate so that the reactor volume is not significantly changed during the test.
2. The feed rate exceeds the maximum substrate utilization rate.
3. The test duration is short and therefore allows simple modeling of biological solids growth.
4. Acclimated activated sludges are used.

A schematic diagram of the FBR is shown in Fig. 6.69. Two liters of mixed liquor are placed in the reactor, and a sample is taken for determination of oxygen utilization rate (OUR) and mixed liquor volatile and total suspended solids prior to the start of the feed flow. The feed is introduced at a flow rate of 100 mL/h, and aliquots of the reactor contents are withdrawn every 20 min for the duration of the 3-h test. The OUR is determined in situ every 30 min during the test. Suspended solids determinations are made every hour during the test.

As discussed under the OECD protocol, the oxygen uptake rate will decrease in the presence of inhibition. As long as there is no inhibition, the OUR will remain constant at a maximum rate. The same limitations apply to the FBR protocol as to the OECD method.

The theoretical responses in a fed batch reactor to both inhibitory and noninhibitory substrates are depicted in Fig. 6.70. In the case where substrate is added at a sufficiently high mass and low volumetric flow rate, the maximum substrate utilization rate will be exceeded, and the change in reactor volume will be insignificant. If the FBR volume

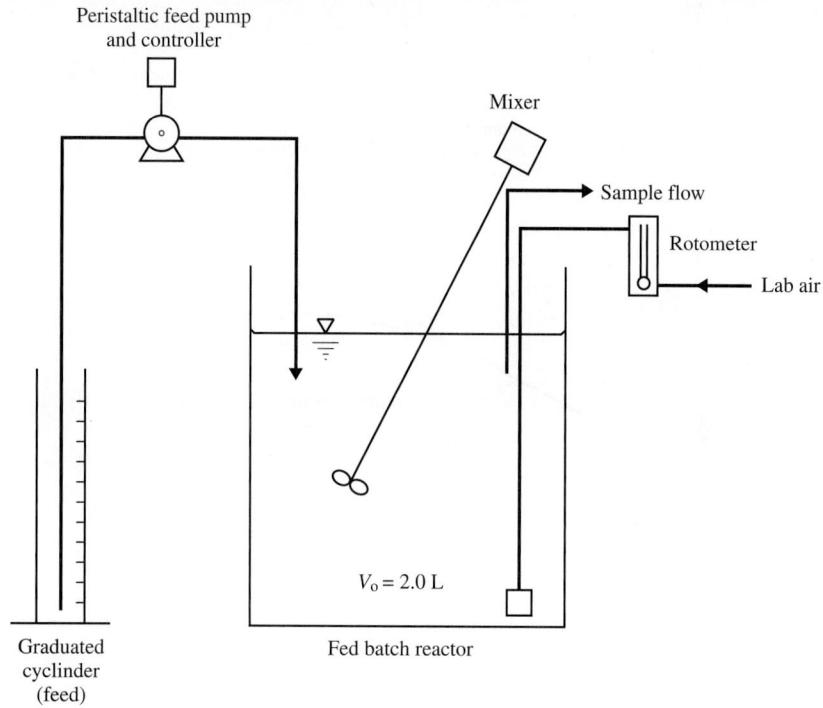

FIGURE 6.69 Fed batch reactor (FBR) configuration.

change is negligible and the mass feed rate exceeds the maximum substrate utilization rate, then a substrate concentration buildup will result in the reactor with time. Noninhibitory substrate response results in a linear residual substrate buildup in the reactor with time. The maximum specific substrate utilization rate q_{max} is calculated as the difference in slopes between the substrate feed rate and the residual substrate buildup rate divided by the biomass concentration. In the case of inhibition, substrate utilization would rapidly decrease, resulting in an upward deflection of the residual substrate concentration curve as shown in Fig. 6.70. As inhibition progresses and acute biotoxicity occurs, the trace of the residual substrate concentration should become parallel to the substrate feed rate. The inhibition constant K_I can be approximated by identifying the inhibitor concentration at the midpoint of the curvilinear portion of the substrate response.

Glucose Inhibition Test

Larson and Schaeffer[45] developed a rapid toxicity test based on the inhibition of glucose uptake by activated sludge in the presence of toxicants. The test was subsequently modified for

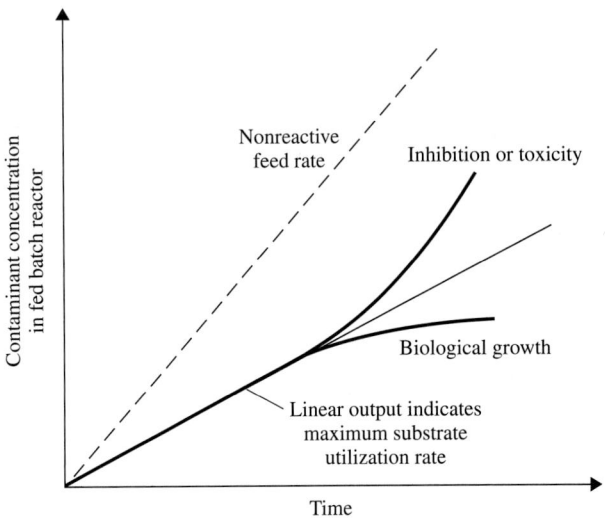

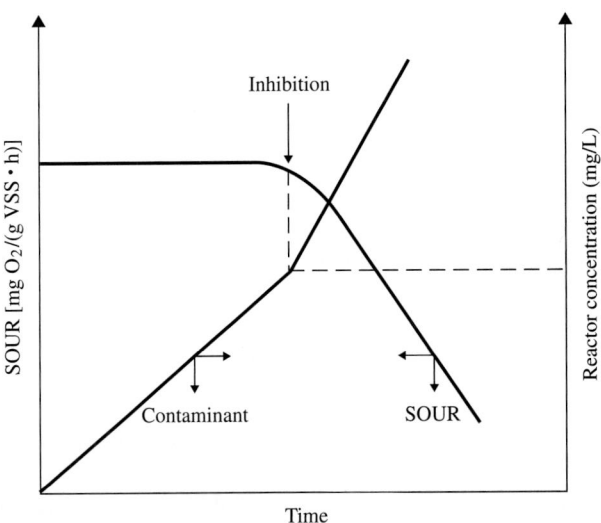

Figure 6.70 Theoretical fed batch reactor output with influent substrate mass flow rate greater than $q_{max} \cdot X_v$ and inhibition effects.

application to a variety of industrial wastewaters. The procedure is as follows:

1. Place 10 mL of sample into a centrifuge tube.
2. Add 10 mL of the stock glucose.
3. Add 10 mL of activated sludge to the centrifuge tube and aerate at a low rate.

4. After 60 min, add two drops of HCl and transfer tubes to the centrifuge.
5. Measure glucose concentration.
6. Sludge control—substitute 10 mL of deionized water for the sample in step 1 and perform steps 2 through 5 as before.
7. Glucose control—place 30 mL of deionized water in a centrifuge tube. Add 1 mL of stock glucose solution. Do not add sludge or aerate. Add two drops of HCl and measure glucose uptake.

The percent inhibition is calculated as follows:

$$\text{Percent inhibition} = \left[\frac{C - C_B}{C_o - C_B}\right]100$$

where C = final glucose concentration in sample solution
C_B = final glucose concentration in sludge control sample
C_o = initial glucose concentration (glucose control)

Inhibition effects using the glucose test are presented in Fig. 6.71 for an organic chemicals wastewater.

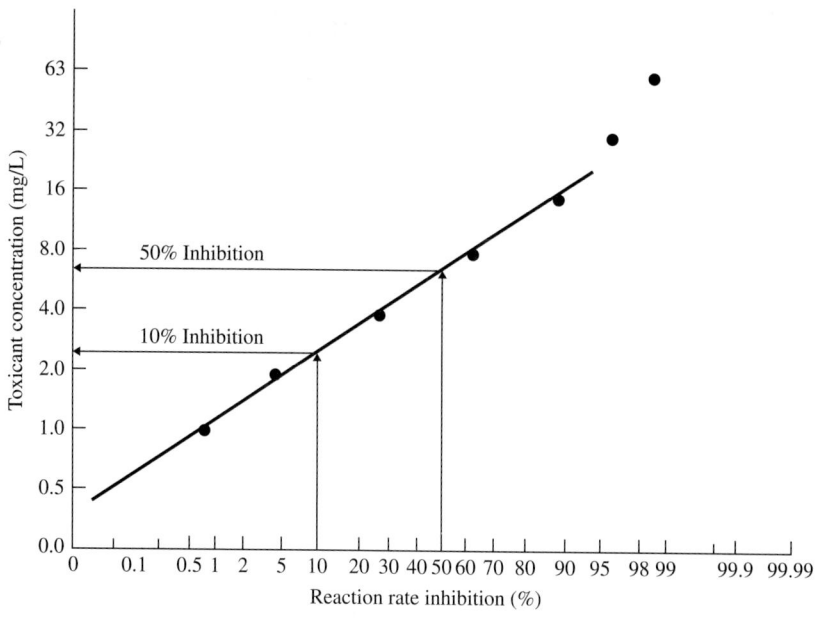

FIGURE 6.71 Determination of bioinhibition effects from the glucose inhibition test.

6.8 Stripping of Volatile Organics

Volatile organic compounds (VOCs) will be stripped from solution by oxygenation in the activated sludge process. Depending on the particular VOC, both air stripping and biodegradation may occur.

The fraction of the VOC that is stripped depends upon several factors. Compound specific factors include the Henry's law constant, the compound's biodegradation rate, and in some cases, the initial concentration of the compound and other substrates. Operational and facility design factors that influence stripping are the method of oxygenation and the power level in the aeration basin. The concentration of nonvolatile organics in the wastewater will affect the composition and mass of the sludge, the operating SRT, and hence, the fraction of VOC that is biodegraded and stripped. Biodegradation and stripping removal data for a range of VOCs are presented in Table 6.26. As a general rule, as more halogen atoms are added to the organic

Compound	Influent Concentration, mg/L	SRT, d	Amount Stripped, %
Toluene	100	3	12–16
	0.1	3	17
	40	3	15
	40	12	5
	0.1	6	22
Nitrobenzene	0.1	6	<1
Benzene	153	6	15
	0.1	6	16
Chlorobenzene	0.1	6	20
1,2-Dichlorobenzene	0.1	6	59
1,2-Dichlorobenzene	83	6	24
1,2,4-Trichlorobenzene	0.1	6	90
o-Xylene	0.1	6	25
1,2-Dichloroethane	150	3	92–96
1,2-Dichloropropane	180	6	5
Methyl ethyl ketone	55	7	3
	430	7	10
1, 1, 1-Trichloroethane	141	6	76

TABLE 6.26 Fate of Selected VOCs in the Activated Sludge Process

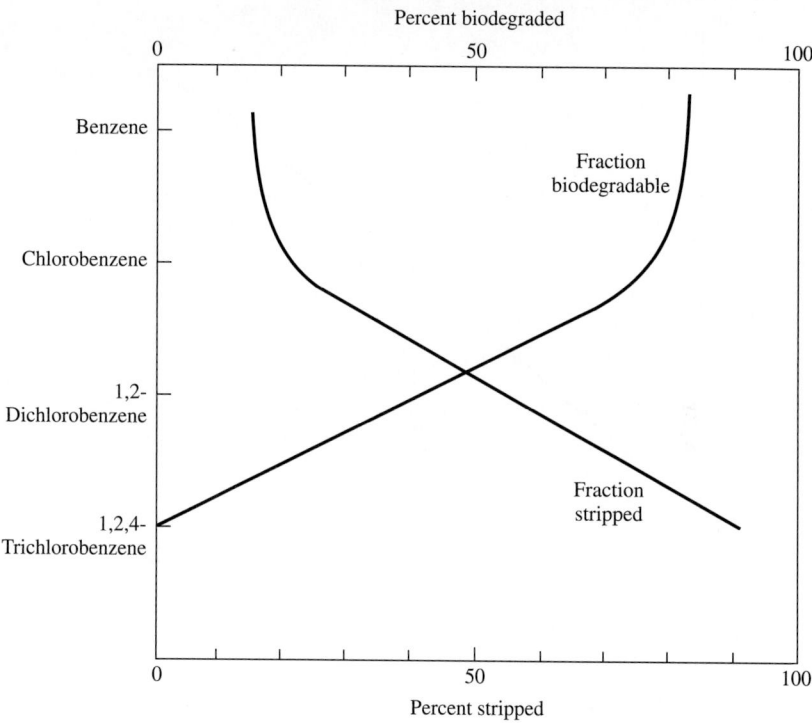

FIGURE 6.72 Relationship between biodegradation and stripping for the chlorinated benzene series. (*Adapted from Weber and Jones, 1983.*)

compound, the rate of biodegradation decreases, and the amount of compound stripped increases. This is shown in Fig. 6.72 for several chlorinated compounds in the benzene series. It should be noted that a greater SRT will usually result in less stripping since a lower power level will be employed and the biomass concentration will be higher. The effects of power level and the type of aeration device on the stripping of benzene are shown in Fig. 6.73, which was developed assuming an influent COD of 250 mg/L and an influent benzene concentration of 10 mg/L. Note that approximately three to four times as much benzene is stripped with surface aeration compared to diffuse aeration at a given power level.

Figure 6.74 represents the volatilization of toluene as a function of air/water for a municipal treatment plant compared to an industrial plant where acclimated biomass is present. Note that significantly more toluene is stripped in the municipal plant, especially at higher air to water ratios. Note that this is a specific case and results are dependent on many factors as discussed in pertinent sections of this book.

In a diffused air oxygenation system, equilibrium between the gas and liquid phases is quickly reached after the bubble is formed.

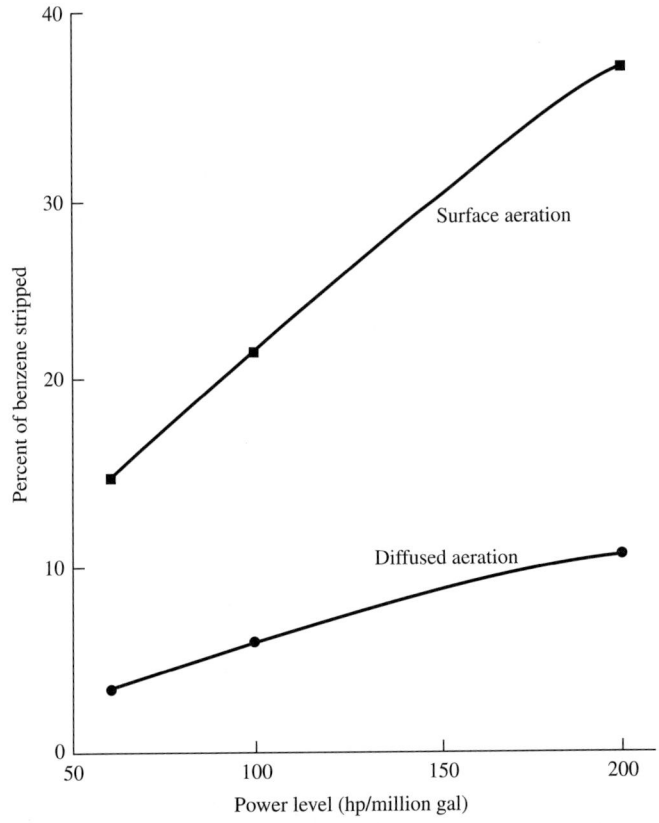

Figure 6.73 Stripping of benzene as related to power level and type of aerator.

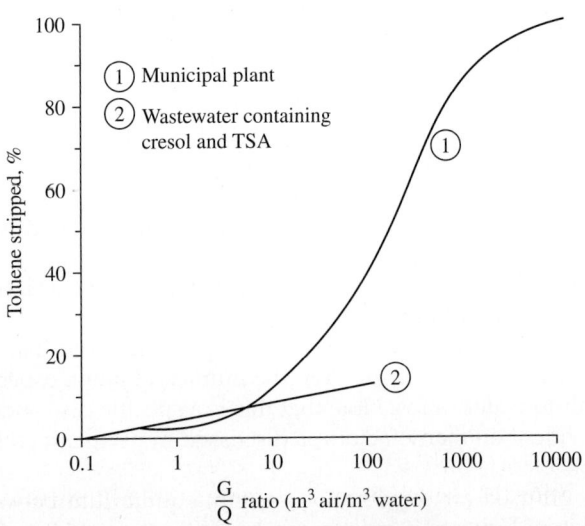

Figure 6.74 Volatilization of toluene from a municipal and an industrial wastewater treatment plant.

Under these conditions, the amount of VOC that is stripped depends primarily on the gas to liquid ratio. Since, under most conditions, the volume of gas is relatively small, stripping is minimal. In a mechanical surface aeration system, however, the volume of gas in contact with the liquid is nearly infinite (i.e., the atmosphere), and hence, a greater quantity of VOC will be stripped.

The fraction emitted to the air of any volatile component entering an activated sludge aeration basin can be derived from the mass balance Eq. (6.47) neglecting the effects of adsorption on the biological floc:

$$Q_o C_{o,i} = Q_o C_{L,i} + r_i + r_{vi} \quad (6.47)$$

where Q_o = flow rate, L/s
$C_{o,i}$ = influent concentration of VOC i, g/L
$C_{L,i}$ = liquid effluent concentration of component i, g/L
r_i = rate of biodegradation of component i, g/s
r_{vi} = rate of volatilization of component i, g/s

For a complete mix system, the fraction of the influent VOC load that is emitted to the air (f_{air}) is given by

$$f_{air} = \frac{r_{vi}}{Q_o C_{L,i} + r_i + r_{vi}} \quad (6.48)$$

When individual compounds, pure or in a mixture, are considered, it is appropriate to use a Monod kinetic model to describe their biodegradation. The concentrations of priority pollutants normally encountered in a complete mix reactor are in the μg/L range in order to comply with current regulations. Under these conditions, the Monod model reduces to a first-order rate expression, and the biodegradation rate of compound i (expressed as COD) is given by

$$r_i = W_i q_m X_{vb} V C_{L,i} / K_S \quad (6.49)$$

where W_i = weighting factor for compound i
q_m = biodegradation rate constant for compound i, g COD/(g VSS · s)
X_v = volatile solids concentration, g/L
V = volume of reactor, L
$C_{L,i}$ = concentration of compound i in the reactor, g/L
K_S = Monod's half saturation constant for component i, g/L

The biomass is considered to be a mixed culture with each organic substrate being the growth-limiting factor for a specific microbial population. Under these circumstances, the fraction of biomass dedicated to a specific substrate would be proportional to the biomass yield associated with this substrate. Hence, the weighting factor would be

$$W_i = a_i C_{B,i} \left[\sum_{i=1}^{n-1} a_i C_{B,i} \right]^{-1} \quad (6.50)$$

where $C_{B,i}$ = concentration of component i that is biodegraded, g/L, and a_i = yield coefficient of component i, g VSS/g COD.

Stripping of VOC is computed by using the method of Roberts et al.[46] This model estimates stripping with surface or diffused aeration on the premise that the transfer rates for volatile solutes are proportional to one another. Since oxygen satisfies the volatility criteria and because a large mass-transfer database for it is available, oxygen has been selected as the reference compound. The overall mass-transfer coefficient for compound i is then proportional to that for oxygen at the operating conditions.

$$(K_L a)_i = \psi (K_L a)_{O_2} \quad (6.51)$$

It has been shown that the proportionality constant (ψ) depends on the liquid-phase diffusivity ratio, D_i/D_{O_2}, and is approximately constant over a wide range of temperature and mixing conditions. Values of ψ are virtually identical in clean water and wastewater, indicating that the transfer rates of dissolved oxygen and organic solutes are inhibited to the same degree as wastewater constituents.

The overall gas transfer coefficient $(K_L a)_{O_2}$ is related to the standard oxygenation rate:

$$(K_L a)_{O_2} = \frac{\text{SOR} \cdot P}{C_s \cdot V} \quad (6.52)$$

where C_s = oxygen solubility in clean water, g/L
P = aeration basin power level, hp
V = aeration basin volume, L
SOR = standard oxygenation rate, g O_2/hp $\cdot$ h

For surface aeration, the stripping of VOCs can be computed from Eq. (6.52):

$$r_{vi} = \psi (K_L a)_{O_2} \cdot C_{L,i} V \quad (6.53)$$

and the fraction of VOC emitted to the air becomes

$$f_{air} = \frac{\psi K_L a_{O_2} V}{Q_o + (W_i q_m X_v V / K_s) + \psi K_L a_{O_2} V} \quad (6.54)$$

For diffused air, assuming the exit air is in equilibrium with the liquid, the fraction of VOC remaining in the liquid phase is

$$r_{vi} = Q_{air} H_c C_{L,i} \qquad (6.55)$$

in which $(H_c)_i$ = Henry's constant for compound i. The fraction of VOC emitted using diffused aeration at an airflow rate of Q_{air} is

$$f_{air} = \frac{Q_{air} H_c}{Q_o + (W_i q_m X_{vb} V / K_s) + Q_{air} H_c} \qquad (6.56)$$

Example 6.13 illustrates the calculation of VOC emissions.

Example 6.13. Determine the fraction of benzene emitted to the air under the following loading conditions for mechanical aeration and diffused aeration. The mechanical aeration system uses 60 hp/million gal and has a $(K_L a)_{O_2} = 1.52/h$, and the diffused aeration system has an airflow rate of 2.16 m³/s.

Solution

$V = 3846$ m³ (136,000 ft³)
$Q_o = 0.178$ m³/s (2820 gal/min)
$C_o = 10$ mg/L
$X_{vb} = 3000$ mg/L
$S_o = 250$ mg COD/L

$$r_i = \frac{W_i q_{mi} X_v V C_{L,i}}{K_{si}}$$

$$W_i = \frac{a_i f_{bio,i} C_{o,i}}{a(S_o - S)} \quad \frac{3.08 \text{ mg COD}}{\text{mg benzene}}$$

Assuming $a_i = a$ and $f_{bio,i} = 0.87$,

$$W_i = \frac{0.87 \times 10 \times 3.08}{(250 - 20)} = 0.117$$

then for mechanical aeration,

$$f_{air,i} = \frac{0.6 \times (1.52/3600) \times 3846}{0.178 + \dfrac{0.117 \times 5.78 \times 10^{-6} \times 3000 \times 3846}{1} + 0.974}$$

$$f_{air,i} = \frac{0.974}{0.178 + 7.803 \times 0.974} = \frac{0.974}{8.955} = 0.109 = 11\%$$

Check $f_{bio,i}$:

$$f_{bio,i} = \frac{7.803}{8.955} = 0.871 = 87\% \quad \text{Checks}$$

For diffused aeration, assume 93 percent biodegradation.

$$W_i = \frac{0.93 \times 10 \times 3.08}{(250-20)} = 0.125$$

$$f_{air,i} = \frac{2.16 \times 0.225}{0.178 + \dfrac{0.125 \times 5.78 \times 10^{-6} \times 3000 \times 3846}{1} + 0.486}$$

$$f_{air,i} = \frac{0.486}{0.178 + 8.336 + 0.486} = \frac{0.486}{9.0} = 0.054 = 5.4\%$$

$$f_{bio,i} = \frac{8.336}{9.0} = 0.926 = 93\% \quad \text{Checks}$$

$S_e = 20$ mg COD/L
$\psi = 0.6$
$q_m = 5.78 \times 10^{-6}/s$
$K_S = 1.0$ mg/L
$H_c = 0.225$

Treatment of VOC Emissions

If the off-gas from the aeration basin contains high concentrations of VOC or ammonia, air quality standards may require off-gas collection and treatment. Several treatment technologies are available, including thermal and catalytic incineration, carbon adsorption, macroreticular resin adsorption, and biological degradation. The removal performance of several of these technologies is shown in Fig. 6.75 as a function of the gas phase VOC concentration. In most cases of activated sludge treatment, the gas phase VOC concentration will be low (< 50 ppmv), and incineration would be required to provide the highest removal efficiency.

Biological treatment of the off-gas is a less costly alternative to incineration and can be done with biofilters which consist of static beds of compost material such as peat or corn silage and bulking agents such as wood chips to maintain porosity and allow sufficient air flow. Granular activated carbon can be added to the bed material to adsorb poorly degradable VOCs. The beds are lightly seeded with biomass and maintained at proper pH, temperature, and moisture content to support biological activity. The off-gas is passed through the bed at surface flow rates of 1 to 10 ft³/(min · ft²)(0.31 to 3.1 m³/min · m²) to provide 2 to 12 min of bed contact time. The biomass concentration depends on the type of bed material and organic loading rates. Bioscrubbers may also be employed in which the off gases are passed through an acclimated suspended growth biological reactor.

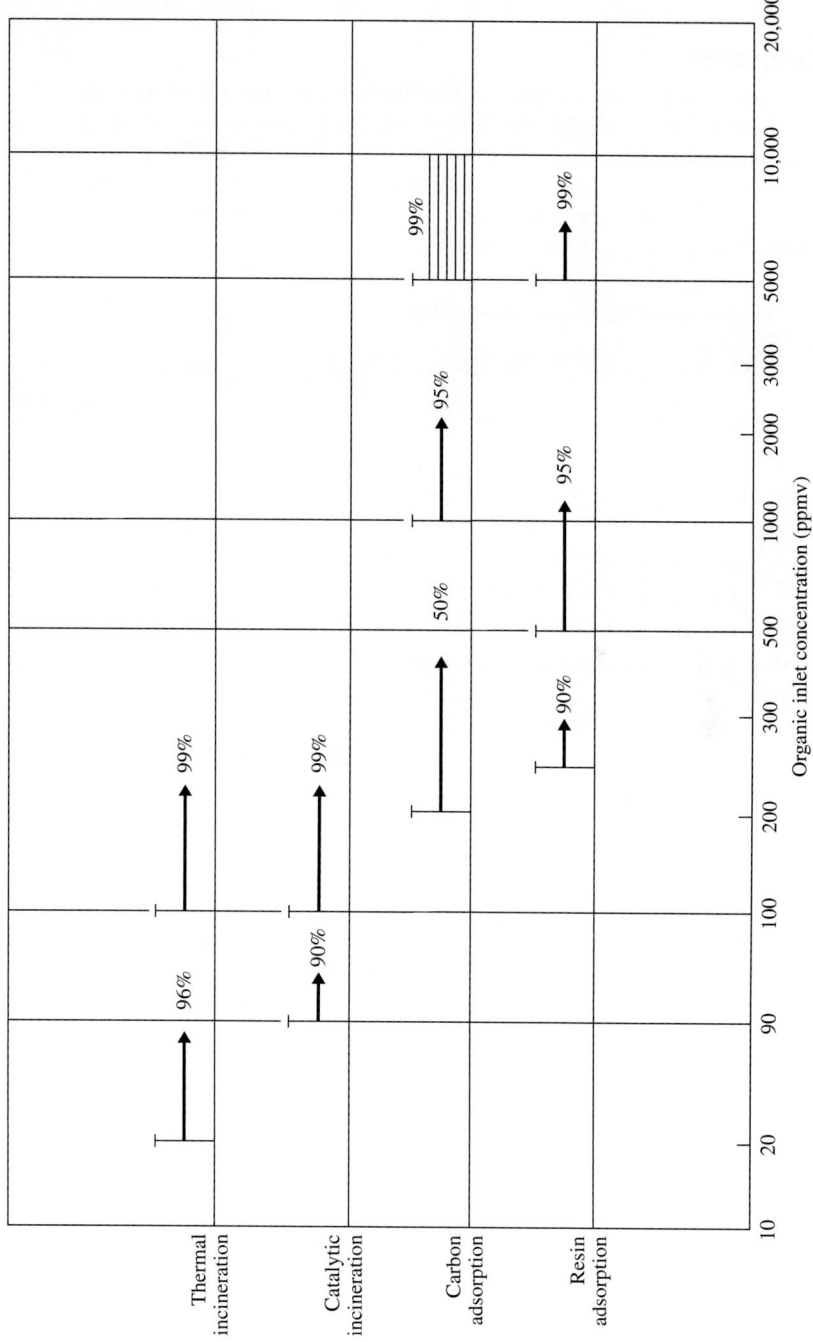

FIGURE 6.75 Performance capabilities of VOC control devices.

6.9 Nitrification and Denitrification

Nitrification

Nitrogen changes in a biological-treatment process are shown in Figs. 6.76 and 6.77. Nitrification is the biological oxidation of ammonia to nitrate with nitrite formation as an intermediate step. The microorganisms involved are the autotrophic species *Nitrosomonas* and *Nitrobacter*, which carry out the reaction in two steps. Nitrogen transformation equations are described as follows:

$$\text{Organic nitrogen} \longrightarrow NH_3\text{-}N$$

$$2NH_4^+ + 3O_2 \xrightarrow{\textit{Nitrosomonas}} 2NO_2^- + 4H_2O + 4H^+ + \text{new cells}$$

$$2NO_2^- + O_2 \xrightarrow{\textit{Nitrobacter}} 2NO_3^- + \text{new cells}$$

For 1 g NH_3-N oxidized

4.33 g of O_2 are consumed
7.15 g of alkalinity (as $CaCO_3$) are destroyed
0.15 g of new cells are formed
0.08 g of inorganic carbon are consumed

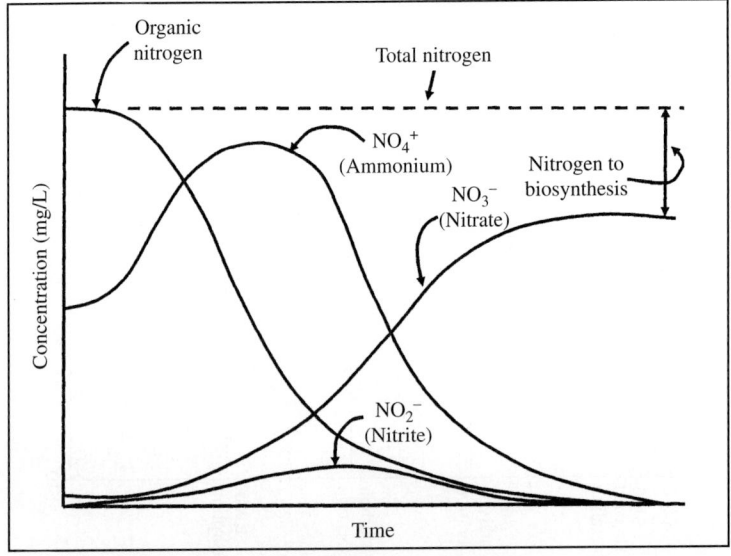

Figure 6.76 Biological nitrogen conversion process.

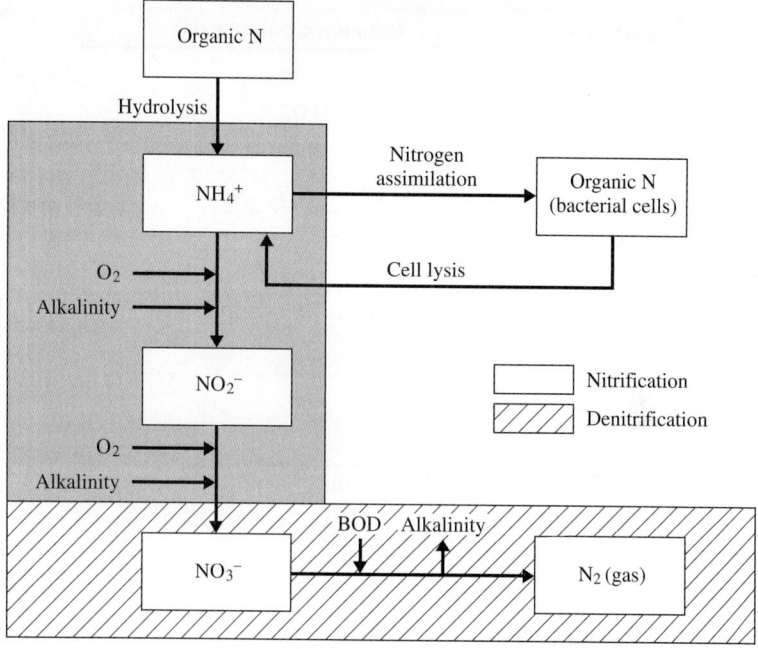

FIGURE 6.77 Nitrogen transformations.

The cell yield for *Nitrosomonas* has been reported as 0.05 to 0.29 mg VSS/mg NH_3-N and for *Nitrobacter* 0.02 to 0.08 mg VSS/mg NH_3-N. A value of 0.15 mg VSS/mg NH_3-N is usually used for design purposes. It is generally accepted that the biochemical reaction rate of *Nitrobacter* is faster than the reaction rate of *Nitrosomonas* and hence there is no accumulation of nitrite in the process and the reaction rate of *Nitrosomonas* will control the overall reaction. Poduska[47] reviewed the results on the effect of micronutrients on the growth of nitrifying bacteria in pure culture as shown in Table 6.27.

Nitrification Kinetics

In order to maintain a population of nitrifying organisms in a mixed culture of activated sludge, the minimum aerobic sludge age $(\theta_c)_{min}$ must exceed the reciprocal of the nitrifiers' net specific growth rate:

$$(\theta_c)_{min} \geq \frac{1}{\mu_{N_T} - b_{N_T}} \tag{6.57}$$

Compound	Concentration, µg/L
Calcium	0.5
Copper	0.005–0.03
Iron	7.0
Magnesium	12.5–0.03
Molybdenum	0.001–1.0
Nickel	0.1
Phosphorus	310.0
Zinc	1.0

Source: Adapted from Poduska, 1973.

TABLE 6.27 Stimulatory Concentration of Micronutrients for Nitrifying Bacteria

where μ_{N_T} = nitrifiers' specific growth rate, d^{-1}, and b_{N_T} = nitrifiers' endogenous decay rate, g $\Delta VSS_N/(g\ VSS_N \cdot d)$. If the SRT is insufficient at operating conditioning, the nitrifiers will be washed from the system. This is demonstrated by Fig. 6.78. Nitrification relative to SRT for an organic chemical wastewater is shown in Fig. 6.79. The nitrifiers' specific growth rate is related to the specific nitrification rate

$$\mu_{N_T} = a_N q_N \tag{6.58}$$

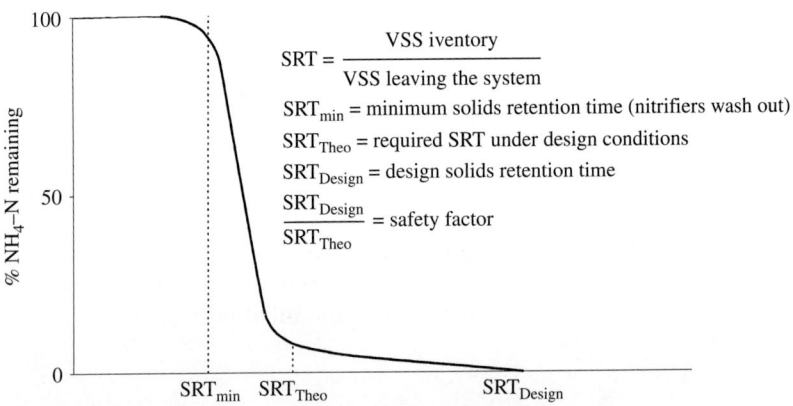

FIGURE 6.78 Relationship of ammonia removal and solids retention time in an activated sludge system.

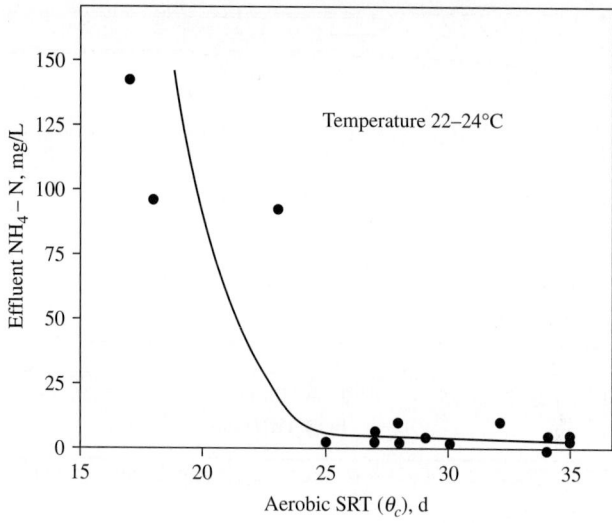

FIGURE 6.79 Nitrification relative to the aerobic SRT for an organic chemicals wastewater.

in which q_N = specific nitrification rate, d^{-1}, and a_N = sludge yield coefficient for nitrifiers.

The ratio of q_N/q_{Nmax} as a function of SRT is presented in Fig. 6.80. The ratio of BOD/TKN and temperature affects the sludge age required for nitrification at a given temperature as shown in Fig. 6.81. The BOD/TKN ratio also affects the nitrification rate at various

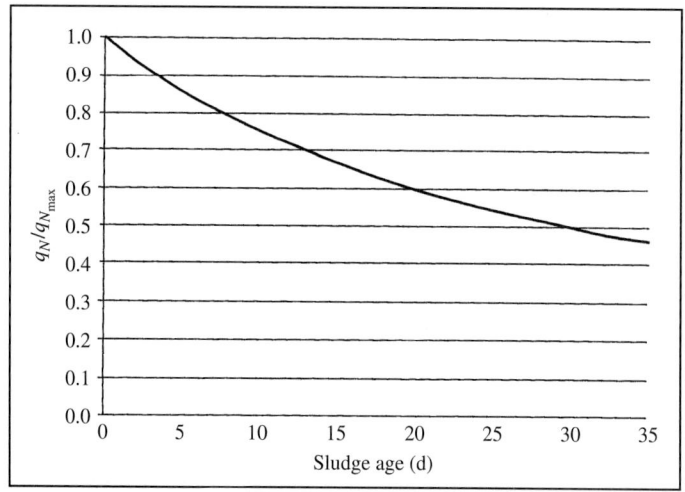

FIGURE 6.80 Effect of sludge age on nitrification rate.

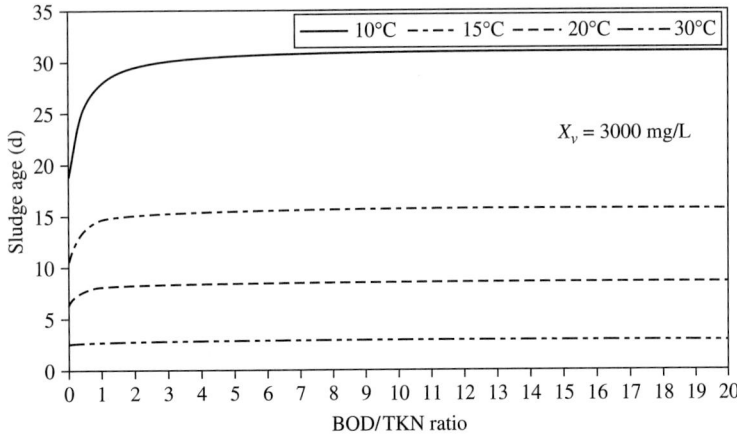

Figure 6.81 Effect of BOD/TKN ratio on sludge age for nitrification.

temperatures as demonstrated by Fig. 6.82. The specific nitrification rate in an activated sludge system also depends on the concentrations of ammonia nitrogen in the effluent and dissolved oxygen as well as the pH. The effects of dissolved oxygen and effluent ammonia are defined as follows:

$$q_N = q_{N_M} \cdot \frac{NH_3\text{-}N}{K_N + NH_3\text{-}N} \cdot \frac{DO}{K_O + DO} \qquad (6.59)$$

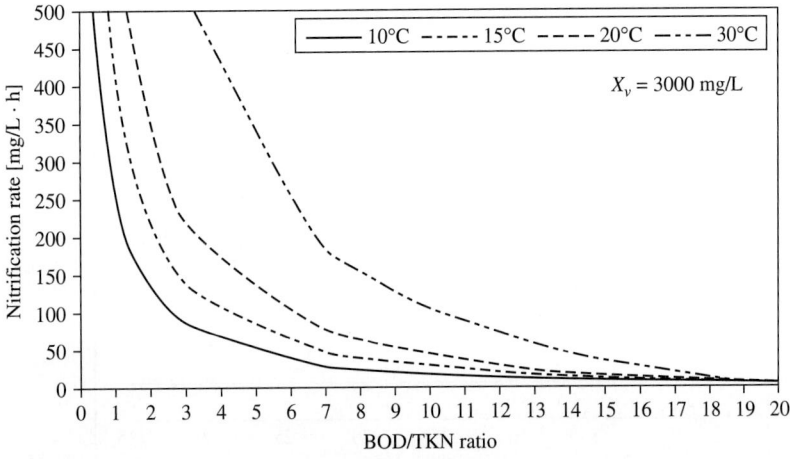

Figure 6.82 Effect of BOD/TKN ratio on nitrification rate.

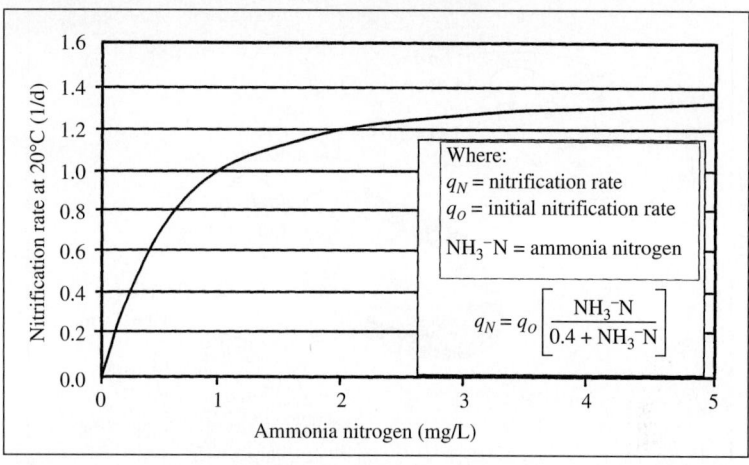

FIGURE 6.83 Effect of ammonia nitrogen conecentration on nitrification rate at 20°C.

where K_N and K_O are the half saturation coefficients for nitrogen and oxygen, respectively. A typical value for K_N is 0.4; K_O may vary from 0 to 1.0 depending on F/M and aeration power level. The effect of ammonia concentration on nitrification is given in Fig. 6.83. As indicated, as the ammonia concentration drops below about 1 mg/L, the nitrification rate decreases significantly.

The influence of mixed liquor dissolved oxygen on the nitrification rate has been somewhat controversial, partly because the bulk liquid concentration is not the same as the concentration within the floc where the oxygen is being consumed. Increased bulk liquid dissolved oxygen concentrations will increase the penetration of oxygen into the floc, thereby increasing the rate of nitrification. At a decreased SRT and higher F/M, the oxygen utilization rate due to carbon oxidation increases, thereby decreasing the penetration of oxygen. Conversely, at a high SRT and lower F/M, the low oxygen utilization rate permits higher oxygen levels within the floc, and consequently, higher nitrification rates occur. Therefore, to maintain the maximum nitrification rate, the bulk mixed liquor dissolved oxygen concentration must be increased as the SRT is decreased. This is reflected in the coefficient K_O and is depicted by Fig. 6.84. The effect of dissolved oxygen on nitrification rates is given by Fig. 6.85 and can be described as follows:

$$q_N = q_{N_{max}} \frac{DO}{K_O + DO}$$

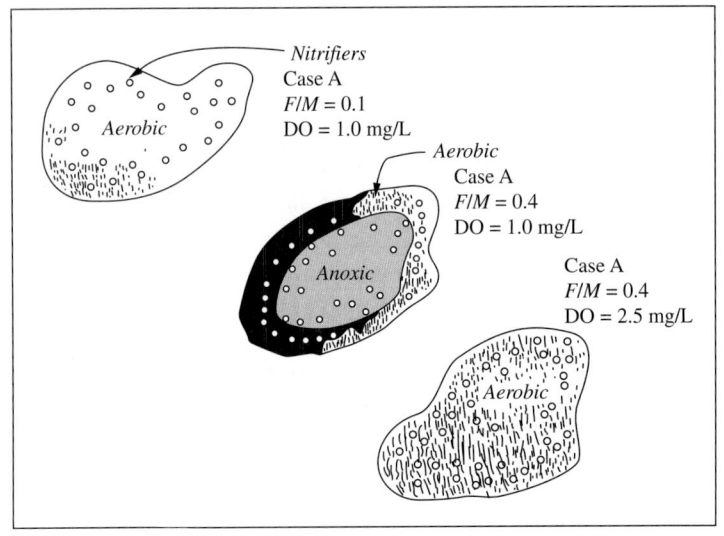

FIGURE 6.84 Effect of F/M on nitrification.

where q_N = nitrification rate
 $q_{N\max}$ = maximum nitrification rate
 DO = dissolved oxygen, mg/L
 K_O = dissolved oxygen correction coefficient

The effect of pH on nitrification rate is shown in Fig. 6.86.

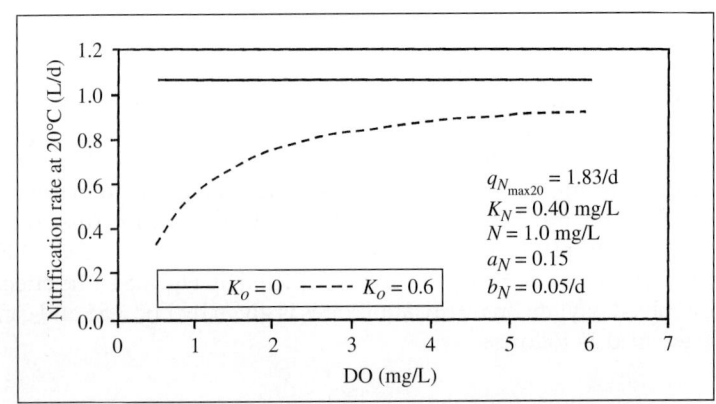

FIGURE 6.85 Effect of dissolved oxygen on nitrification rate at 20°C.

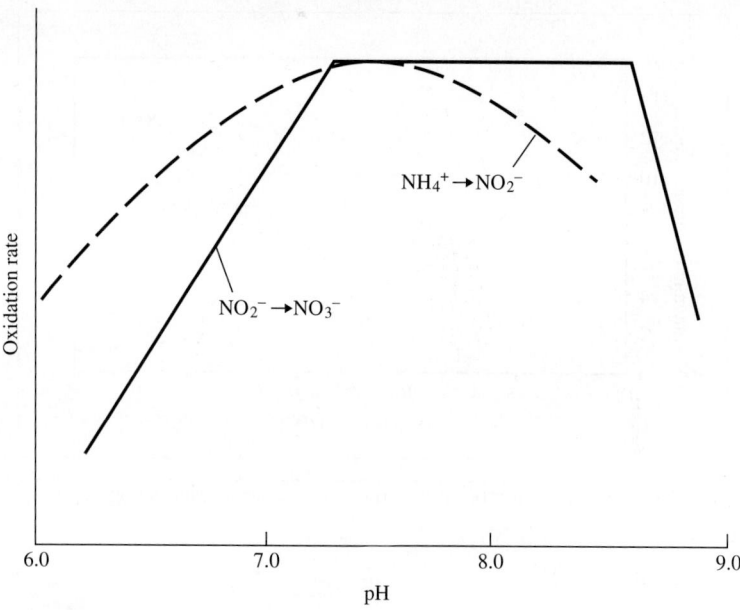

FIGURE 6.86 Effect of pH on ammonia oxidation. (*Adapted from Wong Chong and Loehr, 1975.*)

In treating industrial wastewaters that may inhibit nitrification, the maximum specific nitrification rate must be experimentally determined. The temperature dependence on the specific nitrification rate is given by

$$q_{N(T)} = q_{N(20°C)} \cdot 1.068^{T-20} \qquad (6.60)$$

The effect of temperature on the minimum sludge age required for nitrification is shown in Fig. 6.87.

The endogenous decay coefficient b_N has a temperature coefficient of 1.04:

$$b_N = b_{N(20°C)} \cdot 1.04^{T-20} \qquad (6.61)$$

The preceding factors may be combined into a nitrification rate determination equation:

$$N_R = 1.82 \frac{1}{1 + 0.033 \theta_c} \frac{N_e}{0.4 + N_e} \frac{DO}{K_O + DO} 1.068^{(T-20)} \qquad (6.61a)$$

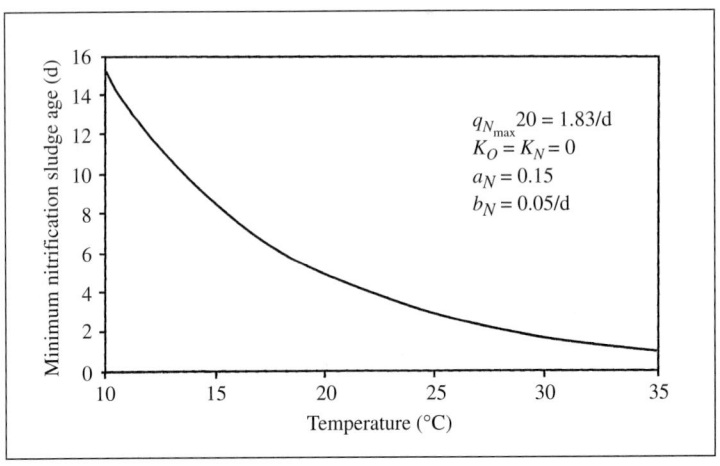

FIGURE 6.87 Effect of temperature on minimum nitrification sludge age.

where N_R = nitrification rate
 θ_c = sludge age, d
 N_e = effluent ammonia concentration
 DO = dissolved oxygen
 K_O = dissolved oxygen correction coefficient
 T = temperature

Nitrification of High-Strength Wastewaters

Wastewaters containing high ammonia concentration and negligible BOD can be treated by biological nitrification. For example, wastewater from a fertilizer manufacturing complex was treated by the activated sludge process. The NH_4^-N content of the influent wastewater varied from 339 to 420 mg/L and inorganic suspended solids varied from 313 to 598 mg/L. The TDS was 6300 mg/L. Because of the high inert suspended solids, the mixed liquor was only 20 percent volatile with a sludge volume index (SVI) of 30 to 40 mL/g. A small fragile floc was generated, which provided an effluent TSS of 55 mg/L. Alkalinity was supplied to the system in the form of sodium bicarbonate. The relationship between nitrification rate and mixed liquor temperature is shown in Fig. 6.88. The temperature correction coefficient, θ, was 1.13. This value is significantly higher than for typical domestic wastewater indicating that the nitrification rate was more sensitive to the operating mixed liquor temperature. The alkalinity requirements showed considerable variation (Fig. 6.89), which was attributed to the presence of alkalinity in the influent suspended solids.

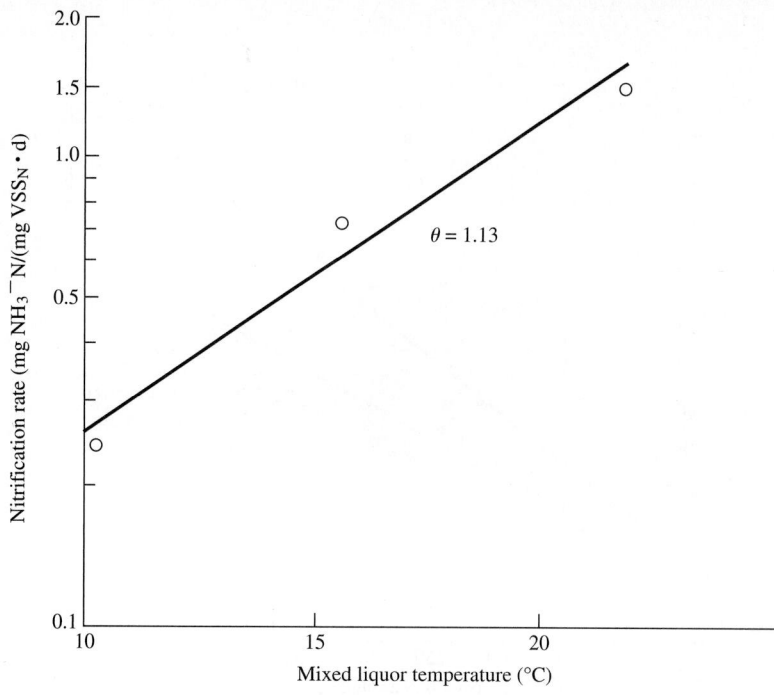

FIGURE 6.88 Relationship between nitrification rate and temperature for a fertilizer wastewater.

Inhibition of Nitrification

In treating industrial wastewaters, nitrification is frequently inhibited, or in some instances prevented, by the presence of toxic organic or inorganic compounds. This is shown in Fig. 6.90, which shows the nitrification results for treatment of an organic chemicals wastewater. These data show that a minimum aerobic SRT of 25 d was required to obtain complete nitrification at 22 to 24°C. The minimum SRT required for complete nitrification of municipal wastewater at these temperatures is approximately 4 d. An SRT of 55 to 60 d was required for complete nitrification of this wastewater at 10°C versus approximately 12 d for municipal wastewater. It was also shown that, at a mixed liquor temperature of 10°C, the nitrifiers were less tolerant to variations in influent composition and temperature than were the heterotrophic organisms responsible for BOD removal and denitrification. Similar results were obtained for a wastewater from a coke plant in which the nitrification rate was approximately one order of magnitude less than that for municipal wastewater, as shown in Fig. 6.91.

Blum and Speece[49] have characterized the toxicity of a variety of organic compounds to nitrification, as summarized in Table 6.28.

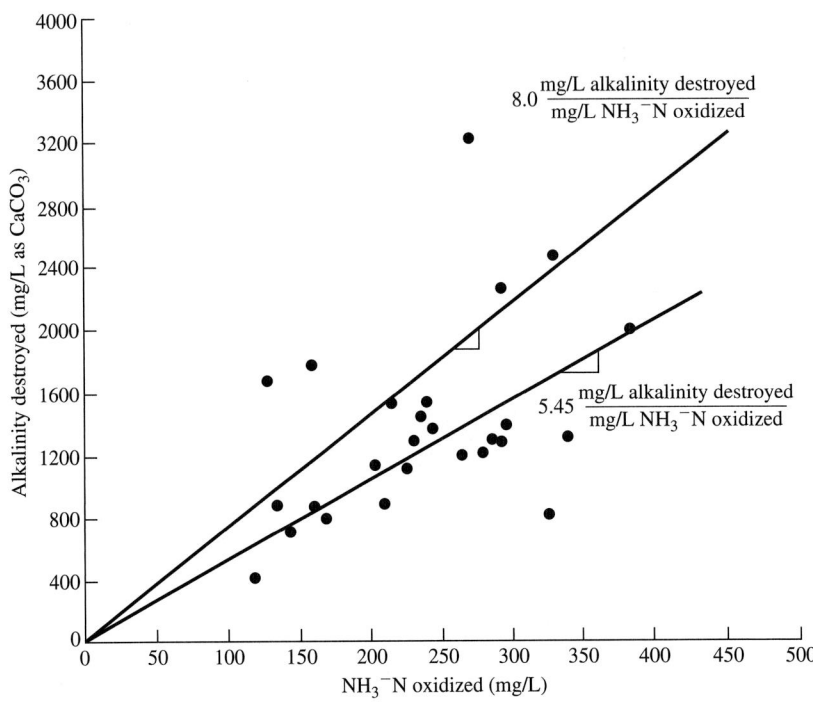

FIGURE 6.89 Alkalinity utilization in the treatment of a fertilizer wastewater.

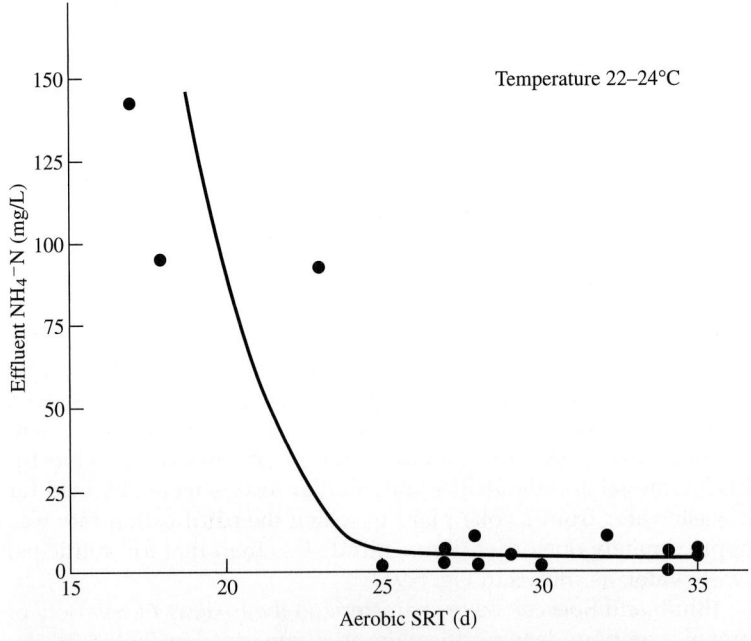

FIGURE 6.90 Nitrification relative to the aerobic SRT for an organic chemicals wastewater. (*Adapted from Anthoisen, 1976.*)

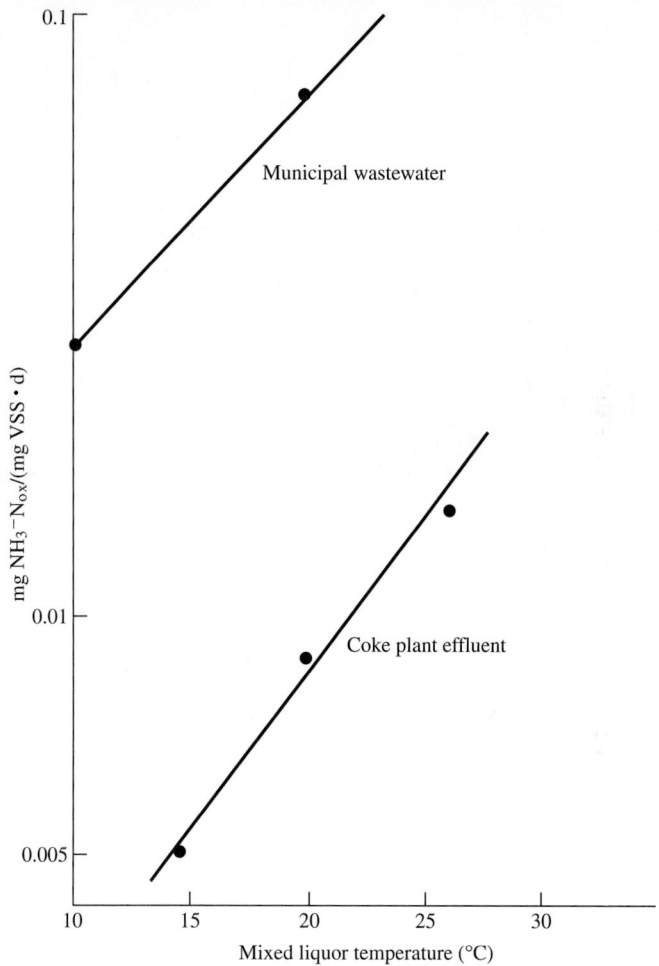

FIGURE 6.91 Relationship between nitrification rate and temperature for municipal wastewater and a coke plant effluent.

The effect of salt levels on nitrification were discussed by Henning and Kayser.[50] They found that fluoride concentrations of 100 mg/L reduced the nitrification rate by 80 percent. Sulfate had no effect at concentrations up to 50 g/L. Chlorides, however, showed significant inhibition, as shown in Fig. 6.92. They showed that nitrification rates were reduced up to 60 percent at NO_2^-N concentrations of several hundred milligrams per liter at pH 8.0.

In cases where nitrification is significantly reduced or totally inhibited, the application of powdered activated carbon (PAC) to

Compound	Biodegradability	Rate of Biodegradation, mg COD/gVSS · h	EC$_{50}$, mg/L	
			Nitrosomonas	Heterotrophs
Cyclohexane	A	—	97	29
Octane	A	—	45	—
Decane	C	—	—	—
Dodecane	D	—	—	—
Methylene chloride	D	—	1.2	320
Chloroform	D	—	0.48	640
Carbon tetrachloride	—	—	51	130
1,1-Dichloroethane	—	—	0.91	620
1,2-Dichloroethane	—	—	29	470
1,1,1-Trichloroethane	—	—	8.5	450
1,1,2-Trichloroethane	—	—	1.9	240
1,1,1,2-Tetrachloroethane	—	—	8.7	230
1,1,2,2-Tetrachloroethane	—	—	1.4	130
Pentachloroethane	—	—	7.9	150
Hexachloroethane	—	—	32	—
1-Chloropropane	D	—	120	700
2-Chloropropane	—	—	110	440
1,2-Dichloropropane	—	—	43	—
1,3-Dichloropropane	C	—	4.8	210
1,2,3-Trichloropropane	—	—	30	290
1-Chlorobutane	D	—	120	230

1-Chloropentane	D	—	99	68
1,5-Dichloropentane	—	—	13	—
1-Chlorohexane	D	—	85	83
1-Chloroctane	—	—	420	52
1-Chlorodecane	D	—	—	40
1,2-Dichloroethylene	D	—	—	—
Trans-1,2-dichloroethylene	—	—	80	1700
Trichloroethylene	A	—	0.81	130
Tetrachloroethylene	—	—	110	1900
1,3-Dichloropropene	—	—	0.67	120
5-Chloro-1-pentyne	—	—	0.59	86
Methanol	A	26	880	20,000
Ethanol	A	32	3900	24,000
1-Propanol	A	71	980	9600
1-Butanol	A	84	—	3900

All *Nitrosomonas* and aerobic heterotroph data corrected for pKa (ionization) and H (gas/liquid partitioning).

$A = \dfrac{\text{BOD}}{\text{TOD}} > 50\%$; readily biodegradable $C = \dfrac{\text{BOD}}{\text{TOD}} < 10\text{--}25\%$; refractory

$B = \dfrac{\text{BOD}}{\text{TOD}} > 25\text{--}25\%$; moderately biodegradable $D = \dfrac{\text{BOD}}{\text{TOD}} < 10\%$; nondegradable

Source: Adapted from Blum and Speece, 1990.

TABLE 6.28 Biodegradability and Biotoxicity Data

Compound	Biodegradability	Rate of Biodegradation, mg COD/gVSS · h	EC_{50}, mg/L	
			Nitrosomonas	Heterotrophs
1-Pentanol	A	—	520	—
1-Hexanol	A	—	—	—
1-Octanol	A	—	67	200
1-Decanol	B	—	—	—
1-Dodecanol	B	—	140	210
2,2,2-Trichloroethanol	—	—	2.0	—
3-Chloro-1,2 propanediol	D	—	—	—
Ethylether	C	—	—	17,000
Isopropylether	D	—	610	—
Acetone	B	—	1200	16,000
2-Butanone	—	—	790	11,000
4-Methyl-2-pentanone	—	—	1100	—
Ethyl-acrylate	—	—	47	—
Butyl-acrylate	—	—	38	470
2-Chloropropionic-acid	A	24	0.04	0.18
Trichloroacetic-acid	D	0	—	—
Diethanolamine	A	16	—	—
Acetonitrile	A	—	73	7500
Acrylonitrile	A	—	6.0	52
Benzene	A	—	13	520
Toluene	A	—	84	110

Xylene	A	—	100	1000
Ethylbenzene	B	—	96	130
Chlorobenzene	D	—	0.71	310
1,2-Dichlorobenzene	—	—	47	910
1,3-Dichlorobenzene	D	—	93	720
1,4-Dichlorobenzene	D	—	86	330
1,2,3-Trichlorobenzene	—	—	96	—
1,2,4-Trichlorobenzene	D	—	210	7700
1,3,5-Trichlorobenzene	—	—	96	—
1,2,3,4-Tetrachlorobenzene	—	—	20	—
1,2,4,5-Tetrachlorobenzene	D	—	9	—
Hexachlorobenzene	D	—	4	350
Benzyl alcohol	A	—	390	2100
4-Chloroanisole	—	—	—	902

All *Nitrosomonas* and aerobic heterotroph data corrected for pKa (ionization) and H (gas/liquid partitioning).

$A = \dfrac{BOD}{TOD} > 50\%$; readily biodegradable $C = \dfrac{BOD}{TOD} < 10\text{–}25\%$; refractory

$B = \dfrac{BOD}{TOD} > 25\text{–}25\%$; moderately biodegradable $D = \dfrac{BOD}{TOD} < 10\%$; nondegradable

Source: Adapted from Blum and Speece, 1990.

TABLE 6.28 Biodegradability and Biotoxicity Data (*Continued*)

Compound	Biodegradability	Rate of Biodegradation, mg COD/gVSS · h	EC$_{50}$, mg/L Nitrosomonas	EC$_{50}$, mg/L Heterotrophs
2-Furaldehyde	B	37	—	—
Benzonitrile	B	—	32	470
m-Tolunitrile	—	—	0.88	290
Nitrobenzene	A	14	0.92	370
2,6-Dinitrotoluene	—	—	183	—
I-Nitronaphthalene	—	—	—	380
Naphthalene	A	—	29	670
Phenanthrene	C	—	—	—
Benzidine	D	—	—	—
Pyridine	A	—	—	—
Quinoline	A	8.5	—	—
Phenol	A	80	21	1100
m-Cresol	A	—	0.78	440
p-Cresol	A	—	27	260
2,4-Dimethylphenol	—	28.2	—	—
3-Ethylphenol	—	—	—	144
4-Ethylphenol	—	—	14	—
2-Chlorophenol	—	—	2.7	360
3-Chlorophenol	—	—	0.20	160
4-Chlorophenol	A	39.8	0.73	98
2,3-Dichlorophenol	—	—	0.42	210

2,4-Dichlorophenol	10.5	0.79	—
2,5-Dichlorophenol	—	0.61	180
2,6-Dichlorophenol	—	8.1	410
3,5-Dichlorophenol	—	3.0	—
2,3,4-Trichlorophenol	—	52	7.8
2,3,5-Trichlorophenol	—	3.9	—
2,3,6-Trichlorophenol	—	0.42	14
2,4,5-Trichlorophenol	—	3.9	23
2,4,6-Trichlorophenol	—	7.9	—
2,3,5,6-Tetrachlorophenol	—	1.3	1.5
Pentachlorophenol	—	6.0	—
2-Bromophenol	—	0.35	—
4-Bromophenol	B	0.83	120
2,4,6-Tribromophenol	—	7.7	—
Pentabromophenol	—	0.27	—
Resorcinol	A	57.5	7.8

All *Nitrosomonas* and aerobic heterotroph data corrected for pKa (ionization) and H (gas/liquid partitioning).

$A = \dfrac{BOD}{TOD} > 50\%$; readily biodegradable $C = \dfrac{BOD}{TOD} < 10\text{--}25\%$; refractory

$B = \dfrac{BOD}{TOD} > 25\text{--}25\%$; moderately biodegradable $D = \dfrac{BOD}{TOD} < 10\%$; nondegradable

Source: Adapted from Blum and Speece, 1990.

TABLE 6.28 Biodegradability and Biotoxicity Data (*Continued*)

Compound	Biodegradability	Rate of Biodegradation, mg COD/gVSS · h	EC$_{50}$, mg/L	
			Nitrosomonas	Heterotrophs
Hydroquinone	B	54.2	—	—
2-Aminophenol	—	21.1	0.27	0.04
4-Aminophenol	—	16.7	0.07	—
2-Nitrophenol	—	14.0	11	11
3-Nitrophenol	—	17.5	—	—
4-Nitrophenol	A	16.0	2.6	160
2,4-Dinitrophenol	—	6.0	—	—

All *Nitrosomonas* and aerobic heterotroph data corrected for pKa (ionization) and H (gas/liquid partitioning).

$A = \dfrac{BOD}{TOD} > 50\%$; readily biodegradable $C = \dfrac{BOD}{TOD} < 10\text{–}25\%$; refractory

$B = \dfrac{BOD}{TOD} > 25\text{–}25\%$; moderately biodegradable $D = \dfrac{BOD}{TOD} < 10\%$; nondegradable

Source: Adapted from Blum and Speece, 1990.

TABLE 6.28 Biodegradability and Biotoxicity Data (*Continued*)

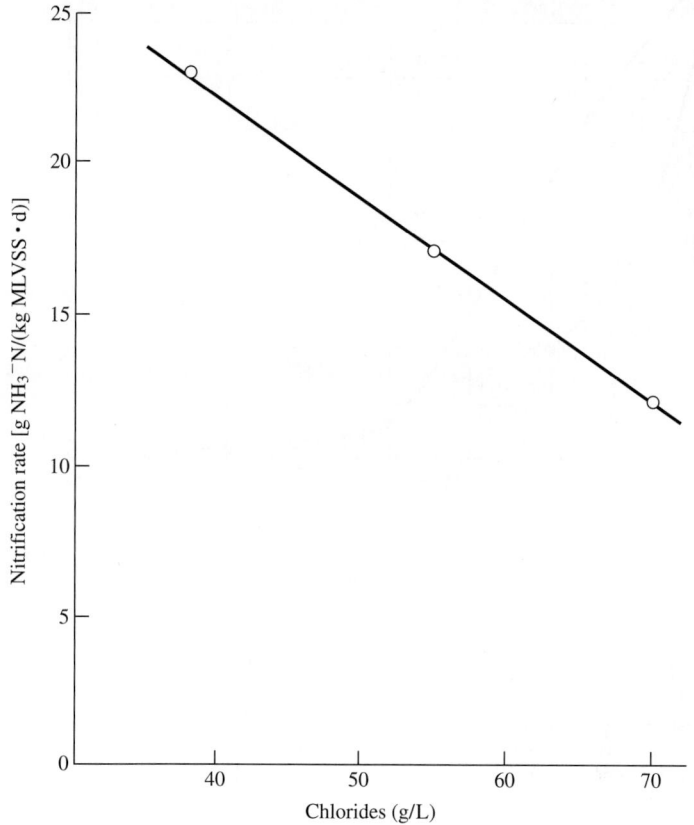

FIGURE 6.92 Nitrification kinetics at different chloride concentrations.

adsorb the toxic agents may enhance nitrification. However, in some cases, excessive quantities of PAC are required to achieve single-stage nitrification. In some cases a second-stage nitrification step can be successfully employed after a first-stage biological process for removal of carbonaceous material and reduction of toxicity.

Metals have been found to be toxic to growing *Nitrosomonas* culture (Skinner and Walker[51]) with complete inhibition for the following metals and concentrations: nickel 0.25 mg/L; Cr, 0.25 mg/L; and Cu, 0.1 to 0.5 mg/L.

Cyanide toxicity to nitrifiers is shown in Fig. 6.93.[52,53]

Unionized ammonia (NH_3) inhibits both *Nitrosomonas* and *Nitrobacter*, as shown in Fig. 6.94.[54] Since the unionized fraction increases with pH, a high pH combined with a high total ammonia concentration will severely inhibit or prevent complete biological nitrification. Since *Nitrosomonas* is less sensitive to ammonia toxicity than

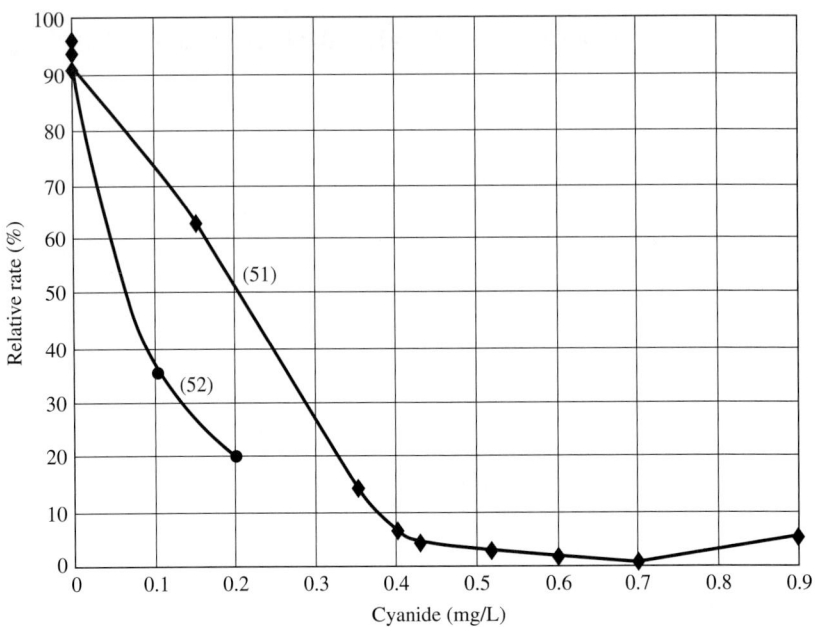

FIGURE 6.93 Relative rate of nitrification as function of cyanide. (*Adapted from Sadick et al., 1996, and Zacharias and Kayser, 1995.*)

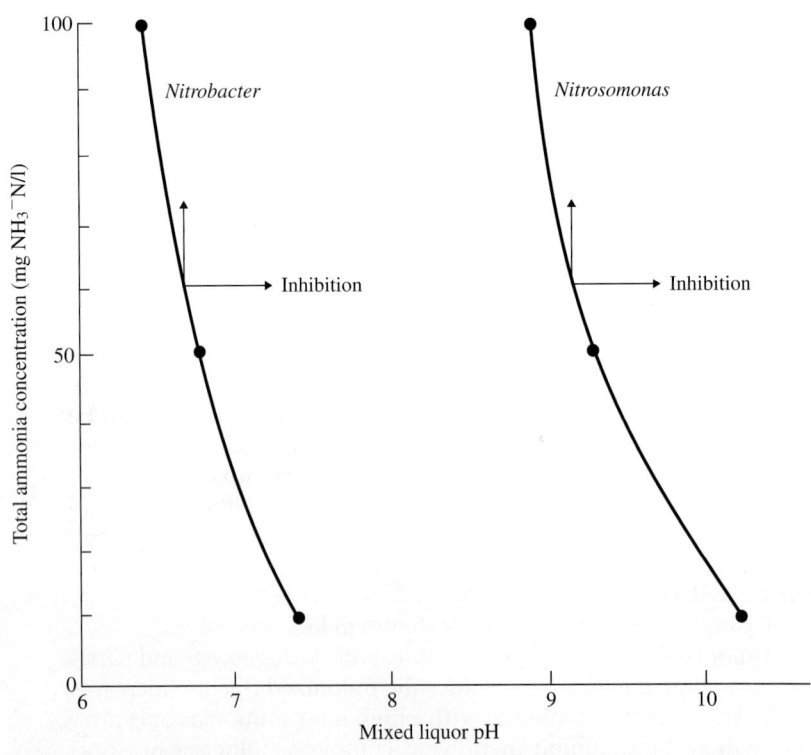

FIGURE 6.94 Ammonia inhibition in the activated sludge process. (*Adapted from Anthoisen, 1976.*)

Nitrobacter, the nitrification process may only be partially complete and result in accumulation of nitrite ion (NO_2^{2-}). This can have severe consequences since NO_2^{2-} is strongly toxic to many aquatic organisms whereas NO_3^--N is not. Ammonia toxicity to activated sludge biomass is rarely a problem in treating municipal wastewaters, since the concentration of total ammonia is low and the mixed liquor pH is near neutral. Industrial wastewaters with high ammonia levels and the potential for high pH excursions, however, may cause biotoxicity and loss of the nitrification process. Under these conditions, it is necessary to control the mixed liquor pH to avoid biotoxicity due to an ammonia spill or shock load. In extreme cases, two stages operated at different pH values may be required to separate the *Nitrosomonas* and *Nitrobacter* and allow complete nitrification. The effect of pH resulting in ammonia and nitrite inhibition to nitrification is presented in Fig. 6.95.

Unfortunately, many industrial wastewaters contain these and other compounds, which alone or together exert a greater but undetermined inhibition effect on the nitrification process. It is essential, therefore, to determine the specific nitrification rate q_N and the $(\theta_c)_{min}$ required to achieve nitrification under actual operating conditions. The value of q_N can be determined using either a batch activated sludge (BAS) test or the semicontinuous fed batch reactor method.

Batch Activated Sludge Nitrification

In the BAS test procedure, a wastewater sample is aerated in the presence of an actively nitrifying sludge. The sludge can be obtained from

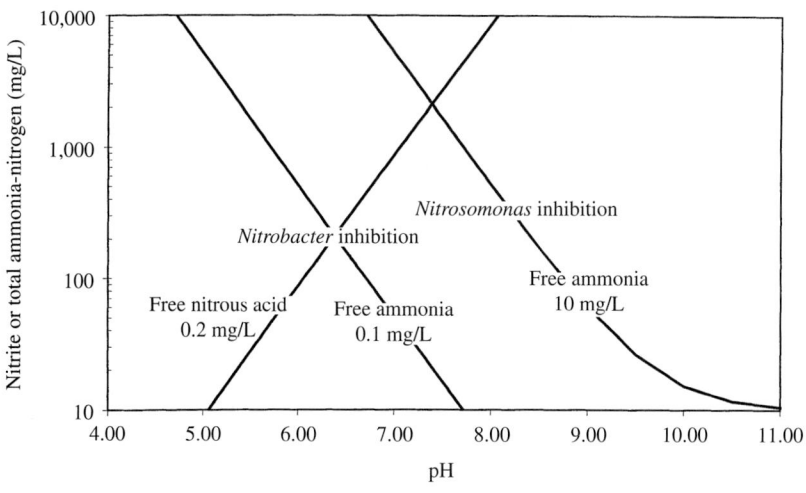

FIGURE 6.95 Ammonia and nitrite inhibition to nitrification.

either a municipal activated sludge plant with negligible industrial load or developed separately from a commercially available nitrifier culture. Regardless of the sludge source, it must be possible to quantify the mass of nitrifiers in the bulk MLVSS concentration since q_N is expressed per unit of nitrifier mass (VSS_N). The initial NH_3^-N concentration in the BAS test should be between 20 and 50 mg/L to eliminate substrate-induced toxicity. If the wastewater contains organic nitrogen, then TKN should also be measured to account for biohydrolysis during the test. Finally, the alkalinity should be adjusted with sodium bicarbonate ($NaHCO_3$) to provide 7.15 mg alkalinity as $CaCO_3$ mg TKN plus a 50 mg/L residual. A control test should be run using the same initial NH_3^-N concentration. The wastewater and control sample should be vigorously aerated and aliquots withdrawn over time for analysis. Inhibition to nitrification can be determined using this test procedure by comparing nitrate production to a control with the same initial ammonia content as shown in Fig. 6.96.

The results of a BAS nitrification test conducted at $T = 21°C$ are shown in Fig. 6.97 for a mixed liquor from a municipal wastewater treatment plant. On the basis of the plant's historical operating data and Eq. (6.61), the fraction of nitrifiers (f_N) in the MLVSS was 0.0245 mg VSS_N/mg VSS. Initially, the wastewater had negligible NO_3^-N and organic nitrogen and an NH_3^-N concentration of 48 mg/L. After 24 h of aeration, 38 mg/L NO_3^-N was produced.

The overall specific NH_3^-N oxidation rate was

$$\frac{38 \text{ mg}/(L \cdot d)}{1200 \text{ mg VSS}/L} = 0.032 \text{ mg } NH_3^-N/(mg \text{ MLVSS} \cdot d)$$

The nitrifier specific nitrification rate was

$$q_N = 0.032/0.0245 = 1.3 \text{ mg } N/(mg \text{ VSS}_N \cdot d)$$

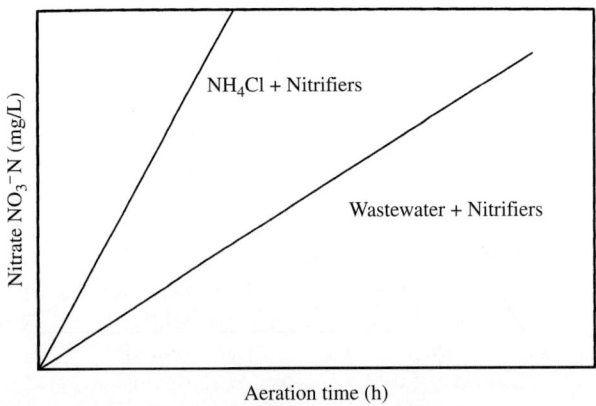

FIGURE 6.96 Test procedure for nitrification rate determination.

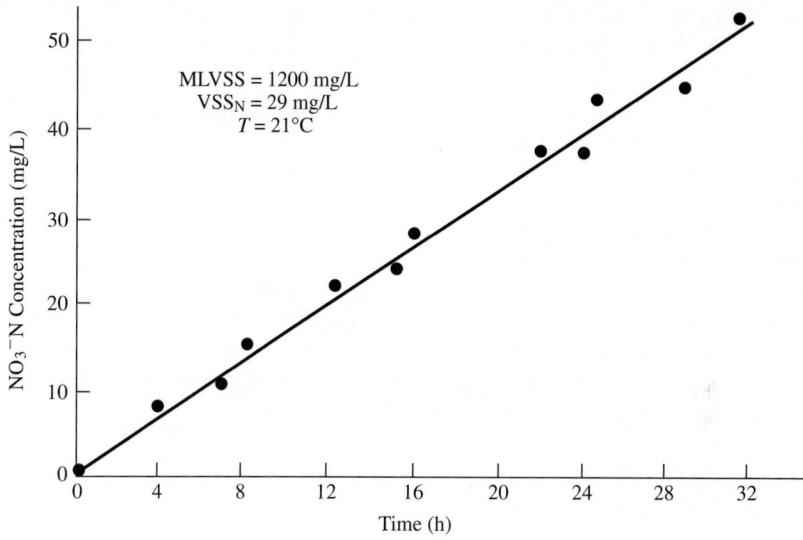

FIGURE 6.97 Results of batch nitrification test.

The nitrifier specific growth rate for $a_N = 0.15$ mg/mg was 0.195 d^{-1}. Neglecting temperature effects, and with $b_N = 0.05$ d^{-1} the $(\theta_c)_{min}$ is

$$(\theta_c)_{min} = \frac{1}{0.195 - 0.05} = 6.9 \text{ d}$$

The $(\theta_c)_{min}$ for municipal wastewater at $T = 21°C$ is approximately 4 d, indicating a two fold inhibition effect on nitrification by the industrial wastewater. The θ_c design would be determined using $(\theta_c)_{min}$ and an appropriate safety factor.

It should be noted that the BAS nitrification test used an existing sludge with a known f_N and VSS$_N$ concentration to determine q_N. The actual mixed liquor established for treatment of this wastewater will have a different f_N value depending on its NH$_3$-N and BOD concentrations. This nitrifier fraction and the measured q_N should be used to determine the hydraulic retention time required in the aeration basin.

Fed Batch Reactor Nitrification Test

The FBR procedure described on p. 322 can also be used to determine the nitrification rate. The requirements for sludge characterization and alkalinity addition are the same as for the BAS nitrification test. The wastewater is added to the reactor at a constant rate, and aliquots are withdrawn over time for analysis. Nitrate (NO$_3^-$N) and nitrite (NO$_2^-$N) production is the preferred method of expressing the test

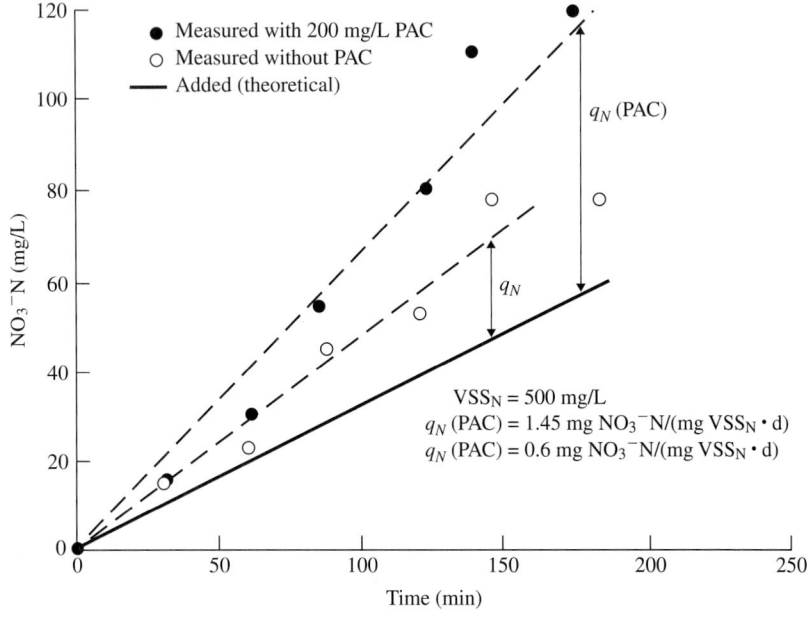

FIGURE 6.98 Determination of nitrification rate using FBR procedure with and without PAC addition.

results and determining q_N, since it eliminates adjustments needed to account for biohydrolysis and cell synthesis. If NH_4^-N removal is used, however, then TKN and COD (or BOD) must also be measured to complete the nitrogen balance.

The results of FBR nitrification tests with and without PAC addition (200 mg/L) are presented in Fig. 6.98. The mixed liquor had a $VSS_N = 500$ mg/L, and the wastewater sample initially had significant concentrations of BOD, organic nitrogen, and NO_3^-N. The nitrification rates were calculated as the difference between the slopes of the two linear traces. They indicate that addition of 200 mg/L of PAC increased the nitrification rate from 0.6 to 1.45 mg $NO_3^-N/$(mg $VSS_N \cdot$ d). The $(\theta_c)_{min}$ can be calculated as in the BAS nitrification test procedure.

Denitrification

Some industrial wastewaters such as those from fertilizer, explosive/propellant manufacture, and the synthetic fibers industry contain high concentrations of nitrates, while others generate nitrates by nitrification. Since biological denitrification generates one hydroxyl ion while nitrification generates two hydrogen ions, it may be advantageous to couple the nitrification and denitrification processes to provide "internal" buffering capacity. While many organics inhibit biological nitrification, this is not generally true for denitrification. Sutton et al.[55] showed that denitrification rates for an organic chemicals plant wastewater were comparable to those observed using nitrified municipal wastewater; however, biological nitrification of the organic chemicals wastewater was severely inhibited. Denitrification uses BOD as a carbon source for synthesis and energy and nitrate as an oxygen source.

$$NO_3^- + BOD \rightarrow N_2 + CO_2 + H_2O + OH^- + \text{new cells}$$

The denitrification process consumes approximately 3.7 g COD per g NO_3^--N reduced and produces 0.45 g VSS and 3.57 g alkalinity per g NO_3^--N reduced. This amounts to one-half the alkalinity that is consumed during nitrification. Some of this alkalinity, however, is lost by reaction with the CO_2 generated by microbial respiration.

Orhon et al.[56] compared the sludge yield under aerobic conditions to anoxic conditions as shown in Table 6.29. McClintock et al.[57] showed that the biomass yield under anoxic conditions was 54 percent that under aerobic conditions and that the endogenous coefficient was 51 percent that under aerobic conditions. Comparative sludge yield will be a function of SRT.

Factors affecting denitrification include:

- Temperature
- Dissolved oxygen
- Substrate biodegradability
- Sludge age

Results of denitrification of an organic chemicals plant wastewater are shown in Fig. 6.99. The temperature effect on denitrification using continuous flow and batch reactors is shown in Fig. 6.100. The temperature coefficient, θ_{DN}, varies from 1.07 to 1.20. The effect of temperature on denitrification rate for $\theta = 1.09$ is shown in Fig. 6.101.

Wastewater	Aerobic	Anoxic, g Cell COD/g COD
Domestic sewage	0.63	0.50
Meat processing	0.64	0.51
Dairy	0.65	0.52
Confectionery	0.72	0.61

*Design conditions
[a]anoxic ~ 0.75 [a]aerobic
[b]anoxic ~ 0.75 [b]aerobic
Source: Adapted from Orhon et al., 1996.

TABLE 6.29 Sludge Yield Coefficient (based on COD)*

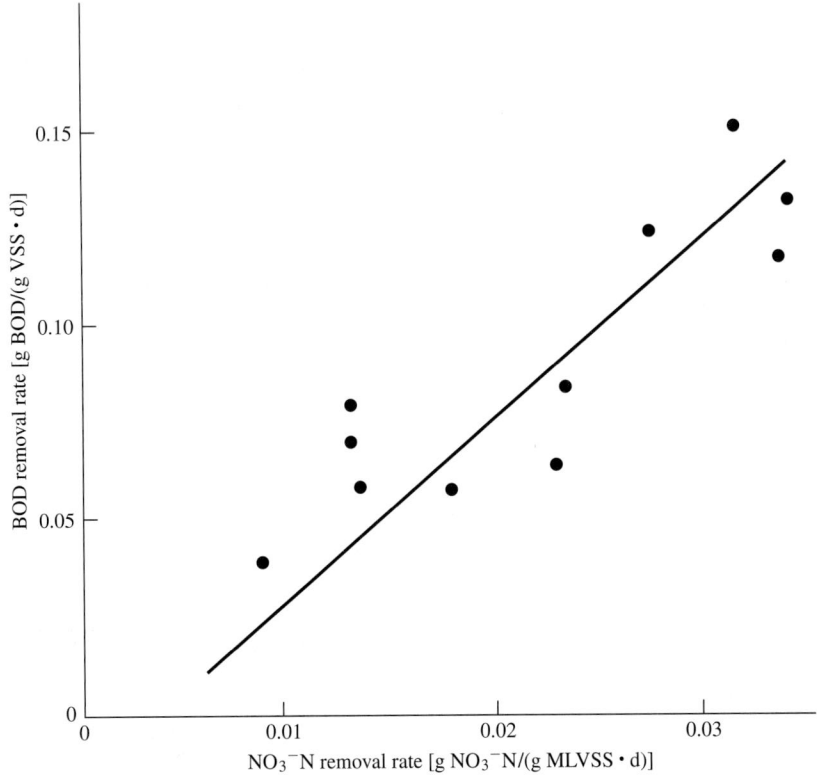

FIGURE 6.99 Relationship between nitrate reduction and BOD removal for an organic chemicals wastewater.

While oxygen inhibits denitrifying facultative bacteria depending on plant operating conditions, flocs may contain anoxic zones where denitrification will occur even when the liquid contains dissolved oxygen. This is illustrated by Fig. 6.102.

OH and Silverstein[58] showed that the IWA relationship was more applicable at high dissolved oxygen levels

$$q_{DN} = q_{DN\,(max)}\left(\frac{1}{1+DO/k}\right) \quad (6.62)$$

A suggested value for k is 0.38 mg/L. It is assumed that k will be a function of floc size and hence power level in the aeration basin.

Field experience has shown that as much as 10 to 25 percent denitrification can occur under aerobic conditions in aeration basins. The effect of DO on nitrification and denitrification is shown in Figs. 6.103 and 6.104.

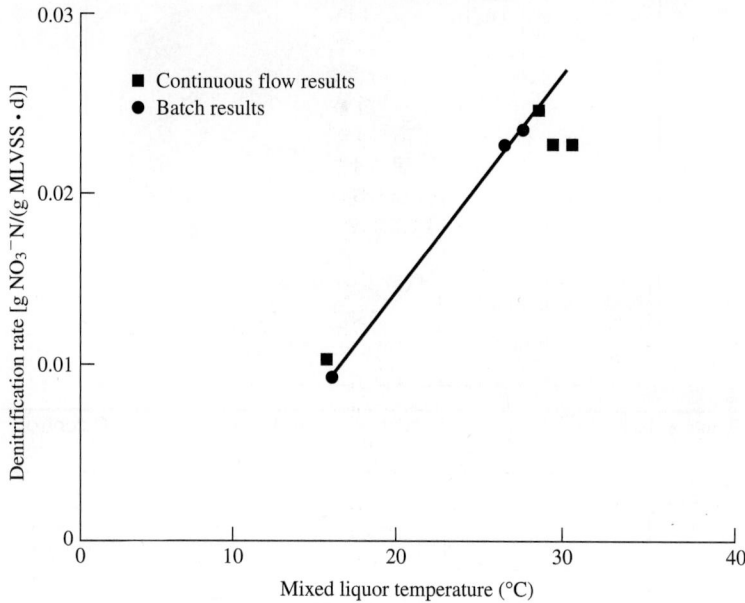

FIGURE 6.100 Relationship between denitrification rate and temperature for an organic chemicals wastewater.

The denitrification rate will depend on the biodegradability of the organics in the wastewater and the concentration of active biomass under aeration similar to the aerobic process. This, in turn, is related to both the SRT or F/M and the presence of inert solids in the sludge. As the F/M increases, the concentration of active biomass and the rate of denitrification increase. Although denitrification can occur

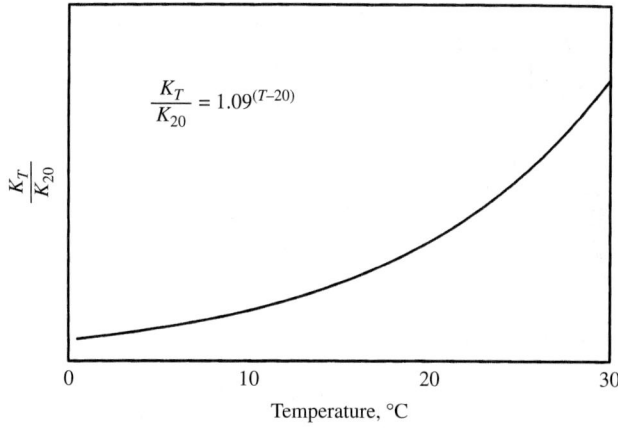

FIGURE 6.101 Effect of temperature on the denitrification rate.

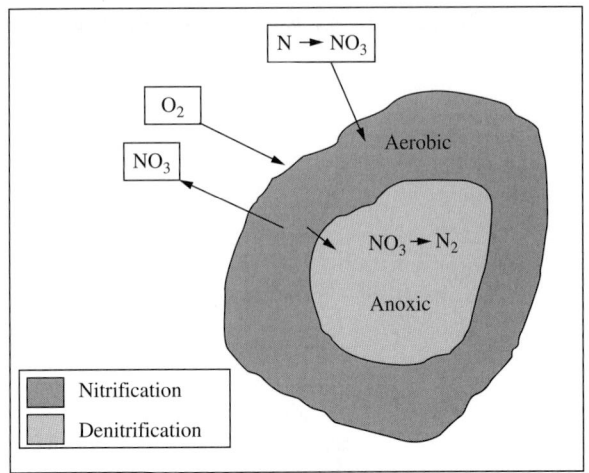

Figure 6.102 Simultaneous nitrification/denitrification through DO control.

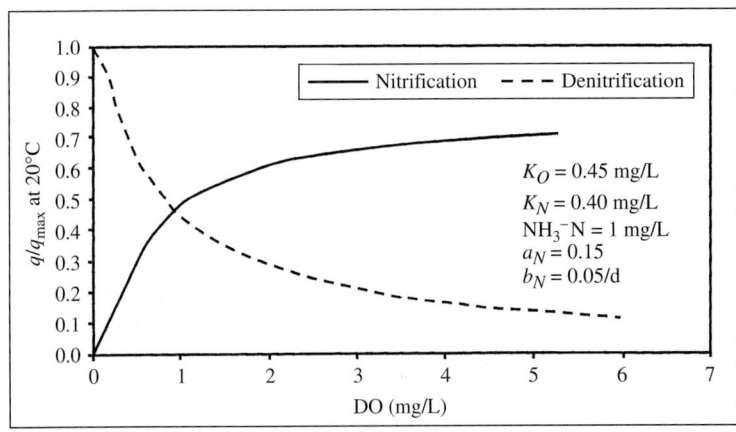

Figure 6.103 Effect of dissolved oxygen on nitrification and denitrification rates at 20°C.

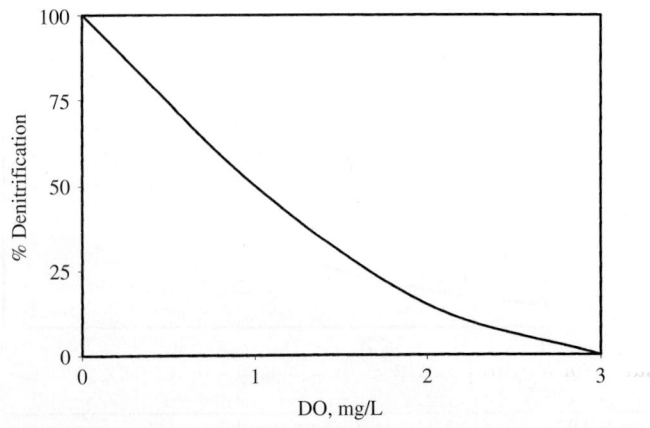

Figure 6.104 Simultaneous nitrification/denitrification.

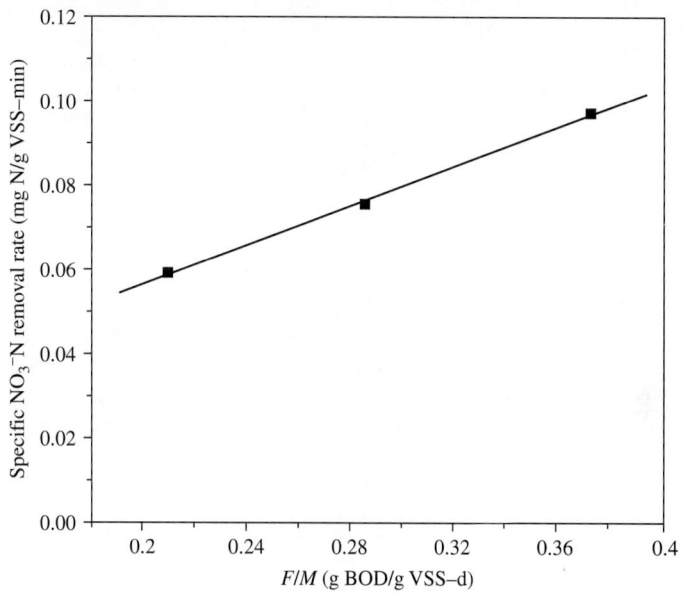

FIGURE 6.105 Relationship between F/M and denitrification rate.

under endogenous conditions (low F/M), using internal biomass reserves, it is very slow and requires long hydraulic retention times. Fig. 6.105 illustrates how the rate of nitrate removal increases with increasing F/M.

Since the rate of denitrification is affected by both wastewater characteristics and process design parameters, it is usually necessary to determine the rate by experimental means. A batch denitrification test should be conducted in which sludge and wastewater are mixed under anoxic conditions [oxidation reduction potential (ORP) = −100 mV], and the residual NO_3^--N concentration is determined with time. Depending on the organic composition of the wastewater, one of several removal rate relationships may be obtained. For a complex wastewater in a plug flow system, a pseudo first-order relationship may exist. In a complete mix system the rate will be proportional to the fraction of organics remaining in solution. Lie and Welander[59] have shown that the denitrification rate can be related to the ORP as shown in Fig. 6.106.

The denitrification rate for a wastewater can also be estimated from the oxygen uptake rate. In this case, the wastewater–anoxic sludge mixture is aerated and the SOUR determined over time. Correlation of R_{DN} and SOUR indicated that 1.0 mg NO_3^--N is equivalent to approximately 3.0 mg O_2, which is in good agreement with the theoretical value of 2.86 mg NO_3^--N/mg O_2.

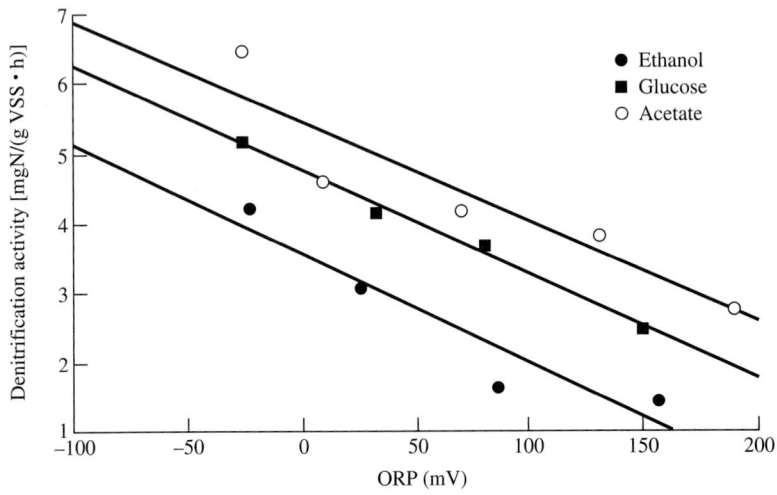

Figure 6.106 Denitrification activity versus ORP for activated sludge from Sjölunda pilot plant with addition of different carbon sources.

In cases where a carbon source is not available in the wastewater, methanol has been used as a carbon source. Various industrial effluents can also be used as a carbon source. Baumann and Krauth[60] have summarized various carbon sources as shown in Table 6.30.

It has been found that, in general, the kinetics of denitrification under anoxic conditions follows the same relationship as aerobic oxidation. This is shown by the batch oxidation of an industrial wastewater under anoxic and aerobic conditions presented in Fig. 6.107.

The BOD to NO_3 ratio for anoxic degradation has been found to vary from 3 to 4 mg BOD/mg NO_3^-N Under complete mix conditions, aerobic oxidation can be defined as:

$$\frac{S_r}{X_d x_v t} = K \frac{S_e}{S_o} \quad (6.63)$$

where S_r = BOD removal, mg/L
 x_v = mixed liquor volatile suspended solids, mg/L
 t = detention time, d
 K = reaction coefficient, d^{-1}
 S_e = effluent BOD, mg/L
 S_o = influent BOD, mg/L

Under anoxic conditions, Eq. (6.63) becomes:

$$\frac{3(NO_3 - N)}{X_d x_v t} = K_{DN} \frac{S_e}{S_o} \quad (6.64)$$

where NO_3^-N = nitrate nitrogen, mg/L
 K_{DN} = denitrification rate coefficient, d^{-1}

Industry	BOD$_5$, mg/L	COD, mg/L	DN Rate, mg NO$_3^-$N/ (g MLSS · h)
Chemical Industry			
Deicing agent (airfield)	65,000	118,000	1.98–3.06
Glue production I	148,500	282,400	0.96–1.26
Glue production II	1,080,000	1,340,000	1.14–2.12
Pharmaceutical industry I	136,000	188,100	4.08
Pharmaceutical industry II	163,000	320,000	1.14–1.53
Photographic industry	126,000	686,000	1.59–1.70
Food Industry			
Alcohol production	3,780	7,300	
Fusel oil	1,320,000	1,780,000	2.79–3.18
Milk processing industry	4,880	7,440	
Plant and vegetable processing	20,650	26,050	4.29
Slaughterhouse	183,000	246,000	1.44
Wine industry	173,100	211,100	5.40
Yeast industry	26,900	28,770	2.79–3.18
Common Substrates			
Acetic acid		1,056,000	3.35
Endogenous			0.26–0.65

Temperature: 13–16°C
Source: Adapted from Baumann and Krauth, 1996.

TABLE 6.30 List of Possible Industrial Wastes or Waste by-products for Denitrification

The aerobic and anoxic kinetic relationships are described by Fig. 6.108. The nitrification rate is a function of the BOD/NO$_3^-$N ratio as shown in Fig. 6.109. It is assumed that 1 mg NO$_3^-$N will consume 3 mg/L BOD. Simultaneous nitrification/denitrification will also occur in the aeration basin. This may vary from 10% at a DO of 2 mg/L to 50% at DO of 1 mg/L.

It is assumed in this equation that the BOD to nitrate ratio is 3. Anoxic and aerobic kinetic coefficients are compared in Table 6.31.

Denitrification in final clarifiers causes floating sludge and increased effluent suspended solids. The nitrogen gas rate production depends on the carbon source available for denitrification, the SRT, temperature, and sludge concentration. Henze et al. estimated that 6 to 8 and 8 to 10 mg/L NO$_3^-$N needs to be denitrified in the sludge blanket to cause sludge flotation at 10 and 20°C, respectively.

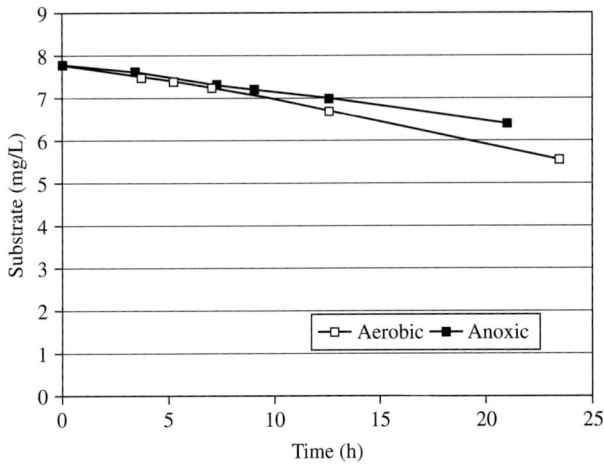

FIGURE 6.107 Comparison between aerobic and anoxic degradable substrate removal.

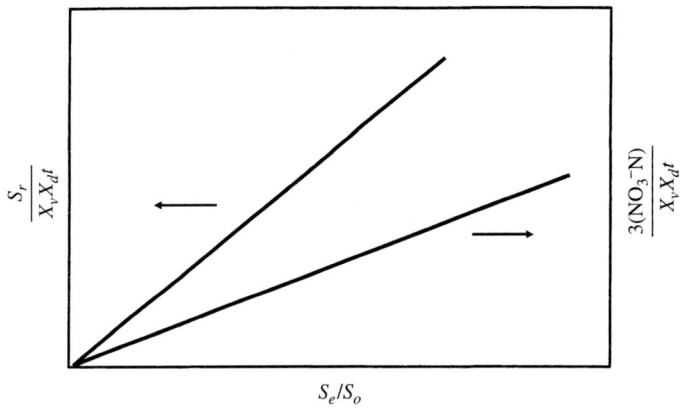

FIGURE 6.108 Aerobic and anoxic kinetic relationships.

	Anoxic	Aerobic
Pharmaceutical	9.2	21.0
Endogenous	4.4	6.3
Pulp and paper	6.0	—

TABLE 6.31 Comparison of Aerobic and Anoxic Kinetic Coefficients (d^{-1})

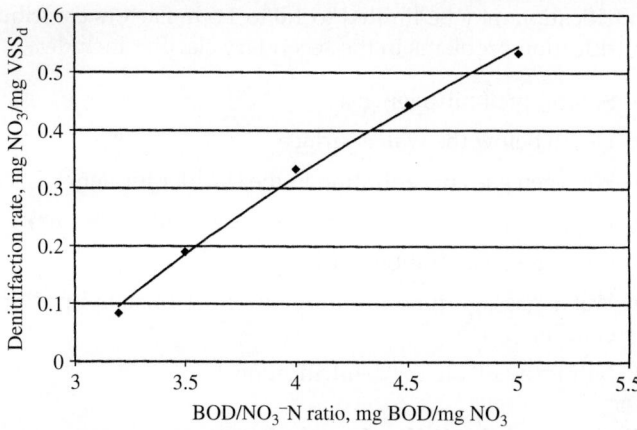

FIGURE 6.109 Denitrification rate versus BOD/ammonia ratio for a given K.

Most denitrification results from endogenous respiration and the utilization of adsorbed slowly degradable organics. This in turn is related to the active mass, which is a function of SRT. A relationship between effluent suspended solids and effluent nitrate concentration is shown in Fig. 6.110. The thickening time in secondary clarifiers

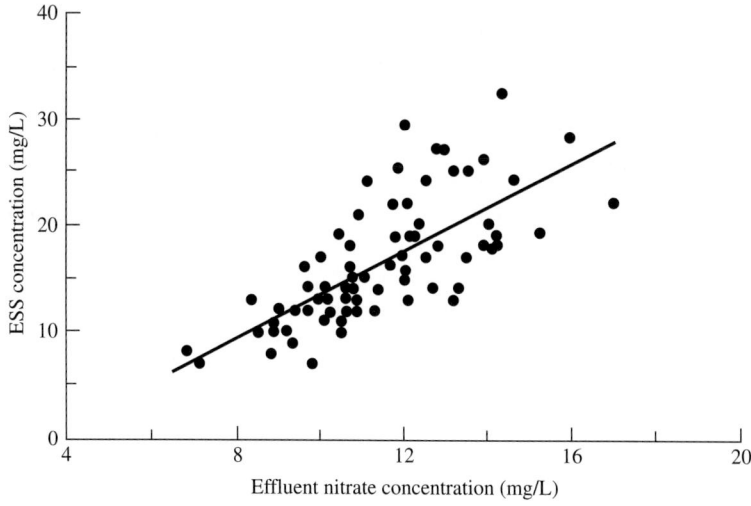

FIGURE 6.110 Effluent suspended solids concentration as a function of effluent nitrate concentration at a large nitrification/denitrification plant. (*Adapted from Sutton et al., 1979.*)

with nitrification may be limited to 1.0 to 1.5 h. Factors contributing to denitrification problems in the secondary clarifier include:

- Solubility of nitrogen gas
- Depth below the water surface
- Nitrogen gas concentration in the clarifier influent
- Nitrogen gas production rate (i.e., denitrification rate)
- Oxygen concentration in the clarifier influent
- Water passage time from clarifier inlet to actual position in clarifier
- Nitrate available for denitrification

Nitrification and Denitrification Systems

A number of alternative treatment systems are available to achieve nitrification and denitrification, in which some form of aerobic-anoxic sequencing is provided. The systems differ in whether they utilize a single sludge or two sludges in separate nitrification and denitrification reactors. The single-sludge system uses one basin and clarifier and the raw wastewater or endogenous reserves as the carbon and energy sources for denitrification.

The two-sludge system uses two basins with separate clarifiers to isolate the sludges. A supplemental carbon source such as methanol (CH_3OH) is provided to the second stage for the carbon and energy source. The simplest configuration of the single-sludge system enables carbonaceous oxidation, nitrification, and denitrification to occur in a single basin by positioning the return sludge and aeration equipment to maintain defined aerobic-anoxic zones in different basin sections. An alternative single-sludge system utilizes a single basin for both aeration and sedimentation by providing intermittent aeration and nonaeration cycles to yield aerobic and anoxic phases of sufficient duration to permit nitrate reduction. Two process flow configurations for single-sludge nitrification and denitrification are shown in Fig. 6.111. In the oxidation ditch (Fig. 6.111a), an aerobic zone exists in the vicinity of the aerator. As the mixed liquor passes away from the aerator, the dissolved oxygen is depleted. Anoxic conditions then exist, and denitrification occurs. This sequence is repeated around the ditch at each aerator installation.

In the internal recycle single-sludge process (Fig. 6.111b), nitrification occurs under aerobic conditions in the second basin. The second basin may be a separate tank (without intermediate clarification) or a single tank with internal baffles to isolate the aerobic and anoxic zones without short-circuiting. Each of these zones can be plug flow or CMAS. Design procedures followed by design examples for nitrification and denitrification follow.

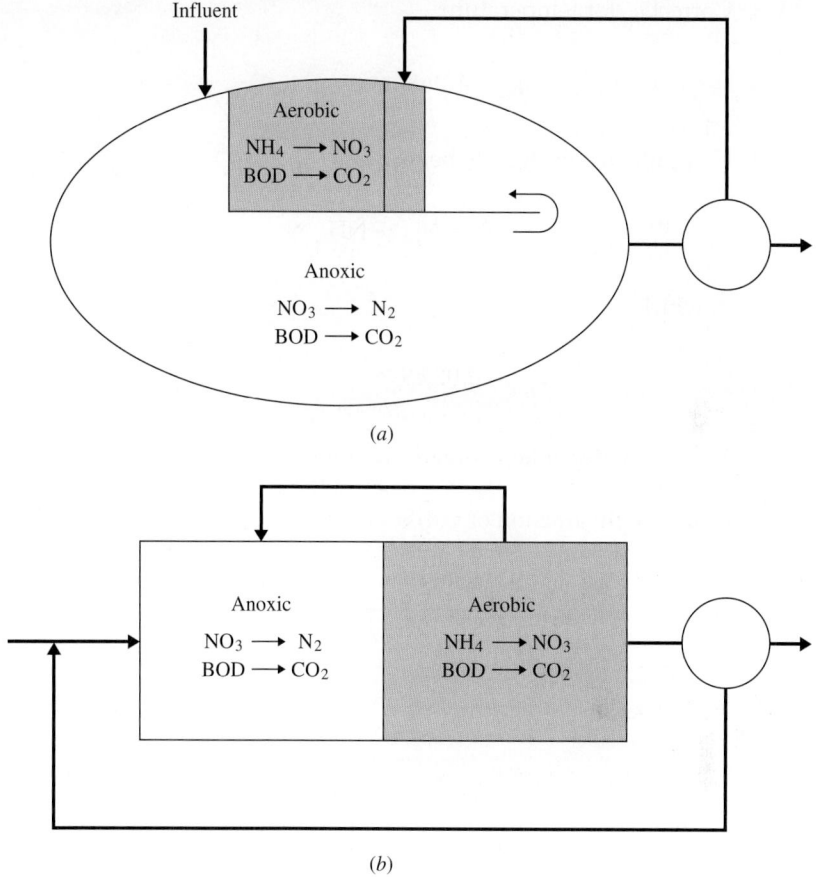

FIGURE 6.111 Alternative single-stage nitrification-denitrification systems.

Nitrification Design Procedure

1. Determine the maximum specific nitrification rate $q_{N(max)}$.
2. Correct $q_{N(max)}$ for the effluent NH_3^-N, DO, and θ_c:

$$q_N = q_{N(max)} \cdot \frac{NH_3^-Ne}{0.4 + NH_3^-N_e} \cdot \frac{DO}{K_o + DO} \cdot f_{a_N}$$

where K_o may vary from 0.2 to 1.

3. Correct q_N for temperature:

$$q_{N_T} = q_{N_{20°C}} 1.068^{(T-20)}$$

4. Compute the nitrogen to be oxidized

$$N_{OX} = TKN - SON - N_{syn} - NH_3^- N_e$$

in which

$$N_{syn} = 0.08\, aS_r$$

SON = nondegradable organic nitrogen

5. Compute the fraction of nitrifiers

$$f_N = \frac{0.15\, N_{OX}}{aS_r + 0.15 N_{OX}}$$

6. Compute the overall nitrification rate

$$R_N = q_N \cdot f_N \cdot X_V$$

7. Compute the required detention time

$$t_N = \frac{N_{OX}}{R_N}$$

8. Compute the WAS

$$\Delta X_V = (aS_r + 0.15\, N_{OX}) - b\, X_d X_V t_N$$

9. Compute the SRT

$$\theta_c = \frac{X_V t_N}{\Delta X_V}$$

10. Compute the oxygen requirements for organics removal

$$O_2/\text{mg/L} = a'S_r + 1.4\, b\, X_d X_V t_N$$

Principles of Aerobic Biological Oxidation

11. Compute the oxygen requirements for nitrification

$$O_2/d = 4.33 \cdot N_{OX}$$

12. Compute the alkalinity consumed

$$Alk = 7.15 \cdot N_{OX}$$

Example 6.14. Design a nitrification system to produce an effluent NH_3-No 1 mg/L at 20°C. The following conditions apply:

$TKNo = 50 \frac{mg}{L}$ $\qquad$ $SON_e = 1 \frac{mg}{L}$

$BOD_{50} = 410 \frac{mg}{L}$ $\qquad$ $SBOD_{5e} = 20 \frac{mg}{L}$

$DO = 2 \frac{mg}{L}$ $\qquad$ $NH_3\text{-}Ne = 1 \frac{mg}{L}$

$a_N = 0.15$ $\qquad$ $a_H = 0.6$

$bN_{20} = \frac{0.05}{d}$ $\qquad$ $b_H 20 = \frac{0.1}{d}$

$X_V = 3000 \frac{mg}{L}$ $\qquad$ $q_{N(max)} = \frac{2.3}{d}$

$\qquad\qquad\qquad\qquad\qquad$ $r_{DN(max)} = \frac{0.06}{d}$

Solution Assume: $\theta_c = 8.22$ d

$$fa_N = \frac{1}{1 + 0.2 \cdot b_{N20} \cdot \theta c} \qquad fa_N = 0.92$$

$$q_N = q_N(max) \cdot fa_N \cdot \frac{(NH_3\text{-}Ne)}{0.4 \frac{mg}{L} + (NH_3\text{-}N)e} \cdot \frac{DO}{0.4 \frac{mg}{L} + DO}$$

$$q_N = 1.27 \frac{1}{d}$$

$Nsyn = 0.08 \cdot a_H \cdot (BOD - BOD_{50})$ $\qquad$ $Nsyn = 18.7 \frac{mg}{L}$

$Nox = TKN_o - (NH_3\text{-}N)e - Nsyn - SONe$ $\qquad$ $Nox = 29.3 \frac{mg}{L}$

$f_N = \frac{a_N \cdot Nox}{a_N \cdot Nox + a_H \cdot (BOD - BOD_{50})}$ $\qquad$ $f_N = 0.0184$

$r_N = q_N \cdot f_N \cdot Xv$ $\qquad$ $r_N = 69.9 \frac{mg}{L \cdot d}$

$t_N = \frac{Nox}{r_N}$ $\qquad$ $t_N = 0.42$ d

$xd_N = \frac{0.8}{1 + 0.2 \cdot b_{N20} \cdot \theta c}$ $\qquad$ $xd_H := \frac{0.8}{1 + 0.2 \cdot b_{H20} \cdot \theta c}$

$$\Delta X_{VN} = a_N \cdot \text{Nox} - b_{N20} \cdot f_N \cdot \chi d_N \cdot Xv \cdot t_N$$

$$\Delta X_{VH} = a_H \cdot (\text{BOD}_{50} - \text{SBOD}_5 e) - b_{H20} \cdot (1 - f_N) \cdot \chi d_H \cdot Xv \cdot t_N$$

$$\Delta X_V = \Delta X_{VN} + \Delta X_{VH} \qquad \Delta X_V = 153 \frac{mg}{L}$$

Check: $\theta c = \dfrac{X_V \cdot t_N}{\Delta Xv}$ \qquad $\theta c = 8.22\ d$

How much denitrification would be expected in the aeration basin?

$$r_{DN} = r_{DN}\ \text{max} \cdot \dfrac{0.38 \frac{mg}{L}}{DO + 0.38 \frac{mg}{L}} \qquad r_{DN} = 9.6 \times 10^{-3}\ \frac{1}{d}$$

$$N_{dn} = r_{DN} \cdot Xv \cdot (1 - f_N) \cdot t_N \qquad N_{dn} = 11.8 \frac{mg}{L}$$

Denitrification Design Procedure

1. Wastewater characterization:

    Soluble BOD — 0.45 μ filtered

    Nitrogen — TKN, $NH_3\text{-}N$, $NO_3\text{-}NO_2\text{-}N$

2. Determine nitrogen to be oxidized.

$$\text{TKN} - NH_3\text{-}N_{\text{eff}} + N_{\text{syn}}\ (0.04\ \text{BOD}_R) + \text{SON}$$

3. Compute $NO_3\text{-}N$ to be denitrified.
4. Estimate simultaneous N/DN and recycle $NO_3\text{-}N$.
    a. Simultaneous N/DN is a function of dissolved oxygen in the aeration basin.
    b. The $NO_3\text{-}N$ denitrified in the anoxic basin is a function of the internal recycle plus the return sludge recycle.
5. The detention time required for denitrification is computed from the relationship (see example):

$$\dfrac{3(NO_3\text{-}N)}{X_V X_d t} = K_{DN} \dfrac{S_e}{S_o}$$

In which $NO_3\text{-}N$ is the concentration of nitrate to be denitrified.

Example 6.15. A design example is presented to show the relationship between detention time and nitrate removal. Figure 6.112 presents the flow sheet and input variables for the example. The procedure for calculating the required aerobic and anoxic detention times is as follows:

Principles of Aerobic Biological Oxidation

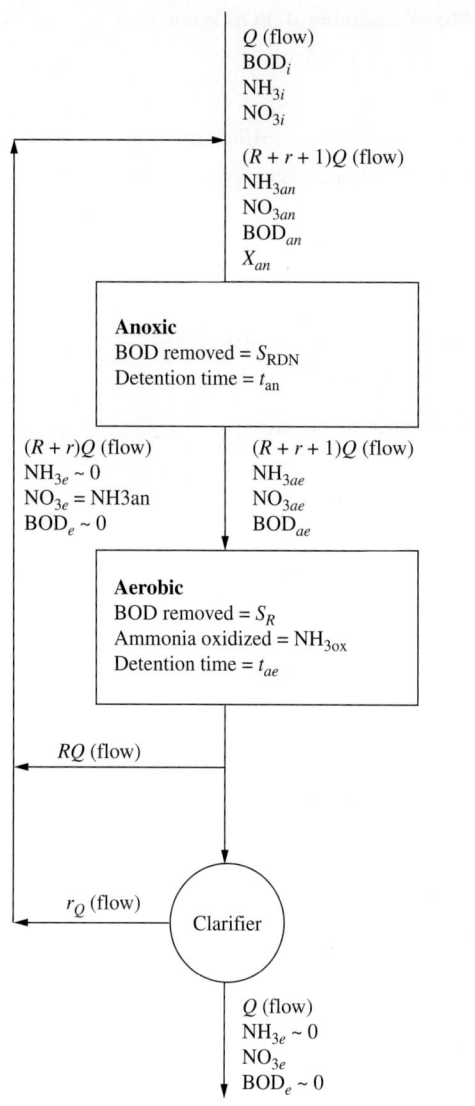

Figure 6.112 Flow sheet and input variables for example.

Assumptions:
1. Most of the BOD is consumed in the process—this is to calculate a conservative detention time. (Therefore, recycle BOD concentration = 0.)
2. Most of the ammonia is consumed in the process—this is to calculate a conservative detention time.
3. The ammonia utilized for biological syntheses is oxidized from the total inlet ammonia at the beginning of the process.
4. Influent viable biomass is very small compared to that in the system.

374 Chapter Six

1. Calculate ammonia to be oxidized, assuming all BOD is removed.
   1.1. N_{OX} = inlet ammonia − (0.04 × BOD removed)
   $N_{OX} = 500 − (0.04 × 1300) = 448$

2. Calculate effluent nitrate and amount to be denitrified in recycle
   2.1. Recycle flow factor = Internal recycle + sludge recycle = $R + r$
   $= 2 + 0.4 = 2.4$
   2.2. Total flow multiplier = $R + r + 1$
   $= 2.4 + 1 = 3.4$
   2.3. Effluent nitrate concentration (NO_{3e}) = N_{OX}/total flow multiplier
   $= 448/3.4 = 131.8$ mg/L
   2.4. Inlet nitrate concentration to anoxic zone = NO_{3an} =
   $$\frac{NO_{3e} \times (R+r) + NO_{3i} \times 1}{(R+r+1)}$$
   $$= \frac{131.8 \times (2.4) + 0 \times 1}{(3.4)} = 93 \text{ mg/L}$$
   2.5. Inlet ammonia to anoxic zone = NH_{3an}
   $$= \frac{NH_{3e} \times (R+r) + N_{ox} \times 1}{(R+r+1)} = 131.76 \text{ mg/L}$$

3. Calculate BOD consumed in the anoxic phase
   3.1. BOD inlet to anoxic zone $= \dfrac{BOD_e \times (R+r) + BOD_i \times 1}{(R+r+1)}$
   $$= \frac{0 \times 2.4 + 1300 \times 1}{3.4} = 382.4 \text{ mg/L}$$
   3.2. $S_{rDN} = 3 \times NO_{3an}$
   $= 3 \times 93 = 279$ mg/L

4. Calculate BOD inlet to the aerobic phase
   4.1. $BOD_i − S_{RDN}$
   $= 382.4 − 279 = 103.4$ mg/L

5. Calculate the detention time for the anoxic phase
   5.1. $t = \dfrac{S_{rDN}}{X_{vb} K_{DN}} \dfrac{BOD_{ae}}{BOD_e}$
   5.2. Volume for anoxic phase = $t \times (R + r + 1) \times Q$

This calculation can be repeated for an internal recycle of 3, 4, 5, and 6. The relationship between detention time and percent removal of nitrate is shown in Fig. 6.113.

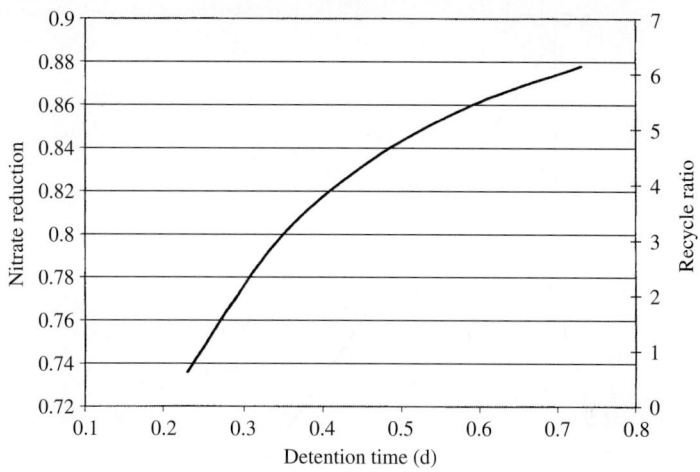

FIGURE 6.113 Nitrate reduction and recycle ratio versus detention time.

6.10 Phosphorus Removal

Phosphorus can be removed from wastewater either chemically or biologically.

Chemical Phosphorus Removal

Chemical phosphorus removal involves precipitation with calcium, iron, or aluminum. Phosphorus is precipitated with calcium salts to low residuals, depending on the pH. The precipitate is a hydroxyapatite, $Ca_5OH(PO_4)_3$:

$$5Ca^{2+} + 7OH^- + 3H_2PO_4^- \rightarrow Ca_5OH(PO_4)_3 + 6H_2O$$

Between pH 9.0 and pH 10.5, precipitation of calcium carbonate competes with calcium phosphate. Unlike aluminum and iron, calcium phosphate solids nucleate and grow very slowly, especially at neutral pH. The addition of seed enhances the reaction, indicating the advantage of solids recycle. Residual soluble phosphorus with respect to pH is shown in Fig. 6.114. The calcium phosphate precipitate is finely divided so that the presence of $Mg(OH)_2$ floc aids in the removal of the calcium phosphate precipitate.

The lime requirements will be dictated by the hardness and the alkalinity, as shown in Fig. 6.115. At high pH levels, low soluble phosphorus levels are achieved but residual particulates may require

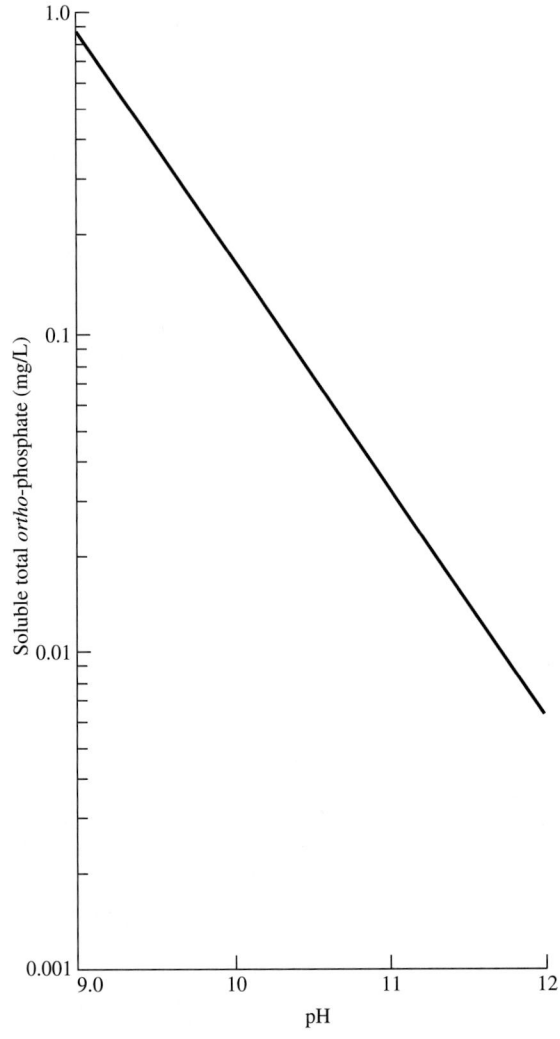

FIGURE 6.114 *Ortho*-phosphate versus pH for lime precipitation.[19]

postfiltration. Recarbonation for pH adjustment after precipitation may redissolve particulate phosphorus if incomplete removal exists prior to pH adjustment.

Iron or aluminum can be employed as a direct precipitation of the metallic phosphate in the case of inorganic wastewater or simultaneous precipitation by addition of the coagulating

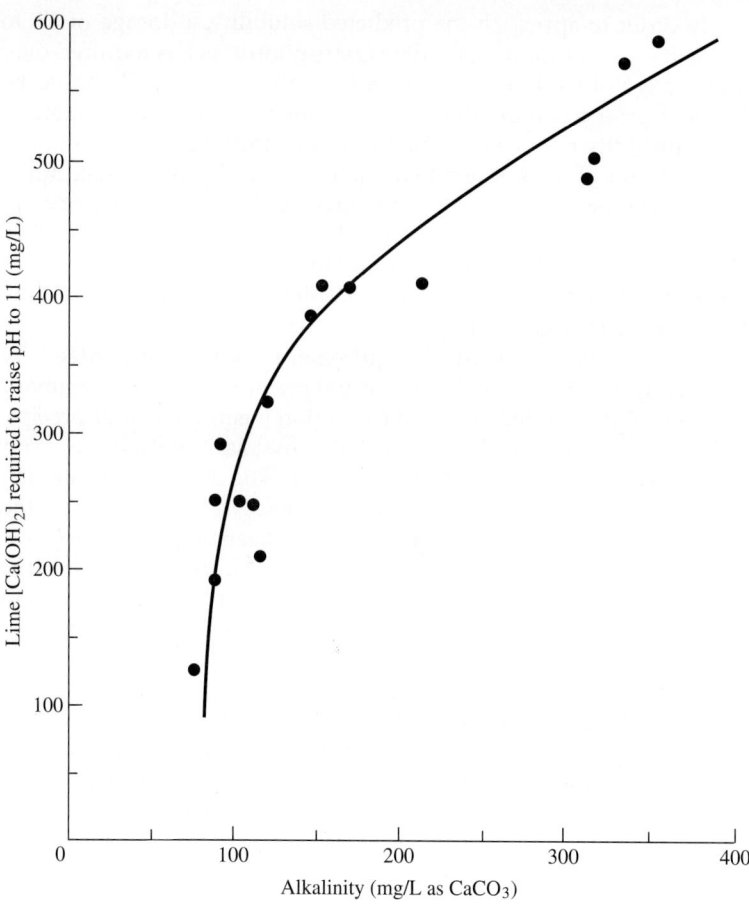

FIGURE 6.115 Lime required to raise the pH to 11 as a function of the watewater alkalinity.[20]

chemicals at the end of the aeration step in the activated sludge process.

$$Al^{+3} + PO_4^{-3} \rightarrow AlPO_4 \downarrow$$

$$Fe^{+3} + PO_4^{-3} \rightarrow FePO_4 \downarrow$$

Theoretical phosphorus residuals from precipitation with aluminum are a function of pH and the Al/P ratio. The precipitate is usually a mixture of $Al(OH)_3$ and $AlPO_4$, although the $AlPO_4$ precipitation is favored over $Al(OH)_3$. The precipitate tends to be amorphous rather than crystalline.

In order to approach the predicted solubility, a dosage of 1.5 to 3.0 moles of aluminum per mole of phosphorus as P is required over a pH range of 6.0 to 6.5. If the water is alkaline, the pH should be lowered prior to alum addition to minimize $Al(OH)_3$ precipitation. Some turbidity may result from the alum addition.

When the alum is added to the activated sludge process, addition should be immediately prior to the final clarifier, in the case of completely mixed systems, or at the end of plug flow aeration basins. This is to avoid phosphorus precipitation in the biological process before microbial utilization and to minimize shear of the chemical flocs in the aeration basin.

In some cases, chemical requirements can be minimized by multiple-point addition: that is, partial precipitation in the primary clarifier or at the head end of the aeration basin with final precipitation at the end of the basin after microbial assimilation.

Iron can be added as $FeSO_4$ or as $FeCl_3$. The dosage is dependent on the dissolved oxygen level, the pH, biological catalysis, and the presence of sulfur and carbonates. Iron has been employed for phosphorus precipitation in biological treatment processes, but has the disadvantage of leaving some iron in the treated effluent. The iron dosage will range from 1.5 to 3.0 mol of iron (Fe^{3+}) per mole of phosphorus (as P). The optimum pH is 5.0, which is too low for conventional biological treatment. Precipitation at neutral pH values may produce a colloidal precipitate requiring a polymer to obtain a minimum total phosphorus residual. A typical representation of soluble phosphorus residual as a function of iron dose is given in Fig. 6.116.

The molar ratio of metallic ion to phosphorus increases as the effluent phosphorus concentration decreases, as shown in Fig. 6.117. These ratios are approximate because side reactions form by-products such as hydroxide and carbonates. The weight ratio Fe^{+3}/influent TP versus effluent TP observed for secondary and tertiary treatment facilities is shown in Fig. 6.118.

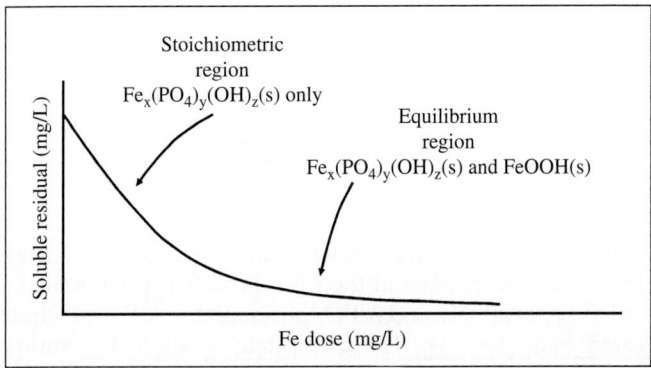

FIGURE 6.116 Typical Fe dose versus soluble P residual curve.

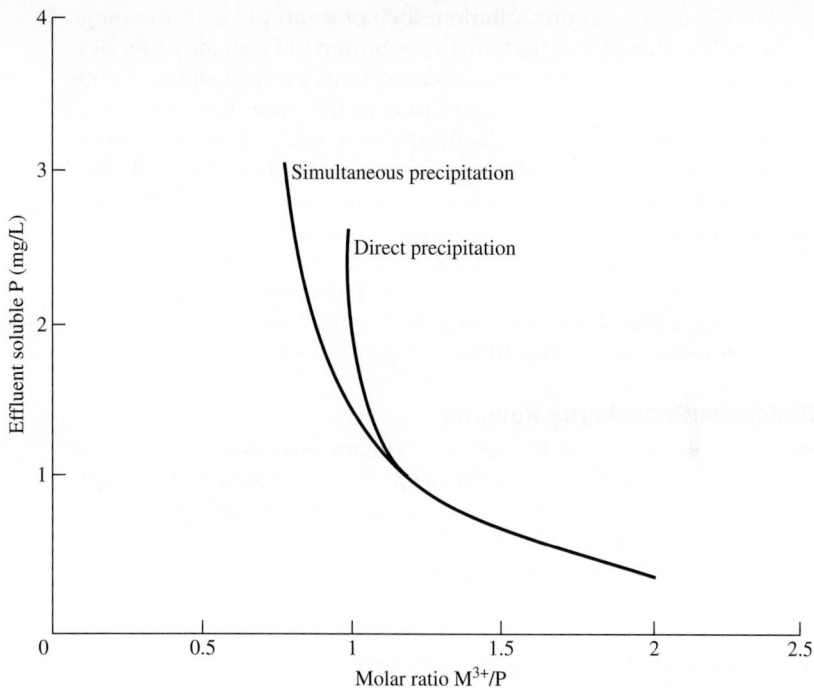

Figure 6.117 Phosphotus removal with aluminum and iron.

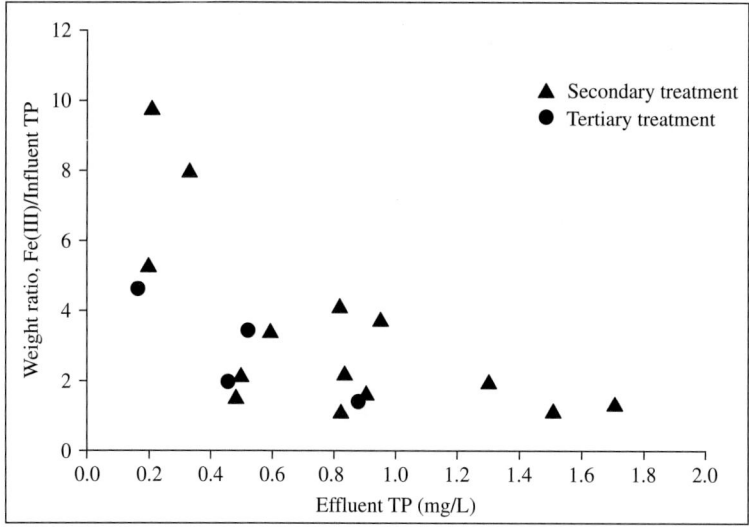

Figure 6.118 Fe(III) to influent TP ratio versus effluent total phosphorus concentration.

For simultaneous precipitation, the optimum pH is in the range of 7 to 8. For direct precipitation, the optimum pH is in the order of 6. For processes with a long precipitation time, such as simultaneous precipitation, the calcium concentration in the water has a beneficial effect with both Al and Fe. Alkalinity has a negative effect on direct precipitation, with short detention times, and the removal of alkalinity is essential for a low-P effluent. This is achieved to a certain extent automatically with the addition of Fe or Al since they behave as acids, although this condition will result in higher chemical consumption and increased sludge production. For low effluent phosphorus concentrations, effluent filtration may be required because of the high phosphorus content of the effluent suspended solids.

Biological Phosphorus Removal

Phosphorus removal in biological treatment processes was initiated by Levin[65] who patented the Phostrip process shown in Fig. 6.119. Return activated sludge (RAS) was fermented in a "stripper" where it released phosphorus. The phosphorus-rich supernatant was treated with lime and the precipitate was removed. The underflow from the stripper was returned to the aeration basin where the phosphorus was removed very efficiently. Today in Germany, this process is making a comeback due to the ability to recover the phosphorus.

Most biological phosphorus removal studies have reported on municipal wastewater experiences. In 1971, Milbury[66] noted that all plants that removed phosphorus were plug-flow systems and were operated not to nitrify, and also, that all these plants experience a release of phosphorus near the inlet zone where no dissolved oxygen could be detected. Barnard[67] was the first to clarify the need for anaerobic contacting between the influent wastewater and activated

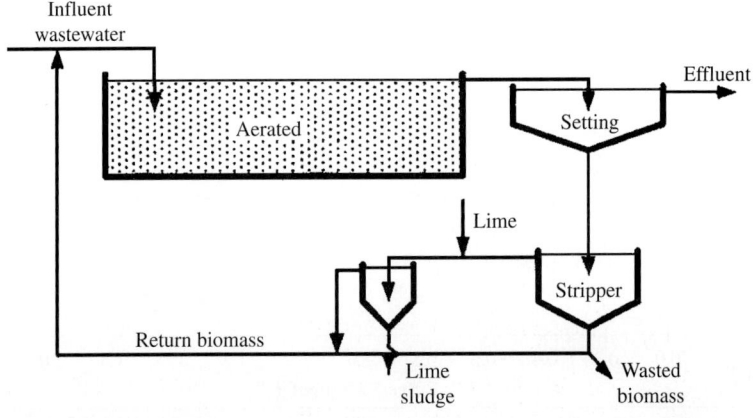

Figure 6.119 Phostrip process for phosphorus removal. (*Adapted from Levin, 1970.*)

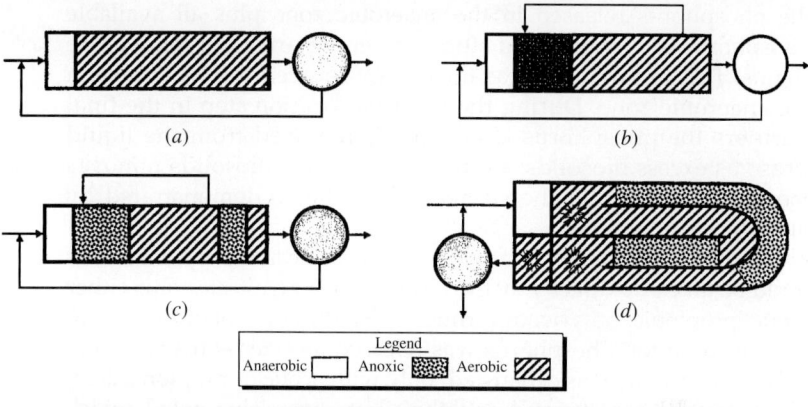

FIGURE 6.120 Phoredox flow sheets. (*Adapted from Barnard, 1974.*)

sludge prior to aerobic degradation for effective biological removal of phosphorus. He observed that nitrates interfered with biological phosphorus removal and noted that a common feature in all plants that removed phosphorus was a release of phosphorus in zones that were either intentionally or inadvertently deprived of oxygen and nitrates, and postulated that a zone free of oxygen and nitrates, followed by an aerated zone, was necessary to obtain excess biological phosphorus removal (EBPR).

The concept was verified by work in South Africa[68] and termed the Phoredox process, variations of which are shown in Fig. 6.120. In 1976 the process was patented only in the United States as the A/O flow sheets.

Variations of the Phoredox flow schemes have been proposed to ensure that nitrates do not enter the anaerobic zone. Practice has shown that any process that prevents nitrates from entering the anaerobic zone will be successful, provided there is sufficient VFA or readily biodegradable carbon measured as COD (rbCOD) in the influent.

Mechanism for Biological Phosphorus Removal

Fuhs and Chen[69] published a landmark paper proposing a theory for biological phosphorus removal by which Acinetobacter and many other organisms, later referred to as phosphate accumulating organisms (PAO) could absorb low-molecular-weight organics (i.e., VFA), especially acetic acid, in the absence of dissolved oxygen and nitrates. These organisms derived energy for the uptake by breaking the phosphate energy bonds of previously stored phosphorus and releasing it to the liquid phase. The VFA taken up in this way is stored as poly-β-hydroxybutyrate (PHB) until the organisms reach the aerobic zone where they metabolize the PHB and use the energy gained to take up the phosphorus released in the anaerobic zone plus all available phosphorus in the feed, and store it as energy-rich polyphosphate chains. This in turn serves as the energy source for VFA uptake in the anaerobic zone. During the solids separation step in the final clarifiers the phosphorus is effectively removed from the liquid phase as excess biosolids. Wasting the surplus biosolids removes the phosphorus from the sludge cycle. This is demonstrated by Fig. 6.121.[70]

Gerber et al.[71] demonstrated that only acetic and propionic acids could be taken up directly by the PAOs by adding nitrates plus either acetic, propionic, butyric, or formic acid and methanol or glucose to the mixed liquor. Phosphorus was released as a zero-order reaction until the substrate was exhausted, only with acetic or propionic acid. Little or no release occurred until the nitrates were eliminated, which indicated a need for fermentation of the other substrates to acetic or propionic acid before uptake by the PAOs. Wentzel et al.[72] concluded that rbCOD can be converted in the anaerobic zone by facilitative organisms to VFA in a first-order reaction in the absence of nitrates or oxygen. This requires staging the anaerobic zone for improved results. Brodisch et al.[73] found that *Aeromonas punctata,* which has the ability to ferment rbCOD to VFA, was present in all plants which removed phosphorus without sufficient VFA in the feed. Various methods have been proposed and evaluated for determination of available rbCOD.[72,74,75]

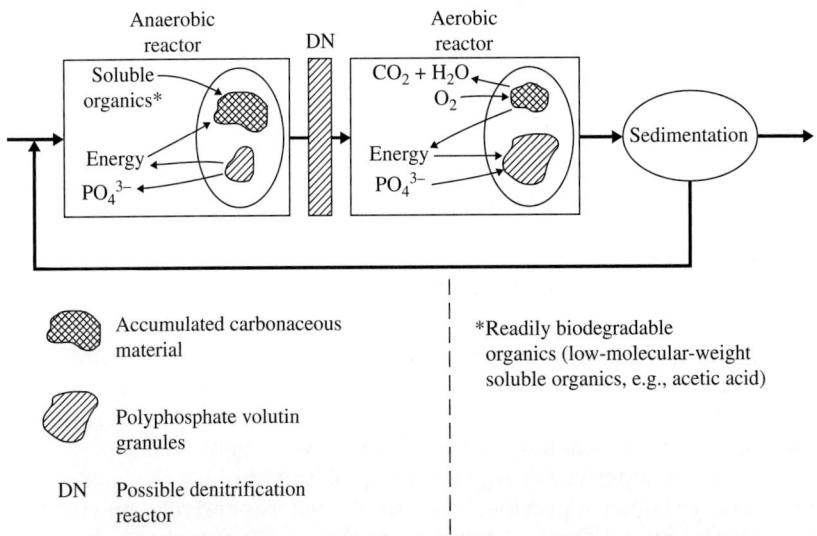

FIGURE 6.121 Biological release and uptake of phosphate, alternating between anaerobic and aerobic conditions. (*Adapted from Grau, 1975.*)

Barnard[76] reported that phosphorus released cannot be removed by subsequent aeration when keeping phosphorus-rich sludge under conditions of no dissolved oxygen, no nitrates, and no supply of VFA since there was no storage of PHB involved. He referred to this phenomenon as secondary release of phosphorus. Additional VFA is required to accumulate phosphorus released in this way. The importance of acid fermentation of primary sludge for production of VFAs for effluent concentrations of less that 0.1 mg/L has been well documented.[77,78]

Glycogen Accumulating Organisms

Glycogen accumulating organisms (GAOs) compete with PAOs for rbCOD in the anaerobic zone of BNR plants by using stored glycogen for their energy source. The VFA taken up in the anaerobic zone is stored as glycogen in the aeration basin. No phosphorus is accumulated by the cells. At temperatures below 29°C and neutral pH values, the PAOs will compete well for food. Rabinowitz et al.[79] reported on three plants where phosphorus reduction was inhibited by GAOs at mixed liquor temperatures above 30°C. Adjustment of feed conditions for acetic and propionic acids favors the growth of PAOs.[80–82]

Biological Phosphorus Removal Design Considerations

Suspended growth biological systems have been shown to remove phosphorus to as low as 0.1 mg/L after filtration. Where there is not sufficient rbCOD in the feed, appropriate carbon sources must be added, either produced through fermentation or from an external source.

A sufficient supply of VFA is key to reliable removal of orthophosphorus to less than 0.1 mg/L. Relying on fermentation of rbCOD in the anaerobic zone may be less effective and may require larger anaerobic zones and a larger mass of carbon. Barnard et al. concluded that when using a mixture of acetic and propionic acid, which is produced by on-site fermentation, the COD/TP ratio in the plant influent should exceed 8. However, when mostly rbCOD is available, it must be fermented in the anaerobic zone and the rbCOD/P ration should be increased to at least 18 to 20. When fermentation of rbCOD is required in the anaerobic zone, the rate of hydrolysis is the rate limiting step, and control of nitrates and dissolved oxygen is essential. Recent developments include the fermentation of a portion of the mixed liquor for VFA production. A SRT in the fermenter of 2 days appears to be optimal.[83]

Generalized plant flow sheets for biological phosphorus removal with and without nitrification-denitrification are shown in Fig. 6.122. Table 6.32 summarizes performance data for a tobacco wastewater treatment plant following the flow sheet shown in Fig. 6.122b.

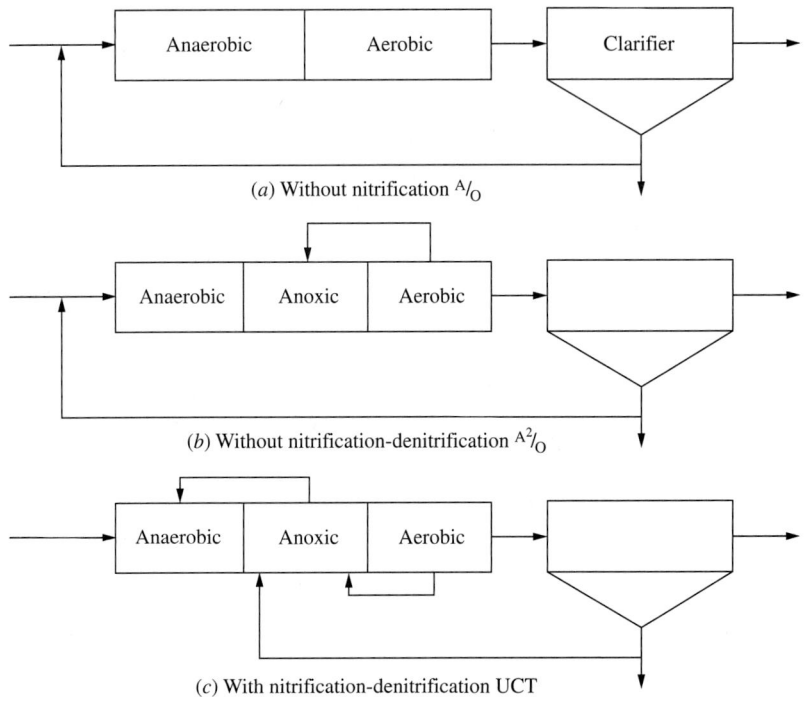

FIGURE 6.122 Biological phosphorus removal with and without nitrification–denitrification.

	Phosphorus, mg/L	NH_3^- N, mg/L	COD mg/L
Primary effluent	9.2	20	683
Anaerobic effluent	19.6	—	—
Aerobic effluent	1.4	3	99

TABLE 6.32 Bio-P Removal for a Tobacco Wastewater Using the Flow Sheet in Fig. 6.121b

The Phoredox (A/O) process and the A^2O process are shown in Fig. 6.122a and b. Nitrification does not occur in the Phoredox process since it is operated at low SRTs of 2 to 3 days at 20°C and 4 to 5 days at 10°C.[84] If nitrification is required, then the process must be modified and excessive amounts of nitrate must be excluded from the anaerobic reactor by RAS recycle. The A^2O process directs the RAS to the anaerobic zone. It operates at an SRT of approximately 8 to 15 days. The UCT process, shown in Fig. 6.122c, directs the RAS to the anoxic zone (1 h HRT). Recycle to the anaerobic zone is from the anoxic zone where nitrate concentration is low. It is used for relatively weak wastewaters where nitrate addition would adversely affect phosphorus removed. The HRT is longer than that used in the

Phoredox process and is generally between 1 and 2 hours. A more detailed description of these and other BPR processes are discussed by Metcalf and Eddy.[85]

Other Mechanisms of Phosphorus Removal

In addition to excess biological P uptake, chemical precipitation may also occur with calcium, magnesium, iron, and aluminum present in the wastewater. Recent data indicate that the physical-chemical removal increases with increasing biological P removal. One reason for the biologically mediated P precipitation may be the high phosphate concentration in the anaerobic reactor created by the biological P release.[86]

Under optimal conditions, the soluble P in the effluent can be reduced to 0.1 mg/L. A combination of BPR with simultaneous precipitation with small dosages of Al or Fe will achieve still lower effluent P concentrations. Chemical phosphorus removal by itself may still be the preferred option for many plants. However, Severn Trent Water has recently changed from using chemicals at all plants to BPR at some plants in order to reduce overall consumption of ferrous sulfate, the supply of which is dwindling.[87] Many of these plants have SRTs which can easily be converted to a Phoredox (AO) phosphorus removal process, shown in Fig. 6.120 by simple portioning of the aeration basin.

Membrane Bioreactors (MBRs)

Membrane bioreactors (MBRs) will be discussed much more fully in Chap. 12. However, here its utilization to enhance nutrient removal will be discussed. MBRs have undergone significant development where it is compatible with the wastewater to be treated. Its primary drawback is the associated power cost.

Barnard[88] discussed the future of the activated sludge process and proposed a flow sheet in which nitrogen and phosphorus removals can be obtained. Nitrogen removal requires both biological nitrification and denitrification that can be accommodated in a MBR. Solids retention time (SRT) in excess of 10 days allows nitrification to very low ammonia concentrations. Denitrification then is necessary to save power and eliminate the need for alkalinity.

Since most existing MBRs operate at high SRTs, effluent ammonia concentrations are quite low, but effluent TN values for domestic sewage are around 8 mg/L. Daily[89] stated that the Cauly Creek plant in Georgia, which was designed in the flow scheme as shown in Fig. 6.123,

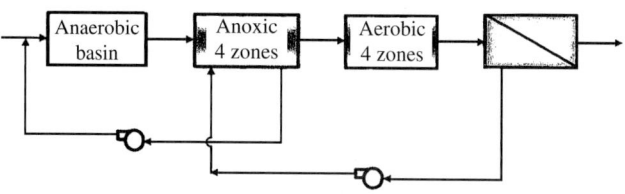

FIGURE 6.123 Flow sheet for MBR plant at Cauly Creek. (*Adapted from* Daily).

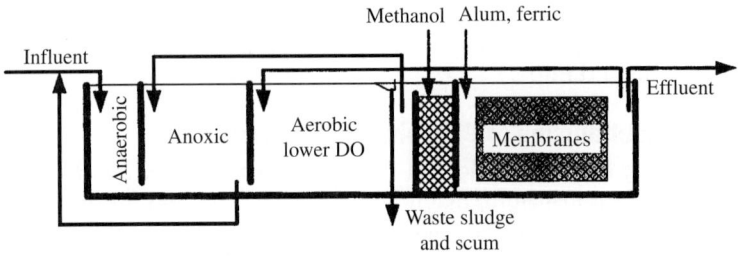

Figure 6.124 MBR configurations with attached growth second anoxic zone for nitrogen and phosphorus removal.

was able to reduce phosphorus concentrations to less than 0.5 mg/L without any chemical addition, and that with the addition of either alum or ferric chloride ($FeCl_3$) to the membrane compartment, average effluent TP of 0.1 mg/L was achieved. The COD/TKN ratio was 11.2 and 81 percent nitrogen removal was achieved. Note that soluble phosphorus may require subsequent reverse osmosis for very low levels (below 0.05 mg/L) to be achieved.

With membrane reactors there is no need for filtration, and the option of using a denitrification filter for removing residual nitrates has been eliminated. To achieve low effluent nitrate levels, configurations should be considered to ensure denitrification to very low levels. Barnard[83] proposed the flow diagram shown in Fig. 6.124 that would include an attached growth section ahead of the membrane basins where methanol-degrading organisms can grow on the media and not be washed out, ensuring less than 1 mg/L of nitrates in the effluent.

6.11 Laboratory and Pilot Plant Procedures for Development of Process Design Criteria

Wastewater Characterization

Wastewater characterization should be based on the equalized wastewater. Depending on the nature of the wastewater and the discharge permit requirements, the following parameters should be evaluated:

- BOD and/or COD or TOC
- Total and volatile suspended solids
- Oil and grease
- Volatile organics
- Priority pollutants
- Toxicity (bioassay)
- Nitrogen forms (TKN, NH_3, NO_2^-, NO_3^-)
- Phosphorus forms ($o\text{-}PO_4$, total P)

For wastewaters that do not contain aquatic toxicity, the following stepwise procedure is applicable to developing the necessary process design data:

1. Adjust the BOD:N:P ratio to 100:5:1, neglecting the wastewater organic nitrogen. Although organic nitrogen may be hydrolyzed to ammonia in the activated sludge process, it is initially neglected in order to ensure adequate nutrients in the experimental phase. The availability of the organic nitrogen will be reevaluated in the final process design.

2. Evaluate the wastewater's potential to promote filamentous bulking. In many cases, if the wastewater is readily degradable ($K > 6$ d^{-1}), filamentous bulking can be expected. If in doubt, evaluate this by operating a complete mix reactor at $F/M \approx 0.4$ d^{-1} for 5 to 8 d to establish the proliferation of filaments.

3. Develop an acclimated mixed liquor. Determine the bioinhibition potential using the FBR procedure. If there is bioinhibition, adjust the initial feed rate of the wastewater to less than 50 percent of the inhibition threshold concentration. Operate the reactor at an F/M of 0.3 d^{-1}. As acclimation proceeds, gradually increase the feed rate until the full waste strength is being treated. For a wastewater with a low bulking potential, use a complete mix reactor. For a wastewater with a high bulking potential, acclimate the mixed liquor in a batch reactor, a sequencing batch reactor, or a biological selector.

Calculations for acclimation of a sludge using a batch fill-and-draw reactor are presented in Example 6.16.

Example 6.16. Determine the operating conditions for a fill-and-draw acclimation procedure using a 20-L reactor volume and the following wastewater characteristics.

BOD = 2500 mg/L
TKN = 12 mg/L
NH_3-N = 2 mg/L
o-PO_4-P = 3 mg/L

Operate the reactor using an $F/M = 0.3$ d^{-1}, with MLVSS = 3000 mg/L.

Solution

$$NH_3 \text{ required} = [2500/(100/5)] - 2$$
$$= 123 \text{ mg N/L}$$

$$o\text{-}PO_4^- \text{ P required} = [2500/(100/1)] - 3$$
$$= 22 \text{ mg P/L}$$

$$t = \frac{S_o}{X_v(F/M)} = \frac{2500}{(3000 \cdot 0.3)}$$
$$= 2.8 \text{ d}$$

Assuming that the wastewater is not bioinhibitory (based on FBR test results), the volume added per feeding cycle is

$$V = \frac{20}{2.8} = 7.14 \text{ l/d}$$

If the FBR results indicate that the wastewater is inhibitory, then the feed should be diluted with a weak, but readily degradable, substrate to provide an S_o = 2500 mg/L. The wastewater contribution to the blended feed volume should be gradually increased (about 10 percent per feed cycle) while maintaining S_o = 2500 mg/L. The SOUR and SVI should be measured daily and the effluent SCOD concentration measured at the end of each feed cycle.

Reactor Operation

At least three reactors should be operated in parallel over an SRT range of 3 to 12 d for a readily degradable wastewater and 10 to 40 d for a less degradable wastewater. Since floc size is related to power level, which in turn affects the reaction coefficient K, the power level in the pilot units should approximate field conditions. The SRT should be maintained by daily wasting of an appropriate mixed liquor volume (i.e., for a 10-d SRT, one-tenth of the reactor volume is wasted daily). The waste sludge mass is computed as the VSS in the wasted reactor volume, plus the VSS in the reactor effluent. The sampling and analytical schedule for the reactors is summarized in Table 6.33. Note that three SRTs are generally required prior to pseudo steady state conditions. If the

Analysis	Frequency
BOD	3/week
COD or TOC	Daily
O_2 uptake rate	Daily
MLVSS	Daily
Dissolved oxygen	Daily
pH	Daily
Temperature	Daily
Nitrogen*	2/week
Phosphorus	2/week
Bioassay†	Weekly
Specific pollutants	Weekly

*The case of a nitrogen-deficient wastewater. If nitrification is desired, TKN, NH_3-N, and NO_3^--N should be run 3 times per week.
†The test species will depend on permit conditions.

TABLE 6.33 Biological Treatment Process Design Study Sampling and Analysis Schedule

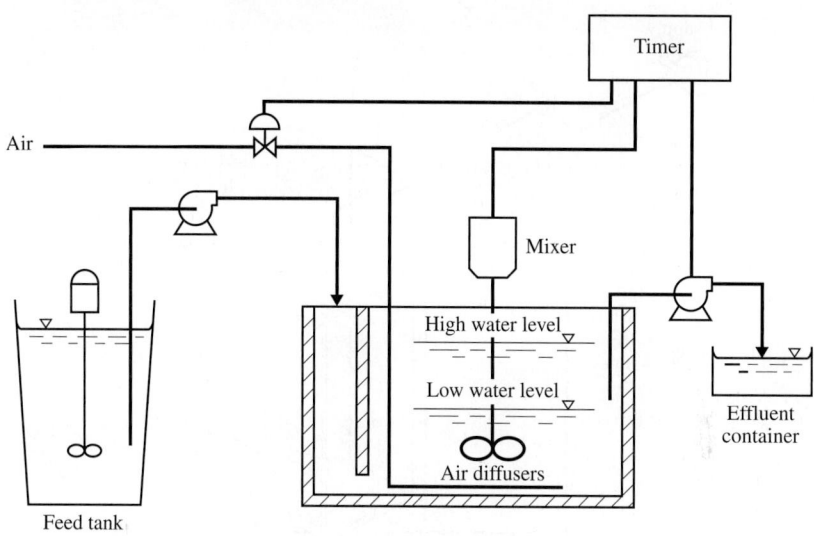

FIGURE 6.125 Pilot plant activated sludge system (SBR modification).

wastewater is sorbable and readily degradable, a sequencing batch reactor (SBR) or a selector should be used to generate a nonfilamentous sludge (Fig. 6.125). For a more refractory wastewater, a complete mix reactor as shown in Fig. 6.126 should be used.

At the end of the treatability study, the degradable fraction X_d and the endogenous decay coefficient b are determined. Sludge from each reactor is washed and aerated, and the concentration of VSS is measured every 2 to 3 d until there is no further reduction in VSS. The degradable fraction and the endogenous decay coefficient can be calculated as shown in Fig. 6.127.

Figure 6.128 presents the methods for graphical analysis of the treatability reactor data. The rate coefficient K is determined by plotting $S_o(S_o - S_e)/X_d X_v t$ versus S_e. If the influent wastewater has a variable organic composition, K will not be constant. The oxygen coefficient a' is determined as the slope of the plot of $O_2/X_d X_v t$ versus $S_r/X_d X_v t$, and the endogenous respiration coefficient b' is the intercept. The sludge yield coefficient a is determined from a plot of $\Delta X_v / X_d X_v t$ versus $S_r/X_d X_v t$. It should be noted that ΔX_v consists of the sludge wasted per day, plus that in the reactor effluent. The design criteria for the final clarifier are determined from zone settling velocity measurements and batch flux analysis of the sludge.

Volatile Organic Carbon

The present emphasis on volatile emissions from wastewater treatment plants requires that stripping be considered in activated sludge process design where volatiles are present in the influent wastewater. There are several factors to be considered in the experimental design:

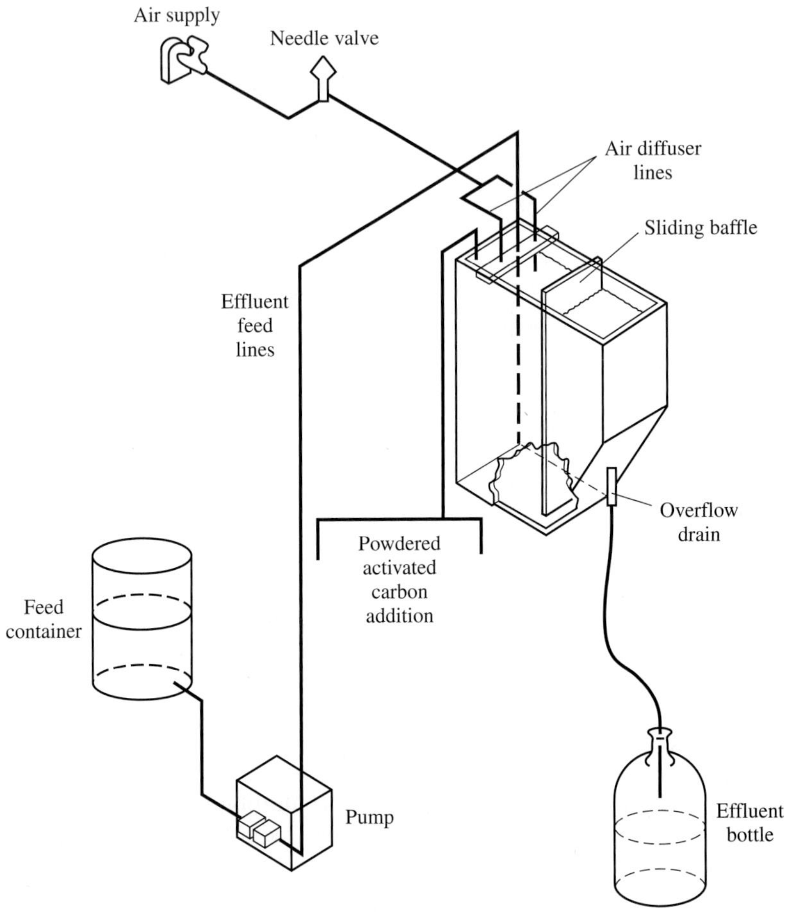

390 Chapter Six

FIGURE 6.126 Biological reactor with alternative PAC addition.

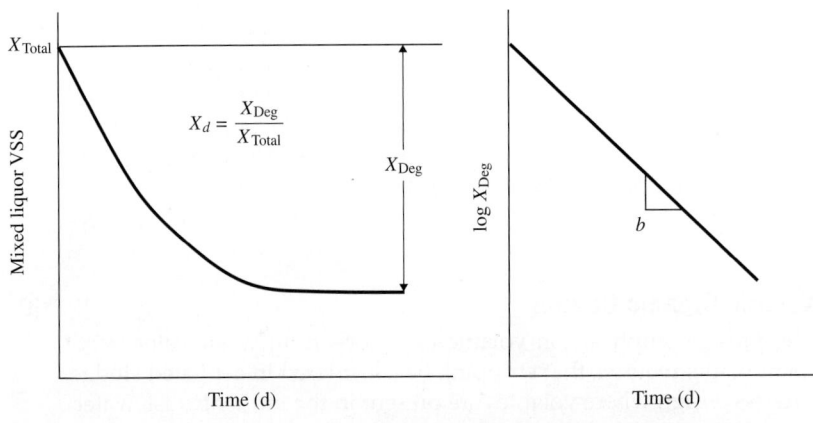

FIGURE 6.127 Determination of the degradable fraction and the endogenous coefficient.

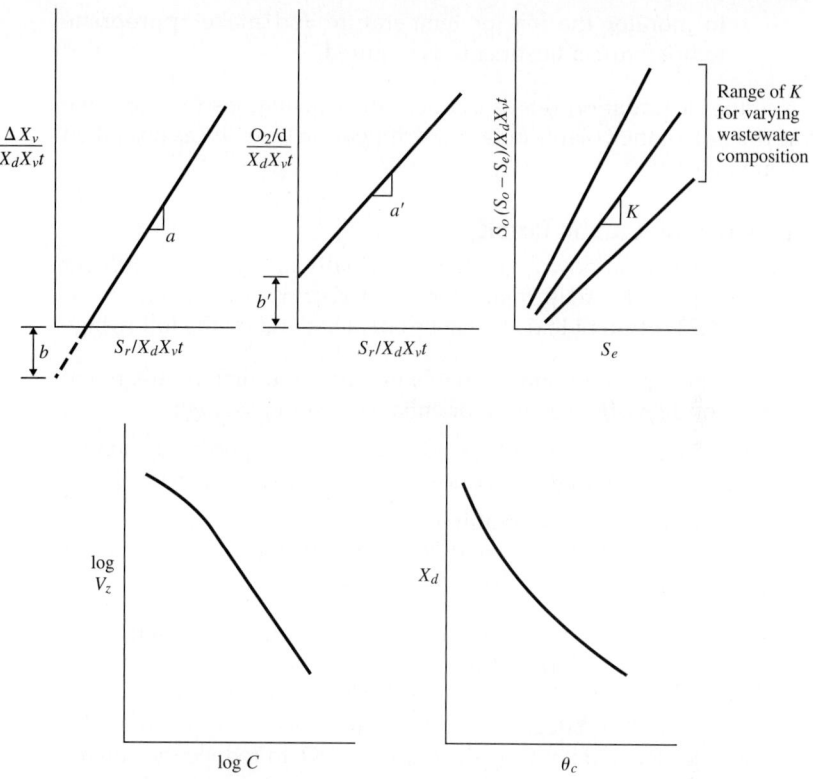

FIGURE 6.128   Parameter correlation plots.

- Both the power level in the aeration basin and the type of aerator (i.e., diffused or mechanical) significantly influence stripping.
- The maximum expected concentration of each particular volatile should be employed. The degradation rate of specific volatiles will be related to both the composite wastewater composition and the process operating conditions, that is, the SRT. Therefore, these variables must be fixed prior to volatile stripping and degradation studies.
- It has been shown that off-gas capture and recirculation will significantly enhance biodegradation of VOCs. Therefore, this process modification should be included in the pilot studies if VOC emission control is required for the plant.
- Covered aeration basins may result in a significant temperature rise due to the exothermic reaction with high-strength wastewaters. Basin temperatures in excess of approximately 38°C may result in floc dispersion. In these cases, it is necessary

to monitor the reactor temperature and make appropriate temperature adjustments as required.

Once the degradation rate is defined, the calculations to determine the fraction of the volatile organics stripped can be made as described in on page 330.

Reduction of Aquatic Toxicity

Aquatic toxicity is now regulated in virtually all industrial effluent discharge permits. Acute toxicity of selected compounds is presented in Table 6.34. Effluent toxicity can be characterized by the following:

- Caused by a nondegradable organic. Requires pretreatment for detoxification and/or enhanced biodegradability.
- Caused by a degradable toxic organic. Requires adjustment of the sludge age to reduce the organic to nontoxic levels.
- Caused by SMP or multiple organics or by-products. Requires application of powdered activated carbon (PAC) or effluent polishing to remove the toxic organics.

One of the first steps in a treatability study, therefore, is to determine the applicability of biological treatment to the wastewater.

Two cases will be considered. In the first, there is a known toxic compound that is biodegradable. A series of reactors at varying SRT are run in order to determine the required SRT to reduce the aquatic toxicity and meet permit requirements. An example of this using nonylphenol from a surfactant manufacturing wastewater is shown in Fig. 6.44. In the second, more common case, toxicity is caused by an unknown mixture of organics or is generated or enhanced by the production of SMP during biooxidation. In this case toxicity may be correlated to effluent COD as shown in Fig. 6.45 for a petroleum refinery wastewater. The testing protocol illustrated in Fig. 6.69 has been developed to define treatment options in this case. The biodegradability of the wastewater is determined by an FBR test. It is important that the biological sludge used in the test is fully acclimated to the wastewater. If the wastewater proves to be nondegradable and toxic, pretreatment must be considered. The primary objective of pretreatment would be detoxification and enhanced biodegradability. If the wastewater is degradable as defined by the FBR procedure, then a long-term oxidation test (e.g., 48 h aeration) is conducted to remove all degradable components. This effluent is then subjected to a bioassay. If this effluent is still toxic, then alternative pretreatment methods or tertiary treatment should be considered. Since it is not possible to distinguish between wastewater component toxicity and SMP toxicity, the wastewater should be pretreated for detoxification. A subsequent long-term oxidation test should then determine the source of the toxicity.

TABLE 6.34 Acute Toxicity of Selected Compounds (96-h LC_{50})

Organics*	Units	Fathead Minnow	Daphnia	Rainbow Trout
Benzene	mg/L	42.70	35.20	38.70
Carbon tetrachloride	mg/L	17.30	15.20	14.50
Chlorobenzene	mg/L	13.20	11.60	11.10
1,1-Dichloroethane	mg/L	120.00	96.40	113.00
1,1,2-Trichloroethane	mg/L	88.70	72.60	81.10
2-Chlorophenol	mg/L	21.60	18.60	18.40
1,4-Dichlorobenzene	mg/L	3.72	3.46	2.89
1,2-Dichlorobenzene	mg/L	87.40	71.10	80.50
2,4-Dinitrophenol	mg/L	5.81	5.35	4.56
4,6-Dinitro-o-cresol	mg/L	2.79	2.65	2.10
Pentachlorophenol	mg/L	170.00	—	—
Ethylbenzene	mg/L	11.00	9.97	9.47
Methylene Chloride	mg/L	326.00	249.00	325.00
Toluene	mg/L	31.00	26.00	27.40
Trichloroethylene	mg/L	55.40	46.20	49.50

Metals[†]	Units	Fathead Minnow	Daphnia	Rainbow Trout
Phenol	mg/L	39.60	33.00	35.40
1,4-Dinitrobenzene	mg/L	1.68	1.61	1.24
2,4,6-Trichlorophenol	mg/L	5.91	5.45	4.62
2,4-Dichlorophenol	mg/L	9.27	8.35	7.40
Naphthalene	mg/L	5.57	5.07	4.44
Nitrobenzene	mg/L	118.00	95.40	110.00
1,1,2,2-Tetrachloroethane	mg/L	31.10	26.70	26.70
Arsenic	mg/L	15,600	5,278	13,340
Chromium, hexavalent	mg/L	43,600	6,400	69,000
Cadmium	mg/L	38.2	0.29	0.04
Copper	mg/L	3.29	0.43	1.02
Lead	mg/L	158.00	4.02	158.00
Mercury	mg/L	—	5.00	249.00
Nickel	mg/L	440.00	54.00	—
Selenium	mg/L	1,460.00	710.00	10,200
Silver	mg/L	.012	.00192	.023
Zinc	mg/L	169.00	8.89	26.20

Inorganics			
Unionized ammonia	pH 7.0	0.093 (23)	0.093 (23)
Total ammonia ( )	pH 8.5	0.260 (6.8)	0.260 (6.8)

*From EPA/Montana state QSAR System.
†EPA, Duluth, 1980.
Note: Highly variable depending on pH an temperature (Federal Reg. Vol. 50, No. 185, Monday, July 29, 1985, pp. 30, 784 to 30,786). Data represent criteria to protect aquatic life at pH 7.0 and 20°C and pH 8.5 and 20°C, 1 h average, mg/L.

TABLE 6.34 Acute Toxicity of Selected Compounds (96-h LC_{50}) (*Continued*)

Chapter Six

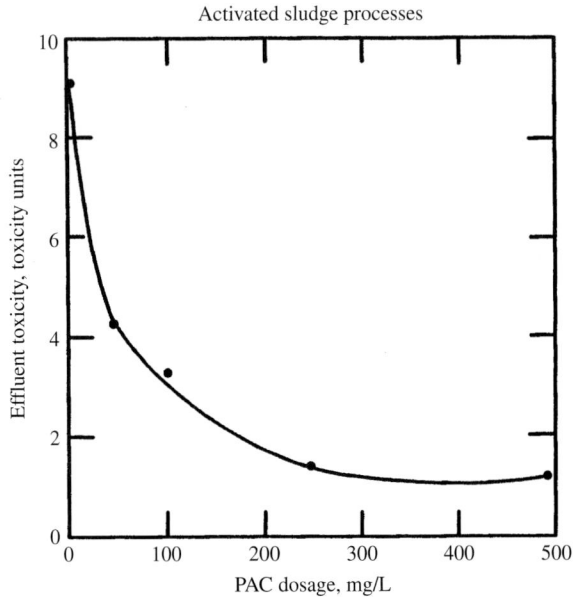

FIGURE 6.129 Toxicity reduction in an organic chemicals wastewater using PAC.

If the toxicity is caused by SMP, tertiary treatment with carbon should be evaluated. It is shown in Chap. 8 that, although laboratory adsorption isotherms will establish the applicability of carbon, they cannot be directly used for process design. A series of reactors should be run at various carbon dosages using the SRT required to remove degradable organics and/or priority pollutants. Each reactor should be preloaded with carbon that has been brought into equilibrium with the reactor effluent. The equilibrium carbon concentration in the reactor mixed liquor is calculated from Eq. (8.12). Toxicity reduction in an organic chemicals wastewater using PAC is illustrated in Fig. 6.129. If the PAC-treated wastewater still exhibits aquatic toxicity, then at-source treatment or elimination of the specific toxic wastewater streams should be considered.

6.12 Problems

6.1. The following data were developed from a pilot plant study at 20°C:

(a) Determine the K for this wastewater.

(b) If $\theta = 1.06$, what is the K at 10°C?

Principles of Aerobic Biological Oxidation

F/M, d⁻¹	X_v, mg/L	S_o, mg/L	S_e, mg/L
0.3	2000	1640	26
0.48	1980	1640	44
0.72	2000	1640	70
0.96	2056	1640	100
1.18	2050	1640	167
2.68	2100	1640	333

6.2. A complete mix activated sludge plant treating an organic chemicals wastewater yielded the following performance data:

$TOD_{influent} = 830$ mg/L

$TOD_{effluent} = 50$ mg/L

Oxygen uptake rate (OUR) = 0.49 mg/(L · min)

MLVSS = 3200 mg/L

$Q = 3.2$ million gal/d (12,100 m³/d)

$t = 0.8$ d

A sample of mixed liquor was aerated in the laboratory to determine the endogenous oxygen uptake rate where

OUR = 0.3 mg/(L · min)

MLVSS = 4700 mg/L

Calculate the endogenous coefficient b' and the coefficients a and a'.

6.3. Determine the operating F/M, MLVSS, and f_b for an activated sludge process under the following operating conditions:

$a = 0.45$

$b = 0.1$ at 20°C

$\theta_c = 10$ d

$X_i = 200$ mg/L

$f_x = 0$

$S_o = 1000$ mg/L

$S_e = 20$ mg/L

$t = 0.9$ d

6.4. Using the data from Example 6.7, compute the detention time required if the influent nondegradable VSS is 100 mg/L. If we wish to maintain the same detention time of 0.52 d, what must the MLVSS be increased to in order to maintain the same effluent quality?

6.5. The average operating and performance conditions for an activated sludge process are illustrated below.

Influent

   $SCOD_o = 1018$ mg/L

   $X_i = 51$ mg/L

   $f_x = 0$

   $Q_o = 5.19$ million gal/d

Effluent

   $SCOD_e = 248$ mg/L

   $VSS_e = 34$ mg/L

Waste Sludge

   $Q_w = 0.08$ million gal/d

   $VSS_w = 6080$ mg/L

Aeration Basin

   $V = 1$ million gal/d

   MLVSS = 2295 mg/L

   $b = 0.18$ d^{-1}

   $t = 0.19$ d

   OUR = 146 mg/(l-h)

Develop a process material balance.

6.6. An untreated industrial wastewater has a BOD of 935 mg/L, negligible organic nitrogen, NH_3-N of 8 mg/L, and a flow rate of 1.5 million gal/d. The wastewater treatment plant operates at an SRT of 15 d with MLVSS of 3000 mg/L. The effluent soluble BOD is 15 mg/L. The sludge yield coefficient is 0.6, b is 0.1 d^{-1}, and the hydraulic retention time is 1.4 d. Compute the nitrogen that must be added.

6.7. A complete mix activated sludge plant is to be designed for the following conditions:

   $Q = 3.5$ million gal/d (1.32×10^4 m^3/d)

   $So = 650$ mg/L

   S (soluble) = 20 mg/L

   $X_v = 3000$ mg/L

   $a = 0.50$

   $a' = 0.52$

   $b = 0.1$/d at 20°C

   $u = 1.065$

   $K = 6.0$/d at 20°C

   $b' = 0.14$/d

Compute:

(a) The aeration volume.
(b) The F/M.
(c) The sludge yield.
(d) The oxygen requirements.
(e) The nutrient requirements.
(f) The effluent quality at 10°C.

6.8. Develop a relationship between the $NH_3\text{-}N/BOD$ ratio and the aerobic volume percentage for a nitrification-denitrification channel. Consider a BOD of 200 mg/L and a range of $NH_3\text{-}N$ of 10 to 70 mg/L. Compute the oxygen requirements for each case. Use the following parameters:

R_N = half the maximum rate
K_{DN} = 0.06 mg $NO_3\text{-}N$/(mg VSS · d) at 20°C
X_v = 3000 mg/L
a' = 0.55
$K_{BOD,\ aerobic}$ = 8.0/d at 20°C
$BOD/NO_3\text{-}N$ = 3.0
a = 0.5
b = 0.1 d^{-1}
X_d = 0.45
S_e = 10 mg/L

References

1. Weber, W. J., and B. E. Jones: EPA NTIS PB86-182425/AS, 1983.
2. Matter-Mutter et al.: *Prop. Water Tech.*, vol. 12, pp. 299–313, 1980.
3. Namkung, J., and Rittman: *J. WPCF*, vol. 59, no. 7, p. 670, 1987.
4. Branghman and Pariss: *Critical Review in Microbiology*, vol. 8, p. 205, 1981.
5. Dobbs, Wang, and Govind: *Environ. Sci. & Tech.*, vol. 23, no. 9, p. 1092, 1989.
6. Kincannon, D. F., and E. L. Stover: EPA Report CR-806843-01-02, 1982.
7. Ford, D. L.: Activated Sludge, University of Texas at Austin Advanced Wastewater Pollution Control Short Course, August 2007.
8. Grady C. P. L., et al.: *Water Research*, vol. 30, p. 742, 1996.
9. Tabak, H. H., and E. F. Barth: *J. WPCF*, vol. 50, p. 552, 1978.
10. Gaudy, A. F.: *J. WPCF*, vol. 34, pt. 2, p. 124, February 1962.
11. Englebrecht, R. S., and R. E. McKinney: *Sewage Ind. Wastes*, vol. .29, pt. 12(l), p. 350, December 1957.
12. McWhorter, T. R., and H. Heukeleklan: *Advances in Water Pollution Research*, vol. 2, Pergamon, New York, 1964.
13. Quirk, T., and W. W. Eckenfelder: *J. WPCF*, vol. 58, pt. 9, p. 932, 1986.
14. Sawyer, C. N.: *Biological Treatment of Sewage and Industrial Wastes*, vol. 1, Reinhold, New York, 1956.
15. Gellman, I., and H. Heukelekian: *Sewage Ind. Wastes*, vol. 25, pt. 10(l), p. 196, 1953.

16. Busch, A. W., and N. Myrick: *Proc. 15th Ind. Waste Conf.*, 1960, Purdue University.
17. Pipes, W.: *Proc. 18th Ind. Waste Conf.*, 1963, Purdue University.
18. Eckenfelder, W. W.: *Principles of Water Quality Management*, CBI, Boston, 1980.
19. Shamas, J. Y., Englande; A. J., *Water Science & Technology*, vol. 1, 1992.
20. Helmers, E. N, J. P. Frame, A. F. Greenbert, and C. N. Sawyer: *Sewage Ind. Wastes*, vol. 23, pt. 7, p. 834, 1951.
21. Wuhrmann, K.: *Biological Treatment of Sewage and Industrial Wastes*, vol. 1, Reinhold, New York, 1956.
22. Tischler, L. F., and W. W. Eckenfelder: *Advances in Water Pollution Research*, vol. 2, Pergamon, Oxford, England, 1969.
23. Gaudy, A. F., K. Komoirit, and M. N. Bhatla: *J. WPCF*, vol. 35, pt. 7, p. 903, July 1963.
24. Grau, P.: *Water Res.*, vol. 9, p. 637, 1975.
25. Adams, C. E., W. W. Eckenfelder, and J. Hovious: *Water Res.*, vol. 9, p. 37, 1975.
26. Van Niekerk et al.: *Wat. Sci. Tech.*, vol. 19, p. 505, 1987.
27. Argaman, Y.: *Water Research*, vol. 25, p. 1583, 1991.
28. Rickard, M. D., and Gaudy, A. F.: *J. WPCF*, vol. 49, R129, 1968.
29. Zahradka., V., *Advances in Water Pollution Research*, vol. 2, Water Pollution Control Federation, Washington, DC, 1967.
30. Philbrook, D. M., and Grady, P. L.: *Proc. 40th Industrial Waste Conf.,* Purdue University, 1985.
31. Hoover, P.: M. S. Dissertation, Vanderbilt University, 1989.
32. Nevalainen, J., et al.: *Wat. Sci. Tech.*, vol. 24, no. 3–4, 1991.
33. Hall, E. R., and Randall, W. G.: *Wat. Sci. Tech.*, vol. 26, no. 1–12, 1992.
34. Higgins, M. J., and J. T. Novak: *Wat. Env. Res.*, vol. 69, no. 2, 1997.
35. Palm, J. C., D. Jenkins, and P. S. Parker: *J. WPCF*, vol. 52, pt. 2, p. 484, 1980.
36. Richard, M. G.: *WEF Ind. Waste Tech. Conf.*, New Orleans, 1997.
37. Eckenfelder, W. W., and J. Musterman: *Activated Sludge Treatment of Industrial Wastewaters*, Technomic Publishing, 1995.
38. Marten, W., and G. Daigger: *Water Env. Research*, vol. 69, no. 7, p. 1272, 1997.
39. Wanner, J.: *Activated Sludge Bulking and Foaming Control*, Technomic Publishing, 1994.
40. Jenkins, D., M. Richard, and G. Daigger: *Manual on the Causes and Control of Activated Sludge Bulking and Foaming Water,* Research Committee, Pretoria, S.A., 1984.
41. Pitter, J., and J. Chudoba: *Biodegradability of Organic Substances in the Aquatic Environment*, CRC Press, Boca Raton, Fla., 1990.
42. Chudoba, J.: *Water Res.*, vol. 19, no. 2, p. 197, 1985.
43. Volskay, V. T., and P. L. Grady: *J. WPCF*, vol. 60, no. 10, p. 1850, 1988.
44. Watkin, A.: Ph.D. dissertation, Vanderbilt University, 1986.
45. Watkin, A., and W. W. Eckenfelder: *Water Sci. Tech.*, vol. 21, p. 593, 1988.
46. Larson, R. J., and S. L. Schaeffer: *Water Res.*, vol. 16, p. 675, 1982.
47. Roberts, P. V., et al.: *J. WPCF*, vol. 56, no. 2, p. 157, 1984.
48. Poduska, R. A.: Ph.D. thesis, Clemson University, 1973.
49. Wong Chong, G. M., and R. C. Loehr: *Water Res.*, vol. 9, p. 1099, 1975.
50. Blum, J. W., and R.A. Speece: *Database of Chemical Toxicity to Bacteria and Its Use in Interspecies Comparisons and Correlations,* Vanderbilt University, 1990.
51. Henning, A., and R. Kayser, *42nd Purdue Ind. Waste Conf.*, p. 893, 1986.
52. Skinner, F. A., and N. Walker: *Arch. Mikrobiol.* pp. 38–339, 1961.
53. Sadick, T. E., et al.: *Proc. WEF*, vol. 1, Dallas, 1996.
54. Zacharias, B., and R. Kayser: *50th Purdue Ind. Waste Conf. Proc.*, Ann Arbor Press, 1995.
55. Anthoisen, A. C.: *J. WPCF*, vol. 48, p. 835, 1976.
56. Sutton et al.: First Workshop, Canadian-German Cooperation, Wastewater Technology Center, Burlington, Ontario, Canada, 1979.
57. Orhon, S. et al.: *Water Sci. Tech.*, vol. 34, no. 5, p. 67, 1996.
58. McClintock, S. A. et al.: *J. WPCF*, vol. 60, no. 3, 1988.

59. OH, J., and J. Silverstein: *Water Res.*, vol. 33, no. 8, p. 1925, 1999.
60. Lie, J., and R. Welander: *Water Sci. Tech.*, vol. 30, no. 6, p. 91, 1994.
61. Baumann, P., and Kh. Krauth: *Proc 2nd Specialized Conference on Pretreatment of Industrial Wastewaters,* Athens, Greece, 1996.
62. Henze H. et al.: *Water Res.*, vol. 27, no. 2, p. 231, 1993.
63. "Secondary Settling Tanks," scientific and technical report no. 6, IAWQ, 1998.
64. Zoltek, J.: *J. WPCF*, vol. 48, p. 179, 1976.
65. Tshobanoglous, G.: *Proc. 12th Sanitary Engineering Conf.*, University of Illinois, 1970.
66. Levin, G. V.: U. S. Patent No. 3,654,147. U.S. Patent Office, Washington D.C., 1970.
67. Milbury, W. F., D. McCaluley, and C. H. Hawthorne.: Operation of conventional activated sludge for maximum phosphorus removal. *J. WPCF,* 43(9), pp. 1890–1901, 1971.
68. Barnard, J. L.: Cut P and N without chemicals. *Water and Wastes Engineering,* Part 1, 11(7), 33-36; Part 2, (11(8), 41–43, 1974.
79. Barnard, J. L.: A review of biological phosphorous removal in activated sludge process. *Water*
70. Fuhs, G. W., and M. Chen: Microbiological basis of phosphate removal in the activated sludge process for the treatment of wastewater. *Microbiol. Ecol.*, 2(2), 119–138, 1975.
71. Arvin, E.: Biological Removal of Phosphorus from wastewater. *CRC Critical Review J.*, vol. 15, pp. 25–64, 1985.
72. Gerber, A., et al.: The effect of acetal and other short-chain compounds on the kinetics of biological nutrient removal processes, *Water SA,* 12, pp. 7–12, 1986.
73. Wentzel, M. C., et al.: Enhanced polyphosphate organism cultures in activated sludge systems, Part 1: Enhanced culture development. *Water SA,* 14, pp. 81–92, 1988.
74. Brodisch, K. E. U., and S. J. Joyner: The role of microorganisms other than *Acinetobacter* in biological phosphate removal in activated sludge process. *Water Sci. Technol.*, vol. 15, pp. 87–103, 1983.
75. WERF Report 99-WWF-3 Methods for Wastewater Characterization in Activated Sludge Modeling. Co-published by IWA Publishing and the Water Environment Federation, 2003.
76. Mamais, D., D. Jenkins, and P. Pitt: A rapid physical-chemical method for the determination of readily biodegradable soluble COD in municipal wastewater. *Water Research,* 27, p. 195, 1993.
77. Barnard, J. L.: Activated primary tanks for phosphate removal. *Water SA,* 10(3), p. 121, 1984.
78. Gu, A. Z., et al.: Investigation of PAOs and GAOs and Their Effects on EBPR Performance at Full-Scale Wastewater Treatment Plants in U.S., *Proc. WEFTEC 2005,* Washington D.C., 2005.

CHAPTER 7
Biological Wastewater Treatment Processes

7.1 Introduction

Biological treatment is designed to accelerate the natural degradation processes and stabilize the wastewater and residuals prior to disposal or reuse. Alternative methods include: aerated lagoons and stabilization basins; activated sludge processes; fixed film processes (trickling filters, rotating biological contactors); and anaerobic treatment. The chosen alternative will depend on characteristics of the raw wastewater, quality of effluent required, land requirements, residuals produced, and costs. The various types of methods, principles of operation, factors affecting performance, design methods, and example problems for each are discussed in this chapter.

7.2 Lagoons and Stabilization Basins

Stabilization basins are a common method of organic wastewater treatment where sufficient land area is available and where groundwater pollution from toxic organics or heavy metals is not a problem.

Stabilization basins can be divided into two classifications: the impounding and adsorption lagoon and the flow-through lagoon. In the impounding-adsorption lagoon, either there is no overflow or there is intermittent discharge during periods of high stream flow. The volumetric capacity of the basin is equal to the total waste flow less losses by evaporation and percolation. If there is intermittent discharge, the required capacity is related to the stream flow characteristics. In view of the large area requirements, impounding lagoons are usually limited to industries discharging low daily volumes of waste or to seasonal operations such as the canning industry.

The flow-through lagoon can be classified into three categories based on the dominant types of biological activity.

Type I. Facultative Ponds

Facultative ponds are divided by loading and thermal stratification into an aerobic surface and an anaerobic bottom. The aerobic surface layer will have a diurnal variation, increasing in oxygen content during the daylight hours due to algal photosynthesis and decreasing during the night, as shown in Fig. 7.1. Sludge deposited on the bottom will undergo anaerobic decomposition, producing methane and other gases. Odors will be produced if an aerobic layer is not maintained. Depths will vary from 3 to 6 ft (0.9 to 1.8 m).

Because oxygen generation by photosynthesis depends on light penetration, highly colored wastewaters such as textile and pulp and paper cannot be treated by this technology.

Type II. Anaerobic Ponds

Anaerobic ponds are loaded to such an extent that anaerobic conditions exist throughout the liquid volume. The biological process is the same as that occurring in anaerobic digestion tanks, that is, primarily organic acid formation followed by methane fermentation. The depth of an anaerobic pond is selected to give a minimum surface area/volume ratio and thereby provide maximum heat retention during cold weather. Depths of 15 ft (4.6 m) are common.

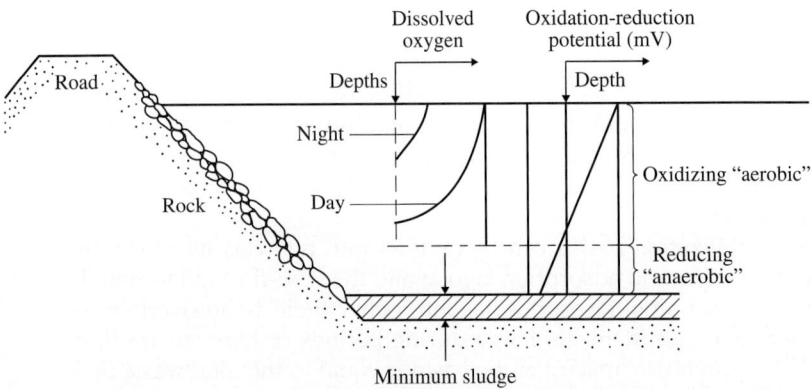

FIGURE 7.1 Waste stabilization pond—facultative type. (*Adapted from Gloyna, 1965.*)[1]

Type III. Aerated Lagoons

These range from a few days' to 2 weeks' detention, depending on the BOD removal efficiency desired. Oxygen is supplied by diffused or mechanical aeration systems, which also cause sufficient mixing to induce a significant amount of surface aeration. Depths from 6 to 15 ft (1.8 to 4.6 m) are common.

Lagoon Applications

For some industrial waste applications, aerobic ponds have been used after anaerobic ponds to provide high-degree treatment. Stabilization basins are also used to polish effluents from biological-treatment systems such as trickling filters and activated sludge.

In aerobic ponds, the amount of oxygen produced by photosynthesis can be estimated from

$$O_2 = CfS \qquad (7.1)$$

where O_2 = oxygen production, lb/(acre · d) or kg/(m² · d)
$C = 0.25$ if O_2 is in lb/(acre · d) or 2.8×10^{-5} if O_2 is in kg/(m² · d)
f = light conversion efficiency, %
S = light intensity, cal/(cm² · d)

If the light conversion efficiency is estimated as 4 percent, then $O_2 = S$. S is a function of latitude and the month of the year and may be expected to vary from 100 to 300 cal/(cm² · d) during winter and summer for a latitude 30°. This, in turn, would imply maximum loadings of 100 to 300 lb BOD_u/(acre · d) [0.011 to 0.034 kg/(m² · d)] in order to maintain any aerobic activity in the pond.

The depth of oxygen penetration in a facultative pond has been estimated as a function of surface loadings, as shown in Fig. 7.2. It should be noted that the data for Fig. 7.2 were developed from oxidation ponds treating domestic wastewater in California. Appropriate adjustments to this curve would have to be made for other types of wastewater being treated in other climatic conditions.

The typical green algae in waste stabilization basins are *Chlamydomonas*, *Chlorella*, and *Euglena*. Common blue-green algae are *Oscillatoria*, *Phormidium*, *Anacystic*, and *Anabaena*. Algae types in a pond will vary seasonally.

In the treatment of highly colored or turbid wastewaters such as from kraft pulp and paper mills, light penetration will be minimal and oxygen input will generate primarily from surface reaeration. Gellman and Berger[3] estimated oxygen input from reaeration at 45 lb O_2/(acre · d) [0.005 kg/(m² · d)]. The performance of stabilization basins in the pulp and paper industry from their data is shown in Fig. 7.3.

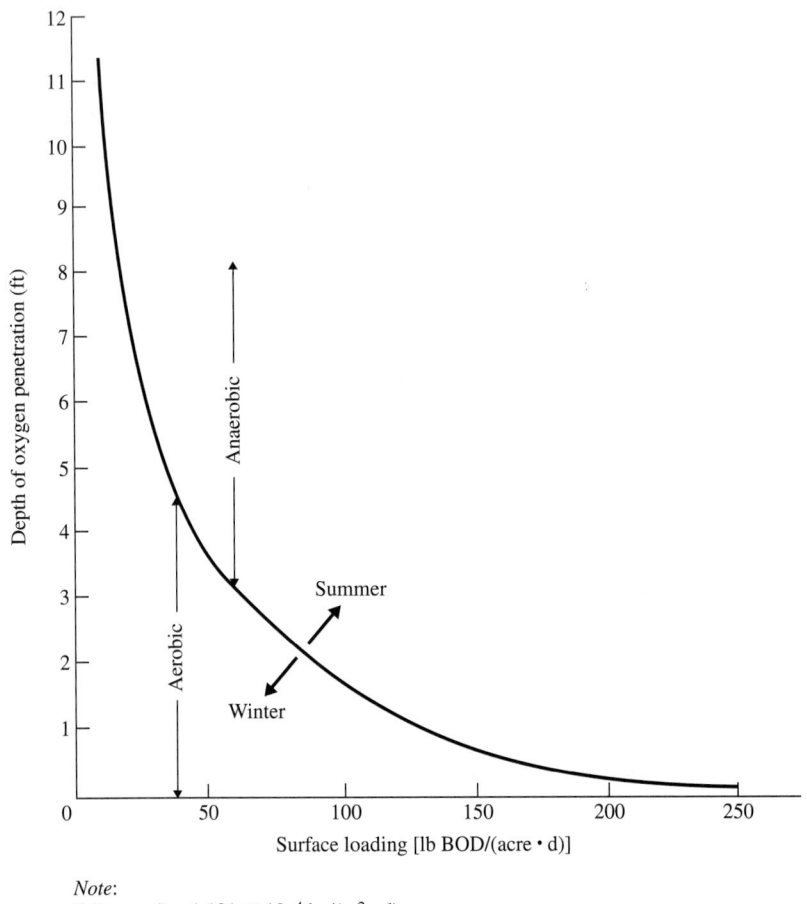

Figure 7.2 Depth of oxygen penetration in facultative ponds. (*Adapted from Oswald, 1968.*)[2]

Several concepts have been employed for the design of facultative and anaerobic ponds. Functional equations for anaerobic and facultative ponds have been developed as follows:

For a single well-mixed pond:

$$\frac{S}{S_o} = \frac{1}{1+kt} \quad (7.2)$$

For multiple ponds:

$$\frac{S}{S_o} = \frac{1}{(1+k_1 t_1)(1+k_2 t_2) \cdots (1+k_n t_n)} \quad (7.2a)$$

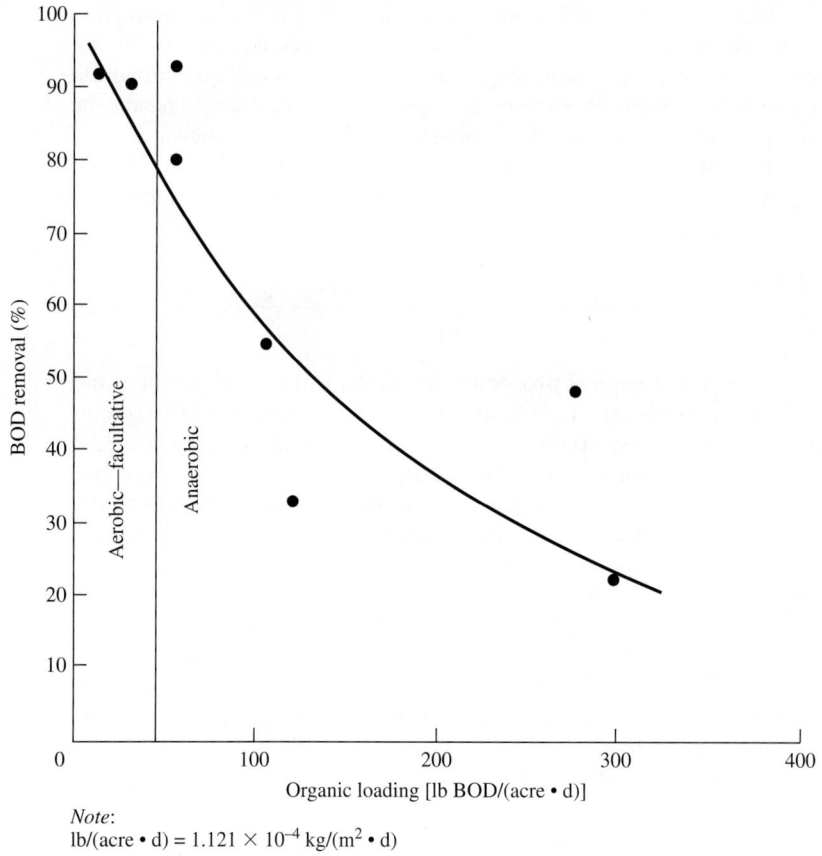

FIGURE 7.3 Waste stabilization pond performance in the pulp and paper industry.

For an infinite number of ponds or for a plug flow pond:

$$\frac{S}{S_o} = e^{-k_n t_n} \tag{7.3}$$

When variable influent concentrations are considered, Eq. (7.2) can be modified to

$$\frac{S}{S_o} = \frac{1}{1 + kt/S_o} \tag{7.4}$$

The equation is functionally the same as that employed for aerated lagoons and activated sludge, except that the rate coefficient k includes the effect of biomass concentration. This is because it is generally impractical or impossible to effectively measure the biomass concentration (VSS) in waste stabilization basins.

When multiple ponds are to be considered [Eq. (7.3)], k is assumed to be the same for all ponds. For complex wastewaters, this is probably not true, since the more readily degradable compounds will be removed in the initial ponds. For these cases, an experimental study would need to be conducted for the wastewater in question to define the change in k.

Data from an organic chemicals plant showed k to be 0.05 d^{-1} under anaerobic conditions at 20°C and 0.5 d^{-1} under aerobic conditions.

In a stabilization basin, the average k_m can be computed:

$$k_m = \frac{k_{aerobic} \times D_{aerobic} + k_{anaerobic} \times D_{anaerobic}}{D_{total}} \tag{7.5}$$

As in all biological processes, the biological activity in the pond will be a function of temperature, and the rate coefficient k can be corrected through the application of Eq. (6.43). Evaluation of the 30-d average performance for a kraft pulp and paper mill showed a θ value of 1.053, as shown in Fig. 7.4. During winter operation in colder climates, the pond will ice over, resulting in anaerobic conditions and reduced performance. (Note that an ice cover will act as an insulator, maintaining higher temperature in the liquid.)

Certain design considerations are important to the successful operation of stabilization basins. These have been discussed by Hermann and Gloyna[4] and Marais.[5] Embankments should be constructed of impervious material with maximum slope between 3:1 and 4:1 and minimum slope of 6:1. A minimum freeboard of 3 ft (0.91 m) should be maintained in the basin. Provision should be made for protecting the bank from erosion. Multicells are recommended for design to reduce short circuiting, enhance kinetics, and reduce erosion. Note that in these systems where there is no recycle, the SRT is equivalent to the HRT. Wind action is important for pond mixing and is effective with a fetch of 650 ft (198 m) in a pond with a depth of 3 ft (0.91 m).

Meat waste has been treated in shallow aerobic ponds only 18 in (0.66 m) deep at loadings of 214 lb BOD/(acre · d) [0.024 kg/(m² · d)]. The waste was presettled and had a concentration of 175 mg/L BOD. BOD reductions of 96 percent in the summer and 70 percent in the winter were obtained.

Use of an aerobic pond treating meat-packing waste after the anaerobic contact process yielded 80 percent BOD removal at a loading of 410 lb BOD/(acre · d) [0.046 kg/(m² · d)]. The concentration of BOD entering the pond was 129 mg/L. Removals at various loadings have been summarized by Steffen.[6]

Packing-house waste loadings to anaerobic ponds 8 to 17 ft (2.4 to 5.2 m) deep have been reported to vary from 0.011 to 0.015 lb BOD/(ft³ · d) [0.176 to 0.240 kg/(m³ · d)], with high BOD removals. Treatment of corn waste with an initial BOD of 2936 mg/L in a lagoon with a 9.6-d detention yielded 59 percent BOD reduction. The organic loading was 0.184 lb BOD/(ft³ · d) [2.95 kg/(m³ · d)].

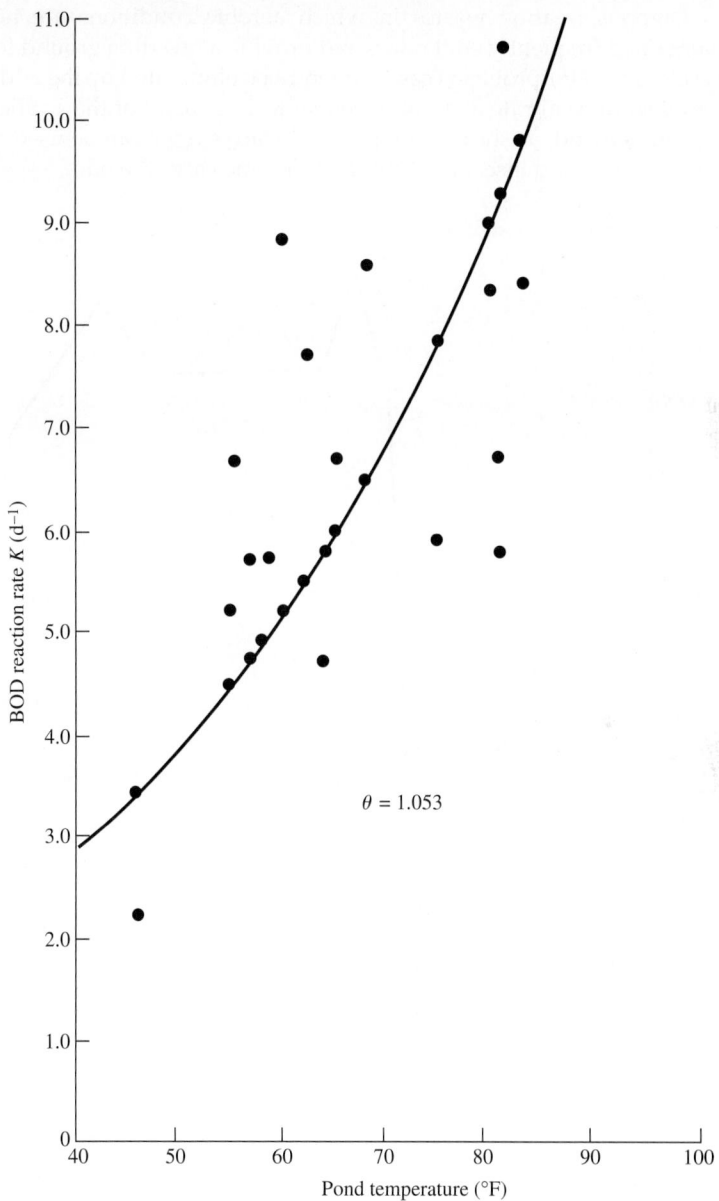

FIGURE 7.4 Temperature effect on 30-d average performance of a pond treating a pulp and paper mill effluent.

In order to improve the efficiency of lagoon operation, nitrogen and phosphorus must be added to nutrient-deficient wastes. The requirements for aerobic and anaerobic processes are discussed on pages 254 and 505, respectively.

Lagoons treating wastes in which aerobic conditions are not maintained frequently emit odors and provide a breeding ground for insects. The odor problems can frequently be eliminated by the addition of sodium nitrate at a dosage equal to 20 percent of the applied oxygen demand, as shown in Fig. 7.5. Surface sprays can be used to reduce the fly and insect nuisance and in some cases the odor.

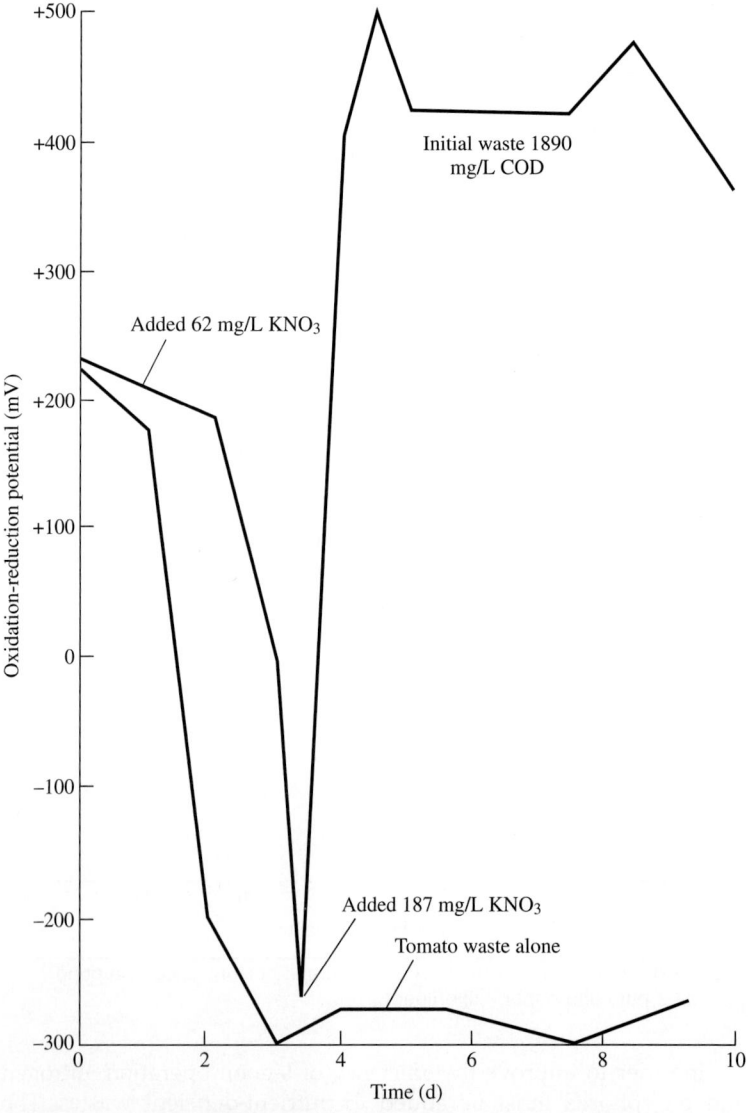

Figure 7.5 Oxidation-reduction potential control of nitrate addition to pond treatment of tomato wastewaters.

Biological Wastewater Treatment Processes

High efficiencies can frequently be attained by the use of anaerobic ponds followed by aerobic ponds. An anaerobic pond with a 6-d detention and 14-ft (4.3 m) depth loaded at 0.014 lb BOD/(ft³ · d) [0.224 kg/(m³ · d)], followed by a 3-ft- (0.9-m-)deep aerobic pond with a 19-d detention loaded at 50 lb BOD/(acre · d) [0.0056 kg/(m² · d)], yielded an overall reduction in BOD from 1100 to 67 mg/L.[7] Performance data for aerobic facultative and anaerobic ponds are summarized for various industrial wastewaters in Table 7.1. The design of facultative ponds is illustrated by Example 7.1.

Example 7.1. An industrial wastewater with a BOD of 500 mg/L is to be treated in a pond or series of ponds with a total retention time of 50 d with a depth of 6 ft. The anaerobic k at 20°C is 0.05 d^{-1} and the aerobic k is 0.51 d^{-1}. Assume the oxygen relationships as shown in Fig. 7.2 apply and the pond temperature is 20°C.

Solution

1. For one pond the applied loading is:

$$\text{Loading} = \frac{2.7 D S_o}{t}$$

$$= \frac{2.7(6)(500)}{50}$$

$$= 162 \text{ lb BOD}/(\text{acre} \cdot \text{d})$$

Dissolved oxygen will exist to a depth of 0.8 ft.

The mean k can be computed:

$$k = \frac{(0.8)(0.5) + 5.2(0.05)}{6} = 0.11 \text{ d}^{-1}$$

The effluent BOD is:

$$S_e = \frac{S_o}{1 + kt}$$

$$= \frac{500}{1 + 0.11(50)}$$

$$= 77 \text{ mg/L}$$

2. For four ponds in series, the retention time in each pond will be 12.5 d. The loading to the first pond is:

$$\frac{2.7(6)(500)}{12.5} = 648 \text{ lb BOD}/(\text{acre} \cdot \text{d})$$

and is anaerobic.

The effluent from the first pond is:

$$S_e = \frac{500}{1 + 0.05 \, (12.5)} = 308 \text{ mg/L}$$

Summary of Average Data from Aerobic and Facultative Ponds

Industry	Area, Acres	Depth, ft	Detention, d	Loading, lb/(acre · d)	BOD Removal, %
Meat and poultry	1.3	3.0	7.0	72	80
Canning	6.9	5.8	37.5	139	98
Chemical	31	5.0	10	157	87
Paper	84	5.0	30	105	80
Petroleum	15.5	5.0	25	28	76
Wine	7	1.5	24	221	
Dairy	7.5	5.0	98	22	95
Textile	3.1	4.0	14	165	45
Sugar	20	1.5	2	86	67
Rendering	2.2	4.2	4.8	36	76
Hog feeding	0.6	3.0	8	356	
Laundry	0.2	3.0	94	52	
Miscellaneous	15	4.0	88	56	95
Potato	25.3	5.0	105	111	

Summary of Average Data from Anaerobic Ponds

Industry	Area, Acres	Depth, ft	Detention, d	Loading, lb/(acre · d)	BOD Removal, %
Canning	2.5	6.0	15	392	51
Meat and poultry	1.0	7.3	16	1260	80
Chemical	0.14	3.5	65	54	89
Paper	71	6.0	18.4	347	50
Textile	2.2	5.8	3.5	1433	44
Sugar	35	7.0	50	240	61
Wine	3.7	4.0	8.8		
Rendering	1.0	6.0	245	160	37
Leather	2.6	4.2	6.2	3000	68
Potato	10	4.0	3.9		

TABLE 7.1 Performance of Lagoon Systems

Summary of Average Data from Combined Aerobic-Anaerobic Ponds

Industry	Area, Acres	Depth, ft	Detention, d	Loading, lb/(acre · d)	BOD Removal, %
Canning	5.5	5.0	22	617	91
Meat and poultry	0.8	4.0	43	267	94
Paper	2520	5.5	136	28	94
Leather	4.6	4.0	152	50	92
Miscellaneous industrial wastes	140	4.1	66	128	

Note:
ft = 0.3048 m
lb/(acre · d) = 1.121 × 10⁻⁴ kg/(m² · d)
acre = 4.0469 × 10³ m²

TABLE 7.1 *(Continued)*

The loading to the second pond is:

$$\frac{2.7(6)(308)}{12.5} = 399 \text{ lb BOD}/(\text{acre} \cdot \text{d})$$

$$S_e = \frac{308}{1+(0.05)(12.5)} = 190 \text{ mg/L}$$

The loading to the third pond is 246 lb BOD/(acre · d) and the effluent is 117 mg/L.

The loading to the fourth pond is 152 lb BOD/(acre · d) and the aerobic depth is 1 ft.

The adjusted k is:

$$\frac{0.5\,(1) + 0.05\,(5)}{6} = 0.125 \text{ d}^{-1}$$

The effluent is:

$$S_e = \frac{117}{1+(0.125)\,(12.5)} = 45.7 \text{ mg/L}$$

Four ponds in series with the same total retention time will produce a superior effluent. This assumes the reaction rate k does not change through the series of basins. This is probably not true in many cases and would have to be determined experimentally.

7.3 Aerated Lagoons

An aerated lagoon is a basin of significant depth, for example, 8 to 16 ft (2.4 to 4.9 m) deep, in which oxygenation is accomplished by mechanical or diffused aeration units and through induced surface aeration.

There are two types of aerated lagoons:

1. The aerobic lagoon in which dissolved oxygen and suspended solids are maintained uniformly throughout the basin.

2. The aerobic-anaerobic or facultative lagoon, in which oxygen is maintained in the upper liquid layers of the basin, but only a portion of the suspended solids is maintained in suspension. These basin types are shown in Fig. 7.6. A photo of a typical aerated lagoon is shown in Fig. 7.7.

In the aerobic lagoon, all solids are maintained in suspension, and this system may be thought of as a "flow-through" activated sludge system, that is, without solids recycle. Thus, the effluent suspended solids concentration will be equal to the aeration basin solids concentration, and the sludge age is equal to the hydraulic retention time (HRT).

In the facultative lagoon, a portion of the suspended solids settle to the bottom of the basin, where they undergo anaerobic decomposition. The anaerobic by-products are subsequently oxidized in the upper aerobic layers of the basin. The facultative lagoon can also be modified to yield a more highly clarified effluent by the inclusion of a separate postsettling pond or a baffled settling compartment.

Aerobic and facultative lagoons are primarily differentiated by the power level employed in the basin. In the aerobic lagoon, the power level is sufficiently high to maintain all solids in suspension and may vary from 14 to 20 hp/million gal (2.8 to 3.9 W/m^3) of basin volume, depending on the nature of the suspended solids in the influent wastewater. Field data demonstrated that 14 hp/million gal (2.8 W/m^3) was generally sufficient to maintain pulp and paper solids in suspension, while 20 hp/million gal (3.9 W/m^3) was required for domestic wastewater treatment.

In the facultative lagoon, the power level employed is sufficient only to maintain dispersion and mixing of the dissolved oxygen.

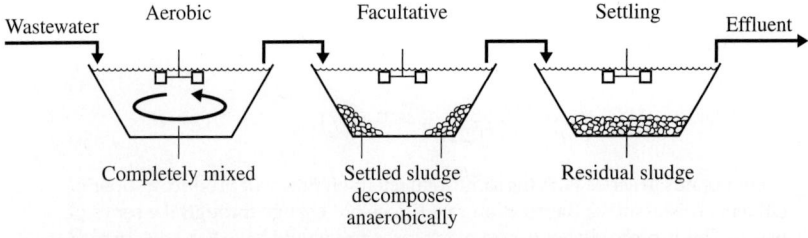

FIGURE 7.6 Aerated lagoon types.

FIGURE 7.7 Aerated lagoon treating pulp and paper mill wastewater.

Experience in the pulp and paper industry showed that a minimum power level employing low-speed mechanical surface aerators is 4 hp/million gal (0.79 W/m³). The use of other kinds of aeration equipment might require different power levels to maintain uniform dissolved oxygen in the basin.

Aerobic Lagoons

At a constant basin detention time, the equilibrium biological solids concentration and the overall rate of organic removal can be expected to increase as the influent organic concentration increases. For a soluble industrial wastewater, the equilibrium biological solids concentration X_v can be predicted from the relationship:

$$X_v = \frac{aS_r}{1+bt} \qquad (7.6)$$

When nondegradable volatile suspended solids are present in the wastewater, Eq. (7.6) becomes

$$X_v = \frac{aS_r}{1+bt} + X_i \qquad (7.7)$$

in which X_i = influent volatile suspended solids not degraded in the lagoon. Combining Eq. (7.6) with the kinetic relationship [Eq. (6.29)], the effluent soluble organic concentration can be computed:

$$\frac{S}{S_o} = \frac{1+bt}{aKt} \qquad (7.8)$$

From Eq. (7.8), it can be concluded that the fraction of effluent soluble organic concentration remaining is independent of the influent organic concentration. For lagoons with fixed detention times, this conclusion is justified because higher influent organic concentrations result in higher equilibrium biological solids levels and, therefore, higher overall BOD removal rates. This relationship is shown in Fig. 7.8 for several wastewaters.

The specific organic reaction rate coefficient K is temperature-dependent, and can be corrected for temperature by Eq. (6.43). The

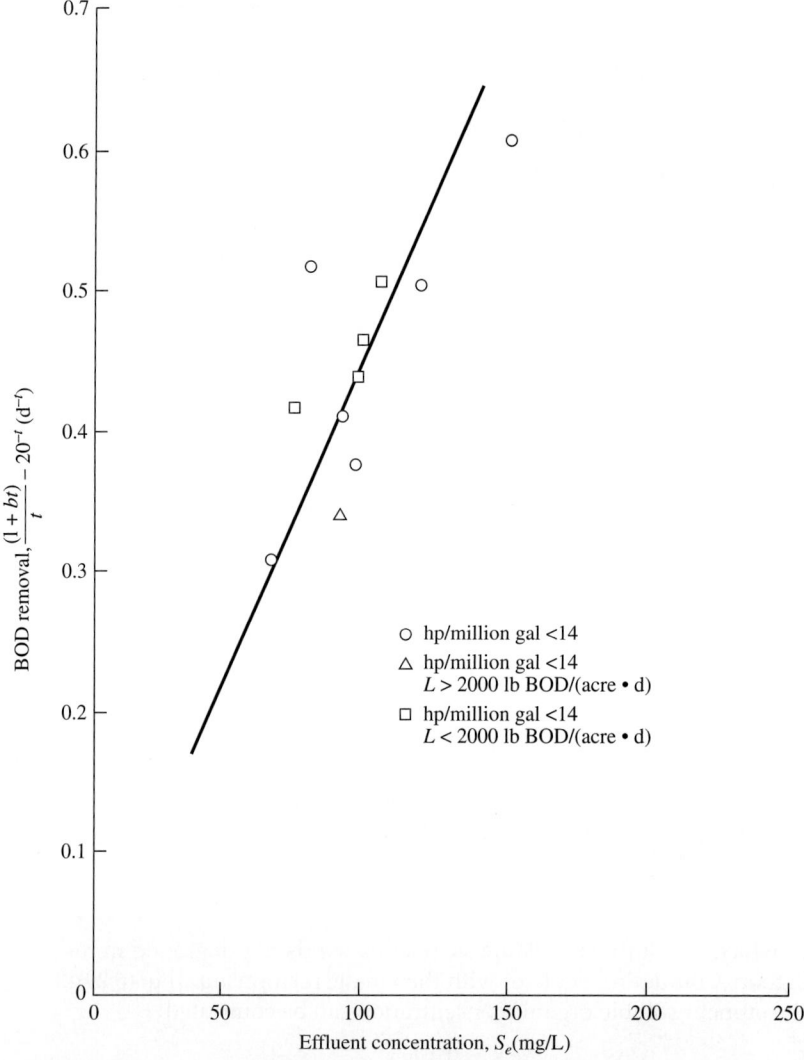

Figure 7.8 Kinetic relationships in aerobic lagoons.

oxygen requirements for an aerobic lagoon are computed using the same relationship employed for activated sludge [Eq. (6.20)]. Aerobic lagoons are employed for pretreatment of high-strength industrial wastewaters prior to discharge to a joint or municipal treatment system, or as the first basin in a two-basin aerated lagoon series, followed by a facultative basin. It should be noted that, while in an aerobic lagoon the soluble organic content is reduced, there is an increase in the effluent suspended solids through synthesis. The relationship obtained for an aerobic lagoon pretreating a brewery wastewater is shown in Fig. 7.9. The relationship between BOD removal and VSS production due to cell synthesis is demonstrated in Fig. 7.10.

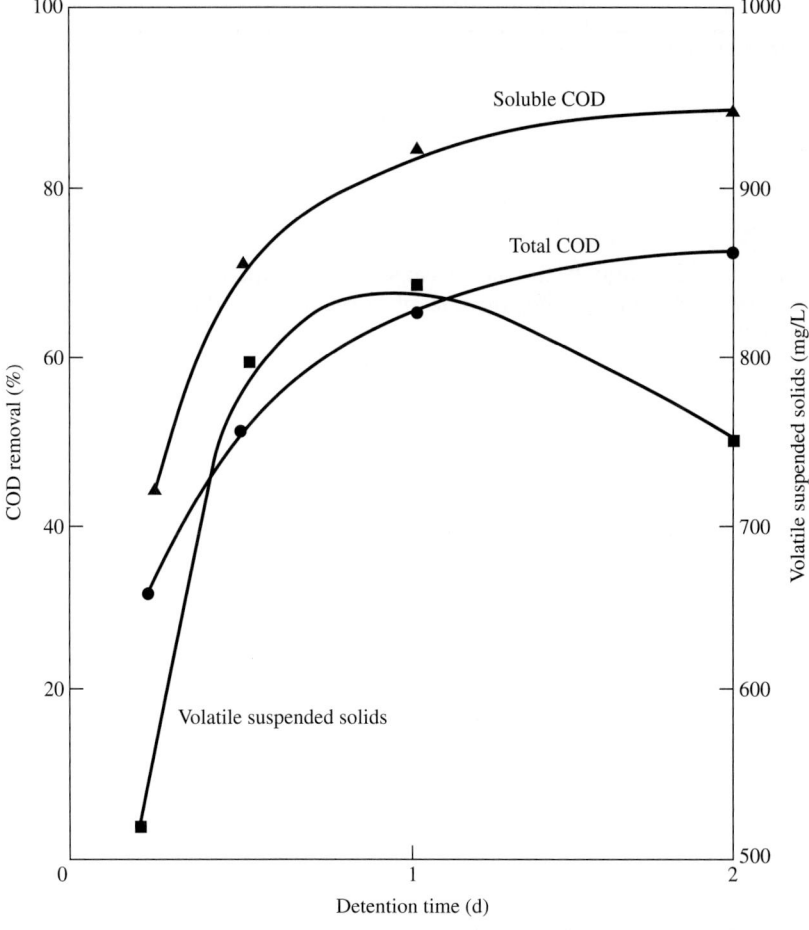

Figure 7.9 COD removal from brewery wastewater through an aerobic lagoon.

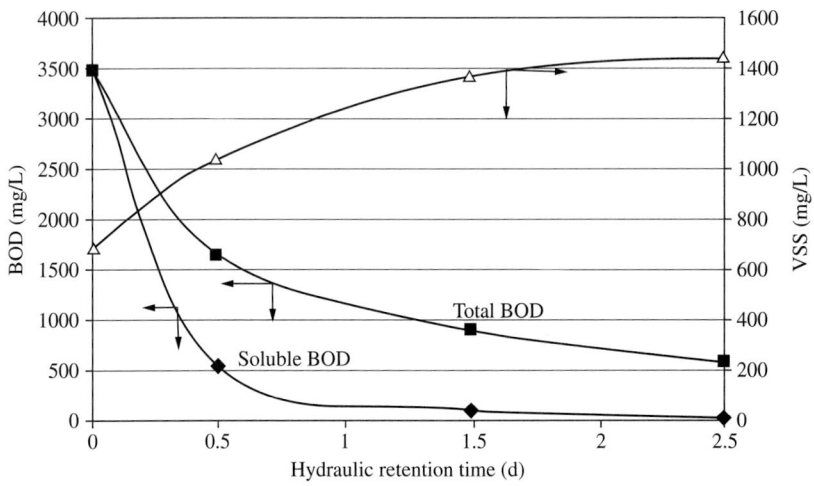

FIGURE 7.10 Bench-scale aerobic aerated lagoon test results for dairy wastewater.

Power requirements should generally be designed for summer operation, since the rates of organic removal will be the greatest during this period of operation.

Facultative Lagoons

In a facultative lagoon, the biological solids level maintained in suspension is a function of the power level employed in the basin. Results from the pulp and paper industry are shown in Fig. 7.11. Average operating values of a chemical industrial wastewater by an aerated lagoon is presented in Table 7.2. Solids deposited in the bottom of the facultative lagoon will undergo anaerobic degradation, which results

Parameter	Value	Unit
Influent TCBOD	8,320	mg/L
Influent TCOD	16,500	mg/L
Effluent TCBOD	480	mg/L
Effluent TCOD	2,300	mg/L
Effluent 1.5 μm TSS	1,500	mg/L
Effluent 0.45 μm TSS	1,950	mg/L
Total dissolved solids	57,480	mg/L
Hydraulic retention time	10	days
Temperature	22	°C

TABLE 7.2 Average Values Observed During Operation of an Aerated Lagoon Treating Chemical Industry Wastewater

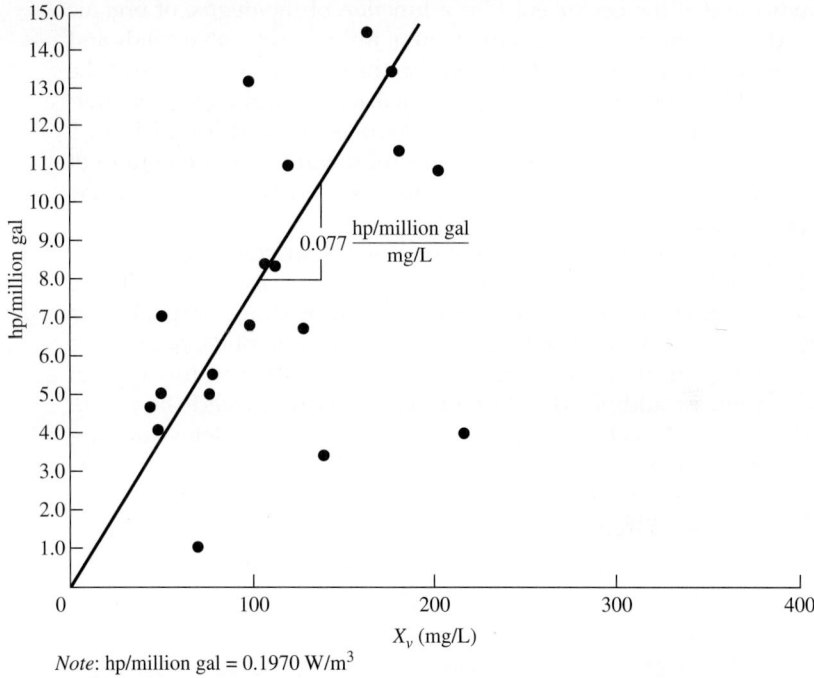

Note: hp/million gal = 0.1970 W/m³

FIGURE 7.11 Correlation for mixing power input versus MLSS concentrations.

in a feedback of soluble organics to the upper aerobic layers. Under these conditions Eq. (6.29) should be modified to

$$\frac{S}{FS_o} = \frac{FS_o}{FS_o + KX_v t} \tag{7.9}$$

in which F is a coefficient that accounts for organic feedback due to anaerobic activity in the deposited sludge layers. The degree of anaerobic activity is highly temperature-dependent and the coefficient F may be expected to vary from 1.0 to 1.4 under winter and summer conditions, respectively, depending on the geographical location of the plant.

In a facultative lagoon, biological solids are maintained at a lower level than in an aerobic lagoon, and soluble organics are fed back to the liquid as anaerobic degradation products. It is therefore not possible to directly compute the oxygen requirements by employing Eq. 6.20. In this case, the oxygen requirements can be related empirically to organic removal and estimated from

$$R_r = F'S_r \tag{7.10}$$

in which F' is an overall oxygen-utilization coefficient for facultative lagoons. Results obtained for various industrial wastewaters would indicate that the coefficient F' is a function of the degree of organic feedback, which in turn is a function of influent settleable solids and temperature. In general, depending on the geographic location of the plant, F' can be estimated to vary from 0.8 to 1.1 during winter operation when anaerobic activity in the basin is low, and from 1.1 to 1.5 during summer operation when anaerobic activity in the bottom of the basin is at a maximum. The value selected will depend on the geographical location of the plant.

Nutrient requirements in aerated lagoons are computed in a similar way to the activated sludge process. In the case of a facultative lagoon, however, anaerobic decomposition of sludge deposited in the bottom of the basin will feed back nitrogen and phosphorus. This is usually sufficient for organic removal occurring in this type of basin, and no additional nitrogen and phosphorus needs to be added. The relationships for the design of aerated lagoon systems are summarized in Table 7.3.[9]

Temperature Effects in Aerated Lagoons

The performance of aerated lagoons is significantly influenced by changes in basin temperature. In turn, the basin temperatures are

	Aerobic	Facultative	Settling
Relationship	hp/million gal 14 to 20	hp/million gal > 4 < 10	
Kinetics	$\dfrac{S_1}{S_0} = \dfrac{1+bt}{aKt}$	$S_2 = \dfrac{F^2 S_1^{2a}}{KX_v t^c + FS_1}$	$t < 2d^b$
Oxygen requirements	$O_2/d = a'S_r + b'X_v t$	$O_2/d = \begin{cases} 0.8 - 1.1\ S_r\ (\text{winter})^d \\ 1.1 - 1.4\ S_r\ (\text{summer}) \end{cases}$	
Nutrient requirements	$N = 0.11\ \Delta X_v$ $P = 0.02\ \Delta X_v$	None required[e]	
Sludge yield	$X_v = \dfrac{aS_r}{1+bt}$	$0.1\ aS_r + \text{inerts}^f$ 3 to 7 percent weight concentration	

[a] BOD feedback from benthal decomposition varies from 1.0 in winter to 1.4 in summer.
[b] To minimize algae growth.
[c] Depends on power level in basin (see Fig. 7.11).
[d] Factors employed depend on geographical location.
[e] Benthal decomposition feeds back N and P.
[f] Most of biovolatile solids will break down through benthal decomposition.
Note: hp/million gal = 1.98×10^{-2} kW/m^3

TABLE 7.3 Aerated Lagoon Design Relationships

influenced by the temperature of the influent wastewater and the ambient air temperature. Although heat is lost through evaporation, convection, and radiation, it is gained by solar radiation. While several formulas have been developed to estimate the temperature in aerated lagoons, the following equation will usually give a reasonable estimate for engineering design purposes:

$$\frac{t}{D} = \frac{T_i - T_w}{f(T_w - T_a)} \tag{7.11}$$

where t = basin detention time, d
  D = basin depth, ft
  T_i = influent wastewater temperature, °F
  T_a = mean air temperature, °F (usually taken as the mean weekly temperature)
  T_w = basin temperature, °F

Equation (7.11) was developed for surface aeration and hence cannot be used for subsurface aeration.

The coefficient f is a proportionality factor containing the heat-transfer coefficients, the surface area increase from aeration equipment, and wind and humidity effects. The approximate value of f is 90 for most aerated lagoons employing surface aeration equipment.

A general temperature model has been developed by Argaman and Adams[8] that considers an overall heat balance in the basin, including heat gained from solar radiation, mechanical energy input, biochemical reaction, and heat lost by long-wave radiation, evaporation from the basin surface, conduction from the basin surface, evaporation and conduction from the aerator spray, and conduction through the basin walls. Their final equation is

$$T_w = T_a + \left[\frac{Q}{A}(T_i - T_a) + 10^{-6}(1 - 0.0071 C_C^2) H_{s,o} + 6.95(\beta - 1)\right.$$

$$+ 0.102(\beta - 1)T_a - e^{0.0604 T_a}\left(1 - \frac{f_a}{100}\right) 1.145 A^{-0.05} V_w$$

$$+ \frac{126 NFV_w}{A} + \frac{10^{-6} H_m}{A} + \frac{1.8 S_r}{A}\right]$$

$$\bigg/ \left[\frac{Q}{A} + 0.102 + (0.068 e^{0.0604 T_a} + 0.118) A^{-0.05} V_w\right.$$

$$+ \frac{4.32 NFV_w}{A}(3.0 + 1.75 e^{0.0604 T_a}) + \frac{10^{-6} UA_w}{A}\right] \tag{7.12}$$

where T_w = basin water temperature, °C
T_a = air temperature, °C
T_i = influent waste temperature, °C
Q = flow rate, m³/d
A = surface area, m²
C_c = average cloud cover, tenths
$H_{s,o}$ = average daily absorbed solar radiation under clear sky conditions, cal/(m² · d)
U = heat-transfer coefficient, cal/(m² · d · °C)
β = atmospheric radiation factor
f_a = relative humidity, percent
N = number of aerators
F = aerator spray vertical cross-sectional area, m²
V_w = wind speed at treetop, m/s
$H_m = 15.2 \times 10^6 p$, where p = aeration power, hp
S_r = organic removal rate, kg COD removed/d
A_w = effective wall area, m²

Equation (7.12) can be used to predict the temperature of diffused air systems by substituting

$$NFV_W = 2Q_A$$

where Q_A = air flow, m³/s.

Aerated Lagoon Systems

Multiple basins may be most effectively employed in aerated lagoon systems under proper conditions. A comparison of single-stage and multistage operation for a pulp and paper mill is shown in Fig. 7.12. As can be seen, multistage operation is more efficient in terms of organic removal. In addition, series operation may be desired where land availability is a concern. In considering a thermal balance, a minimum total basin volume can be obtained by employing two basins in series. The first basin volume is minimized to maintain a high temperature, a high biological solids level, and a resulting high BOD reaction rate in an aerobic lagoon. The second basin is a facultative basin at lower power (mixing) levels that permit solids settling and decomposition in the bottom of the basin. An optimization procedure can be employed to determine the smallest total basin volume and the lowest aeration horsepower for a specified effluent quality. Where low effluent suspended solids are desired, a final settling basin can be employed. The settling basin should have

1. A sufficiently long detention period to effect the desired suspended solids removal
2. Adequate volume for sludge storage
3. Minimal algal growth
4. Minimal odors from anaerobic activity

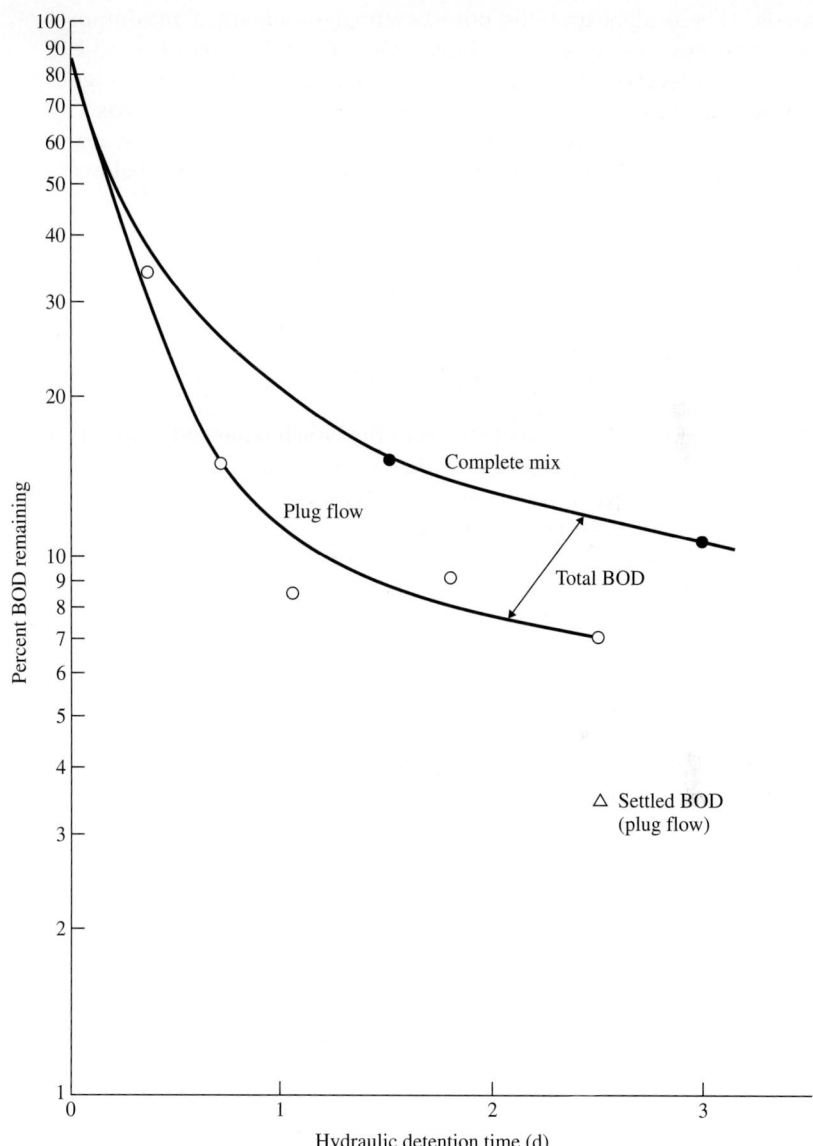

Figure 7.12 Pilot plant results for single- and multistage operation.

Unfortunately, these design objectives are not always compatible. Frequently short retention times are required to inhibit algal growth, which are too short for proper settling. Also, adequate volume must remain above the sludge deposits at all times to prevent the escape of the odorous gases of decomposition.

In order to achieve these objectives, a minimum detention time of 1 d is usually required to settle the majority of the settleable suspended solids. Where algal growths pose potential problems, a maximum detention time of 3 to 4 d is recommended. For odor control, a minimum water level of 3 ft (0.9 m) should be maintained above the sludge deposits at all times. Parker[10] has shown that a scale-up factor needs to be applied in aerated lagoon process design. The effect of scale on aerated lagoon treatment of a petrochemical wastewater is shown below:

Scale	t, d	K_{20}
Bench	0.83	10.0
Pilot	0.83	5.3
Full	1.3	2.8

Parker[10] attributes this to wall effects in the bench scale and to differences in fluid shearing intensity. In any event, caution should be exercised in scaling up a design from laboratory data.

One problem with aerated lagoons, as well as other technologies, is the effect of shock loads on effluent quality. In one pulp and paper mill effluent, a dissolved oxygen probe at the head end of the basin was used to trigger a diversion to a spill pond as shown in Fig. 7.13. In this case when the dissolved oxygen dropped below 2 mg/L, indicating a shock load, a portion of the wastewater was diverted. The spill basin contents were then pumped at a controlled rate to the aerobic lagoon under dissolved oxygen control. The variability of one aerated lagoon performance is shown in Fig. 7.14.[11]

The design of an aerated lagoon system is illustrated by Example 7.2. Where temperature is not a consideration, a simplified procedure can be employed as shown in Example 7.3.

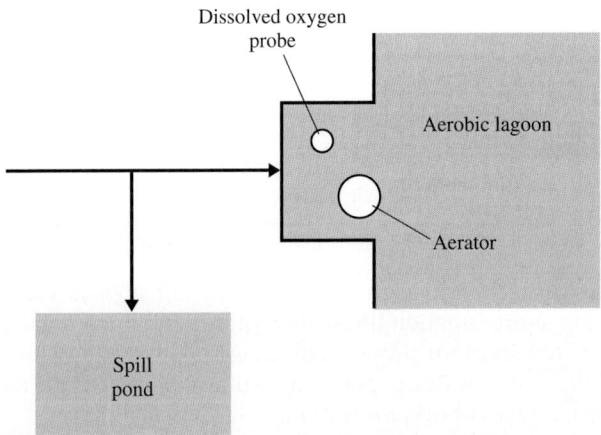

FIGURE 7.13 Spill diversion control.

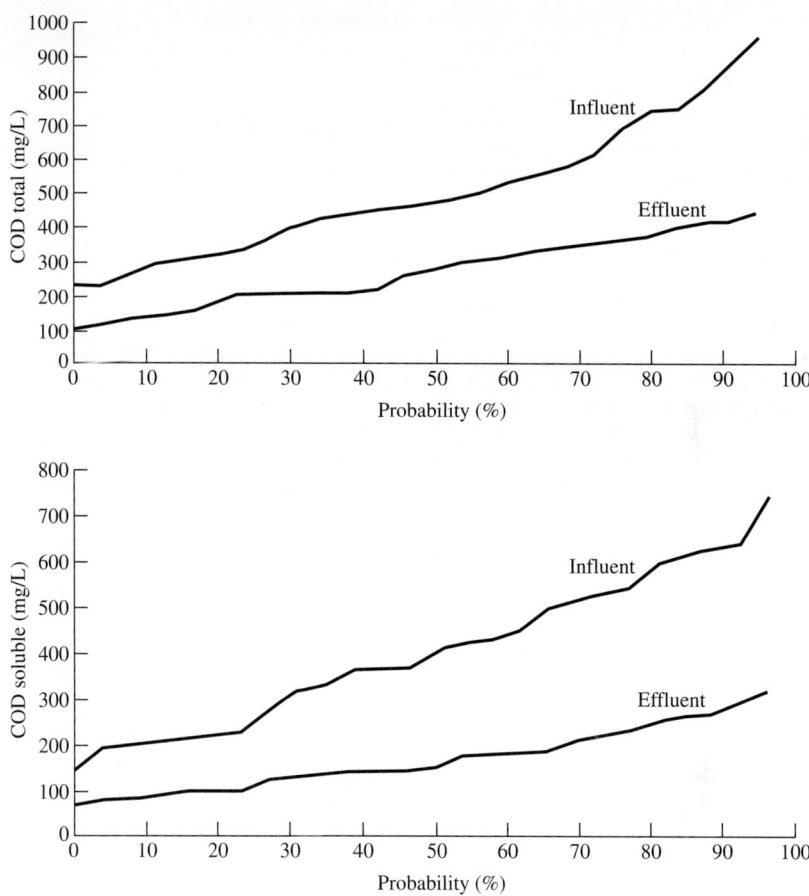

FIGURE 7.14 Performance of aerated ponds—COD removal. (*Adapted from Galil et al., 1996.*)

Example 7.2. Design a two-stage aerated lagoon system of 12 ft (3.66 m) depth to treat an 8.5 million gal/d (32,170 m³/d) industrial wastewater with the following characteristics:

Influent: BOD_5 = 425 mg/L

Temperature = 85°F (29°C)

SS = 0 mg/L

Ambient temperature: summer = 70°F (21°C)

winter = 34°F (1°C)

Kinetic variables: K = 63/d at 20°C
a = 0.5
b = 0.2/d at 20°C
a' = 0.52
b' = 0.28/d at 20°C
θ = 1.035 for BOD reaction rate

$\theta = 1.024$ for oxygen-transfer efficiency
$F = 1.0$ (winter)
$ = 1.4$ (summer)
$F' = 1.5$ (summer)
$N_o = 3.2$ lb $O_2/$(hp $\cdot$ h) [1.95 kg/(kW $\cdot$ h)]
$\alpha = 0.85$
$\beta = 0.90$
$C_L = 1.0$ mg/L

The final effluent should have a maximum soluble BOD_5 of 20 mg/L for the summer and 30 mg/L for the winter.

Solution
In general, the required detention time of the lagoon system for reaching a terminal BOD_5 is controlled by winter temperatures while the oxygen requirements and power level will usually be controlled by summer conditions.

(a) *Design of basin based on minimum detention time*

1. Compute basin volume by assuming a detention time.
   For $t = 2$ d:
   $$V = Qt$$
   $$= 8.5 \times 2$$
   $$= 17 \text{ million gal} \quad (64{,}350 \text{ m}^3)$$

2. Calculate winter water temperature in basin:

   $$\frac{t}{D} = \frac{T_i - T_w}{f(T_w - T_a)}$$

   or

   $$T_w = \frac{DT_i + ftT_a}{D + ft}$$

   $$= \frac{12 \times 85 + 1.6 \times 2 \times 34}{12 + 1.6 \times 2}$$

   $$= 74.3°F \quad \text{or} \quad 23.5°C$$

3. Correct the BOD reaction rate for winter conditions:

   $$K_{(T_2)} = K_{(T_1)} \theta^{T_2 - T_1}$$

   $$K_{23.5} = 6.3 \times 1.035^{23.5-20}$$

   $$= 7.11/\text{d}$$

4. Calculate the winter effluent soluble BOD_5:

   $$\frac{S_e}{S_o} = \frac{1 + bt}{aKt}$$

   $$S_e = \frac{1 + 0.2 \times 2}{0.5 \times 7.11 \times 2} \, 425$$

   $$= 83.7 \text{ mg/L}$$

5. Repeat steps 2 through 4 for summer conditions:

$$T_w = \frac{12 \times 85 + 1.6 \times 2 \times 70}{12 + 1.6 \times 2}$$

$$= 81.8°F \text{ or } 27.7°C$$

$$K_{27.7} = 6.3 \times 1.035^{27.7-20}$$

$$= 8.21/d$$

$$S_e = \frac{1 + 0.2 \times 2}{0.5 \times 8.21 \times 2} \, 425$$

$$= 72.5 \text{ mg/L}$$

6. Calculate average volatile suspended solids concentration under summer conditions:

$$X_v = \frac{aS_r}{1 + bt}$$

$$= \frac{0.5(425 - 72.5)}{1 + 0.2 \times 2}$$

$$= 126 \text{ mg/L}$$

7. Calculate oxygen requirement:

$$R_r = a'S_r + b'X_v$$

$$= 0.52 \times (425 - 72.5) \times 8.5 \times 8.34 + 0.28 \times 126 \times 8.5 \times 2 \times 8.34$$

$$= 17{,}996 \text{ lb/d} \quad (8170 \text{ kg/d})$$

8. Compute horsepower requirement:

$$N = N_o \frac{\beta C_s - C_L}{C_{s(20)}} \, a\theta^{T-20}$$

$$= 3.2 \, \frac{0.90 \times 7.96 - 1.0}{9.2} \, 0.85 \times 1.024^{27.7-20}$$

$$= 2.19 \text{ lb/(h·p)} \quad (1.33 \text{ kg/kW·h})$$

$$\text{hp} = \frac{R_r}{N}$$

$$= \frac{17{,}996}{2.19 \times 24} = 342 \text{ hp} \quad (257 \text{ kW})$$

9. Check power level:

$$PL = \frac{\text{hp}}{V}$$

$$= \frac{342}{17} = 20.1 \text{ hp/million gal} \quad (0.40 \text{ kW/m}^3)$$

For a conservative design, the minimum power level should be 14 hp/million gal (0.28 kW/m³). If the power level is significantly less than 14 hp/million gal, then 14 hp/million gal should be used.

10. Repeat steps 1 through 9 using different detention times. The results can be tabulated as follows:

t, d	Winter or Summer	T_w °F	T_w °C	K, d^{-1}	S_e, mg/L	X_v, mg/L	R_r, lb/d	hp Required	PL, hp/million gal
1.0	W	79.0	26.1	7.77	131				
	S	83.2	28.4	8.41	121	127	13,727	261	30.7
2.0	W	74.3	23.5	7.11	83.7				
	S	81.8	27.7	8.21	72.5	126	17,996	342	20.1
3.0	W	70.4	21.3	6.59	68.8				
	S	80.7	27.1	8.04	56.4	115	20,436	389	153
4.0	W	67.3	19.6	6.21	61.6				
	S	79.8	26.6	7.91	48.4	105	22,219	476 (423)*	14 (12.4)*
5.0	W	64.6	18.1	5.90	57.6				
	S	79.0	26.1	7.77	43.8	95	23,480	595 (449)	14 (10.6)
6.0	W	62.3	16.8	5.64	55.3				
	S	78.3	25.7	7.66	40.7	87	24,528	714 (469)	14 (9.2)

*Minimum power level applied.

(b) *Design of basin based on minimum detention time*

It is assumed that the minimum power level in a facultative lagoon should be 4 hp/million gal (0.079 kW/m³) to maintain 50 mg/L of volatile suspended solids in suspension. The design of this basin is based on selecting an effluent BOD concentration from the first basin at given detention times to be the influent to the second basin.

1. Calculate the temperature in this basin and adjust the BOD reaction rate by assuming a detention time. For example, the effluent BOD concentrations at a 2-d detention time for the first basin are 83.7 and 72.5 mg/L for winter and summer, respectively. Basin temperature:

$$T_w = \frac{DT_i + ftT_a}{D + ft}$$

Corrected influent concentration:

$$S'_o = FS_o$$

Corrected BOD reaction rate:

$$K_{20} = \frac{6.3}{425} S'_o$$

Assume $t = 5$ d.

Winter:
$$T_w = \frac{12 \times 74.3 + 1.6 \times 5 \times 34}{12 + 1.6 \times 5}$$
$$= 58.2°F \quad \text{or} \quad 14.6°C$$

$$K_{20} = \frac{6.3}{425} \, 1.0 \times 83.7$$
$$= 1.24/d$$

$$K_{14.6} = 1.24 \times 1.035^{14.6-20}$$
$$= 1.03/d$$

Summer:
$$T_w = \frac{12 \times 81.8 + 1.6 \times 5 \times 70}{12 + 1.6 \times 5}$$
$$= 77.1°F \quad \text{or} \quad 25.1°C$$

$$K_{20} = \frac{6.3}{425} \, 1.4 \times 72.5$$
$$= 1.50/d$$

$$K_{25.1} = 1.50 \times 1.035^{25.1-20}$$
$$= 1.79/d$$

2. Compute the detention time required to reduce the soluble BOD to the prescribed level:
$$S'_o = FS_o$$
$$t = \frac{S'_o(S'_o - S_e)}{KX_v S_e}$$

Winter: $S'_o = 83.7 \text{ mg/L}$
$$t = \frac{83.7(83.7 - 30)}{1.03 \times 50 \times 30}$$
$$= 2.91 \text{ d}$$

Summer: $S'_o = 1.4 \times 72.5$
$$= 102 \text{ mg/L}$$
$$t = \frac{102(102 - 20)}{1.79 \times 50 \times 20}$$
$$= 4.67 \text{ d}$$

Now summer conditions are controlling.

3. Repeat steps 1 and 2 until the computed detention time in step 2 is close enough to the assumed value in step 1. Final results are as follows:

$t = 4.67$ d, $V = 39.7$ million gal (150,300 m^3)

Winter: $T_W = 58.8°F$ or $14.9°C$

$K_{14.9} = 1.04/d$

$S_e = 21.5$ mg/L

Summer: $T_W = 77.3°F$ or $25.2°C$

$K_{25.2} = 1.79/d$

$S_e = 20.0$ mg/L

4. Calculate oxygen requirement:

$$R_r = F'S_r$$

$$= 1.5 \times (1.4 \times 72.5 - 20.0) \times 8.5 \times 8.34$$

$$= 8666 \text{ lb/d} \quad (3934 \text{ kg/d})$$

5. Calculate horsepower requirement:

$$N = N_o \frac{\beta C_s - C_L}{C_s(20)} a\theta^{T-20}$$

$$= 3.2 \frac{0.90 \times 8.36 - 1.0}{9.2} 0.85 \times 1.024^{25.2-20}$$

$$= 2.18 \text{ lb/(hp·h)} \quad [1.33 \text{ kg/(kW·h)}]$$

$$\text{hp} = \frac{R_r}{N}$$

$$= \frac{8666}{2.18 \times 24}$$

$$= 166 \text{ hp} \quad (125 \text{ kW})$$

6. Check power level:

$$PL = \frac{\text{hp}}{V}$$

$$= \frac{166}{39.7}$$

$$= 4.2 \text{ hp/million gal} \quad (0.083 \text{ kW/m}^3)$$

7. Repeat steps 1 through 6 using different detention times in the first basin with the adequate detention time in the second basin to meet the desired effluent quality. The results are tabulated below:

S_o, mg/L	T_w °F	T_w °C	K, d^{-1}	t, d	S_e, mg/L	X_v, mg/L	R_r, lb/d	hp Required	PL, hp/million gal
131	55.0	12.8	1.51	8.57	22.1	50			
121	76.2	24.6	2.94	8.57	20.0	50	10,740	291 (205)*	4.0 (2.8)*
83.7	58.8	14.9	1.04	4.67	21.5	50			
72.5	77.3	25.2	1.79	4.67	20.0	50	8,666	166	4.2
68.8	59.2	15.1	0.862	3.33	22.3	50			
56.4	77.4	25.2	1.40	3.33	20.0	50	6,274	120	4.2
61.6	58.4	14.7	0.761	2.72	23.0	50			
48.4	77.2	25.1	1.19	2.72	20.0	50	5,083	97	4.2
57.6	57.3	14.1	0.697	2.34	23.8	50			
43.8	76.9	24.9	1.08	2.34	20.0	50	4,392	84	4.2
55.3	56.1	13.4	0.653	2.12	24.6	50			
40.7	76.5	24.7	0.993	2.12	20.0	50	3,934	75	4.2

*Minimum power level applied.

(c) *Optimum lagoon system*

This two-stage lagoon system can be optimized to minimize either the total detention time or the total horsepower to be installed.

The following figure shows the summarized design results of the two-stage lagoon system. The minimum total required detention time is 6.33 d,

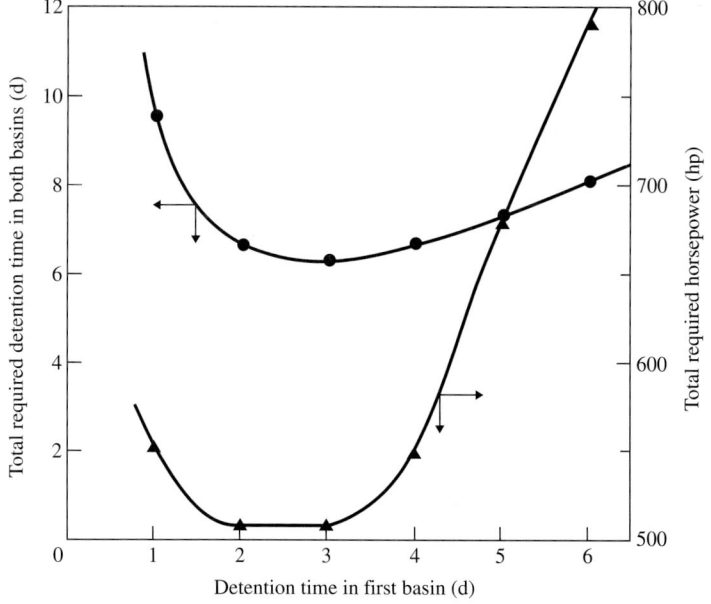

Effect of detention time in first basin on total required time and horsepower.

with a corresponding total installed horsepower of 509 hp (382 kW). The minimum total horsepower is 508 hp (381 kW) with a total detention time of 6.67 d. The second alternative may be the optimum system since the increased basin size will be more tolerable to the influent fluctuation in both flow rate and constituent concentration. It will also be justified by an economic gain realized by operating with less power input, although the difference in the horsepower installed is minimal. In this alternative, the detention time in the first basin is 2 d and the detention time in the second basin 4.67 d. The results are tabulated below:

	t, d			hp Installed		
First Basin	Second Basin	Total	First Basin	Second Basin	Total
1.0	8.57	9.57	261	291	552
2.0	4.67	6.67	342	166	508
3.0	3.33	6.33	389	120	509
4.0	2.72	6.72	476	97	573
5.0	2.34	7.34	595	84	679
6.0	2.12	8.12	714	75	789

Example 7.3. Design an aerated lagoon system consisting of an aerobic and a facultative lagoon for the following conditions:

Flow = 4 million gal/d
$S_o = 450$ mg/L
$S_e = 20$ mg/L (soluble)
$K = 6$ d^{-1}
$b = 0.1$ d^{-1}
$a = 0.6$
$a' = 0.5$

Solution Design the aerobic lagoon:

$$\frac{S_e}{S_o} = \frac{1+bt}{aKt}$$

t, d	S_e/S_o	S_e
1	0.3	135
2	0.17	77
3	0.12	54
4	0.097	44

Use a 2-d detention time in the aerobic lagoon.

$$X_v = \frac{aS_r}{1+bt}$$

$$= \frac{0.6 \cdot 373}{1.2} = 187 \text{ mg/L}$$

Oxygen requirement:

$$O_2 = a'S_r + 0.14 \, X_v t$$

$$= 0.5 \cdot 373 + 0.14 \cdot 187 \cdot 2$$

$$= 239 \text{ mg/L}$$

lb O_2/d $= 239 \cdot 4 \cdot 8.34 = 7968$

$$\text{hp} = \frac{7968}{24 \cdot 1.5 \text{ lb } O_2/(\text{hp} \cdot \text{h})} = 22$$

$$\text{hp/million gal} = \frac{221}{8} = 27.7$$

Nutrient requirements:

$$N = 0.11 \, \Delta X_v = 0.11 \cdot 187 = 20.6 \text{ mg/L}$$

$$= 20.6 \cdot 4 \cdot 8.34 = 687 \text{ lb/d}$$

$$P = 0.02 \, \Delta X_v = 0.02 \cdot 187 = 3.74 \text{ mg/L}$$

$$= 3.74 \cdot 4 \cdot 8.34 = 125 \text{ lb/d}$$

Facultative lagoon: Reduce K to consider the lower reaction rate in the facultative lagoon [Eq. (6.31)].

$$K_1 = K \frac{S_1}{S_o} = 6 \cdot \frac{77}{450}$$

$$= 1.02 \text{ d}^{-1}$$

Use a feedback factor of 1.2. Assume $X_v = 100$ mg/L.

$$\frac{(FS_1)^2 - S_e(FS_1)}{S_e K X_v} = t$$

$$t = 3.27 \text{ d}$$

Oxygen requirement:

$$(77 \cdot 1.2 - 20) \cdot 4 \cdot 8.34 = 2400 \text{ lb/d}$$

$$\text{hp} = \frac{2400}{24 \cdot 1.5 \text{ lb } O_2/(\text{hp} \cdot \text{h})} = 67$$

$$\text{hp/million gal} = \frac{67}{3.27 \cdot 4} = 5$$

7.4 Activated Sludge Processes

The objective of the activated sludge process is to remove soluble and insoluble organics from a wastewater stream and to convert this material into a flocculent microbial suspension that is readily settleable and will permit the use of gravitational solid-liquid separation techniques. A number of different modifications or variants of the activated sludge process have been developed since the original experiments of Arden and Lockett in 1914. These variants, to a large extent, have been developed out of necessity or to suit particular circumstances that have arisen. For the treatment of industrial wastewater, the common generic flowsheets, as discussed in Chap. 6, are shown in Fig. 6.56. The nature of the wastewater will dictate the process type. Figure 7.15 illustrates the types of reaction that will occur and hence dictate the candidate process.

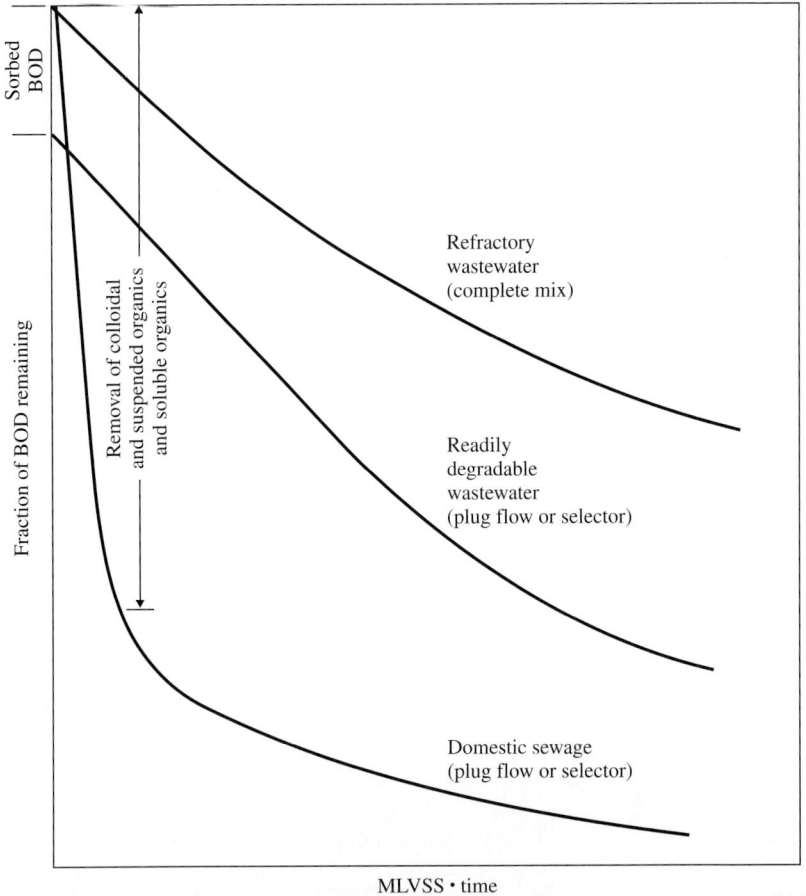

Figure 7.15 Removal of BOD in the activated sludge process.

Plug Flow Activated Sludge

The plug flow activated sludge process uses long, narrow aeration basins to provide a mixing regime that approaches plug flow. Wastewater is mixed with a biological culture under aerobic conditions. The biomass is then separated from the liquid stream in a secondary clarifier. A portion of the biological sludge is wasted and the remainder returned to the head of the aeration tank with additional incoming waste. The rate and concentration of activated sludge returned to the basin determine the mixed liquor suspended solids concentration. As discussed in Chap. 6, a plug flow regime promotes the growth of a well-flocculated, good settling sludge. If the wastewater contains toxic or inhibiting organics, they must be removed or equalized prior to entering the head end of the aeration basin. The oxygen utilization rate is high at the beginning of the aeration basin and decreases with aeration time. Where complete treatment is achieved, the oxygen utilization rate approaches the endogenous level toward the end of the aeration basin.

The dispersion number N_D can be used to express the degree of longitudinal mixing in an activated sludge plant. The dispersion number is a dimensionless number

$$N_D = \frac{D}{UL} \qquad (7.13)$$

where N_D = dispersion number
U = mean velocity of flow, m/s
L = total length of aeration tank, m
D = coefficient of axial dispersion, m²/s

A value of 0.068 m²/s for D was determined by Boon et al.[12] in 24 activated sludge plants. Values of N_D less than about 0.1 should ensure good plug flow hydraulics in practice.

Equation (7.13) can be reexpressed:

$$N_D = \frac{Dt}{L^2} \qquad (7.13a)$$

in which t = hydraulic detention time based on $Q + R$. The axial dispersion coefficient N_D is related to the air flow, increasing by a factor of 2 with an air flow increase from 20 to 100 std ft³/min per 1000 ft³ of tank volume.[13]

Modification of the way in which wastewater and recycle sludge are brought into contact in a plug flow system can have a number of benefits. For readily degradable wastewaters, a baffled inlet section will ensure sorption. The provisions of a separate zone at the inlet, with a volume of about 15 percent of the total aeration volume, together with a low-energy subsurface mechanical stirrer,

FIGURE 7.16 Plug flow activated sludge process.

can achieve controlled anoxic conditions such that nitrate associated with the recycled sludge fed to the zone can partially satisfy the BOD fed to the zone. In cases where nitrification occurs, recycle of nitrified mixed liquor from the end of the aeration basin to the anoxic zone at the head end can achieve significant denitrification. A typical plug flow process is shown in Fig. 7.16.

Complete Mix Activated Sludge

To obtain complete mixing in the aeration tank, proper choices of tank geometry, feeding arrangement, and aeration equipment are required. Through the use of complete mixing, with either diffused or mechanical aeration, it is possible to establish a constant oxygen demand as well as a uniform mixed liquor suspended solids (MLSS) concentration throughout the basin volume. Hydraulic and organic load transients are dampened in these systems, giving a process that is very resistant to upset from shock loadings. Influent wastewater and recycled sludge are introduced to the aeration basin at different points. An activated sludge plant is shown in Fig. 7.17. Readily degradable wastewaters such as food-processing wastes will tend toward filamentous bulking in a complete mix system, as has previously been discussed.

FIGURE 7.17 Complete mix activated sludge plant.

Such conditions may be minimized by the inclusion of a precontacting zone to effect a high level of substrate availability to the recycled mixed liquor. The precontacting zone should have a retention time in the order of 15 min to maximize biosorption. Design parameters for this contacting zone appear to be waste-specific and require bench-scale trials for their assessment. By contrast, complex chemical wastewaters do not support filamentous growth and complete mix processes work very effectively in them. Performance data from the treatment of several industrial wastewaters are shown in Table 7.4.

Extended Aeration

In this process, sludge wasting is minimized. This results in low growth rates, low sludge yields, and relatively high oxygen requirements by comparison with the conventional activated sludge processes. The trade-off is between a high-quality effluent and less sludge production. Extended aeration is a reaction-defined mode rather than a hydraulically defined mode, and can be nominally plug flow or complete mix. Design parameters typically include a food/microorganisms (F/M) ratio of 0.05 to 0.15, a sludge age of 15 to 35 days, and mixed liquor suspended solids concentrations of 3000 to 5000 mg/L. The extended aeration process can be sensitive to sudden increases in flow due to resultant high MLSS loadings to the final clarifier, but is relatively insensitive to shock

Wastewater	Influent BOD, mg/L	Influent COD, mg/L	Effluent BOD, mg/L	Effluent COD, mg/L	T, °C	F/M BOD, d^{-1}	F/M COD, d^{-1}	SRT, d	MVLSS, mg/L	HRT, d	SVI, mL/g	ZSV, ft/h
Pharmaceutical	2950	5840	65	712	10.4	0.11	0.19		4970	5.4		
	3290	5780	23	561	20.8	0.11	0.18		5540	5.4		
Coke and by-products chemical plant	1880	1950	65	263		0.18	0.21		2430	4.1	42.4	26
Diversified chemical industry	725	1487	6	257	21	0.41	0.71		2874	0.61	119	4.54
Tannery	1020	2720	31	213	21	0.18	0.45	16	1900	3		
	1160	4360	54	561	21	0.15	0.49	20	2650	3		
Alkylamine manufacturing	893	1289	12	47		0.146	0.21		1977	3.1	133	4.2
ABS	1070	4560	68	510	33.5	0.24	0.94	6	2930	1.5	23	28.7
Viscose rayon	478	904	36	215		0.30	0.47		2759	0.57	117	4.7
Polyester and nylon fibers	207	543	10	107	13.1	0.18	0.40		1689	0.664	116	7.9
	208	559	4	71	22.4	0.20	0.48		1433	0.712	144	8.6
Protein processing	3178	5355	10	362	10	0.054	0.08		2818	21	180	2.9
	3178	5355	5.3	245	26.2	0.100	0.16		2451	12.7	215	2.7
Propylene oxide	532	1124	49	289	20	0.20	0.31		2969	1	51	12.5
	645	1085	99	346	37	0.19	0.25		2491	1.4	32	3.7
Paper mill	375	692	8	79	9.3	0.111	0.19	18.9	1414	2.38	63	22
	380	686	7	75	23.3	0.277	0.45	5.2	748	1.83	504	10
Vegetable oil	3474	6302	76	332		0.57	1.00		1740	3.5	49.2	30
Organic chemicals	453	1097	3	178	20.3	0.10	0.21		2160	2.02	111	6.9
Cotton fiber pulping	1540	—	17	—	—	0.40	—	20	1200	3.4	—	—

TABLE 7.4 CMAS Treatment Performance for Selected Industrial Wastewaters

loads in concentration due to the buffering effect of the large biomass volume. While the extended aeration process can be used in a number of configurations, a significant number are installed as loop-reactor systems where aerators of a specific type provide oxygen and establish a unidirectional mixing to the basin contents. The use of loop-reactor systems and modifications thereof in wastewater treatment have been significant over recent years.

Oxidation Ditch Systems

A number of loop-reactor or ditch system variants are now available. In any ditch system it is necessary to adequately match basin geometry and aerator performance in order to yield an adequate channel velocity for mixed liquor solids transport. The key design factors in these systems relate to the type of aeration that is to be provided. It is normal to design for a 1 ft/s (0.3 m/s) midchannel velocity in order to prevent solids deposition. The ditch system is particularly amenable to those cases where both BOD and nitrogen removal are desired. Both reactions can be achieved in the same basin by alternating aerobic and anoxic zones, as shown in Fig. 7.18.

A typical oxidation ditch aeration basin consists of a single channel or multiple interconnected channels, as shown in Fig. 7.19.

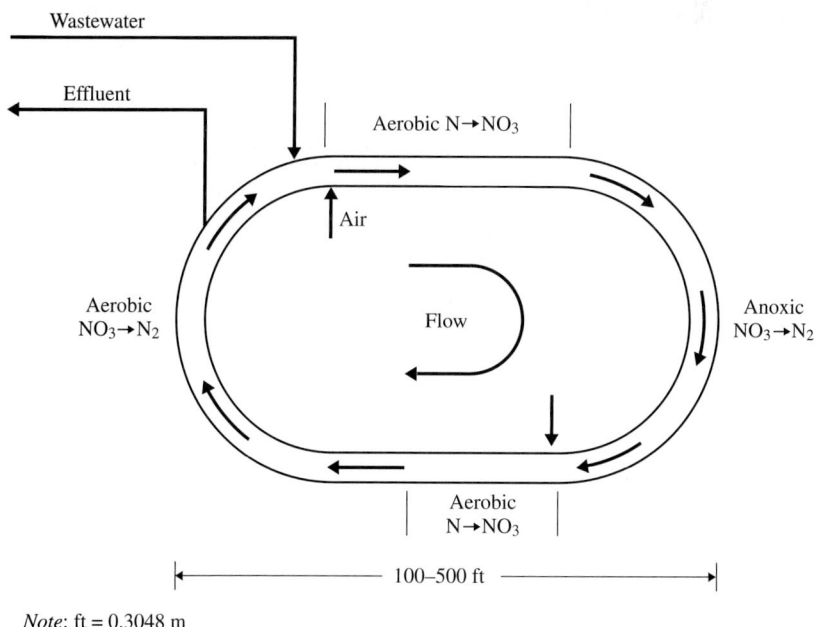

Note: ft = 0.3048 m

FIGURE 7.18 Oxidation ditch with nitrification and denitrification.

FIGURE 7.19 Oxidation ditch.

Sequencing Batch Reactor

Sequencing batch reactor (SBR) systems are experiencing increasing popularity because of costs and operating flexibility for particular types of wastewaters. The SBR utilizes two or more basins operating in such a manner that one is filling while the other is emptying. It involves the fill aeration sequence followed by the sludge settling sequence. The wastewater is added over a short period of time to maximize biosorption and flocculant sludge growth. Aeration is continued for a selected period of time followed by a quiescent settling phase and decantation of the treated effluent. Signification biological nitrification, denitrification, and phosphorous removal can be achieved by operation modification and recycle time adjustment. The SBR operates on timed cycles—fill, react, settle, decant—as shown in Fig. 7.20. The feed rate can be adjusted to a batch mode in the case of readily degradable wastewaters to avoid filamentous bulking. The treatment cycle for a pulp and paper mill wastewater is shown in Fig. 7.21. Denitrification is achieved through anoxic cycles. Performance data are shown in Tables 7.5, 7.6, and 7.7. An SBR plant is shown in Fig. 7.22. A design example is given in Example 7.4.

Example 7.4. Design an SBR to treat a wastewater having a flow rate of 0.50 million gal/d and a BOD of 500 mg/L. Assume an $X_v t$ of 1250 (mg · d)/L and a feed plus aeration period of 10 h, with 1 h settling and 1 h effluent decant periods (12 h total cycle time).

Biological Wastewater Treatment Processes 441

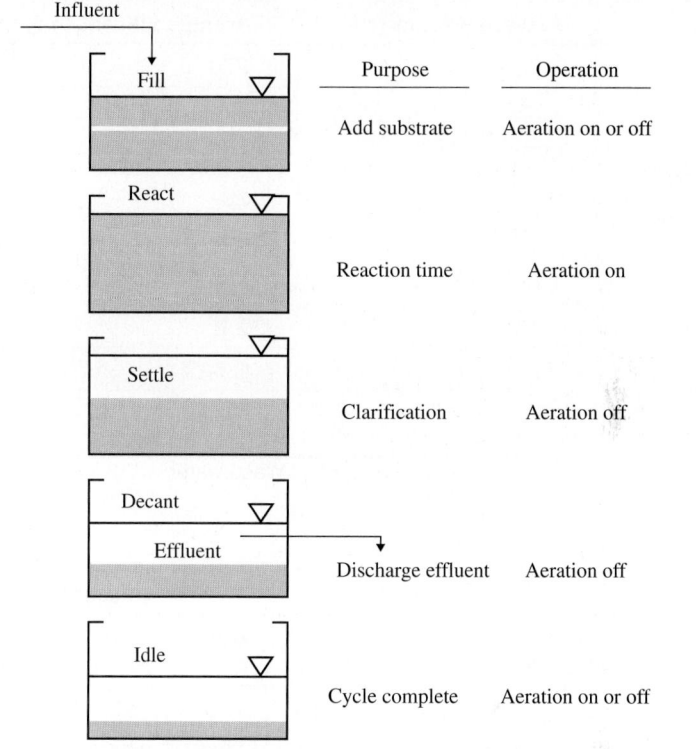

FIGURE 7.20 SBR operation sequence.

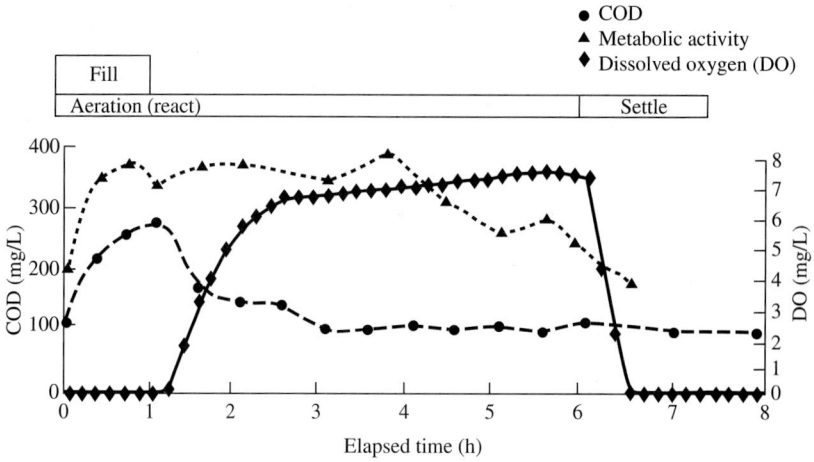

FIGURE 7.21 COD and metabolic activity of the activated sludge during an SBR cycle of reactor for pulp and paper mill wastewater (20-d sludge age, 8-h cycle time, 6-h aerated react time). (*Adapted from Franta and Wilderer, 1997.*)[14]

Parameter	Average Influent (mg/L)	Average Effluent (mg/L)
TCOD	2,240	276
SCOD	1,810	224
TBOD	955	16
SBOD	844	13
TSS	219	20
VSS	199	11
NH_3-N	0.15	2.2
O-PO_4	2.4	1.7
T(°C)	18	–

TABLE 7.5 SBR Treatment of a Recycle Paper Mill Wastewater

	Influent			Effluent		
	Flow, million gal/d	COD, mg/L	TSS, mg/L	COD, mg/L	BOD, mg/L	TSS, mg/L
Mean	0.094	2400	315	106	15	21
Standard deviation	0.023	903	164	26	104	17
Minimum	0.033	870	73	68	5	3
Maximum	0.133	5636	1030	172	70	80

TABLE 7.6 Treatment of Dairy Wastewater by an Intermittent Activated Sludge System

Sample Point	TOC	TOX	Phenol	Benzoic Acid	o-CBA[†]	m-CBA	p-CBA
Feed, mg/L	8135	780	1650	2475	840	240	285
Effluent, mg/L	409	240	<1	7	3	<2	6

[*]24-h cycle; MLSS = 10,000 mg/L with HRT = 10 d (10% feeding over 4-h fill period).
[†]Chlorobenzoic acid.

TABLE 7.7 SBR Treatment of a Chemical Wastewater[*]

FIGURE 7.22 Sequencing batch reactor (SBR) plant.

Solution

$$X_v = \frac{1250}{10/24} = 3000 \text{ mg/L}$$

The volume treated during one cycle is 0.25 million gal, and the MLVSS is $(0.25)(3000)(8.34) = 6255$ lb.

At an SVI of 150 mL/g, the volume required for storage of the settled sludge is

$$\left(150 \frac{\text{mL}}{\text{g}}\right)\left(454 \frac{\text{g}}{\text{lb}}\right)\left(3.53 \times 10^{-5} \frac{\text{ft}^3}{\text{mL}}\right) = 2.4 \text{ ft}^3/\text{lb MLSS}$$

Correcting for MLVSS/MLSS = 0.8, the volume required is

$$V = (6255)(1/0.8)(2.4)(7.48)/10^6 = 0.141 \text{ mg}$$

Provide 3 ft of freeboard between the settled sludge blanket and the supernatant water level at the end of the decant cycle.

The volume for aeration and settled sludge is

$$0.25 \text{ mg} + 0.141 \text{ mg} = 0.391 \text{ mg}$$

Select a side water depth (SWD) of 16 ft. The area is

$$\frac{391,000}{(7.48)(16)} = 3267 \text{ ft}^2$$

and the tank diameter is 65 ft. Use a total tank depth of 19 ft to include freeboard.

The MLVSS under aeration will be

$$\frac{6255}{(3267 \cdot 16 \cdot 7.48/10^6)(8.34)} = 1920 \text{ mg/L}$$

$$\text{MLSS} = \frac{1}{0.8} \times 1920 = 2400 \text{ mg/L}$$

Figure 7.23 is a diagrammatic representation of the operating sequences for a continuous-flow intermittently aerated and decanted activated sludge system.[15] Each sequence (t_0–t_1, t_1–t_2, t_2–t_3) of the cycle t_0–t_3 is initiated by a time-base controller. The treatment cycle begins after the end of decantation from the previous cycle. Aeration begins at time t_0 and continues until time t_1, during which time influent wastewater increases the volume of mixed liquor for aeration. At time t_1, aeration stops and is followed by a nonaeration sequence in which the mixed liquor undergoes settlement and anoxic processes can occur. Following the settlement/anoxic sequence t_1–t_2, treated effluent is discharged during the period t_2–t_3; at the completion of t_3, the same sequence of events is repeated. Operational sequences are developed to optimize specific functions within the main process. For example, a denitrification cycle will require sufficient aeration to provide for total carbonaceous and nitrogenous oxidation within the time period t_0–t_1, and the period t_1–t_3 must be adequate to effect the reduction of nitrate.

An important feature of these plants is their ability to accept prolonged high-flow conditions without loss of mixed liquor solids. The hydraulic capacity of conventional continuous systems is limited by the operational capacity of the secondary settlement unit.

The decanting device in these systems is located at the vessel end opposite the inlet. The movable weir of the unit is positioned out of the mixed liquor during aeration and settlement. During the decant sequence, a hydraulic ram is activated, which drives the weir through the surface layer of the vessel to the design bottom water level. In this

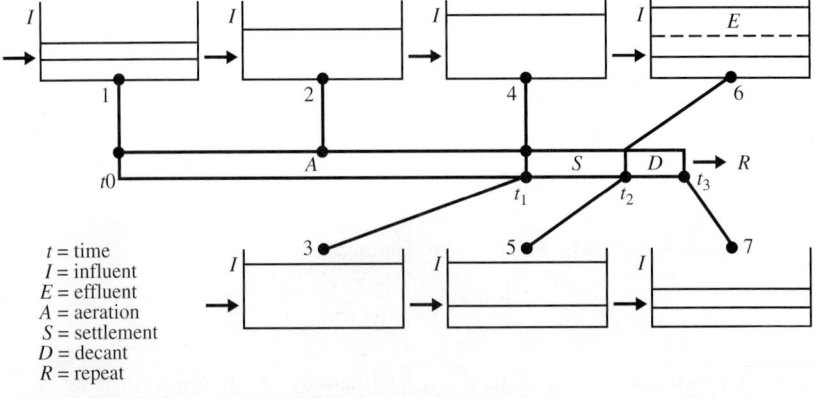

FIGURE 7.23 Diagrammatic representation of continuous-flow sequentially aerated activated sludge.

way, a surface layer of treated effluent is skimmed continuously during the decant sequence from the vessel and discharged out of the vessel by gravity via the carrier system of the decanter.

Plants can be designed for an average F/M ratio of 0.05 to 0.20 lb BOD/(lb MLSS · d) [0.05 to 0.20 kg BOD/(kg MLSS · d)], depending on the quality of effluent that is specified, at a bottom water level suspended solids concentration of up to 5000 mg/L. In calculating the volume occupied by the sludge mass, an upper sludge volume index of 150 mL/g is used. To ensure solids are not withdrawn during decant, a buffer volume is provided, the depth of which is generally in excess of 1.5 ft (0.5 m) between bottom water level and top sludge level after settlement.

Batch-Activated Sludge

Batch-activated sludge is similar to the intermittent system, except that it is usually employed for high-strength, low-volume industrial wastewaters. Wastewater is added over a short time period to maximize biosorption and flocculent sludge growth. Aeration is then continued for up to 20 h. The mixed liquor is then settled and the treated effluent decanted. A typical batch-activated sludge system is shown in Fig. 7.24. Performance data for a high-strength chemical wastewater is shown in Table 7.8. A design example is shown by Example 7.5.

Example 7.5. Design a batch-activated sludge plant for the following wastewater:

$Q = 50{,}000$ gal/d (190 m³/d)
$BOD(S_o) = 500$ mg/L
$TKN = 2$ mg/L
$a = 0.6$
$a' = 0.55$
$b = 0.1$ d⁻¹
$F/M = 0.1$ d⁻¹
$S = 10$ mg/L (soluble)

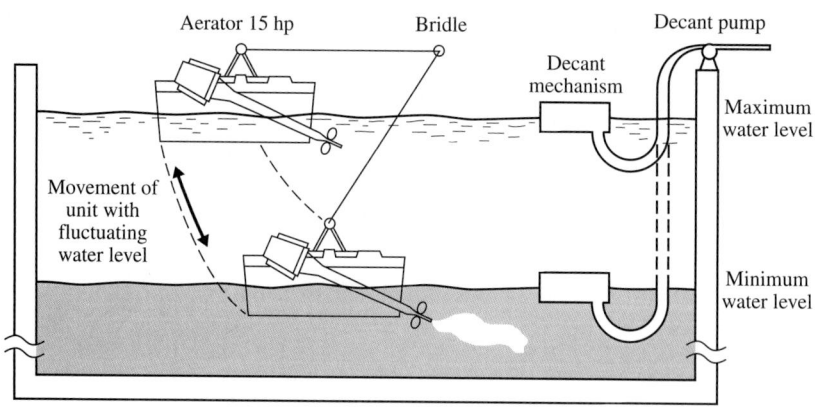

Note: hp = 0.7456 kW

FIGURE 7.24 Batch-activated sludge system.

TABLE 7.8 Batch-Activated Sludge Treatment Performance for a Specialty Chemicals Wastewater

Parameter	Monthly Average Value							
	March	April	May	June	July	August	September	October
Influent TBOD, mg/L	5,734	5,734	5,734	5,734	7,317	7,317	7,317	7,317
Effluent SBOD, mg/L	43	57	49	119	39	156	91	391
Influent COD, mg/L	10,207	10,207	10,207	10,207	15,242	15,242	15,242	15,242
Effluent COD, mg/L	920	1,992	1,456	2,067	705	2,023	1,682	2,735
Effluent TSS, mg/L	386	828	640	940	250	657	640	700
MLSS, mg/L	9,246	2,430	5,520	3,108	10,300	2,025	9,761	4,572
HRT, d	16.8	16.7	15.6	7.7	14.9	14.6	6.7	6.5
SRT, d	50	50	50	50	30	30	30	30
PAC, mg/L	1,500	—	500	—	2,000	—	2,000	—
Feed time, h	4	4	4	4	4	4	4	4
Aeration time, h	23	23	23	23	23	23	23	23
Settling time, h	1	1	1	1	1	1	1	1
SVI (mL/g)	19	157	32	74	65	75	19	74
F/M (COD basis)	0.11	0.3	0.19	0.19	0.21	0.60	0.51	0.68

The plant will be operated with 20-h aeration, 2-h sedimentation, and 2-h decant.

Solution
The BOD removal will be

$$S_r Q = (500-10) \text{ mg/L } (8.34) \frac{\text{lb/million gal}}{\text{mg/L}} (0.05) \frac{\text{million gal}}{d}$$

$$= 204 \text{ lb/d } (93 \text{ kg/d})$$

At an F/M of 0.1 d^{-1}, the required MLVSS is

$$X_v V = \frac{QS_o}{F/M} = \frac{0.05 \times 500 \times 8.34}{0.1}$$

$$= 2085 \text{ lb VSS } (946 \text{ kg})$$

and at 85 percent volatile solids the MLSS is

$$\frac{2085}{0.85} = 2453 \text{ lb SS } (1113 \text{ kg})$$

Assuming the sludge has an SVI of 100 mL/g, the sludge volume will be

$$\frac{100 \text{ mL}}{\text{g SS}} \times \frac{454 \text{ g}}{\text{lb}} \times 3.53 \times 10^{-5} \frac{\text{ft}^3}{\text{mL}} = 1.6 \text{ ft}^3/\text{lb}$$

and the volume required for the settled sludge is

$$2453 \text{ lb} \times 1.6 \text{ ft}^3/\text{lb} = 3925 \text{ ft}^3 \text{ or } 29,360 \text{ gal } (111 \text{ m}^3)$$

If sludge is to be wasted twice per month, storage must be provided for accumulated sludge.

At an estimated degradable fraction of the VSS of 0.4, the daily accumulation of VSS is

$$\Delta X_v = aS_r Q - b X_d X_v V$$

$$= 0.6(204) - 0.1 \times 0.4 \times 2085$$

$$= 39 \text{ lb VSS/d } (18 \text{ kg/d})$$

Storage for 15 d will be

$$\frac{39 \text{ lb VSS/d}}{0.85 \text{ VSS/SS}} \times 15 \text{ d} \times 1.6 \text{ ft}^3/\text{lb SS} = 1100 \text{ ft}^3 \text{ or } 8240 \text{ gal } (31 \text{ m}^3)$$

The total volume of basin (excluding freeboard) will be 50,000 + 29,360 + 8240 = 87,600 gal (332 m³). This will be a basin 35.25 ft (10.7 m) in diameter and 12 ft (3.7 m) deep. If 3 ft (0.9 m) is provided for freeboard, the operational basin dimensions will be 35.25 ft (10.7 m) in diameter by 15 ft (4.6 m) depth.

The oxygen requirements can be calculated:

$$O_2/d = a'S_r Q + 1.4b \times X_d X_v V$$

$$= 0.55(204) + 1.4 \times 0.1 \times 0.4 \times 2085$$

$$= 229 \text{ lb/d or } 9.54 \text{ lb/h } (4.3 \text{ kg/h})$$

The required power at 1.5 lb O_2/(hp · h) is

$$\frac{9.54}{1.6} = 6.4 \text{ hp (use 7.5 hp)} \quad [0.91 \text{ kg}/(\text{kW} \cdot \text{h})]$$

This is equivalent to 86 hp/million gal (17 W/m³) of basin volume, which should provide adequate mixing. The nutrient requirement will be

$$N = 0.123 \frac{X_d}{0.8} \Delta X_v + 0.07 \left(\frac{0.8 - X_d}{0.8}\right) \Delta X_v$$

$$= 0.123 \times \frac{0.4 \times 39}{0.8} + 0.07 \times \frac{0.8 - 0.4}{0.8} \times 39$$

$$= 3.8 \text{ lb/d as N} \quad (1.7 \text{ kg/d})$$

$$P = 0.026 \times \frac{0.4 \times 39}{0.8} + 0.01 \times \frac{0.8 - 0.4}{0.8} \times 39$$

$$= 0.7 \text{ lb/d as P} \quad (0.32 \text{ kg/d})$$

Oxygen-Activated Sludge

The high-purity oxygen system is a series of well-mixed reactors employing concurrent gas-liquid contact in a covered aeration tank, as shown in Fig. 7.25. The process has been used for the treatment of municipal, pulp and paper mill, and organic chemical wastewaters. Feed wastewater, recycle sludge, and oxygen gas are introduced into the first stage. Gas-liquid contacting can be done by submerged turbines, jet aeration, or surface aeration.

Oxygen gas is automatically fed to either system on a pressure demand basis with the entire unit operating, in effect, as a respirator; a restricted exhaust line from the final stage vents the essentially odorless gas to the atmosphere. Normally the system will operate most economically with a vent gas composition of about 50 percent oxygen. For economic considerations, about 90 percent of oxygen utilization with on-site oxygen generation is desired. Oxygen may be generated by a traditional cryogenic air-separation process for large installations (75 million gal/d) (2.8 × 10⁵ m³/d) or a pressure-swing adsorption (PSA) process for smaller installations. The power requirements for surface or turbine aeration equipment vary from 0.08 to 0.14 hp/thousand gal (0.028 kW/m³). At peak load conditions, the oxygen system is usually designed to maintain 6.0 mg/L dissolved oxygen in the mixed liquor.

Since high dissolved oxygen concentrations are maintained in the mixed liquor, the system can usually operate at high F/M levels (0.6 to 1.0) without filamentous bulking problems. The maintenance of an aerobic floc with high zone settling velocities also permits high MLSS concentrations in the aeration tank. Solids levels will usually range from 4000 to 9000 mg/L, depending on the BOD of the wastewater and design volume of the system.

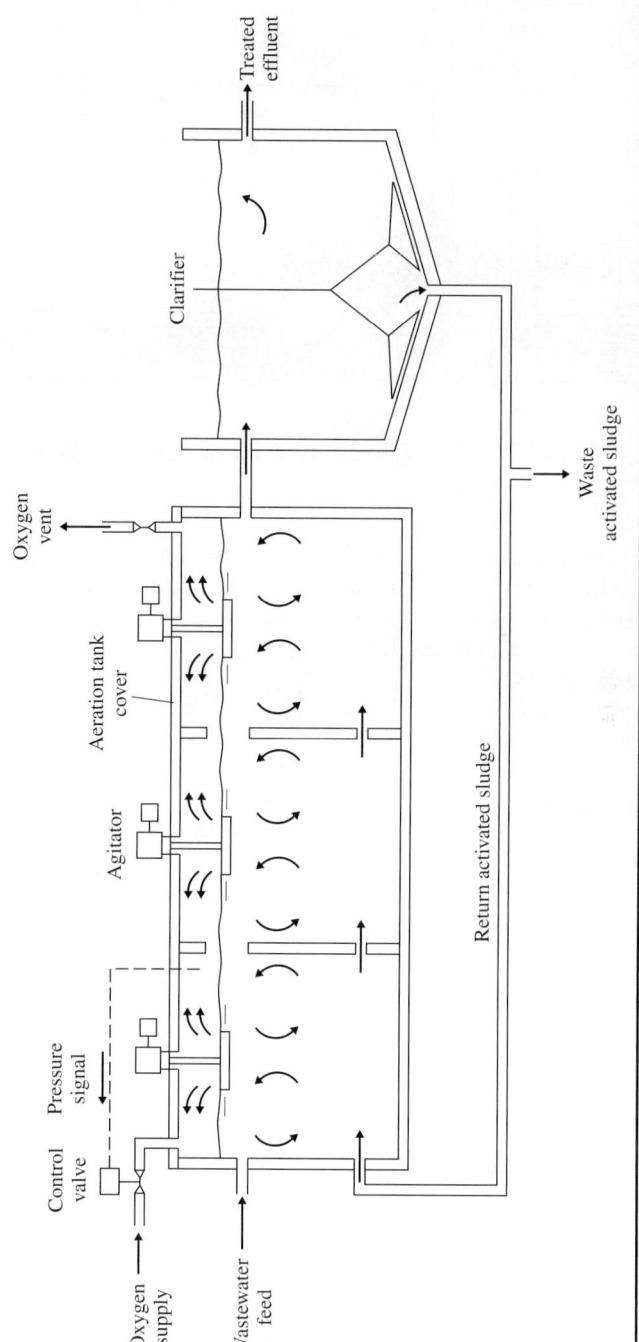

Figure 7.25 Schematic diagram of three-stage oxygen system.

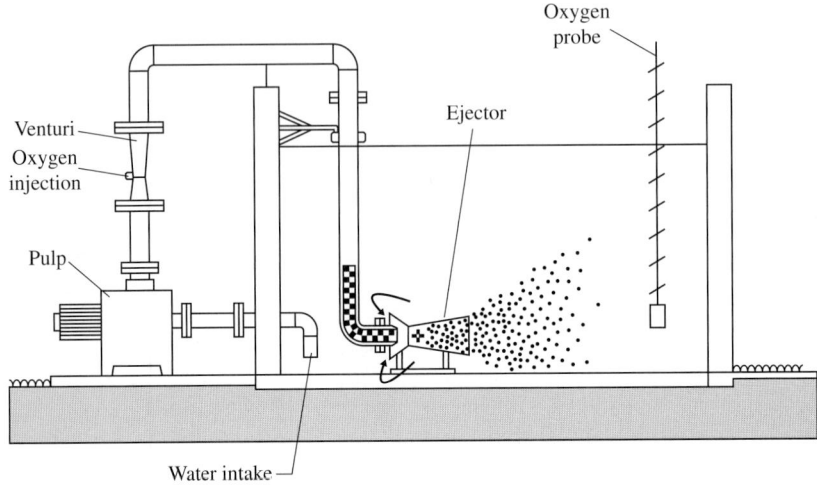

Figure 7.26 OXY-DEP process for biological wastewater treatment using pure oxygen.

Pure oxygen also is employed in open aeration basins in which oxygen under high pressure is mixed with the influent wastewater. When introduced into the aeration basin, the supersaturated gas comes out of solution in the form of microscopic bubbles. This process is shown in Fig. 7.26.

A membrane filtration unit, shown in Fig. 7.27, permits operation at high MLSS (10,000 to 40,000 mg/L) and is relatively independent of sludge quality. It also produces a high-quality effluent that can be readily disinfected and is frequently suitable for process recycle. This type of system with pure oxygen can have an exceptionally small footprint.

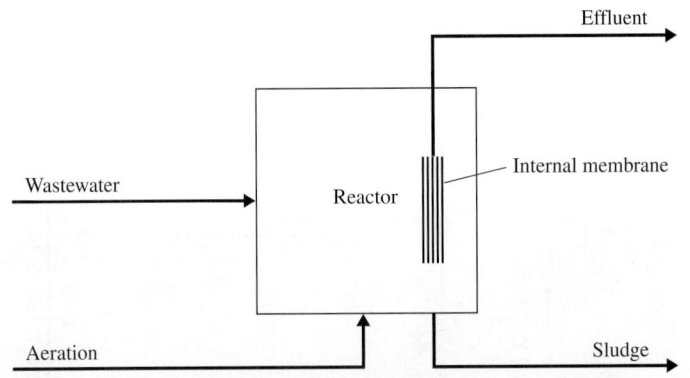

Figure 7.27 Diagram of membrane biological reactor with ZeeWeed membranes.

A more extensive discussion of MBR design and performance is provided in Sec. 12.5.

Deep-Shaft-Activated Sludge

The deep-shaft-activated sludge process operates at an F/M of 1 to 2 d^{-1} (BOD basis) using a mixing energy level of 800 to 1500 hp/million gal of aeration basin volume. The shaft depth varies from 150 to 400 ft. The operating mixed liquor dissolved oxygen concentration varies from 10 mg/L to 20 mg/L since the increasing shaft depth increases the saturation concentration. The MLSS levels vary from 8000 to 12,000 mg/L. Solids-liquid separation is provided by dissolved air flotation at high MLSS concentrations (greater than 10,000 mg/L) and by vacuum degasification and conventional gravity clarification at lower MLSS levels. A block flow diagram for the deep-shaft process is shown in Fig. 7.28. Performance data for a brewery wastewater is shown in Table 7.9.

Biohoch Process

The Biohoch reactor consists of an aeration section divided by a perforated plate into lower and upper zones and a cone-shaped final clarifier surrounding the aeration section. The air is supplied to the reactor by means of radial flow jets installed at the reactor bottom.

The untreated wastewater is pumped into the reactor through the radial flow jets or through separate pipes. Turbulence in the lower zone is sufficient to provide completely mixed conditions. In the upper chamber, the stabilizing and degassing zone, bubbles

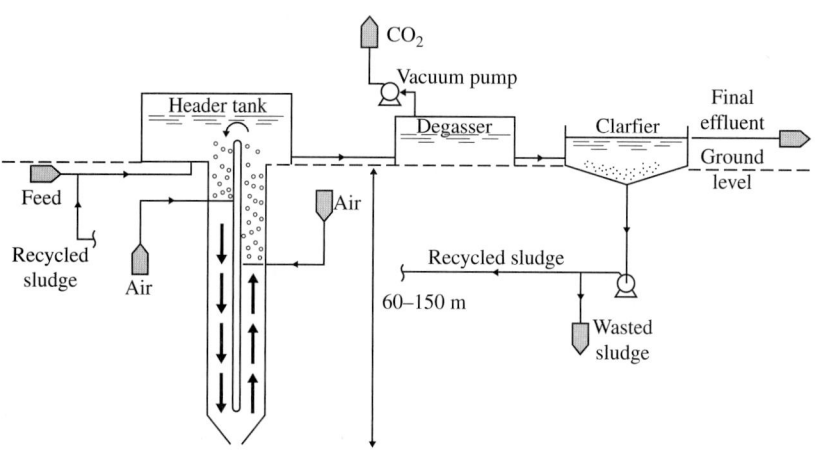

FIGURE 7.28 Flowsheet for deep-shaft activated sludge process.

Parameter	Performance
Average flow	0.65 million gal/d
Average BOD_5	2,400 mg/L
MLSS	12,000 mg/L
MLVSS	7,920 mg/L
F/M	1.51
Hydraulic detention time	0.2 d
Clarifier loading	618 gal/(d · ft²)
Recycle solids	4 percent
Effluent BOD_5	78 mg/L
TSS	91 mg/L

Source: Adapted from Cuthbert and Pollock, 1995.

TABLE 7.9 Brewery Wastewater Treatment in the Deep-Shaft Process[16]

adhering to the activated sludge are removed, since they impede sedimentation in the final clarifier. The depth of the aeration zone is approximately 65 ft.

The Biohoch reactor is illustrated in Fig. 7.29 and performance data for treatment of an organic chemicals wastewater are shown in Table 7.10.

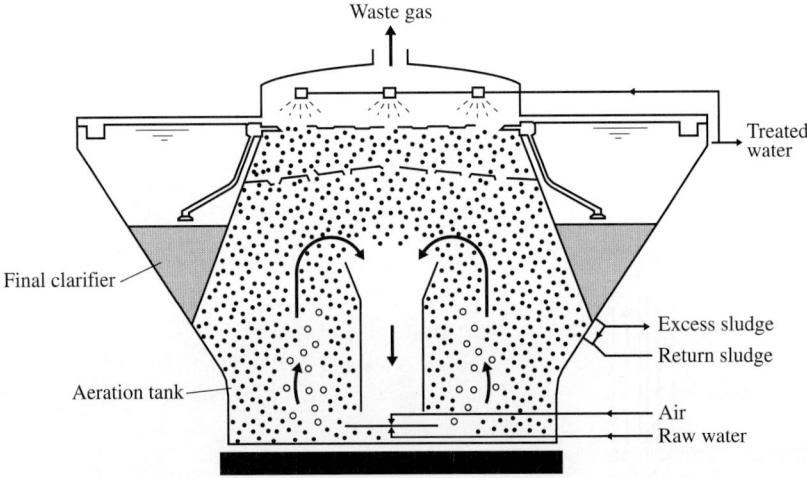

FIGURE 7.29 Flow scheme of a Biohoch reactor.

Parameter	Value
Flow rate	0.63 million gal/d
Influent BOD	5,000 mg/L
Effluent BOD	40 mg/L
Influent COD	6,000 mg/L
Effluent COD	750 mg/L
F/M (BOD basis)	0.43 1b/(1b MLSS · d)
MLSS	3,500 mg/L
SRT	5.4 d
HRT	80 h
Temperature	95°F

TABLE 7.10 Treatment of an Organic Chemicals Wastewater in the Biohoch Reactor

Integrated Fixed Film Activated Sludge

An innovation that has been implemented in recent years is the integration of fixed film media into activated sludge reactors to improve performance and in some cases to minimize expansion of existing facilities. In plants where nitrification and denitrification is practiced, nitrification is usually the rate-limiting step and the media is placed in the aerobic zone to enhance nitrification at low temperatures. The most common media is Ringlace ropelike media and small floating plastic sponges such as Captor and Linpur. The floating media are kept in place by screens.

Randall[17] reviewed performance data of IFAS systems. In one case, using Ringlace nitrification was three times greater than the control reactor and denitrification was 2.5 times greater. The integration of media also decreased the SVI and the solids loading to the final clarifier. Floating sponges have been used in several ways. In one case the sponges were placed before the activated sludge zone. Sponges have also been used after the activated sludge aeration basin for enhanced nitrification. Randall added sponges to the activated sludge aeration basin and reduced the volume requirements for nitrification by 20 percent. Pilot plant work at Broomfield, CO, Cheyenne, WY, and Mamaroneck, NY, indicates that a 50 percent reduction of SRT could continue to achieve acceptable nitrification.[18] In one New England plant sponge media provided greater than 80 percent total nitrogen removal by simultaneous nitrification and denitrification.[19] Sponges may also be used for refractory organics requiring long sludge ages as a second stage treatment.

Thermophilic Aerobic Activated Sludge

Thermophilic aerobic activated sludge offers the advantages of rapid degradation rates and low sludge yield. The optimal temperature for thermophilic oxidation is 55 to 60°C, but common terminology generally includes any process operating at temperatures of 45°C or higher. The reaction rate has been reported to be 3 to 10 times greater than mesophilic operation and the endogenous rate 10 times greater, thereby greatly reducing the net sludge yield. A COD removal of 20,000 to 40,000 mg/L coupled with an oxygen transfer efficiency of 10 to 20 percent are necessary for autoheating to thermophilic temperatures. One disadvantage is that thermophilic bacteria fail to flocculate, making biomass separation in the effluent a problem.

Final Clarification

The final clarifier is a key element in the activated sludge process, and failure to consider major design concepts will result in process upset. Factors affecting secondary clarifier performance include:

- Rising solids from denitrification
- Settled solids scoured from high sludge blankets (i.e., thickening overloads)
- Nonsettleable solids resulting from flocculation problems
- Settleable solids loss resulting from flow short-circuiting or high velocity currents

A primary operating parameter for clarifier control is the height of the sludge blanket. Too high a blanket may result in poor clarification efficiency. Factors affecting high sludge blankets include:

- Secondary effluent flow rate
- RAS flow rate
- MLSS concentration
- Total surface area available for clarification
- Sludge settling characteristics

Activated sludge exhibits zone settling, which is discussed on p. 106. Batch settling of activated sludge can be characterized by the relationship:

$$V_z = V_o e^{-KX} \qquad (7.14)$$

where V_z = zone settling velocity
X = sludge concentration at V_z
V_o and K = empirical constants

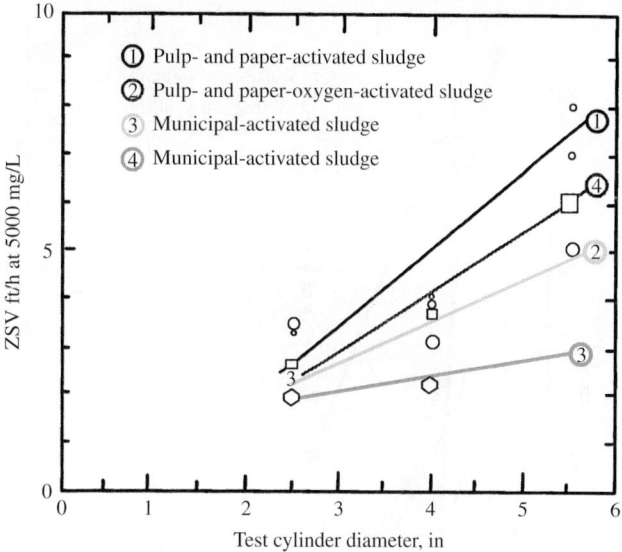

FIGURE 7.30 Effect of test cylinder diameter on stirred ZSV.

Zone settling velocities (ZSVs) are typically determined by column testing. It is important to recognize that the ZSV will be influenced by test cylinder diameter as shown for several different activated sludges in Fig. 7.30. ZSV will also be impacted by solids concentration, decreasing with increasing MLVSS as shown by Fig. 7.31

The settling properties of activated sludge are commonly related to the sludge volume index (SVI). Daigger and Roper,[20] among others, showed that the zone settling velocity (Eq. 7.14) of municipal activated sludges could be related to SVI:

$$V_Z = 7.80 e^{-(0.148+0.0021 SVI)X} \qquad (7.14a)$$

In Eq. (7.14a), V_z, X, and SVI are in meters per hour, grams per liter, and milliliters per gram, respectively. The importance of good settling, as reflected by lower SVI in terms of required clarifier diameter and cost, is reflected by Fig. 7.32. The clarifier functions primarily as a thickener in which the flux to the clarifier is related to the MLSS concentration, the influent flow rate, the underflow rate, the available clarifier surface area, and the sludge settling characteristics. A state point of analysis is used to determine the operational control of the final clarifier. The essential components are as follows:

- *State point.*—the intersection of the overflow rate and underflow rate operating lines
- *Overflow rate operating lines.*—having a slope equal to the secondary clarifier(s) surface overflow rate

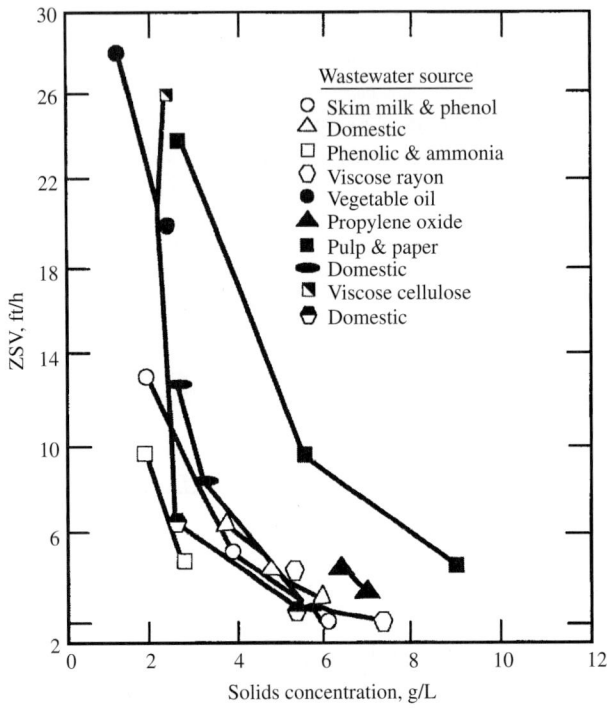

FIGURE 7.31 Zone-setting velocity characteristic of sludges treating various industrial wastewaters.

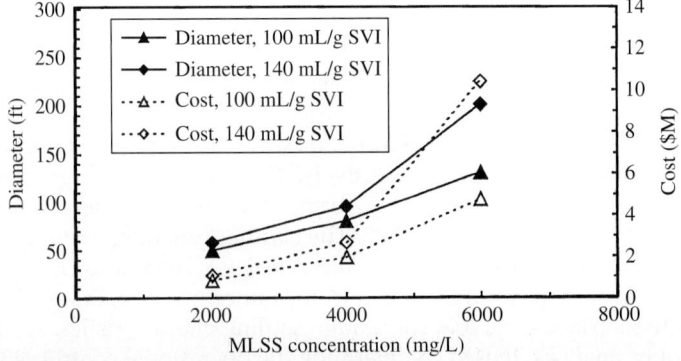

FIGURE 7.32 Effect of settling characteristics (SVI) and MLSS concentration on sizing and cost of three equally-sized secondary clarifiers (Plant flow = 15 mgd, RAS flow = 15 mgd).

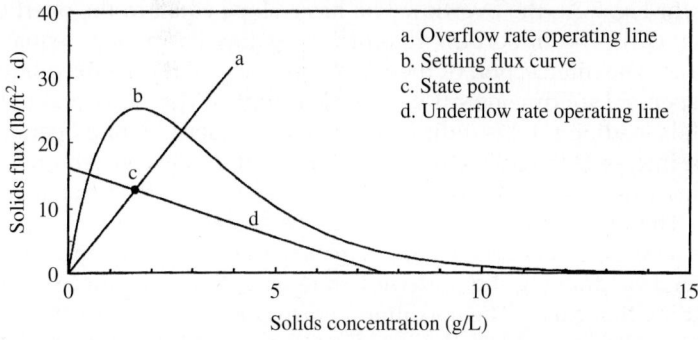

FIGURE 7.33 Identification of essential components of a state point analysis.

- *Underflow rate operating lines.*—having a slope equal to the negative of the RAS flow rate divided by the total secondary clarifier surface area
- *Settling flux curve.*—defined by the settling characteristics of the activated sludge; either SVI or the Vesilind settling parameters, V_o and K

Figures 7.33 and 7.34 illustrate these relationships.

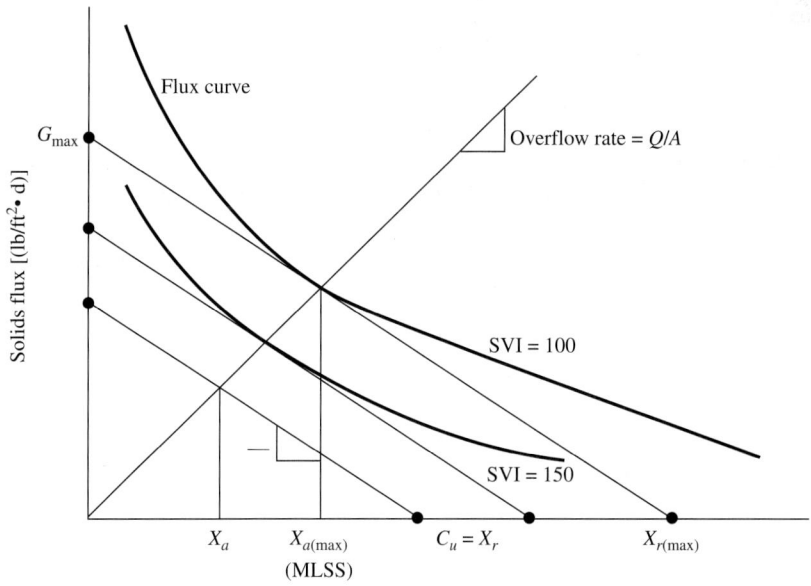

Note: lb/(ft² · d) = 4.89 kg/(m² · d)

FIGURE 7.34 Final clarifier relationships.

In Fig. 7.34, the overflow line has a slope equal to the overflow rate, Q/A. The underflow rate operating line has a slope equal to $-R/A$. The intersection of these two lines occurs at the mixed liquor suspended solids concentration X_a. The applied flux G_A (equal to the solids loading rate) is indicated where the underflow rate operating line intersects the solids flux axis. The return sludge concentration X_r is indicated where this line intersects the solids concentration axis.

The maximum flux G_{max} that can be successfully transmitted to the bottom of the clarifier and out in the return sludge stream is determined by drawing the underflow rate operating line tangent to the settling flux curve. The maximum return sludge solids concentration that can be achieved, $X_{r(max)}$, is indicated where this line intersects the solids concentration axis. When the underflow rate operating line is tangent to the settling flux curve, the system is said to be underloaded with respect to thickening. Operation below this limit is desired; prolonged operation above this limit will fill the clarifier with solids, which may result in a gross loss of solids in the effluent.

The amount of solids that can be processed through a clarifier is dictated by the sludge settling characteristics. This relationship is shown in Fig. 7.34: as the sludge settleability decreases (SVI rises from 100 to 150 mL/g), the maximum flux decreases ($G_{max1} < G_{max2}$); the maximum return sludge solids concentration also decreases ($X_{r(max)1} < X_{r(max)2}$). Repeating the calculation procedure for various underflow concentrations results in a series of clarifier underflow, that is, recycle, concentrations with their corresponding maximum solids flux rates in which each represents a limiting design or operating condition. Figure 7.35 represents these calculations for a range of SVI sludges for municipal wastewater. Acceptable operating points on Fig. 7.35 are those which lie below the line for the particular SVI. The dashed lines or operating lines are derived from clarifier mass balances assuming effluent suspended solids and sludge wastage are small relative to the clarifier solids feed. The state point on a given operating line is defined by the solids flux G and the corresponding clarifier underflow or recycle solids concentration X_u. The clarifier will not be solids-limiting, provided the state point lies below the limiting solids flux line for the particular SVI. This is denoted as $X_{a(max)}$ in Fig. 7.34. This analysis is illustrated by Example 7.6. A clarifier design procedure is shown in Example 7.7.

Example 7.6. A final clarifier is operating under the following conditions:

Surface area A = 2000 ft^2 (186 m^2)
Influent flow Q = 1.2 million gal/d (4540 m^3/d)
Sludge recycle R = 0.6 million gal/d (2270 m^3/d)
Mixed liquor suspended solids X_a = 3000 mg/L

Compute:

The overflow rate Q/A
The underflow rate R/A
The solids flux G

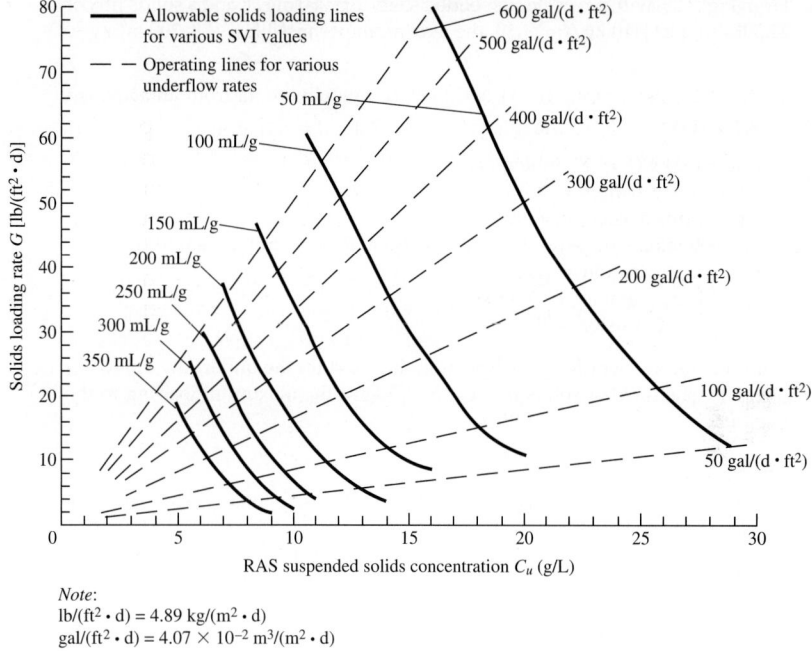

Figure 7.35 Clarifier design and operation diagram. (*After Daigger and Roper*[20]).

The recycle suspended solids X_r
The maximum SVI for clarifier operation

Solution
The overflow rate is

$$\frac{1.2 \times 10^6}{2000} = 600 \text{ gal}/(d \cdot ft^2) \quad [24.4 \text{ m}^3/(m^2 \cdot d)]$$

The underflow rate is

$$\frac{0.6 \times 10^6}{2000} = 300 \text{ gal}/(d \cdot ft^2) \quad [12.2 \text{ m}^3/(m^2 \cdot d)]$$

The solids flux is

$$G = \frac{X_a(Q+R)}{A} = \frac{3000(1.2+0.6)8.34}{2000} = 22.5 \text{ lb}/(ft^2 \cdot d) \quad [110 \text{ kg}/(m^2 \cdot d)]$$

The recycle suspended solids is

$$\frac{R}{Q} = \frac{X_a}{X_r - X_a}$$

or

$$X_r = \frac{X_a(1+R/Q)}{R/Q} = \frac{3000(1+0.6/1.2)}{0.6/1.2} = 22.5 \text{ lb}/(ft^2 \cdot d) \quad [110 \text{ kg}/(m^2 \cdot d)]$$

From Fig. 7.35, with a recycle suspended solids of 9000 mg/L and a solids flux of 22.5 lb/(ft² · d) [110 kg/(m² · d)], the maximum permissible SVI is 200 mL/g.

Example 7.7. State point analysis. Given the following data from an activated sludge plant:

Influent flow $Q = 12$ million gal/d
Number of clarifiers $= 2$
Surface area of each, $A = 9000$ ft²
Raw activated sludge (RAS) flow from each, $R = 3.5$ million gal/d
MLSS concentration $= 3$ gal/L
V_o settling parameter $= 564$ ft/d
K settling parameter $= 0.41/g$

Compute (a) the overflow rate operating line and (b) the underflow operating line. (c) Construct the solids flux curve. (d) Find the maximum loading to the clarifier.

Solution

(a) The slope of the overflow rate operating line equals the surface overflow rate Q/A.

The overflow rate is

$$\frac{Q}{A} = \frac{12 \text{ million gal/d}}{2 \times 9000} = 666 \text{ gal/(d·ft}^2)$$

Converting Q/A to ft/d gives

$$\frac{666}{7.48} = 89 \text{ ft/d}$$

But the overflow rate in ft/d is equal to

$$\frac{16 \cdot G \text{ lb/(ft}^2 \cdot \text{d)}}{\text{MLSS (g/L)}}$$

Therefore, for an MLSS of 3 g/L, G can be computed:

$$\frac{16 \cdot G}{3} = 89$$

$$G = 16.7 \text{ lb/(ft}^2 \cdot \text{d)}$$

In Fig. 7.36, the overflow rate operating line is the line passing through the origin and the point defined by $G = 16.7$ lb/d, MLSS $= 3$ g/L.

(b) The slope of the underflow rate operating line equals the underflow rate.

The underflow rate is

$$\frac{2R}{2A} = \frac{2 \times 3.5 \text{ million gal/d}}{2 \times 9000 \text{ ft}^2} = 389 \text{ gal/(d·ft}^2)$$

Converting the underflow rate to ft/d gives

$$\frac{389}{7.48} = 52 \text{ ft/d}$$

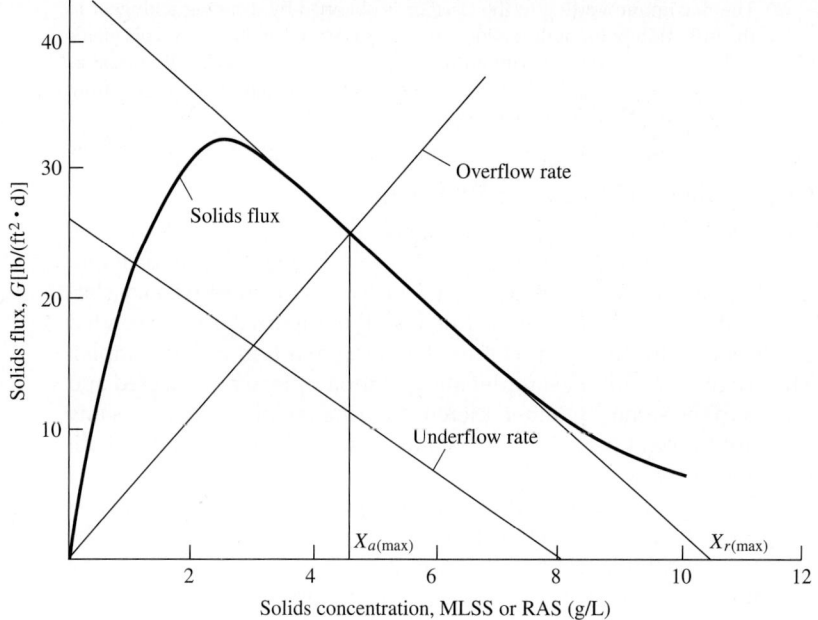

FIGURE 7.36 Plot of overflow and underflow operating lines and solids flux curve.

The underflow rate in ft/d is equal to

$$\frac{16G \; (\text{lb/d} \cdot \text{ft}^2)}{\text{RAS (g/L)}}$$

The flux G to the clarifier is

$$G = \frac{(Q+R)X_a}{2A}$$

$$= \frac{(12+7)3 \cdot 8.34 \cdot 10^3}{18{,}000} = 26.4 \; \text{lb/(ft}^2 \cdot \text{d)}$$

The RAS concentration is computed

$$\frac{16 \cdot 26.4 \; \text{lb/(ft}^2 \cdot \text{d)}}{\text{RAS(g/L)}} = 52 \; \text{ft/d}$$

$$\text{RAS} = 8.1 \; \text{g/L}$$

The underflow rate operating line is now plotted on Fig. 7.36 as the line extending from 8.1 g/L on the RAS axis to 26.4 lb/(ft² · d) on the G axis.

(c) The solids flux curve can be constructed by modifying Eq. (7.14)

$$V_z X = G = V_o X e^{-KX}$$

For $V_o = 564$ ft/d and $K = 0.4$ L/g, the solids flux curve is plotted on Fig. 7.36.

(d) The maximum loading to the clarifier is obtained by drawing a tangent to the inflection point of the solids flux curve as shown in Fig. 7.36. This yields a flux of 44 lb/(ft^2 · d) with an underflow RAS of 10.4 g/L. The MLSS as shown in Fig. 7.32 is 4500 mg/L. The underflow rate is increased from 389 gal/(d · ft^2) to 563 gal/(d · ft^2).

Flocculation and Hydraulic Problems

Wahlberg et al.[21] devised a series of tests, called the *dispersed suspended solids* (DSS) and flocculated suspended solids (FSS) tests, to differentiate between flocculation and hydraulic problems in secondary clarifiers. DSS is operationally defined as the supernatant suspended solids concentration after 30 min of settling in a Kemmerer sampler. The use of a Kemmerer sampler allows the sample to be collected and settled in the same container, thereby sparing the biological flocs any unquantifiable aggregation or breakup effects during an intermediate transfer step.

FSS is operationally defined as the supernatant suspended solids concentration after 30 min of flocculation and 30 min of settling. The FSS concentration is particularly useful because it is a measure of the effluent TSS concentration that would be possible if the mixed liquor entering a secondary clarifier is optimally flocculated and the flow characteristics in the clarifier are ideal. Differences between the FSS concentration and the effluent TSS concentration reflect any inefficiencies caused by poor flocculation or poor hydraulics. Most activated sludges will produce an effluent less than 10 mg/L in TSS if properly flocculated and ideally settled.

The existence of density currents in secondary clarifiers is well known. These currents result from the underwater waterfall that results when the solids-laden mixed liquor enters the relatively solids free secondary clarifier. High velocity jets also can occur as a result of improperly designed secondary clarifier inlets that fail to dissipate the kinetic energy of the influent flow. Poor secondary clarifier designs resulting in chronic hydraulic problems have plagued the wastewater treatment industry. Hydrodynamic models are being used successfully to identify hydraulic problems and to design modifications to correct them. Increasingly, these models are being used to design new secondary clarifiers as well.

Treatment of Industrial Wastewaters in Municipal-Activated Sludge Plants

Municipal wastewater is unique in that a major portion of the organics are present in suspended or colloidal form. Typically, the BOD in municipal sewage will be 50 percent suspended, 10 percent colloidal, and 40 percent soluble. By contrast, most industrial wastewaters are almost 100 percent soluble. In an activated sludge plant treating municipal wastewater, the suspended organics are rapidly enmeshed in the flocs, the colloids are adsorbed on the flocs, and a portion of the soluble organics are adsorbed. These reactions occur in the first few minutes of aeration contact. By contrast, for readily degradable wastewaters, that is, food processing, a portion of the BOD is rapidly sorbed and the remainder removed as a function of time and biological solids concentration. Very little sorption occurs in refractory wastewaters. These phenomena are shown in Fig. 7.15. The kinetics of the activated sludge process will therefore vary, depending on the percentage and type of industrial wastewater discharged to the municipal plant, and must be considered in the design calculations.

The percentage of biological solids in the aeration basin will also vary with the amount and nature of the industrial wastewater. For example, municipal wastewater without primary clarification will yield a sludge that is 47 percent biomass at a 3-d sludge age. Primary clarification will increase the biomass percentage to 53 percent. Increasing the sludge age will also increase the biomass percentage as volatile suspended solids undergo degradation and synthesis. Soluble industrial wastewater will increase the biomass percentage in the activated sludge.

As a result of these considerations, there are a number of phenomena that must be considered when industrial wastewaters are discharged to municipal plants:

1. *Effect on effluent quality.* Soluble industrial wastewaters will affect the reaction rate K, as shown in Table 6.8. Refractory wastewaters such as tannery and chemical will reduce K, while readily degradable wastewaters such as food processing and brewery will increase K.

2. *Effect on sludge quality.* Readily degradable wastewaters will stimulate filamentous bulking, depending on basin configuration, while refractory wastewaters will suppress filamentous bulking.

3. *Effect of temperature.* An increased industrial wastewater input, that is, soluble organics, will increase the temperature coefficient θ, thereby decreasing efficiency at reduced operating temperatures.

4. *Sludge handling.* An increase in soluble organics will increase the percentage of biological sludge in the waste sludge mixture. This generally will decrease dewaterability, decrease cake solids, and increase conditioning chemical requirements. An exception is pulp and paper mill wastewaters, in which pulp and fiber serves as a sludge conditioner and will enhance dewatering rates.

It should be noted that most industrial wastewaters are nutrient deficient; that is, they lack nitrogen and phosphorus. Municipal wastewater, with a surplus of these nutrients, will provide the required nutrient balance. However, wastewaters from industries such as poultry processing can be very high in both nitrogen and phosphorus relative to organic matter, and substantially increase the costs of nutrient removal at municipal plants.

Pretreatment of an industrial wastewater in a municipal plant is shown in Example 7.8.

Example 7.8. A wastewater is to be pretreated in an activated sludge process and then blended with a domestic wastewater for final treatment in a second activated sludge basin as shown below:

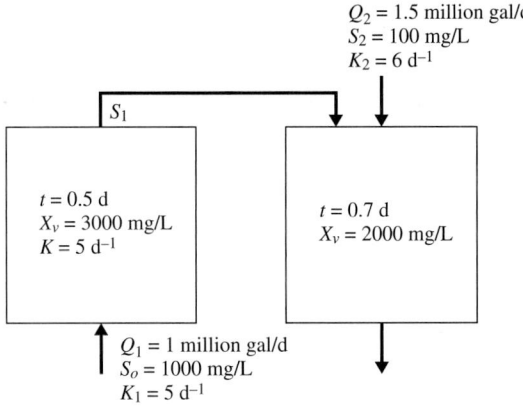

Solution
The effluent SBOD from the first aeration basin is, from Eq. (6.29),

$$(S_e)_1 = \frac{1000^2}{1000 + (5)(3000)(0.5)}$$

$$= 118 \text{ mg/L}$$

The reaction rate K_2 of the pretreated industrial wastewater in the second basin is, from Eq. (6.30),

$$K_2 = 5\left(\frac{118}{1000}\right) = 0.59 \text{ d}^{-1}$$

The influent concentration to the second basin is

$$(S_o)_2 = \frac{(118)(1.0) + (100)(1.5)}{(1.0 + 1.5)}$$

$$= 107 \text{ mg/L}$$

The average rate coefficient, $\overline{K}$, after blending the two wastewaters is, from Eq. (6.31),

$$\frac{1}{\overline{K}} = \frac{\frac{1}{0.59}(118)(1.0) + \frac{1}{6}(100)(1.5)}{(118)(1.0) + (100)(1.5)}$$

$$\overline{K} = 1.2 \text{ d}^{-1}$$

The SBOD from the second basin will be

$$(S_e)_2 = \frac{107^2}{107 + (1.2)(2000)(0.7)}$$

$$= 6.5 \text{ mg/L}$$

Effluent Suspended Solids Control

The importance of effluent solids control is underscored by the increasingly stringent effluent quality requirements. In addition to the solids themselves, they also contribute BOD, nutrients, potentially sorbed metals and priority pollutants. Since kinetic equations determine soluble BOD, the BOD associated with the solids fraction must be considered. At higher F/M operation more TBOD/TSS is observed due to the higher degradable fraction of biomass. This is illustrated by Fig. 7.37.

Carryover of suspended solids in the secondary clarifier effluent can have several causes:

- Floc shear due to high aeration basin power levels
- Poor clarifier hydraulics
- High wastewater TDS concentration
- Low or high mixed liquor temperature
- Rapid change in mixed liquor temperature
- Low mixed liquor surface tension

High mixed liquor turbulence levels created by turbine-type or mechanical surface aerators can cause floc breakup that results in

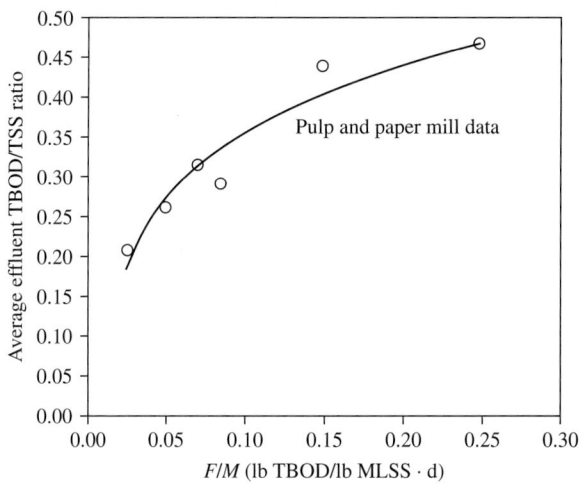

FIGURE 7.37 Calculation of BOD contributed by effluent suspended solids.

Flocculation Time, min	Settled TSS,* mg/L at	
	690 hp/million gal	360 hp/million gal
0	81	64
1	30	28
3	28	22
5	26	19
7	27	21

*Supernatant TSS following flocculation and 15-min settling period.

TABLE 7.11 Floc Shear Test Results for Pulp and Paper Mill Wastewaters

high-effluent suspended solids. This problem can frequently be solved by reducing the aeration basin power level and/or by installing a flocculation zone between the aeration basin and the final clarifier. Results of flocculation of mixed liquor from an activated sludge plant treating a pulp and paper mill wastewater at two aeration basin power levels are shown in Table 7.11. These data indicate that flocculation times of 1 to 3 min were effective at reducing the settled effluent TSS concentration. The effect of turbulence on effluent suspended solids when fine bubble diffusers are used is shown in Fig. 7.38.

High-effluent suspended solids levels will frequently result from a poor clarifier hydraulic design that causes density currents and/or short-circuiting. These conditions result in an upwelling of floc solids at the clarifier peripheral weir. This problem can be minimized by installing a Stamford baffle, which redirects the upflow of solids away from the effluent weir. A Stamford baffle and its performance are illustrated by Fig. 7.39.

In the case of industrial wastewaters, however, high effluent suspended solids of a dispersed character can result from one of several causes:

1. High total dissolved salt (TDS) may cause an increase in nonsettleable suspended solids. While the specific cause of floc dispersion has not been defined, increasing the TDS has generally resulted in an increase in nonsettleable, dispersed solids. High TDS will also increase the specific gravity of the liquid, thereby reducing the settling rate of the biological sludge. The salt content appears to have little effect on the kinetics of the process under acclimated conditions.

2. Dispersed suspended solids increase as the aeration basin temperature decreases. For example, an aerated sludge plant in West Virginia treating an organic chemicals wastewater had

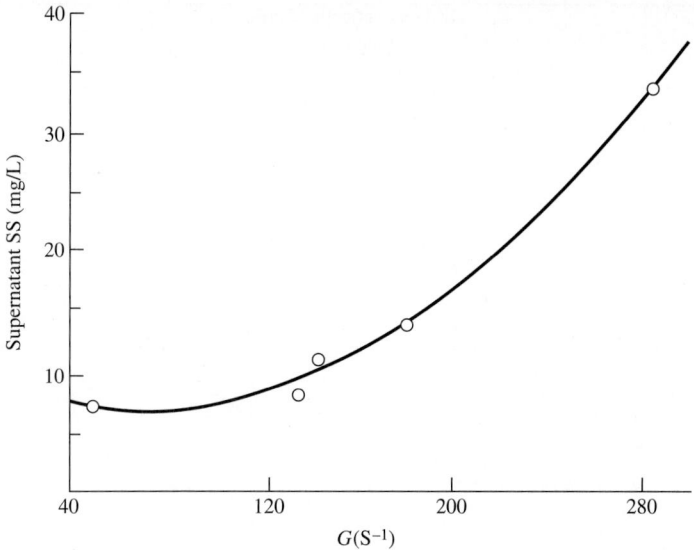

G is defined as the mean velocity gradient, which is a measure of the mixing intensity. For example, what is G for an air flow of 20 scfm/1000 ft³ in an aeration basin of 0.5 mg, a 26 ft SWD at 20°C

$$G = \sqrt{\frac{Q_a \gamma h}{V \mu}}$$

in which Q_a = air flow, m³/sec
γ = liquid specific weight, N/m³
h = liquid depth, m
V = aeration basin volume, m³
μ = absolute viscosity, N sec/m²

$$G = \sqrt{\frac{0.63 \text{ m}^3/\text{sec} \cdot 9790 \text{ N/m}^3 \cdot 7.47 \text{ m}}{1894 \text{ m}^3 \cdot 1.003 \times 10^{-3} \frac{\text{N} \cdot \text{sec}}{\text{m}^2}}}$$

$= 155 \text{ sec}^{-1}$

FIGURE 7.38 The effect of G on effluent SS concentration for fine-bubble aerated plants. (Adapted from Parker et al., 1992.)[22]

an effluent suspended solids concentration of 42 mg/L during summer operation and 104 mg/L during winter operation. The change in coagulant dosage with respect to temperature at a Tennessee organic chemicals plant is shown in Fig. 7.40.

3. Dispersed suspended solids increase with a decrease in surface tension. At one deinking mill the effluent suspended solids were directly related to the surfactant usage in the mill.

4. The nature of the organic content may increase the effluent suspended solids. While this effect is ill-defined, some plants will consistently generate high effluent suspended solids.

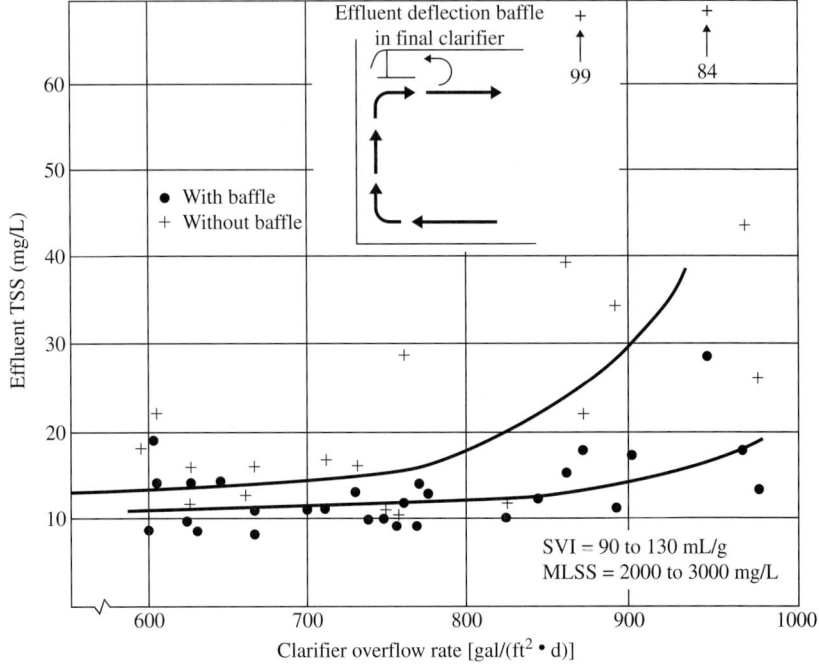

FIGURE 7.39 Effect of Stamford baffle on clarifier effluent TSS concentration.

Effect of TDS on effluent TSS concentrations and activated sludge treatment of agricultural chemicals wastewater is shown in Tables 7.12 and 7.13, respectively. The effluent suspended solids can be reduced by the addition of a coagulant prior to the final clarifier. It is important that there be sufficient time for flocculation to occur. This can be achieved by a flocculation chamber between the aeration basin and the clarifier or by a flocculation well within the clarifier, as shown in Fig. 7.41.

Cationic polyelectrolytes, or alum or iron salts, may be used as a coagulant. Choice of coagulant depends on a test program to select the most economical solution. Data reported by Paduska[23] for a cationic polyelectrolyte are shown in Fig. 7.40. It is important when using a cationic polymer to avoid overdosing, which will cause a charge reversal and a redispersion of the solids. Results of ferric chloride coagulation of activated sludge effluents as a function of dose is presented in Table 7.14. The temperature effect as previously noted is also shown in Fig. 7.40. An activated sludge design is illustrated by Example 7.9.

Example 7.9. Given the following data:

$Q = 4$ million gal/d (15,140 m³/d)
$S_o = 610$ mg/L
$S_e = 40$ mg/L

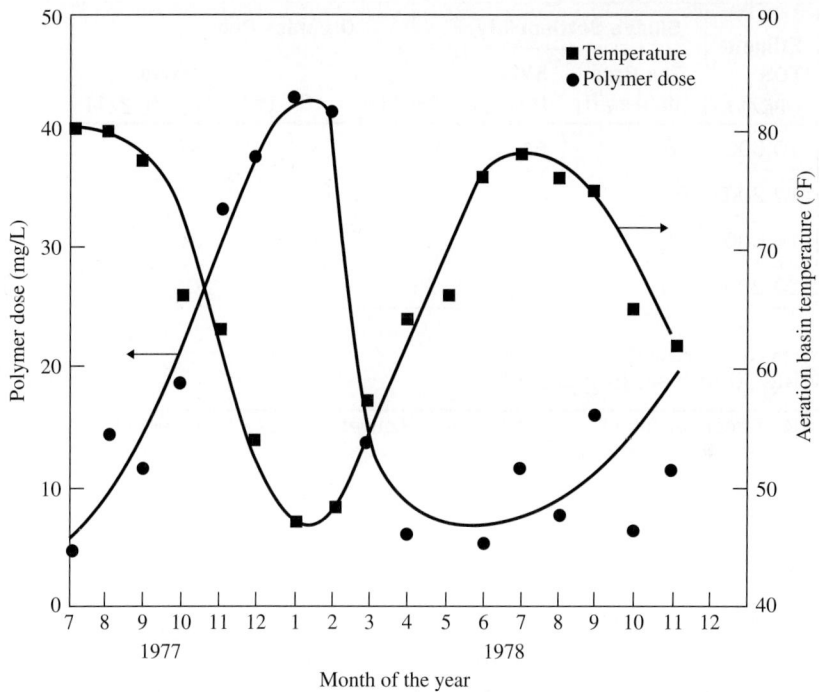

FIGURE 7.40 Polymer addition for effluent suspended solids control. (*Adapted from Paduska, 1979.*)

Wastewater	Aeration Basin		Effluent TSS (mg/L)
	F/M (mg BOD/ mg MLVSS · d)	TDS (%)	
Agricultural chemicals and specialty organic chemicals (1987)	0.17	1.3	34
	0.17	1.6	45
	0.18	2.0	109
Contaminated groundwater (1992)	0.40	1.2	20
	0.40	3.5	130
	0.40	7.0	250
Specialty organic chemicals (2000)	0.14	2.0	69
	0.17	3.0	94

TABLE 7.12 Effect of Mixed Liquor TDS Concentration on Effluent TSS Concentrations

Unit No.	Effluent TDS (mg/L)	Sludge Settleability		Organics Removal		
		Flux Rate (lb/d-sq ft)	SVI (mL/g)	BOD (%)	TOC (%)	Settled TSS† (mg/L)
1	10,600	48	61	94	55	32
2	13,200	51	49	96	53	34
3	15,600	51	47	96	57	38
4	20,200	55	46	93	53	101

*Units operated at 25°C and $F/M = 0.2/d$.
†Following 30-min settling period.

TABLE 7.13 Effects of Total Dissolved Solids on Activated Sludge Treatment of Agricultural Chemicals Wastewater*

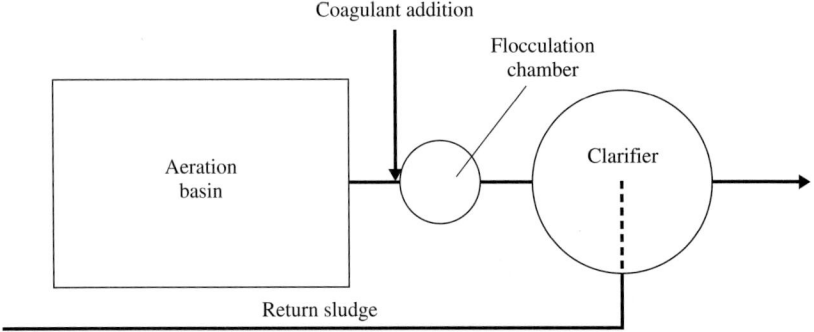

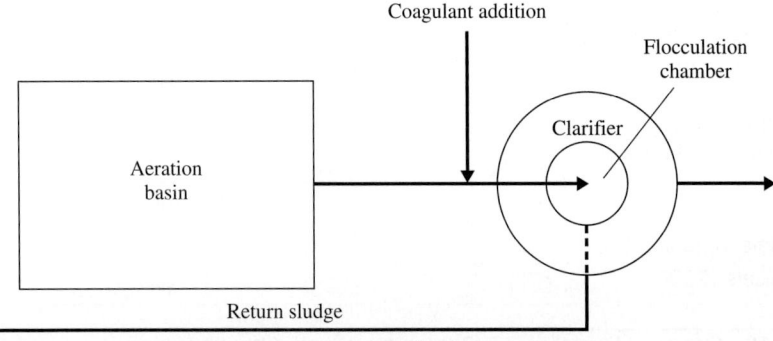

FIGURE 7.41 Coagulant addition for suspended solids control.

Dose* (mg/L)	Settled TSS† (mg/L)	pH (units)
0	175	7.7
100	114	7.6
200	54	7.4
400	11	6.9
600	5	6.7

*Dose expressed as $FeCl_3 \cdot 6H_2O$.
†Following 30-min settling period.

TABLE 7.14 Ferric Chloride Coagulation of Activated Sludge Effluent

$SS_{effl} = 40$ mg/L
mg BOD/mg SS = 0.3
$K = 3.0$ d^{-1} at 20°C
$b = 0.1$ d^{-1} at 20°C
$X_v = 3000$ mg/L
$a = 0.55$
$a' = 0.50$
$\theta_b = 1.04$
$\theta_K = 1.065$

Compute:

(a) The sludge age and F/M to meet this effluent quality
(b) The N and P requirements at 20°C
(c) The MLVSS at 10°C to meet the same effluent quality
(d) The excess sludge at 10°C
(e) The oxygen requirements at 30°C

Solution

(a) Food to microorganisms ratio F/M and sludge age θ_c:
F/M: The effluent soluble BOD is

$$S = S_e - 0.3SS_{effl} = 40 - 0.3 \times 40 = 28 \text{ mg/L}$$

The detention time is obtained by rearranging Eq. (6.28):

$$t = \frac{(S_o - S)S_o}{KX_vS} = \frac{(610-28) \text{ mg/L} \times 610 \text{ mg/L}}{3.0/\text{d} \times 3000 \text{ mg/L} \times 28 \text{ mg/L}}$$

$$= 1.141 \text{ d}$$

$$F/M = \frac{S_o}{(X_v t)} = \frac{610 \text{ mg/L}}{3000 \text{ mg/L} \times 1.41 \text{ d}} = 0.144/\text{d}$$

Sludge age: The BOD removed is

$$S_r = S_o - S = 610 - 28 = 582 \text{ mg/L}$$

The degradable fraction is given by

$$X_d = \frac{aS_r + bX_v t - [(aS_r + bX_v t)^2 - (4bX_v t)(0.8aS_r)]^{0.5}}{2bX_v t}$$

$$= 0.47$$

The sludge age can be calculated:

$$\theta = \frac{X_v t}{aS_r - bX_d X_v t}$$

$$= \frac{3000 \text{ mg/L} \times 1.41 \text{ d}}{0.55 \times 582 \text{ mg/L} - 0.1/\text{d} \times 0.47 \times 3000 \text{ mg/L} \times 1.41 \text{ d}}$$

$$= 34.9 \text{ d} \approx 35 \text{ d}$$

Note that θ_c could have been calculated first using the following equation [obtained by the above equation with Eq. (6.7)]:

$$\theta_c = \frac{-(aS_r - bX_v t) + [(aS_r - bX_v t)^2 + 4(abX_n' S_r)(X_v t)]^{0.5}}{2abX_n' S_r}$$

with $X_n' = 0.2$, this equation gives $\theta_c = 35$ d; then X_d can be computed from Eq. (6.7):

$$X_d = \frac{X_d'}{(1 + bX_n' \theta_c)}$$

$$= \frac{0.8}{1 + 0.1 \times 0.2 \times 35} = 0.47$$

(b) Nitrogen N and phosphorus P requirements:
The excess volatile sludge produced at 20°C is computed from Eq. (6.12):

$$\Delta X_{v20} = (aS_r - bX_d X_v t)Q$$

$$= (0.55 \times 582 \text{ mg/L} - 0.1/\text{d} \times 0.47 \times 3000 \text{ mg/L}$$

$$\times 1.41 \text{ d}) (4 \text{ million gal/d}) \times [8.34 \text{ (lb/million gal)}/(\text{mg/L})]$$

$$= 4046 \text{ lb/d} \quad (1837 \text{ kg/d})$$

The nitrogen and phosphorus requirements are given by Eqs. (6.22) and (6.23), therefore

$$N = 0.123 \left(\frac{X_d}{0.8}\right) \Delta X_v + 0.07 \left(\frac{0.8 - X_d}{0.8}\right) \Delta X_v$$

$$= \left[0.123 \left(\frac{0.47}{0.8}\right) + 0.07 \left(\frac{0.8 - 0.47}{0.8}\right)\right] (4046) \text{ lb/d}$$

$$= 409 \text{ lb/d} \quad (186 \text{ kg/d})$$

$$P = 0.026 \left(\frac{X_d}{0.8}\right) \Delta X_v + 0.01 \left(\frac{0.8 - X_d}{0.8}\right) \Delta X_v$$

Biological Wastewater Treatment Processes

$$= \left[0.026\left(\frac{0.47}{0.8}\right) + 0.01\left(\frac{0.8-0.47}{0.8}\right)\right](4046) \text{ lb/d}$$

$$= 79 \text{ lb/d} \quad (36 \text{ kg/d})$$

(c) Mixed liquor volatile suspended solids at 10°C, X_{v10}:
The kinetic coefficients at 10°C are computed:

$$K_{10°C} = K_{20°C}\theta_K^{10-20} = 3.0/d \times 1.065^{-10}$$

$$= 1.6/d$$

$$b_{10°C} = b_{20°C}\theta_b^{10-20} = 0.1/d \times 1.04^{-10}$$

$$= 0.068/d$$

From Eq. (6.29), rearranged,

$$X_{v10} = \frac{S_r S_o}{K_{10°C} St} = \frac{582 \text{ mg/L} \times 610 \text{ mg/L}}{1.6/d \times 28 \text{ mg/L} \times 1.41 \text{ d}}$$

$$= 5624 \text{ mg/L}$$

(d) Excess sludge at 10°C, ΔX_{v10}:
With the new values of K, b, and X_v in Eq. (6.14), the degradable fraction at 10°C is

$$X_{d10} = 0.40$$

The excess sludge is computed from Eq. (6.12):

$$\Delta X_{v10} = (0.55 \times 582 - 0.068 \times 0.40 \times 5624 \times 1.41) \times (4)(8.34) \text{ lb/d}$$

$$= 3483 \text{ lb/d} \quad (1581 \text{ kg/d})$$

(e) Oxygen requirements at 30°C, R_{30}
The kinetic coefficients at 30°C are

$$K_{30°C} = 3.0/d \times 1.065^{30-20}$$

$$= 5.6/d$$

$$b_{30°C} = 0.1/d \times 1.04^{10}$$

$$= 0.15/d$$

Then

$$X_{v30} = \frac{582 \times 610}{5.63 \times 28 \times 1.41}$$

$$= 1597 \text{ mg/L}$$

and

$$X_{d30} = 0.54$$

$$R_{30} = (a'S_r + 1.4b_{30°C} \, X_{d30} X_{v30} t)Q$$

$$= (0.50 \times 582 \text{ mg/L} + 1.4 \times 0.15/\text{d} \times 0.54 \times 1597 \text{ mg/L}$$

$$\times 1.41 \text{ d})(4 \text{ million gal/d})[8.34(\text{lb/million gal})/\text{mg/L}]$$

$$= 18{,}226 \text{ lb/d} \quad (8275 \text{ kg/d})$$

7.5 Trickling Filtration

A trickling filter is a packed bed of media covered with slime growth over which wastewater is passed. As the waste passes through the filter, organic matter present in the waste is removed by the biological film.

Plastic packings are employed in depths up to 40 ft (12.2 m), with hydraulic loadings as high as 4.0 gal/(min · ft²) [0.16 m³/(min · m²)]. Depending on the hydraulic loading and depth of the filter, BOD removal efficiencies as high as 90 percent have been attained on some wastewaters. In one industrial plant, a minimum hydraulic loading of 0.5 gal/(min · ft²) [0.02 m³/(min · m²)] was required in order to avoid the generation of filter flies (psychoda). Figure 7.42 shows an installation of a plastic-packed filter.

Theory

As wastewater passes through the filter, nutrients and oxygen diffuse into the slimes, where assimilation occurs, and by-products and CO_2 diffuse out of the slime into the flowing liquid. As oxygen diffuses

Figure 7.42 Plastic-packed trickling filter.

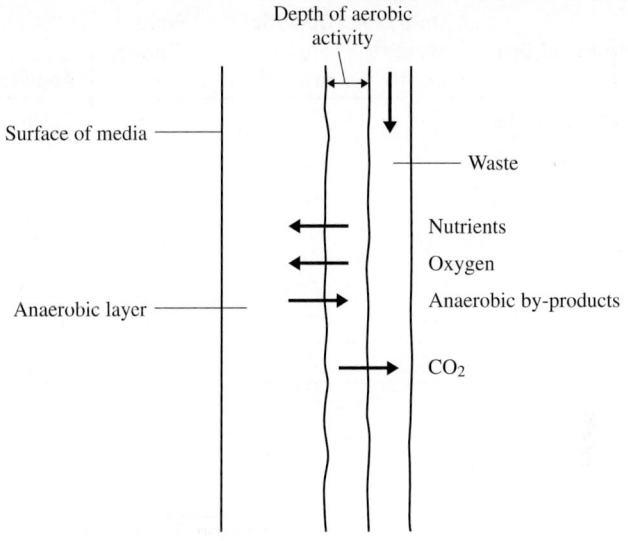

FIGURE 7.43 Mode of operation of a trickling filter.

into the biological film, it is consumed by microbial respiration, so that a defined depth of aerobic activity is developed. Slime below this depth is anaerobic, as shown in Fig. 7.43.

As in BOD removal by activated sludge, BOD removal through a trickling filter is related to the available biological slime surface and to the time of contact of wastewater with that surface.

The mean time of contact of liquid with the filter surface is related to the filter depth, the hydraulic loading, and the nature of the filter packing:

$$t = \frac{CD}{Q^n} \tag{7.15}$$

where t = mean detention time
D = filter depth, ft
Q = hydraulic loading, gpm/ft^2
C and n = constants related to the specific surface and the configuration of the packing

There are a number of commercial packings available. These include vertical-flow, random-packed, and cross-flow media. The properties of these media are summarized in Table 7.15. Investigations of several types of packings showed that

$$C = C'A_v^m \tag{7.16}$$

in which A_v is the specific surface expressed in square feet per cubic foot (*note:* ft^2/ft^3 = 3.28 m^2/m^3). For spheres, rock, and Polygrid plastic

Medium Type	Nominal Size, in	Unit Weight, lb/ft³	Specific Surface area, ft²/ft³	Void Space, %	Application
Bundle (sheet)	24 × 24 × 48	2–5	27–32	>95	C, CN, N
	24 × 24 × 48	4–6	42–45	>94	N
Rock	1–3	90	19	50	N
Rock	2–4	100	14	60	C, CN, N
Random (Dump)	Varies	2–4	25–35	>95	C, CN, N
	Varies	3–5	42–50	>94	N
Wood	48 × 48 × $1\frac{7}{8}$	10.3	14		C, CN

Note: $C = CBOD_{5R}$
$CN = CBOD_{5R}$ and NOD_R
$N =$ tertiary NOD_R

1 in = 25.4 min
1 lb/ft³ = 16.05 kg/m³
1 ft²/ft³ = 0.305 m²/m³

TABLE 7.15 Comparative Physical Properties of Common Trickling Filter Media

media without slime, C' has a value of 0.7 and m a value of 0.75. This relationship will vary for others that have a different general configuration.

The exponent n in Eq. (7.15) has been observed to decrease with decreasing specific surface A_v. Equations (7.15) and (7.16) can be combined to yield a general expression for the mean retention time through any type of filter packing:

$$\frac{t}{D} = \frac{C' A_v^m}{Q^n} \qquad (7.17)$$

The mean retention time increases considerably in the presence of filter slime, in some cases being as much as 4 times that prevailing with the nonslimed surface.

In order to avoid filter plugging, a maximum specific surface of 30 ft²/ft³ (98 m²/m³) is recommended for the treatment of carbonaceous wastewaters. Specific surfaces in excess of 100 ft²/ft³ (328 m²/m³) can be used for nitrification because of the low yield of biological cellular material.

Recent data have shown improved performance with cross-flow media with the same specific surface as other media because of the increased time of contact of the wastewater passing through the filter.

In Chap. 6, it is shown that soluble BOD removal in the plug flow activated sludge process could be expected to follow the relationship

$$\frac{S}{S_o} = e^{-K_b X_v t / S_o}$$

These equations may be analogously applied to trickling filters under the following conditions:

1. The specific surface must remain constant. This is true for any specific filter medium but varies from medium to medium.
2. The medium must have a uniform, thin slime cover. This does not always occur, particularly in the case of rock filters. It is essential therefore, that the entire surface be wetted. A minimum hydraulic loading of 0.50 gal/(min · ft^2) is recommended.

The second condition also requires uniform distribution of the hydraulic loading to the filter. Albertson[24] has shown the dosing cycle should be optimized to ensure media wetting and flushing of excessive slime growth. Albertson and Eckenfelder[25] have shown that, with adequate hydraulic loading for media wetting, the BOD removal efficiency is independent of depth and that the value of K is depth-dependent as follows:

$$K_2 = K_1 \left(\frac{D_1}{D_2}\right)^{0.5} \tag{7.18}$$

This condition is violated when heavy slime buildup short-circuits the filter.

For the retention time as defined by Eq. (7.15) and assuming that the available bacterial surface to be proportional to the specific surface A_v,

$$\frac{S}{S_o} = e^{-KA_vD/Q^nS_o} \tag{7.19}$$

In Eq. (7.19) the hydraulic loading of Q includes both the forward flow and the recirculated flow. It has been shown that in many cases BOD removal can be increased by recirculation of filter effluent around the filter. The recirculated flow serves as a diluent to the influent waste. When recirculation is used, the applied BOD to the filter, S_o, becomes

$$S_o = \frac{S_a + NS}{1 + N} \tag{7.20}$$

where S_o = BOD of waste applied to the filter after mixing with the recirculated flow
S_a = BOD of the influent
S = BOD of filter effluent
N = recirculation ratio, R/Q

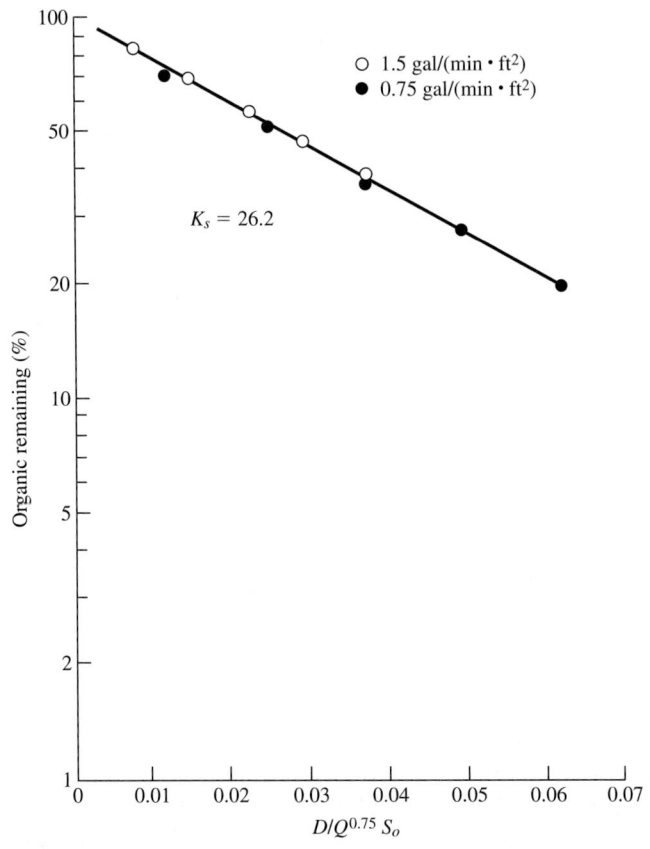

Figure 7.44 Treatment of dilute black liquor (S_o = 400 mg/L) on plastic packing.

Filter performance in accordance with Eq. (7.19) is shown in Fig. 7.44. Combining Eqs. (7.19) and (7.20) yields the BOD removal relationship:

$$\frac{S}{S_a} = \frac{e^{-KA_v D/Q^n S_o}}{(1+N) - Ne^{-KA_v D/Q^n S_o}} \tag{7.21}$$

When the BOD of the waste exhibits a decreasing removal rate with decreasing concentration, the BOD in the recirculated flow will be removed at a lower rate than that prevailing in the influent. In this case, one must apply a coefficient of retardation to the recirculated flow.

It has been shown that BOD removal through the filter can be related to the applied organic loading expressed as lb BOD/(1000 ft³ · d):

$$\frac{S}{S_o} = e^{-kA_v/L} \tag{7.22}$$

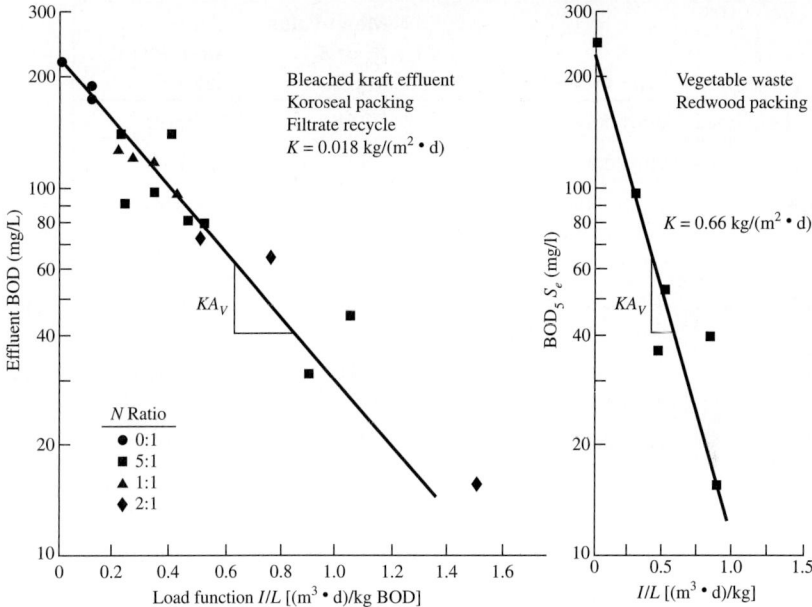

FIGURE 7.45 Data correlation for two wastewaters in accordance with Eq. (7.22).

in which L is expressed as lb BOD/(1000 ft³ · d) (kg of BOD/m³ · d). Data correlated in accordance with Eq. 7.22 are shown in Fig. 7.45. It should be noted that when the exponent n in Eqs. (7.22) is unity, Eqs. (7.22) and (7.19) become mathematically the same. Filter performance characteristics for various wastewaters, in accordance with Eq. (7.22), are shown in Table 7.16.

Oxygen Transfer and Utilization

Oxygen is transferred from air passing through the filter to the films of liquid passing over the slimes. Since the rate of oxygen transfer is related to fluid mixing and turbulence, it can be expected that hydraulic loading and media configuration will influence the transfer rate. Experimentation on various types of filter media has shown that the following relationship applies:[28]

$$\frac{-dC}{dD} = K_o(C_s - C_L) \qquad (7.23)$$

where D = filter depth
   C_L = concentration of dissolved oxygen in the liquid passing over the filter
   K_o = transfer rate coefficient

Type of Waste	Type of Medium	Mean Value, S_o or S_a, mg/L	Rate Coefficient, kg/m²
Pharmaceutical	Vinyl core	5248	0.2160
Phenolic	Vinyl core	340	0.0210
Sewage	6 parallel: basalt, slag, Surfpac I and II, Flocor, Cloisonyle	280	0.0480
Sewage	4 parallel: slag 6 in and 4 in, Flocor, Surfpac	332	0.0480
Sewage	4 parallel: slag 6 in and 4 in, Flocor, Surfpac	215	0.0500
Kraft mill sludge recycle	Vinyl core	210	0.0160
Kraft mill filtrate recycle	Vinvl core	220	0.0180
Vegetable	Del Pak	235	0.0660
Fruit canning	Surfpac	2200	0.0930
Kraft mill	Surfpac	130	0.005
Fruit processing	Surfpac	3200	0.001
Pulp and paper	Vinyl core	280	0.016

TABLE 7.16 Trickling Filter Performance

This integrates to

$$\frac{(C_s - C_L)_1}{(C_s - C_L)_2} = e^{-K_o(D_2 - D_1)} \tag{7.24}$$

The transfer rate coefficient K_o is related to the hydraulic loading to the filter. The oxygen transfer relationship is shown in Fig. 7.46.

The oxygen transfer can be expressed in mass of oxygen per hour per unit volume of filter medium by the following relationship

$$N = 5.0 \times 10^{-4} K_o (C_s - C_L) Q \tag{7.25}$$

where $N = $ lb $O_2/(\text{ft}^3 \cdot \text{h})$
$C_s = $ oxygen saturation, mg/L
$C_L = $ dissolved oxygen concentration, mg/L
$Q = $ hydraulic loading, gal/(min · ft²)
$K_o = $ transfer rate coefficient, ft⁻¹

or

$$N = 0.06 K_o (C_s - C_L) Q$$

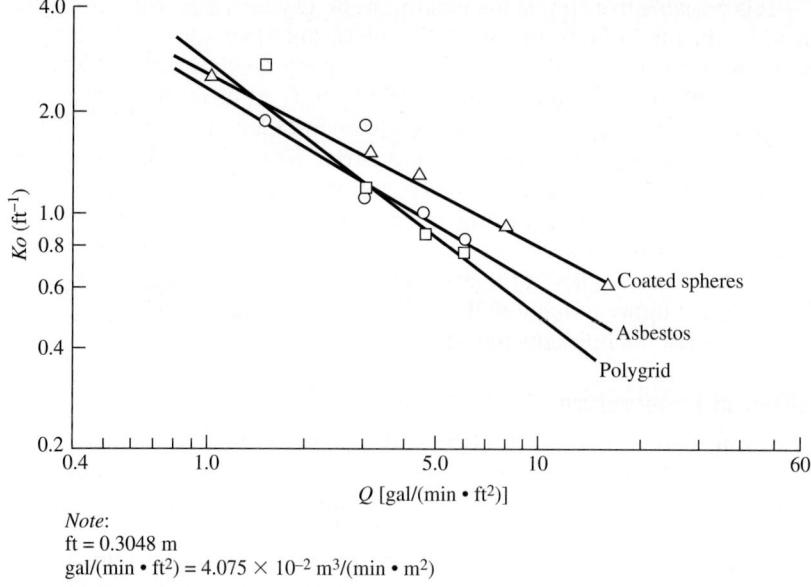

FIGURE 7.46 Relationship between the oxygen transfer rate coefficient and hydraulic loading.

where $N = \text{kg O}_2/(\text{m}^3 \cdot \text{h})$
$Q = \text{m}^3/(\text{m}^2 \cdot \text{min})$
$K_o = \text{m}^{-1}$

The ability of a filter to remove BOD is limited by the aerobic activity of the filter film. This in turn is limited by the amount of oxygen transferred to the slime from the flowing liquid.

The activity of the filter films is measured by the surface oxygen-utilization rate. British studies[29] of the treatment of domestic sewage showed an oxygen-utilization rate of 0.028 mg $\text{O}_2/(\text{cm}^2 \cdot \text{h})$. Studies of the treatment of dilute black liquor at a filter loading of 400 lb BOD/(1000 ft³ · d) [6.41 kg/(m³ · d)] yielded a utilization rate of 0.0434 mg $\text{O}_2/(\text{cm}^2 \cdot \text{h})$.[27]

The total quantity of aerobic film can be estimated from the relationship:

$$h = \sqrt{\frac{2D_L C_L}{K_r \rho}} \tag{7.26}$$

where h = depth of film to which oxygen penetrates
D_L = diffusivity of oxygen through the filter film
C_L = concentration of dissolved oxygen in the liquid passing the film surface
ρ = density of the filter film
K_r = unit oxygen-utilization rate of the filter film

It is possible to estimate the maximum BOD which can be assimilated by the filter film from Eqs. (7.25) and (7.26). Albertson[24] has estimated that oxygen is not limiting up to organic loadings of 160 lb O_2/(1000 ft^3 · d) for media containing 30 ft^3/ft^3. In order to ensure adequate oxygen ventilation, fans should be installed.

In treating high-concentration industrial wastewaters, a maximum influent concentration must be maintained in order to avoid anaerobic conditions prevailing in the filter slimes and resultant escape of anaerobic products and odors to the atmosphere. Depending on the degradability of the wastewater this may vary from 600 to 1200 mg/L. Higher influent BOD concentrations require recirculation for dilution of the influent strength.

Effect of Temperature

The performance of trickling filters will be affected by changes in the temperature of the filter films and the liquid passing over the films. It is usually assumed that these two temperatures will be essentially the same when only the aerobic portion of the film is considered. A decrease in temperature results in a decrease in respiration rate, a decrease in oxygen-transfer rate, and an increase in oxygen saturation. The combined effect of these factors results in an increase of aerobic film at a lower activity level, yielding a somewhat reduced efficiency at lower temperatures. The relationship of efficiency and temperature can be expressed as[26]

$$E_T = E_{20} \times 1.035^{T-20} \tag{7.27}$$

where E = filter efficiency and T = temperature, °C

Trickling Filter Applications

In most cases, the reaction rate K for soluble industrial wastewaters is relatively low, and hence filters are not economically attractive for high treatment efficiency (85 percent BOD reduction) of such wastewaters. Plastic-packed filters, however, have been employed as a pretreatment for high-strength wastewaters in which BOD removals in the order of 50 percent have been achieved at hydraulic and organic loadings of greater than 4 gal/(min · ft^2) [0.16 m^3/(min · m^2)] and 500 lb BOD/(thousand ft^3 · d) [8.0 kg/(m^3 · d)]. Performance characteristics for several industrial wastewaters are shown in Table 7.17 and Fig. 7.47. Example 7.10 presents a trickling filter design.

Example 7.10. One million gallons per day (3785 m^3/d) of an industrial wastewater with a BOD of 900 mg/L is to be pretreated to an effluent BOD of 300 mg/L. A plastic tower with a specific surface of 30 ft^2/ft^3 (98 m^2/m^3) and a depth of 20 ft (6.1 m) is to be used. Determine the hydraulic loading, the recirculation ratio, and the filter area required. The maximum BOD applied to the filter is 600 mg/L. The reaction rate coefficient for this wastewater is 1.0. The media has a coefficient n of 0.5.

Waste	Hydraulic Loading, MGAD	Depth, ft	Raw BOD	Recycle Ratio	BOD Removal (clarified), %	Temperature	BOD Loading, lb/1000 ft³
Sewage	126	21.6	145	3	88		54
	252		131	3	82		110
	252		175	1	70		250
	63		173	0	78		95
	126	10.8	152	0	76		67
	189		166	0	45		549
	135		165	0	57		390
	95		185	0	51		304
Citrus	72	21.6	542	3	69		199
	189		464	2	42		612
Citrus and sewage	189	21.6	328	2	53		384
Kraft mill	365	18.0	250	0	10	34	
	185	18.0	250	0	24	36	
	200	21.6	250	0	23	40	
	90	21.6	250	0	31	33	
Black	47	18	400	0	73	24	200
Liquor	95	18	400	0	58	29	380
	189	18	400	0	58	35	780

Note: MGAD = 0.9354 m³/(m²·d)
ft = 0.3048 m
lb/1000 ft³ = 16 × 10⁻³ kg/m³

TABLE 7.17 High-Rate Trickling Filtration Performance

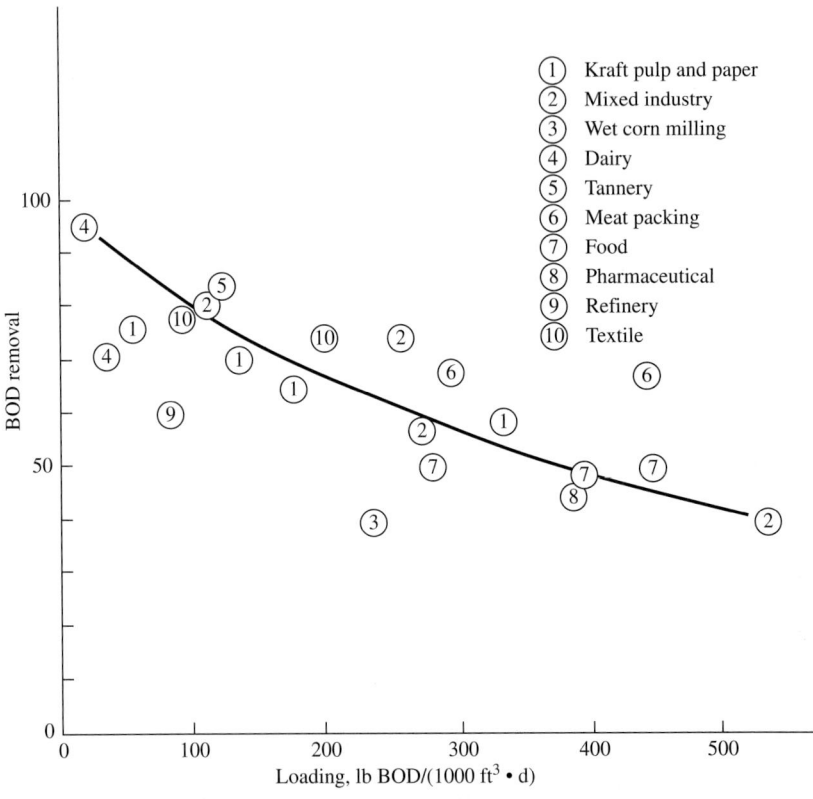

Figure 7.47 Pretreatment of organic wastewater by high-rate trickling filters using plastic media ($A_v = 30$ ft²/ft³).

Solution Calculate the required recycle ratio:

$$S_o = \frac{S_a + NS_e}{1+N}$$

$$600 = \frac{900 + 300N}{1+N}$$

$$N = 1.0$$

Calculate the hydraulic loading:

$$\frac{S}{S_o} = e^{-KA_v D/Q^n S_o}$$

$$\frac{300}{600} = e^{-(1 \times 30 \times 20)/(Q_o^{0.5} \times 600)} = e^{-x}$$

$$x = 0.69 = \frac{1}{Q_o^{0.5}}$$

$$Q_o = 2 \text{ gal/(min} \cdot \text{ft}^2) \quad [0.082 \text{ m}^3/(\text{min} \cdot \text{m}^3)]$$

Since $N = 1.0$, $R = Q$, and the wastewater flow is 1.0 gal/(min $\cdot$ ft^2) [0.041 m^3/(min $\cdot$ m^2)],

$$A = \frac{10^6 \text{ gal/d} \times 6.94 \times 10^{-4} \text{ d/min}}{1.0 \text{ gal/(min} \cdot \text{ft}^2)} = 694 \text{ ft}^2 \quad (64.5 \text{ m}^2)$$

The filter volume $V = AD = 694 \times 20 = 13{,}880 \text{ ft}^3$ (393 m^3).

Check for oxygen requirements:
The BOD removed through the filter is

$$2 \text{ million gal/d} \times (600 - 300)(\text{mg/L}) \times 8.34 \frac{\text{lb/million gal}}{\text{mg/L}}$$

$$= 5000 \text{ lb/d} \quad (2270 \text{ kg/d})$$

Assuming 0.8 lb O$_2$/lb BOD removed, the oxygen requirement is 4000 lb/d (1816 kg/d).

From Fig 7.46 at a hydraulic loading of 2 gal/(min $\cdot$ ft^2), the oxygen-transfer rate coefficient K_o is 2.0/ft.

The oxygen transfer can be calculated from Eq. 7.25:

$$N = 5.0 \times 10^{-4} \times 2.0/\text{ft} \times (9-1) \text{ mg/L} \times 2 \text{ gal/(min} \cdot \text{ft}^2)$$

$$= 160 \times 10^{-4} \text{ lb O}_2/(\text{ft}^3 \cdot \text{h}) \quad [0.26 \text{ kg/(m}^3 \cdot \text{h})]$$

Assuming an α value of 0.85,

$$N = 160 \times 10^{-4} \times 0.85 = 136 \times 10^{-4} \text{ lb O}_2/(\text{ft}^3 \cdot \text{h}) \quad [0.22 \text{ kg/m}^3 \cdot \text{h}]$$

lb O$_2$/d = 136×10^{-4} lb/(ft$^3 \cdot$ h) $\times 24$ h/d $\times 13{,}880$ ft^3 = 4341 lb/d (2055 kg/d)

Nitrification in a trickling filter is related to the organic loading. Since heterotropic growth is considerably greater than that of nitrifiers, the nitrifiers compete very poorly for film growth. Therefore, only under low organic loadings will significant nitrification occur. This is shown in Fig. 7.48.

Tertiary Nitrification

Nitrification in a trickling filter in the absence of carbonaceous organics follows a zero-order reaction for effluent $NH_3\text{-}N$ concentrations of 3.5 to 10 mg/L. Results, reported by Jiumm et al.[30] are shown in Fig. 7.49 for a synthetic wastewater activated sludge effluent. Maximum ammonia removal rates for municipal wastewater have been reported to vary from 1.2 to 1.8 g $NH_3\text{-}N/(m^2 \cdot d)$. This rate must be corrected for industrial effluents. Reported rates

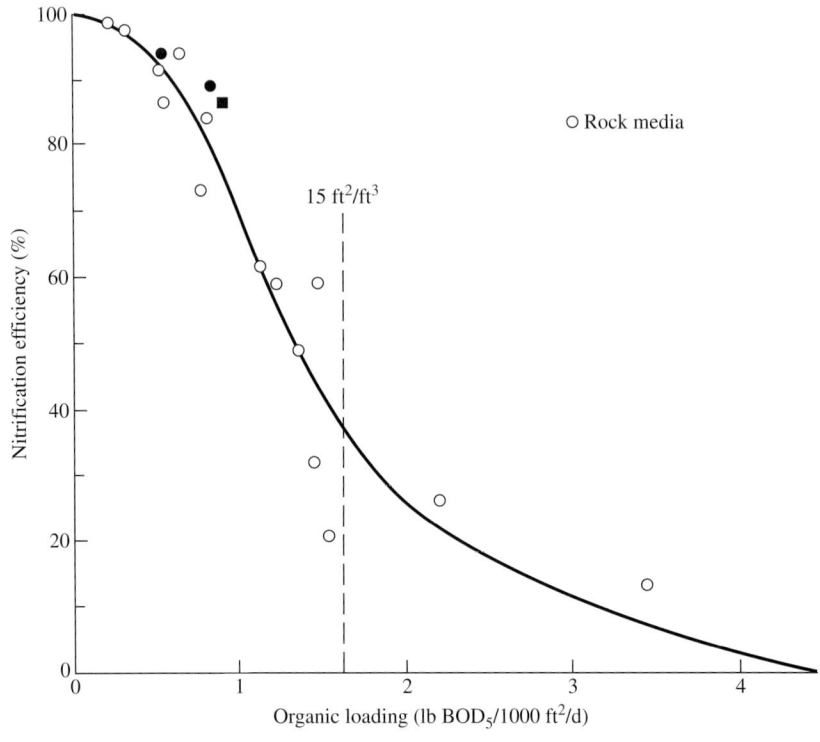

Figure 7.48 Combined carbon oxidation–nitrification performance.

are based on a soluble BOD applied to the filter of less than 20 mg/L. In order to ensure wetting of the filter surface, hydraulic loadings should exceed 0.8 gal/(min · ft²). Jiumm et al.[30] have shown that alkalinity requirements are less than the theoretical value of 7.2 mg alkalinity per mg of N oxidized. They showed a variation of 6.7 mg alkalinity/mg N oxidized at low nitrogen loadings to 4.6 mg/L at high loadings. This decrease in alkalinity requirement may be attributed to denitrification in the biofilm, which generates some alkalinity. Tertiary nitrification in a trickling filter is illustrated by Example 7.11.

Example 7.11. It is desired to reduce NH_3^-N from 100 to 20 mg/L through a 20-ft-deep trickling filter using media with a specific surface area of 44 ft²/ft³. No inhibition is expected. The wastewater flow is 1 million gal/d.

Compute:

The filter volume required
The hydraulic loading on the filter
The alkaline loading on the filter

Use a NH_3^-N removal rate of 0.28 lb NH_3^-N/(1000 ft² · d)

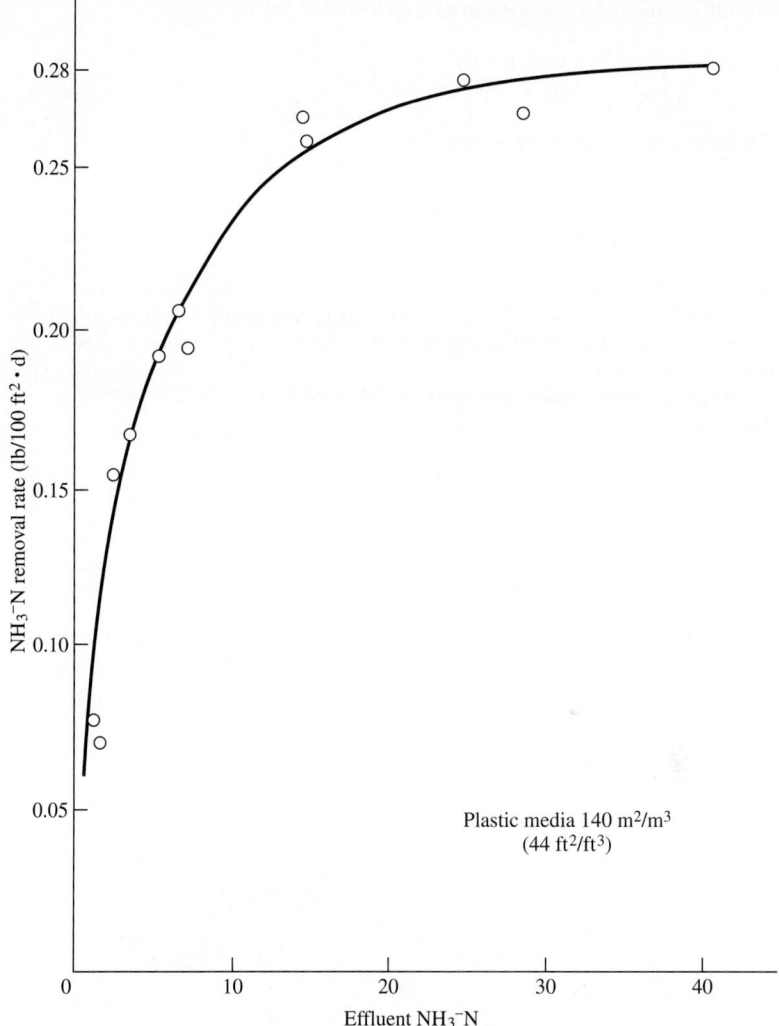

FIGURE 7.49   Tertiary nitrification through a trickling filter.

Solution   The NH_3^-N removed is

$$(100-20)(1)(8.34) = 667 \text{ lb/d}$$

The packing area required is

$$\frac{667 \text{ lb } NH_3^-N/d}{0.28 \text{ lb } NH_3^-N/(1000 \text{ ft}^2 \cdot d)} = 2383 \times 10^3 \text{ ft}^2$$

The filter volume, for a specific surface area of 44 ft²/ft³, is

$$\frac{2383 \times 10^3 \text{ ft}^2}{44 \text{ ft}^2/\text{ft}^3} = 54 \times 10^3 \text{ ft}^3$$

The surface area of the filter is therefore

$$\frac{54 \times 10^3}{20 \text{ ft}} = 2700 \text{ ft}^2$$

and the filter diameter is thus 52 ft.

The hydraulic loading is based on the need to maintain 0.8 gal/(min · ft²) across the filter surface. This will require 2160 gal/min, or a recycle flow of 2160 − 690 = 1470 gal/min.

For the alkalinity requirement, assume 7.14 lb alkalinity/lb NH_3^-N removed. Then

$$(667)(7.14) = 4762 \text{ lb/d}$$

7.6 Rotating Biological Contactors

The rotating biological contactor consists of large-diameter plastic media mounted on a horizontal shaft in a tank, as shown in Fig. 7.50. The contactor is slowly rotated with approximately 40 percent of the

FIGURE 7.50 Rotating biological contactor. (*Courtesy of Envirex Inc.*)

surface area submerged. A 1- to 4-mm layer of slime biomass is developed on the medium. (This would be equivalent to 2500 to 10,000 mg/L in a mixed system.) As the contactor rotates, it carries a film of wastewater through the air, resulting in oxygen and nutrient transfer. Additional removal occurs as the contactor rotates through the liquid in the tank. Shearing forces cause excess biomass to be stripped from the media, as in a trickling filter. This biomass is removed in a clarifier. The attached biomass is shaggy with small filaments, providing a high surface area for organic removal to occur. The present medium consists of high-density polyethylene with a specific surface of 37 ft^2/ft^3 (121 m^2/m^3). Single units are up to 12 ft (3.7 m) in diameter and 25 ft (7.6 m) long, containing up to 100,000 ft^2 (9290 m^2) of surface in one section.

The primary variables affecting treatment performance are:

1. Rotational speed
2. Wastewater retention time
3. Staging
4. Temperature
5. Disk submergence

In the treatment of low-strength wastewaters (BOD to 300 mg/L), performance increased with rotational speed up to 60 ft/min (18 m/min), with no improvement noted at higher speeds. Increasing rotational speed increases contact, aeration, and mixing, and would therefore improve efficiency for high-BOD wastewaters. However, increasing rotational speed rapidly increases power consumption, so that an economic evaluation should be made of the trade-off between increased power and increased area.

In the treatment of domestic wastewater, performance increases with liquid volume to surface area ratios up to 0.12 gal/ft^2 (0.0049 m^3/m^2). No improvement was noted above this value.

In many cases, significant improvement was observed by increasing from two to four stages with no significant improvement with greater than four stages. Several factors could account for these phenomena. The reaction kinetics would favor plug flow or multistage operation. With a variety of wastewater constituents, acclimated biomass for specific constituents may develop in different stages. Nitrification will be favored in the later stages, where low BOD levels permit a higher growth of nitrifying organisms on the media. In the treatment of industrial wastewaters with high BOD levels or low reactivity, more than four stages may be desirable. For high-strength wastewaters, an enlarged first stage may be employed to maintain aerobic conditions. An intermediate clarifier may be employed where high solids are generated to avoid anaerobic conditions in the contactor

	SCN⁻, mg/L	CN_t^-, mg/L	CN_c^-, mg/L	Cu, mg/L	NH_3^-N, mg/L
Influent	51.9	6.94	4.60	0.99	5.5
Effluent	<1.0	0.35	0.05	0.03	0.5
Results Across Four Stages:					
			Stage		
	Influent	1	2	3	4
SCN⁻, mg/L	51.9	20.3	3.0	<1.0	<1.0
NH_3^-N, mg/L	2.2	8.4	3.4	0.9	0.3

CN_c^- = cyanide amenable to chlorination
CN_t^- = total cyanide

TABLE 7.18 RBC Performance on Goldmine Tailings Wastewater

basins. RBC performance treating minewater tailings is shown in Table 7.18.

Design for industrial wastewaters will usually require a pilot plant study. A kinetic model similar to that developed for activated sludge may be employed:

$$\frac{Q}{A}(S_o - S) = kS \qquad (7.28)$$

where Q = flow rate
  A = surface area
  S_o = influent substrate concentration
  S = effluent substrate concentration
  k = reaction rate

Or, for wastewaters with a highly variable influent strength,

$$\frac{Q}{A}(S_o - S) = K\frac{S}{S_o} \qquad (7.29)$$

Equation (7.26) for several industrial wastewaters is shown in Fig. 7.51. For high-BOD wastewaters, performance can be improved by surrounding the media with enriched oxygen to enhance oxygen transfer and BOD removal.

Several factors become apparent from Fig. 7.51. The maximum BOD removal rate $Q/A(S_o - S)$ for a given operating condition (rotational speed, gas oxygen content, etc.) will relate to both the concentration of influent BOD and the biodegradability of the wastewater.

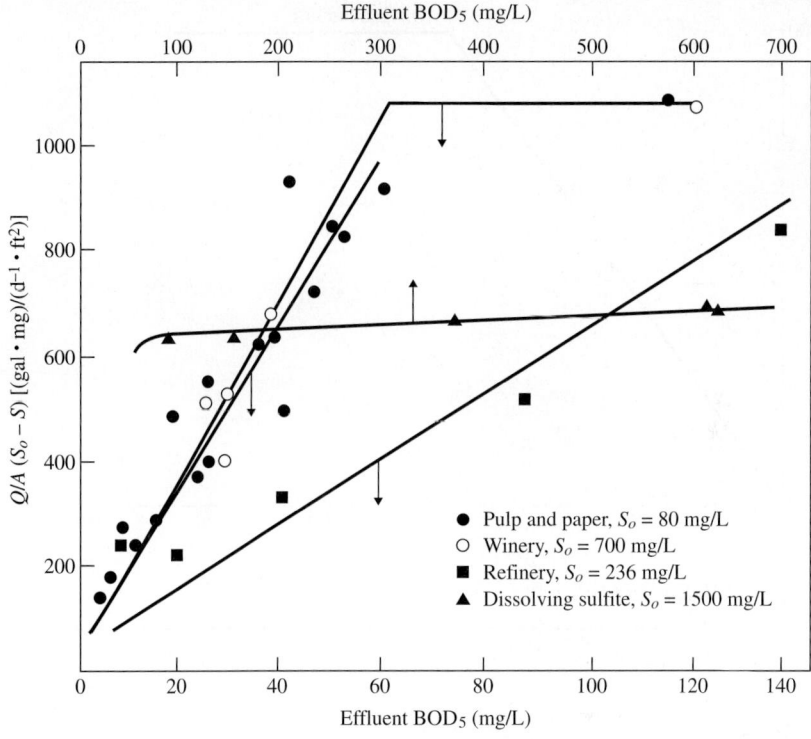

FIGURE 7.51 BOD removal characteristics of an RBC treating industrial wastewaters.

The performance of multiple contactors in series can be defined by the relationship:

$$\frac{S}{S_o} = \left(\frac{1}{1+kA/Q}\right)^n \quad (7.30)$$

in which n is the number of stages. Since at some loading oxygen will be limiting, it is important that the loading for each stage be checked for a multistage system.

As previously noted, the economics of the alternatives should be evaluated for each application. A process design is shown in Example 7.12.

Example 7.12. An RBC is to be designed for an effluent soluble BOD of 20 mg/L for a wastewater with an initial soluble BOD of 300 mg/L. The flow is 0.5 million gal/d (1900 m³/d). Compute the total number of stages and the required area. The performance relationship is shown in Fig. 7.52.

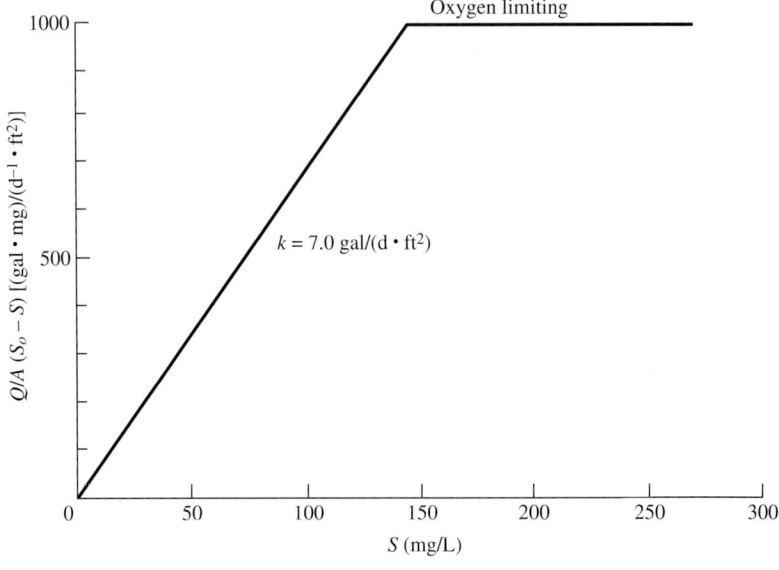

FIGURE 7.52 BOD removal relationship.

Solution

The maximum loading to an RBC occurs at that point where oxygen becomes limiting, as shown in Fig. 7.52. The maximum hydraulic loading can be computed:

$$\frac{Q}{A} = \frac{[(Q/A)S_r]_{max}}{S_o - [(Q/A)S_r]_{max}/k}$$

The performance of multiple contactors in series can be defined by the relationship:

$$\frac{S_e}{S_o} = \left(\frac{1}{1+kA/Q}\right)^n$$

$$\frac{Q}{A} = \frac{k}{(S_o/S_e)^{1/n} - 1}$$

The maximum hydraulic loading can be computed:

$$\left(\frac{Q}{A} S_r\right)_{max} = 1000 \text{ gal}/(d \cdot ft^2) \cdot mg/L$$

$$\frac{Q}{A} = \frac{1000}{300 - 1000/7} = 6.37 \text{ gal}/(d \cdot ft^2) \ [0.26 \text{ m}^3/(d \cdot m^2)]$$

The required area for various numbers of stages can be computed as tabulated below.

Number of Stages, n	Q/A, gal/ (d · ft²)	$(Q/A)/n$, gal/ (d · ft² · Stage)	$A \times 10^{-3}$, ft²/ Stage
1	0.5	0.5	1000
2	2.44	1.22	410
3	4.83	1.61	310
4	7.29	1.82	275
5	10.0	2.0	250

$$\frac{Q}{A} = \frac{7.0}{(S_o/S_e)^{1/n} - 1}$$

The hydraulic loading (Q/A) can be plotted versus the number of stages, as shown in Fig. 7.53. In order not to exceed the maximum loading to the first stage, the maximum number of stages would be 3.5. In this case, the plant would be designed for three stages.

The calculations can also be made graphically, as shown in Fig. 7.54.

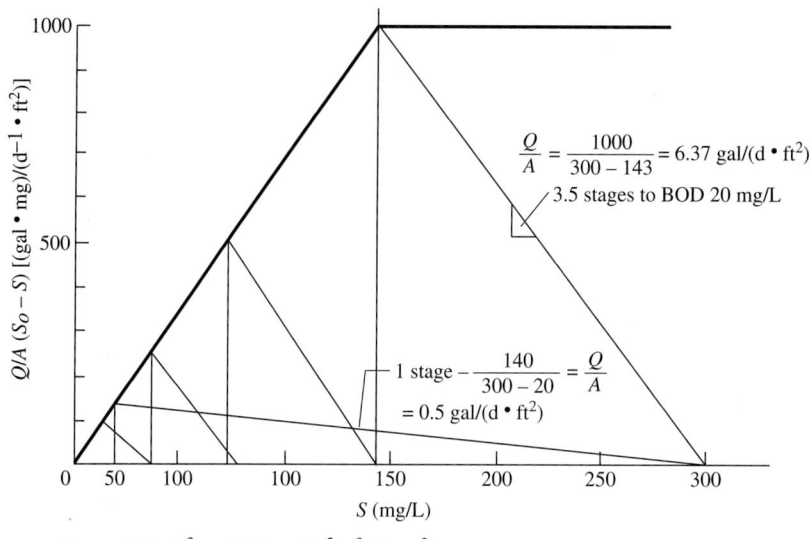

FIGURE 7.53 Graphical design solution.

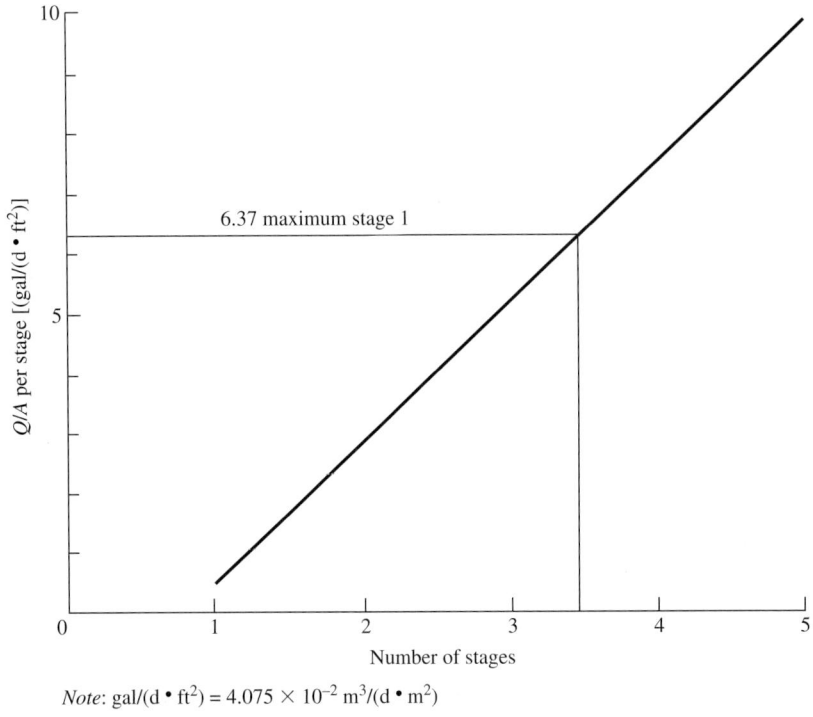

FIGURE 7.54 Calculation of number of stages.

7.7 Anaerobic Treatment Processes

Anaerobic decomposition involves the breakdown of organic waste to gas (methane and carbon dioxide) in the absence of oxygen. During the last 25 years, anaerobic treatment has become a more attractive alternative for stabilization of specific waste types. This is due to its relative low total cost (as much as half as aerobic systems), beneficial use of methane produced, low sludge production, and low energy and nutrient requirements. A new generation of reactors has been designed to retain biomass on static (upflow anaerobic filters, downflow stationary fixed film reactors) or moving (fluidized bed reactors) supports, and also by promoting the growth of good settling flocculating biomass (anaerobic contactor, anaerobic sludge blanket reactors, expanded granular sludge bed reactors). These modifications have resulted in reactor SRT becoming independent of HRT, thus allowing for operation at short HRT (6 h to 1 week) and higher organic loading rates (4 to 40 kg COD/m^3 reactor/d). Smaller reactors and a more stable operation have resulted.

Now over 850 anaerobic reactors are in operation worldwide. Approximately 75 percent of these treat wastewaters from the food and related industries. Presently, at least 63 anaerobic treatment systems are treating chemical and petro-chemical wastewater globally. This listing has been summarized by Macarie.[31] With appropriate pretreatment to increase biodegradability, reduce toxicity or inhibition, or enhance the environment (i.e., reduce salinity), the number of applications of this technology is expected to increase significantly in the future.

Process Alternatives

The anaerobic process is operated in one of several ways. Some of these are shown in Fig. 7.55.

1. The *anaerobic filter reactor* establishes growth of the anaerobic organisms on a packing medium. The filter may be operated upflow, as shown in Fig. 7.55, or downflow. The packed filter medium, while retaining biological solids, also provides a mechanism for separating the solids and the gas produced in the digestion process. Jennett and Dennis,[35] treating a pharmaceutical wastewater, were able to achieve a 97 percent removal of COD at a loading of 3.5 kg COD/($m^3 \cdot d$) using plastic media (37°C). Sachs et al.[36] achieved an 80 percent COD reduction at a loading of 0.56 kg COD/($m^3 \cdot d$) (35°C) and a 36-h HRT for a synthetic organic chemicals wastewater. Obayashi and Roshanravan[37] obtained a 70 percent COD reduction at a loading of 2 kg COD/($m^3 \cdot d$) on a rendering wastewater. Start-up periods may vary from 3 to 9 months, depending on substrate and OLR differences (Colleran et al.).[38]

2. The *anaerobic contact process*[32,33] provides for separation and recirculation of seed organisms, thereby allowing process operation at retention periods of 6 to 12 h. A degasifier is usually needed to minimize floating solids in the separation step. For high-degree treatment, the solids retention time has been estimated at 10 d at 90°F (32°C); the estimate doubles for each 20°F (11°C) reduction in operating temperature. Steffen and Bedker reported a full-scale anaerobic contact process (30 to 35°C) that achieved 90 percent COD removed while treating a meat-packing wastewater at a loading of 2.5 kg COD/($m^3 \cdot d$) and an HRT of 13.3 h. The SRT was about 13.3 d. Speece[34] has shown that with a 6 percent rate of synthesis, the days required for a twofold and a tenfold increase in biomass is 12 and 40 d, respectively.

3. In the fluidized-bed reactor (FBR) wastewater is pumped upward through a sand bed on which microbial growth has been developed. Biomass concentrations exceeding 30,000 mg/L

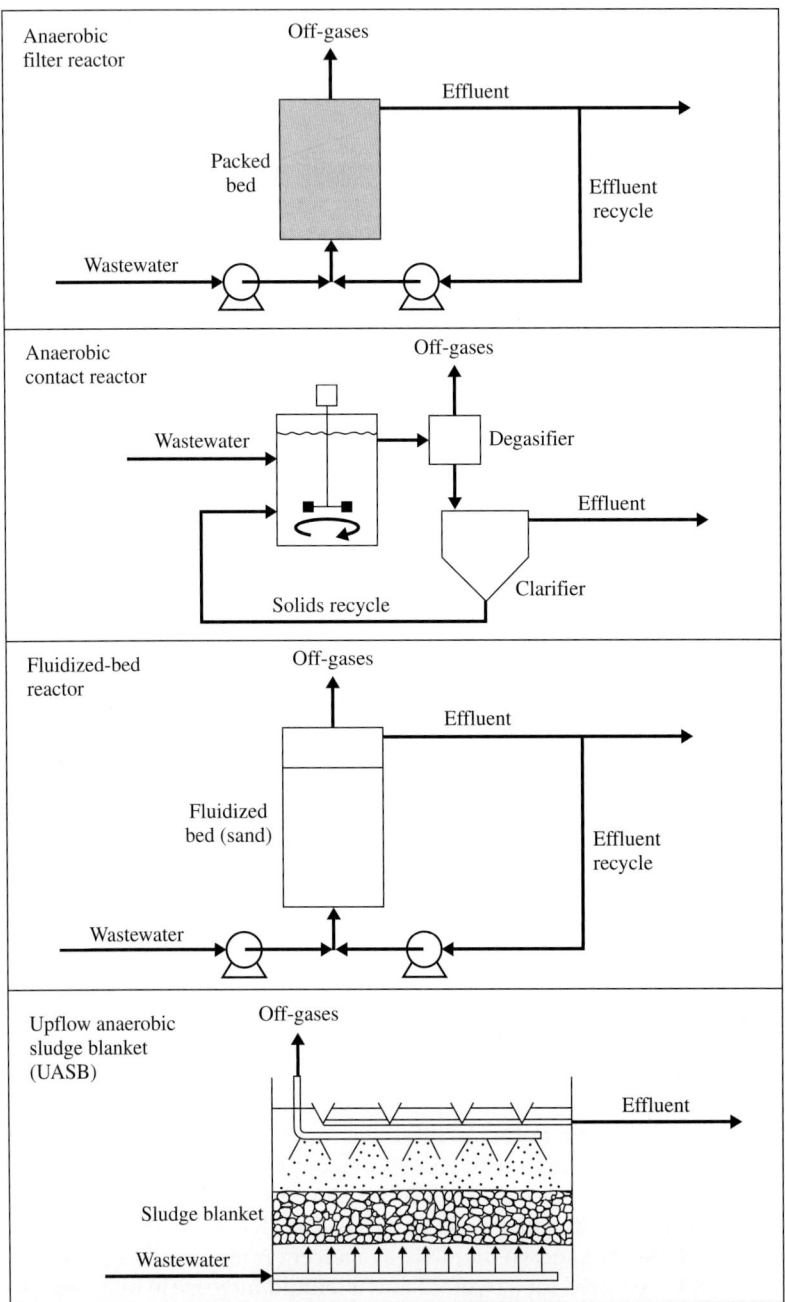

FIGURE 7.55 Anaerobic wastewater treatment processes.

have been reported. Effluent is recycled to mix with the feed in quantities dictated by the strength of the wastewater and the fluidization velocity. Organic removal efficiencies of 80 percent were achieved at loadings of 4 kg COD/(m$^3 \cdot$ d) on dilute wastewaters.[41]

4. In the *upflow anaerobic sludge blanket* (UASB) process, wastewater is directed to the bottom of the reactor, where it must be distributed uniformly. The wastewater flows upward through a blanket of biologically formed granules which consume the waste as it passes through the blanket. Methane and carbon dioxide gas bubbles rise and are captured in the gas dome. Liquid passes into the settling portion of the reactor, where solids-liquid separation takes place. The solids return to the blanket area while the liquid exits over the weirs. Formation of granules and their maintenance is extremely important in the operation of the process. Palns et al.[39] hypothesized that favorable granule formation requires a plug flow reactor configuration with a neutral pH, a zone of high H_2 partial pressure, a nonlimiting NH_3^-N source, and a limited source of cysteine, and a substrate that produces H_2 as an intermediate. With carbohydrate wastewaters the alkalinity requirement is 1.2 to 1.6 g alkalinity as $CaCO_3$/g influent COD, which is sufficient to maintain the pH above 6.6. Guiot et al.[40] found that the addition of trace metals enhanced biomass activity. To keep the blanket in suspension, an upflow velocity of 2 to 3 ft/h (0.6 to 0.9 m/h) has been used. Waste stabilization occurs as the waste passes through the sludge bed, with solids concentrations in the sludge bed having been reported to reach as high as 100 to 150 g/L. Loadings up to 600 lb COD/(100 ft$^3 \cdot$ d) [96 kg/(m$^3 \cdot$ d)] have been used successfully for certain wastewaters. In pilot studies, organic loadings of 15 to 40 kg COD/(m$^3 \cdot$ d) at liquid residence times of 3 to 8 h successfully treated high-strength wastewaters. A full-scale plant treating beet sugar wastewater achieved 80 percent removal with loadings of 10 kg COD/(m$^3 \cdot$ d) and an HRT of 4 h.

5. The ADI-BVF process is a low-rate anaerobic reactor with intermittent mixing and sludge recycle. The reactor has two zones: a reaction zone at the inlet end and a clarification zone at the outlet end. The reactor can be an aboveground tank or a lined earthen basin with a floating insulated membrane cover for gas recovery and temperature and odor control. Because of the large reactor volume, equalization requirements are minimized. A typical unit is shown in Fig. 7.56.

6. The *expanded granular sludge bed* (EGSB) process was developed and marketed by Biothane (Biobed Process) and Paques (IC Reactor) in the 1990s. A schematic of both processes is

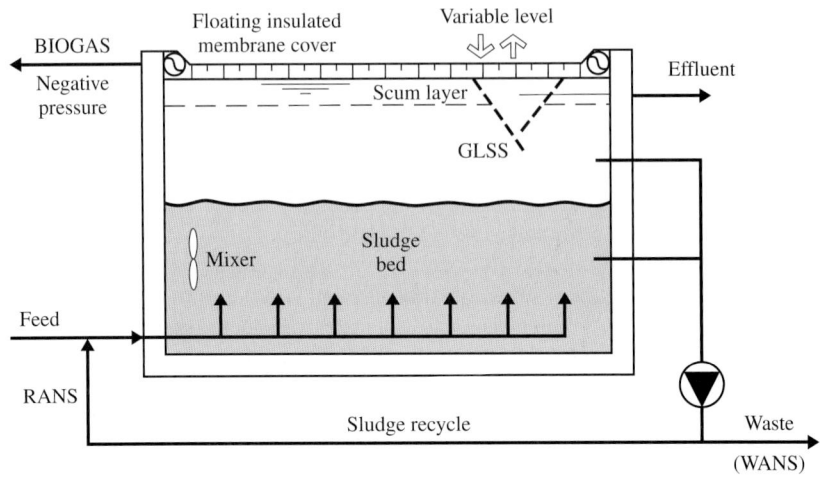

FIGURE 7.56 ADI-BVF® reactor. (*Courtesy of ADI Systems Inc.*)

presented in Fig. 7.57. These are operated at temperature ranges of 25 to 35°C, a HRT of 3 to 18 h, and a loading rate of 8 to 24 kg COD/m³/d. A 90 percent COD reduction has been experienced at a thermoplastics plant in the Netherlands at an organic load of 10 kg COD/m³ · d.[42]

Performance data for the various anaerobic processes are shown in Table 7.19. Performance of the ADI-BVF process treating a variety of industrial wastewaters is shown in Table 7.20.

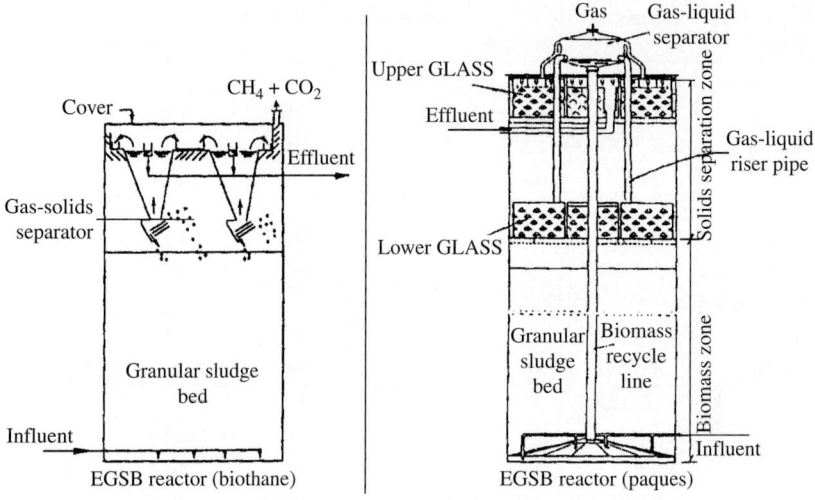

FIGURE 7.57 Expanded granular sludge bed (EGSB) processes developed and marketed by biothane (biobed process) and paques (IC Reactor).

Mechanism of Anaerobic Fermentation

Although the process kinetics and material balances are similar to those of aerobic systems, certain basic differences require special consideration. The conversion of organic acids to methane gas yields little energy; hence, the rate of growth is slow and the yield of organisms by synthesis is low. The kinetic rate of removal and the sludge yield are both considerably less than in the activated sludge process. The quantity of organic matter converted to gas will vary from 80 to 90 percent. Because there is less cell synthesis, nutrient requirements are correspondingly less than in the aerobic systems. High process efficiency requires elevated temperatures and the use of heated reaction tanks. Some of this heat may be provided by the methane produced by the process.

In anaerobic fermentation, roughly four groups of microorganisms sequentially degrade organic matter. Hydrolytic microorganisms degrade polymer-type material such as polysaccharides and proteins to monomers. This reduction results in no reduction of COD. The monomers are then converted into fatty acids (VFA) with a small amount of H_2. The principal acids are acetic, propionic, and butyric with small quantities of valeric. In the acidification stage there is minimal reduction of COD. Should a large amount of H_2 occur, some COD reduction will result. This seldom exceeds 10 percent. All acids higher than acetic acid are converted into acetate and H_2 by acetogenic microorganisms. The conversion of propionic acid is

$$C_3H_6O_2 + 2H_2O \rightarrow C_2H_4O_2 + CO_2 + 3H_2$$

500 Chapter Seven

Wastewater	Loading, kg/(m³ · d)	HRT, d	Temperature, °C	Removal, %	Reference
Anaerobic contact process:					
Meat packing	3.2 (BOD)	12	30	95	33
Meat packing	2.5 (BOD)	13.3	35	95	32
Keiring	0.085 (BOD)	62.4	30	59	40
Slaughterhouse	3.5 (BOD)	12.7	35	95.7	40
Citrus	3.4 (BOD)	32	34	87	—
Upflow filter process:					
Synthetic	1.0 (COD)	—	25	90	43
Pharmaceutical	3.5 (COD)	48	35	98	36
Pharmaceutical	0.56 (COD)	36	35	80	35
Guar gum	7.4 (COD)	24	37	60	37
Rendering	2.0 (COD)	36	35	70	44
Landfill leachate	7.0 (COD)	—	25	89	45
Paper mill foul condensate	10–15 (COD)	24	35	77	46
Fluidized bed reactor process:					
Synthetic	0.8–4.0 (COD)	0.33–6	10–3	80	47
Paper mill foul condensate	35–48 (COD)	8.4	35	88	46
USAB process:					
Skimmed milk	71 (COD)	5.3	30	90	47
Sauerkraut	8–9 (COD)	—	—	90	47
Potato	25–45 (COD)	4	35	93	42
Sugar	22.5 (COD)	6	30	94	42
Champagne	15 (COD)	6.8	30	91	35
Sugar beet	10 (COD)	4	35	80	47
Brewery	95 (COD)	—	—	83	48
Potato	10 (COD)	—	—	90	48
Paper mill foul condensate	4–5 (COD)	70	35	87	46
ADI-BFV process:					
Potato	0.2 (COD)	360	25	90	—
Cornstarch	0.45 (COD)	168	35	85	—
Dairy	0.32 (COD)	240	30	85	—
Confectionary	0.51 (COD)	336	37	85	—

TABLE 7.19 Performance of Anaerobic Processes

Wastewater	Raw Wastewater				Anaerobic Effluent			
	COD, mg/L	BOD, mg/L	BOD/COD	SS, mg/L	COD, mg/L	BOD, mg/L	BOD/COD	SS, mg/L
Potato processing	4,263	2,664	0.62	1,888	144	32	0.22	70
Yeast, cane molasses	13,260	6,630	0.50	1,086	4,420	600	0.14	883
Brewery and municipal	9,750	2,790	0.29	4,146	332	179	0.54	168
Clam processing	3,813	1,895	0.50	856	594	337	0.57	130
Corn processing and municipal	5,780				1,210			136
Hardboard mill effluent	12,930	5,990	0.46	486	2,590	740	0.29	507
Dairy wastewater	13,076	7,204	0.55	1,919	596	173	0.29	260
Semichemical pulp mill	6,826	2,221	0.32	851	3,822	524	0.14	881
Brewery	2,692	1,407	0.52	778	295	122	04.1	201
Alcohol stillage—1	90,000	23,000	0.26					
Alcohol stillage—2	120,000	40,000	0.33		57,000	4,700	0.08	
Alcohol stillage—3	98,000	31,000	0.32		54,000	6,000	0.11	
Alcohol stillage—4	80,000	24,000	0.30		36,000	4,100	0.11	
Dairy	3,250	1,970	0.61	252	372	111	0.30	55
Potato processing	1,890	1,090	0.58	341	165	98	0.59	50
Kraft foul condensates	13,960	6,710	0.48	10	1,076	660	0.61	190
Molasses stillage	65,000	25,000	0.38	5,000	15,000	1250	0.08	500
Corn wet milling	3,510	1,700	0.48	1,080	410	133	0.32	64

TABLE 7.20 Anaerobic Treatment of Industrial Wastewaters COD, BOD, and SS Values for Different Wastewaters

Wastewater	Raw Wastewater				Anaerobic Effluent		
	COD, mg/L	BOD, mg/L	BOD/COD	SS, mg/L	COD, mg/L	BOD, mg/L	BOD/COD	SS, mg/L
Pulp and paper	5,349	2,287	0.43	3,792	965	308	0.32	199
Dairy	25,541	20,575	0.81	974	737	190	0.26	337
Dairy	19,200	10,400	0.54	3,400	770	130	0.17	500
Brewery	4,011	2,786	0.69	139	510	306	0.60	105
Industrial and domestic	3,000	1,620	0.54	550	300	105	0.35	120
Dairy	8,830	7,890	0.89	1,670	150	86	0.57	53
Potato processing	8,356	5,300	0.63	5,250	1,113	486	0.44	708
Apple processing	3,994	2,441	0.61	2,573	174	87	0.50	54
Olive processing	13,395	5,550	0.41	289	2,332	786	0.34	212
Beans and pasta processing	2,604	1,200	0.46		1,285	528	0.41	
Pharmaceutical	9,200	4,000	0.43	2,400	3,300	850	0.26	350
Pharmaceutical	7,100	3,300	0.46	1,000	1,490	460	0.31	170
Confectionary	10,560	6,550	0.62	1,050	320	70	0.22	180
Potato processing	12,489	5,978	0.48	9,993	4,692	1573	0.34	2200
Ethanol corn processing	1,155	743	0.64	20	397	204	0.51	162

TABLE 7.20 Anaerobic Treatment of Industrial Wastewaters COD, BOD, and SS Values for Different Wastewaters (*Continued*)

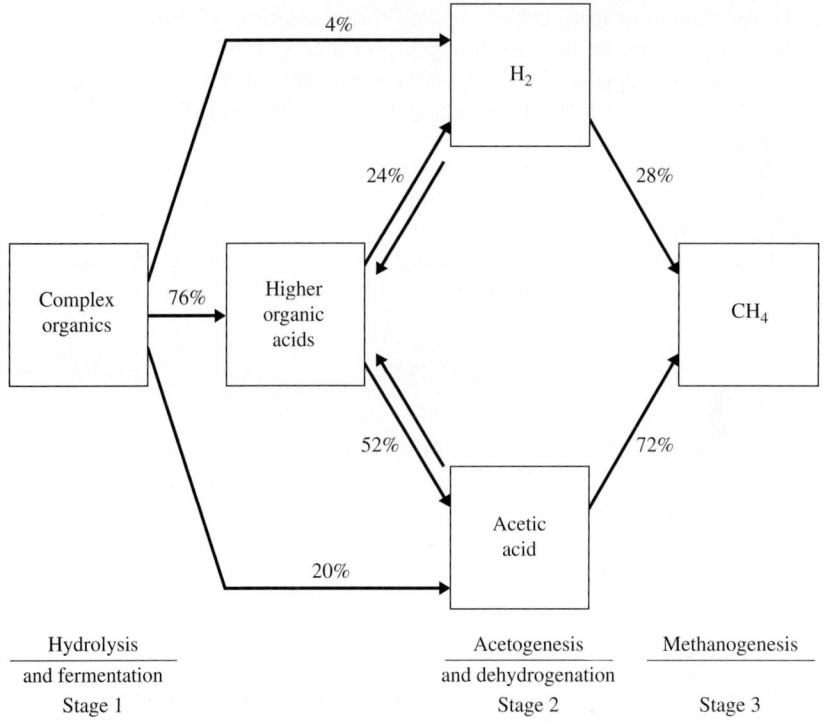

FIGURE 7.58 Three stages of methane fermentation.

In this reaction, COD reduction does occur in the form of H_2. This reaction will occur only if the concentration of H_2 is very low. The breakdown of organic acids to CH_4 and CO_2 is shown in Fig. 7.58. The acetic acid and H_2 are converted to CH_4 by methanogenic organisms.

Acetic acid: $$C_2H_4O_2 \rightarrow CO_2 + CH_4$$

$$CH_3COO^- + H_2O \rightarrow CH_4 + HCO_3^-$$

Hydrogen: $$HCO_3^- + 4H_2 \rightarrow CH_4 + OH^- + 2H_2O$$

The specific activity of typical anaerobic processes treating soluble industrial wastewaters is approximately 1 kg COD utilized/(kg biomass · d). There are two classes of methanogens that convert acetate to methane, namely *Methanothrix* and *Methanosarcina*. *Methanothrix* has a low specific activity, so it will predominate in systems with a low steady-state acetate concentration. In highly loaded systems, *Methanosarcina* will predominate with a higher specific activity (3 to 5 times as high as *Methanothrix*) if trace nutrients are available. These trace nutrients are iron, cobalt, nickel, molybdenum, selenium, calcium, magnesium, and microgram per liter levels of vitamin B.[34]

Speece[34] has reported trace mineral requirements as Ca 0.018 mg/g Ac, Fe 0.023 mg/g Ac, Ni 0.004 mg/g Ac, Co 0.003 mg/g Ac, and Zn 0.02 mg/g Ac.

Zehnder et al.[50] found that optimal methanogen growth and the specific rate of methane production required between 0.001 and 1.0 mg/L sulfur as S.

The kinetic relationship commonly employed for anaerobic degradation is the Monod relationship:

$$\frac{ds}{dt} = \frac{q_m S X}{K_s + S} \tag{7.31}$$

where ds/dt = substrate utilization rate, mg/(L · d)
q_m = maximum specific substrate utilization rate, g COD/(g VSS · d)
S = effluent concentration, mg/L
X = biomass concentration, mg/L
K_s = half saturation concentration, mg/L

Typical values for the coefficients are (Lawrence and McCarty)[51]:

Temperature, °C	q_{max}, d^{-1}	K_s, mg/L
35	6.67	164
25	4.65	930
20	3.85	2130

Reactor staging will frequently result in a reduced volume: 1 lb (0.454 kg) of COD or ultimate BOD removed in the process will yield 5.62 ft³ (0.16 m³) of methane at 0°C and 6.3 ft³ CH_4 at 35°C.

The quantity of cells produced during methane fermentation will depend on the strength of the waste, the character of the waste, and the retention of the cells in the system. As in an aerobic system, a portion of the cells produced will be destroyed by endogenous metabolism. The data of McCarty and Vath[52] are shown in Fig. 7.59. A relationship similar to Eq. (6.12) can be used to estimate cell yield. The relationships obtained by McCarty and Vath are:

Amino and fatty acids: $A = 0.054F - 0.038M$
Carbohydrates: $A = 0.46F - 0.088M$
Nutrient broth: $A = 0.076F - 0.014M$

where A = biological solids accumulated, mg/L
M = mixed liquor volatile suspended solids, mg/L
F = COD utilized, mg/L

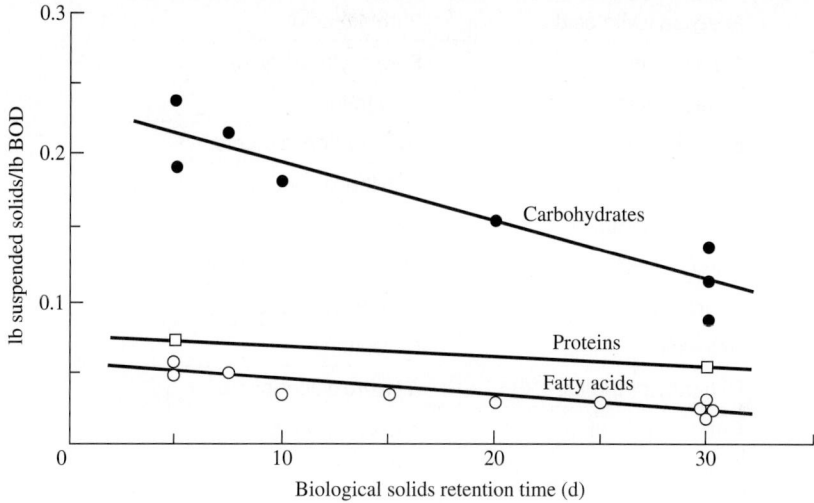

FIGURE 7.59 Biological solids production resulting from methane fermentation. (*After McCarty and Vath, 1962.*)

McCarty[49] has estimated the cell composition as $C_5H_9NO_3$, in which the nitrogen requirement is 11 percent of the net cell weight. The phosphorus requirement has been estimated as 2 percent of the biological cell weight. The COD of the cells is 1.21 mg/mg volatile suspended solids.

Biodegradation of Organic Compounds under Anaerobic Conditions

Anaerobic processes can break down a variety of aromatic compounds. It is known that anaerobic breakdown of the benzene nucleus can occur by two different pathways: photometabolism and methanogenic fermentation. It has been shown that benzoate, phenylacetate, phenylpropionate, and annamate are completely degraded to CO_2 and CH_4. Lower fatty acids were detected as reaction intermediates. While long acclimation periods were required to initiate gas production, the time required could be reduced by adapting the bacteria to an acetic acid substrate before adapting them to the aromatic.

Chmielowski et al.[52–53] showed that phenol, p-cresol, and resorcinol were completely converted to CH_4 and CO_2. A tabulation of organic compounds mineralized under anaerobic conditions is shown in Table 7.21.[50] As described by Speece,[34] *fortuitous metabolism* is defined as biodegradation of refractory organics by microbes from which they derive no energy. If a cosubstrate is added along with a refractory organic such as chloroform, the biomass is able to metabolize it fortuitously. Cosubstrates include sugar, methanol, and lactate.

Acetylsalicylic acid	Phthalic acid
Acrylic acid	Polyethylene glycol
p-Anisic acid	Pyrogallol
Benzoic acid	p-Aminobenzoic acid
Benzyl alcohol	Butylbenzylphthalate
2,3-Butanediol	4-Chloroacetanilide
Catechol	m-Chlorobenzoic acid
m-Cresol	Diethylphthalate
p-Cresol	Geraniol
Di-n-butylphthalate	4-Hydroxyacetinilide
Dimethylphthalate	p-Hydroxybenzyl alcohol
Ethyl acetate	2-Octanol
2-Hexanone	Propionanilide
o-Hydroxybenzoic acid	Butylbenzylphthalate
p-Hydroxybenzoic acid	m-Chlorobenzoic acid
3-Hydroxybutanone	m-Methoxyphenol
1-Octanol	o-Nitrophenol
Phenol	p-Nitrophenol
Phloroglucinol	

Source: Adapted from Shelton and Tiedjc, 1984.[54]

TABLE 7.21 Organics Mineralized Under Anaerobic Conditions

As in aerobic processes, soluble microbial products (SMP) are also generated under anaerobic conditions. Kuo et al.[55] found that SMP production ranged from 0.2 to 1.0 percent for acetate and 0.6 to 2.5 percent for glucose.

Food-processing and brewery wastes are readily broken down anaerobically with BOD removals in the range of 85 to 95 percent. In the anaerobic treatment of distillery waste, special precautions must be taken to reduce the inhibitory tendency of sulfates contained in the waste by either dilution or adjustment of loading rates. While animal wastes are readily treated anaerobically, ammonia toxicity may be a problem in dealing with fresh waste that contains a large amount of urine. Vath[56] showed that linear anionic and nonionic ethoxylated surfactants underwent degradation as observed by a loss of surfactant properties. Numerous investigators have shown that a variety of pesticides including lindane and isomers of benzene hexachloride will degrade under anaerobic conditions.

Wastewater	BOD, mg/L	COD, mg/L
Sugar	50–500	250–1500
Dairy	150–500	250–1200
Maize starch	—	500–1500
Potato	200–300	250–1500
Vegetable	100	700
Wine	3,500	—
Pulp	350–900	1,400–8000
Fibre board	2500–5500	8800–14900
Paper mill	100–200	280–300
Landfill leachate	—	500–4000
Digester supernatant	400	800–1400
Brewery	—	200–350
Distillery	—	320–400

TABLE 7.22 BOD/COD Characteristics of Effluents from Anaerobic Treatment of Various Wastewaters

Many of the higher-molecular-weight hydrocarbons that make up oil are decomposable by anaerobic bacteria. Shelton and Hunter[57] demonstrated that anaerobic decomposition of oil occurs naturally in bottom deposits.

An excellent review of the anaerobic treatment of industrial wastewaters by anaerobic processes has been presented by Obayashi and Gorgan.[58] The BOD/COD characteristics of effluents from anaerobic treatment of various wastewaters are shown in Table 7.22. Macarie has summarized experiences of anaerobic processes treating chemical and petrochemical wastewaters.[31]

Factors Affecting Process Operation

The anaerobic process will effectively function over two temperature ranges: the mesophilic range of 85 to 100°F (29 to 38°C) and the thermophilic range of 120 to 135°F (49 to 57°C). Although the rates of reaction are much greater in the thermophilic range, the maintenance of higher temperatures is usually not economically justifiable.

The methane organisms function over a pH range of 6.6 to 7.6 with an optimum near pH 7.0. When the rate of acid formation exceeds the rate of breakdown to methane, a process unbalance results in which the pH decreases, gas production falls off, and the CO_2 content of the gas increases.

pH control is therefore essential to ensure a high rate of methane production. Lime is commonly used to raise the pH of an anaerobic system when there is a process imbalance. Caution must be taken, since excess application of lime will result in precipitation of calcium carbonate. Sodium bicarbonate can alternatively be used for pH adjustment. It is desirable to have a bicarbonate alkalinity in the range of 2500 to 5000 mg/L in order to provide a buffer capacity to handle the increase in volatile acids with a minimal decrease in pH. If sufficient amounts of alkalinity are not present in the feed, then the alkalinity can be controlled by reducing the feed rate or adding alkalinity to the wastewater.

Alkalinity is required to neutralize the volatile fatty acids (VFA) generated and the H_2CO_3 formed, which results from the high partial pressure of CO_2 in the reactor. These alkalinity requirements are shown in Fig. 7.60.

In a well-operated complete mix reactor, the VFA will be in the range of 20 to 200 mg/L. In plug flow systems, however, elevated concentrations of VFA in the inlet region will reduce additional alkalinity. In order to avoid a decrease in pH, sufficient alkalinity must be present to compensate for the high CO_2 content. At a 30 percent CO_2 content in the digester gas, 1500 mg/L of alkalinity is necessary.

Inorganic salts in low concentration may provide stimulation; in high concentration, they may be toxic.[59] In some cases, adaptation will increase the tolerance level of the organisms. Table 7.23 summarizes some chemicals inhibitory to anaerobic process.[34]

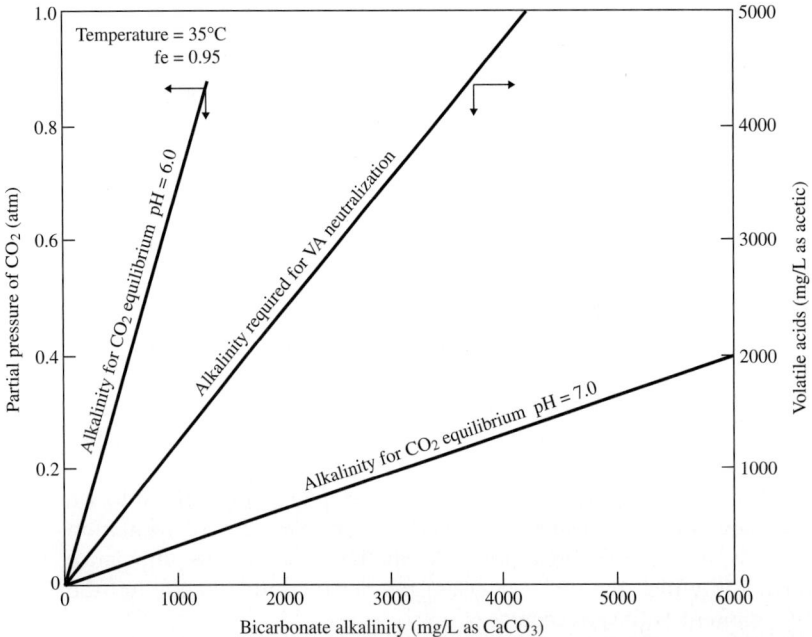

Figure 7.60 Design diagram for alkalinity requirements.

Toxicity of Some Organic Chemicals		Toxicity of Some Inorganic Chemicals		
	50% Inhibition		Mild Inhibition	Strong Inhibition
Acetaldehyde	440 mg/L	Sodium	3,500–5,500	8,000
Acrolein	10	Potassium	2,500–4,500	12,000
Bacitracin	20	Calcium	2,500–4,500	8,000
Chloroform	15	Magnesium	1,000–1,500	3,000
Creosote	1	Chromium (VI)		50–70 total 3 soluble
Cyanide	1	Chromium (II)		200–600 total
Dinitrophenol	40	Sulfide:	50% @ 6% H_2S in gas	
Ethylbenzene	340	(approximate)	25% @ 4% H_2S in gas	
Freons	1		10% @ 2% H_2S in gas	
Formaldehyde	70			
Long chain fatty acids	500			
Monensin (Rumensin)	2			

Source: Adapted from speece, 1996.

TABLE 7.23 Chemical Inhibition to Anaerobic Processes

Toxicity of Some Organic Chemicals		Toxicity of Some Inorganic Chemicals	
	50% Inhibition	Mild Inhibition	Strong Inhibition
Nitrobenzene	10		
Tannins	700		
Virginiamycin	10		
Quaternary ammonium compounds	25		
Tertiary amines	50		
Diamines	100		
Chlorine or equivalent	80		
Peroxyacetic acid	12		

Source: Adapted from speece, 1996.

TABLE 7.23 Chemical Inhibition to Anaerobic Processes (*Continued*)

The presence of antagonistic ions may sharply reduce the inhibitory effect of specific cations. Kugelman and McCarty[60] have shown that 300 mg/L potassium will reduce by 80 percent the inhibitory effect of 7000 mg/L sodium. The inhibitory effect will be completely eliminated by the addition of 150 mg/L calcium, but the calcium without the potassium will exert no beneficial effect. Ammonia is toxic at concentrations in excess of 3000 mg/L and inhibitory at levels greater than 1500 mg/L. The toxicity is related to the pH, which indicates how much ammonia is present in gaseous form. Gaseous ammonia is more inhibitory than the ammonium ion.

The maximum nontoxic concentration of soluble sulfide in an anaerobic system is 200 mg/L. Since sulfide is present as H_2S (gaseous), HS^-, and precipitated sulfide, the total sulfide concentration may be considerably higher. The loss of sulfide in the gas would also permit higher concentrations of either sulfate or sulfide in the waste feed to the system.

The heavy metals, such as copper, zinc, and nickel, are toxic at low concentrations. In the presence of sulfides, however, some of these metals are precipitated. The toxicity to organisms is related to their concentration in the digester itself.

When toxic materials are present, they must be removed before treatment or diluted below the toxic levels. The proper approach will depend principally on the overall waste composition.

7.8 Laboratory Evaluation of Anaerobic Treatment

Anaerobic testing requires a slightly different approach than that used for aerobic evaluation. Anaerobic treatability is typically assessed by using methane production as a primary indicator of waste stabilization. This is supplemented by measurements of COD removal and biomass levels.

Inhibition can be quantified by the manner in which the toxicant influences the maximum specific substrate utilization rate, q_m, and the half-saturation coefficient, K_s. Young and Cowan[61] have combined these effects on rate of substrate removal as

$$R_s = \frac{q_m [q^*] SX_a}{q_m [q^*] SX_a / K_{so}[K_s^*] + S}$$

where R_s = rate of substrate conversion, mg COD/L – h
q_m = intrinsic maximum specific substrate utilization rate in the absence of inhibitory substances
S = substrate concentration
X_a = active biomass
K_{so} = intrinsic half-velocity coefficients in the absence of inhibitory substances

q_m and K_s* are inhibition terms.

Han and Levenspield[62] presented a generalized equation to determine the inhibition terms:

$$q* = [1 - I/I*]^n$$
$$K_s* = [1 - I/I*]^{-m}$$

where $I*$ = toxicant concentration at complete inhibition

If I is low or $I*$ is high, the kinetic coefficients are reduced to the intrinsic values for nontoxic environments, or have q_m and K_{so}. The coefficients describing the effect of toxic organic chemicals can be determined using respirometric techniques as shown by Young and Cowan.[61]

As toxicant concentrations increase, a decrease in COD removal efficiency or increasing effluent COD and reduced biomass activity is experienced. Operation at longer SRT will produce several benefits. These include:

- Improved resistance to environmental factors—toxicity, loading variability, temperature, and so on
- Enhanced treatment efficiency
- Less waste sludge production

The SRT can be increased by:

- Increasing inventory of biomass
- Biomass recycle
- Employing fixed media to increase biomass inventory
- Increasing reactor volume

Some design considerations regarding wastewater characteristics are:[63]

- Typically COD should be > 3,000 mg/L.
- Influent suspended solids < 10 percent of COD.
- Fats, oil, and grease (FOG) may limit design.
- Calcium < 1 percent of COD.
- Sulfates < 5 percent of COD.
- Ammonia can be toxic at > 1,500 mg/L.
- TKN < 10 percent of COD or < 1,500 mg/L.
- Pretreatment for metals precipitation.
- Source control for organic toxicants.
- Addition of iron to bind sulfides.

Owen et al.[64] developed an anaerobic biotransformation assay termed the biochemical methane potential (BMP). This assays the concentration of organic pollutants in a wastewater which can be anaerobically converted to CH_4 and evaluates the potential anaerobic process efficiency. The test procedure has been described by Speece[34] as follows.

The procedure for implementing the BMP assay involves placing an aliquot of the effluent sample, normally 50 mL, in a 125-mL serum bottle with an anaerobic inoculum. In many cases the reactor effluent already contains an adequate inoculum. In other cases an acclimated inoculum can be taken directly from an anaerobic reactor.

The headspace in the serum bottle should be purged with CO_2 at 30 to 50 percent CO_2 composition for pH control and N_2 or CH_4. The serum bottle is then incubated at 35°C and CH_4 production recorded after a prescribed number of days (usually 5 days). The gas production is measured by inserting a hypodermic needle, connected to a calibrated fluid reservoir, through the serum cap. At this temperature, 395 mL of CH_4 production is equivalent to 1 g COD reduction, a stoichiometric relationship which allows calculation of the COD reduction in the liquid phase.

It is important that CO_2 production be excluded because CO_2 does not represent COD reduction under anaerobic conditions. For example, if 2000 mg/L (COD equivalent) of biodegradable organic pollutant remains in the effluent, a BMP assay would indicate that after a period of time 39.5 mL of CH_4 net gas production would result from a 50 mL sample of effluent.

A cardinal rule of the BMP assay, as with the BOD, is that the biomass must be acclimated to the pollutants. Care must be exercised to run a control with only the anaerobic inoculum to ensure adequate time and acclimation for the biomass to metabolize the pollutant. Whereas a 20-day BOD is considered to represent the ultimate demand aerobically, the BMP may be extended to 30 or 60 days to accommodate acclimation of the biomass to toxic and/or unusual pollutants occurring in some industrial wastewaters.

Since COD conversion is normally proportional to the product of biomass and time, the relative amount of biomass inoculum will affect the rate of conversion, but not the net ultimate value. A BMP analysis of a degradable wastewater is shown in Fig. 7.61. The corresponding COD reduction is shown in Fig. 7.62.

Owen et al.,[64] as described by Speece,[34] also developed a very useful and simple assay procedure to evaluate the potential toxicity of a wastewater sample to the anaerobic biomass, the anaerobic toxicity assay (ATA). The biomass to be evaluated is placed in a serum bottle and gassed with 50 percent CO_2 and 50 percent CH_4; then the wastewater sample is injected in increasing volumes into successive bottles. This procedure results in a range of dilution of the wastewater with the initial inocula of biomass. Excess substrate is also added initially to the serum bottles to avoid substrate limitation. If there is toxicity in the wastewater sample, it will be reflected in a reduced initial rate of gas production in proportion to the volume of wastewater added.

Because the aceticlastic methanogens are commonly the most sensitive to toxicity in the consortium, this characteristic can be

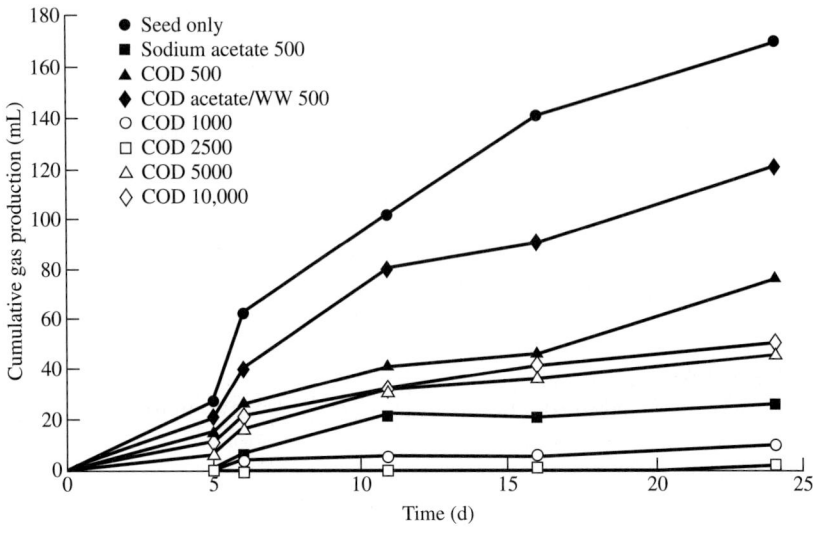

FIGURE 7.61 Cumulative gas production for nontoxic wastewaters.

assayed by adding a surplus of acetate (10,000 mg/L of calcium acetate salt is recommended). More complex substrates such as glucose, ethanol, propionate, or other complex substrates can be added in excess to assay toxicity to members in the consortium other than methanogens. Biomass activity can be determined using specific methanogenic activity (SMA). This test requires measurement of methane production from acetic acid under kinetically saturated condition at 35°C. Adequate mixing to avoid diffusion gradients in

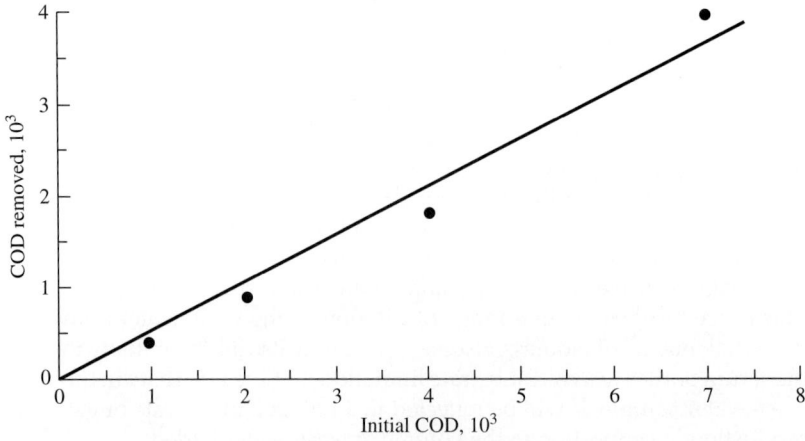

FIGURE 7.62 COD removal relative to initial COD.

granules or biofilms is required. SMA units are g COD/g VSS/d. Methane generated is converted to COD units by the following conversion: 0.39 L of methane/d is equivalent to 2.53 g COD/d.[61]

The significant difference between the BMP and the ATA assays is that the ATA is flooded with acetate (or other simple substrate noted above) as well as the wastewater sample, whereas the BMP is not. Also it must be borne in mind that in the ATA assay, the initial rate of gas production is of primary interest, while in the BMP it is the total amount of gas production which is important. Acclimation phenomena can be observed in all three assays as the biomass demonstrates the ability to acclimate to the toxicity. In the BMP, if gas production rate (corrected for the control) per unit volume of wastewater decreases as the amount injected into the bottles increases, this change is also an indication of inherent toxicity in the wastewater, as shown in Fig. 7.63. A detailed discussion regarding anaerobic bioassays and treatability testing is presented by Young and Cowan.[61]

Methane production can be estimated by the relationship

$$G = 5.62 (S_r - 1.42 \Delta X_v) \qquad (7.32)$$

where $G = CH_4$ produced per day, ft^3/d
S_r = BOD removed, lb/d
ΔX_v = VSS produced, lb/d

or

$$G = 0.351 (S_r - 1.42 \Delta X_v) \qquad (7.33)$$

where G is in m^3/d and S_r and ΔX_v are in kg/d. One cubic foot (0.0283 m^3) of methane has a net heating value of 960 Btu (1.01 × 10^6 J).

A design example is shown in Example 7.13.

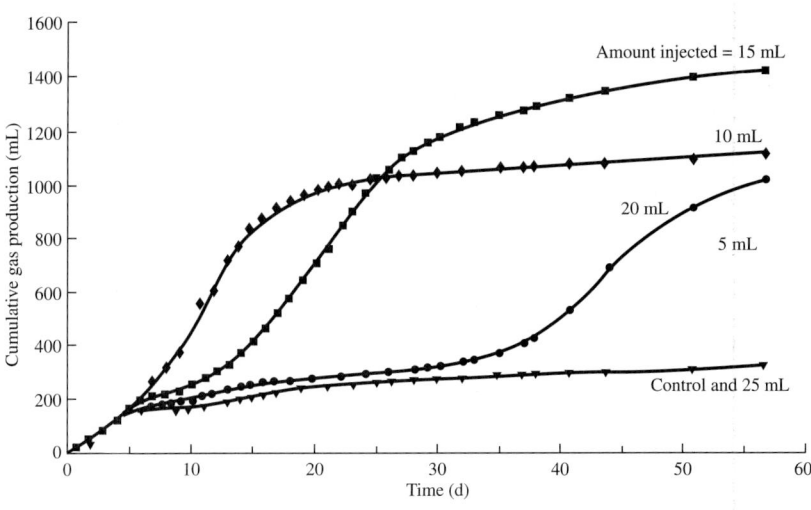

FIGURE 7.63 Cumulative gas production for inhibitory wastewater.

Example 7.13. Design an anaerobic contact process to achieve 90 percent removal of COD from a wastewater flow of 100,000 gal/d (379 m³/d).

Total influent COD = 10,300 mg/L
Nonremovable COD = 2200 mg/L
Removable COD (COD_R) = 8100 mg/L
COD to be removed = 90 percent

Process parameters:

SRT = 20 days minimum
Temperature = 35°C
a = 0.136 mg VSS/mg COD_R
b = 0.021 mg VSS/(mg VSS · d)
k = 3.24 1/(mg · d)
X_v = 5000 mg/L

Solution
Digester volume
From the kinetic relationship:

$$t = \frac{S_o S_r}{X_v kS}$$

$$= \frac{(8100)(7290)}{(5000)(3.24)(810)} = 4.5 \text{ d}$$

The digester volume is therefore

(4.5 d)(0.1 million gal/d) = 0.45 million gal   (1700 m³)

Check SRT (sludge age):

$$SRT = \frac{X_v t}{\Delta X_v} = \frac{X_v t}{aS_r - bX_v t}$$

$$= \frac{(5000)(4.5)}{(0.136)(7290) - (0.021)(5000)(4.5)} = 43.4 \text{ d}$$

This is in excess of the recommended SRT of 20 d to ensure the growth of methane formers.

A vacuum degasifier or flash aerator should be provided between the digester and the clarifier to purge gas from the sludge. A flash aerator inhibits further methane production in the clarifier with resulting floating sludge.

Sludge yield
The sludge yield from the process is

$$\Delta X_v = aS_r - bX_v t$$

$$= (0.136)(7290) - (0.021)(5000)(4.5) = 519 \text{ mg/L}$$

$$\Delta X_v = 519 \text{ mg/L} \times 0.1 \text{ million gal/d} \times 8.34 \ \frac{\text{lb/million gal}}{\text{mg/L}}$$

$$= 433 \text{ lb/d} \quad (196 \text{ kg/d})$$

Gas production

$$G = 5.62(S_r - 1.42 \Delta X_v)$$

$$= 5.62[(7290)(0.1)(8.34) - (1.42)(433)]$$

$$= 30{,}700 \text{ ft}^3 \text{ CH}_4/\text{d} \ (869 \text{ m}^3/\text{d})$$

Heat requirements

These can be estimated by calculating the energy required to raise the influent wastewater temperature to 95°F (35°C) and allowing 1°F (0.56°C) heat loss per day of detention time.

$$\text{Average wastewater temperature} = 75°F \ (23.9°C)$$

$$\text{Heat-transfer efficiency} = 50 \text{ percent}$$

$$\text{Btu} = \frac{W(T_i - T_e)}{E} \times (\text{specific heat})$$

$$\text{Btu}_{\text{required}} = \frac{(0.1 \text{ million gal/d})(8.34 \text{ lb/gal})(95° + 4.5° - 75°)}{0.5} \times \left(\frac{1 \text{ Btu}}{\text{lb} \cdot °F}\right)$$

$$= 40{,}900{,}000 \text{ Btu/d} \ (4.3 \times 10^{10} \text{ J/d})$$

The heat available from gas production is

$$\text{Btu}_{\text{available}} = (30{,}700 \text{ ft}^3 \text{ CH}_4/\text{d})(960 \text{ Btu/ft}^3 \text{ CH}_4)$$

$$= 29{,}500{,}000 \text{ Btu/d} \quad (3.1 \times 10^{10} \text{ J/d})$$

External heat of $40{,}900{,}000 - 29{,}500{,}000 = 11{,}400{,}000$ Btu/d (1.2×10^{10} J/d) should be supplied to maintain the reactor at 95°F (35°C).

Nutrient requirements
The nitrogen requirement is

$$N = 0.12 \Delta X_v = 0.12 \times 433 \text{ lb/d}$$

$$= 52 \text{ lb/d} \ (23.6 \text{ kg/d})$$

The phosphorus requirement is

$$P = 0.025 \Delta X_v = 0.025 \times 433 \text{ lb/d}$$

$$= 11 \text{ lb/d} \quad (5 \text{ kg/d})$$

7.9 Problems

7.1. A wastewater of 3 million gal/d (11,350 m³) has a BOD of 2100 mg/L and a VSS content of 400 mg/L. It is to be pretreated in an aerobic lagoon with a retention period of 1.2 d. The K is 8.0/d. Compute:

(a) The effluent soluble BOD in mg/L
(b) The effluent VSS in mg/L
(c) The oxygen required in mass/d if
   $a = 0.5$
   $a' = 0.55$
   $b = 0.15/d$

7.2. An industrial facility is to treat its waste in a series of aerated lagoons. The final effluent should have a maximum soluble BOD_5 of 20 mg/L for the summer months and 30 mg/L for the winter months.

Flow = 8.5 million gal/d (32,170 m³/d)
$BOD_{5\ inf} = 425$ mg/L
Temperature of waste = 95°F (35°C)
$K = 6.3$ d⁻¹ at 20°C
$\theta = 1.065$
$a = 0.5$
$a' = 0.52$
$b = 0.20$ d⁻¹
$\alpha = 0.88$
$C_1 = 1.0$ mg/L
$T_a = 30°F$ (winter)
   $= 65°F$ (summer)

Design the lagoons and optimize the system with respect to area. Design and specify the aeration system. Present a diagram showing basin geometries and aerator locations.

7.3. Design a two-stage aerated lagoon to operate at 20°C for the following conditions:

Flow = 2 mgd
BOD, $S_o = 500$ mg/L
$S_{e\ soluble} = 20$ mg/L
$K = 10^{-1}$ d
$a = 0.6$
$b = 0.1$ d⁻¹
$a' = 0.55$

Assume the detention period in the aerobic lagoon is 1 d.

7.4. The parameters for an activated sludge process are

Wastewater

Flow = 150 m³/h (0.95 million gal/d)
$COD_T = 450$ mg/L

$COD_D = 250$ mg/L
$BOD_5 = 125$ mg/L
TSS = 42 mg/L
Eff. $BOD_T = 20$ mg/L
$BOD_S = 10$ mg/L
$COD_D = 30$ mg/L

Process
$a'COD = 0.55$
$aVSS/COD = 0.32$
$aVSS/BOD = 0.55$
$\theta_c = 8$ d
$F/M_{BOD} = 0.35$
$K_{BOD} = 5$ d^{-1}

Assume the 42 mg/L influent TSS are 80 percent volatile and are nondegradable. Assume X_v to be 3000 mg/L and the sludge age to be 8 d.

Compute the following:

(a) Waste activated sludge

(b) Oxygen requirements

(c) Nutrient requirements

(d) Sludge recycle

(e) Biological selector

(f) Final clarifier area for a flux of 20 lb/(ft² · d)

7.5. A batch-activated sludge plant is to be designed for the following conditions:

$Q = 40{,}500$ gal/d (153 m³/d)
$S_o = 975$ mg/L
$S_e = 10$ mg/L
Period of wastewater discharge = 11.0 h

A laboratory bench-scale study yielded the following operational parameters:

$(F/M)_{av} = 0.2/$d
$a = 0.53$
$b = 0.16/$d
$a' = 0.43$
Eff. SS = 50 mg/L (85% volatile)
Equilibrium sludge density = 0.8 lb sludge/ft³ (12.8 kg/m³)

The sludge is to be stored for a period of 30 d when the reactor will be emptied and cleaned. A value of 45 lb BOD removed/(hp · d) [27.4 kg/(kW · d)] was also found to be in effect.

(a) Calculate the following:

  (i) Volume of active sludge required to oxidize each day's accumulation of degradable solids

  (ii) The 30 days sludge storage volume

(iii) The daily volume for the waste
(iv) The freeboard at 5 percent of the waste volume
(v) The total volume and basin dimensions

(b) Calculate the amount of oxygen per day required.

(c) Calculate the effluent BOD total.

(d) Estimate the aeration power requirements.

7.6. A 5 million gal/d (18,930 m³/d) municipal-activated sludge plant is presently operating at an F/M of 0.3 d^{-1}. The plant plans to accept 0.5 million gal/d (1890 m³/d) of a brewery wastewater with a BOD of 1200 mg/L. Determine what changes need to be made to avoid filamentous bulking. Use the F/M dissolved oxygen relationship shown in Fig. 6.51. The municipal sewage has a BOD of 200 mg/L. The effluent soluble BOD is 10 mg/L.

7.7. A municipal plant receives fruit-processing wastewaters and has a combined BOD of 600 mg/L. The return sludge has a concentration of 8000 mg/L SS. What recycle is required for a contactor with a floc load of 150 mg BOD/g VSS?

7.8. Compute the required clarifier area for a wastewater flow of 1.0 million gal/d (3785 m³/d), an MLSS of 3000 mg/L, an SVI of 150 mL/g, and an R/Q of 0.5. What is the maximum MLSS if the SVI increases to 250 mL/g and the recycle rate stays the same?

7.9. A trickling filter presently achieves 52 percent BOD reduction from a brewery wastewater under the following conditions:

$N = 3{:}1$
$Q = 3 \text{ gal}/(\text{min} \cdot \text{ft}^2) \; [0.122 \text{ m}^3/(\text{m}^2 \cdot \text{min})]$
$d = 20 \text{ ft } (6.1 \text{ m})$
$S_o = 850 \text{ mg/L}$
Flow = 0.8 million gal/d (3030 m³/d)
$n = 0.5$

In-plant changes will reduce the flow to 0.5 million gal/d and the BOD to 700 mg/L. For the same recycle, what will the new BOD removal efficiency be?

7.10. The following data were obtained on a rotating biological contactor (RBC) for a bleached pulp-mill effluent. The test unit had four stages.

Average Flow, gal/(d · ft²)	Influent, mg/L	Stages, mg/L			
		1	2	3	4
2.0	76	26	10	7	6
4.0	72	41	24	16	12
6.0	81	42	25	15	10
8.0	78	52	38	25	18
10.0	83	60	47	36	30

Correlate the data and design a system for 80 percent BOD removal, assuming an influent BOD of 85 mg/L.

7.11. Consider Example 7.9. At present, an air-activated sludge plant follows the anaerobic process. The activated sludge plant operates under the following conditions to produce an effluent with a COD of 30 mg/L (degradable):

$T = 20°C$
$F/M = 0.3\ d^{-1}$
$X_v = 2500\ mg/L$

The load to the anaerobic plant is to be increased to 120,000 gal/d (454 m³/d) with the same removable COD.

(a) Compute the new effluent from the anaerobic plant assuming X_v remains the same. What will the new gas production be?

(b) What modifications to the aerobic plant must be made to maintain the same effluent quality? Assume the sludge settling characteristics are the same as originally. Compute the following:
  (i) The new sludge recycle and the new MLSS. The volatile content of the sludge is 74 percent.
  (ii) The new F/M.
  (iii) Calculate the increase in power required assuming 1.5 lb O_2/(hp · h) [0.91 kg/kW · h]

The process parameters are:

$a = 0.5$
$a' = 0.6$
$b = 0.15$ at $20°C$

References

1. Gloyna, E. F.: *Waste Stabilization Pond Concepts and Experiences,* Division of Environmental Health, World Health Organization, Geneva, 1965.
2. Oswald, W. J.: *Advances in Water Quality Improvement* (E. F. Gloyna and W. W. Eckenfelder, eds.), vol. 1, University of Texas Press, Austin, Texas, 1968.
3. Gellman, I., and H. F. Berger: *Advances in Water Quality Improvement* (E. F. Gloyna and W. W. Eckenfelder, eds.), vol. 1, University of Texas Press, Austin, Texas, 1968.
4. Hermann, E. R., and E. F. Gloyna: *Sewage Ind. Wastes,* vol. 30, p. 963, 1958.
5. Marais, G. V., and V. A. Shaw: *Trans. S. Afr. Inst. Civ. Engrs.,* vol. 3, p. 205, 1961.
6. Steffen, A. J.: *J. WPCF,* vol. 35, pt. 4, p. 440, 1963.
7. Cooper, R. C.: *Dev. Appl. Microbiol.,* vol. 4, pp. 95–103, 1963.
8. Argamon, Y., and C. Adams: *Proc. 8th International Conf. on Water Pollution Research,* Pergamon Press, Oxford, England, 1976.
9. Eckenfelder, W. W., C. Adams, and S. McGee: *Advances in Water Pollution Research,* vol. 2, Pergamon Press, Oxford, 1972.
10. Parker, D.: Personal communication.
11. Galil, N. et al.: *Proc. 2nd Specialized Conf. on Pretreatment of Industrial Wastewaters,* Athens, Greece, 1996.
12. Boon, A. G. et al.: Report from Water Resources Research Center, Stevenage, U.K., 1983.

13. Murphy, K. L., and P. L. Timpany: *J. San. Eng. Div.*, ASCE, SA 5, October, 1967.
14. Franta, J. R., and P. A. Wilderer: *Water Sci. Tech.*, vol. 35, no. 1, p. 67, 1997.
15. Goronszy, M.: *J. WPCF*, vol. 41, pt. 2, p. 274, 1979.
16. Cuthbert, D. C., and D. C. Pollock: *Wat. Env. Assoc. of Texas*, May 1995.
17. Randall, C. W.: *Water and Env. Management*, vol. 12, no. 5, p. 375, 1998.
18. Johnson, T. L., J. P. McQuarrie, and A. R. Shaw: Integrated Fixed-Film Activated Sludge (IFAS): The New Choice for Nitrogen Removal Upgrades in the United States, *Proc. WEFTEC*, 2004.
19. Masterston, T., J. Federico, G. Hedman, and S. Duerr: *Upgrading for Total Nitrogen Removal with a Porous Media IFAS System*, BETA Group, Inc., Lincoln, Rhode Island.
20. Daigger, G. T., and R. E. Roper: *J. WPCF*, vol. 57, p. 859, 1985.
21. Wahlberg, E. J. et al: *Proc. Wat. Env. Fed. 68th Ann. Conf.*, vol. 1, p. 435, 1995.
22. Parker, D. S. et al.: *Wat. Sci. Tech.* vol. 25, no. 6, p. 301, 1992.
23. Paduska, R. A.: *Proc 24th Industrial Waste Conf.*, Purdue University, Lafayette, Ind., 1979.
24. Albertson, O. E.: *WPCF Operations Forum*, p. 15, 1989.
25. Albertson, O. E., and Eckenfelder, W. W.: *Proc. 2nd International Conf. on Fixed Film Biological Processes*, Washington, D.C., 1984.
26. Howland, W. E.: *Proc. 12th Ind. Waste Conf.*, Purdue University, vol. 94, p. 435, 1958.
27. Eckenfelder, W. W., and E. L. Barnhart: *J. WPCF*, vol. 35, p. 535, 1963.
28. Eckenfelder, W. W.: *Proc. ASCE SA*, vol. 4, pt. 2, pp. 33, 860, July 1961.
29. Department of Scientific and Industrial Research: "Water Pollution Research, 1956" (British), Her Majesty's Stationery Office, London, 1957.
30. Jiumm, M. H. et al.: *Proc. 1st International Conf. on Fixed Film Biological Processes*, Kings Island, Ohio, 1982.
31. Macarie, H.: *Overview on the Application of "Anaerobic Digestion to Chemical and Petrochemical Wastewaters," Waste Minimization and End of Pipe Treatment in Chemical and Petrochemical Industries*, Water Science & Technology, Oxford, UK: Elsevier Science Ltd., vol. 42, No. 5-6, 2000.
32. Steffen, A. J., and M. Bedker: *Proc. 16th Ind. Waste Conf.*, Purdue University, 1961.
33. Schropfer, G. J. et al.: *Sewage Ind. Wastes*, vol. 27, p. 460, 1955.
34. Speece, R. F.: *Anaerobic Biotechnology for Industrial Wastewaters*, Archae Press, Nashville, Tenn., 1996.
35. Jennett, J. C., and N. D. Dennis: *J. WPCF*, vol. 47, p. 104, 1975.
36. Sachs, E. F. et al.: *Proc. 33rd Ind. Waste Conf.*, Purdue University, 1978.
37. Obayashi, A. W., and M. Roshanravan: Unpublished report, Illinois Institute of Technology, Chicago, 1980.
38. Colleran, E. S. et al.: *Proc. 7th International Symposium on Anaerobic Digestion*, South Africa, p. 160, 1994.
39. Palns, S. S. et al.: *Water SA IT*, pp. 47–56, pp. 1991.
40. Guiot, S. et al: *Wat. Sci. Tech.*, vol. 25, p. 1, 1992.
41. Jewell, W. J. et al.: *J. WPCF*, vol. 53, p. 482, 1981.
42. Constable, S. W. C., and R. Kras: Selection, Start-up and Operation of an Anaerobic Pretreatment System for Wastewater from a Thermoplastic Production Facility. In: *Proc. 71st Annual Water Environment Federation Conf.*, Orlando, Florida, 1998.
43. Young, J. C., and P. L. McCarty: *Technical Report 87*, Department of Civil Engineering, Stanford University, 1968.
44. Witt, E. R. et al.: *Proc. 34th Ind. Waste Conf.*, Purdue University, 1979.
45. Dewalle, F. B., and E. S. K. Chian: *Biotech. Bioengineering*, vol. 18, p. 1275, 1976.
46. Donovan, G.: Personal communication.
47. Lettinga, G., and W. de Zeeuw: *Proc. 35th Ind. Waste Conf.*, Purdue University, 1980.
48. Pette, K. C. et al.: *CSM Suiker*, Amsterdam, Netherlands, 1986.
49. McCarty, P. L.: *Progress in Wat. Tech.*, vol. 7, p. 157, 1975.
50. Zehnder, A. J. et al.: *Anaerobic Digestion*, Elsevier, Amsterdam, 1982.

51. Lawrence, A. W., and McCarty, P. L.: *J. WPCF,* vol. 41, pp. R1–R17, 1969.
52. McCarty, P. L., and C. A. Vath: *Int. J. Air Water Pollution,* vol. 6, p. 65, 1962,
53. Chmielowski, J. et al.: *Zesz. Nauk,* Politech Slaska Inz. (Polish), vol. 8, p. 97, 1965.
54. Shelton, D. R., and J. M. Tiedjc: *Applied and Env. Microbiol.,* vol. 47, pp. 850–857, 1984.
55. Kuo, W. C. et al.: *Water Env. Research,* 1995.
56. Vath, C. A.: *Soap and Chem. Specif.,* March 1964.
57. Shelton, T. B., and J. B. Hunter: *J. WPCF,* vol. 47, pt. 9, p. 2257, 1975.
58. Obayashi, A. W., and J. M. Gorgan: *Management of Industrial Pollutants by Anaerobic Processes,* Lewis Publishers, Chelsea, Mich., 1985.
59. McCarty, P. L. et al.: *J. WPCF,* vol. 35, pt. 1, p. 501, 1963.
60. Kugelman, I. J., and P. L. McCarty: *Proc. 19 Ind. Waste Conf.,* Purdue University, 1964.
61. Young, J. C., and P. M. Cowan: *Respirometry for Environmental Science and Engineering,* S. J. Enterprises, Springdale, Arkansas, July 2004.
62. Han, K., and O. Levenspiel: "Extended MondKinetics for Substrate, Product and Cell Inhibition," *Biotechnol. Bioeng.,* 32, pp. 430-437, 1998.
63. Young, J.: "'*Anaerobic Treatment of Industrial Wastewaters,*' presented at Biotechnology in Waste Management for Sustainable Development," Chiang Mai, Thailand, August 2002.
64. Owen, W. R. et al.: *Water Res.,* vol. 13, p. 485, 1979.

CHAPTER 8
Adsorption

8.1 Introduction

Many industrial wastes contain organics which are refractory and which are difficult or impossible to remove by conventional biological treatment processes. Examples are ABS and some of the heterocyclic organics. These materials can frequently be removed by adsorption on an active-solid surface. The most commonly used adsorbent is activated carbon.

While other adsorbents can be used, activated carbon is able to remove a broad range of adsorbates, including many synthetic organic chemicals and inorganics such as heavy metals. The process may be of particular consideration when a BAT effluent quality is required and/or toxicity removal is an issue. This chapter presents the basic concepts of adsorption, laboratory evaluation techniques, regeneration requirements, and system design. Powdered activated carbon to enhance activated sludge performance is also discussed.

8.2 Theory of Adsorption

A solid surface in contact with a solution tends to accumulate a surface layer of solute molecules because of the unbalance of surface forces. Chemical adsorption results in the formation of a monomolecular layer of the adsorbate on the surface through forces of residual valence of the surface molecules. Physical adsorption results from molecular condensation in the capillaries of the solid. In general, substances of the highest molecular weight are most easily adsorbed. There is a rapid formation of an equilibrium interfacial concentration, followed by slow diffusion into the carbon particles. The overall rate of adsorption is controlled by the rate of diffusion of the solute molecules within the capillary pores of the carbon particles. The rate varies reciprocally with the square of the particle diameter, increases with increasing concentration of solute, increases with increasing temperature, and decreases with increasing molecular weight of the solute. Morris and Weber[1] found the rate of adsorption to vary as the square root of the time of contact, as shown in Fig. 8.1.

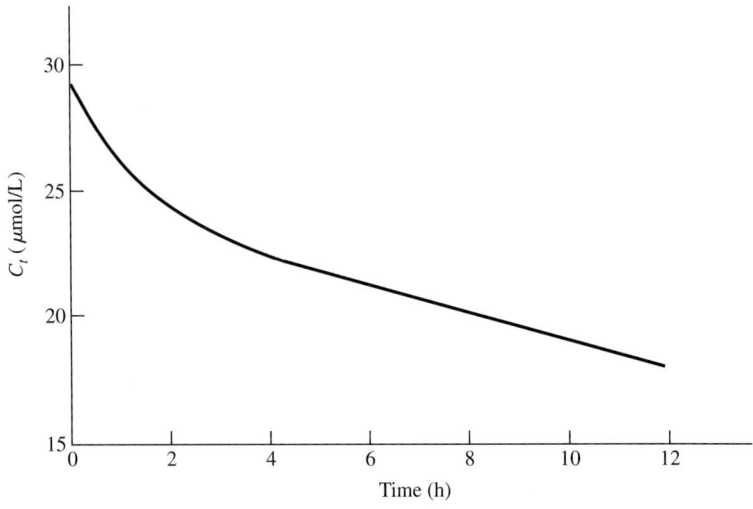

Figure 8.1 Rate of adsorption of 2-dodecyl benzene sulfonate by Columbia carbon (30°C, 75 mg/L, 0.273 mm diameter). (*After Morris and Weber, 1964.*)

The adsorptive capacity of a carbon for a solute will likewise be dependent on both the carbon and the solute.

Most wastewaters are highly complex and vary widely in the adsorbability of the compounds present. Molecular structure, solubility, etc., all affect the adsorbability. These effects are shown in Table 8.1. The relative adsorbability of organics on carbon is shown in Table 8.2.

Formulation of Adsorption

The degree to which adsorption will occur and the resulting equilibrium relationships have been correlated according to the empirical relationship of Freundlich and the theoretically derived Langmuir relationship. For practical application, the Freundlich isotherm usually provides a satisfactory correlation. The Freundlich isotherm is expressed as

$$\frac{X}{M} = kC^{1/n} \qquad (8.1)$$

where X = weight of substance adsorbed
M = weight of adsorbent
C = concentration remaining in solution

and k and n are constants depending on temperature, the adsorbent, and the substance to be adsorbed. The Freundlich constants for several priority pollutants are shown in Table 8.3.[2]

1. An increasing solubility of the solute in the liquid carrier decreases its adsorbability.
2. Branched chains are usually more adsorbable than straight chains. An increasing length of the chain decreases solubility.
3. Substituent groups affect adsorbability:

Hydroxyl	Generally reduces adsorbability. Extent of decrease depends on structure of host molecule.
Amino	Effect similar to that of hydroxyl but somewhat greater. Many amino acids are not adsorbed to any appreciable extent.
Carbonyl	Effect varies according to host molecule. Glyoxylic acid more adsorbable than acetic but similar increase does not occur when introduced into higher fatty acids.
Double bonds	Variable effects as with carbonyl.
Halogens	Variable effects.
Sulfonic	Usually decreases adsorbability.
Nitro	Often increases adsorbability.

4. Generally, strong ionized solutions are not as adsorbable as weakly ionized ones; that is, undissociated molecules are in general preferentially adsorbed.
5. The amount of hydrolytic adsorption depends on the ability of the hydrolysis to form an adsorbable acid or base.
6. Unless the screening action of the carbon pores intervenes, large molecules are more sorbable than small molecules of similar chemical nature. This is attributed to more solute carbon chemical bonds being formed, making desorption more difficult.
7. Molecules with low polarity are more sorbable than highly polar ones.

TABLE 8.1 Influence of Molecular Structure and Other Factors on Adsorbability

The Langmuir equation is based on an equilibrium between condensation and evaporation of adsorbed molecules, considering a monomolecular adsorption layer:

$$\frac{X}{M} = \frac{abC}{1+aC} \qquad (8.2)$$

This can be reexpressed in linear form as

$$\frac{1}{X/M} = \frac{1}{b} + \frac{1}{ab}\frac{1}{C} \qquad (8.2a)$$

where b = amount adsorbed to form a complete monolayer on the surface and a = constant which increases with increasing molecular size.

Compound	Molecular Weight	Aqueous Solubility, %	Concentration, mg/L		Adsorbability, g Compound/g Carbon	Percent Reduction
			Initial C_o	Final C_f		
Alcohols						
Methanol	32.0	∞	1000	964	0.007	3.6
Ethanol	46.1	∞	1000	901	0.020	10.0
Propanol	60.1	∞	1000	811	0.038	18.9
Butanol	74.1	7.1	1000	466	0.107	53.4
Aldehydes						
Formaldehyde	30.0	∞	1000	908	0.018	9.2
Acetaldehyde	44.1	∞	1000	881	0.022	11.9
Propionaldehyde	58.1	22	1000	723	0.057	27.7
Butyraldehyde	72.1	7.1	1000	472	0.106	52.8
Aromatics						
Benzene	78.1	0.07	416	21	0.080	95.0
Toluene	92.1	0.047	317	66	0.050	79.2
Ethyl benzene	106.2	0.02	115	18	0.019	84.3
Phenol	94	6.7	1000	194	0.161	80.6

TABLE 8.2 Amenability of Selected Organic Compounds to Activated Carbon Adsorption

Compound	K, mg/g	1/n
Hexachlorobutadiene	360	0.63
Anethole	300	0.42
Phenyl mercuric acetate	270	0.44
p-Nonylphenol	250	0.37
Acridine yellow	230	0.12
Benzidine dihydrochloride	220	0.37
n-Butylphthalate	220	0.45
N-Nitrosodiphenylamine	220	0.37
Dimethylphenylcarbinol	210	0.33
Bromoform	200	0.83
β-Naphthol	100	0.26
Acridine orange	180	0.29
α-Naphthol	180	0.31
α-Naphthylamine	160	0.34
Pentachlorophenol	150	0.42
p-Nitroaniline	140	0.27
1-Chloro-2-nitrobenzene	130	0.46
Benzothiazole	120	0.27
Diphenylamine	120	0.31
Guanine	120	0.40
Styrene	120	0.56
Dimethyl phthalate	97	0.41
Chlorobenzene	93	0.98
Hydroquinone	90	0.25
p-Xylene	85	0.16
Acetophenone	74	0.44
1,2,3,4-Tetrahydronaphthalene	74	0.81
Adenine	71	0.38
Nitrobenzene	68	0.43
Dibromochloromethane	63	0.93

TABLE 8.3 Summary of Freundlich Parameters at Neutral pH

Since most wastewaters contain more than one substance which will be adsorbed, direct application of the Langmuir equation is not possible. Morris and Weber[1] have developed relationships from the Langmuir equation for competitive adsorption of two substances:

$$\frac{X_A}{M} = \frac{a_A b_A C_A}{1 + a_A C_A + a_B C_B} \tag{8.3a}$$

$$\frac{X_B}{M} = \frac{a_B b_B C_B}{1 + a_A C_A + a_B C_B} \tag{8.3b}$$

More complex relationships could similarly be developed for multi-component mixtures. It should be noted that although the equilibrium capacity for each individual substance adsorbed in a mixture is less than that of the substance alone, the combined adsorption is greater than that of the individuals alone. In industrial application, contact times of less than 1 h are usually used. Equilibrium is probably closely realized when high carbon dosages are employed, since the rate of adsorption increases with carbon dosage.

8.3 Properties of Activated Carbon

Activated carbons are made from a variety of materials including wood, lignin, bituminous coal, lignite, and petroleum residues. Granular carbons produced from medium volatile bituminous coal or lignite have been most widely applied to the treatment of wastewater. Activated carbons have specific properties depending on the material source and the mode of activation. Property standards are helpful in specifying carbons for a specific application. In general, granular carbons from bituminous coal have a small pore size, a large surface area, and the highest bulk density. Lignite carbon has the largest pore size, least surface area, and the lowest bulk density. Adsorptive capacity is the effectiveness of the carbon in removing desired constituents such as COD, color, phenol, etc., from the wastewater. Several tests have been employed to characterize adsorptive capacity. The phenol number is used as an index of a carbon's ability to remove taste and odor compounds. The iodine number relates to the ability of activated carbon to adsorb low-molecular-weight substances (micropores having an effective radius of less than 2 μm), while the molasses number relates to the carbon's ability to adsorb high-molecular-weight substances (pores ranging from 1 to 50 μm). In general, high iodine numbers will be most effective on wastewaters with predominantly low-molecular-weight organics, while high molasses numbers will be most effective for wastewaters with a dominance of high-molecular-weight organics. Properties of commercial carbons are shown in Table 8.4.[3]

Physical properties	NORIT (lignite)	Calgon Filtrasorb 300 (8 × 30) (bituminous)	Westvaco Nuchar WV-L (8 × 30) (bituminous)	Witco 517 (12 × 30) (bituminous)
Surface area, m^2/g (BET)	600–650	950–1050	1000	1050
Apparent density, g/cm^3	0.43	0.48	0.48	0.48
Density, backwashed and drained, lb/ft^3	22	26	26	30
Real density, g/cm^3	2.0	2.1	2.1	2.1
Particle density, g/cm^3	1.4–1.5	1.3–1.4	1.4	0.92
Effective size, mm	0.8–0.9	0.8–0.9	0.85–1.05	0.89
Uniformity coefficient	1.7	1.9 or less	1.8 or less	1.44
Pore volume, cm^3/g	0.95	0.85	0.85	0.60
Mean particle diameter, mm	1.6	1.5–1.7	1.5–1.7	1.2
Specifications				
Sieve size (U.S. standard series)				
Larger than No. 8 (max. %)	8	8	8	*
Larger than No. 12 (max. %)	*	*	*	5
Smaller than No. 30 (max. %)	5	5	5	5
Iodine No.	650	900	950	1000
Abrasion No., minimum	†	70	70	85
Ash, %	†	8	7.5	0.5
Moisture as packed (max. %)	†	2	2	1

*Not applicable to this size carbon.
† No available data from the manufacturer.
Note: $lb/ft^3 = 16\ kg/m^3$.
Source: Adapted from U.S. EPA, 1973.

TABLE 8.4 Properties of Commercially Available Carbons

Laboratory Evaluation of Adsorption

In order to evaluate the feasibility and economics of adsorption, a laboratory adsorption study should be conducted. If granular carbon is to be evaluated, the carbon must first be ground to pass a 325-mesh screen. Grinding the carbon will not significantly affect its adsorptive capacity, but will increase the rate of adsorption. The time of contact required to approach equilibrium should first be evaluated. A carbon dosage of 500 mg/L is mixed with waste for various periods of time, and the degree of adsorption is determined at selected time intervals. A mixing time sufficient to achieve 90 percent or more of equilibrium should be used for subsequent studies. Usually, a 2-h contact is sufficient to attain greater than 90 percent of equilibrium, though in some cases a longer contact time is required. The initial testing should include a 24-h contact time. If the equilibrium value after 2 h is greater than 90 percent of the 24-h value, the 2-h test can be used.

Various dosages of carbon are then mixed with waste for the time interval selected. The carbon is then filtered off and the concentration remaining in solution is measured. These data are then plotted in accordance with the Freundlich isotherm to determine the adsorption characteristics. This is shown in Fig. 8.2. Note that in most wastewaters there will be a nonsorbable residual. A crude estimate of the carbon required can be made from this plot for any required removal.

Depending on the characteristics of the wastewater, one type of carbon may be superior to another (Fig. 8.3) since the capacity is greater at equilibrium effluent concentrations. For TOC, carbon-2 is better for carbon column operation, since the capacity is greater in equilibrium with the influent (C_o) at column exhaustion, while carbon-1 would be better for batch treatment.

Continuous Carbon Filters

Although batch laboratory adsorption studies provide useful information on the application of adsorption to the removal of specific waste constituents, continuous carbon filters provide the most practical application of this process in waste treatment. The reasons for this are:

1. High capacities in equilibrium with the influent concentration rather than the effluent concentration can be approached, as shown in Fig. 8.4.

2. Biological activity in the presence of degradable organics will affect the apparent carbon capacity.

The carbon filter can be considered as a non–steady-state process in which, as an increasing quantity of water is passed through the bed, the adsorbent is removed in an increasing amount. Consider the system shown in Fig. 8.5. As water initially passes through the uppermost layers, rapid adsorption occurs in equilibrium with the effluent concentration. As this water passes through the bed, the equilibrium

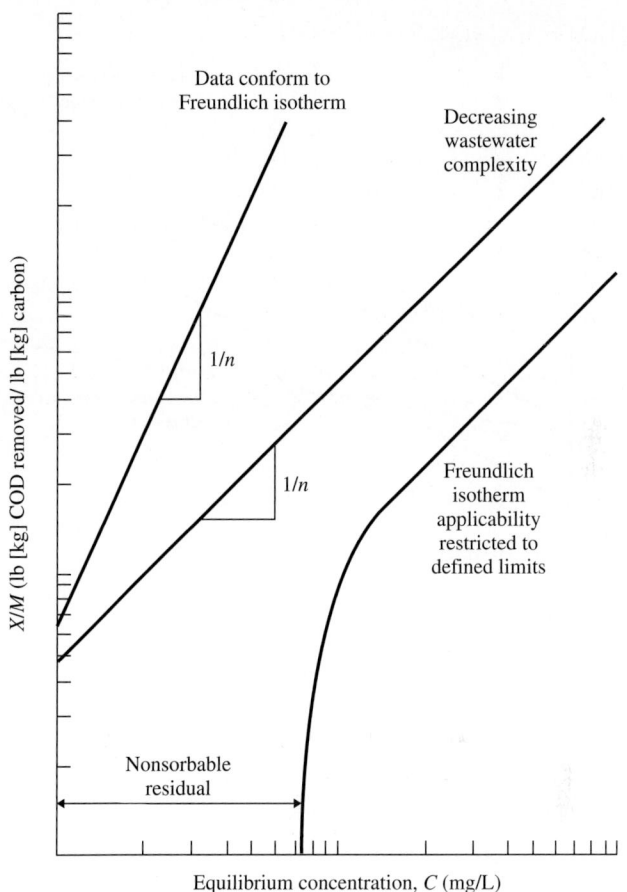

FIGURE 8.2 Freundlich adsorption isotherm.

shifts with the decreasing concentration of remaining solute; this results in a substantially solute-free effluent. With continuing flow of water, the adsorption zone, in equilibrium with the influent concentration, moves downward in the bed. As this zone approaches the bottom of the bed, the concentration of solute in the effluent increases. The *breakpoint* is defined as the volume of water passed through the bed before a maximum effluent concentration is reached. As the adsorption zone falls to the bottom of the bed, the effluent concentration increases until it equals the influent concentration.

The breakthrough point:

1. Decreases with decreased empty bed contact time (EBCT)
2. Decreases with increasing particle size of adsorbent
3. Decreases with increased initial solute concentration

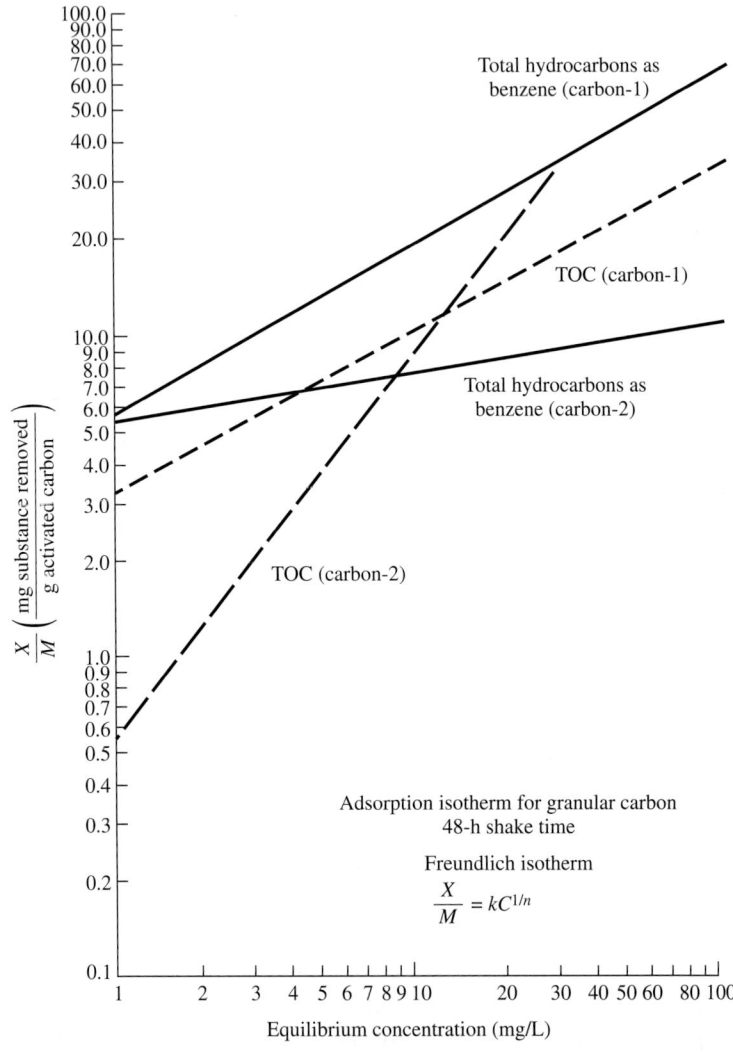

Figure 8.3 Freundlich isotherms for total hydrocarbons and total organic carbon.

A granular carbon pilot plant for the development of breakthrough curves is shown in Fig. 8.5.

Carbon Regeneration

It is generally feasible to regenerate spent carbon for economic reasons. In the regeneration process, the object is to remove from the carbon pore structure the previously adsorbed materials. The modes of regeneration are thermal, steam, or solvent extraction; acid or base treatment; and chemical oxidation. The methods other than thermal

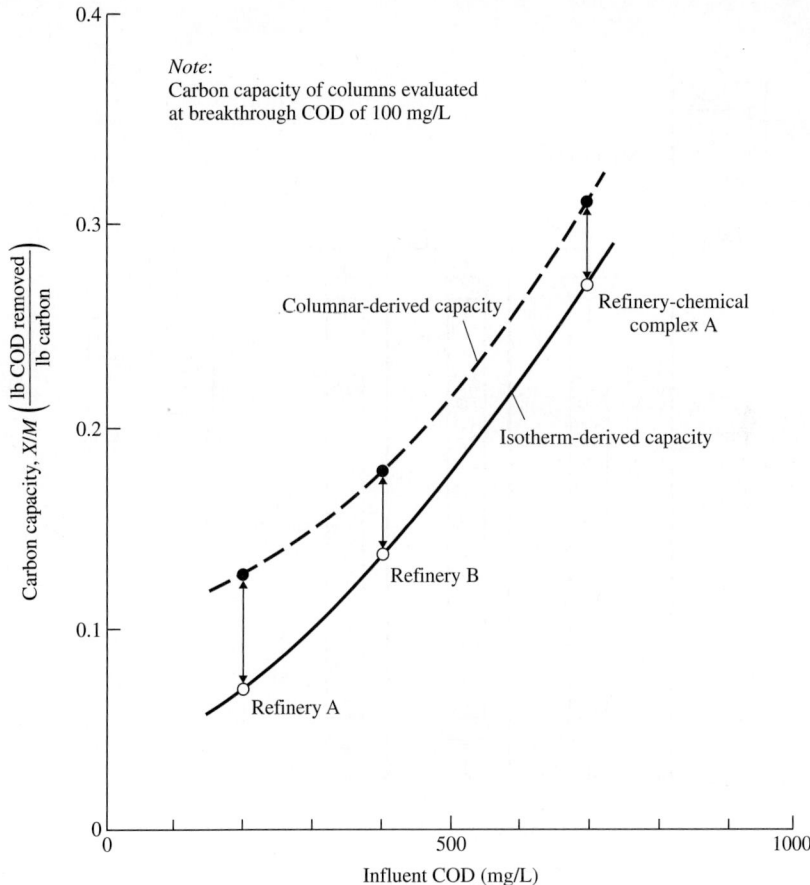

FIGURE 8.4 Carbon capacity from batch and column systems.

are usually to be preferred when applicable, since they can be accomplished in situ. The difficulty arises that adsorption from multicomponent wastewaters usually does not lend itself to high-efficiency regeneration by these methods. An exception is phenol, which can be treated with caustic to convert it to the more soluble phenate and a single chlorinated hydrocarbon, which can be removed with steam. In most wastewater cases, however, thermal regeneration is required. *Thermal regeneration* is the process of drying, desorption, and high-temperature heat treatment (1200 to 1800°F; 650 to 980°C) in the presence of limited quantities of water vapor, flue gas, and oxygen. Multiple-hearth furnaces or fluidized-bed furnaces can be used.

Weight losses of carbon result from attrition and carbon oxidation. Depending on the type of carbon and furnace operation, this usually amounts to 5 to 10 percent by weight of the carbon regenerated.

536 Chapter Eight

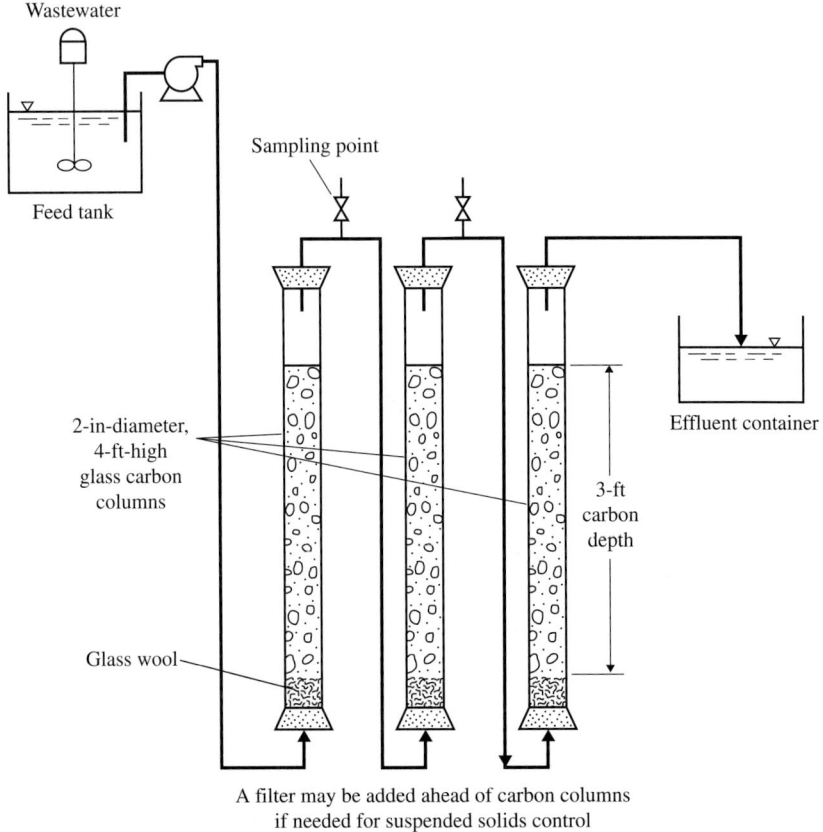

Figure 8.5 Granular activated carbon (GAC) columns schematic (laboratory scale).

There is also a change in carbon capacity through regeneration that may be caused by a change in pore size (usually an increase resulting in a decrease in iodine number) and a loss of pores by deposition of residual materials. In the evaluation of carbons for a wastewater-treatment application, the change in capacity through successive regeneration cycles should be evaluated. In most cases, three to six regeneration cycles will define the maximum capacity loss. This is shown in Fig. 8.6.[4]

Adsorption System Design

In granular carbon columns, the carbon capacity at breakthrough as related to exhaustion is a function of the waste complexity, as shown in Fig. 8.7.[5] A single organic such as dichloroethane will yield a sharp breakthrough curve such that the column is greater than 90 percent

Adsorption 537

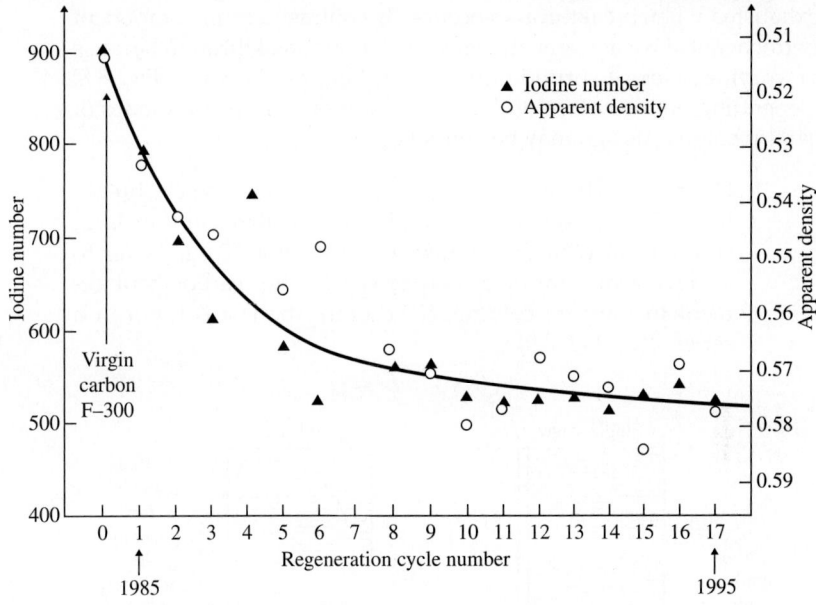

FIGURE 8.6 Regenerated carbon quality trends. (*Adapted from Roll and Crocker, 1996.*)

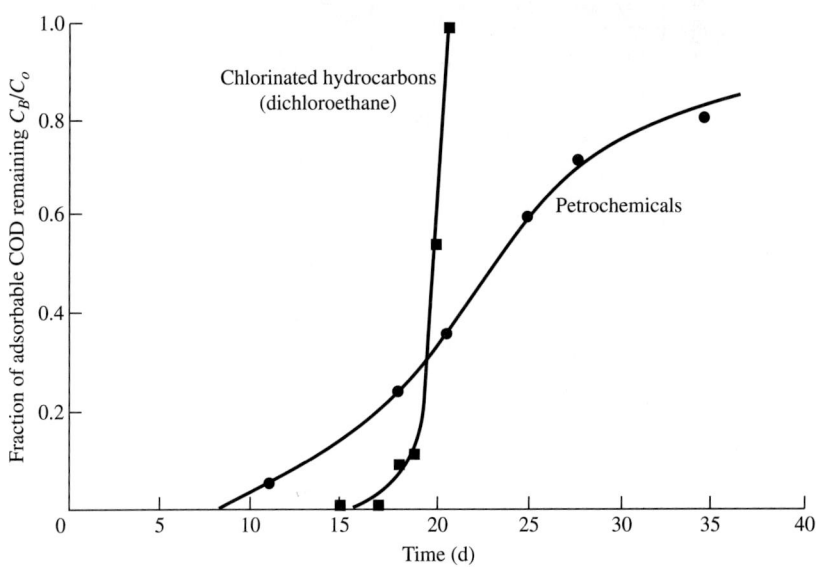

FIGURE 8.7 Continuous carbon column breakthrough curves. (*Adapted from Argamon and Eckenfelder, 1976.*)

exhausted when breakthrough occurs. By contrast, a multicomponent petrochemical wastewater shows a drawn-out breakthrough because of varying rates of sorption and desorption, as shown in Fig. 8.8. Depending on the nature of the wastewater one of several modes of carbon column design may be employed:

1. *Downflow.* These are fixed beds in series. When breakthrough occurs in the last column, the first column is in equilibrium with the influent concentration (C_o) in order to achieve a maximum carbon capacity. After carbon replacement in the first column, it becomes the last column in a series, etc., (Fig. 8.9).

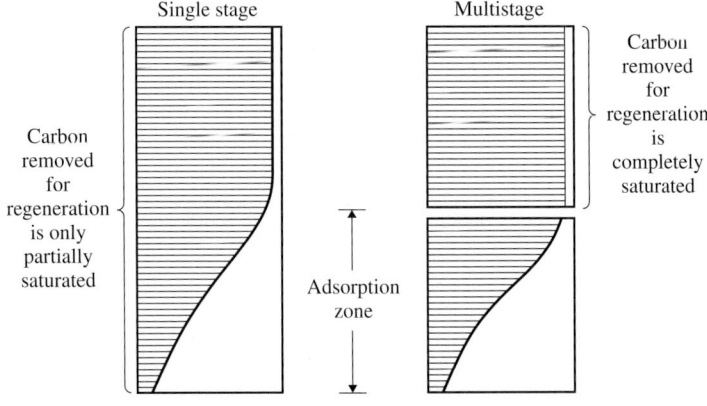

(*a*) Deep adsorption zone

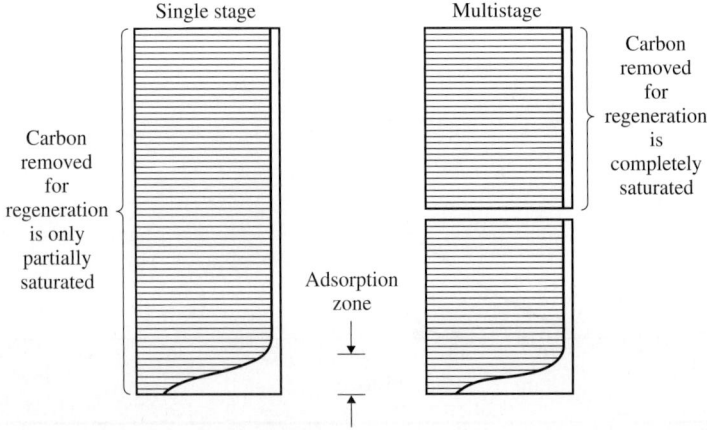

(*b*) Shallow adsorption zone

Figure 8.8 Adsorption zones for single- and multicomponent wastewaters.

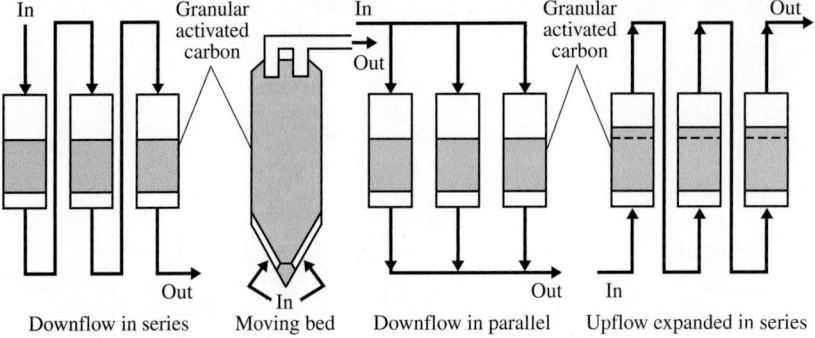

FIGURE 8.9 Types of GAC column design. (*Courtesy of Calgon Carbon Corporation.*)

2. *Multiple units.* These are operated in parallel with the effluent blended to achieve the final desired quality. The effluent from a column ready for regeneration or replacement, which is high in COD, is blended with the other effluents from fresh carbon columns to achieve the desired quality (Fig. 8.9). This mode of operation is most adaptable to waters in which the capacity at breakthrough/capacity at exhaustion ratio is near 1.0, as in the dichloroethane case previously mentioned.

3. *Upflow.* Expanded beds are used when suspended solids are present in the influent or when biological action occurs in the bed (Fig. 8.9).

4. *Continuous counterflow.* These are column or pulsed beds with the spent carbon from the bottom (in equilibrium with influent solute concentration) sent to regeneration. Since this design cannot be backwashed, residual biodegradable organic content in the influent should be very low to avoid plugging. Regenerated and makeup carbon is fed to the top of the reactor (Fig. 8.9). A granular carbon system is shown in Fig. 8.10.

5. *Upflow-downflow.* This concept provides a countercurrent two-bed series system. The two beds are arranged so that the gravity, open-top structures are operated in a series upflow "roughing" contactor and a downflow "polishing" contactor. Once breakthrough occurs, the pair of columns are taken off line, the spent upflow column is regenerated, and the unused capacity of the downflow column is used by reversing the flow and employing it as the upflow reactor, using the former upflow column containing regenerated carbon as the downflow polishing unit.

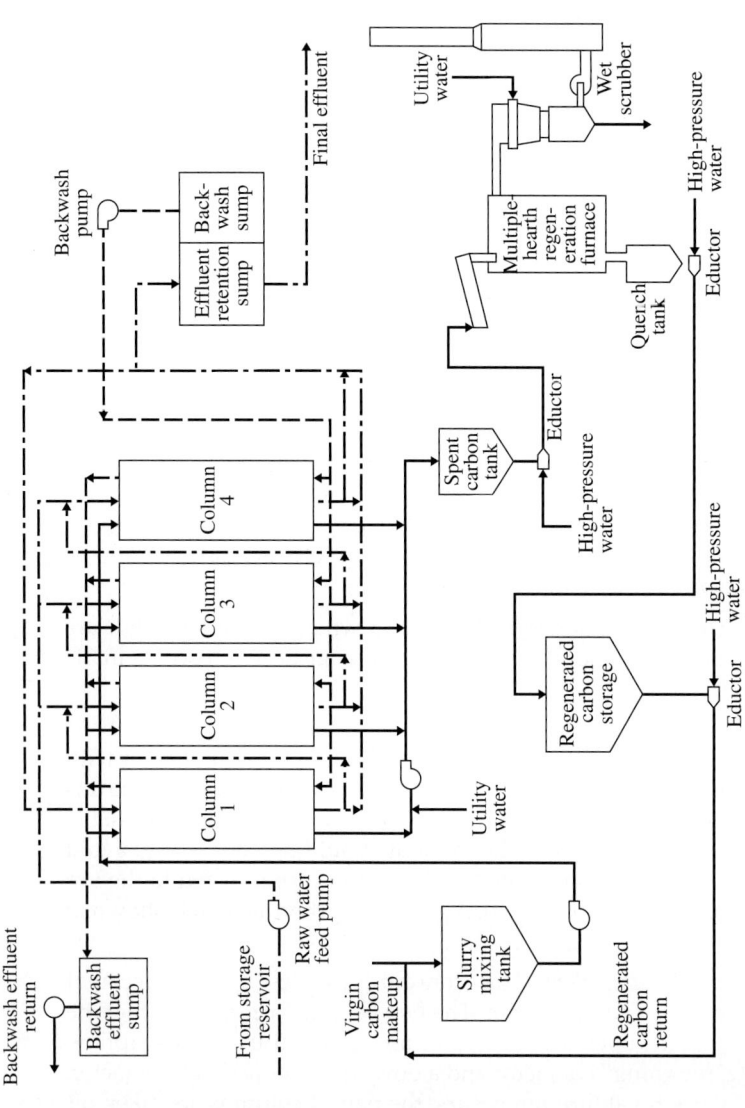

Figure 8.10 GAC process flowsheet.

Bohart and Adams[6] developed a relationship, based on a surface-reaction-rate theory, which can be used to predict the performance of continuous carbon columns:

$$\ln\left(\frac{C_o}{C_B} - 1\right) = \ln(e^{KN_oX/v} - 1) - KC_o t \tag{8.4}$$

Since $e^{KN_oX/v}$ is much greater than unity, Eq. (8.4) can be simplified to

$$t = \frac{N_o}{C_o v}\left[X - \frac{v}{KN_o}\ln\left(\frac{C_o}{C_B} - 1\right)\right] = \frac{N_o}{C_o}\left[\text{EBCT} - \frac{1}{KN_o}\ln\left(\frac{C_o}{C_B} - 1\right)\right] \tag{8.5}$$

where t = service time
    v = linear flow rate
    X = depth of bed
    K = rate constant
    N_o = adsorptive capacity
    C_o = influent concentration
    C_B = allowable effluent concentration
    EBCT = empty bed contact time, X/v

The bed depth that is theoretically just sufficient to prevent penetration of concentration in excess of C_b at zero time is defined as the *critical depth* and determined from Eq. (8.5) when $t = 0$:

$$X_o = \frac{v}{KN_o}\ln\left(\frac{C_o}{C_B} - 1\right) \tag{8.6}$$

The critical EBCT (EBCT_o) is:

$$\text{EBCT}_o = \frac{1}{KN_o}\ln\left(\frac{C_o}{C_B} - 1\right) \tag{8.6a}$$

From Eq. (8.5) it can be shown that the adsorptive capacity N_o can be determined from the slope of a linear plot of t versus X or EBCT. The rate constant K is then calculated from the intercept of this plot:

$$b = -\frac{1}{C_o K}\ln\left(\frac{C_o}{C_B} - 1\right) \tag{8.7}$$

The application of the Bohart-Adams equation to the removal of ABS[4] in continuous carbon columns is shown in Example 8.1.

Chapter Eight

Example 8.1. Experiments in a carbon column 1 in (2.54 cm) in diameter were conducted on a 10-ppm ABS solution.[5] The data obtained are given in Table 8.5.

Compute: (a) the coefficients in the Bohart-Adams equation for this carbon and (b) the carbon required per year to treat 100,000 gal (380 m³) of water per week with an ABS content of 10 mg/L to a residual of 0.5 ppm.

The tower will be 5 ft (1.5 m) in depth and 2 ft (0.6 m) in diameter; compute the adsorption efficiency of the carbon.

Solution

(a)
$$t = \frac{N_o}{C_o v}\left[X - \frac{v}{KN_o}\ln\left(\frac{C_o}{C_B} - 1\right)\right]$$

where $N_o = M\ \text{ABS}/L^3\ C$
$X = L$
$C_o = M/L^3$
$v = L/T$
$K = L^3/MT$
$t = T$

Area of column A_c, (1-in diameter) $= 0.00545\ \text{ft}^2\ (5.06\ \text{cm}^2)$
Calculation of N_o and K

$$C_o = 10\ \text{mg/L} \times 62.4 \times 10^{-6}\ \text{lb/ft}^3 = 0.000624\ \text{lb/ft}^3$$

Flow Rate, gal/(min · ft²)	Bed Depth, ft	Throughput Volume, gal
2.5	2.5	363
	5.0	1216
	7.5	2148
5.0	2.5	141
	5.0	730
	10.0	2190
10.0	5.0	332
	10.0	1380
	15.0	2760

Note: gal/(min · ft²) = 4.07×10^{-2} m³/(min · m²)
ft = 0.3048 m
gal = 3.785×10^{-3} m³

TABLE 8.5 Column Capacity Data for The Adsorption of ABS from a 10-ppm Aqueous Solution (capacity to 0.5 ppm breakpoint)

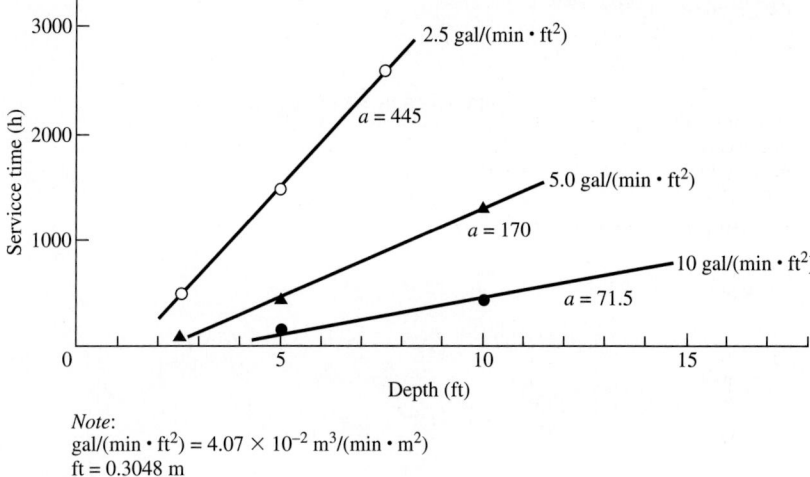

Note:
gal/(min · ft²) = 4.07 × 10⁻² m³/(min · m²)
ft = 0.3048 m

FIGURE 8.11 Correlation of service time vs. column depth

From the equation for t above, N_o is computed from the slope of the t versus X plot (see Fig. 8.11 and Table 8.6). To prepare Table 8.6 the following considerations apply:

$$v(\text{ft}/\text{h}) = q[\text{gal}/(\text{min} \cdot \text{ft}^2)] \times 8.02(\text{ft}/\text{h})/[\text{gal}/(\text{min} \cdot \text{ft}^2)]$$

$$V(\text{ft}^3) = \text{throughput (gal)} \times 0.134 \, (\text{ft}^3/\text{gal})$$

$$t = \frac{V}{vA_c}$$

| Flow | | | Volume | |
q, gal/(min · ft²)	v, ft/h	Depth, ft	Treated V, ft³	t, h
2.5	20	2.5	49	440
		5.0	162	1480
		7.5	268	2620
5.0	40	2.5	19	87
		5.0	98	440
		10.0	290	1340
10	80	5.0	44	102
		10.0	184	420
		15.0	370	835

Note: gal/(min · ft²) = 8.02 ft/h = 4.07 × 10⁻² m³/(min · m²)

TABLE 8.6 Service Time Determination for Various Hydraulic Loading Rates

At 2.5 gal/(min · ft²):

$$N_o = C_o v_a$$
$$= 0.000624 \times 20 \times 445 = 5.55 \text{ lb/ft}^3 \ (89 \text{ kg/m}^3)$$

At 5.0 gal/(min · ft²):

$$N_o = 0.000624 \times 40 \times 170 = 4.24 \text{ lb/ft}^3 \ (67.9 \text{ kg/m}^3)$$

At 10 gal/(min · ft²):

$$N_o = 0.000624 \times 80 \times 71.5 = 3.57 \text{ lb/ft}^3 \ (57.2 \text{ kg/m}^3)$$

$$b = -\frac{1}{C_o K} \ln\left(\frac{C_o}{C_B} - 1\right) \quad \text{and} \quad K = -\frac{1}{C_o b} \ln\left(\frac{C_o}{C_B} - 1\right)$$

From Fig. 8.12 the values of b are taken, and K can be calculated giving the values presented in Table 8.7.

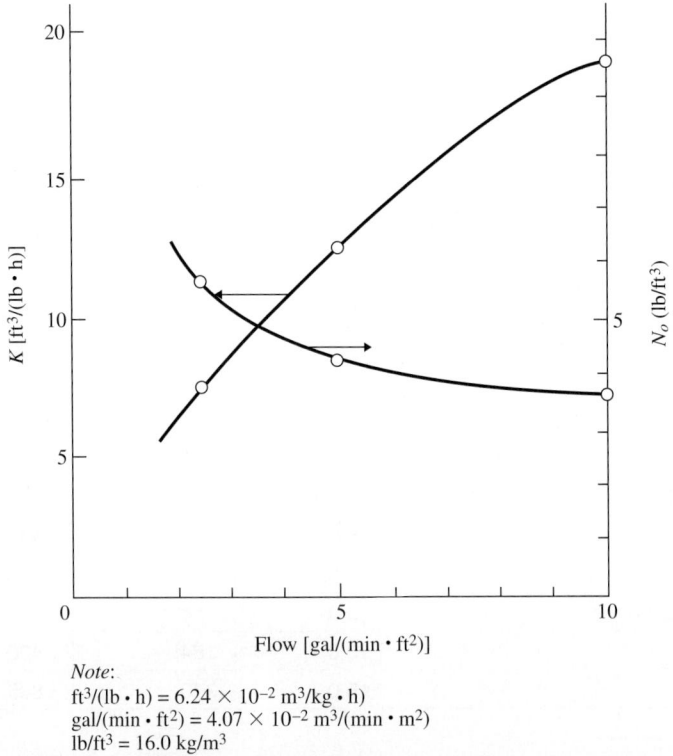

Note:
ft³/(lb · h) = 6.24 × 10⁻² m³/kg · h)
gal/(min · ft²) = 4.07 × 10⁻² m³/(min · m²)
lb/ft³ = 16.0 kg/m³

Figure 8.12 Determination of rate constant K as a function of adsorptive capacity N_o.

Flow, gal/(min · ft²)	b, h	K, ft³/(lb · h)
2.5	−630	7.5
5.0	−370	12.7
10.0	−250	18.8

Note: ft³/(lb · h) = 6.24 × 10⁻² m³/(kg · h)
gal/(min · ft²) = 4.07 × 10⁻² m³/(min · m³)
lb/ft³ = 16.0 kg/m³

TABLE 8.7 Determination of Rate Constant K as a Function of Hydraulic Loading

Calculation of critical layer [Eq. (8.6)]

$$X_o = \frac{v}{KN_o} \ln\left(\frac{C_o}{C_a} - 1\right)$$

$$X_o = \frac{20}{7.5 \times 5.55} \cdot 2.95 = 1.41 \text{ ft } (0.43 \text{ m}) \quad \text{for } 2.5 \text{ gal/(min} \cdot \text{ft}^2)$$

In this manner

$$X_o = 2.19 \text{ ft } (0.67 \text{ m}) \quad \text{for } 5.0 \text{ gal/(min} \cdot \text{ft}^2)$$

$$X_o = 3.51 \text{ ft } (1.07 \text{ m}) \quad \text{for } 10.0 \text{ gal/(min} \cdot \text{ft}^2)$$

(b) Using a tower of 24 in diameter (0.61 m), 5 ft deep (1.53 m), 3.15 ft² (0.29 m²), 15.75 ft³ (0.45 m³),

$$\frac{100{,}000 \text{ gal/week}}{5 \text{ d/week} \times 1440 \text{ min/d}} = 13.9 \text{ gal/min } (0.05 \text{ m}^3/\text{min})$$

or 4.4 gal/(min · ft²)[0.18 m³/(min · m²)]

At 4.4 gal/(min ·ft²), $v = 35$ ft/h, $K = 11.5$ ft³/(lb · h), and $N_o = 4.2$ lb/ft³,

$$t = \frac{N_o}{C_o v}\left[X - \frac{v}{KN_o} \ln\left(\frac{C_o}{C_B} - 1\right)\right]$$

$$= \frac{4.2}{0.000624 \times 35}\left(5 - \frac{35}{11.5 \times 4.2} \cdot 2.95\right)$$

$$= 551 \text{ h}$$

At a flow rate of 834 gal/h the total volume treated before breakthrough is (551)(834) = 460,000 gal. For a yearly volume of 5.2 million gal, 11 carbon charges are required.

Bed efficiency

Total ABS adsorbed = 460,000 × 8.34 × 9.5 = 36.4 lb

Total capacity = 4.2 lb ABS/ft³ C × 15.7 ft³ C = 66 lb

$$\% \text{ efficiency} = \frac{36.4}{66} \times 100 = 55\%$$

Efficiency computed from X_o:
Using Eq. (8.6),

$$X_o = \frac{35}{11.5 \times 4.2} \ln\left(\frac{10}{0.5} - 1\right) = 2.13 \text{ ft}$$

$$\% \text{ efficiency} = \frac{X - X_o}{X} \times 100 = \frac{5 - 2.13}{5} \times 100 = 57\%$$

Hutchins[7] presented a modification of the Bohart-Adams equation which required only three column tests to collect the necessary data. This is called the bed depth service time (BDST) approach. The Bohart-Adams equation can be expressed as

$$t = aX + b$$

$$= a'\text{EBCT} + b \tag{8.8}$$

where a = slope = $N_o/C_o v$
b = intercept = $-1/KC_o[\ln(C_o/C_B - 1)]$
$a' = va$

If a value of a is determined for one flow rate, values for other flow rates can be computed by multiplying the original slope by the ratio of the original and new flow rates. The b value change is insignificant with respect to changing flow rates. No adjustment for flow rate is needed if the EBCT versus t plot is used. Adjustment for changing initial concentration can be made as follows:

$$a_2 = a_1 \frac{Q_1}{Q_2}$$

$$b_2 = b_1 \frac{C_1}{C_2} \frac{\ln(C_2/C_F - 1)}{\ln(C_1/C_B - 1)}$$

in which C_F and C_B are the effluent concentrations at C_2 and C_1, respectively. In order to develop a BDST correlation, a number of pilot columns of equal depth are operated in series and breakthrough curves plotted for each, as shown in Fig. 8.13. These data are then used to

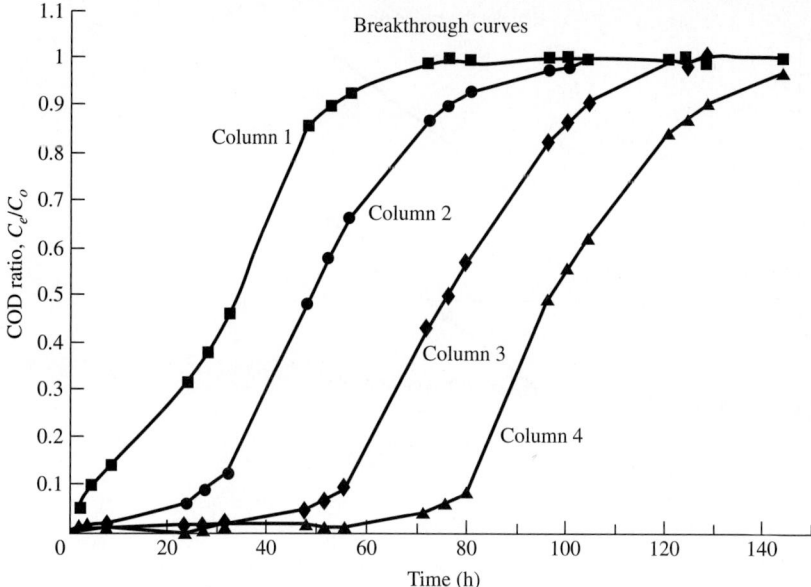

FIGURE 8.13 Column breakthrough curves.

plot a BDST correlation by recording the operating time required to reach a certain removal at each EBCT. A BDST plot of the data from Fig. 8.13 is shown in Fig. 8.14.

The slope of the BDST line is equal to the reciprocal velocity of the adsorption zone and the X intercept is the critical depth defined as the minimum bed depth required to obtain the desired effluent quality at time zero.

If the adsorption zone is arbitrarily defined as the carbon layer through which the liquid concentration varies from 90 to 10 percent of the feed concentration, then this zone is defined by the horizontal distance between these two lines in the BDST plot.

In order to design an adsorption system with maximum carbon utilization, the carbon removed should be near saturation in equilibrium with the influent concentration. In multistage columns, the first run of three columns will be made with all fresh carbon. When the third column breaks through, the first column should be exhausted and a fourth column with fresh carbon is placed at the end of the train. This process is repeated each time the last column breaks through.

The BDST curve should be developed from the breakthrough curves after the third column breaks through. Assuming that 90 percent removal represents exhaustion, the horizontal distance between 90 and the desired breakthrough concentration is taken as the depth of the adsorption zone. This is also the minimum in a pulsed or moving-bed system. For a multistage system, the number of stages

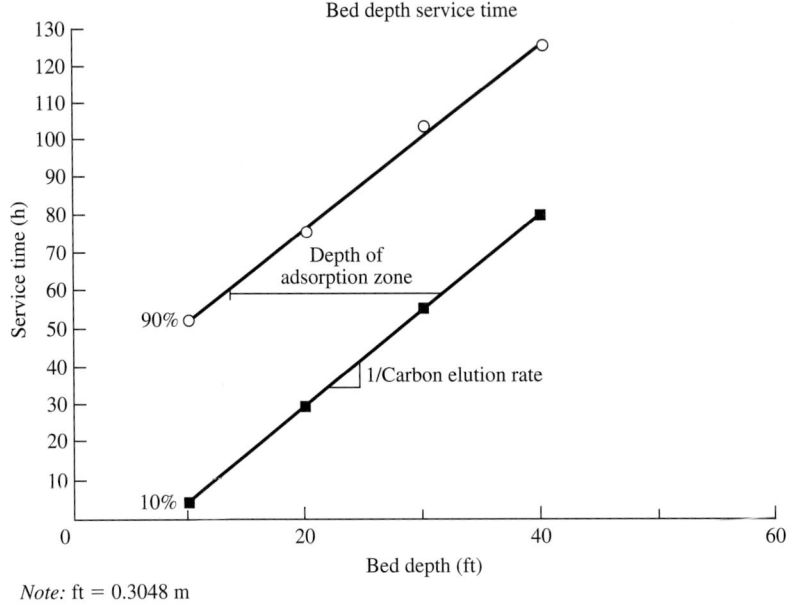

FIGURE 8.14 BDST design curves.

and the bed depth in each stage are related to the depth of the adsorption zone:

$$n = \frac{D}{d} + 1 \qquad (8.9)$$

where n = number of stages in series
D = depth of adsorption zone
d = depth of single stage

Selection of d should be based on practical considerations and should be an integer fraction of D. Selecting a small d will result in small-size equipment with lower carbon inventory, but a high number of stages and consequently more costly equipment. Example 8.2 illustrates a BDST design.

Example 8.2. A petrochemical washwater with a flow of 85,000 gal/d (322 m³/d) has to be treated to an effluent standard of 50 mg/L. A four-column pilot plant was operated with a carbon that had a density of 30 lb/ft³ (481 kg/m³). The columns were 10 ft (3 m) long and loaded at a hydraulic rate of 5 gal/(min · ft²) [0.20 m³/(min · m²)]. The pilot plant is operated in series: the effluent from column 1 is passed to the top of column 2 and sequentially to columns 3 and 4.

Calculate the depth of the adsorption zone, the required number of columns, the time required to exhaust a column, the column diameter, the daily carbon use, and the solution carbon adsorption loading.

Solution The data should be plotted as sequential breakthrough curves (Fig. 8.13) of the bed effluent concentration as a ratio of the influent concentration of the system (630 mg/L).

The breakthrough curves for the four columns are symmetrical. The *adsorption zone* is defined as the carbon layer through which the concentration varies from 10 percent to 90 percent of the feed concentration. The breakthrough times are plotted as the service times and are a function of total carbon bed depth (Fig. 8.11).

In this case, the lines are almost parallel and the depth of the adsorption zone is between 18 and 19 ft. The total bed depth will be the adsorption zone plus an additional column:

$$\text{Number of columns} = \frac{19}{10} + 1 = 2.9, \text{ round up to } 3$$

The bed depth service time (Fig. 8.14) data fit well in the Bohart-Adams equation:

$$t = \frac{N_o}{C_o v}(X) - \frac{1}{C_o K} \ln\left(\frac{C_o}{C_e} - 1\right)$$

At 10 percent,

$$t(h) = 2.57(X) - 21.5$$

At 90 percent,

$$t(h) = 2.50(X) + 27.0$$

$$\text{Adsorption velocity} = \frac{1}{\text{slope}} = \frac{1}{2.57 \text{ h/ft}}$$

$$= 0.39 \text{ ft/h or } 9.34 \text{ ft/d} \quad (2.85 \text{ m/d})$$

The rest of the design and operation can be calculated from the above determinations:

$$\text{Time to exhaust a column} = \frac{10 \text{ ft}}{9.34 \text{ ft/d}} \times 24 \text{ h/d} = 26 \text{ h}$$

$$\text{Area required} = \frac{Q}{A} = \frac{85,000 \text{ gal/d}}{5 \text{ gal/(min} \cdot \text{ft}^2)} \times \frac{d}{1440 \text{ min}} = 11.8 \text{ ft}^2 \text{ (1.1 m}^2\text{)}$$

Diameter = 3.88 ft (1.2 m)

Carbon use = 11.8 ft² × 9.34 ft/d × 39 lb/ft³ = 3300 lb/d (1500 kg/d)

$$\text{Carbon adsorption loading} = \frac{(630-50) \text{ mg/L} \times Q}{\text{carbon use}}$$

$$= \frac{580 \text{ mg/L} \times 85,000 \text{ gal/d}}{3300 \text{ lb/d}} \times \frac{8.34 \text{ lb/gal}}{10^6 \text{ mg/L}}$$

$$= 0.125 \text{ lb COD/lb carbon}$$

GAC Small Column Tests

The two dominant factors that control the breakthrough in GAC columns are the adsorption capacity and adsorption kinetics. As previously discussed, pilot columns utilize the same reliable predictors of breakthrough behavior in full-scale columns in terms of both capacity and rate of adsorption. However, this approach may require time-consuming and expensive studies. Rapid methods to design GAC columns from small columns have been developed to reduce the study time and cost. Examples of methods using small columns are the short fixed bed, the minicolumns, the high-pressure minicolumns, the dynamic minicolumn adsorption technique, the accelerated column tests, the small-scale columns, and the rapid small-scale column test (RSSCT). The use of the RSSCT, which does not require the use of complicated models, will be illustrated.

The RSSCT method, which was developed by Frick[8] and improved and applied by Crittenden and coworkers,[9,10] is a scaled-down version of a pilot- or full-scale GAC column. If the RSSCT and the GAC column use carbons with the same bulk density and capacity, similitude is maintained through the use of dimensional analysis for the relationship between particle size, column length or empty bed contact time (EBCT), and operation time. These relationships may be summarized by Eqs. (8.10) and (8.11).

$$\frac{\text{EBCT}_S}{\text{EBCT}_L} = \frac{t_S}{t_L} \qquad (8.10)$$

where EBCT_S = EBCT of small-particle column
EBCT_L = EBCT of large-particle column
t_S = operation time of small-particle column
t_L = operation time of large-particle column

Equal Reynolds numbers (Re) for the RSSCT and the full-scale GAC column assure hydraulic similarity with equal length of the mass transfer zone (MTZ) to length of the column ratio. This equality in the Re numbers is expressed by the following equation:

$$\text{Re}_S = \frac{d_S}{\nu} \frac{v_S}{\varepsilon_S} = \frac{d_L}{\nu} \frac{v_L}{\varepsilon_L} \qquad (8.11)$$

This is illustrated in Example 8.3.

Example 8.3. Design a 100 gal/min (379 L/min) full-scale GAC column using the following information from a small-column study:
$d_S = 0.06$ cm (>0.0077 cm)
$\varepsilon_S = 0.40$ (0.36 to 0.48)
$\rho_b = 0.42$ g/mL (0.32 to 0.42 g/mL)

$ID_S = 1.5$ cm ($ID_S/d_S > 25$–50)
$L_S = 5$ cm
$Re_S = 1$
$t_{bS} = 2$ d (time to breakthrough or $C = C_{to}$)
$t_{eS} = 3.5$ d (time to exhaustion or $C = 0.95\ C_o$)
C_o = influent generation
C_{to} = treatment objective
C = effluent concentration
$T_w = 20°C$

Solution

1. Small column

    a. Dimensions:

       Surface Area

       $$A_S = \frac{\pi ID^2}{4} = \frac{\pi (1.5)^2}{4} = 1.77 \text{ cm}^2$$

       Volume of carbon bed

       $$V_{bS} = A_S \cdot L_S = 1.77 \times 5 = 8.84 \text{ cm}^3$$

    b. Mass of carbon in the column

       $$M_{cS} = V_{bS} \cdot \rho b = 8.84 \times 0.42 = 3.71 \text{ g C}$$

    c. Hydraulic loading rate (approaching velocity) using $Re_S = 1$

       $$V_S = \frac{Re_S\,\varepsilon}{d_S} = \frac{1 \times 0.602 \text{ cm}^2/\text{min} \times 0.4}{0.06 \text{ cm}} = 4.01 \frac{\text{cm}}{\text{min}} \quad \left(\frac{0.98 \text{ gal/min}}{\text{ft}^2}\right)$$

    d. Flow rate

       $$Q_S = V_S A_S = 4.01 \times 1.77 = 7.10 \text{ cm}^3/\text{min}$$

    e. $\text{EBCT}_S = \dfrac{V_{bS}}{Q_S} = \dfrac{8.84}{7.10} = 1.25$ min

    f. Length of mass transfer zone

       $$\text{MTZ}_S = L_S \left[\frac{t_{eS} - t_{bS}}{t_{eS}}\right] = 5 \left[\frac{3.5 - 2}{3.5}\right] = 2.1 \text{ cm } (42\% \text{ of } L_S)$$

2. GAC adsorber

    a. Select GAC

       $d_L = 0.1$ cm

       $\varepsilon_L = 0.4$

       $\rho_b = 0.42$ g/cm^3 = 420 kg/m^3

b. Dimensions for an $EBCT_L$ of 15 min and an HLR_L = 4 gal/(min · ft²) = 16.3 cm/min = V_L

$$V_{bL} = Q_L \times EBCT_L = 100 \times 15 = 1500 \text{ gal} = 5.68 \text{ m}^3$$

$$A_L = \frac{Q_L}{HLR} = \frac{100}{4} = 25 \text{ ft}^2 = 2.33 \text{ m}^2$$

$$ID_L = \left(\frac{4A_L}{\pi}\right)^{0.5} = \left(\frac{4 \times 2.33 \text{ m}^2}{\pi}\right)^{0.5} = 1.72 \text{ m} \quad (5.65 \text{ ft})$$

$$L_L = \frac{V_{bL}}{A_L} = \frac{5.68}{2.33} = 2.44 \text{ m} \quad (8.0 \text{ ft})$$

c. Time to breakthrough

$$t_{bL} = t_{bS} \frac{EBCT_L}{EBCT_S} = 2 \times \frac{15}{1.25} = 24 \text{ d}$$

d. Mass of carbon in the column

$$M_{cL} = V_{bL} \times \rho_b = 5.68 \text{ m}^3 \times \frac{420 \text{ kg}}{\text{m}^3} = 2390 \text{ kg} \quad (5260 \text{ lb})$$

e. Estimate bed volumes of wastewater treated to breakthrough:

$$BV_b = \frac{t_{bL}}{EBCT_L} = \frac{24}{15} \times 1440 = 2300$$

3. Estimate the volume of wastewater treated per unit mass of carbon (specific volume):

$$V_{sp} = \frac{Q_L \, t_{bL}}{M_{cL}} = \frac{0.379 \times 24}{2390} \times 1440 = \frac{5.5 \text{ m}^3}{\text{kg}} \quad \left(\frac{660 \text{ gal}}{\text{lb}}\right)$$

4. Estimate the carbon utilization rate:

$$CU_r = \frac{1}{V_{sp}} = 0.000182 \, \frac{\text{kg}}{1} = 182 \, \frac{\text{mg}}{1} \quad \left(\frac{0.00152 \text{ lb}}{\text{gal}} = \frac{1.52 \text{ lb}}{1000 \text{ gal}}\right)$$

5. Estimate the length of the MTZ:

$$MTZ_L = L_L \times 0.42 = 2.44 \times 0.42 = 1.02 \text{ m} \quad (3.4 \text{ ft})$$

6. Determine configuration of GAC system:

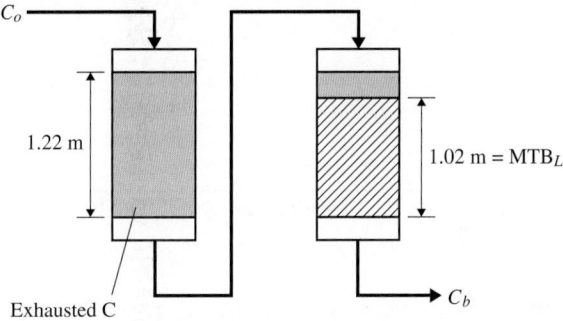

Using a safety factor of 1.5, use two columns in series with:

$$ID_L = 1.72 \text{ m}$$

$$L_L = 1.83 \text{ m}$$

$$V_{bL} = \frac{\pi \, 1.72^2}{4} \times 1.83 = 4.25 \text{ m}^3/\text{column}$$

$$EBCT_L = 2 \times \frac{4.25}{0.379} = 22 \text{ min} \quad (11 \text{ min per column})$$

$$t_b = 2 \times \frac{22}{1.25} = 35 \text{ d}$$

Performance of Activated Carbon Systems

Activated carbon columns are employed for the treatment of toxic or nonbiodegradable wastewaters and for tertiary treatment following biological oxidation.

When degradable organics (BOD) are present in the wastewater, biological action provides biological regeneration of the carbon, thus increasing the apparent capacity of the carbon. Biological activity may be an asset or a liability. When the applied BOD is in excess of 50 mg/L, anaerobic activity in the columns may cause serious odor problems while aerobic activity may cause plugging due to the biomass generation by aerobic activity.

Most heavy metals are removed through carbon columns as shown in Table 8.8 for a petroleum refinery wastewater. In order to avoid reduced capacity after regeneration, the carbon should be acid-washed prior to reuse. The effectiveness of activated carbon in the treatment of various industrial wastewaters is shown in Table 8.9.

Parameter	API Separator, mg/L	Carbon-Treated, mg/L
Chromium	2.2	0.2
Copper	0.5	0.03
Iron	2.2	0.3
Lead	0.2	0.2
Zinc	0.7	0.08

TABLE 8.8 Heavy Metals Removal on Activated Carbon from a Petroleum Refinery Wastewater

High-molecular-weight compounds such as SMP are strongly adsorbed on carbon and will replace weakly adsorbed compounds, as shown in Fig. 8.15.

8.4 The PACT Process

Powdered activated carbon (PAC) can be added to the activated sludge process for enhanced performance (the PACT process). The flowsheet for this process is shown in Fig. 8.16. The addition of PAC has several process advantages, namely, decreased variability in effluent quality and removal by adsorption of nondegradable organics (principally color), reduction of inhibition in industrial wastewater treatment, and removal of refractory priority pollutants. PAC can be integrated into existing biological treatment facilities at minimum capital cost. Since the addition of PAC enhances sludge settleability, conventional secondary clarifiers will usually be adequate, even with high carbon dosages. In some industrial waste applications, nitrification is inhibited by the presence of toxic organics. The application of PAC has been shown to reduce or eliminate this inhibition. Batch isotherm screening tests are used on the biological effluent in order to select the optimal carbon. Bench-scale continuous reactors, as shown in Fig. 6.70, can be used to develop process design criteria; several reactors are run in parallel: a control with no PAC and several with varying dosages of PAC.

The PAC dosage and the PAC mixed liquor solids concentration are related to the sludge age:

$$X_p = \frac{X_i \theta_c}{t} \qquad (8.12)$$

where X_p = equilibrium PAC MLSS content
X_i = PAC dosage
t = hydraulic retention time

Type of Industry	Initial TOC (or phenol), mg/L	Initial Color OD	Average Reduction, %	Carbon Exhaustion Rate, lb/1000 gal
Food and kindred products	25–5300	—	90	0.8–345
Tobacco manufacturers	1030	—	97	58
Textile mill products	9–4670	—	93	1–246
	—	0.1–5.4	97	0.1–83
Apparel and allied products	390–875	—	75	12–43
Paper and allied products	100–3500	—	90	3.2–156
	—	1.4	94	3.7
Printing, publishing, and allied industries	34–170	—	98	4.3–4.6
Chemicals and allied products	19–75,500	—	85	0.7–2905
	(0.1–5325)	—	99	1.7–185
	—	0.7–275	98	1.2–1328
Petroleum refining and related industries	36–4400	—	92	1.1–141
	(7–270)	—	99	6–24
Rubber and miscellaneous plastic products	120–8375	—	95	5.2–164
Leather and leather products	115–9000	—	95	3–315
Stone, clay, and glass products	12–8300	—	87	2.8–300
Primary metal industries	11–23,000	—	90	0.5–1857
Fabricated metal products	73,000	—	25	606

Note: lb/1000 gal = 0.120 kg/m^3

TABLE 8.9 Results from Adsorption Isotherms on Various Industrial Wastewaters

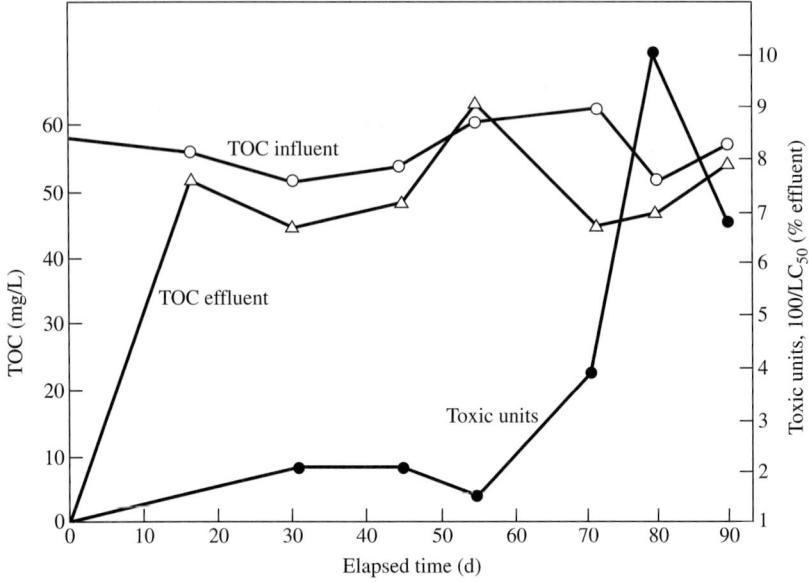

FIGURE 8.15 TOC and toxicity reduction by granular carbon columns.

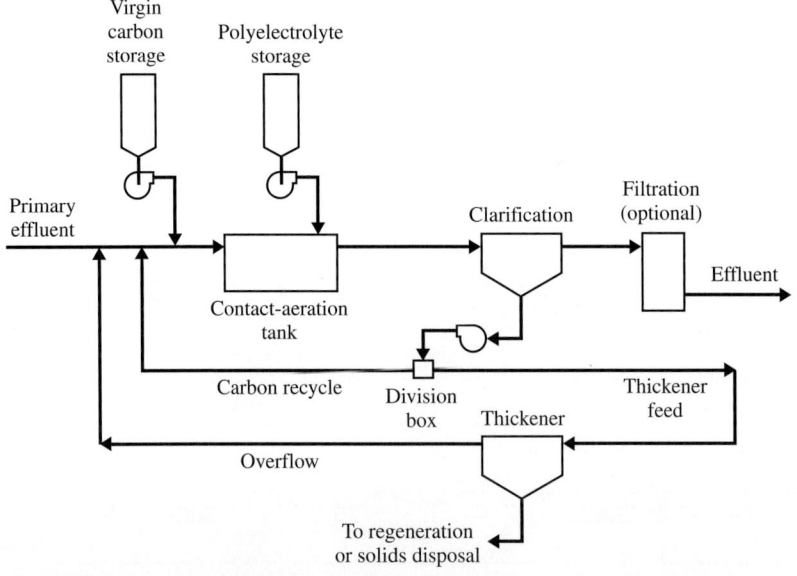

FIGURE 8.16 PACT wastewater-treatment system general process diagram.

The sludge age affects the PAC efficiency with higher sludge ages enhancing the organic removal per unit of carbon; affects the molecular configuration of the adsorbate, depending on varying biological uptake patterns and end products; and establishes the equilibrium biological solids level in the aeration basin. There is some evidence that the attached biomass degrades some of the low-molecular-weight compounds that are adsorbed, as demonstrated by superior TOC removal rates for PAC when added to an aeration basin as opposed to isotherm predictions of adsorption capacity. Figure 8.17 was developed by evaluating the difference in TOC removal for biological units operated in parallel with and without PAC. As can be seen, the performance of the carbon in the bioreactor was significantly greater than predicted by the isotherm only. The mechanisms felt to be responsible for this phenomena include:

1. Additional biodegradation of organics due to decreased biological toxicity or inhibition via activated carbon.
2. Degradation of normally nondegradable substances due to increased exposure time to the biomass through adsorption on the carbon. The carbon with adsorbed material remains in the system for one sludge age, typically 10 to 30 d, while without

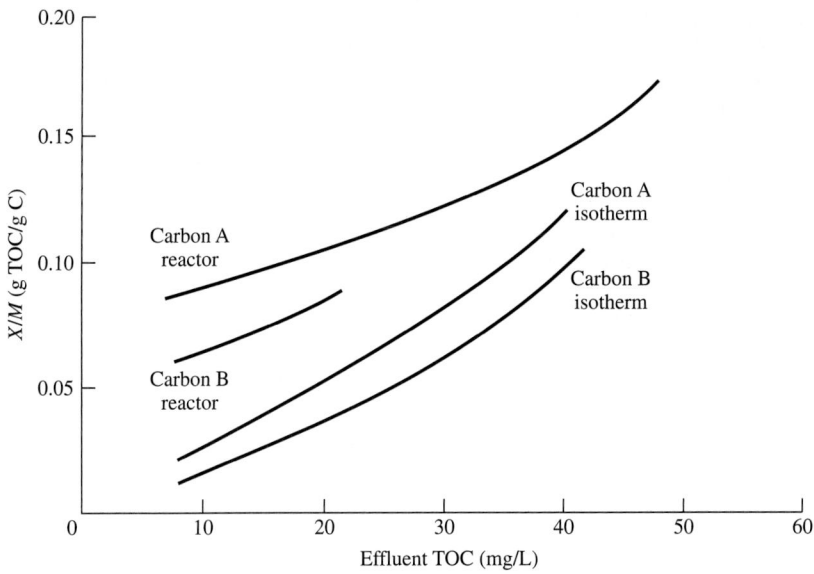

Figure 8.17 Performance relationship of PACT reactors with isotherm data.

PAC feed, mg/L	SRT, d	TOC, mg/L	TKN, mg/L	NH_3^-N, mg/L	NO_2^-N, mg/L	NO_3^-N, mg/L
0	40	31	72	68	4.0	0
33	30	20	6.3	1	4.0	9.0
50	40	26	6.4	1	1.0	13.0

Influent conditions: TOC = 535 mg/L, TKN = 155 mg/L, NH_3^-N = 80 mg/L; pH = 7.5.

TABLE 8.10 Effect of PAC on Nitrification of Coke Plant Wastewaters

carbon the substances would remain in the system for only one hydraulic retention time, typically 6 to 36 h.

3. Substitution/adsorption phenomena, replacement of low-molecular-weight compounds with high-molecular-weight compounds, resulting in improved adsorption efficiency and lower toxicity.

Elimination of nitrification inhibition by the addition of PAC for a coke plant wastewater is shown in Table 8.10.

When there is a small or intermittent application of PAC, the carbon is disposed of with the excess sludge. Continuous application at larger plants, however, requires regeneration of the carbon. This can be accomplished by the use of wet air oxidation (WAO).

In the WAO process, the biological carbon-sludge mixture is treated in a reactor at 450°C and 750 lb/in^2 (51 atm) for 1 h in the presence of oxygen. The biological sludge is oxidized and solubilized under these conditions and the carbon regenerated. The exothermic reaction will provide energy for the reaction, provided that the influent solids content exceeds 10 percent. The decant liquor from the reactor will contain 5000 mg/L BOD, which is recycled back to the aeration basins. In some cases, there is an ash buildup which must be removed from the system. Depending on the characteristics of the wastewater and the type of carbon used, there may be significant losses in carbon capacity through regeneration. This phenomenon should be evaluated by a pilot plant study for any specific application.

Carbon dosages may vary from 20 to 200 mg/L, depending on the results desired. Since the carbon is abrasive, equipment selection should consider this fact. The effect of PAC dosage on the removal of chlorinated benzenes is shown in Fig. 8.18.[11] Performance of the PACT process for an organic chemicals wastewater is shown in Table 8.11. Activated sludge process performance with and without PAC is summarized in Table 8.12.

Adsorption

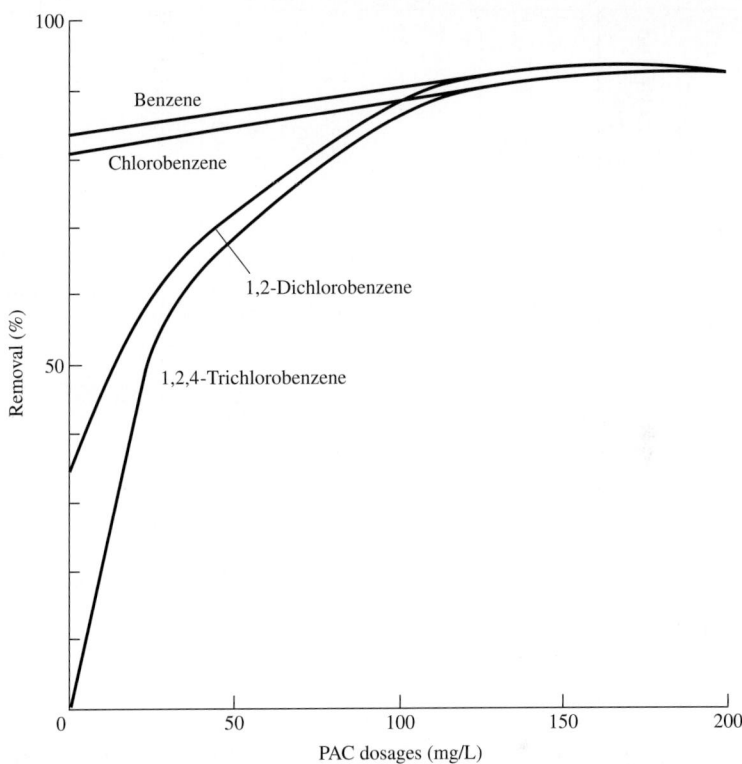

FIGURE 8.18 Effect of PAC dosage on chlorinated benzenes. (*Adapted from Weber and Jones, 1983.*)

	Wastewater Composition, mg/L							Bioassay* LC_{50}
	BOD	TOC	TSS	Color	Cu	Cr	Ni	
Influent	320	245	70	5365	0.41	0.09	0.52	
Biotreatment	3	81	50	3830	0.36	0.06	0.35	11
+ 50 mg/L PAC	4	68	41	2900	0.30	0.05	0.31	25
+ 100 mg/L PAC	3	53	36	1650	0.18	0.04	0.27	33
+ 250 mg/L PAC	2	29	34	323	0.07	0.02	0.24	>75
+ 500 mg/L PAC	2	17	40	125	0.04	<0.02	0.23	>87

*Percentage of wastewater in which 50 percent of aquatic organisms survive for 48 h.

TABLE 8.11 Larger Doses of Powdered Activated Carbon Result in Greater Removal of Organic Carbon, Color, and Heavy Metals

	Industrial					
	Organic Chemicals[13]		Textile Finishing[13]		Berndt-Polkowski[12]	
Operating Conditions	PACT*	Activated Sludge	PACT	Activated Sludge	PACT*	Air Activated Sludge
Aeration, d time	6	6	2.4	2.4	3.2–4.5	†
SRT, time	25	45*				
Temperature, °C	25	25			13	16
Performance results					20	20
Influent characteristics, mg/L						
BOD_5	4,035	4,035	660	Unknown		
TOC	2,965	2,965			134	128
COD	10,230	10,230	1362	1590	364	320
TKN	120	120	Unknown	106	39.4	32.0
$NH_3\text{-}N$	76	76	74	31	19.5	19.8
Chlorinated hydrocarbons	5–67	5				
Phenol	8.1	8.1	5	26		
Effluent characteristics, mg/L						
BOD_5	11	17			1.2	24
TOC	25	65			49.8	63

COD	102	296	116	270	
TKN	4	—	6	29	
NH$_3$-N	0.8	—	3.6	15	
Chlorinated hydrocarbons	0.1	0.9			
Phenol	0.01	0.22	0.44	1.6	
Color, APHA	94	820	240	600	
Detergents			0.8	11.4	

*Includes wet oxidation carbon regeneration.
†Contact stabilization facility; aeration time = 4.2 h, stabilization time = 6.2 h.

TABLE 8.12 Suspended Growth Systems Comparison (with and without PAC)

8.5 Problems

8.1. Given the following data for the adsorption of napthlene on Filtersorb 400 carbon:

Carbon dose, mg/L	Effluent C_f, mg/L
0	9.94
11.2	5.3
22.3	3.0
56.1	0.71
168.3	0.17
224.4	0.06

(a) Develop the adsorption isotherm and the Freundlich parameters.

(b) What is the adsorption capacity at an initial concentration of 10 mg/L?

(c) What is the carbon dose required to reduce the concentration from 1 to 0.01 mg/L?

8.2. The breakthrough curves for a pilot plant operation are shown in Fig. P8.2. The pilot columns were 3.5 ft (1.07 m) long. Design a carbon column system to a breakthrough TOC concentration of 20 mg/L. The influent wastewater has a flow of 3100 gal/min (11.7 m³/min) and an influent TOC of 60 mg/L. Use a hydraulic loading of 4 gal/ (min · ft²) [0.163 m³/(min · m²)] and a carbon density of 28 lb/ft³ (449 kg/m³).

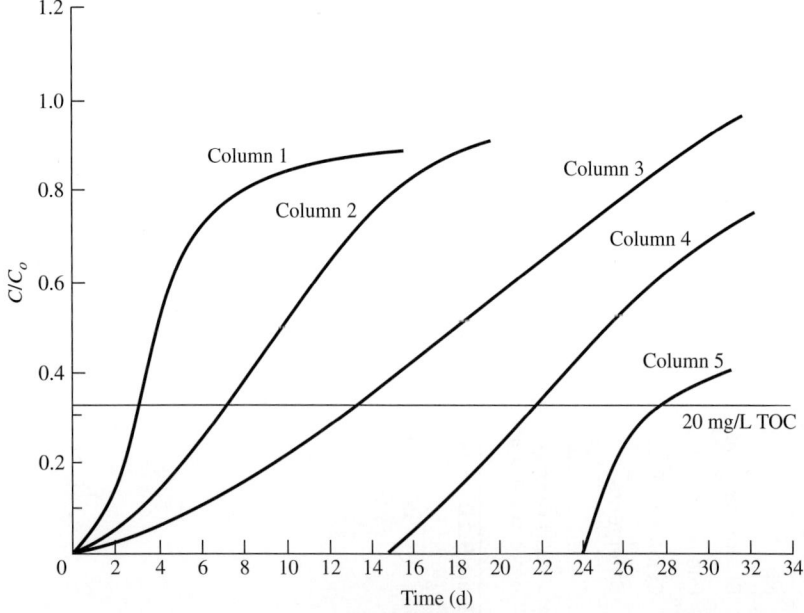

FIGURE P8.2 Breakthrough curves for GAC columns.

References

1. Morris, J. C., and W. J. Weber: "Adsorption of Biochemically Resistant Materials from Solution," Environmental Health Series AWTR-9, May 1964.
2. EPA: "Carbon Adsorption Isotherms for Toxic Organics," EPA-600/8-80-023, April 1980.
3. U.S. EPA: *Process Design Manual for Carbon Adsorption,* Technology Transfer, 1973.
4. Roll, R. R., and Crocker, D. N.: *Proc. WEF,* vol. 1, Dallas, 1996.
5. Argamon, Y., and W. W. Eckenfelder: *Water 1975 Symposium,* Series 72, p. 151, Association of Industrial Chemical Engineers, New York, 1976.
6. Dale, J., J. Malcolm, and I. M. Klotz: *Ind. Eng. Chem.,* vol. 38, pt. 1, p. 289, 1946.
7. Hutchins, R. A.: *Chem. Engineering,* vol. 80, pt. 19, p. 133, 1973.
8. Calgon Carbon Corp., Pittsburgh.
9. Crittenden, J. et al.: *J. Env. Eng. ASCE,* vol. 113, no. 2, p. 243, 1987.
10. Crittenden, J. et al.: *J. AWWA,* vol. 77, p. 87, 1991.
11. Weber, W. J., and B. E. Jones: EPA NTIS PB86-182425/AS, 1983.
12. Berndt, C., and L. Polkowski: "A Pilot Test of Nitrification with PAC," *50th Ann. Control States WPCA,* 1977.
13. US Filter/Zimpro, Inc. Rothchild, Wisc.

CHAPTER 9
Ion Exchange

9.1 Introduction

Ion exchange can be used for the removal of undesirable anions and cations from a wastewater. In the process, ions of a given species are displaced from an insoluble exchange material, which may be natural or synthetic media by ions of a different species in solution. Typically, cations are exchanged for hydrogen or sodium and anions for hydroxyl ions. The process can be used to remove heavy metals, total dissolved solids and nitrogen (ammonia and nitrate). Once the exchange capacity of the resin is depleted, it must be regenerated prior to reuse. Present-day synthetic media are highly specific for target ions and are very efficient. Recovery of removed constituents offers opportunity for reclamation and reuse. The theory of ion exchange, experimental procedures for determining its applicability and design requirements, and particular application to plating wastewater treatment are presented herein.

9.2 Theory of Ion Exchange

Ion exchange resins consist of an organic or inorganic network structure with attached functional groups. Most ion exchange resins used in wastewater treatment are synthetic resins made by the polymerization of organic compounds into a porous three-dimensional structure. The degree of crosslinking between organic chains determines the internal pore structure, with higher crosslink density giving smaller pore sizes. From a kinetic viewpoint, a low degree of crosslinking would enhance diffusion of ions through larger pores. However, physical strength decreases and swelling in water increases as crosslink density is lowered. The functional ionic groups are usually introduced by reacting the polymeric matrix with a chemical compound containing the desired group. Exchange capacity is determined by the number of functional groups per unit mass of resin.

Ion exchange resins are called *cationic* if they exchange positive ions and *anionic* if they exchange negative ions. Cation exchange resins have acidic functional groups, such as sulfonic, whereas anion exchange resins contain basic functional groups, such as amine. Ion exchange resins are often classified by the nature of the functional group as strong acid, weak acid, strong base, and weak base. The strength of the acidic or basic character depends upon the degree of ionization of the functional groups, as with soluble acids or bases. Thus, a resin with sulfonic acid groups would act as a strong-acid cation exchange resin. The most common strong-acid ion exchange resin is prepared by copolymerizing styrene and divinylbenzene followed by sulfonation of the copolymer. The degree of crosslinking is controlled by the fraction of divinylbenzene in the initial mixture of monomers.

The types of ion exchange resins are:

1. *Strong-acid cation resins.* Strong-acid resins are so named because their chemical behavior is similar to that of a strong acid. The resins are highly ionized in both the acid ($R\text{-}SO_3H$) and salt ($R\text{-}SO_3Na$) form, over the entire pH range.

2. *Weak-acid cation resins.* In a weak-acid resin, the ionizable group is a carboxylic acid (—COOH) as opposed to the sulfonic acid group (SO_3H^-) used in strong-acid resins. These resins behave like weak organic acids that are weakly dissociated.

3. *Strong-base anion resins.* Like strong-acid resins, strong-base resins are highly ionized and can be used over the entire pH range. These resins are used in the hydroxide (OH) form for water deionization.

4. *Weak-base anion resins.* Weak-base resins are like weak-acid resins, in that the degree of ionization is strongly influenced by pH.

5. *Heavy-metal selective chelating resins.* Chelating resins behave like weak-acid cation resins but exhibit a high degree of selectivity for heavy-metal cations. Chelating resins tend to form stable complexes with the heavy metals. In fact, the functional group used in these resins is an EDTA compound. The resin structure in the sodium form is expressed as R-EDTA-Na.

The reactions that occur depend upon chemical equilibria situations in which one ion will selectively replace another on the ionized exchange site. Cation exchange on the sodium cycle can be illustrated by the following reaction:

$$Na_2 \cdot R + Ca^{2+} \rightleftharpoons Ca \cdot R + 2Na^+ \qquad (9.1)$$

where R represents the exchange resin. When all the exchange sites have been substantially replaced with calcium, the resin can be regenerated by passing a concentrated solution of sodium ions through the bed. This reverses the equilibrium and replaces the calcium with sodium.

A 5 to 10 percent brine solution is usually used for regeneration:

$$2Na^+ + Ca \cdot R \rightleftharpoons Na_2 \cdot R + Ca^{2+} \qquad (9.2)$$

Similar reactions occur for cation exchange on the hydrogen cycle:

$$Ca^{2+} + H_2 \cdot R \rightleftharpoons CaR + 2H^+ \qquad (9.3)$$

Regeneration with 2 to 10 percent H_2SO_4 yields

$$Ca \cdot R + 2H^+ \rightleftharpoons H_2 \cdot R + Ca^{2+} \qquad (9.4)$$

Anion exchange similarly replaces anions with hydroxyl ions:

$$SO_4^{2-} + R \cdot (OH)_2 \rightleftharpoons R \cdot SO_4 + 2OH^- \qquad (9.5)$$

Regeneration with 5 to 10 percent sodium hydroxide will renew the exchange sites:

$$R \cdot SO_4 + 2OH^- \rightleftharpoons R \cdot (OH)_2 + SO_4^{2-} \qquad (9.6)$$

In addition to the factors of concentration, the nature of the exchanger, and the exchanging ions, such factors as temperature and the particle size of the exchanger are also of considerable importance to the kinetics of ion exchange. The degree of exchange depends on several factors:

1. The size and valence (charge) of the ions entering into the exchange
2. The concentration of ions in the water or solution
3. The nature (both physical and chemical) of the ion exchange substance
4. The temperature

The following sequence shows the selectivity and ease of exchange of cations (Clifford et al.[1]): $Ra^{2+} > Ba^{2+} > Sr^{2+} > Ca^{2+} > Ni^{2+} > Cu^{2+} > Co^{2+} > Zn^{2+} > Mn^{2+} > UO_2^{2+} > Ag^+ > Cs^+ > K^+ > NH_4^+ > Na^+ > Li^+$. Thus, radium is the most preferred and lithium is the least preferred cation. The following sequence shows the selectivity of exchange on anions: $HCRO_4^- > CrO_4^{2-} > ClO_4^- > SeO_4^{2-} > SO_4^{2-} > NO_3^- > Br^- > HPO_4^-, HA_sO_4^-, SeO_3^{2-} > CO_3^{2-} > CN^- > NO_2^- > Cl^- > H_2PO_4^-, H_2AsO_4^-, HCO_3^- > OH^- > CH_3COO^- > F^-$. The least preferred anion has the shortest retention time and appears first in the effluent, whereas the most preferred anion has the longest retention time and is eluted last.

In certain cases a nonpreferred ion can be converted into a polyvalent complex that is highly preferred by the resin, thereby making ion exchange a feasible decontamination process. For example, the removal of UO_2^{2+} from acid solutions is generally not practical because of competition from the highly preferred polyvalent cations Fe^{3+} and Al^{3+}. However, UO_2^{2+} ion undergoes a stepwise formation of anion complexes with sulfate as follows (Dorfiner[2]):

$$UO_2^{2+} + nSO_4^2 \rightleftharpoons UO_2(SO_4)n^{2-2n} \quad (n = 1, 2, 3)$$

In this case, the resulting two-valent ($n = 2$) and four-valent ($n = 3$) complex anions are highly preferred by strong-base anion exchangers.

The performance and economics of ion exchange are related to the capacity of the resin to exchange ions and to the quantity of regenerant required. Since exchange occurs on an equivalent basis, the capacity of the bed is usually expressed as equivalents per liter of bed volume. In some cases capacity has been expressed as kilograms of $CaCO_3$ per unit of bed volume or as mass of ions per unit volume of bed. In like manner, the quantity of ions to be removed in the wastewater is expressed as equivalents per liter of wastewater to be treated.

In a fixed-bed exchanger, there is a relationship between the operating capacity of the bed and the quantity of regenerant employed. Resin utilization is defined as the ratio of the quantity of ions removed during treatment to the total quantity of ions that could be removed at 100 percent efficiency. The regenerant efficiency is the quantity of ions removed from the resin compared to the quantity of ions present in the volume of regenerant used. The resin utilization will increase as the regenerant efficiency decreases. Typical performance curves for a cation exchange resin are shown in Fig. 9.1. The shape of these curves will vary, depending upon the characteristics of the resin and the concentration of the regenerant used.

Treatment of a wastewater by ion exchange involves a sequence of operating steps. The wastewater is passed through the resin until the available exchange sites are filled and the contaminant appears in the effluent. This process is defined as the *breakthrough*. At this point treatment is stopped and the bed is backwashed to remove dirt and to regrade the resin. The bed is then regenerated. After regeneration, the bed is rinsed with water to wash out residual regenerant. The bed is then ready for another treatment cycle.

The treatment and regeneration cycle is shown in Fig. 9.2. In this figure, area *ABHG* is the quantity of ions in the volume of solution treated before breakthrough. Area *ABC* is the quantity of ions leaking through the column and area *ACHG* is the quantity of ions removed by the exchange resin. The resin utilization is therefore area *ACHG/K*, where *K* is the ultimate capacity of the resin. Area *BCDEF* is the quantity of ions removed from the bed during regeneration. The regeneration efficiency is therefore area *BCDEF/R*, where *R* is equal to the concentration of the regenerant times its volume.

Ion Exchange   569

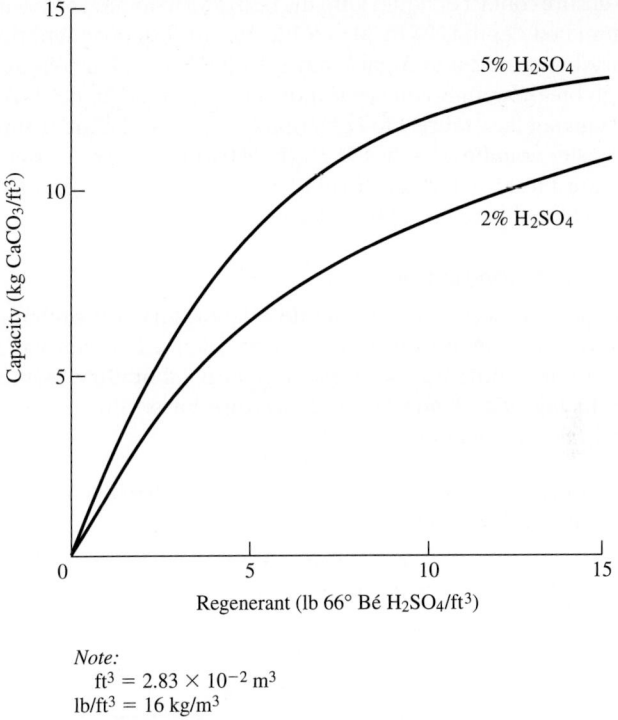

Note:
ft^3 = 2.83 × 10^{-2} m^3
lb/ft^3 = 16 kg/m^3

FIGURE 9.1   Performance of a cation exchange resin.

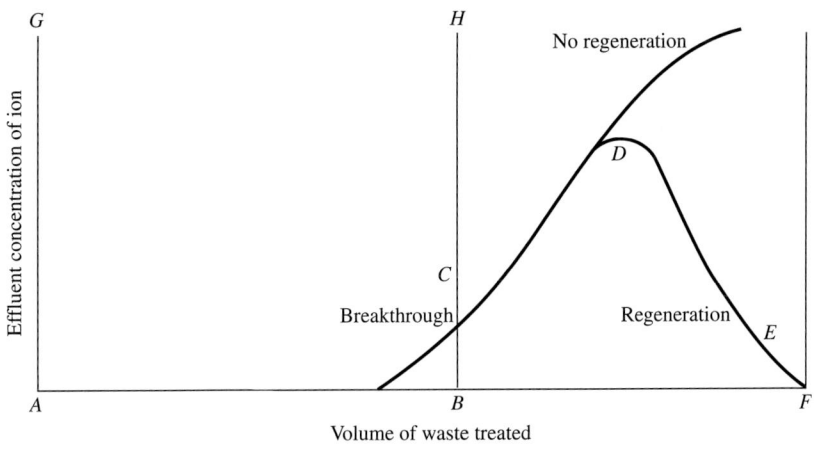

FIGURE 9.2   Treatment and regeneration cycle of an ion exchange resin.

To ensure contact of liquid with the resin and to minimize leakage, the minimum bed depth is 24 to 30 in (61 to 76 cm). The treatment flow rate can vary between 2 and 5 gal/(min · ft³) [0.27 to 0.67 m³/(min · m³)], although breakthrough will occur more quickly at the higher flow rates. The regenerant flow rate is 1 to 2 gal/(min · ft³) [0.13 to 0.27 m³/(min · m³)]. A rinsewater volume of 30 to 100 gal/ft³ (4.0 to 13.4 m³/m³), applied at a flow rate of 1 to 1.5 gal/(min · ft³) [0.13 to 0.20 m³/(min · m³)], will usually be sufficient to flush a bed of residual regenerant.

Experimental Procedure

It is frequently necessary to operate a laboratory ion exchange column to develop the necessary design criteria for the removal of ions from complex industrial wastes. A typical laboratory assembly is shown in Fig. 9.3. A suggested procedure based on the apparatus shown in Fig. 9.3 is detailed as follows:

1. Rinse the column for 10 min with deionized water at a flow rate of 50 mL/min.
2. Adjust the flow rate to the column to 50 mL/min of solution containing the waste to be treated.

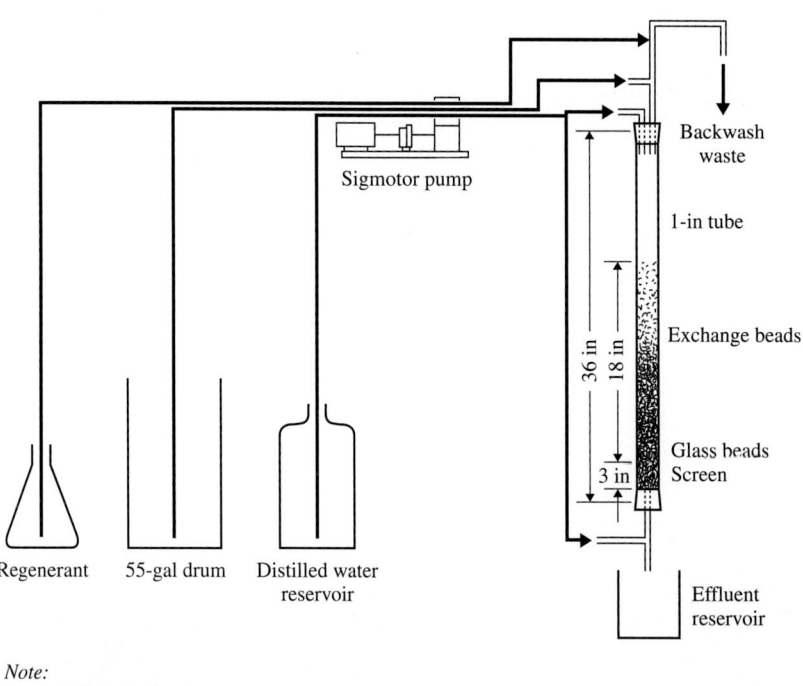

Note:
gal = 3.79 × 10⁻³ m³
in = 2.54 cm

Figure 9.3 Laboratory ion exchange column.

3. Measure the initial volume of solution to be treated.
4. Start the treatment cycle. Develop the breakthrough curve until the ion concentration reaches the maximum effluent limit.
5. Backwash to 25 percent bed expansion for 5 to 10 min. (Use distilled water for the backwash operation.)
6. Regenerate at a flow rate of 6 mL/min, using the concentration and volume recommended for the resin. Collect the spent regenerant and measure the recovered ions.
7. Rinse the column with distilled water.

When several runs are made, it is possible to develop a relationship between resin utilization and regenerant efficiency and to select the optimum operating level for the system.

Macroreticular resins are employed for the removal of specific nonpolar organic compounds. These resins are highly specific and can be formulated to remove one compound or a class of compounds. The resins are solvent-regenerated. Treatment results of selected compounds using macrorecticular resins are shown in Table 9.1.

Compound	Influent, $\mu g/L$	Effluent, $\mu g/L$
Carbon tetrachloride	20,450	490
Hexachloroethane	104	0.1
2-Chloronaphthalene	18	3
Chloroform	1,430	35
Hexachlorobutadiene	266	<0.1
Hexachlorocyclopentadiene	1,127	1.5
Napthalene	529	<3
Tetrachloroethylene	34	0.3
Toluene	2,360	10
Aldrin	84	0.3
Dieldrin	28	0.2
Chlordane	217	<0.1
Endrin	123	1.2
Heptachlor	40	0.8
Heptachlor epoxide	11	<0.1

TABLE 9.1 Macroreticular Resin Treatment of Selected Compounds

Arsenic (v) has been removed with a strong base anion exchange resin.[3] As (v) is present as the divalent anion $HAsO_4^{2-}$ appears to be preferred on strong-base resins over monovalent anions in wastewaters.

Selenium can be removed by ion exchange under the following conditions:

Oxidation of all aqueous to selenate SeO_4^{2-} anion.

Strong-base anion exchange removal of selenate anion. The amount of selenate that can be removed is dependent on sulfate and nitrate concentrations in the water.[3]

Ammonia can be removed by ion exchange using a natural inorganic zeolite clinoptilolite, which has an unusual selectivity for ammonium ions.[3] This unusual selectivity, which makes it attractive for ammonium ions, is caused by structurally related ion sieve properties. While the total exchange capacity of clinoptilolite is somewhat less than that of synthetic organic resins, its selectivity for the ammonium ion compensates. Regeneration is accomplished using a 3 to 6 percent NaCl, which is reused following NH_3 removal from the spent regenerant by air stripping or break-point chlorination.

9.3 Plating Waste Treatment

One of the major applications of ion exchange in industrial waste treatment has been in the plating industry, where chrome recovery and water reuse have often resulted in considerable savings.[4-7]

For recovery of the spent chromic acid in plating baths, the chromic acid is passed through a cation exchange resin to remove other ions (Fe, Cr^{3+}, Al, etc.). The effluent can be returned to the plating bath or to storage. Since the maximum concentration of CrO_3 that can be passed through some resins to avoid deterioration is 14 to 16 oz/gal (105 to 120 kg/m³) as CrO_3, the bath may require dilution and the recovered solution may require makeup to strength.

The rinsewaters are first passed through a cation exchanger to remove metal ions. The effluent from this unit is passed through an anion exchanger to remove chromate and to obtain demineralized makeup water. It is desirable to pass the rinsewater through the cation unit first to avoid precipitation of metal hydroxides on the exchange resin. The anion exchanger is regenerated with sodium hydroxide; this results in a mixture of Na_2CrO_4 and NaOH in the spent regenerant. This mixture is passed through a cation exchanger to recover H_2CrO_4, which is returned to the plating bath. The recovered chromic acid from the spent regenerant will average 4 to 6 percent concentration. The spent regenerant from the cation exchanger will require neutralization and possibly precipitation of metallic ions before it is discharged to the sewer.

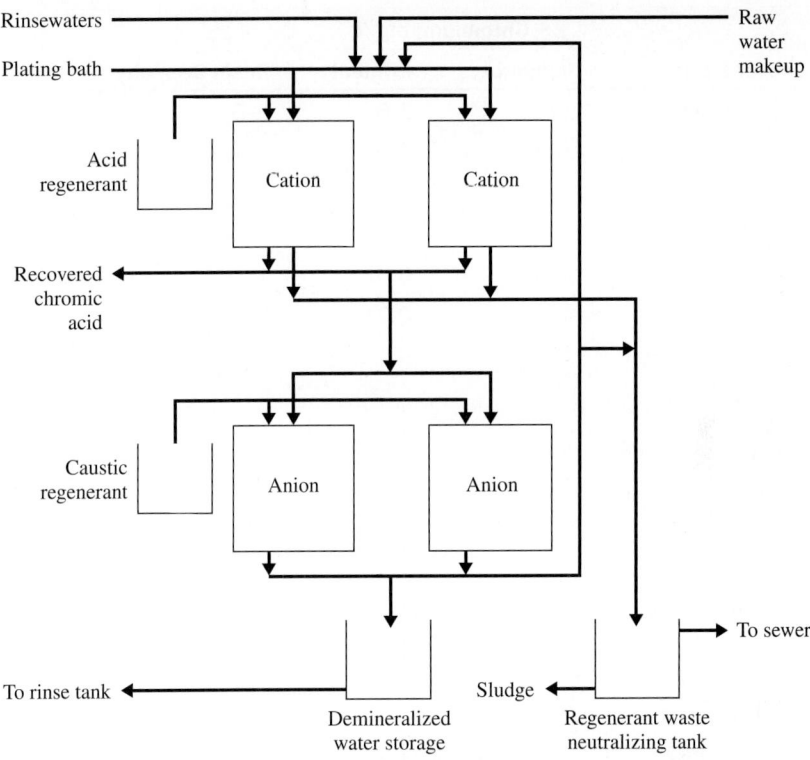

FIGURE 9.4 Ion exchange system for chromate removal and water reuse.

Since most of the metal ions are eluted in the first 70 percent of the regenerant volume, neutralization requirements can be reduced by reuse of the last portion of the acid regenerant for the subsequent regeneration.[4] In like manner, the last portion of the caustic regeneration can be used for neutralization of the spent cation regenerant.

In cation units, the regeneration requirements are higher than in water purification because of the competition with the H^+ present in the waste solution. For water reuse, 4 to 5 lb/ft³ (64 to 80 kg/m³) H_2SO_4 may be required for regeneration, but for recovery of H_2CrO_4 as much as 25 lb/ft³ (400 kg/m³) H_2SO_4 may be required to reduce leakage of sodium ions. A flow diagram of an ion exchange process for a plating plant is shown in Fig. 9.4. Ion exchange performance is shown in Table 9.2.[8]

Example 9.1. A general plating plant operates 16 h/d, 5 d/week. The total discharge of rinsewaters has the following characteristics:

Copper   22 mg/L as Cu
Zinc   10 mg/L as Zn
Nickel   15 mg/L as Ni
Chromium 130 mg/L as CrO_3

Wastewater Source	Chromium, mg/L		Resin Capacity*
	Influent	Effluent	
Cooling tower blowdown	17.9	1.8	5–6
	10.0	1.0	2.5–4.5
	7.4–10.3	1.0	—
	9.0	0.2	2.5
Plating rinsewater	44.8	0.025	1.7–2.0
	41.6	0.01	5.2–6.3
Pigment manufacture	1210	<0.5	—

*lb. chromate/ft^3 resin.
Source: Adapted from Patterson, 1985.

TABLE 9.2 Ion Exchange Performance in Hexavalent Chromium Removal

	Exchanger	
	Cation	Anion
Regenerant	H_2SO_4	NaOH
Dosage, lb/ft^3	12	4.8
Concentration, %	5	10
Flow rate	0.5 gal/(min · ft^3)	
Operating capacity	1.5 equiv wt/L	3.8 lb CrO_3/ft^3

Note: lb/ft^3 = 16.0 kg/m^3
gal/(min · ft^3) = 0.134 m^3/(min · m^3)

TABLE 9.3 Typical Ion Exchange Operating Characteristics

The rate of flow is 50 gal/min (0.19 m^3/min), and in-plant separation is not feasible. Design an exchanger system to include water and chromium recovery. The operating characteristics of the cation exchanger are given in Table 9.3.

Solution

Anion exchanger
In the anion exchanger, CrO_3 is exchanged for OH.

$$130 \text{ mg/L} \times 50 \text{ gal/min} \times 60 \text{ min/h} \times 16 \text{ h/d} \times 8.34$$

$$\times 10^{-6} \frac{\text{lb/gal}}{\text{mg/L}} = 52 \text{ lb/d}$$

For a resin capacity of 3.8 lb CrO_3/ft^3 at a regeneration level of 4.8 lb NaOH/ft^3 and a daily regeneration,

$$\text{Volume of resin} = \frac{52}{3.8} = 13.7 \text{ ft}^3 \quad (0.39 \text{ m}^3)$$

Treatment flow rate is 3.6 gal(min · ft³) [0.48 m³/(min · m³)], for a resin depth of 30 in (0.76 m), 2 units, 2 ft (0.61 in) diameter by 30 in (0.76 in) deep, plus 50 percent for bed expansion.

Regeneration

$$\text{NaOH required} = 4.8 \times 13.7 = 66 \text{ lb/reg} \quad (30 \text{ kg/reg})$$

$$\text{Regenerant tank volume} = 66 \text{ lb NaOH} \times \frac{1}{0.10} \times \frac{1}{9.6} \text{ lb reg/gal}$$

$$= 68 \text{ gal} \quad (0.26 \text{ m}^3)$$

$$\text{Rinse requirement at } 100 \text{ gal/ft}^3 = 1370 \text{ gal} \quad (5.2 \text{ m}^3)$$

Cation exchanger
The cations to be removed are:

$$\text{Zn} \quad \frac{10 \text{ mg/L}}{32.7 \text{ mg/meq}} = 0.306 \text{ meq/L}$$

$$\text{Cu} \quad \frac{22 \text{ mg/L}}{31.8 \text{ mg/meq}} = 0.693 \text{ meq/L}$$

$$\text{Ni} \quad \frac{15 \text{ mg/L}}{29.4 \text{ mg/meq}} = 0.511 \text{ meq/L}$$

In the cation unit, Cu, Zn, and Ni are exchanged for H^+.
The total daily equivalents are

$$(0.306 + 0.693 + 0.511) \times 10^{-3} \times 50 \times 60 \times 16 \times 3.78 \text{ l/gal} = 273 \text{ equiv wt/d}$$

For an operating capacity of 1.5 equiv wt/L at a regeneration level of 12 lb H_2SO_4/ft³ (5 percent), the resin required for a 2-d regeneration is

$$\frac{273 \times 2}{1.5 \times 28.3 \text{ L/ft}^3} = 13.0 \text{ ft}^3 \text{ resin} \quad (0.36 \text{ m}^3)$$

The treatment flow rate is 3.8 gal/(min · ft³) [0.51 m³/(min · m³)]. Use 2 units, 2 ft (0.61 m) diameter by 30 in (0.76 m) deep plus 50 percent for bed expansion.

Regeneration
Using 5 percent H_2SO_4 at 12 lb/ft³, H_2SO_4 required is

$$12 \times 13 = 156 \text{ lb} \quad (71 \text{ Kg})$$

$$\text{Regenerant tank} = 156 \times \frac{1}{0.05} \times \frac{1}{1.0383 \times 8.34 \text{ lb/gal}}$$

$$= 360 \text{ gal} \quad (1.36 \text{ m}^3)$$

where 1.0383 is the specific gravity of 5% H_2SO_4.

$$\text{Rinse requirement} = 120 \text{ gal/ft}^3 \times 13 \text{ ft}^3 = 1560 \text{ gal} \quad (5.9 \text{ m}^3)$$

Anion regenerant capacity for chromium recovery:

$$\text{Sodium} = \frac{66 \text{ lb NaOH} \times 453 \text{ g/lb}}{40 \text{ g/equiv wt}} = 750 \text{ equiv wt}$$

If it is assumed that 70 percent of anion exchanger regenerant will pass through the cation unit, 525 equiv wt must be exchanged, which is compatible with the capacity of the cation units.

9.4 Problem

9.1. A metal finishing plant operates 8 h/d, 6 d/week. The total discharge of reuse waters has the following characteristics:

Zn	15 mg/L
Ni	12 mg/L
Cr	190 mg/L as CrO_3

The flow rate is 100 gal/min (0.38 m³/min). Design an ion exchange system to include water and chromium recovery. The operating characteristics of the resins are given in Example 9.1.

References

1. Clifford, D. et al.: *J. ES&T*, vol. 20, no. 11, p. 1072, 1986.
2. Dorfner, K.: *Ion Exchange Properties and Applications*, Ann Arbor Science Pub., 1973.
3. Bin, Luo et al.: *Toxicity Reduction: Evaluation and Control*, Technomic Pub. Co., Lancaster, Pa., 1998.
4. Fadgen, T. J.: *Proc. 7th Ind. Waste Conf.*, 1952, Purdue University.
5. Paulson, C. F.: *Proc. 7th Ind. Waste Conf.*, 1952, Purdue University.
6. Rich, L. G.: *Unit Processes of Sanitary Engineering*, John Wiley, New York, 1963.
7. Keating, R. J., R. Dvorin, and V. J. Calise: *Proc. 9th Ind. Waste Conf.*, 1954, Purdue University.
8. Patterson, J.: *Industrial Wastewater Treatment Technology*, 2nd ed., Butterworth Pub., Stoneham, Mass., 1985.

CHAPTER 10
Chemical Oxidation

10.1 Introduction

Chemical oxidation generally refers to the use of oxidizing agents such as ozone (O_3), hydrogen peroxide (H_2O_2), permanganate (MnO_4^-), chloride dioxide (ClO_2), chlorine (Cl_2 or $HOCl$), or even oxygen, (O_2), without the need for microorganisms for the reactions to proceed. These reactions frequently require one or more catalysts in order to increase the rate of reaction to acceptable levels. Catalysts include simple pH adjustment, UV light, transition metal cations, enzymes, and a variety of proprietary catalysts of unreported composition.

Chemical oxidation is typically applied to situations where organic compounds are nonbiodegradable (refractory), toxic, or inhibitory to microbial growth. However, chemical oxidation is also effective for the destruction of many inorganic compounds and the elimination of odorous compounds, for example, oxidation of sulfides ($H_2S \rightarrow SO_4^{2-}$).

While oxygen serves as a readily available and extremely economical oxidant for biological treatment processes, other chemical oxidants are relatively expensive and cannot compete economically with aerobic biological treatment. However, it is not necessary to carry chemical oxidation to the fullest extent of reaction (conversion of organic carbon to CO_2). Partial oxidation of compounds may be sufficient to render specific compounds, such as priority pollutants, more amenable to subsequent biological treatment. On a general basis, the oxidation of specific compounds may be characterized by the extent of degradation of the final oxidation products:[1]

1. *Primary degradation.* A structural change in the parent compound.
2. *Acceptable degradation (defusing).* A structural change in the parent compound to the extent that toxicity is reduced.
3. *Ultimate degradation (mineralization).* Conversion of organic carbon to inorganic CO_2.
4. *Unacceptable degradation (fusing).* A structural change in the parent compound resulting in an increase in toxicity.

With the exception of unacceptable by-products, chemical oxidation can result in reduced toxicity or inhibitory behavior, and greatly increased biodegradability of the parent compounds at dosages far less than required for ultimate degradation, that is, the "stoichiometric dosage." Therefore, coupled chemical/biological oxidation processes, with chemical oxidation used for pretreatment of difficult wastes is frequently considered as a treatment alternative.[2] This chapter will consider stoichiometrics, properties, and applicability of commonly used oxidants and hydrothermal processes.

10.2 Stoichiometry

It is important to define the stoichiometric relationship between an oxidant and the compounds to be treated so that required oxidant dosages may be estimated, and experiments may be designed within reasonable limits. A general approach may be taken so that one can easily convert the stoichiometry for one particular compound from one oxidant to another. For convenience, the half-reactions for each oxidant can be expressed in terms of "free reactive oxygen," $O^{\cdot}$, derived for each oxidant, or, using simple oxygen as an example:

$$O_2 \rightarrow 2O^{\cdot} \qquad (10.1)$$

This may be developed from the electrochemical half-reaction for any oxidant by balancing the electrons with the equivalent free reactive oxygens based on water, or:

$$H_2O \rightarrow O^{\cdot} + 2e^- + 2H^+ \qquad (10.2)$$

where the free reactive oxygen is half a diatomic oxygen.

Then, as an example we may consider one of the possible half-reactions for permanganate:

$$MnO_4^- + 2H_2O + 3e^- \rightarrow MnO_2 + 4OH^- \qquad (10.3a)$$

Balancing electrons with Eq. (10.2) gives

$$\begin{array}{r} 2MnO_4^- + 4H_2O + 6e^- \rightarrow 2MnO_2 + 8OH^- \\ + \quad 3H_2O \rightarrow 3O^{\cdot} + 6e^- + 6H^+ \\ \hline MnO_4^- + H_2O \rightarrow 2MnO_2 + 3O^{\cdot} + 2OH^- \end{array} \qquad (10.3b)$$

Equation (10.3b) is the equivalent half-reaction to Eq. (10.3a), except the electrons are replaced by the free reactive oxygens. The same procedure can be completed for any oxidant using Eq. (10.2) and the appropriate half-reaction. Then, any oxidation reaction may be expressed in terms of the equivalent free oxygen and related back to a specific oxidant by the stoichiometry of the half-reaction. The half-reactions for a variety of oxidants are presented in Table 10.1.

Half-Reaction	Equivalent Reactive Oxygens	
	Mol [O] per mol Oxidant (n)	Mol [O] per kg Oxidant
Chlorine:		
$Cl_2 + H_2O \rightarrow O^\bullet + 2Cl^- + 2H^-$	0.5	14.1
$HOCl \rightarrow O^\bullet + Cl^- + H^-$	1.0	19.0
Chlorine dioxide:		
$2ClO_2 + H_2O \rightarrow 5O^\bullet + 2Cl^- + 2H^+$	2.5	37.0
Hydrogen peroxide:		
$H_2O_2 \rightarrow O^\bullet + H_2O$	1.0	29.4
Permanganate:		
pH < 3.5		
$2MnO_4^- + 6H^+ \rightarrow 2Mn^{2+} + 5O^\bullet + 3H_2O$	2.5	15.8
3.5 < pH < 7.0		
$2MnO_4^- + 2H^- \rightarrow 2MnO_2 + O^\bullet + 3H_2O$	1.5	9.5
7.0 < pH < 12.0		
$2MnO_4^- + H_2O \rightarrow 3O^\bullet + 2MnO_2 + 2OH^-$	1.5	9.5
12.0 < pH < 13.0		
$2MnO_4^- + H_2O \rightarrow 2MnO_4^{-2} + O^\bullet + 2H^+$	0.5	3.2
Ozone:		
high pH		
$O_3 \rightarrow O^\bullet + O_2$	1.0	20.8
low pH		
$O_3 \rightarrow 3O^\bullet$	3.0	61.4

TABLE 10.1 Oxidant Half-Reactions

For the ultimate conversion of an organic compound to CO_2 and H_2O, a general stoichiometric equation may be derived for reactive oxygens:

$$C_aH_bO_c + dO^{\cdot} \rightarrow aCO_2 + (b/2)H_2O \tag{10.4}$$

where $d = 2a + b/2 - c$. Then, the equation may be balanced for any oxidant by adding the half-reaction for the oxidant times the number of free reactive oxygens required (d) divided by the stoichiometric number of free reactive oxygens produced (see half-reactions in Table 10.1).

Example 10.1. For phenol (C_6H_5OH), balance the half-reactions with peroxide (H_2O_2) and permanganate (MnO_4^-) as oxidants.

Solution
From Eq. (10.4),

$$C_6H_5OH + 14O^{\cdot} \rightarrow 6CO_2 + 3H_2O$$

H_2O_2:

$$C_6H_5OH + 14O^{\cdot} \rightarrow 6CO_2 + 3H_2O$$
$$\underline{+ \;(14/1)\;(H_2O \rightarrow O^{\cdot} + H_2O)}$$
$$= C_6H_5OH + 14H_2O_2 \; 6CO_2 + 17H_2O$$

MnO_4^-:

$$C_6H_5OH + 14O^{\cdot} \rightarrow 6CO_2 + 3H_2O$$
$$\underline{+ \;(14/3)\;(2MnO_4^- + H_2O \rightarrow 3O^{\cdot} + 2MnO_2 + 2OH^-)}$$
$$= C_6H_5OH + (28/3)^{\cdot} MnO_4^- + (5/3)\;H_2O\;6CO_2 + (28/3)\;MnO_2 + (28/3)\;OH^-$$

In many cases, a wastewater will consist of a wide variety of compounds rather than one or a few specific known compounds. Therefore, the use of the theoretical stoichiometry cannot be applied to specific compounds without some error introduced by the background oxidizable compounds. However, the general approach developed using the free reactive oxygens can be adapted to the surrogate chemical oxygen demand, COD. This readily measurable parameter can be converted to the total stoichiometric requirement for an arbitrary wastewater as follows:

$$\text{Oxidant demand (mg oxidant/L)} = (2/n)\;(MW/32)\;COD \tag{10.5}$$

where n = moles O per mol oxidant (see Table 10.1)
        MW = molecular weight of oxidant (g/mol)
        COD = chemical oxygen demand (mg O_2/L)

For H_2O_2 and $KMnO_4$ as examples:

H_2O_2 stoichiometric demand = (2/1) (34/32) COD = 2.13 COD

$KMnO_4$ stoichiometric demand = (2/1.5) (158/32) COD = 6.58 COD

For the stoichiometric requirements and the yield of free reactive oxygens, that is, moles O per kg (from Table 10.1), the cost for the stoichiometric dosage of each oxidant may be determined (based on $/kg oxidant). Then, a ranking of candidate oxidants based on the potential oxidant cost can be established, and their actual effectiveness and cost determined from laboratory data.

10.3 Applicability

Apart from oxygen, most oxidants are expensive and not competitive with biological wastewater treatment for high-strength, large-volume wastewaters. However, chemical oxidation processes are generally designed for wastewaters that are not amenable to biological treatment, that is, toxic, inhibitory, and/or refractory compounds. In addition, significant savings can be obtained through the coupling of chemical oxidation with biological treatment by pretreating toxic/refractory wastes to improve biological treatment performance. The partial oxidation of many wastes can be economically achieved, yielding a variety of readily biodegradable organic acids.

The average oxidation state (OX) of organic carbon in the waste mixture may be expressed as:[3]

$$OX = \frac{4\,(TOC - COD)}{TOC} \qquad (10.6)$$

The COD/TOC ratio can be used as a primary parameter to determine the extent of the overall reaction. Figure 10.1 shows a typical case. The initial TOC before reaction is point A. Point B identifies the initial reaction products, and point C identifies the products after the reaction is substantially complete. An evaluation for toxicity and for biodegradability is made at points B and C to determine both the dosage and the time of contact.

The efficiency of treatment is based on the ability of the oxidant to produce an acceptable organic by-product without significant ultimate conversion of organic carbon to carbon dioxide. This can be expressed as:

$$f = \frac{OD_T}{OX_T} \qquad (10.7)$$

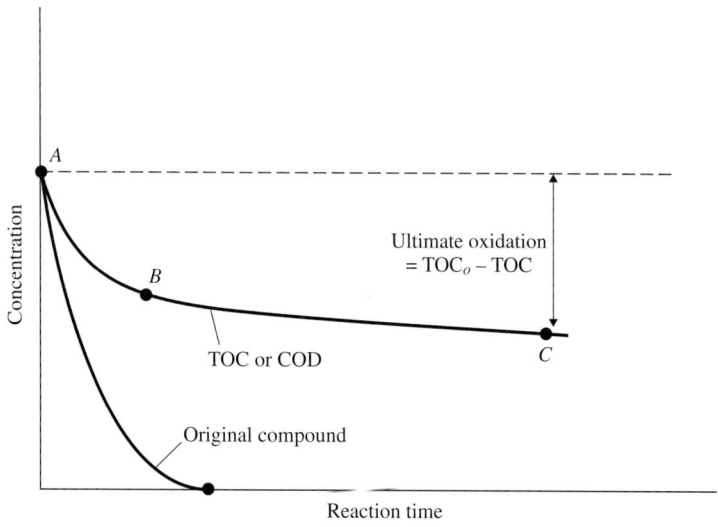

FIGURE 10.1 Change in original compound and TOC or COD over the course of reaction with an oxidant.

where f = fraction of ultimate oxidant demand
OD_T = oxidant consumed, mg/L
OX_T = total stoichiometric oxidant demand (can be based on COD), mg/L

Results for the oxidation of 2,4-dichlorophenol (DCP) and 2,4-dinitro-*ortho*-cresol (DNOC) are shown in Fig. 10.2. The results show that H_2O_2 was not used for ultimate conversion of the organic carbon, but rather the original compounds were drastically altered, leaving by-products in which the mean oxidation state of the carbon was more highly oxidized (positive instead of negative). DNOC exhibited the greatest change (OX = +2.22) while DCP showed the least (OX = +0.867).

10.4 Ozone

Ozone, O_3, is a powerful oxidant (E_H > 2.0 V) that is commonly used for disinfection and wastewater treatment. Ozone is a metastable gas at normal temperatures and pressures and must be generated on site. At high pressures decomposition is rapid. Therefore, generation and mass transfer operations are carried out at low pressure, typically less than 20 lb/in² gauge.[4]

Ozone generation is the key to economical operation. Generators operate on electrical current and produce ozone from either air or pure oxygen gas streams. Moisture retards the process and air must be dehumidified prior to ozone generation (dew point < −60°C).

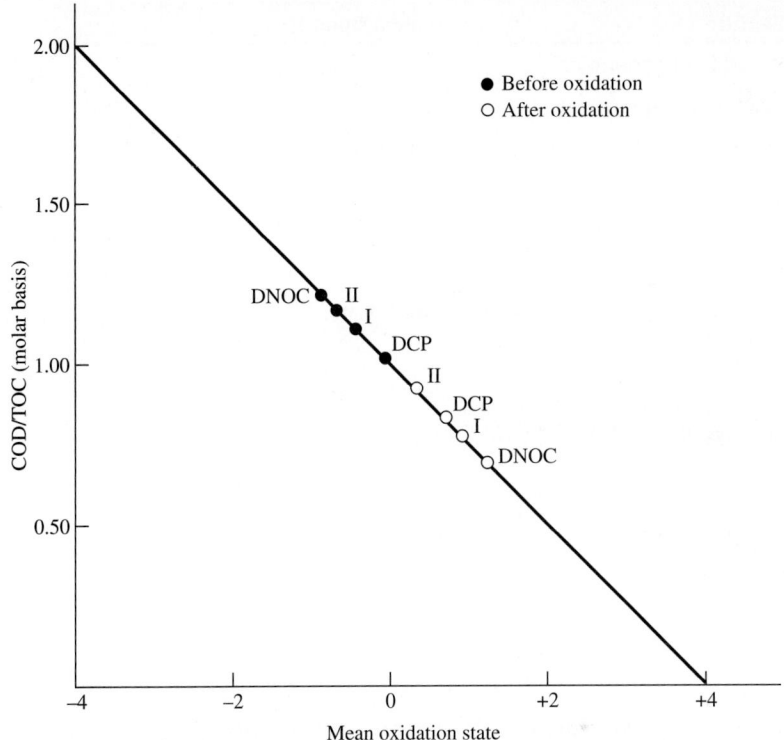

FIGURE 10.2 The COD/TOC ratio versus the mean oxidation state of carbon in the system. Note that a +4 oxidation state corresponds to inorganic carbon (CO_2) and a −4 oxidation state corresponds to methane. I and II refer to different times of oxidation.

Recent advances in pure oxygen generation have greatly reduced the cost of ozone production. A comparison of air and pure O_2 for ozone production is presented in Table 10.2.

Ozone decomposes in water, especially at high pH values, to produce free radicals: $O_3 \rightarrow O^{\cdot} + O_2$ ($n = 1$, see Table 10.1). The rate of decomposition can be written as a function of pH (represented by OH^-) in terms of a first-order decay:[5]

$$\frac{dO_3}{dt} = r_d = -9.811 \times 10^7 \, [OH^-]^{0.123} [O_3] \exp(-5606/T) \qquad (10.8)$$

where
r_d = rate of O_3 self-decomposition, M/min
$[OH^-], [O_3]$ = concentrations of OH^- and O_3 in aqueous solution, M
T = temperature, K

Feature	Typical Operation
Principle of generation	Corona discharge
Electrical frequency	Low: 50 or 60 Hz
	Median: 60–1000 Hz
	High: > 1000 Hz
Operating pressure	7 to 14 lb/in² gauge (low frequency)
	20 lb/in² gauge (high frequency)
Cooling	Water or oil
Ozone output	Air: 1–6%
	Pure O_2: 8–14%
Energy consumption, kWh/lb	Air: 6.4–9.1
	Pure O_2: 3.2–6.8

Source: Adapted from Langlais, Rekhow, and Brink, 1991.

TABLE 10.2 Typical Systems for Generating Ozone

The ozone reaction mechanism is then uniquely dependent on the rate of mass transfer into an aqueous system, and three operating regions of reactivity exist:

Region 1: High pH and/or low ozone dose rate, implying that the major reaction mechanism with solutes entails ·OH. Little selectivity will be exhibited.

Region 2: Approximately equal rates of decomposition and mass transfer. Both direct O_3 and ·OH reactions are important.

Region 3: Low pH and/or extremely high ozone dose rate, implying that direct oxidation of solutes by O_3 is the controlling reaction. Ozone may exhibit a high degree of selectivity.

The mechanisms of ozone oxidation of organics are:

1. Oxidation of alcohols to aldehydes and then to organic acids:

$$RCH_2OH \xrightarrow{O_3} RCOOH$$

2. Substitution of an oxygen atom onto an aromatic ring
3. Cleavage of carbon double bonds

Ozonation can be employed for the removal of color and residual refractory organics in effluents. In one case, while there was a decrease in TOC in the final filtered effluent, the soluble BOD increased from 10 to 40 mg/L because of the conversion from long-chain biologically refractory organics to biodegradable compounds. Similar results were obtained from the ozonation of a secondary effluent from low- and high-rate activated sludge units treating a tobacco-processing wastewater, as shown in Table 10.3. TOC will not be reduced until the organic carbon has been oxidized to CO_2, while COD will generally be reduced with any oxidation.

Oxidation of unsaturated aliphatic or aromatic compounds causes a reaction with water and oxygen to form acids, ketones, and alcohols. At a pH greater than 9.0 in the presence of redox salts such as Fe, Mn, and Cu, aromatics may form some hydroxyaromatic structures (phenolic) which may be toxic. Many of the by-products of ozonation are readily biodegradable.

For many of the priority pollutants there is a rapid, first-order reduction in COD to some level followed by a slow or even zero removal, indicating the nonreactive nature of the by-products.

COD reduction by ozonation of an aerated lagoon effluent treating an organic chemicals wastewater is shown in Fig. 10.3.

Phenol can be oxidized with ozone, producing as many as 22 intermediate products between phenol and CO_2 and H_2O. The reaction is first order with respect to phenol and proceeds optimally over a pH of 8 to 11. The ozone consumption is 4 to 6 mol O_3 consumed/mole phenol oxidized. This requires in the order of 25 mole O_3/mol phenol to be generated in the gas phase.

Ozone may be used in combination with UV radiation to catalyze the oxidation of nonreactive organics, such as saturated hydrocarbons and highly chlorinated organics. Ozone reacts with UV light (253.7 nm) in aqueous systems to produce hydrogen peroxide:[6]

$$O_3 + H_2O \xrightarrow{uv} O_2 + H_2O_2 \qquad (10.9)$$

Parameter	Low F/M (0.15) Time, min*		High F/M (0.60) Time, min*	
	0	60	0	60
BOD, mg/L	27	22	97	212
COD, mg/L	600	154	1100	802
pH	7.1	8.3	7.1	7.6
Org-N, mg/L	25.2	18.9	40	33
NH_3-N, mg/L	3.0	5.8	23	25
Color (Pt-Co)	3790	30.0	5000	330

*Loading of 155 mg O_3/min.

TABLE 10.3 Ozonation Results on Secondary Clarifier Effluent

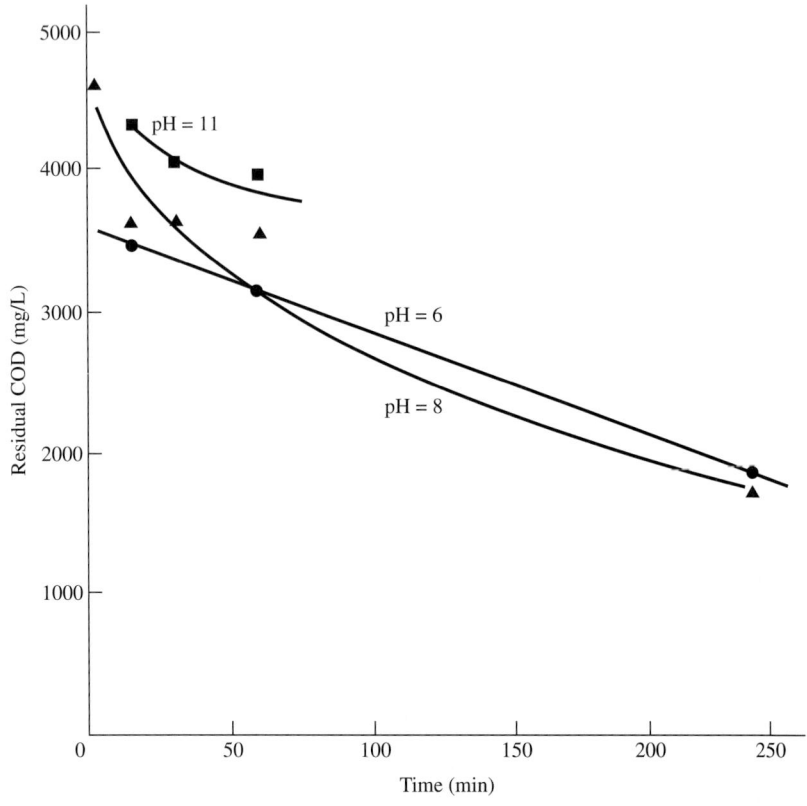

FIGURE 10.3 Effect of pH on COD removal by ozonation-aerated lagoon effluent.

These two oxidants, O_3 and H_2O_2, react with one another, providing O_3 and hydroxyl radical, ˙OH, at low pH values. In some cases, H_2O_2 may be added directly to complement the coupled oxidation process, in advanced oxidation processes (AOPs). In addition, UV light may react directly with some of these organics, further promoting reaction of the first by-products with O_3.

10.5 Hydrogen Peroxide

Hydrogen peroxide, H_2O_2, is commercially available in a variety of grades, with 30 or 50 percent (by weight) solutions being most common for wastewater applications. Inhibitors, typically phosphates, are added to prolong storage times. Hydrogen peroxide has a long history of use for oxidation of sulfides in sewer lines and wastewater treatment plants, and more recently has been widely applied to toxic and refractory organics. Sulfide oxidation with H_2O_2 is summarized in Table 10.4.

Acidic or neutral pH
$H_2O_2 + H_2S \rightarrow 2H_2O + S$
-Reaction time: 15–45 min
Catalyst: Fe^{2+}
pH: 6.0–7.5
Reaction time: sec
Basic pH
$4H_2O_2 + S^{2-} \rightarrow SO_4^{2-} + 4H_2O$
Reaction time: 15 min

TABLE 10.4 Sulfide Oxidation with Hydrogen Peroxide

Alkaline peroxidation (pH 9.5) will oxidize formaldehyde: $2CH_2O + H_2O_2 + 2OH^- \rightarrow 2HCOO^- + H_2 + 2H_2O$.

Alkaline peroxidation (pH 10 to pH 12) is an effective means of providing total cyanide destruction. The reactions are $CN^- + H_2O_2 \rightarrow OCN^- + H_2O$, $OCN^- + 2H_2O \rightarrow NH_4^+ + CO_3^{2-}$. This process was evaluated for treatment of a wastestream discharged from specialty polymers manufacturing. The wastewater was discharged at 40°C and pH 7.5 and contained 4700 mg/L COD and 96 mg/L CN. Hydrogen peroxide was added to the wastestream and allowed to react at pH 10.7 (sodium hydroxide addition was required) for 6 h. Following 1 h of reaction, 5600 mg/L H_2O_2 had been consumed and the residual CN concentration was 8.3 mg/L. Following 6 h of reaction time, 6000 mg/L H_2O_2 had been consumed and the residual CN concentration was 2.5 mg/L.

Reactions with H_2O_2 alone are slow, and a catalyst is generally required. A wide variety of reaction schemes are possible, using high pH (alkaline catalysis); metals such as ferrous sulfate (Fenton's reagent), complexed Fe (Fe-EDTA or Heme), Cu, or Mn; or natural enzymes such as horseradish peroxidase. By far, the most common catalyst is ferrous iron ($FeSO_4$ or Fenton's reagent), at a pH of about 3.5. A series of reactions with Fe are hypothesized leading to the generation of free radicals, $\cdot OH$ and $HO_2\cdot$ and regeneration of Fe (II):

$$Fe^{2+} + H_2O_2 \rightarrow Fe^{3+} + OH^+ + \cdot OH \qquad (10.10a)$$

and

$$Fe^{3+} + H_2O_2 \rightarrow Fe^{2+} + HO_2\cdot + H^+ \qquad (10.10b)$$

A chain reaction then occurs between the hydroxyl radical and an organic compound R:

$$RH + {}^{\bullet}OH \rightarrow R^{\bullet} + H_2O \qquad (10.11a)$$

$$R^{\bullet} + O_2 \rightarrow ROO^{\bullet} \qquad (10.11b)$$

$$ROO^{\bullet} + RH \rightarrow ROOH + R^{\bullet} \qquad (10.11c)$$

The reaction proceeds optimally at pH between 2.0 and 4.0, primarily so that the Fe (III) produced stays in solution as Fe^{3+} rather than precipitating as the ferrihydroxides, $Fe(OH)_3$ or FeOOH. Some processes make use of this, precipitating the Fe (III) after the reaction is complete and recycling the ferrihydroxides back to the process for reuse. A typical catalyzed H_2O_2 system is shown in Fig. 10.4.

Other catalysts include UV radiation (UV-H_2O_2). It is hypothesized that the H_2O_2 molecule can be split directly into hydroxyl radicals by UV light:

$$H_2O_2 \xrightarrow{UV} 2\,{}^{\bullet}OH \qquad (10.12)$$

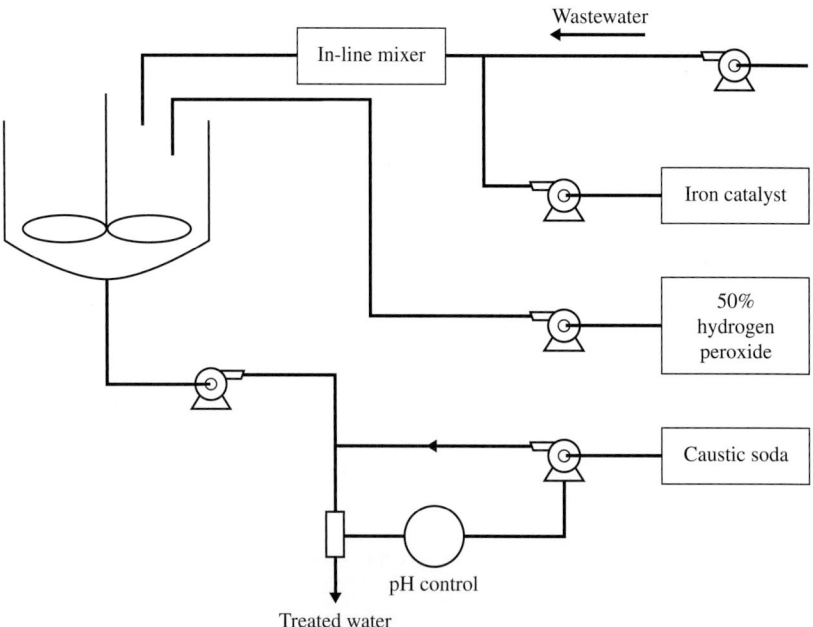

Note: A pH of ≤ 5 is required for this reaction.

Figure 10.4 Wastewater treatment system with catalyzed hydrogen peroxide.

Ultraviolet light is derived from lamps that have been developed for a high quantum yield in the appropriate wavelengths for hydroxyl radical generation, based on decomposition of ferrioxalate as an indicator. Additionally, heat generated by the lamps can increase the rates of reaction significantly. The UV-H_2O_2 process is generally applied to aqueous wastes of lower color, turbidity, and concentration, such as contaminated groundwaters. However, proprietary additives are available for higher-strength wastes. Many compounds are effectively treated by UV-H_2O_2 including benzene, toluene, xylene, trichloroethylene, and perchloroethylene, but some are refractory to treatment, including chloroform, acetone, trinitrobenzene, and *n*-octane. Performance is shown in Fig. 10.5.

Several catalysts are compared in Table 10.5. It should be noted that catalysts must be evaluated experimentally for each wastewater and/or specific pollutant.

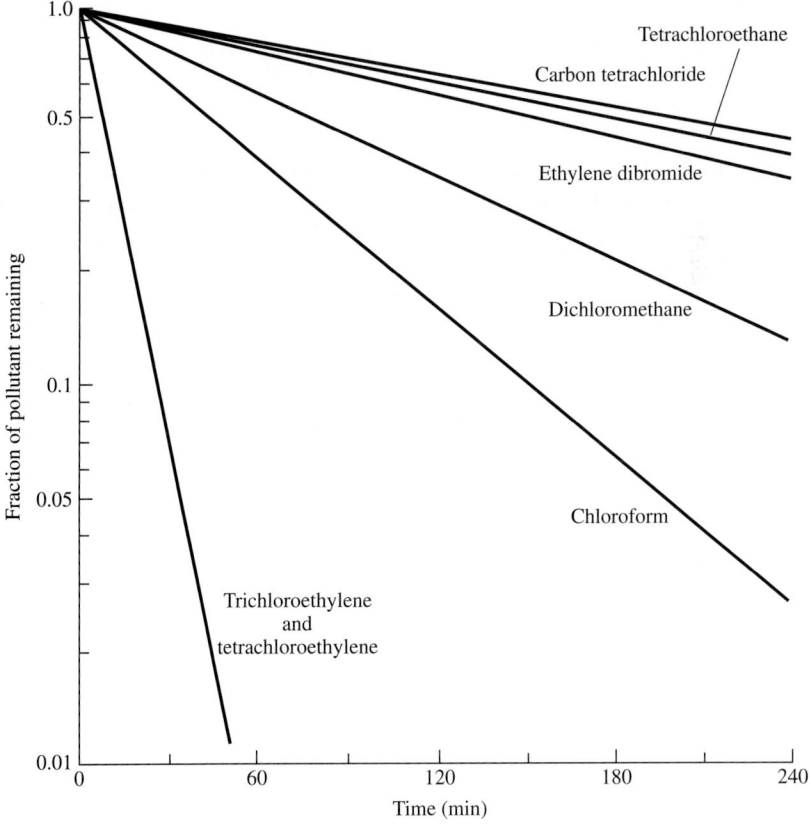

FIGURE 10.5 Comparison of rates of reaction of halogenated aliphatics at 20°C with UV and hydrogen peroxide. (*Adapted from Sundstrom et al., 1994.*)

Catalyst	Optimum pH	Conditions Required	Comments
Fenton's reagent: $FeSO_4$	2–4	H_2O_2:Fe $\approx$ 10:1	Generation of Fe$(OH)_3$ sludge
Alkaline pH	> 9	—	—
Complexed iron:			
Fe-EDTA,* Heme	5–10	H_2O_2:Fe $\approx$ 5:1	Complexed iron remains in solution
UV, radiation	2–10[†]	Low turbidity Low UV absorbance Low organic contaminant	Electrical costs dominate

*As ferrous or ferric ethylenediaminetetraacetate.
[†]pH plays little role.

TABLE 10.5 Catalysts for Hydrogen Peroxide Treatment of Wastes

Hydrogen peroxide has been shown to be useful in many cases for removal of specific undesirable pollutants such as priority pollutants, reduction of toxicity, and/or improvements in biodegradability (both rate and extent of degradation). Table 10.6 summarizes the results of tests on a variety of aromatic compounds oxidized which indicate reactivity with these compounds, COD and TOC reduction, and a significant improvement in toxicity (except for 2,6-dichlorophenol). Oxidation with H_2O_2 can then be coupled with biological treatment to further treat the COD to an acceptable level. In many cases oxidation of refractory COD will result in an increase of BOD as shown in Fig. 10.6.[7]

10.6 Chlorine

Chlorine has a long history of use as an oxidant in water and wastewater treatment, and has been especially successful for color removal where organic dyes are present. However, recent concerns regarding the formation of chlorinated by-products, such as chloroform, have greatly reduced the applications of chlorine in wastewater. Chlorination of an activated sludge effluent for color removal is shown in Fig. 10.7. While chlorine is highly successful for color removal, chlorinated organics must be accounted for. Similar effluents have had difficulty meeting discharge permit requirements where zero- and low-flow receiving streams were involved.

Compound	% Destruction	% COD Reduction	% TOC Reduction[†]	Toxicity (EC_{50})[§]	
				Original Compound	After Oxidants
Nitrobenzene	>99	72.4	37.3	62	217
Benzoic acid	>99	75.8	48.8	289	>292
Phenol	>99	76.1	44.1	69	NT
o-Cresol	>99	75.0	55.6	29	NT
m-Cresol	>99	73.3	38.2	16	NT
p-Cresol	>99	71.8	40.0	5	NT
o-Chlorophenol	>99	75.1	47.9	52	—
m-Chlorophenol	>99	75.0	41.3	18	NT
p-Chlorophenol	>99	75.7	21.7	3	NT
2,3-Dichlorophenol	>99	70.2	52.6	10	NT
2,4-Dichlorophenol	>99	68.9	50.3	6	—
2,5-Dichlorophenol	>99	74.2	41.8	18	NT
2,6-Dichlorophenol	>99	61.1	32.5	54	26[¶]
3,5-Dichlorophenol	>99	69.1	48.9	5	NT
2,3-Dinitrophenol	>99	80.1	50.7	86	NT
2,4-Dinitrophenol	>99	72.5	51.0	22	NT
Aniline	>99	76.5	43.4	394	NT

[*]Stoichiometric dose of H_2O_2, 5×10^{-3} M pollutant, 50 mg/L $FeSO_4$ (as Fe) for catalyst; samples reacted in batch for 24 h at 20°C (±2°).
[†]Based on original compound only.
[‡]Corresponds to ultimate conversion.
[§]EC_{50} = concentration which causes a 50 percent reduction in activity. Based on Microtox toxicity; values are as mg DOC/l; NT = nontoxic.
[¶]Shows increased toxicity.

TABLE 10.6 Oxidation of Various Aromatic Compounds with H_2O_2[*,†]

Chlorine is still frequently applied for the oxidation of cyanides in metal finishing operations. The oxidation of cyanide by chlorine proceeds through several reactions that are highly pH-dependent, and a two-step process is typically practiced:

Step 1
Reaction 1: Reaction of cyanide with hypochlorite to form cyanogen chloride (all pH values):

$$CN^- + OCl^- + H_2O \xrightarrow{\text{slow}} CNCl + 2OH^- \qquad (10.13a)$$

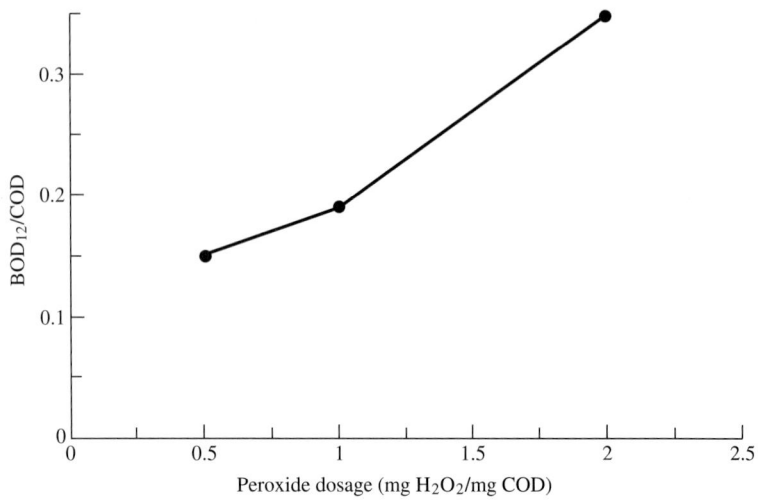

Figure 10.6 BOD_{12}/COD ratio at different peroxide dosages. (*Adapted from Albers and Kayser, 1998.*)

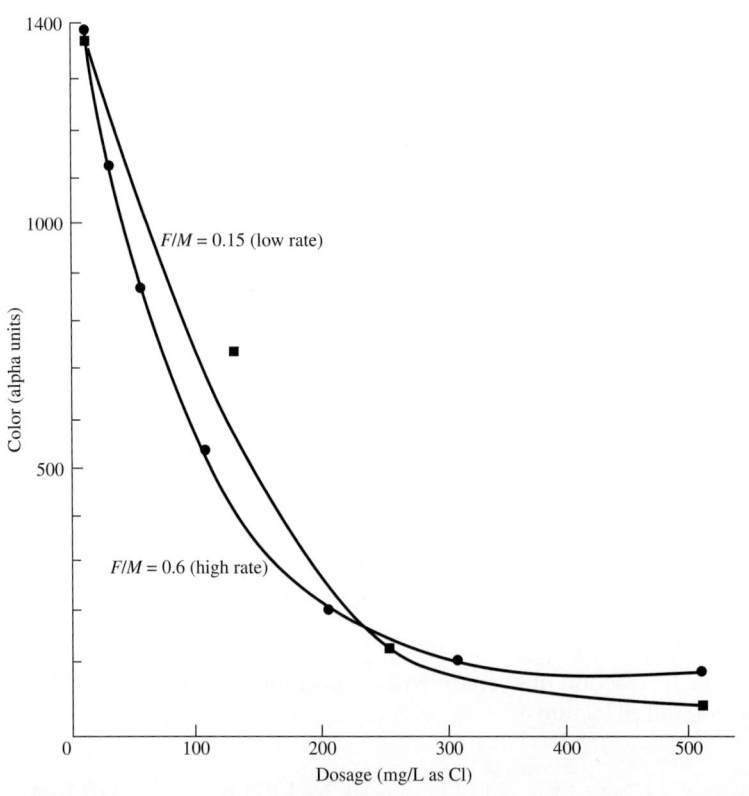

Figure 10.7 Color removal by chlorination.

Reaction 2: Hydrolysis of cyanogen chloride to form cyanate, CNO^- (minimum pH 9 to 10; pH 11.5 recommended):

$$CNCl + 2OH^- \xrightarrow{fast} CNO^- + Cl^- + H_2O \quad (10.13b)$$

Step 2
Reaction 3: Oxidation of cyanate to form bicarbonates and nitrogen gas (pH 8.0 to 8.5):

$$2CNO^- + 3HOCl \rightarrow 2HCO_3^- + N_2 + 3Cl^- + H^+ \quad (10.13c)$$

Cyanogen chloride is a toxic gas and must be eliminated immediately. It is not stable and hydrolyzes rapidly to CNO^- at high pH. Therefore, reactions 1 and 2 are carried out simultaneously in step 1 at the recommended pH of 11.5.

At this point, CNO^- is about 1000 times less toxic than CN^-, and frequently the reaction is stopped at this point. However, complete treatment requires step 2, where CNO^- is oxidized to HCO_3^- and N_2 (gas) at pH 8 to 8.5. While a lower pH (< 7.6) would be preferred, the potential for HCN evolution in the event of a process malfunction places a lower limit on pH in this process.

In practice, the dosages of chlorine depend heavily on the identity of the metals complexed with CN^- and the presence of other background constituents in the wastewater. Since these concentrations change continually, the system is usually operated on the basis of oxidation-reduction potential (ORP). A summary of typical dosages and operating parameters is presented in Table 10.7. Chlorine is typically applied as sodium hypochlorate (15 percent NaOCl) for operation at under 20 gal/min and chlorine gas (Cl_2) for larger operations. As an example problem consider the example on the next page.

Reaction Step	Chlorine Dose*	Operating ORP[†], mV
Step 1: $CN^- \rightarrow CNO^-$	Stoichiometric (Cu, Zn), 2 × Stoichiometric (Fe, Co, Ni)	~350[†]
Step 2: $CNO^- \rightarrow HCO_3^- + N_2$	Stoichiometric	500–800

*This must be tuned experimentally for each wastewater.
[†]Stoichiometric dosages based on Eqs. (10.10a) and (10.10c).

TABLE 10.7 Summary of Cyanide Oxidation Practice with Chlorine

Example 10.2. How much Cl_2 must be supplied to oxidize 130 mg/L cyanide (as CN) given a flow of 10,000 L/d?

(a) Consider oxidation to CNO^-.

(b) Consider complete oxidation to HCO_3^- and N_2.

First, consider that Cl_2 reacts with water to produce HOCl and HCl:

$$Cl_2 + H_2O \leftrightarrow HOCl + HCl$$

Therefore, only half of our chlorine is effective. Next, all calculations must be done on a molar basis, or molecular weight of CN = 12 + 14 = 26 g/mole. Then

$$\frac{130 \text{ mg/L cyanide}}{(26 \text{ g/mol})(1000 \text{ mg/g})} = 5.0 \times 10^{-3} \text{ M}$$

Solution

(a) Combine reactions 1 and 2 [Eqs. (10.10a) and (10.10b)] to give an overall reaction of CN^- to CNO^-:

$$\text{Reaction 1} + \text{Reaction 2} = CN^- + OCl^- \rightarrow CNO^- + Cl^-$$

Therefore, 1 mol of OCl^- is required for each CN^-:

$$5 \times 10^{-3} \text{ MCN}^- \left(\frac{1 \text{ mol OCl}^-}{1 \text{ mol CN}^-} \right) = 5 \times 10^{-3} \text{ M OCl}^- \text{ required}$$

$$= \frac{\text{stoichiometric dosage}}{5 \times 10^{-3} \text{ M Cl}_2 \text{ required}}$$

Finally, for a 10,000 L/d flow:

$$(5 \times 10^{-3} \text{ mol Cl}_2/\text{L})(10{,}000 \text{ L/d}) = 50 \text{ mol Cl}_2/\text{d}$$

then Cl_2 = 71 g/mol, or

$$\frac{(50 \text{ mol Cl}_2/\text{d})(71 \text{ g/mol})}{1000 \text{ g/kg}} = 3.55 \text{ kg Cl}_2/\text{d}$$

and, since dosage is 1 to 2 times stoichiometric, depending on the metal complexed with CN^- (see Table 10.5), the actual required dose = 3.55 to 7.10 kg Cl_2/d.

(b) Consider reaction 3 [Eq. (10.13c)]:

$$2CNO^- + 3HOCl \rightarrow 2HCO_3^- + N_2 + 3Cl^- + H^+$$

Note that 3 moles HOCl produced per 2 mol CNO⁻ and 1 mole CNO⁻ produced per mole CN⁻ are initially present (100 percent conversion). Therefore,

$$5\times 10^{-3}\ \text{MCNO}^- = \left(\frac{3\ \text{mol HOCl}}{2\ \text{mol CNO}^-}\right)(10{,}000\ \text{l/d})\left(\frac{71\ \text{g/mol}}{1000\ \text{g/kg}}\right) = 5.33\ \text{kg Cl}_2/\text{d}$$

and the total requirement is step 1 + step 2, or

$$\text{Cl}_2 = (3.55\ \text{to}\ 7.10\ \text{kg/d}) + 5.33\ \text{kg/d} = 8.88\ \text{to}\ 12.43\ \text{kg Cl}_2/\text{d}$$

Electrode potential relationships in the alkaline chlorination of cyanide wastewater are shown in Fig. 10.8.

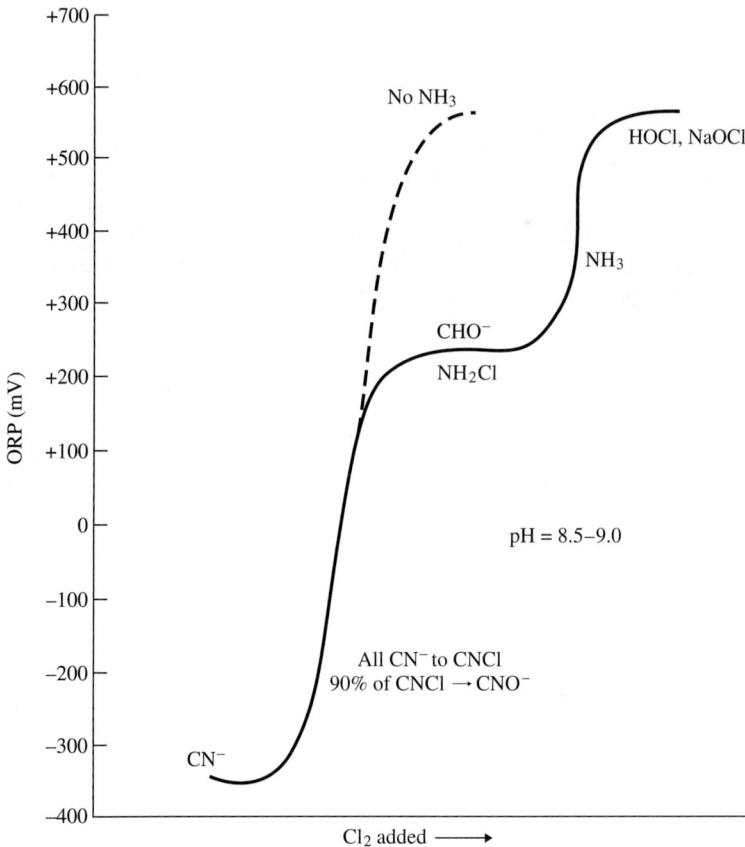

FIGURE 10.8 Electrode potential relationships in the alkaline chlorination of cyanide waste.

10.7 Potassium Permanganate

Permanganate, MnO_4^-, is a powerful oxidizing agent (E_h = 1.68 V) that is reactive over a wide pH range (although rates are reportedly faster at high pH) for a variety of organic and inorganic compounds. This is supplied in stable form as a solid (96.5 to > 99 percent purity) or in concentrated aqueous form. Applications of $KMnO_4$ traditionally include odor control (oxidation of inorganic and organic sulfides) and textile, tannery, steel processing, metal finishing, pulp and paper, and oil refinery wastes. Oxidation of sulfides proceeds differently under acidic or alkaline conditions:

Acidic

$$3H_2S + 2KMnO_4 \rightarrow 3S° + 2H_2O + 2KOH + 2MnO_2(s)$$

Alkaline

$$3H_2S + 8KMnO_4 \rightarrow 3K_2SO_4 + 2H_2O + 2KOH + 8MnO_2(s)$$

Unlike other oxidants, a solid reaction by-product, $MnO_2(s)$, is formed and must be disposed of as a waste sludge, along with other precipitates/solids already understood as part of the waste itself. This sludge can be considerable for concentrated wastewaters. Data show that $KMnO_4$ can be effective in the destruction of specific compounds and in toxicity reduction for phenolics and other aromatic compounds, and even certain chlorinated aliphatics, such as trichloroethylene, perchloroethylene, and trichloroethane.

10.8 Oxidation Overview

The chemical oxidation of various aromatic compounds by H_2O_2, $KMnO_4$, and O_3 is shown in Table 10.8.[8] The toxicity before and after oxidation of these compounds is shown in Table 10.9.

Application of chemical oxidation for the treatment of different types of dyes is shown in Table 10.10.

Based on numerous studies, combined AOPs have been shown to be more effective than single oxidant applications. Typically, use is restricted to relatively low COD wastewaters due to oxidant cost. As previously mentioned, complete oxidation to terminal end products is not generally necessary, thus reducing oxidant dosage and cost.

10.9 Hydrothermal Processes

Hydrothermal processes refer to aqueous treatment of wastewaters at elevated temperatures and pressures. There are three basic operating regimes that are in use or have been investigated in the laboratory:

Chemical Oxidation

Compound	Initial Oxidation State	TOC Removal, %[a,b,c]			COD Removal, %[d]		
		H_2O_2	$KMnO_4$	O_3	H_2O_2	$KMnO_4$	O_3
Pyrrolidine	−1.76	34.9	NR[e]	32.1	72.1	NR[e]	58.5
Sulfanilic acid	−0.84	46.3	NR	57.5	74.9	NR	57.4
Naphthalene	−0.80	46.2	NR	0.0	80.4	NR	>99.0
Diphenylamine	−0.66	69.4	NR	30.6	87.7	NR	90.0
Skatole	−0.66	0.0	NR	0.0	39.0	NR	38.1
Benzaldehyde	−0.57	78.6	67.6	74.4	93.5	79.1	74.2
Indole	−0.50	62.3	60.3	60.9	95.5	91.0	77.2
Catechol	−0.33	57.0	52.2	22.0	80.5	66.3	30.7
Hydroquinone	−0.33	30.7	27.3	17.2	78.5	71.2	45.0
Resorcinol	−0.33	56.5	27.8	29.1	79.8	73.1	50.1
Vanillin	−0.25	70.3	53.4	63.6	87.8	55.2	63.6
Pyrogallol	0.00	45.4	22.1	28.5	75.1	78.2	48.5
Salicylic acid	0.00	28.6	31.6	31.2	74.6	49.8	41.6
Coumarin	+0.22	25.9	NR	NR	65.3	NR	NR
Phthalic acid	+0.25	37.0	NR	31.1	71.2	NR	52.0
Average[f]	−0.44	45.9	42.8	36.8	77.1	70.5	55.9

[a] Arranged in order of lowest oxidation state of organic carbon.
[b] Mean oxidation state of the organic carbon calculated from molecular structure.
[c] Corresponds to the percentage converted to CO_2, i.e., ultimate oxidation.
[d] All compounds were initially at a concentration of 5×10^{-3} M (as the original compound) unless limited by solubility (naphthalene and diphenylamine).
[e] NR = nonreactive (no measurable reduction in COD).
[f] Average based on reactive compounds only.
Source: Adapted from Bowers, 1997.

TABLE 10.8 Chemical Oxidation of Various Aromatic Compounds Using H_2O_2, $KMnO_4$, and O_3

1. *Wet air oxidation (WAO).* A commonly used process for more concentrated wastes, especially those that are toxic and/or biologically refractory. These processes use an oxidant, primarily O_2 from air, to partially oxidize organics, yielding a variety of low-molecular-weight organic acids (readily biodegradable). Usual temperatures range from 150 to 320°C and pressures from 150 to 3000 lb/in² gauge (1.0 to 20.7 Mpa). This process is shown in Fig. 10.9.

2. *Hydrothermal hydrolysis.* Hydrolysis of organic compounds can occur at elevated temperatures and pressures, for example, $CN^- \rightarrow HCOO^-$ or $CCl_4 \rightarrow HCl$. To date, all

Compound	Original EC$_{50}$, %*	EC$_{50}$ After Chemical Oxidation, %*,†		
		H$_2$O$_2$	KMnO$_4$	O$_3$
Pyrrolidine	NT‡	NT	NR§	NT
Sulfanilic acid	14.0 (49.6)	62.0 (117.8)	NR	70.0 (107.1)
Naphthalene	41.0 (5.3)	NT (7.0)¶	NR	NT (14.0)¶
Diphenylamine	14.0 (5.0)	83.0 (9.1)	NR	88.0 (22.0)
Skatole	0.9 (2.6)	4.2 (12.6)	NR	8.0 (24.2)
Benzaldehyde	4.8 (20.2)	54.0 (48.6)	11.0 (15.0)	39.0 (42.1)
Indole	1.3 (6.2)	56.0 (100.8)	25.0 (47.5)	45.0 (84.6)
Catechol	8.1 (28.4)	30.0 (45.3)	49.5 (95.5)	39.0 (108.0)
Hydroquinone	0.3 (1.0)	1.6 (3.9)	63.0 (172.0)	48.0 (143.0)
Resorcinol	55.0 (194.7)	64.0 (98.6)	42.0 (107.9)	48.0 (120.5)
Vanillin	22.0 (103.8)	53.0 (74.2)	52.0 (236.4)	65.0 (113.1)
Pyrogallol	2.8 (9.4)	61.0 (111.6)	50.0 (130.5)	42.0 (102.1)
Salicylic acid	27.0 (110.4)	25.0 (73.0)	26.0 (90.2)	28.0 (98.3)
Coumarin	2.0 (12.4)	NT (460.0)¶	NR	NR
Phthalic acid	NT	NT	NR	NT

*Based on Microtox toxicity; % volume (or concentration) to reduce the light produced from luminescence by 50%.
†Values in parentheses represent the EC$_{50}$ on a TOC basis (as mg/L).
‡NT = nontoxic, i.e., no detectable light reduction.
§NR = nonreactive compound for this particular oxidant.
¶Value in parentheses for nontoxic compounds represents the highest TOC value at no dilution.
Source: Adapted from Bowers, 1997.

TABLE 10.9 Toxicity of Compounds Before and After Oxidation

Type of Dye	Fenton (FSR)	UV/H$_2$O$_2$	O$_3$/H$_2$O$_2$
Reactive	+	+	+
Direct	+	+	+
Metal complex	+	+	+
Pigment	+	−	−
Disperse	−	+	+
Vat	0	−	−
Mixtures	0	+	0

Recommended +
Not recommended −
Applicable 0

TABLE 10.10 Process Recommendations for Different Types of Dyes

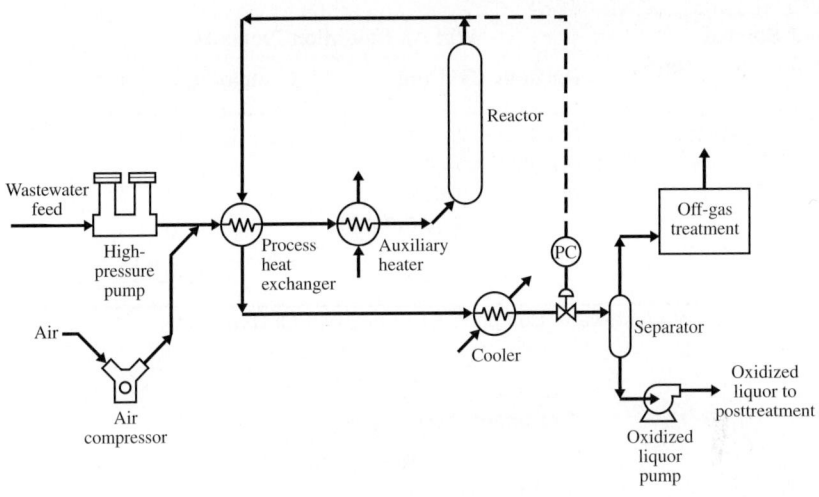

FIGURE 10.9 Wet air oxidation flow scheme.

investigations have been on a laboratory scale. Temperatures proposed for hydrothermal hydrolysis range from 200 to 374°C and pressure from 220 to 3200 lb/in^2 gauge.

3. *Supercritical water oxidation (SCWO).* Aqueous oxidation of organics takes place to completion, at even higher temperature and pressures than wet air oxidation. SCWO takes place beyond the critical point of water (about 374°C and 218 atm). Typical operating conditions are 400 to 650°C and 3500 to 5000 lb/in^2 gauge (24.1 to 34.5 Mpa). At these temperatures and pressures, materials of construction become critical and solubility of salts can decrease dramatically, causing fouling.

To date, only wet air oxidation has achieved any degree of commercial success. Hydrothermal hydrolysis and SCWO have been mainly demonstrated in the laboratory, and only a few operations have been run continuously in the field. However, WAO is a commercial process, with at least three commercial suppliers of equipment (U.S. Filter/Zimpro, Kenox Corporation, and Nippon Petrochemical) and over 250 units sold to customers. Field units operate a roughly 2.5 to 300 gal/min capacity and treat a variety of wastes from spent caustic solutions to pharmaceuticals. The fate of wastewater components is shown in Table 10.11. A summary treatment performance for several case studies is presented in Table 10.12. These systems all operate with air as the source of oxygen. The high temperatures, pressures, and corrosivity of these wastes require annual maintenance and inspection as well as frequent cleaning to remove scale deposit from the boilers and heat exchangers. With regular maintenance, a

Influent Soluble, Colloidal, or Suspended Species	Wet Air Oxidation Products	
	Partially Oxidized	Completely Oxidized
Complex organics	Low-molecular-weight organics (carboxylic acids, aldehydes, ketones, hydrocarbons)	CO_2, H_2O, HX
Inorganics	NH_4, N_2, SO_3	SO_4^{2-}, PO_4^{3-}, NO_3^{-}

TABLE 10.11 Fate of Wastewater Components During Wet Air Oxidation

	Treatment Parameters		
Waste Compound[†]	Temperature, °C	Residence Time, min	Destruction, %
Acenaphthene	275	60	99.99
Acenaphthene	275	60	99.0
Carbon tetrachloride	275	60	99.7
Chloroform	275	60	99.5
Dibutylphthalate	275	60	99.5
Malathion	250	60	99.9
Mercaptons	200	—	>99.99
4-Nitrophenol	275	—	99.6
Phenols	200	—	97.7–98.2

*Data from U.S. Filter/Zimpro, Rothchild, Wisconsin.
[†]Influent concentrations ranged from 287 to 11,800 mg/L.

TABLE 10.12 Summary of Operating Conditions and Destruction Efficiencies at WAO Installations*

WAO unit is expected to be in service 80 to 85 percent of the time, while having secondary pumps and compressors available can increase in-service time to 90 to 95 percent.

10.10 Problem

10.1. A wastewater has a flow of 40,000 gal/d (151 m³/d) containing 35 mg/L CN^-. Compute the chemical requirements for the alkaline chlorination of this wastewater for CN^- removal.

References

1. Lyman, W. J., W. F. Reehl, and D. H. Rosenblatt: *Handbook of Chemical Property Estimation Methods*, American Chemical Society, Washington D.C., 1990.
2. Lankford, P. W., and W. W. Eckenfelder: *Toxicity Reduction in Industrial Effluents*, Van Nostrand Reinhold, New York, 1990.
3. Stumm, W., and J. J. Morgan: *Aquatic Chemistry*, John Wiley & Sons, New York.
4. Langlais, B., D. A. Reckhow, and D. R. Brink, eds.: *Ozone in Water Treatment: Application and Engineering*, Lewis Publishers, Chelsea, Mich., 1991.
5. Sullivan, D. E., and J. A. Roth: "Kinetics of Ozone Self-Decomposition in Aqueous Solution," *Water 1979*, vol. 76, no. 197; pp. 142–149.
6. Sundstrom, D. W. et al.: presented at the Association of Industrial Chemical Engineers Spring Meeting, April 1994.
7. Albers, H., and Kayser, R.: *42nd Purdue Ind. Waste Conf.*, p. 893, 1988.
8. Bowers, A.: *Chemical Oxidation*, Technomic Pub. Co., Lancaster, Pa., 1997.

CHAPTER 11
Sludge Handling and Disposal

11.1 Introduction

Handling and disposal of residuals produced during wastewater treatment is difficult and expensive. Costs can be as high as 50 to 60 percent of the total system cost. The solids and biosolids (collectively termed *sludge* herein) resulting from waste stabilization are usually liquid or semisolid liquid ranging from 0.25 to12 percent solids by weight. These residuals are typically organic and may become offensive if mismanaged. Hence, the objectives of residuals management are severalfold:

- Render stable sludge
- Reduce sludge volume
- Destroy pathogens
- Facilitate reuse
- Minimize costs

The large quantities of residuals produced offer the opportunity for reuse and innovative processes for "value added" product development. Consequently, the design and operation of the entire wastewater treatment system should be consistent with this goal. Criteria for reuse include:

- Nonhazardous/nontoxic (regulatory consideration)
- Stable (regulatory consideration)
- Noninfectious (regulatory consideration)
- Product must meet end-use specifications

Process selection for solids handling depends on:

- Sludge characteristics
- Ultimate disposal/reuse options

603

- Land availability
- Climate
- Costs
- Plant size

Considerations to any residual management program include:

- Nature of residuals
- Minimize sludge quantity (increase percent solids) and capture fines
- Sludge handling and transportation
  a. Decrease volume
  b. Increase stability
  c. Decrease odors
  d. Decrease moisture
  e. Decrease percent volatile solids
  f. Increase floc strength
- Dewatering characteristics and conditioning
- Ultimate disposal and reuse options
  a. Liability issues
  b. Marketing survey
  c. Outreach programs
  d. Permitting issues

Most of the treatment processes normally employed in industrial water pollution control yield a sludge from a solids-liquid separation process (sedimentation, flotation, etc.) or produce a sludge as a result of a chemical coagulation or a biological reaction. These solids usually undergo a series of treatment steps involving thickening, dewatering, and final disposal. Organic sludges may also undergo treatment for reduction of the organic or volatile content prior to final disposal. Sludges contain free water, capillary water, and bound water. Free water is removed by thickening of the sludge. Capillary water is removed through dewatering. Bound water can be removed only by chemical or thermal means. These mechanisms are shown in Fig. 11.1.

In general, gelatinous-type sludges such as alum or activated sludge yield lower concentrations, whereas primary and inorganic sludges yield higher concentrations in each process sequence.

Conventional sludge handling alternatives are shown in Fig. 11.2. The processes selected depend primarily on the nature and characteristics of the sludge and on the final disposal method employed or on the selected reuse option. For example, activated sludge is more effectively concentrated by flotation than by gravity thickening. Final disposal by incineration desires a solids content that supports its own combustion. In some cases, the process sequence is apparent from experience with similar sludges or by geographical or economic

Sludge Handling and Disposal

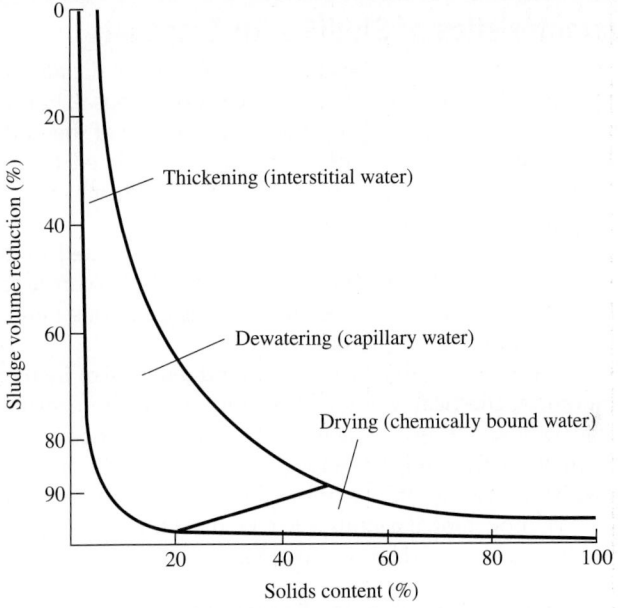

FIGURE 11.1 Mechanisms of sludge dewatering.

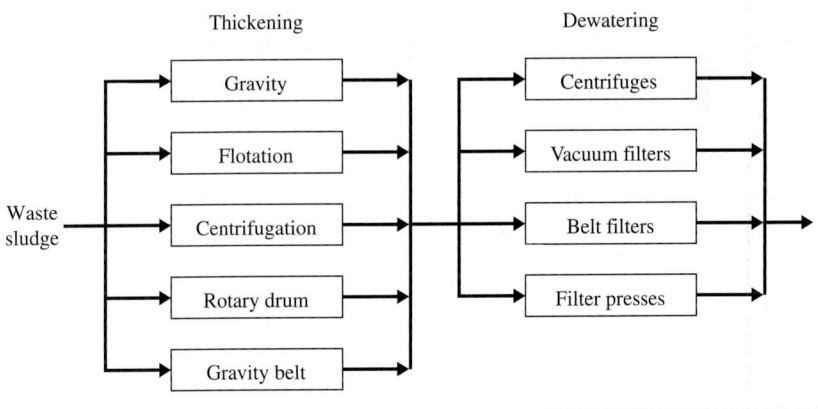

FIGURE 11.2 Sludge handling process alternatives.

constraints. In other cases, an experimental program must be developed to determine the most economical solution to a particular problem.

This chapter discusses sludge characterization and stabilization techniques and methods for sludge handling and disposal, including thickening, dewatering, and final treatment/deposition methods. Process descriptions, design criteria, and examples are included.

11.2 Characteristics of Sludges for Disposal

The physical and chemical characteristics of sludges dictate the most technically and economically effective means of disposal. For thickening, the concentration ratio C_u/C_o (the concentration of the underflow divided by the concentration of the influent) is related to the mass loading [lb solids/(ft² · d) or kg/(m² · d)], which indicates the feasibility of gravity thickening.

The dewaterability of a sludge by filtration is related to the specific resistance. While the specific resistance of a sludge can be reduced by the addition of coagulants, economic considerations may dictate alternative dewatering methods.

Ultimate disposal usually considers land disposal or incineration. In incineration, the heat value of the sludge and the concentration attainable by dewatering dictate the economics of the operation. Land disposal may use the sludge as a fertilizer or soil conditioner, as in the case of waste activated sludges or in a confined landfill for hazardous sludges. It is important, if a sludge is to be used for land disposal, that heavy metals be removed by pretreatment.

Leaching Tests to Characterize Residuals

Sustainable waste management requires that attention be given to reuse of residuals produced during processing and treatment. To be usable, these residuals must be rendered nonhazardous and meet end-use criteria. Characterization of residuals with respect to toxicity, therefore, is a critical issue. Residuals, including sludges containing toxics, if improperly managed may pose a substantial hazard to human health or the environment.

The toxicity characteristic leaching procedure (TCLP) is a protocol developed by the U.S. Environmental Protection Agency (USEPA) (Method 1311) to evaluate the propensity of a waste material to release hazardous constituents in a mismanagement scenario, such as placement of a hazardous waste into a municipal solid waste landfill. The procedure involves chemical analyses of an extract of the waste. If the waste contains less than 0.5 percent filterable solids, the filtrate from the Method 1311 protocol is considered to be the extract. Wastes with higher solids contents are leached with an acetic acid solution to produce an extract. A specific liquid to solid ratio (L/S) of 20:1 is employed as is a zero head extraction is used to account for volatiles. The 100-g sample (<40 mesh) and extraction fluid is rotated for 18 ± 2 h at 30 rpm. The extraction fluid depends on the alkalinity of the waste. Very alkaline wastes are leached with a fixed amount of glacial acetic acid at pH 2.88 ± 0.05. Other wastes are leached at a pH of 4.93 ± 0.05. If any of the extract concentrations from a waste exceed the regulatory levels established by USEPA, that waste is considered to exhibit the hazardous waste characteristic of toxicity. At present, USEPA has promulgated regulatory levels for 40 elements and compounds, including heavy metals, volatile and semivolatile organics, pesticides, and herbicides (see 40 CFR 261.24), as shown in Table 11.1.

Contaminant	Regulatory Level (mg/L)
Arsenic	5.0
Barium	100.0
Benzene	0.5
Cadmium	1.0
Carbon tetrachloride	0.5
Chlordane	0.03
Chlorobenzene	100.0
Chloroform	6.0
Chromium	5.0
o-Cresol	200.0*
m-Cresol	200.0*
p-Cresol	200.0*
Cresol	200.0*
2,4-D	10.0
1,4-Dichlorobenzene	7.5
1,2-Dichloroethane	0.5
1,1-Dichloroethylene	0.7
2,4-Dinitrotoluene	0.13†
Endrin	0.02
Heptachlor (and its epoxide)	0.008
Hexachlorobenzene	0.13†
Hexachlorobutadiene	0.5
Hexachloroethane	3.0
Lead	5.0
Lindane	0.4
Mercury	0.2
Methyoxychlor	10.0

*If o-, m-, and p-cresol concentrations cannot be differentiated, the total cresol concentration is used. The regulatory level of total cresol is 200 mg/L.

†Quantification limit is greater than the calculated regulatory level. The quantification limit therefore becomes the regulatory level.

TABLE 11.1 Maximum Concentration of Contaminants for the Toxicity Characteristic

Contaminant	Regulatory Level (mg/L)
Methyl ethyl ketone	200.0
Nitrobenzene	2.0
Pentachlorophenol	100.0
Pyridine	5.0
Selenium	1.0
Silver	5.0
Tetrachloroethylene	0.7
Toxaphene	0.5
Trichloroethylene	0.5
2,4,5-Trichlorophenol	400.0
2,4,6-Trichlorophenol	2.0
2,4,5-TP (Silvex)	1.0
Vinyl chloride	0.2

*If *o*-, *m*-, and *p*-cresol concentrations cannot be differentiated, the total cresol concentration is used. The regulatory level of total cresol is 200 mg/L.

†Quantification limit is greater than the calculated regulatory level. The quantification limit therefore becomes the regulatory level.

TABLE 11.1 Maximum Concentration of Contaminants for the Toxicity Characteristic (*Continued*)

Regulatory levels are based on a dilution attenuation factor (DAF). The amount of dilution and attenuation will depend on hydrogeologic conditions, the types of pollutants and the rate of discharge. The TCLP determines the allowable limit based on primary drinking water standards. A DAF of 100 is often used to determine the TCLP TC regulatory level for metals.

Note that the TCLP is not required if total analyses indicate compounds are not present or present at levels which could not exceed the regulatory limit. Bulk density testing can be used to determine these levels.

Another batch leaching test is used to simulate the effect of acid rain on land-disposed waste. The USEPA Synthetic Precipitation Leaching Procedure (SPLP) is performed similar to the TCLP test but the extraction fluid consists of two inorganic acids, nitric and sulfuric. Extraction for samples east of the Mississippi River are run at pH 4.22 ± 0.05 to reflect industrialization and high population effects; west of the Mississippi it is conducted at pH 5.0. Leachate concentrations are compared to drinking water standards and groundwater clean-up target levels. A DAF of 20 is usually employed. If a waste is not hazardous in regard to TC, as defined by the TCLP leaching test, the potential risk to human health and the environment must be evaluated when disposal and reuse options of the waste are considered. The leaching results of the SPLP test may then be compared to risk-based water quality standards or guidelines. If risk due to leaching is of no concern, possible direct exposure should be considered for reuse options based on total concentrations of waste constituents. Additional leaching tests may still be conducted at various conditions of pH, L/S ratio, time, redox conditions, kinetics, and so on.

Another batch extraction procedure employed by USEPA is the multiple extraction procedure (MEP) currently used for delisting hazardous wastes. It follows the same procedure as the TCLP run for 24 h. Following the first extraction, at least eight additional extractions are repeated using an inorganic mixture of nitric and sulfuric acids at pH 3.01 ± 0.2 to simulate acid rain.

The state of California uses a waste extraction test (WET) to classify wastes as either hazardous or nonhazardous for regulatory purposes. The test is similar to the TCLP but uses a buffered 0.2M sodium citrate pH 5.0 extraction solution over 48 h at a L/S ratio of 10:1. Results are compared to 17 inorganic listed contaminants. WET testing usually results in higher leachate concentrations for some metals as compared to extraction by SPLP.

Regulatory batch leaching tests are also commonly used in Europe. Descriptions and procedures for many of these have been discussed by Van de Sloot et al.[1] An overview of the use of leaching tests for solid and hazardous waste manager is presented by Townsend et al.[2] The TCLP test may not allow for extrapolation of long-term effects and does not address other potential leaching mechanisms. Hence, its use for most risk-based waste management decisions has been questioned.[3]

11.3 Aerobic Digestion

Aerobic digestion when applied to excess biological sludges involves the oxidation of cellular organic matter through endogenous metabolism. The oxidation of cellular organics has been found to follow first-order kinetics when applied to the degradable volatile suspended solids.[4] Under batch or plug flow conditions,

$$\frac{(X_d)_e}{(X_d)_o} = e^{-k_d t} \tag{11.1}$$

where $(X_d)_e$ = degradable volatile solids after time t
$(X_d)_o$ = initial degradable volatile solids
k_d = reaction rate coefficient, d^{-1}
t = time of aeration, d

The kinetic parameters can be determined from a batch oxidation in the laboratory as shown in Fig. 11.3.

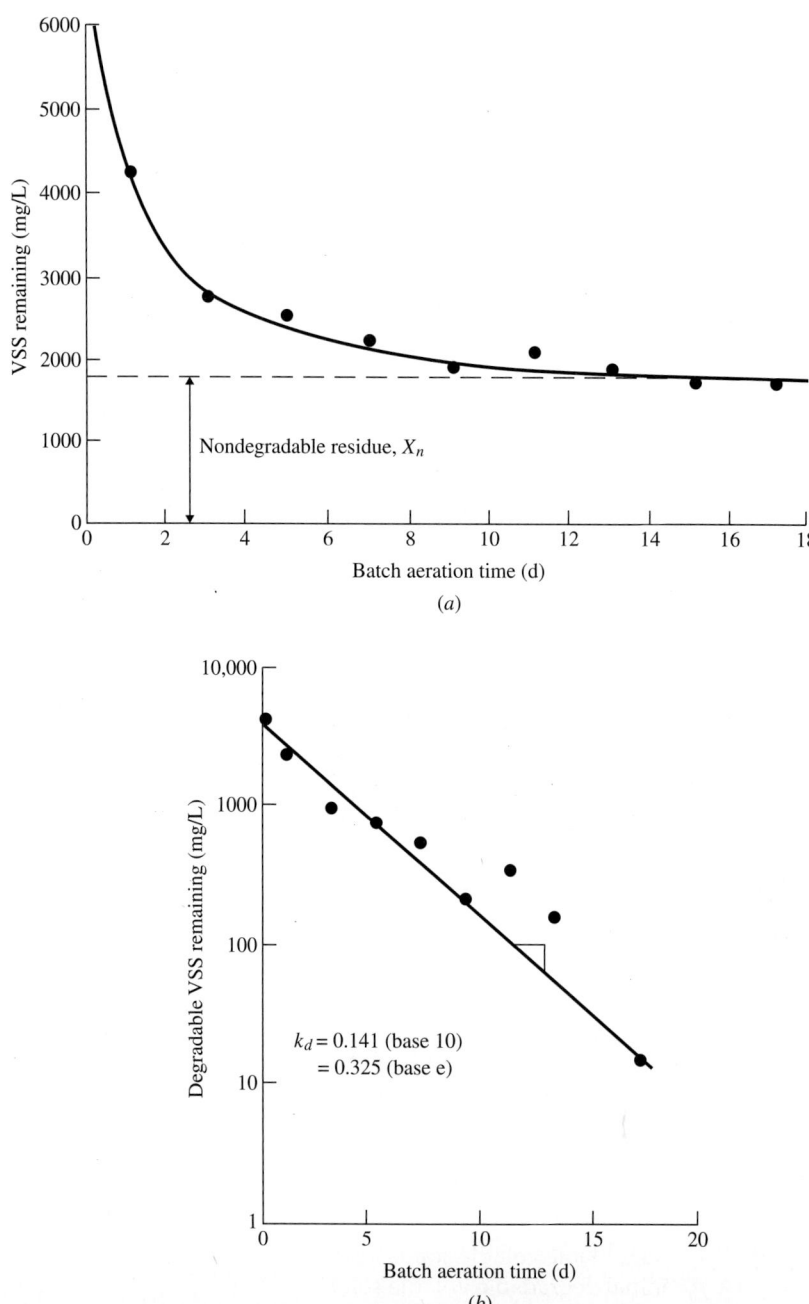

Figure 11.3 Kinetics of aerobic sludge digestion. (a) Chronological destruction of VSS in batch reactor; (b) correlation of degradable VSS with detention time.

Sludge Handling and Disposal

If the total volatile suspended solids are considered, Eq. (11.1) becomes

$$\frac{X_e - X_n}{X_o - X_n} = e^{-k_d t} \tag{11.1a}$$

where X_o = initial VSS
X_e = effluent VSS
X_n = nondegradable VSS

For a completely mixed reactor, the relationship is modified to

$$\frac{X_e - X_n}{X_o - X_n} = \frac{1}{1 + k_d t} \tag{11.2}$$

and the required retention time is

$$t = \frac{X_o - X_e}{k_d (X_e - X_n)} \tag{11.3}$$

For n completely mixed reactors in series,

$$\frac{X_e - X_n}{X_o - X_n} = \frac{1}{(1 + k_d t_n)^n} \tag{11.4}$$

In accordance with the kinetic relationship, mixed reactors in series are more efficient than one mixed reactor. For example, to achieve 90 percent removal of degradable solids at 20°C would require 9.7 days in a single-stage digester and 7.2 days in a three-stage digester.

The oxygen requirements for aerobic digestion can be estimated as 1.4 lb of oxygen consumed for each pound of VSS destroyed (1.4 kg O_2/kg VSS destroyed). Nitrogen and phosphorus will be released by the oxidation process. Under mesophilic digestion conditions, with long sludge ages and a nitrifier seed, nitrification will usually occur. If nitrification does not occur there will be a buildup of ammonia and nondegradable COD. Oxygen and alkalinity must be available for this oxidation. Under thermophilic operation, nitrification will be inhibited.

Temperature will affect the rate coefficient k_d. The temperature relationship is shown in Fig. 11.4. Conventional aerobic digestion design employs secondary clarifier underflow (0.5 to 1.5 percent solids) in one or more completely mixed aeration basins. Power levels of 15 to 20 standard ft³/(min · thousand ft³) [15 to 20 std m³/(min · thousand m³)] using diffused air or 100 hp/million gal (0.02 kW/m³) using surface mechanical aerators are usually adequate for providing both mixing and oxygen requirements. Prethickening the sludge offers a number of advantages, in particular, reducing the basin volume requirements and increasing the temperature due to the exothermic heat of reaction.

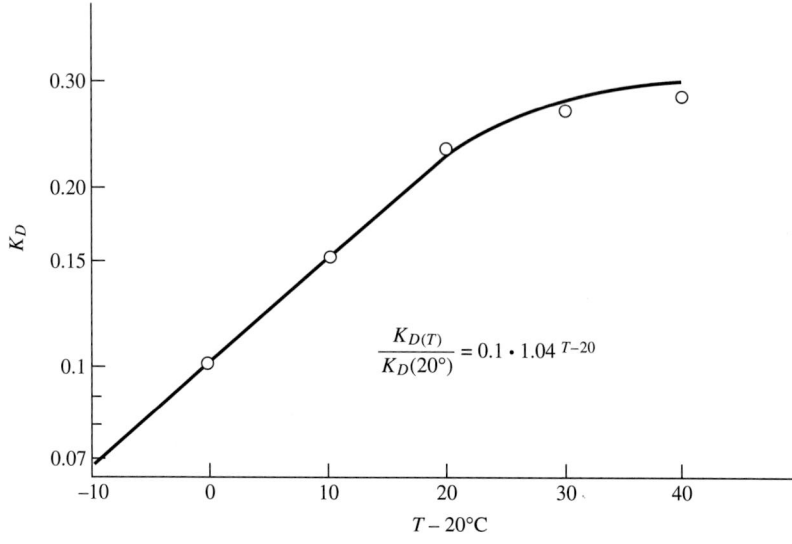

FIGURE 11.4 Temperature effect on aerobic digestion.

Andrews and Kambhu[5] have estimated the heat of combustion as 9000 Btu/lb (2.1×10^7 J/kg) VSS destroyed. Aerobic digestion designs are illustrated in Example 11.1. Aerobic digestion requirements will depend on the operating sludge age in the aeration process. As the sludge age is increased, more of the degradable biomass is oxidized in the aeration basin and hence less will be oxidized in the aerobic digester. Consider two cases as shown in Example 11.1.

Example 11.1. Design an aerobic digester to yield a final degradable fraction of 0.37 for a waste activated sludge from a system with a 10-d SRT and a 30-d SRT. The following data apply for the wastewater:

$S_r = 690$ mg/L
$Q = 5$ million gal/d (18,925 m³/d)
$a = 0.6$
$b = 0.1$ d^{-1}

Solution Equation (6.11) can be rearranged to calculate ΔX_v.

$$\Delta X_v = \frac{aS_r}{1 + bX_d \theta_c}$$

For the 10-d SRT, $X_d = 0.67$.

$$\Delta X_v = \frac{0.6 \cdot 690}{(1 + 0.1 \cdot 0.67 \cdot 10)}$$

$= 248$ mg/L   (10,337 lb/d; 4693 kg/d)

For the 30-d SRT, $X_d = 0.5$.

$$\Delta X_v = \frac{0.6 \cdot 690}{(1 + 0.1 \cdot 0.5 \cdot 30)}$$

$$= 166 \text{ mg/L} \quad (6905 \text{ lb/d; } 3135 \text{ kg/d})$$

A degradable fraction of 0.37 yields a degradable VSS of 2000 lb/d (908 kg/d).

	10-d SRT	30-d SRT
X_o	10,337 lb/d	6905 lb/d
X_D	6926 lb/d (3144 kg/d)	3494 lb/d (1586 kg/d)
X_N	3411 lb/d	3411 lb/d (1549 kg/d)
X_{DN}	2000 lb/d	2000 lb/d
X_e	5411 lb/d	5411 lb/d (2457 kg/d)

Required detention time:

$$t = \frac{X_o - X_e}{k_d(X_e - X_n)}$$

$$k_d = 0.155 \text{ d}^{-1}$$

For a 10-d SRT:

$$t = \frac{10{,}337 - 5411}{0.155(5411 - 3411)}$$

$$= 15.9 \text{ d}$$

For a 30-d SRT:

$$t = \frac{6905 - 5411}{0.155(5411 - 3411)}$$

$$= 4.8 \text{ d}$$

Oxygen requirements:
For VSS destruction,

$$\text{Oxygen} = (X_D - X_{DN}) \cdot 1.4$$

For a 10-d SRT,

$$\text{Oxygen} = (6926 - 2000) \cdot 1.4$$

$$= 6896 \text{ lb/d} \quad (3131 \text{ kg/d})$$

For a 30-d SRT,

$$\text{Oxygen} = (3494 - 2000) \cdot 1.4$$

$$= 2092 \text{ lb/d} \quad (950 \text{ kg/d})$$

Nitrification (see Fig. 6.11):

It is assumed that the NH_3^-N released through endogenous metabolism is nitrified.

$$N_{ox} = N_o X_o - N_e X_e$$

For a 10-d SRT,

$$N_{ox} = 0.091 \cdot 10{,}337 - 0.072 \cdot 5411$$
$$= 551 \text{ lb/d} \quad (250 \text{ kg/d})$$

For a 30-d SRT,

$$N_{ox} = 0.076 \cdot 6905 - 0.072 \cdot 5411$$
$$= 135 \text{ lb/d } (61 \text{ kg/d})$$

The oxygen requirements for nitrification are

$551 \cdot 433 = 2386$ lb/d   (1083 kg/d)   10-d SRT

$135 \cdot 433 = 585$ lb/d   (266 kg/d)   30-d SRT

The total oxygen requirements are

6896 lb/d + 2386 lb/d = 9282 lb/d   (4214 kg/d)   10-d SRT

2092 lb/d + 585 lb/d = 2677 lb/d   (1215 kg/d)   30-d SRT

The alkalinity requirements are

$551 \cdot 7.14 = 3934$ lb/d   (1786 kg/d)   10-d SRT

$135 \cdot 7.14 = 964$ lb/d   (438 kg/d)   30-d SRT

Denitrification:

Assume sludge at 10,000 mg/L. For the 10-d SRT system at 10,337 lb/d, the sludge flow is 0.124 million gal/d. The volume of the digester is

$$0.124 \text{ million gal/d} \cdot 15.9 \text{ d} = 19.7 \text{ million gal}$$

The oxygen-uptake rate for VSS oxidation is

$$\frac{6896 \text{ lb/d}}{1.97 \cdot 8.34 \cdot 24} = 17.5 \text{ mg/(L} \cdot \text{h)}$$

The denitrification rate mg $NO_3^-N/(l \cdot h)$ is estimated as

$$17.5 \cdot 0.25 = 4.4 \text{ mg } NO_3/(L \cdot h)$$

$$NO_3^-N = 551 \text{ lb/d} \quad (250 \text{ kg/d})$$

or 33.5 mg/L in the aeration volume. Then

$$t_{DN} = 33.5/4.4 = 7.6 \text{ h}$$

Recently a focus has been on autothermal aerobic digestion (ATAD). In this process the exothermic heat of combustion of the volatile solids increases the temperature in the reactor to the thermophilic range, that is, 55°C. Deeny et al.[21] showed that an influent VSS of 3 percent was sufficient to maintain a digester temperature of 55 to 60°C. This is illustrated in Example 11.2.

Example 11.2. Determine the temperature achieved in an aerobic digester with feed solids of 3 percent at 22°C. The daily sludge volume is 50,000 gal (189 m³). Volatile suspended solids reduction is 40 percent and the endogenous rate k_d is 0.3 d⁻¹. The degradable fraction of the sludge is 0.67.

Solution From Eq. (11.3),

$$t = \frac{X_o - X_e}{k_d(X_e - X_n)}$$

in which

$$X_o = 30{,}000 \text{ mg/L}$$
$$X_e = 30{,}000 \cdot 0.6 = 18{,}000 \text{ mg/L}$$
$$X_n = 30{,}000 \cdot 0.33 = 9900 \text{ mg/L}$$
$$t = \frac{30{,}000 - 18{,}000}{0.3(18{,}000 - 9900)} = 4.9 \text{ d}$$

The volume of the digester is

$$50{,}000 \cdot 4.9 = 245{,}000 \text{ gal} \quad (927 \text{m}^3)$$

The VSS reduced is

30,000 mg/L · 0.4 · 0.05 million gal/d · 8.34 = 5004 lb/d   (2272 kg/d)

The heat generated is

9300 Btu/lb · 5004 lb/d = 46.5 × 10⁶ Btu/d   (49.1 kJ/d)

The water per day is

50,000 gal/d · 8.34 lb/gal = 0.417 × 10⁶ lb/d   (189,000 kg/d)

The temperature increase will be

$$46.5 \times 10^6 \text{ Btu/d} = \Delta T \cdot 1 \text{ Btu/(lb} \cdot {}^\circ\text{F)} \cdot 0.417 \times 10^6 \text{ lb/d}$$
$$\Delta T = 111.5 {}^\circ\text{F} \quad (44 {}^\circ\text{C})$$

Considering heat losses, this should be sufficient to maintain a digester temperature in excess of 55°C.

The oxygen requirements are

$$5004 \text{ lb/d} \cdot 1.4 = 7006 \text{ lb/d} \quad (3181 \text{ kg/d})$$

and the oxygen uptake rate will be

$$\frac{7006 \text{ lb/d}}{0.245 \text{ mg} \cdot 8.34 \cdot 24} = 142 \text{ mg/(l·h)}$$

11.4 Gravity Thickening

Gravity thickening is accomplished in a tank equipped with a slowly rotating rake mechanism that breaks the bridge between sludge particles, thereby increasing settling and compaction. Gravity thickening is usually applied to primary and chemical sludges which thicken well by gravity settling. A typical gravity thickener is shown in Fig. 11.5.

The primary objective of a thickener is to provide a concentrated sludge underflow. The area of thickener for a specified underflow concentration is related to the mass loading [lb/(ft² · d), kg/(m² · d)] or to the unit area [ft²(lb/d), m²/(kg/d)].

The mass loading can be computed from a stirred laboratory cylinder test. For municipal sewage, the mass loading might be expected to vary from 4 lb/(ft² · d) [19.5 kg/(m² · d)] for waste activated sludge to 22 lb/(ft² · d) [107 kg/(m² · d)] for primary sludge.

A procedure for the design of gravity thickeners has been developed by Dick.[6] The most important criterion in thickener design and operation is the mass loading or solids flux expressed as pounds of solids fed per square foot per day (or kilograms per square meter

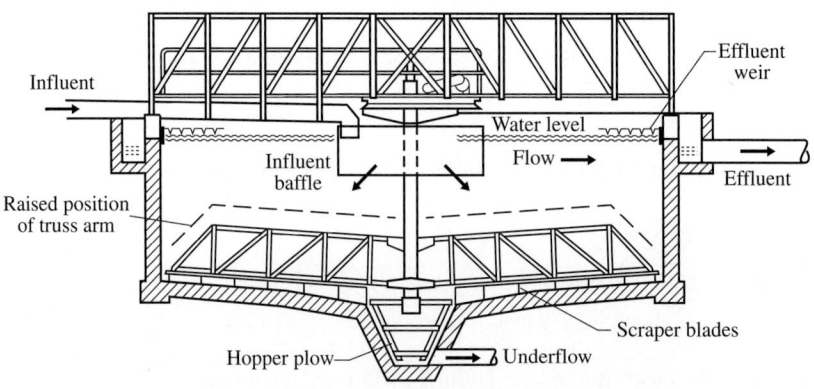

FIGURE 11.5 Gravity thickener (*courtesy of Link Belt, FMC Company*).

per day). The limiting flux that produces the desired underflow for a given area must equal the solids loading rate to the thickener:

$$G_L = \frac{C_o Q_o}{A} = \frac{M}{A} \quad (11.5)$$

where Q_o = influent flow, ft³/d (m³/d)
C_o = influent solids, lb/ft³ (kg/m³)
M = solids loading, lb/d (kg/d)
G_L = limiting solids flux, lb/(ft² · d) [kg/(m² · d)]
A = area, ft² (m²)

The limiting flux can be obtained from the following rationale.
The capacity of a thickener for removing solids under batch conditions is

$$G_B = C_i V_i \quad (11.6)$$

where G_B = batch flux, lb/(ft² · d) [kg/(m²/d)]
C_i = solids concentration, lb/ft³ (kg/m³)
V_i = settling velocity at C_i, ft/d (m/d)

A relationship can be developed between C_i and V_i that is usually linear on a log scale over a wide range of concentrations, as shown for an activated sludge in Fig. 11.6.
In a continuous thickener the solids are removed both by gravity and by the velocity resulting from the removal of sludge from the tank bottom:

$$G = C_i V_i + C_i U \quad (11.7)$$

where G = continuous solids flux, lb/(ft² · d) [kg/(m²/d)], and U = downward sludge velocity due to sludge removal, ft/d (m/d). G can be varied by controlling U since this is determined by the underflow pumping rate. Assuming total solids removal from the bottom:

$$U = \frac{Q_u}{A} = \frac{C_u Q_u}{C_u A} = \frac{M}{C_u A} = \frac{G_L}{C_u} \quad (11.8)$$

where Q_u = underflow, ft³/d (m³/d), and C_u = underflow concentration, lb/ft³ (kg/m²). It is important to note from Eq. (11.8) that by increasing U, the withdrawal rate decreases the underflow concentration C_u. A batch flux curve as shown in Fig. 11.7 can be employed to determine the limiting flux G_L for a given underflow concentration C_u. This is because the slope of any line connecting G_L on the y axis with C_u on the x axis on the batch flux curve, as shown in Fig. 11.7, is obtained by plotting G_B [computed from Eq. (11.6)] against its corresponding concentration C_i.

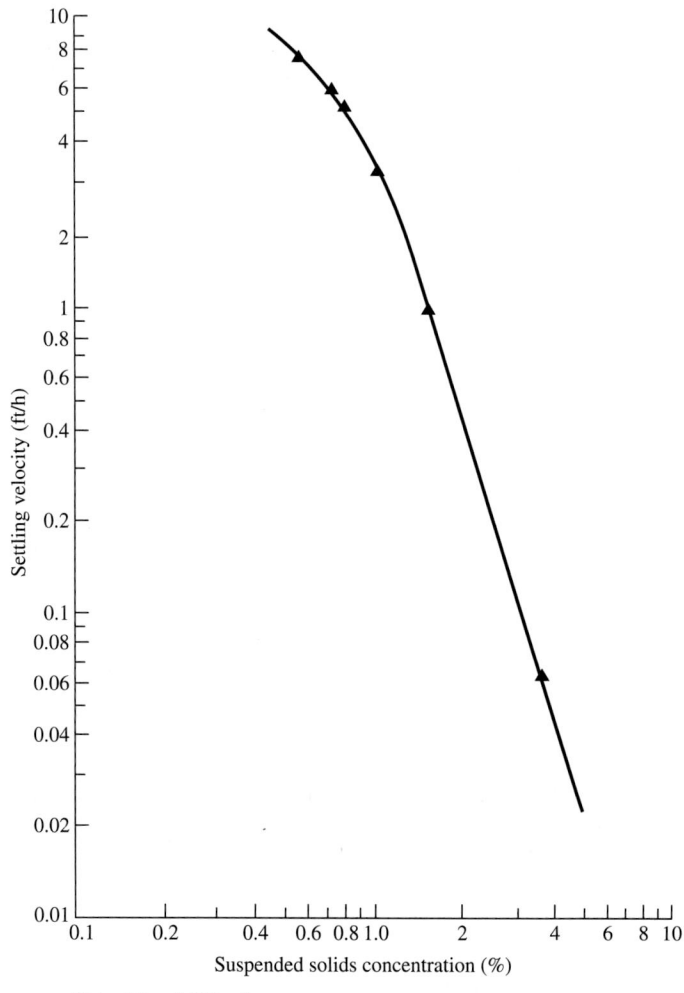

FIGURE 11.6 Sludge settling characteristics.

The required thickener area, A, is then computed from Eq. (11.5). It should be noted that the selected underflow concentration C_u must be less than the ultimate concentration attainable, C_∞. C_∞ is determined from thickening studies as

$$C_\infty = \frac{C_o H_o}{H_\infty} \qquad (11.9)$$

where C_o = initial solids concentration
 H_o = initial height
 C_∞ = final or ultimate concentration
 H_∞ = final height

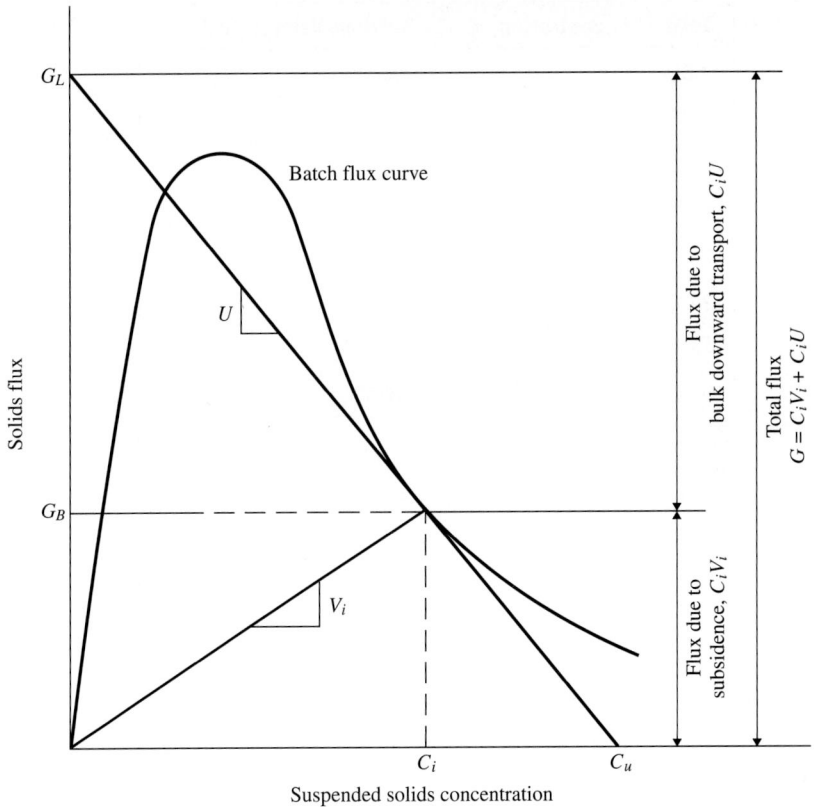

FIGURE 11.7 Batch flux plot illustrating how to determine limiting flux for a continuous thickener.

Example 11.3 illustrates a gravity thickener design.

Example 11.3. Waste sludge from a chemical coagulation process is to be gravity-thickened from 0.5 to 4 percent. The average sludge volume is 550,000 gal/d (2082 m³/d) with a variation of 450,000 to 700,000 gal/d (1703 to 2650 m³/d). Determine the thickener area required and the underflow solids concentration at minimum flow.

Solution The relationship between zone settling velocity and suspended solids concentration is shown in Table 11.2. The batch flux curve is developed by plotting the flux G versus its corresponding concentration. For example, at 2 percent solids,

$$G = 0.02 \times 62.4 \text{ lb/ft}^3 \times 0.50 \text{ ft/h} \times 24 \text{ h/d}$$
$$= 15.0 \text{ lb/(ft}^2 \cdot \text{d)} \quad [73.3 \text{ kg/(m}^2 \cdot \text{d)}]$$

The batch flux curve is shown in Fig. 11.8. For the desired underflow concentration of 4 percent, the limiting flux is found from the batch flux curve by constructing a

Solids Concentration, %	Settling Velocity, ft/h
0.50	7.5
0.75	5.5
1.00	4.2
1.25	3.1
1.50	1.5
2.00	0.50
4.00	0.075
6.00	0.030

Note: ft/h = 0.3048 m/h.

TABLE 11.2 Batch Settling Data

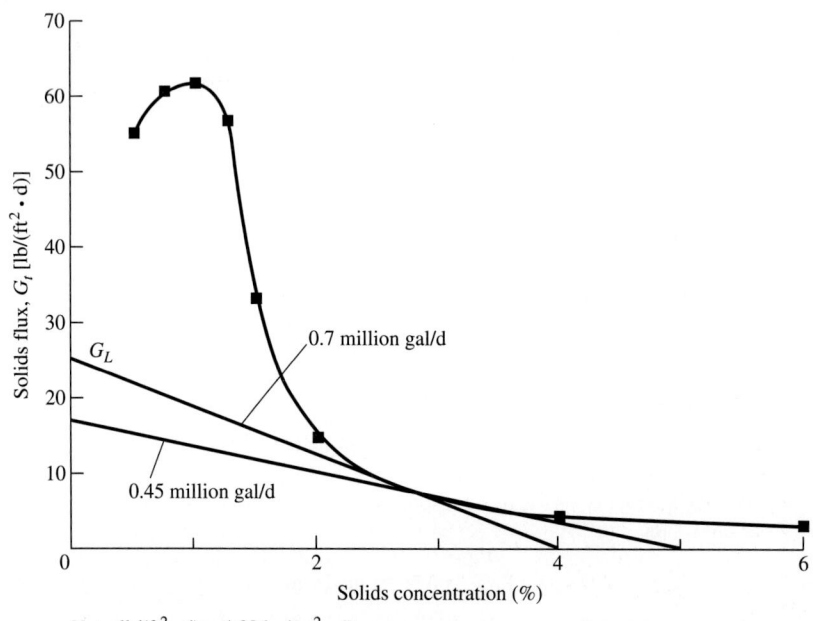

Note: lb/(ft² · d) = 4.89 kg/(m² · d)

FIGURE 11.8 Batch flux curve.

tangent which extends from 4 percent solids to the limiting flux of $G_L = 26$ lb/(ft² · d) [127 kg/(m² · d)]. The required thickener area is calculated:

$$A = \frac{C_o Q_o}{G}$$

$$= \frac{(0.7 \text{ million gal/d})(5000 \text{ mg/L})(8.34)(\text{lb/million gal})(\text{mg/L})}{26 \text{ lb/(ft}^2 \cdot \text{d)}}$$

$$= 1123 \text{ ft}^2 \quad (104 \text{ m}^2)$$

When the sludge flow to the thickener is 0.45 million gal/d, the solids flux will be

$$G = \frac{(0.45 \text{ million gal/d})(5000 \text{ mg/L})(8.34)(\text{lb/million gal})/(\text{mg/L})}{1123 \text{ ft}^2}$$

$$= 16.7 \text{ lb/(ft}^2 \cdot \text{d)} \quad [81.6 \text{ kg (m}^2 \cdot \text{d)}]$$

From the batch flux curve (Fig. 11.8), the underflow concentration at this loading will be 4.9 percent.

11.5 Flotation Thickening

Thickening through dissolved air flotation is particularly applicable to gelatinous sludges such as activated sludge. In flotation thickening, small air bubbles released from solution attach themselves to and become enmeshed in the sludge flocs. The air-solid mixture rises to the surface of the basin, where it concentrates and is removed. The primary variables are recycle ratio, feed solids concentration, air/solids (A/S) ratio, and solids and hydraulic loading rates. Pressures between 50 and 70 lb/in² (345 to 483 kPa or 3.4 to 4.8 atm) are commonly employed. Recycle ratio is related to the A/S ratio and the feed solids concentration. The float solids are related to the A/S ratio, as shown in Fig. 11.9. Typical design criteria are:

Solids loading rate	50 to 100 kg/m² d
Pressurized effluent recycle	15 to 120 percent
Pressure requirement	3 to 5 bars
A/S ratio	0.005 to 0.060 kg/kg

Experience has shown that in some cases dilution of the feed sludge to a lower concentration increases the concentration of the floated solids. A flotation thickener is shown in Fig. 11.10. Performance data for the thickening of excess activated sludge is shown in Table 11.3. The use of polyelectrolytes will usually increase the solids capture.

The quality of the sludge has a significant effect on its ability to thicken by flotation. For example, a filamentous, bulking activated sludge may not achieve 2 percent solids, compared to 4 to 5 percent with a well-flocculated sludge. This in turn will affect all other solids handling operations.

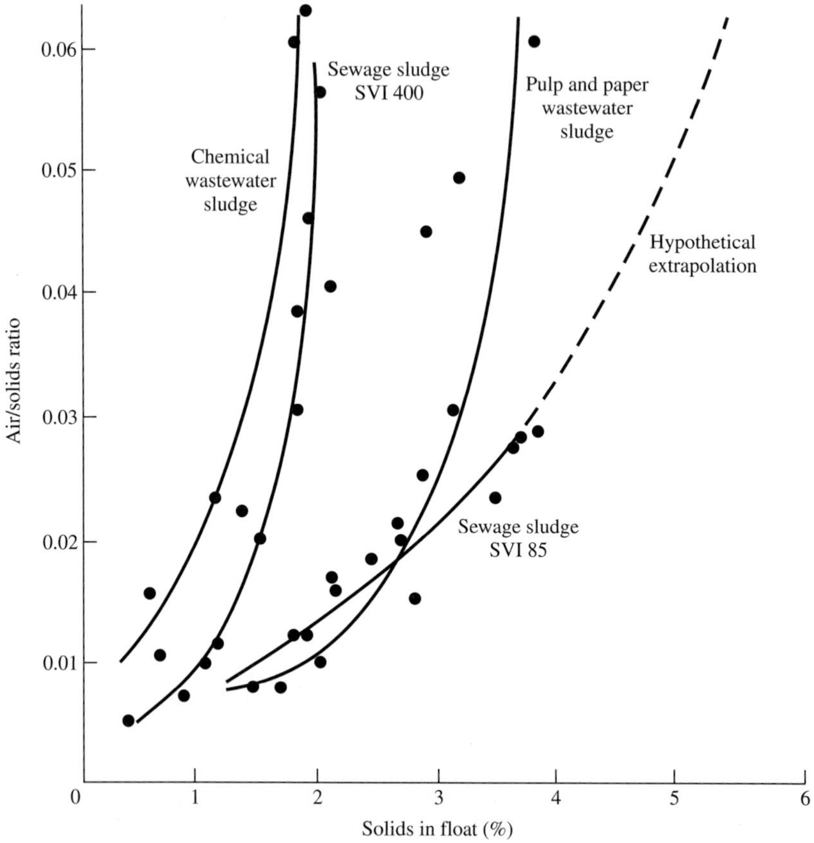

FIGURE 11.9 Influence of air/solids ratio on float solids content.

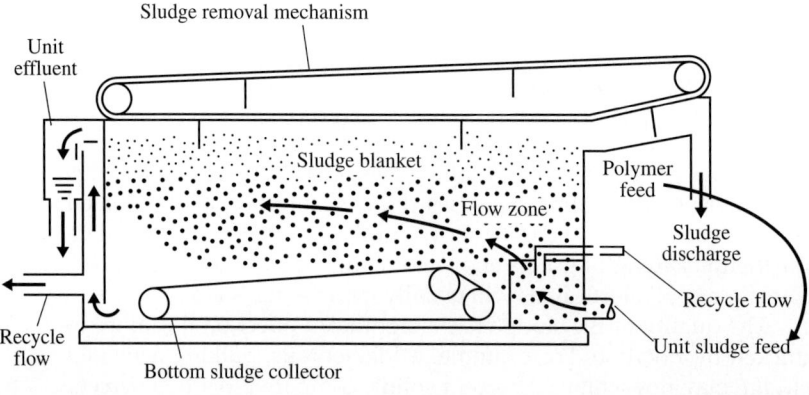

FIGURE 11.10 Flow diagram of a flotation unit (*courtesy of Komline-Sanderson Engineering Company*).

Equipment	Type of Sludge	Loading	Cake Solids, %	Chemicals, lb/ton Polymer	Reference
Thickening					
Gravity	WAS	5–6 lb/(ft² · d)	2.5–3.0	None	12
Gravity	Pulp and paper 53% P, 47% WAS	25	4	None	12
	67% P, 33% WAS	25	6	None	—
	100% P	25	9	None	—
DAF	WAS	2.9–4.5 lb/(ft² · h)	4–5.7	Low dosage	12
Solid bowl centrifuge	WAS	75–100 gal/min	5–7	None	12
Basket centrifuge	Citrus, WAS	25–40 gal/min	9–10	10–20	13
Gravity belt	WAS	315 gal/min	5.5	39	—

Note: lb/(ft² · d) = 4.88 kg/(m² · d)
gal/min = 3.78 × 10⁻³ m³/min
WAS = waste activated sludge
P = primary sludge

TABLE 11.3 Thickening and Dewatering of Wastewater Sludges

Equipment	Type of Sludge	Loading	Cake Solids, %	Chemicals, lb/ton Polymer	Reference
Solid bowl centrifuge	Paper mill, WAS	100 gal/min	11	10	14
Solid bowl centrifuge	Chemical, WAS	—	7–9	5–10	15
Dewatering					
Basket centrifuge	Citrus, WAS	25–40 gal/min	9–10	10–20	13
Basket centrifuge	Paper mill, WAS	60 gal/min	11	5	14
Belt press	Citrus, WAS	40 gal/(min · m)	18	10–20	13
Belt press	Paper mill, WAS	70 gal/(min · m)	16	6.5	14
Belt press	Chemical, WAS	—	13–15	10–20	15
Belt press	Organic chemical, WAS	190 gal/min	15	25	16
Belt press	Deinking primary	500 L/m	37	4	17
Belt press	Bleached and unbleached kraft; 67 percent P, 33 percent WAS	240 L/m	27	12	17
Belt press	Kraft linerboard	75 L/m	19	25	17

Note: lb/(ft^2 · d) = 4.88 kg/(m^2 · d)
gal/min = 3.78 × 10^{-3} m^3/min
WAS = waste activated sludge
P = primary sludge

TABLE 11.3 Thickening and Dewatering of Wastewater Sludges (*Continued*)

11.6 Rotary Drum Screen

A rotary drum screen consists of a stainless steel or nonferrous wire mesh screen cloth. Screen openings typically vary from 6 to 20 mm. The drum revolves at about 4 rev/min around a horizontal axis and operates slightly less than half submerged. The wastewater flows in one end of the drum and outward through the screen cloth. Solids are raised above the liquid level by rotation of the screen and are backflushed into receiving troughs by high-pressure jets. With the finer mesh, effluent may be used for spray water. A typical rotary drum screen is shown in Fig. 11.11.

Typical design criteria for waste activated sludge are

Loading rate	33 L/min/m²
Polymer consumption	5 to 9 kg/ton of solids
Capture rate	95 to 99 percent

Low SVI sludge will thicken to 6 to 10 percent with a filtrate clarity of 100 to 500 mg/L.

11.7 Gravity Belt Thickener

The gravity belt thickener reduces the volume of water in the sludges by using gravity forces to remove water that has been freed by polymer or chemical conditioning. Polymer or chemical is injected into the sludge feed line and mixed by an adjustable in-line mixer. Conditioned sludge enters a stainless steel tank and is then distributed evenly across the width of the belt without shearing the flocculated sludge

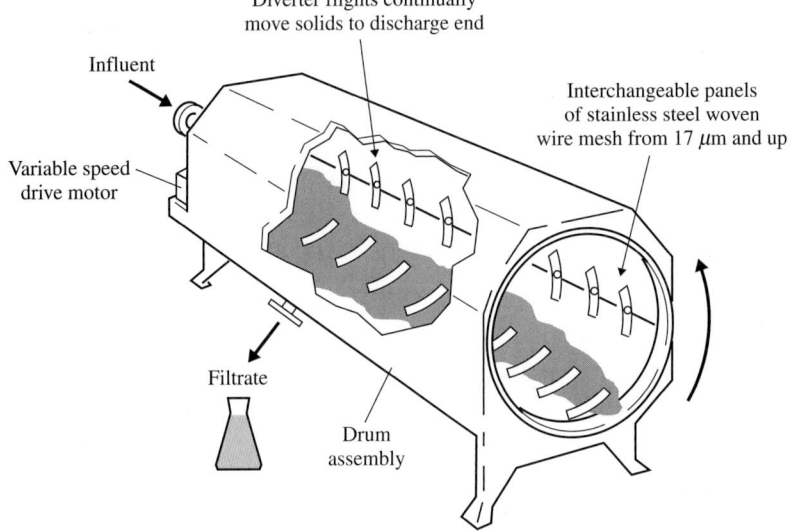

FIGURE 11.11 Rotary drum screen thickener.

particles. A series of free-floating plows furrow and roll the sludge to expose ponded water to open areas on the belt while a low-wear grid system supports the belt and shears capillary water from its underside. An adjustable-angle discharge dam rolls the sludge for maximum thickened solids concentration. Thickened sludge is continuously removed by an adjustable spring-tensioned blade. The drainage belt then passes through a high-pressure/low-volume shower assembly to remove any particles trapped in the belt. Typically a hydraulic loading rate up to 35 $m^3/h/m$ of width is employed. Thickened sludge solids concentration of 4 to 7 percent is achievable. A typical gravity belt thickener is shown in Fig. 11.12.

11.8 Disk Centrifuge

In the disk-nozzle separator, the feed enters at the top and is distributed between a multitude of channels, or spaces between the stacked conical disks. Solid particles settle through the layer of liquid flowing in these channels to the underside of the disk, and then slide down to a sludge compaction zone. The thickened sludge is flushed out of the bowl with a portion of the wastewater, thus limiting the solids concentration to 10 to 20 times the feed rate. The disk-nozzle separator finds its major application in the thickening of activated and similar sludges. They are very efficient in thickening waste activated sludge at high feed rates without the addition of polymers. In an industrial waste treatment plant in Germany, excess activated sludge has been thickened from 1 percent to 8 to 10 percent solids.

11.9 Basket Centrifuge

In the basket-type centrifuge, feed is introduced in the bottom of the basket. At equilibrium, solids settle out of the annular moving liquid layer to the layer that builds up on the bowl wall, while the centrate overflows the lip at the top. When solids have filled the basket, feed is stopped, the basket speed is reduced, and a knife moves into the cake, discharging it from the bottom of the casing. Cycles are automated and cake unloading requires less than 10 percent of the cycle time. Chemical addition is generally not required for high solids recovery. However, the unit operates at low centrifugal forces and has a discontinuous cake discharge and a fairly low solids handling capacity.

Activated sludge can be thickened to 4.5 to 8.0 percent total solids by using up to 2.5 g of cationic polymer per kilogram waste. A low-SVI sludge will yield the higher cake density.

A solid bowl centrifuge is also used for sludge thickening. It can usually obtain over 90 percent capture without the use of polymers.

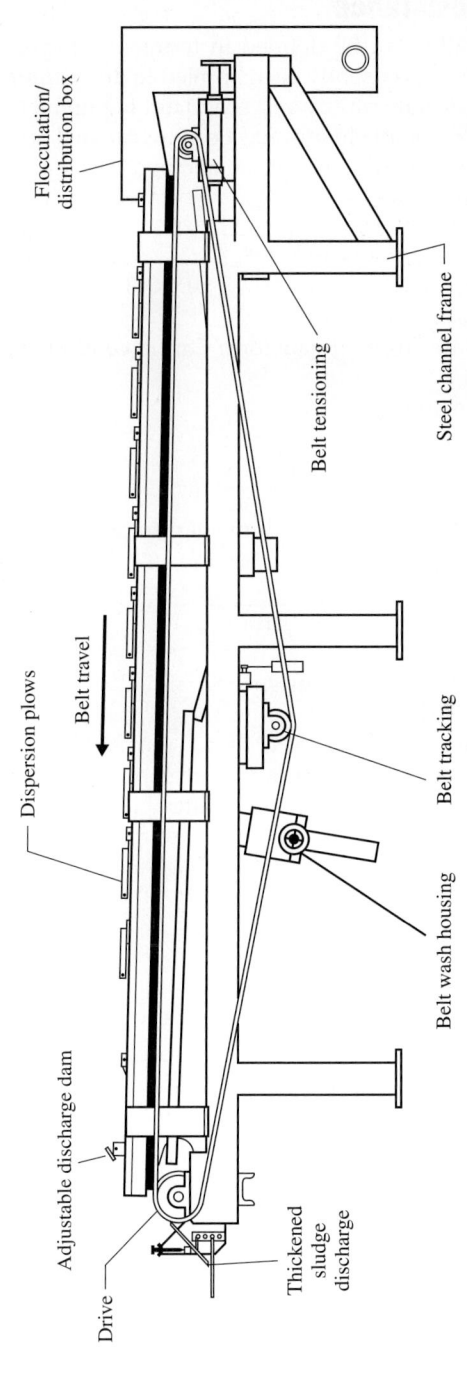

FIGURE 11.12 Gravity belt thickener.

11.10 Specific Resistance

Sludge dewaterability can be defined in terms of its specific resistance. While this test has usually been applied to drum filters, it does give a measure of dewaterability and coagulant requirements.

The rate of filtration of sludges has been formulated according to Poiseuille's and Darcy's laws by Carman and by Coackley and Jones:[7]

$$\frac{dV}{dt} = \frac{PA^2}{\mu(rcV + R_m A)} \quad (11.10)$$

where V = volume of filtrate, mL
t = cycle time (approximate form time in continuous drum filters), s
P = vacuum, in Hg
A = filtration area, cm^2
μ = filtrate viscosity, s^2/g
r = specific resistance, gr · s^2/g^2
c = weight of solids/unit volume of filtrate, g/mL

$$c = \frac{1}{C_i/(100 - C_i) - C_f(100 - C_f)} \text{ units} \quad (11.11)$$

where C_i = initial moisture content, %, and C_f = final moisture content, %.

R_m, the initial resistance of the filter media, can usually be neglected, as it is small compared to the resistance developed by the filter cake. The specific resistance r is a measure of the filterability of the sludge and is numerically equal to the pressure difference required to produce a unit rate of filtrate flow of unit viscosity through a unit weight of cake.

Integration and rearrangement of Eq. (11.10) yields

$$\frac{t}{V} = \left(\frac{\mu rc}{2PA^2}\right)V + \frac{\mu R_m}{PA} \quad (11.12)$$

From Eq. (11.12), a linear relationship will result from a plot of t/V versus V. The specific resistance can be computed from the slope of this plot:

$$r = \frac{2bPA^2}{\mu c} \quad (11.12a)$$

where b = slope of the t/V versus V plot.

Although specific resistance has limited value for the design of a sludge dewatering device, it provides a valuable tool for the evaluation of the relative filterability of sludges. Typical values are given in Table 11.4.

Description		Specific Resistance, $(gr \cdot s^2/g^2) \times 10^{-7}$	Coefficient of Compressibility
Domestic activated sludge		2800	
Activated (digested)		800	
Primary (raw)		1310–2110	
Primary (digested)		380–2170	
Primary (digested)		1350	25.00
Primary (digested)			
Detention time	Stage		
7.5 d	1	1590	
10.0 d	1	1540	
15.0 d	1	1230	
20.0 d	1	530	
30.0 d	1	760	
15.0 d	2	400	
20.0 d	2	400	
30.0 d	2	480	
Activated sludge + 13.5% $FeCl_3$		45	
Activated sludge + 10.0% $FeCl_3$		75	
Activated sludge + 125% (by weight) newsprint		15	
Activated digested sludge + 6% $FeCl_3$ + 10% CaO		5	
Activated digested sludge + 125% newsprint + 5% CaO		4.5	
Vegetable-processing sludge		46	7.00
Vegetable tanning		15	20.00
Lime neutralization acid mine drainage		30	10.50
Alum sludge (waterworks)		530	14.50
Neutralization of sulfuric acid with lime slurry		1–2	

*All values were recorded at 500 gr/cm² pressure.

TABLE 11.4 Specific Resistance of Sludges*

Description	Specific Resistance, $(gr \cdot s^2/g^2) \times 10^{-7}$	Coefficient of Compressibility
Neutralization of sulfuric acid with dolomitic lime slurry	3	0.77
Aluminum processing	3	0.44
Paper industry	6	
Coal (froth flotation)	80	1.60
Distillery	200	1.30
Mixed chrome and vegetable tannery	300	
Chemical wastes (biological treatment)	300	
Petroleum industry (from gravity separators)		
Refinery A	10–100	0.50
Refinery B	100	0.70

*All values were recorded at 500 gr/cm² pressure.

TABLE 11.4 Specific Resistance of Sludges* (*Continued*)

Most wastewater sludges form compressible cakes in which the filtration rate (and the specific resistance) is a function of the pressure difference across the cake:

$$r = r_0 P^s \tag{11.13}$$

where s = coefficient of compressibility. The greater the value of s, the more compressible is the sludge. When $s = 0$, the specific resistance is independent of pressure and the sludge is incompressible.

Some generalizations on filtration characteristics can be made. Filterability is influenced by particle size, shape, and density, and by the electrical charge on the particle. Smaller particles exert a greater chemical demand than larger particles. The larger the particle size, the higher is the filter rate and the lower the cake moisture. Municipal and industrial sludges filter very poorly, and coagulants must be added; lime and ferric salts have been the most common coagulating agents used in the past. Polyelectrolytes have proved effective coagulants in many applications. Frequently, the dual use of anionic and cationic polymers is the most economic and effective procedure. The cationic polymer affects charge neutralization and the anionic affects polymer particle bridging and agglomeration of the particles. Note that excessive coagulant dosages result in a charge reversal and in an increase in specific resistance.

Laboratory Procedures

Filtration characteristics of sludges can be obtained in the laboratory by the Büchner funnel test. The Büchner funnel test can be used to determine specific resistance. It is usually possible, however, to determine the sludge compressibility s and the optimum coagulant dosage from a series of Büchner funnel tests.

The procedure for the Büchner funnel test is as follows (see Fig. 11.13):

1. Prepare the Büchner funnel by placing a wire mesh or screen under the filter paper to ensure drainage.
2. Moisten the filter paper with water and adjust the vacuum to obtain a seal.

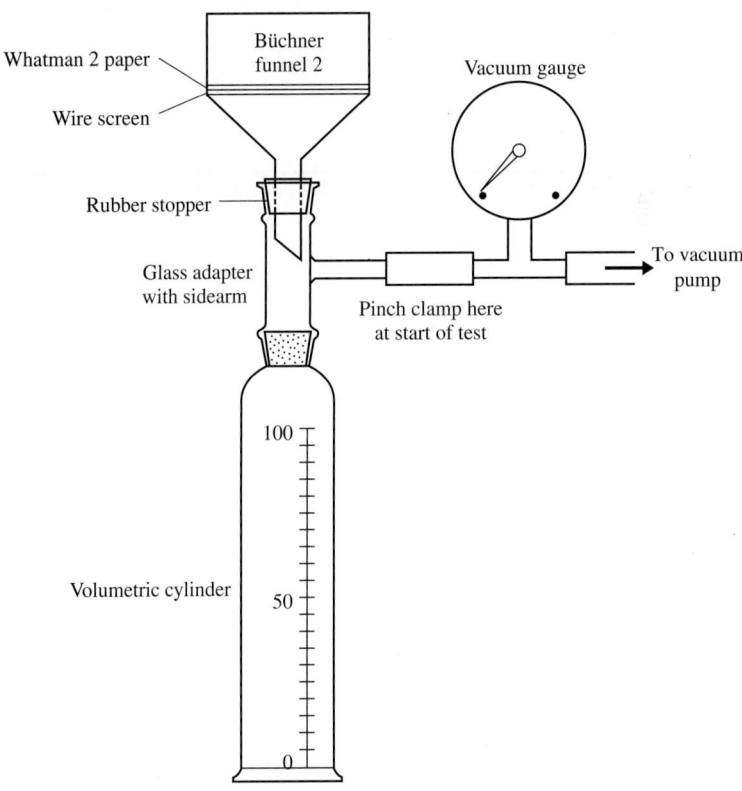

Figure 11.13 Büchner funnel assembly for determination of sludge filterability.

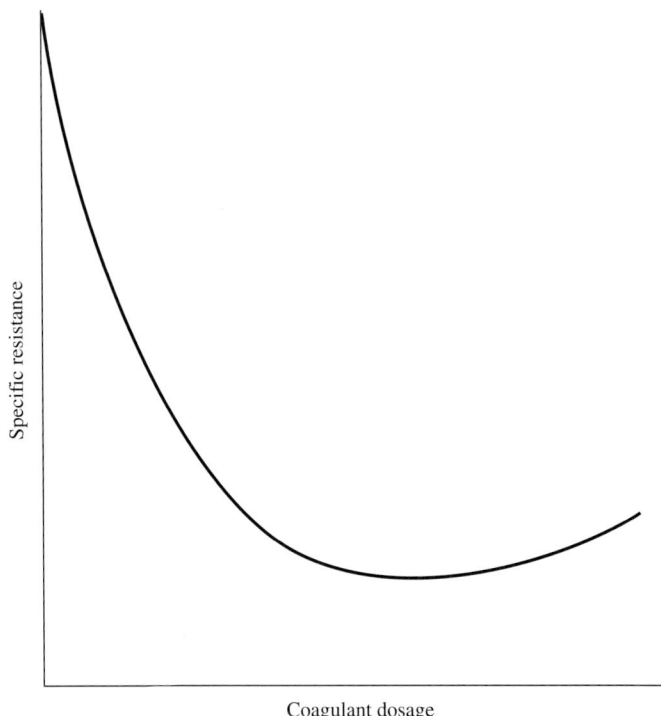

FIGURE 11.14 Relation between specific resistance and coagulant dosage.

3. Condition the sludge if necessary; mix it and permit it to stand 30 s to 1 min; use 200 mL samples. The relationship between specific resistance and coagulant dosage is shown in Fig. 11.14.
4. Transfer it to the Büchner funnel, allow sufficient time for a cake to form (5 to 10 s), and apply the vacuum.
5. Record the milliliters of filtrate after selected time intervals (usually 5 to 10 s).
6. Continue filtration until the vacuum breaks.
7. Determine the initial and final solids in the feed sludge and cake.
8. Record the data obtained and calculate the specific resistance in accordance with Eq. (11.12a).

Calculations for specific resistance are shown in Example 11.4.

Example 11.4. Calculate the specific resistance given the values in the table and the data in Fig. 11.15,

where $A = 176.5$ cm^2
$T = 84°$F (29°C)

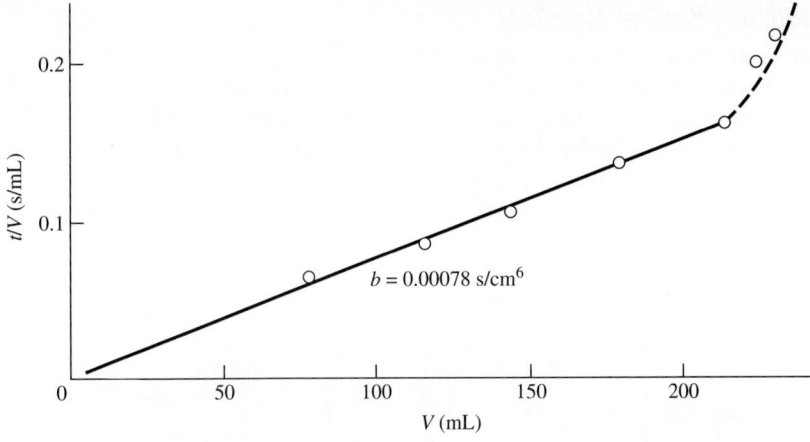

FIGURE 11.15 Determination of variable b for specific resistance calculation.

$P = 20$ in Hg $= 704$ g/cm^2
$C_i = 97.6\%$
$C_j = 77.4\%$
$\mu = 0.01$ s^2/g (viscosity at 29°C)
$b = 0.0007$

Time, s	Volume, mL	t/V
5	78	0.064
10	114	0.088
15	142	0.106
25	178	0.140
35	212	0.165
45	224	0.201
50	228	0.220

Solution

$$r = \frac{2PbA^2}{\mu c}$$

$$c = \frac{1}{C_i/(100-C_i) - C_f/(100-C_f)}$$

$$= \frac{1}{97.6/2.4 - 77.4/22.6} = 0.0269 \text{ g/mL}$$

$$r = \frac{(2)(0.00078)(704)(176.5)^2}{(0.0269)}$$

$$= 12.7 \times 10^7 \text{ gr} \cdot \text{s/g}^2$$

Capillary Suction Time Test

An evaluation technique based on the capillary suction time (CST) has been found to be a rapid, easy, inexpensive, and reproducible method of characterizing the dewaterability of a given sludge. The assembly developed by the Water Research Center (WRC) at Stevanage Laboratory, England, is shown in Fig. 11.16. Filtrate is withdrawn from the sludge sample by capillary adsorbent filter paper.

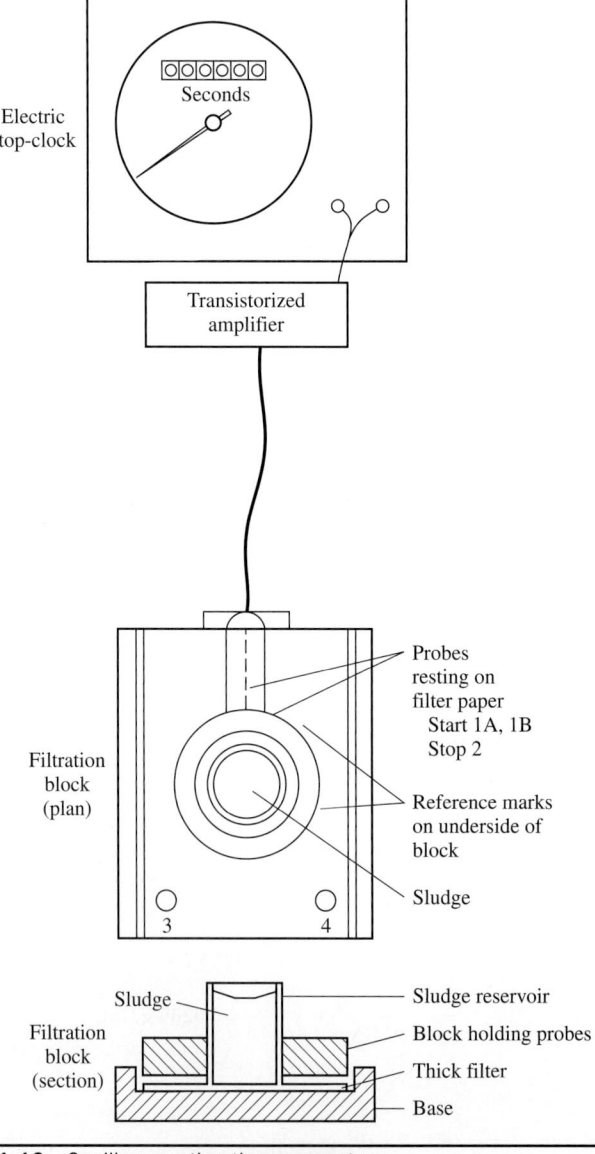

FIGURE 11.16 Capillary suction time apparatus.

Filterability is measured by observing the time for an area of paper to become wetted. The CST can be correlated to specific resistance. The optimum coagulant and its respective dosage can be determined from a CST versus coagulant dosage relationship.

11.11 Centrifugation

Centrifuge performance is affected by both machine and process variables. The significant machine variables for the solid bowl decanter are bowl speed, pool volume, and conveyor speed. Process variables include the feed rate of solids to the machine, solids characteristics, chemical addition, and temperature.

The solid bowl decanter consists of an imperforated cylindrical-conical bowl with an internal helical conveyor as shown in Fig. 11.17. The feed sludge enters the cylindrical bowl through the conveyor discharge nozzles. Centrifugal forces compact the sludge against the bowl wall, and the internal scroll or conveyor, which rotates slightly slower than the bowl, conveys the compacted sludge along the bowl wall toward the conical section (beach area) and out.

When the feed rate to a centrifuge is increased, the retention time in the unit is decreased and the recovery decreases. Flow rates are usually limited to 0.5 to 2.0 gal/(min · hp) [3.65 to 14.6 m^3/(d · kW)] to obtain satisfactory solids recovery. Since the lower recovery results in the removal of only larger particles, a drier cake is produced. Increasing the feed solids concentration reduces the liquid overflow from the machine, resulting in an increased recovery of solids.

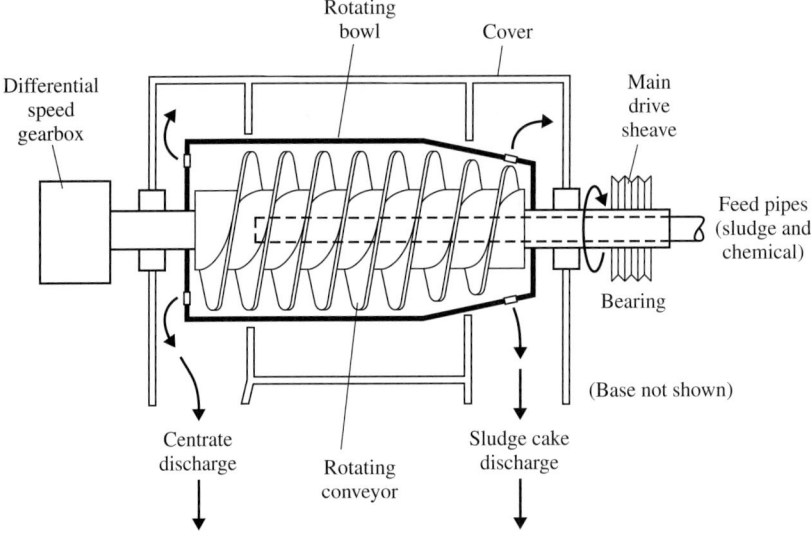

FIGURE 11.17 Continuous countercurrent solid bowl conveyor discharge centrifuge.

Chemical flocculants (polyelectrolytes) are used to increase recovery. The flocculants both increase the structure strength of the solids and flocculate fine particles. Because of the increased removal of the fine particles, chemical addition usually lowers the cake dryness. Centrifuge performance characteristics are shown in Fig. 11.18.

Bernard and Englande[8] correlated centrifuge performance data in accordance with the relationship

$$R = \frac{C_1 (C_2 + P)^m}{Q^n} \qquad (11.14)$$

where R = recovery, percent
   P = polymer dosage, lb/ton dry solids feed (kg/ton)
   Q = feed rate, gal/(min · ft²) [m³/(min · m²)]
   C_1, C_2 = constants
   m, n = exponents

Centrifuge performance in accordance with Eq. (11.14) is shown in Fig. 11.19. Centrifuge design is illustrated in Example 11.5.

Example 11.5. Determine the polymer required and size of centrifuge with respect to surface area for dewatering 10,000 lb/d (4536 kg/d) of sludge previously thickened to 4 percent. The centrifuge will operate over 8 h with a 95 percent solids recovery. The pilot plant data yielded the following relationship:

$$R(\%) = \frac{48(0.47 + P)^{0.37}}{Q^{0.52}}$$

where R = solid recovery efficiency, %
   P = polymer dosage, lb/ton of sludge
   Q = hydraulic loading, gal/(min · ft²)

Solution   For 4 percent solids concentration, the sludge flow is

$$Q' = 10,000 \times \frac{1}{0.04} \times \frac{1}{8.34}$$
$$= 30,000 \text{ gal/d} \qquad (114 \text{ m}^3/\text{d})$$

If the centrifuge will operate for 8 h/d, the total sludge flow to the centrifuge is

$$Q' = 30,000 \times \frac{24}{8}$$
$$= 90,000 \text{ gal/d or } 62.5 \text{ gal/min} \qquad (341 \text{ m}^3/\text{d or } 0.237 \text{ m}^3/\text{min})$$

At 95 percent solids recovery, the relationship between polymer requirement and hydraulic loading can be calculated by

$$95 = \frac{48(0.47 + P)^{0.37}}{Q^{0.52}} \qquad \text{(plot } P \text{ versus } Q \text{ in Fig. 11.20)}$$

Sludge Handling and Disposal 637

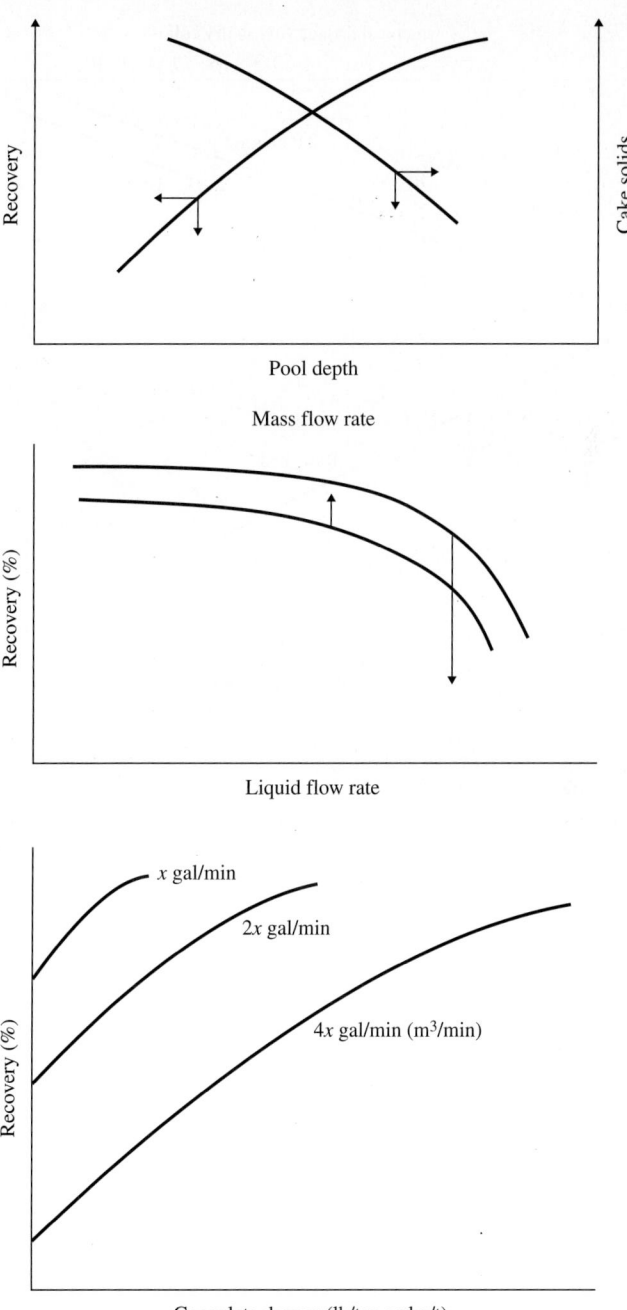

Figure 11.18 Centrifuge operating relationships.

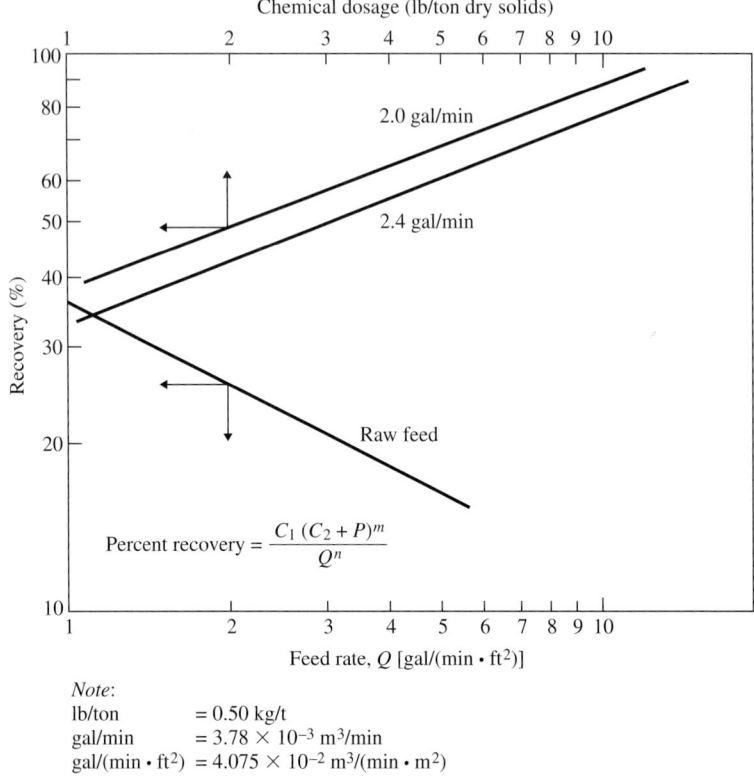

FIGURE 11.19 Solids recovery for digested activated sludge with cationic polymer.

The centrifuge size is computed by

$$A = \frac{62.5}{Q} \text{ ft}^2 \quad \text{(plot } A \text{ versus } Q \text{ in Fig. 11.20)}$$

From Fig. 11.20,

$$A = 46 \text{ ft}^2 \quad (4.3 \text{ m}^2)$$
$$P = 46 \text{ lb/d} \quad (21 \text{ kg/d})$$

11.12 Vacuum Filtration

Vacuum filtration was one of the most common methods for dewatering wastewater sludges. Vacuum filtration dewaters a slurry under applied vacuum by means of a porous medium, which retains the solids but allows the liquid to pass through. Media used include cloth, steel mesh, or tightly wound coil springs.

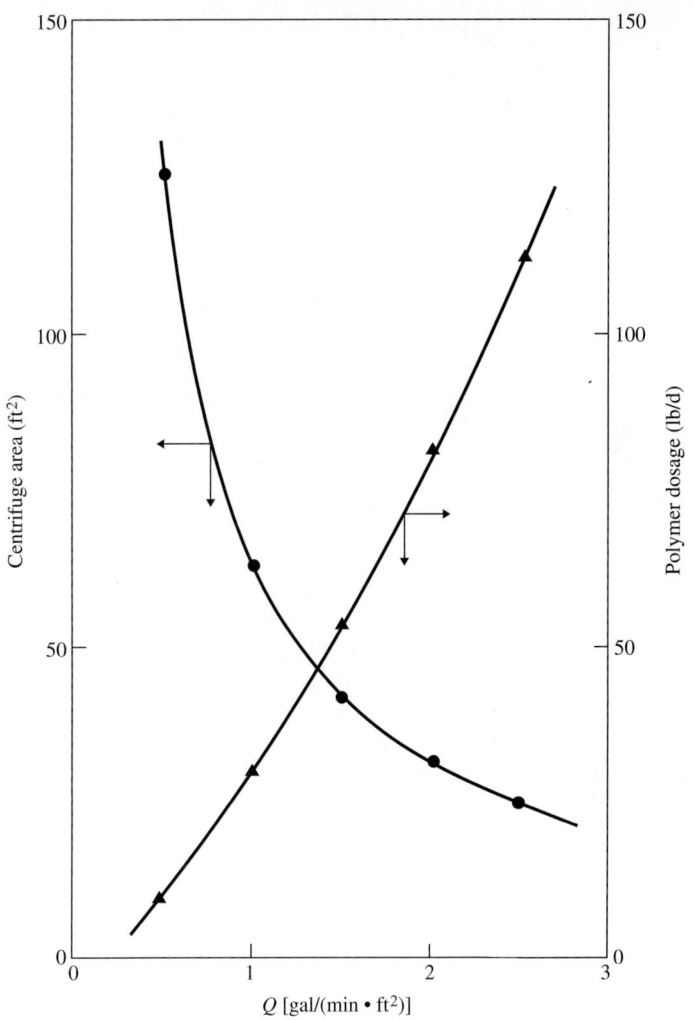

FIGURE 11.20 Required centrifuge area and polymer dosage for various hydraulic loadings.

In vacuum filter operation, a rotary drum passes through a slurry tank in which solids are retained on the drum surface under applied vacuum. The drum submergence can vary from 12 to 60 percent. As the drum passes through the slurry, a cake is built up and water is removed by filtration through the deposited solids and the filter media. The time the drum remains submerged in the slurry is the *form time*, t_f. As the drum emerges from the slurry tank, the deposited cake is further dried by liquid transfer to air drawn through the cake by the applied vacuum. This period of the drum's cycle is called the

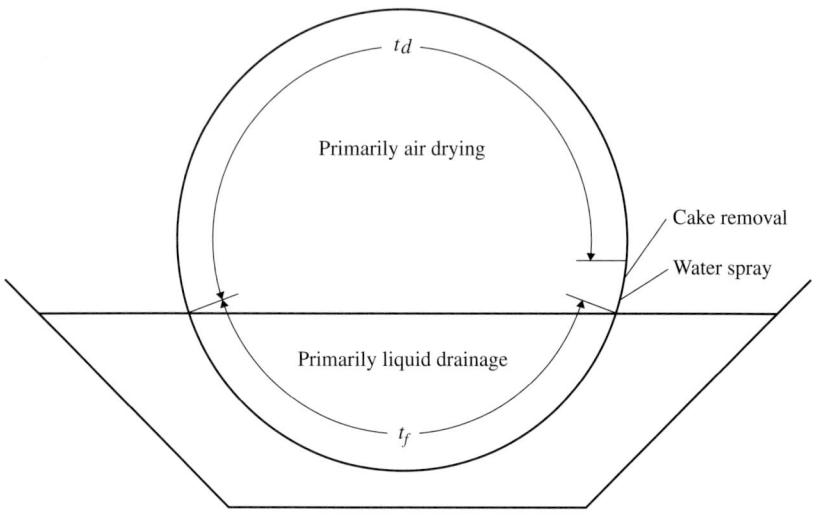

FIGURE 11.21 Mechanism of vacuum filtration.

dry time, t_d. At the end of the cycle, a knife edge scrapes the filter cake from the drum to a conveyor. The filter medium is usually washed with water sprays prior to again being immersed in the slurry tank. A vacuum filter is schematically shown in Fig. 11.21.

The variables that influence the dewatering process are solids concentration, sludge and filtrate viscosity, sludge compressibility, chemical composition, and the nature of the sludge particles (size, shape, water content, etc.).

The filter operating variables are vacuum, drum submergence and speed, sludge conditioning, and the type and porosity of the filter medium. Note that this methodology is energy intensive.

Equation (11.12) can be modified to express filter loading (neglecting the initial resistance of the filter medium):

$$L_f = 35.7 \left(\frac{P^{1-s}}{\mu R_o} \right)^{1/2} \frac{c^m}{t_f^n} \qquad (11.15)$$

where $R_o = r_o \times 10^{-7}$, gr·s²/g²
P = vacuum, lb/in²
c = solids deposited per unit volume filtrate, g/mL
μ = filtrate viscosity, cP
t_f = form cycle time, min
L_f = filter loading, lb/(ft²·h)
m, n = constants related to the sludge characteristics
s = coefficient of compressibility

For routine calculations C_i is used in Eq. (11.15) as c.

Equation (11.15) is in terms of form time and is conventionally converted to cycle time by

$$L_c = L_f \frac{\%\ \text{submergence}}{100} \times 0.8$$

The factor 0.8 compensates for the area of the filter drum where the cake is removed and the medium washed. The total cycle time on a filter may vary from 1 to 6 min. Submergence of the drum may vary from 10 to 60 percent, resulting in a maximum spread of form time of 0.1 to 3.5 min. This also yields a maximum spread of dry time of 2.5 to 4.5 min. In general, the filter yield from highly compressible cakes is unaffected by increases in form vacuum varying from 12 to 17 in (30 to 43 cm) of mercury.

Diatomaceous earth is composed of fossil skeletons of microscopic diatoms and is used as a filter aid in rotary vacuum precoat filtration and pressure filtration. Typically 2 to 6 in (5 to 15 cm) of diatomaceous earth or perlite is applied prior to sludge application. To maintain a porous cake, filter aid may also be added continuously to the sludge slurry prior to vacuum filtration. Collected solids plus a few thousandths of an inch of precoat are scraped off by a knife as the drum advances at rates of 0.5 to 5 min/rev. Depending on the waste stream and filter aid, flow rates can vary from 2 to 50 gal/(h · ft²) [81 to 2035 L/(h · m²)]. Operating costs with diatomaceous earth or perlite precoat are high because of the need to replenish expended filter aid. Precoating is especially required for dewatering of gelatinous-type solids such as alum sludge, clarification of oily wastes, and for sludges characterized by a high percentage of fines. Precoating is also generally employed during pressure filtration. As with the vacuum filter operation, it both protects the filter medium against frequent blinding and provides a thin, nonadherent parting plane between the cake and the medium which minimizes cake discharge difficulties. The optimal type and quantity of precoat must be established by laboratory evaluation. Vacuum filtration design is shown in Example 11.6.

Example 11.6. A combined primary and activated sludge from a pulp and paper mill is to be dewatered. The sludge flow is 100 gal/min (0.38 m³/min) of 6 percent solids. Design a vacuum filter to operate 16 h/d, 7 d/week, using 15 in Hg (381 mm Hg) vacuum (7.35 lb/in²) and 30 percent submergence. Laboratory and pilot studies have shown that:

1. The coefficient $m = 0.25$.
2. The coefficient $n = 0.65$.
3. An optimum cake solids of 28 percent is obtained at a 3-min dry time.
4. The coefficient of compressibility is 0.85.
5. The specific resistance r_o is 1.3×10^7 gr · s²/g².

Solution

$$L_f = 35.7\left(\frac{P^{1-s}}{\mu R_o}\right)^{1/2} \frac{c^m}{t_t^n}$$

The cycle time is $3/0.7 = 4.3$ min.

$$t_f = 4.3 - 3 = 1.3 \text{ min}$$

$$c = \frac{1}{94/6 - 72/28} = 0.077 \text{ g/cm}^3$$

$$L_f = 35.7\left[\frac{7.35^{0.15}}{(1)(1.3)}\right]^{1/2} \frac{0.077^{0.25}}{1.3^{0.65}}$$

$$= 16.3 \text{ lb}/(\text{ft}^2 \cdot \text{h}) \quad [79.1 \text{ kg}/(\text{m}^2 \cdot \text{h})]$$

From Eq. (11.17),

$$L_c = 16.3\left(\frac{30}{100}\right) \times 0.8$$

$$= 3.91 \text{ lb}/(\text{ft2} \cdot \text{h}) \quad [19.1 \text{ kg}/(\text{m}^2 \cdot \text{h})]$$

Sludge to be filtered = 100 gal/min × 8.34 lb/gal × 0.06 × 60 min/h × 24/16

$$= 4500 \text{ lb/h} \quad (2041 \text{ kg/h})$$

Filter area required $= \dfrac{4500}{3.91} = 1151 \text{ ft}^2 \quad (107 \text{ m}^2)$

11.13 Pressure Filtration

Pressure filtration is applicable to almost all water and wastewater sludges. The sludge is pumped between plates that are covered with a filter cloth. The liquid seeps through the cloth, leaving the solids behind between the plates. The medium may or may not be precoated. When the spaces between the plates are filled, the plates are separated and the solids removed. The pressure exerted on the cake during formation is limited to the pumping force and filter closing system design. Filters are designed at pressures ranging from 50 to 225 lb/in^2 (345 to 1550 kPa). As the final filtration pressure increases, a corresponding increase in dry cake solids is obtained. Most municipal sludges can be dewatered to produce 40 to 50 percent cake solids with 225-lb/in^2 (690-kPa) filters. Filtrate quality will vary from 10 mg/L suspended solids with precoat to 50 to 500 mg/L with unprecoated cloth, depending on the medium, type of solids, and type of conditioning. Conditioning chemicals are the same as used in vacuum filtration (lime, ferric chloride, or polymers). Materials such as ash have also been used. It should be noted that where 40 to 50 percent cakes are reached industrial applications usually employ a filter aid

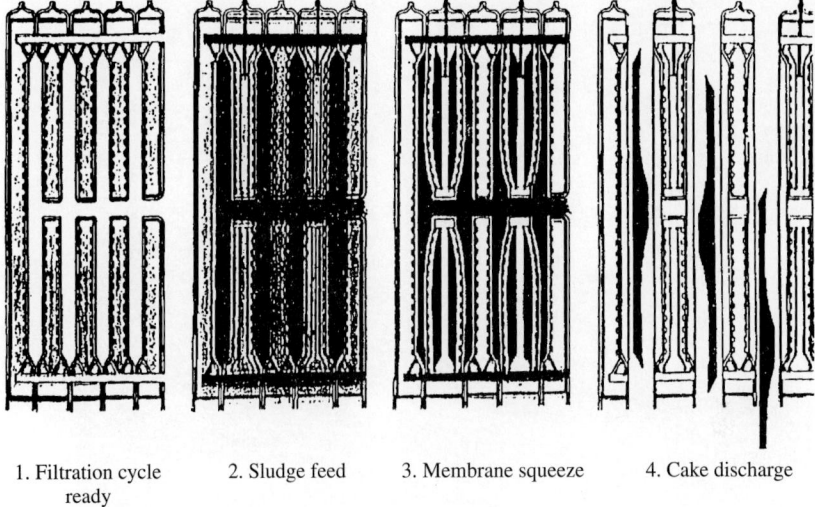

1. Filtration cycle ready 2. Sludge feed 3. Membrane squeeze 4. Cake discharge

FIGURE 11.22 Membrane filter press.

to increase the shear resistance as pressure is applied. Coal fines are used in Germany to increase the fuel value. In the United States, 20 to 30 percent lime and ferric chloride is frequently used, which reduces the Btu content of the cake. If the cake is to be incinerated, filter aids with high Btu content should be considered. Poorly filterable sludges yielding highly compressible cakes can be conditioned with skeleton builders to produce a more porous and incompressible cake. For oily sludges both hydrated lime and fly ash have been successfully used.[22] A membrane filter press operation is shown in Fig. 11.22. A pressure filter is shown in Fig. 11.23. Example 11.7 illustrates a pressure filter design.

Example 11.7. Size a plate and frame pressure filter to dewater sludge using the following data:

Average loading = 13,300 lb/d (6030 kg/d) dry TSS
Maximum loading = 25,000 lb/d (11,340 kg/d) dry TSS
Average sludge concentration = 3.0 percent
Minimum sludge concentration = 2.0 percent

A series of pilot tests were conducted and the following conditions were found:

Total cycle time = 3.5 h*
Average cake solids = 40 percent
Minimum cake solids = 30 percent
Cake density = 70 lb/ft^3 (1120 kg/m^3)
Conditioning = 100 lb FeCL$_3$/ton (50 kg/t) dry solids
             +200 lb lime/ton (100 kg/t) dry solids

---

*Allows sufficient time for cloth washing and cake removal.

FIGURE 11.23 Plate and frame pressure filter.

Size dewatering equipment to treat the average sludge load in 1 shift/d and the maximum load in 2 shifts/d in a 7-day week.

Solution   Calculate the volume of sludge to be treated.

$$\text{Average volume} = \frac{13{,}300 \text{ lb dry SS/d}}{0.03 \text{ lb dry SS/lb sludge} \times 8.34 \text{ lb/gal}}$$

$$= 53{,}000 \text{ gal/d} \quad (200 \text{ m}^3/\text{d})$$

$$\text{Maximum volume} = \frac{25{,}000}{0.02 \times 8.34} = 150{,}000 \text{ gal/d} \quad (570 \text{ m}^3/\text{d})$$

Calculate the dewatered volume.

$$\text{Average} = \frac{13{,}300 \text{ lb/d} + (300 \text{ lb/ton} \times 5 \times 10^{-4} \text{ ton/lb} \times 13{,}300 \text{ lb/d})}{0.4 \text{ lb TSS/lb cake} \times 70 \text{ lb/ft}^3}$$

$$= 545 \text{ ft}^3/\text{d} \quad (15 \text{ m}^3/\text{d})$$

$$\text{Maximum} = \frac{25{,}000 \text{ lb/d} + (300 \times 5 \times 10^{-4} \times 25{,}000)}{0.30 \times 70} = 1369 \text{ ft}^3/\text{d} \quad (39 \text{ m}^3/\text{d})$$

Calculate the number of filter cycles required per day knowing the cycle time of 3.5 h and designing for a single pressure filter.

$$\text{Average number of cycle per day} = \frac{1 \text{ shift/d} \times 8 \text{ h/shift}}{3.5 \text{ h/cycle}} \cong 2 \text{ cycles/d}$$

$$\text{Maximum number of cycles per day} = \frac{2 \times 8}{3.5} \cong 4 \text{ cycles/d}$$

Calculate the volume of dewatered sludge or pressure filter volume required per cycle.

$$\text{Average filter volume per cycle} = \frac{545}{2} = 273 \text{ ft}^3 \ (8 \text{ m}^3/\text{d})$$

$$\text{Maximum filter volume per cycle} = \frac{1369}{4} = 342 \text{ ft}^3 \ (10 \text{ m}^3/\text{d})$$

Select the size for the filter press. The volume per chamber of the press is 3.0 ft³ (0.085 m³). The maximum filter volume per cycle is 342 ft³. Therefore, we need a press with a minimum of 114 (342/3) chambers. A standard filter press should then be selected which meets this criteria.

Hydraulic presses have also been applied to further dewater filter-cake paper mill sludges for incineration. Board mill sludge has been dewatered to 40 percent solids from 30 percent solids at a pressure of 300 lb/in² (2070 kPa) and a pressing time of 5 min.

11.14 Belt Filter Press

In a belt filter press, shown in Fig. 11.24, chemically conditioned sludge is fed through two filter belts and is squeezed by force to drive water through these belts. Variations of this device are successfully used to dewater municipal and industrial sludges.

A belt filter press (shown schematically in Fig. 11.25) employs not only the concept of cake shear with simultaneous application of pressure but also low-pressure filtration and thickening by gravity drainage.

FIGURE 11.24 Belt filter press.

646 Chapter Eleven

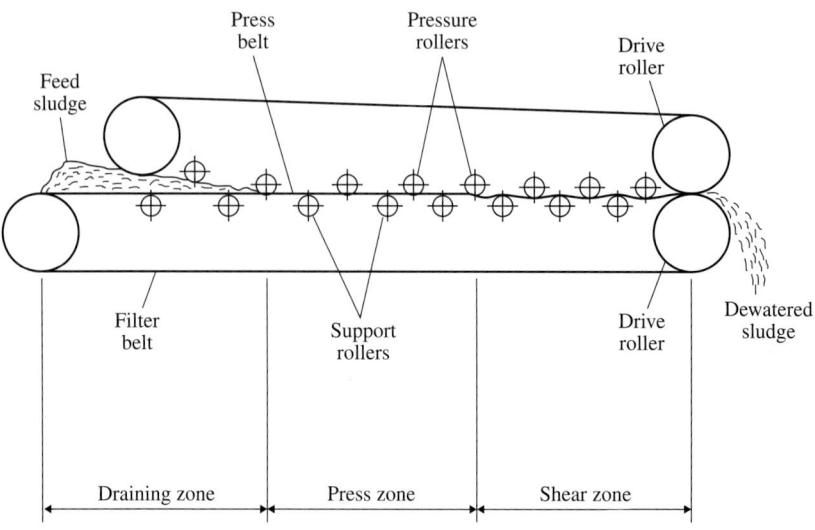

FIGURE 11.25 Schematic of a belt filter press.

An endless filter belt runs over a drive and guide roller at each end like a conveyor belt. The upper side of the filter belt is supported by several rollers. Above the filter bed a press belt runs in the same direction and at the same speed. The drive roller for this belt is coupled with the drive roller of the filter belt.

The press belt can be pressed on the filter belt by means of a pressure roller system whose rollers can be individually adjusted horizontally and vertically. The sludge to be dewatered is fed on the upper face of the filter belt and is continuously dewatered between the filter and press belts. Note how the supporting rollers of the filter belt and pressure rollers of the pressure belt are adjusted in such a way that the belts and the sludge between them describe an S-shaped curve. This configuration induces parallel displacement of the belts relative to each other due to the difference in radius, producing shear in the cake. After dewatering in the shear zone, the sludge is removed by a scraper. Available data indicate that over a range of 0.5 to 12 percent dissolved solids, a filter is relatively insensitive to concentration, but is very sensitive to rate of flux per unit area. On the average belt, washwater flow approximately equals the sludge application rate. A belt filter design is illustrated in Example 11.8.

Example 11.8. Design a belt filter to dewater 86,600 gal/d (330 m^3/d) of 2 percent thickened sludge. The sludge is produced from a pulp and paper wastewater treatment facility and comprises approximately 23 percent waste activated and 77 percent primary sludge. Pilot scale tests were conducted on a 0.5-m belt width pilot press with the values given in Table 11.5.

The following design criteria have been selected to fall in the midrange the operating specification of the full-scale belt press:

Sludge Handling and Disposal

	Run Number				
	1	2	3	4	5
Inlet TS, %	1.76	1.76	2.34	2.34	2.34
Sludge flow, gal/min	15	28	15	20.1	28
Sludge throughput, lb/h	132	247	176	235	328
Cake TS, %	30.2	28.5	31.5	30.9	35.1
Solids capture, %	95.7	94.4	95.2	94.7	94.3
Polymer dosage, lb/ton dry solids	8.3	11.3	5.1	4.6	4.5
Belt speed, ft/min	5	10	5	10	20
Upper belt pressure			5 bar (72.5 lb/in²)		
Lower belt pressure			5 bar (72.5 lb/in²)		
Belt tensions			45 bar (653 lb/in²)		

Note: gal/min = 3.785×10^{-3} m³/min
lb/h = 0.4536 kg/h
lb/ton = 0.5 kg/t
ft/min = 0.3048 m/min

TABLE 11.5 Pilot Belt Filter Testing Results

Cake total solids 30%
Solids capture   95%
Belt speed   10 ft/min (3 m/min)
Throughput   200 lb/(h · 0.5 m) (181 kg/h/m)
Polymer usage   5 lb/ton dry solids (2.5 kg/t)

Solution Design for two 8-h shifts with a total of 14 h of belt press operation 5 d/week as follows:

$$0.0866 \text{ million gal/d} \times 20{,}000 \text{ mg/L} \times 8.34 = 14{,}445 \text{ lb/d} \quad (6550 \text{ kg/d})$$

$$\frac{14{,}445 \text{ lb/d} \times 7 \text{ d}}{14 \text{ h/d} \times 5 \text{ d}} = 1445 \text{ lb/h} \quad (660 \text{ kg/d})$$

$$\text{Required belt width} = \frac{1445 \text{ lb/h}}{200 \text{ lb/(h} \cdot 0.5 \text{ m)}} = \frac{1445 \text{ lb/h}}{400 \text{ lb/(h} \cdot \text{m)}} = 3.6 \text{ m}$$

11.15 Screw Press

A screw press (Fig. 11.26) may be employed to dewater thickened sludge or for further dewatering of conventionally dewatered sludge. The screw press employs tapered shafts to create a gradual forced volume reduction of the material to be dewatered. The moisture is removed through a perforated screen. The press is equipped with a

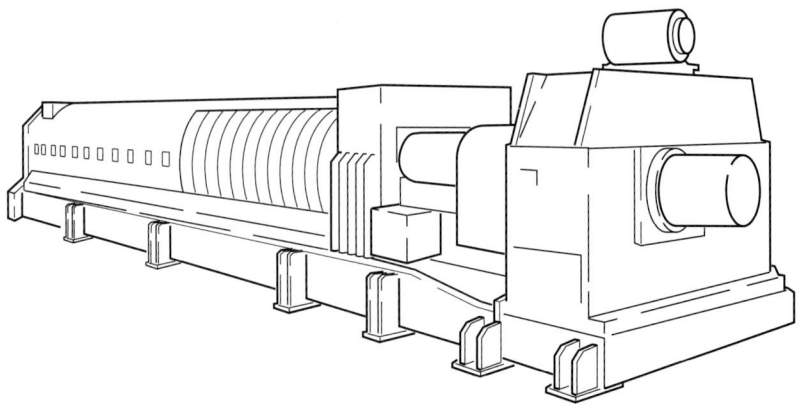

FIGURE 11.26 Screw press (*courtesy of ANDRITZ*).

shaft rotary steam joint allowing for the application of steam pressures up to 100 lb/in^2. The press used for further dewatering is designed to handle predewatered cake with solids content in excess of 15 to 20 percent total solids.

11.16 Sand Bed Drying

For small industrial waste treatment plants, sludge can be dewatered on open or covered sand beds. Drying of the sludge occurs by percolation and evaporation. The proportion of the water removed by percolation may vary from 20 to 55 percent, depending on the initial solids content of the sludge and on the characteristics of the solids. The design and use of drying beds are affected by climatic conditions (rainfall and evaporation). Sludge drying beds usually consist of 4 to 9 in (10 to 23 cm) of sand over 8 to 18 in (20 to 46 cm) of graded gravel or stone. The sand has an effective size of 0.3 to 1.2 mm and a uniformity coefficient less than 5.0. Gravel is graded from $\frac{1}{8}$ to 1 in (0.32 to 2.54 cm). The beds are provided with underdrains spaced from 9 to 20 ft apart (2.7 to 6.1 m). The underdrain piping may be vitrified clay laid with open joints having a minimum diameter of 4 in (10 cm) and a minimum slope of about 1 percent. The filtrate is returned to the treatment plant.

Wet sludge is usually applied to the drying beds at depths of 8 to 12 in (20 to 30 cm). Removal of the dried sludge in a "liftable state" varies with both individual judgment and final disposal means, but usually involves sludge of 30 to 50 percent solid.

In many cases, the bed turnover can be substantially increased by the use of chemicals. Alum treatment can reduce the sludge drying time by 50 percent. The use of polymers can increase the rate of bed dewatering and also increase the depth of application. Bed yield has been reported to increase linearly with polymer dosage.

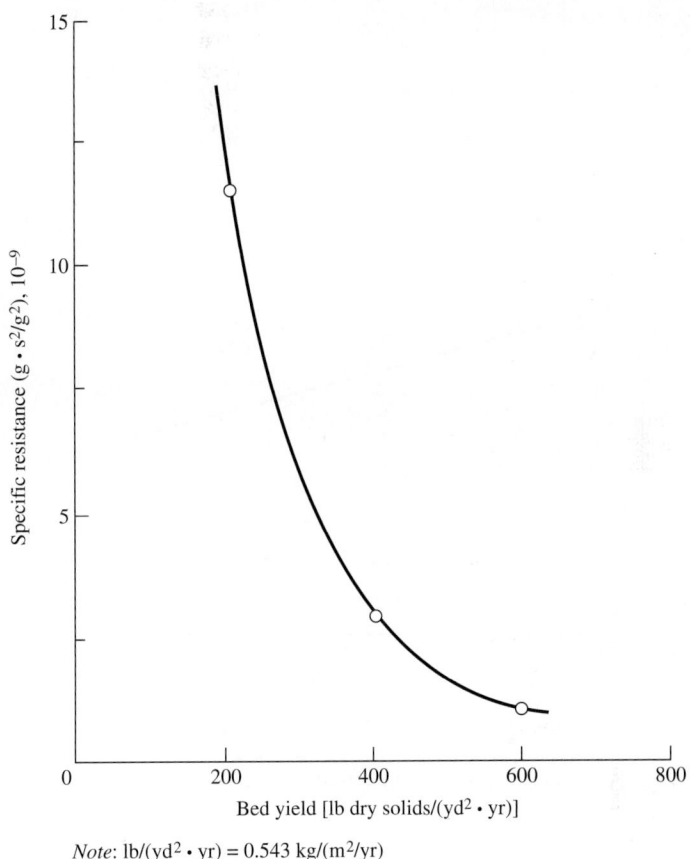

FIGURE 11.27 Relationship between the specific resistance of the sludge and drying bed yield as liftable sludge (*after Swanwick, 1963*).

A rational method has been developed by Swanwick[9] based on the observed dewatering characteristics of a variety of sewage and industrial waste sludges. In this procedure, sludge after drainage (usually 18 to 24 h) is permitted to air dry to the desired consistency. The moisture difference (initial − final) is that which must be evaporated. Depending on the cumulative rainfall and evaporation for the geographical area in question, the time required for various times of the year for evaporation of this moisture is computed. The required bed area may then be determined. Bed yield as related to specific resistance is shown in Fig. 11.27.

11.17 Factors Affecting Dewatering Performance

The sludge mixture (i.e., primary versus activated sludge) will affect both the chemical conditioning requirements and the final cake solids. Data from a pulp and paper mill are shown in

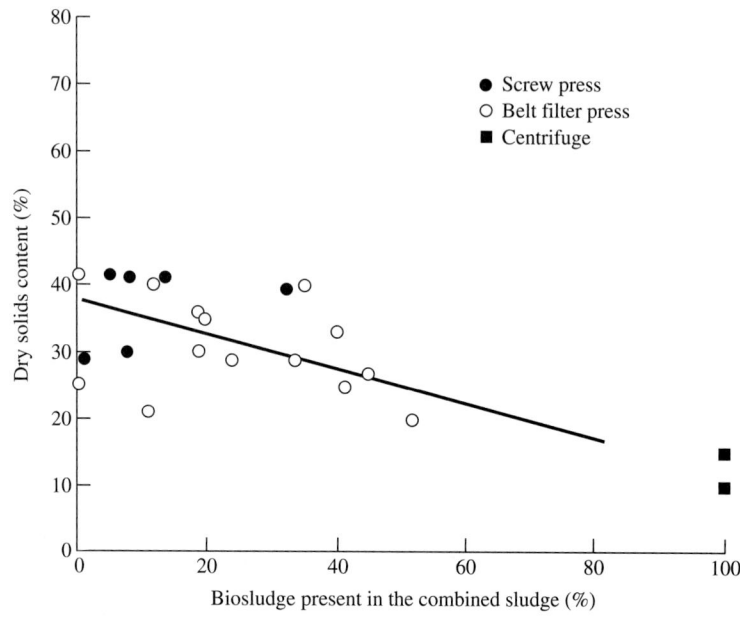

FIGURE 11.28 The results from dewatering of mixed primary and waste activated sludge from a pulp and paper mill with different methods.

Fig. 11.28.[10] Barber and Bullard[11] showed that a well-flocculated activated sludge produced a 15 to 16 percent solids cake on a belt filter press compared to 10 percent with an extracellular polysaccharide bulking sludge. In addition, the flocculant sludge used 57 percent less dewatering polymers. The effect of waste activated sludge and dewatering is shown in Fig. 11.29. Thickening and dewatering of wastewater sludges is shown in Table 11.3.

11.18 Land Disposal of Sludges

Land disposal of wet sludges can be accomplished by lagooning or by application of liquid sludge to land by truck or spray system. Liquid sludge may also be carried by pipeline to a remote agricultural or lagoon site.

Lagooning is commonly employed for the disposal of inorganic industrial waste sludges. Organic sludges usually receive aerobic and anaerobic digestion prior to lagooning to eliminate odors and insects. Lagoons may be operated as substitutes for drying beds, with the sludge periodically removed and the lagoon refilled. In a permanent lagoon, supernatant liquor is removed. When it is filled with solids, the lagoon is abandoned and a new site selected.

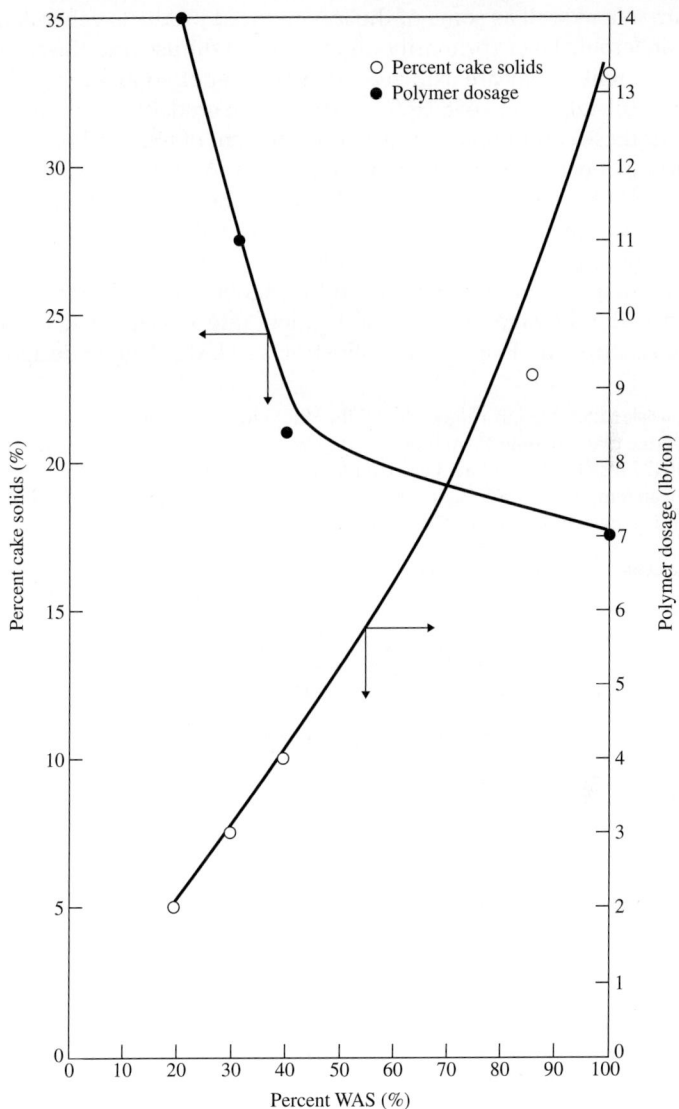

Figure 11.29 Screw press performance for waste sludge from a virgin pulp and recycle mill.

In general, lagoons should be considered where large land areas are available and the sludge will not present a nuisance to the surrounding environment.

Benthal stabilization occurs in a lagoon as a result of anaerobic or a combination of aerobic and anaerobic mechanisms. Below the surface aerobic layer, anaerobic conditions prevail in which methane gas is evolved, as well as other products of anaerobic decomposition.

Ammonia as well as some of the less-reduced products generated in the anaerobic layer (primarily organic acids) diffuse into the aerobic layer in which they are oxidized. Rich[18] has reported an average benthal stabilization rate of 80 g/(m² · d) of biodegradable solids at 20°C. Under these conditions, as much as 63 percent of total carbon stabilization can occur via methane fermentation. Assuming oxidation of all ammonia and BOD released to the water, the oxygen uptake would be 86 g O_2/(m² · d). The sludge should concentrate in the bottom of the lagoon to 2.5 to 3.0 percent solids.

Assuming continuous input to the lagoon and withdrawal once a year, the average annual stabilization rate is estimated as 68 g biomass/(m² · d). Example 11.9 illustrates a sludge lagoon design.

Example 11.9. Design a lagoon to stabilize the sludge generated from an activated sludge plant treating 1.0 million gal/d (3785 m³/d) of a wastewater with a BOD of 425 mg/L. The activated sludge plant operates at a sludge age of 45 d. The mean temperature is 20°C, $a = 0.55$ g, $b = 0.1$/d, $t = 0.71$ d, MLVSS = 3000 mg/L and 80 percent is volatile, $S = 10$ mg/L.

Solution X_d is calculated from Eq. (6.6.):

$$X_d = \frac{0.8}{1+(0.1)(45)(0.2)} = 0.42$$

$$\Delta X_v = [0.55(425-10) - 0.1 \times 0.42 \times 3000 \times 0.71](8.34) \times 1.0$$

$$= 1158 \text{ lb VSS d} \quad (525 \text{ kg/d})$$

or

$$\frac{1158}{0.8} = 1448 \text{ lb SS/d} \quad (657 \text{ kg/d})$$

Assume 75 percent of the VSS will degrade in the lagoon; the area required is

$$A = \frac{R_a}{B_{av}}$$

in which R_a is the loading rate in lb/d of VSS to be degraded and B_{av} is the average annual stabilization rate [600 lb VSS/(acre · d) or 0.067 kg/(m² · d)].

The required area is

$$A = \frac{1158 \text{ lb/d} \times 0.75}{600 \text{ lb/(acre·d)}}$$

$$= 1.45 \text{ acres or } 63{,}162 \text{ ft}^2 \quad (5868 \text{ m}^2)$$

The sludge depth accumulation can be calculated by first finding the residual sludge:

$$\left(\frac{1158}{0.8} - 1158 \times 0.75\right) \text{ lb residue SS/d} \times 365 \text{ d/yr}$$

$$= 211{,}500 \text{ lb residual sludge/yr} \quad (96{,}000 \text{ kg/yr})$$

At 3 percent solids, the volume of sludge is

$$\frac{211{,}500 \text{ lb/yr}}{0.03 \times 62.5 \text{ lb/ft}^3} = 112{,}800 \text{ ft}^3/\text{yr} \quad (31{,}950 \text{ mm}^3/\text{yr})$$

The sludge depth accumulation will then be

$$\frac{112{,}800 \text{ ft}^3/\text{yr}}{1.45 \text{ acre} \times 43{,}560 \text{ ft}^3/\text{acre}} = 1.8 \text{ ft/yr} \quad (0.55 \text{ m/yr})$$

If the oxygen requirements are 760 lb O_2/(acre · d) [0.086 kg/(m^2 · d)], the total O_2 requirement will be 1102 lb/d (760 × 1.45). Using a 10-ft (3.05-m) water column and an aerator transfer efficiency of 10 percent, the required airflow will be

$$\text{Required airflow} = \frac{1102 \text{ lb/d}}{1440 \text{ min/d} \times 0.1 \times 0.232 \text{ lb } O_2/\text{lb air} \times 0.0746 \text{ lb air/ft}^3}$$

$$= 442 \text{ standard ft}^3/\text{min} \quad (12.5 \text{ m}^3/\text{min})$$

If each diffuser operates at 4 standard ft³/min, the number of diffusers will be 110.

In several cases, biological sludges after aerobic or anaerobic digestion have been sprayed on local land sites from tank wagons or pumped through agricultural pipes. This employs multiple applications at low dosages from 100 dry tons/acre (22.4 kg/m²) for average conditions to 300 tons/acre (67.3 kg/m²) in areas of low rainfall.

Excess activated sludge has been disposed of in oxidation ponds in which algal activity maintains aerobic conditions in the overlaying liquids while the sludge undergoes anaerobic digestion. This procedure has been successfully employed for municipal activated sludge at Austin, Texas, and excess activated sludge from a petrochemical plant in Houston, Texas. Lagoon loading rates of 600 lb VSS/(acre · d) [0.0673 kg/(m² · d)] have been employed.

Many organic sludges can be incorporated into the soil without mechanical dewatering. Surface application can be accomplished by spreading from a truck or spraying. Sludge may also be injected into the soil 8 to 10 in (20 to 25 cm) below the surface by a mobile unit, as shown in Fig. 11.30. Injection offers the advantage of minimizing surface runoff and odor problems. An important consideration is the heavy metals content of the sludge. At a pH greater than 6.0, heavy metals will exchange for Ca^{2+}, Mg^{2+}, Na^+, and K^+. This natural ability to exchange heavy metals by the soil is called the *cation exchange capacity* (CEC) and is expressed in milliequivalents per hundred grams of dry soil. The amount of heavy metals from sludge is influenced by such factors as pH and aerobic or anaerobic conditions. The CEC of sandy soil may vary from 0 to 5 while clay soils will have a CEC between 15 and 20. The nutrient content of the sludge will support the growth of plants. The organic portion of the soil will also

FIGURE 11.30 Sludge injection vehicle.

chelate heavy metals. In accordance with USEPA Regulation 503.13, ceiling concentrations of metals for land disposal shall not exceed the concentrations shown in Table 11.6. USEPA annual pollutant loadings are given in Table 11.7.

Prior to incorporation, sludges should receive a minimum degree of stabilization. Chow[19] has recommended aerobic digestion of 15 d to reduce the volatile content to less than 55 percent.

Each crop has a nutrient requirement (N, P, K, etc.). The annual quantity of sludge that can be incorporated depends on the available

Pollutant	Cumulative Pollutant Loading Rate, kg/ha
Arsenic	41
Cadmium	39
Copper	1500
Lead	300
Mercury	17
Nickel	420
Selenium	100
Zinc	2800

TABLE 11.6 Cumulative Pollutant Loading Rates

Pollutant	Annual Pollutant Loading Rate, kg/ha 365-d Period
Arsenic	2.0
Cadmium	1.9
Copper	75
Lead	15
Mercury	0.85
Nickel	21
Selenium	5.0
Zinc	140

TABLE 11.7 Annual Pollutant Loading Rates

nitrogen content of the sludge and the nitrogen uptake of the selected crop. Excess application of sludge can result in oxidation of ammonia to nitrate, which can contaminate the groundwater. Since all the applied organic nitrogen is not available to the crops in the same year, there is a sequential removal of organic nitrogen. Normally, about 40 percent of the organic nitrogen applied in the first year is available for crop growth that year. Subsequently, 20, 10, 5, and, 2.5 percent of the organic nitrogen is available for the second, third, fourth, and fifth years, respectively.

The Illinois EPA[20] recommends a minimum depth of earth cover to the annual water table of 10 ft (3 m) with a permeability rate of 2 to 20 in/h (5 to 51 cm/h). A maximum land slope of 8 percent is recommended. A minimum soil pH of 6.5 should be maintained. The design of a sludge land incorporation system is shown in Example 11.10.

Example 11.10. Design a land incorporation system for an excess activated sludge. The sludge characteristics are

Quantity, gal/d	6560
Amount, lb/d	3500
NH_3^- N, mg/L	235
Org-N, mg/L	865
SS, mg/L	63,000
PO_4, mg/L	30

Note: gal/d = 3.785 × 10³ m³/d
lb/d = 0.4536 kg/d

Solution *Metal analysis of sludge:*

Metal	mg/kg (Dry Solids)
Al	700
Cd	3.0
Ca	105,000
Cr	400
Cu	60
Fe	6000
Pb	30
Ni	150
Zn	120
K	150

For example, consider cadmium. The total Cd in the sludge is

$$3\frac{\text{mg Cd}}{\text{kg sludge}} \times 3500\frac{\text{lb sludge}}{\text{d}} \times \frac{\text{kg}}{2.2\text{ lb}} = 4773\frac{\text{mg.Cd}}{\text{d}}$$

For each acre (0.414 ha), one can apply, without exceeding the allowable yearly loading rate of 1.9 kg/ha (Table 11.7),

$$1.9\frac{\text{kg Cd}}{\text{ha}} \times 0.405\frac{\text{ha}}{\text{acre}} \times 10^6\frac{\text{mg}}{\text{kg}} = 769,500\frac{\text{mg Cd}}{\text{d}}$$

The years to reach the maximum allowable loading of 39 kg/ha (Table 11.6) at the 1.9 kg/ha yearly rate would be

$$39\text{ kg/ha} \times 1.9\text{ kg/(ha} \cdot \text{yr)} = 20.5\text{ yr}$$

But, for the 4773 mg Cd/d in the sludge, a somewhat longer time is available:

$$39\text{ kg/ha} = \frac{39 \times 10^6\text{ mg Cd}}{4773\text{ mg Cd/d}} = 8170\text{ d} = 22.3\text{ yr}$$

The maximum annual loading in pounds per acre is

$$1.9\frac{\text{kg}}{\text{ha} \cdot \text{y}} \times \frac{0.414\text{ ha}}{\text{acre}} \times 2.2\frac{\text{lb}}{\text{kg}} = 1.73\text{ lb/(acre} \cdot \text{yr)}$$

Agronomic loading:

The maximum allowable nitrogen loading for Bermuda grass is 0.0224 kg/(m² · yr). This equates to 200 lb/(acre · yr). For subsurface incorporation, NH_3 availability is 100 percent and org-N availability is 40 percent.

The available N for the first year of application is

$$235 \text{ mg NH}_3^-\text{N/L} + 865 \text{ mg org-N/L} \times 0.4 = 581 \text{ mg N/L}$$
$$= 0.00486 \text{ lb/gal}$$

The sludge loading is therefore

$$\frac{200}{0.00486} = 41{,}152 \text{ gal/(acre} \cdot \text{yr)} \quad [0.0385 \text{ m}^3/(\text{m}^2 \cdot \text{yr})]$$

The acres required are thus

$$\frac{6560 \times 365}{41{,}152} = 58 \text{ acres} \quad (234{,}720 \text{ m}^2)$$

In subsequent years of application, additional organic nitrogen conversion should be considered.

Oily sludges have successfully been disposed of on land. Recent data published indicate the following:

1. The oil degradation rate is directly related to the percentage of oil in the soil.
2. Fertilization improved the degradation rate.
3. Aeration (tilling) frequencies vary (from 1 week to 2 months).
4. Between 380 and 400 m³ (2000 and 2500 bbl) of oil per hectare should be degraded in an 8-month growing season.
5. Sludge farming is about one-fifth as expensive as incineration.

11.19 Incineration

The sludge cake must be disposed of. This can be accomplished by hauling the cake to a land disposal site or by incineration.

The variables to be considered in incineration are the moisture and volatile content of the sludge cake and the thermal value of the sludge. The moisture content is of primary significance because it dictates whether the combustion process will be self-supporting or whether supplementary fuel will be required. The thermal values of sludges may vary from 5000 to 10,000 Btu/lb (1.16 × 10⁷ to 2.33 × 10⁷ J/kg).

Incineration involves drying and combustion. Various types of incineration units are available to accomplish these reactions in single or combined units. In the incineration process, the sludge temperature is raised to 212°F (100°C), at which point moisture is evaporated from the sludge. The water vapor and air temperature are increased to the ignition point. Some excess air is required for complete combustion of the sludge. Self-sustaining combustion is often possible with dewatered waste sludges once the burning of auxiliary fuel raises the incinerator temperature to the ignition point. An autogenous sludge is one in which there is a favorable ratio of water to volatiles. Generally speaking, that ratio must be slightly better than 2 lb H_2O/lb volatiles for the sludge to approach autogenous conditions. The primary end products of combustion are carbon dioxide, sulfur dioxide, and ash.

Incineration can be accomplished in multiple-hearth furnaces in which the sludge passes vertically through a series of hearths. In the upper hearths, vaporization of moisture and cooling of exhaust gases occur. In the intermediate hearths, the volatile gases and solids are burned. The total fixed carbon is burned in the lower hearths. Temperatures range from 1000°F (538°C) at the top hearth to 600°F (316°C) at the bottom. The exhaust gases pass through a scrubber to remove fly ash and other volatile products. This is shown in Fig. 11.31.

In the fluidized bed, sludge particles are fed into a bed of sand fluidized by upward-moving air. A temperature of 1400 to 1500°F (760 to 815°C) is maintained in the bed, resulting in rapid drying and burning of the sludge. Ash is removed from the bed by the upward-flowing combustion gases.

11.20 Problems

11.1. An aerobic digester has a retention time of 10 d at 20°C and the following characteristics:

Reaction rate $k_d = 0.155$/deg
Influent SS = 10,500 mg/L
Volatile = 85 percent
Nondegradable VSS = 32 percent
$X_d = 0.68$
$b = 0.1$/d
Flow = 100,000 gal/d (379 m³/d)

(a) Calculate the composition of the digester effluent, TSS, VSS.

(b) Calculate the oxygen requirements including nitrification.

11.2. A thickener is operating under the following conditions:

Solids flux = 20 lb/(ft² · d) [97.7 kg/(m² · d)]
Maximum underflow concentration = 45 percent
Area = 600 ft² (55.7 m²)

A change in plant operations modifies the characteristics of the sludge to a flow of 180,000 gal/d (681 m³/d) with a concentration of 1 percent. The thickening characteristics are those of Example 11.3. What will be the new maximum underflow concentration?

Sludge Handling and Disposal 659

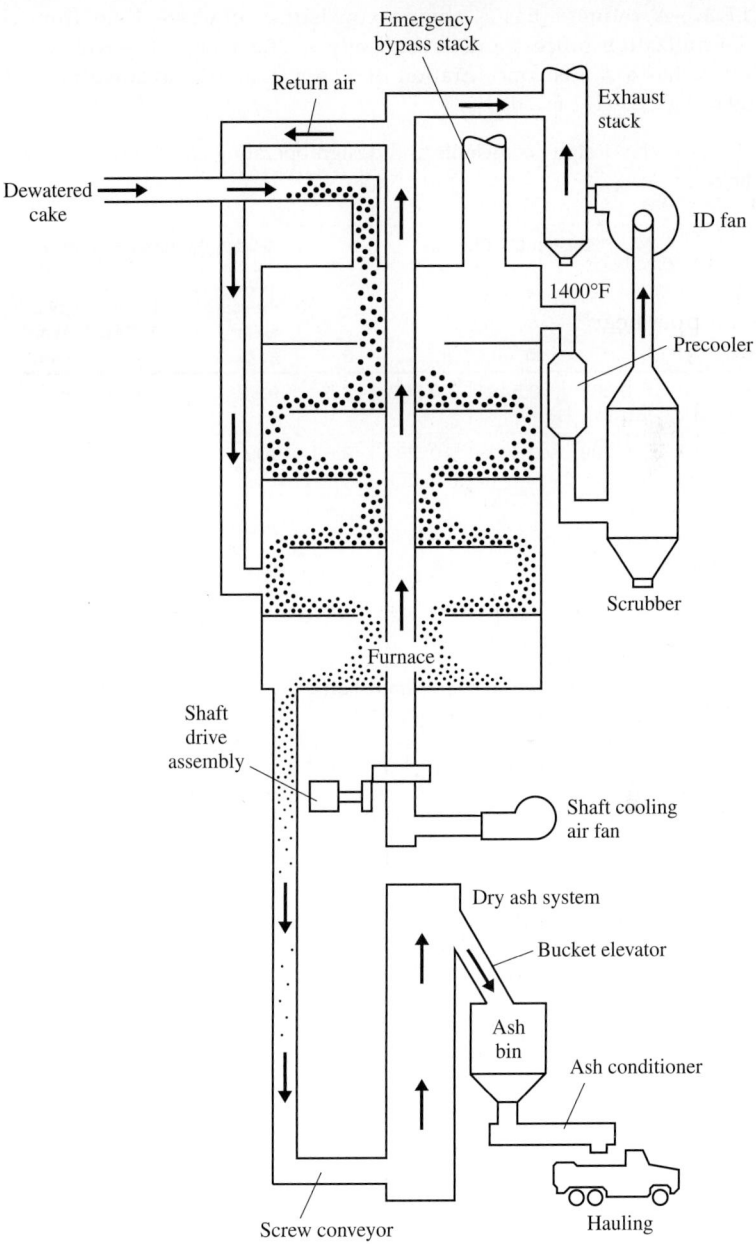

FIGURE 11.31 Multiple-hearth system.

11.3. A refinery has a sludge consisting of waste lime from a neutralization process and several oily wastestreams. The composite waste has a solids concentration of 52.6 g/L and an average flow of 29,600 gal/d (112 m³/d).

Given: The design coefficients and design operating conditions as shown below:

Process	Design Coefficients					Design Operation Conditions		
	μ, cP	$\dfrac{1-s}{2}$	m	n	R_o	Vacuum or Pressure, lb/in² gage	Cake Solids, %	Cycle Time, min
Vacuum filter	1	0.087	0.548	−0.562	3.5	9.8	34	6
Filter press	1	0.299	0.306	−0.559	7.63	100	40	1200

Determine:

(a) The required vacuum filter area based on a 7 d/week operation with 8 h/week for precoating.

(b) The required filter press area based on a 7 d/week operation with 4 h of downtime for every cycle.

Use the relationship:

$$L = \frac{35.7 P^{(1-s)/2}(c^m)(t)^n}{(\mu R_o)^{1/2}}$$

where L = filter loading, lb/(ft² · h)
P = vacuum pressure, lb/in² gage
c = solids deposited per unit volume of filtrate, g/mL
t = cycle time, min
μ = viscosity, cP
R_o = cake resistivity

Note 1: $c = \left[\dfrac{\text{sludge moisture (\%)}}{\text{sludge solids (\%)}} - \dfrac{\text{cake moisture (\%)}}{\text{cake solids (\%)}} \right]^{-1}$

Note 2: The above relationship can be used to determine the loading on a filter press where

P = filter pressure, lb/in² gage

References

1. Van der Sloot, H., L. Heasman, and P. Quevauiller: *Harmonization of Leaching/Extraction Tests*. Elsevier, Amsterdam, The Netherlands, 1997.
2. Townsend, T., Y. C. Jang, and T. Tolaymat: "A Guide to the Use of Leaching Tests in Solid Waste Management Decision Making." Report # 03-01(A) prepared for the Florida Center for Solid and Hazardous Waste Management, University of Florida, March, 2003.
3. U.S. Environmental Protection Agency. "Waste Leachability: The Need for Review of Current Agency Procedures," EPA-SAB-EEC-COM-99-002, Science Advisory Board, Washington, D.C., 1999.
4. Stien, R., C. E. Adams, and W. W. Eckenfelder: *Water Res.*, vol. 8, p. 213, 1974.
5. Andrews, J. F., and K. Kambhu: "Thermophilic Aerobic Digestion of Organic Solid Waste," Final Progress Report, Clemson University, Clemson, S.C., 1970.
6. Dick, R. I.: "Thickening," in *Process Design in Water Quality Engineering*, E. L. Thackston and W. W. Eckenfelder, eds., Jenkins, Austin, Texas, 1972.
7. Coackley, P., and B. R. S. Jones: *Sewage Ind. Waste*, vol. 28, pt. 8; p. 963, 1956.
8. Bernard, J., and A. J. Englande: "Centrifugation," in *Process Design in Water Quality Engineering*, E. L. Thackston and W. W. Eckenfelder, eds., Jenkins, Austin, Texas, 1972.
9. Swanwick, J. D.: *Advances in Water Pollution Research*, vol. II, p. 387, Pergamon Press, New York, 1963.
10. Saunamaki, R.: *Wat. Sci. Tech.*, vol. 35, no. 2-3, p. 235, 1997.
11. Barber, J. B., and Bullard, C. M.: *Proc. WEF Industrial Waste Conf.*, Nashville, 1998.
12. Eckenfelder, W. W.: *Principles of Water Quality Management*, CBI, Boston, Mass., 1980.
13. Bassett, P. J., et al.: *Proc. 33rd Ind. Waste Conf.*, Purdue University, 1978.
14. Dickey, R. O., and R. C. Ward: *Proc. 36th Ind. Waste Conf.*, Purdue University, 1978.
15. Podusks, R. A., et al.: *Proc. 35th Ind. Waste Conf.*, Purdue University, 1980.
16. Leonard, R. J., and J. W. Parrott: *Proc. 33rd Ind. Waste Conf.*, Purdue University, 1978.
17. Miner, R. G.: *J. WPCF*, vol. 52, pt. 9(2), p. 389, 1980.
18. Rich, L.: *Water Res.*, vol. 16, pt. 9(l), p. 399, 1982.
19. Chow, V.: *Sludge Disposal on Land*, 3M Co., St. Paul, Minn., 1979.
20. Illinois EPA: "Design Criteria for Municipal Sludge Utilization on Agricultural Land," Tech. Policy, WPC-3, 1977.
21. Deeny, K., et al.: *Proc. 40th Purdue Industrial Waste Conf.*, Purdue University, 1985.
22. Zall, J., et al.: *J. WPCF*, vol. 59, no. 7, 1987.

CHAPTER 12
Miscellaneous Treatment Processes

12.1 Introduction

Polishing treatment and the final disposal methods for wastewater not considered in previous chapters are discussed herein. Methodologies include: land treatment, deep-well disposal, membrane processes, membrane bioreactors, granular media filtration, and microscreens.

12.2 Land Treatment

A wide variety of food-processing wastewaters, including meat, poultry, dairy, brewery, and winery wastewaters, have been applied successfully to the land. Disposal of industrial wastes by irrigation can be practiced in one of several ways, depending on the topography of the land, the nature of the soil, depth of water table and the characteristics of the waste:

1. Distribution of waste through spray nozzles over relatively flat terrain
2. Distribution of waste over sloping land which runs off to a natural watercourse
3. Disposal through ridge and furrow irrigation channels

Irrigation

Irrigation includes those systems where loading rates are about 2 to 4 in/week (5 to 10 cm/week) and a crop is grown. Methods of application include various sprinkler systems, ridge and furrow, and surface flooding. It is desirable to have spray application periods followed by rest periods in the ratio of approximately 1:4 or greater, for example, 0.5 h spray:2 h rest.

Screened waste is pumped through laterals and sprayed through sprinklers located at appropriate intervals, as shown in Fig. 12.1. The waste percolates through the soil, and during this process the organics undergo biological degradation. The liquid is either stored in the soil layer or discharged to the groundwater. Most spray irrigation systems use a cover crop of grass or other vegetation to maintain porosity in the upper soil layers. The most popular cover crop is reed canary grass (*Phalaris arundinacea*). This grass develops an extensive root system, has a relatively large leaf area, and is tolerant to adverse conditions. There is a net waste loss by evapotranspiration (evaporation to the atmosphere and absorption by the roots and leaves of plants). This may amount to as much as 10 percent of the waste flow.

Loamy well-drained soil is most suitable for irrigation systems; however, soil types from clays to sands are acceptable. A minimum depth to groundwater of 5 ft (1.5 m) is preferred to prevent saturation of the root zone. Underdrain systems have been used successfully to adapt to high groundwater or impervious subsoil conditions.

Water-tolerant perennial grasses have been used most commonly because they take up large quantities of nitrogen, are low in maintenance, and keep the soil infiltration rates as high as possible. Seasonal

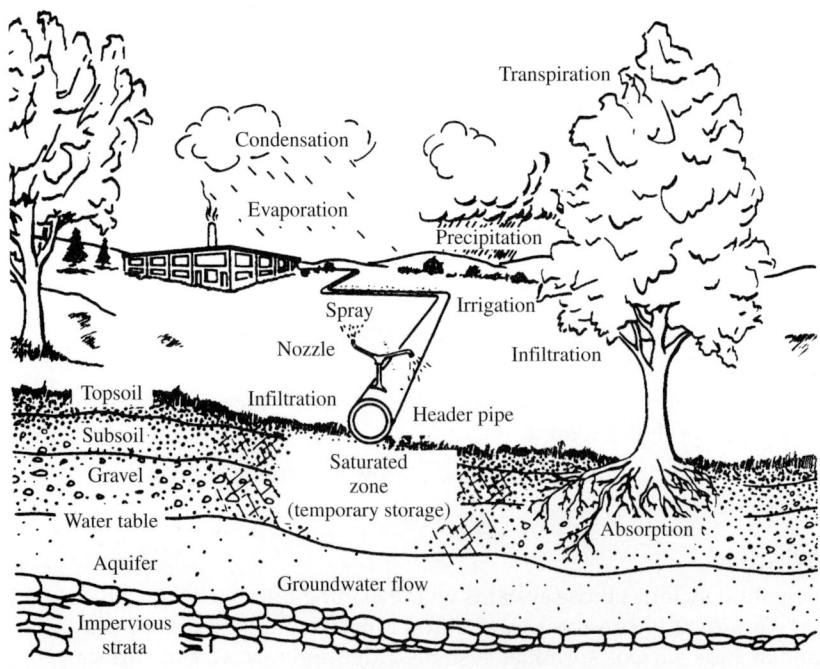

Figure 12.1 Spray irrigation system.

canning wastewaters are often used to irrigate corn or annual forages to coincide with the production of wastewater.

In some cases, wastes have been sprayed into woodland areas. Trees develop a high-porosity soil cover and yield high transpiration rates. A small elm tree may take up as much as 3000 gal/d (11.4 m³/d) under arid conditions.

The principal factors governing the capacity of a site to adsorb wastewater are[1]

1. *Character of the soil.* A sandy type of soil will have a high filtration rate; clay will pass very little water.
2. *Stratification of the soil profile.* Some soils will exhibit clay lenses, which are a barrier to flow.
3. *Depth to groundwater.* The quantity of wastewater that can be sprayed on a given area will be proportional to the depth of soil through which it must travel to the groundwater. Sufficient soil depth must be available, however, to effect biological degradation of the organics.
4. *Initial moisture content.* The capacity of the soil to adsorb water is proportional to the initial moisture content of the soil.
5. *Terrain and ground cover.* A cover crop will increase the quantity of water that can be absorbed by a given area. A sloping site will increase the runoff.

In a spray irrigation field, water absorbed into the ground is held in capillary suspension until approximately 95 percent of saturation. Additional water will flow into the groundwater under a head equal to the distance from the ground surface to the groundwater table. The discharge of wastewater may be at a steady rate or at a short-term rate. The capacity to absorb at a steady rate is proportional to the overall coefficient of permeability for the soil between the ground surface and the groundwater table[1]:

$$Q = 328 \times 10^3 KS \qquad (12.1)$$

where Q = gal/(min · acre)
K = overall coefficient of permeability, ft/min
S = degree of saturation of soil (near 1.0 for steady-rate application)

or

$$Q = 1.00 KS \qquad (12.1a)$$

for Q in m³/(min · m²) and K in m/min.

Description of Fine Component	K, ft/min
Trace fine sand (0.10%)	1.0–0.2
Trace silt (0–10%)	0.8–0.04
Coarse and fine silt (10–20%)	0.012–0.002
Fissured clay soils (50%) and organic soils (50%)	0.0008–0.0004
Dominating clay soils (up to 100%)	<0.0002

Note: ft/min = 0.3048 m/min.

TABLE 12.1 Variation of Permeability with Soil Characteristics[1]

The coefficient of permeability K depends on the soil characteristics, as shown in Table 12.1. The overall coefficient will depend on the variation in soil characteristics for various depths below the surface and can be computed as

$$K = \frac{H}{H_1/K_1 + H_2/K_2 + \cdots + H_n/K_n} \quad (12.2)$$

where H is the total depth to the groundwater in feet; $H_1, H_2, \ldots, H_n$ are the thickness of layers of soil profile; and K_1, K_2, etc., are the average permeability coefficients for each layer. Soil borings are usually necessary to determine the overall coefficient K. Care should be taken to avoid missing clay lenses that may be present. The short-term rate is proportional to the capillarity of the soil and the initial moisture content. It is usually considerably higher than the steady rate.

Rapid Infiltration

Rapid infiltration systems are characterized by percolation of most of the applied wastewater through the soil and into the subsurface. The method is restricted to use with rapidly permeable soils such as sands and sand loams. This type of system is normally thought to be associated with recharge or spreading basins, although in food-processing applications, high-rate sprinkler systems have been used to provide distribution of the wastewater.

In rapid infiltration systems, plants play a relatively minor role in terms of treatment of the applied wastewater. Physical, chemical, and biological mechanisms operating within the soil are responsible for treatment. The more permeable the soil, the farther the wastewater must travel through the soil to receive treatment. In very sandy soils, this minimum distance is considered to be approximately 15 ft (5 m).

Overland Flow

Overland flow is a fixed film biological-treatment process. In overland flow, land treatment wastewater is applied at the upper reaches of the grass-covered slopes and allowed to flow over the vegetated surface to runoff collection ditches. The wastewater is treated by a thin film down the length of the slope. The process is best suited to slowly permeable soils but can also be used on moderately permeable soils that have relatively impermeable subsoils.

Wastewater is usually applied by sprinklers to the upper two-thirds of slopes that are 150 to 200 ft (46 to 61 m) in length. A runoff collection ditch or drain is provided at the bottom of each slope. Treatment is accomplished by bacteria on the soil surface and within the vegetative litter as the wastewater flows down the sloped, grass-covered surface to the runoff collection drains. Ideally, the slopes should have a grade of 2 to 4 percent to provide adequate treatment and prevent ponding or erosion. The system may be used on naturally sloped lands or it may be adapted to flat agricultural land by reshaping the surface to provide the necessary slopes.

The characteristics of land treatment systems are summarized in Table 12.2.

Waste Characteristics

In addition to soil conditions, there are several waste characteristics that require consideration in a spray irrigation system. Suspended solids should be removed from the waste, either by screening or by sedimentation, before it is sprayed. Solids will tend to clog the spray nozzles and may mat the soil surface, rendering it impermeable to further percolation. An excessively acid or alkaline pH will be harmful to the

Feature	Irrigation	Rapid Infiltration	Overland Flow
Hydraulic loading rate, cm/d	0.2–1.5	1.5–30	0.6–3.6
Land required, Ha*	24–150	1.2–24	10–60
Soil type	Loamy sand to clay	Sands	Clay to clay loam
Soil permeability	Moderately slow to moderately rapid	Rapid	Slow

*Field area in hectares not including buffer area, roads, or ditches for 3785 m^3/d (1 million gal/d) flow.

TABLE 12.2 Comparative Characteristics of Land Treatment Systems[2]

cover crop. High salinity will impair the growth of a cover crop and in clay soils will cause sodium to replace calcium and magnesium by ion exchange. This will cause soil dispersion, and as a result drainage and aeration in the soil will be poor. A maximum salinity of 0.15 percent has been suggested to eliminate these problems.[3]

The soil is a highly efficient biological treatment process, and the performance of a system is usually governed by the hydraulic capacity of the soil as opposed to the organic loading rate. Oxygen exchange into soils depends on the air-filled pore spaces. In saturated soils, oxygen transfer will be similar to that in oxidation ponds. In well-drained soils, oxygen exchange of the surface is rapid because of mass flow. Below the first 4 in (10 cm), however, oxygen exchange is slow because it depends on diffusion.

Research by Jewell[4] showed that organic loading rates onto soil can exceed 16,000 lb/(acre · d) [1.79 kg/(m² · d)] on a COD basis without exceeding bacterial capabilities. Adamczyka[5] reported problems with irrigation systems loaded at 2000 to 5000 lb/(acre · d) [0.22 to 0.56 kg/m² · d)] of BOD but no problems at 500 lb/(acre · d) [0.056 kg/(m² · d)]. The problems of the very high BOD loadings can include damage to vegetation, odors, and leaching of undegraded organics into the subsurface. Loadings in the range of 535 lb/(acre · d) [0.060 kg/(m² · d)] are generally acceptable for irrigation and rapid infiltration systems. For overland flow the limiting loading rate has not been defined but is probably 134 to 180 lb/(acre · d) [0.017 to 0.020 kg/(m² · d)]. Nutrient addition in the form of nitrogen or phosphates may be necessary for nutrient-deficient wastes unless the fields are adequately fertilized.

Design of Irrigation Systems

A model has been developed[6] for irrigation that includes mass balances of water and nitrogen and provides estimates of nitrogen losses to the groundwater. The principal mechanism for nitrogen removal is crop growth.

The land application system has four components: pretreatment, wastewater transmission to the irrigation site, a lagoon for storage of the wastewater during time periods in which irrigation is infeasible, and the irrigation site. The principal variables are

Q_m = wastewater flow to land application site (L^3/T)
A = irrigated land area (L^2)
r = average wastewater application rate (L/T)

If Q_m is in million gallons per day, A in acres, and r in inches per week,

$$\frac{Q_m}{A} = \frac{r}{258} \qquad (12.3)$$

If Q_m is in cubic meters per day, A in square meters, and r in centimeters per week,

$$\frac{Q_m}{A} = \frac{r}{700}$$

If T is the number of weeks of the irrigation season and P and ET are irrigation season precipitation and evapotranspiration (in inches), respectively, the amount of water entering the groundwater below the irrigation site is $7Q_mT/A + 0.02715(P - ET)$, in million gallons per acre. If P and ET are in centimeters,

$$\text{Flow to groundwater (m}^3) = 7Q_m(\text{m}^3/\text{d})\frac{T}{A\ (\text{m}^2)} + 0.01(P - ET)$$

Similarly, if n is the nitrogen concentration of the pretreated wastewater (in milligrams per liter) and NC is the nitrogen removal by the growing crop (M/L^2), an estimate of nitrogen entering the groundwater from the irrigation site is

$$7(8.34)nQ_mT/A - NC$$

if NC is in pounds per acre, or

$$7 \times 10^{-3} nQ_m(\text{m}^3/\text{d})\frac{T}{A(\text{m}^2)} - NC$$

if NC is in kilograms per square meter.

If groundwater standards require the average nitrogen concentrations in seepage water to be less than the drinking water standard of 10 mg/L,

$$\frac{7nQ_mT/A - NC/8.34}{7Q_mT/A + 0.02715(P - ET)} < 10 \qquad (12.4)$$

which reduces to

$$\frac{Q_m}{A} < \frac{NC}{58.4T(n-10)} + \frac{0.0388(P - ET)}{T(n-10)} \qquad (12.4a)$$

In metric units

$$\frac{7nQ_mT/A - NC \times 10^3}{7Q_mT/A + 0.01(P - ET)} < 10$$

which reduces to

$$\frac{Q_m}{A} < \frac{143\text{NC}}{T(n-10)} + \frac{(P-\text{ET})}{70T(n-10)}$$

If the nitrogen application in the wastewater is to be equal to or less than the crop requirement, NC, then

$$\frac{7(8.34)nQ_m T}{A} < \text{NC} \tag{12.5}$$

or

$$\frac{Q_m}{A} < \frac{\text{NC}}{58.4nT} \tag{12.5a}$$

or in metric units

$$7 \times 10^{-3} nQ_m \frac{T}{A} < \text{NC}$$

or

$$\frac{Q_m}{A} < \frac{143\text{NC}}{nT}$$

Equations (12.4) and (12.5) constrain the nitrogen loading rate at the land application site. The liquid loading rate will be constrained by the drainage capacity of the soil, $\bar{r}$ (in inches per week or centimeters per week):

$$r < \bar{r} \tag{12.6}$$

These equations will define the land area requirements for a specific wastewater. A design is shown in Example 12.1.

Example 12.1. A food-processing wastewater is to be land-irrigated. Compute the area requirements for the following conditions:
1 million gal/d (3785 m³/d)
500 mg/L BOD
25 mg/L N

Solution Regulations limit the application rate to 3 in/week (7.6 cm/week) and 500 lb BOD/(acre · d) [0.056 kg/(m² · d)]. For this maximum loading rate,

$$\frac{Q_m}{A} = \frac{r}{258 \ (\text{acre} \cdot \text{in/week})/(\text{million gal/d})}$$

$$\frac{1}{A} = \frac{3}{258}$$

$$A = 86 \text{ acres} \quad (35 \text{ ha} = 350{,}000 \text{ m}^2)$$

The BOD in terms of pounds per day is

$$\text{lb BOD/d} = \frac{500 \text{ mg BO}}{L} \times 1 \text{ million gal/d} \times 8.34 \frac{\text{lb/million gal}}{\text{mg/L}}$$

$$= 4170 \text{ lb/d} \quad (1893 \text{ kg/d})$$

For this BOD loading rate, a relatively small area is adequate:

$$A = \frac{4170 \text{ lb/d}}{500 \text{ lb BOD/(acre} \cdot \text{d)}} = 8.34 \text{ acres} \quad (3.40 \text{ ha} = 34{,}000 \text{ m}^2)$$

However, the nitrogen for the crop requirement with reed canary grass is 200 lb/acre (0.022 kg/m²). The spraying period for the cannery is 12 weeks. Therefore,

$$\text{lb N} = \frac{25 \text{ mg N}}{L} \times \text{million gal/d} \times 8.34 \frac{\text{lb/million gal}}{\text{mg/L}} \times 12 \text{ weeks} \times 7 \frac{\text{d}}{\text{week}}$$

$$= 17{,}512 \text{ lb N} \quad (7951 \text{ kg})$$

$$A = \frac{17{,}514 \text{ lb N}}{200 \text{ lb N/acre}} = 88 \text{ acres} \quad (35.9 \text{ ha} = 359{,}000 \text{ m}^2)$$

or

$$\frac{Q_m}{A} = \frac{NC}{58.4nT}$$

$$\frac{1}{A} = \frac{200}{58.4 \times 25 \times 12}$$

$$A = 88 \text{ acres} \quad (35.9 \text{ ha} = 359{,}000 \text{ m}^2)$$

The area requirement is 88 acres, dictated by the crop requirement rather than the regulatory maximum loading or the BOD.

Performance of Land Application Systems

Wastes that have been successfully disposed of by spray irrigation include cannery,[7] pulp and paper,[8] dairy, and tannery wastes. Spent sulfite liquor[9,10] has been applied up to a rate of 116 lb solids/(d · yd²) [63 kg/(d · m²)]; this resulted in up to 95 percent removal of BOD through a 10-ft (3-m) soil layer. Boxboard[8] wastes have been sprayed on a blown silt loam with a gravel underlay and alfalfa cover crop at a short-term rate of 0.7 in/h (1.8 cm/h) and a total daily rate of 0.21 to 0.56 in/d (0.5 to 1.4 cm/d). This is equivalent to 0.2 to 1.0 acres per ton of production per day [0.89 to 4.40 m²/(kg/d)]. Strawboard cook liquor, including beater-washer water and machine water,[8] has been sprayed twice a week for 6 h/d at a rate of 0.5 in/d (1.3 cm/d). A rate of 1.2 in/d (3 cm/d) resulted in ponding. Kraft mill wastes have been found to require 1.5 acres per ton of production (6.7 m²/kg) or more.[8]

Waste	Acres Sprayed	Application Rate, gal/(min · acre)	Duration	Average Loading lb BOD/(acre · d)	lb SS/(acre · d)	Reference
Tomatoes	5.63	178	7.5 h/d	413	364	1
	6.40	86	3.7 h/d	155	139	1
Corn	2.28	153.5	10 h/d	864	500	1
Asparagus and beans	0.90	282.0	5.6 h/d	22.5	356	1
Lima beans	6.65	65	16 h/d	65	46	1
Cherries	2.24	96.5	17 h/d	807	654	1
Paperboard	1.30	77	3 h/5 d			8
Hardboard	100	42	12 h/10 d			8
	300	24	18 h/10 d			
Strawboard	1.5	94	6 h/3 d			8
Kraft	70	98	8 h/week			8

Note: gal/(min · acre) = 9.35 × 10^{-7} m^3/(min · m^2)
lb/(acre · d) = 1.12 × 10^{-4} kg/(m^2 · d)
acre = 4.05 × 10^3 m^2

TABLE 12.3 Spray Irrigation of Industrial Wastes

Runoff of airport deicing fluid (ADF) consisting of propylene glycol plus additives were treated on soil.[11] Waste activated sludge amendments on the order of 955 mg/kg soil, in addition to lime amendments and tilling, enhanced the degradation rate. Diluting ADF solutions in excess of 20 percent by weight prior to soil application is necessary to prevent inhibition. The degradation followed first-order kinetics.

Total petroleum hydrocarbons were degraded in soil.[12] Soil aeration was accomplished by ploughing the soils monthly, and nutrients were incorporated into the soils by adding dry fertilizers. Complete destruction was achieved in 50 d. Typical data for spray irrigation systems are shown in Table 12.3. Results for food-processing wastewaters are shown in Table 12.4. Overland flow systems are summarized in Table 12.5.

12.3 Deep-Well Disposal

Deep-well disposal involves the injection of liquid wastes into a porous subsurface stratum that contains noncommercial brines.[13] The wastewaters are stored in sealed subsurface strata isolated from

Design Flow, million gal/d	Overall Application Rate, in/week	Spray Field Scheduling Application Rate, in/h	Use/Rest Ratio	Spray Time, h	Total Spray Area, acres	Number of Spray Fields	Crop Cover	Soil	Pretreatment	Product
0.2	2.7	0.25	1:3	0.33	23	4	Reed canary grass	Coarse and fine sediments	Activated sludge	Dairy cream, butter, cheese, powdered milk
0.189	0.68	0.037	1:8	5.25	85.6	6	Reed canary grass	Low plastic clay	Lagoons	Beets, cabbage
1.8	5.5	0.25	1:5	20	61	5	Forested with heavy undergrowth	Genesee fine sandy loam	Screening and pH adjustment	Beets, corn, peas
1.0	3.15	0.30	1:7	12	80	4	Reed canary grass, and clover	Ontario loam and Cazenovia silt loam	Aerated lagoons	Corn, beans, peas, carrots, potatoes
0.64	3.53	0.21	1:28	6	73.5	7	Reed canary grass	Williamson and Wallington loam	pH adjustment and aerated lagoons	Beans, cherries, beets, apples

Note: gal/d = 3.79×10^{-3} m^3/d
in/week = 2.54 cm/week
acre = 4.05×10^3 m^2

TABLE 12.4 Spray Irrigation of Food-processing Wastewaters[5]

Design flow, million gal/d	Overall Application Rate, in/week	Spray Field Scheduling Application Rate, in/h	Use/Rest Ratio	Spray Time, h	Total Spray Area, acres	Number of Spray Fields	Crop Cover	Soil	Pretreatment	Product
0.8	2.7	0.15	1:8	8	80	9	Reed canary grass	Schoharie silty loam	Lagoons	Sauerkraut, beans
0.26	3.36	0.2	1:7	12	20	4	Reed canary grass	Gravelly loam	Screening, aerated lagoons	Beans, apples
0.025	2.6	0.11	1:6	18	9	5	Reed canary grass	Chenango gravel	Screening	Cheddar cheese
0.1	1.0	0.1	1:17	12	27	9	Japanese millet	Fine, sandy loam and silt loam	Lagoons	Cherries, prunes, apples
0.047	2.02	0.12	1:9	12	6	6	Reed canary grass	Sandy silt	Screening pH adjustment lagoon	Cider, vinegar

Note: gal/d = 3.79×10^{-3} m^3/d
in/week = 2.54 cm/week
acre = 4.05×10^3 m^2

TABLE 12.4 Spray Irrigation of Food-processing Wastewaters[5] (*Continued*)

Miscellaneous Treatment Processes

Name and Location	Loading Rate, cm/d	Slope Length, m	Type of Wastewater
Campbell Soup Co.,			
Chestertown, Maryland	1.0	53–76	Poultry
Hunt-Wesson Foods,			
Davis, California	2.2	51	Tomato
El Paso, Texas	2.5	62	Meat packing
Middlebury, Indiana	1.5	79	Poultry
Campbell Soup Co.,			
Napoleon, Ohio	1.5	53–61	Tomato
Paris, Texas	1.6–3.6	60–90	Vegetable soup
Sebastopol Co-op,			
Sebastopol, California	1.5	45	Apple
Nabisco, Inc.			
Woodbury, Georgia	0.9–11	80–84	Peanuts, pimientos
Frito-Lay Inc.,			
Wooster, Ohio	0.5–1.0	67	Vegetable

TABLE 12.5 Selected Overland Flow Systems[2]

groundwater or mineral resources. Disposal wells may vary in depth from a few hundred feet (< 100 m) to 15,000 ft (4570 m) with capacities ranging from less than 10 to more than 2000 gal/min (38 to 7550 L/min). Wastes disposed of in wells are usually highly concentrated, toxic, acidic, or radioactive, or wastes high in inorganic content which are difficult or excessively expensive to treat by some other process.

The disposal system consists of the well and pretreatment equipment necessary to prepare the waste for suitable disposal into the well. A casing, generally of steel, is cemented in place to seal the disposal stratum from other strata penetrated during drilling of the well. An injection tube transports the waste to the disposal stratum as shown in Fig. 12.2.

Oil or freshwater is used to fill the annular space between the injection tube and the casing and extends to but is sealed from the injection stratum. Leaks in the injection tube or drainage to the casing can be detected by monitoring the pressure of the fluid.

The system includes a basin to level fluctuations in flow, pretreatment equipment, and high-pressure pumps. Pretreatment

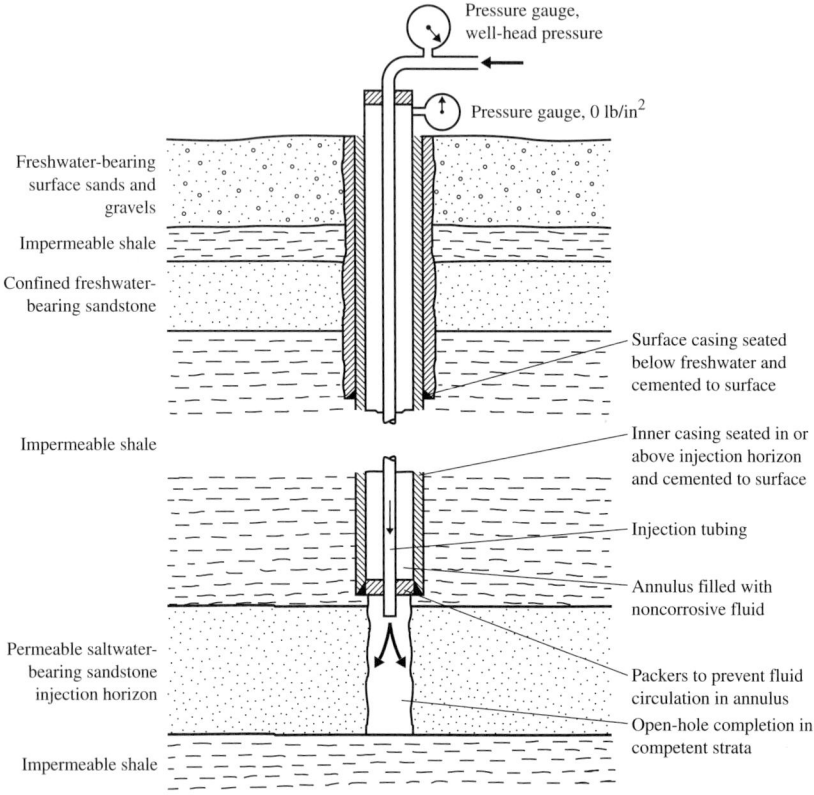

Figure 12.2 Schematic diagram of a waste injection well completed in a competent sandstone. (*After Donaldson, 1964*[14].)

requirements are determined by the characteristics of the wastewater, compatibility of the wastewater and the formation water, and the characteristics of the receiving stratum.

Pretreatment may include the removal of oils and floating material, suspended solids, biological growths, dissolved gases, precipitatable ions, acidity, or alkalinity. A typical system is shown in Fig. 12.3.

The best disposal areas include sedimentary rock in the unfractured state, including sandstones, limestones, and dolomites and unconsolidated sands. Fractured strata should be avoided because, if a vertical fissure exists, groundwater contamination may result.

The well-head pressure is related to the difference between the bottom-hole pressure and the reservoir pressure. Cores of the injection location are needed to evaluate the porosity and permeability of the stratum and any possible reactions between the wastewater and the stratum.

Miscellaneous Treatment Processes

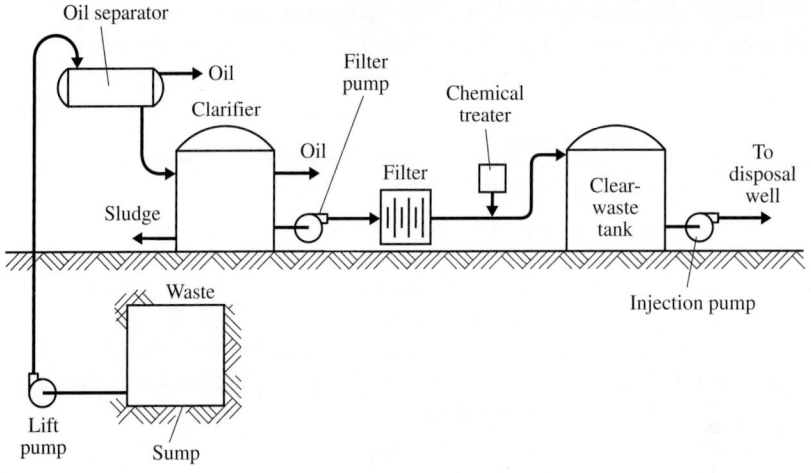

FIGURE 12.3 Typical subsurface waste-disposal system. (*After Donaldson, 1964.*)

Although wastewaters should be generally free of suspended solids, some vugular formations will accept suspended solids without problems or an increase in injection pressure. In some cases, injection can be increased by well stimulation, which involves the injection of mineral acids to dissolve calcium carbonate and other acid-soluble particulates that tend to plug the stratum. Mechanical procedures involve scratching, swabbing, washing, and underreaming the well bore and shotting the uncased stratum with explosives or hydraulic fracturing.

The cost of a well system will be affected by the depth of the injection well, type of formation, geographic location, waste volume, required pretreatment, and injection pressure.

It should be recognized that there is a significant regulatory difference in the deep-well disposal of Resource Conservation and Recovery Act (RCRA) hazardous and nonhazardous wastes. The Land Ban Provisions of the 1984 Hazardous and Solid Waste Amendments (HSWA) prohibited the land disposal of selected wastes as well as the deep-well injection of hazardous wastes without strict standard for pretreatment. A "no migration petition for 10,000 years" is one of the barriers before hazardous wastes can be injected. Title 40 (40 CFR) contains the regulations applicable to these activities.

- Part 146, Subpart G—Underground Injection Control (UIC) Program: Criteria and Standards Applicable to Class I Hazardous Waste Injection Wells, imposes more stringent permit requirements for hazardous waste disposal wells.
- Part 148, Hazardous Waste Injection Restrictions, sets rules for UIC land ban implementation and sets standards and procedures for the petition process.

There is less strict procedure in applying and getting approval to dispose nonhazardous wastes in Class I injection wells. However, there are certain fundamental factors which apply to all injection wells, namely:

- Injection will not be used for hazardous waste disposal in any areas where seismic activity could potentially occur.
- Injected wastes must be compatible with the mechanical components of the injection well system and the natural formation water. The waste generator may be required to perform physical, chemical, biological, or thermal treatment for removal of various contaminants or constituents from the waste to modify the physical and chemical character of the waste to assure compatibility.
- High concentrations of suspended solids (typically >2 ppm) can lead to plugging of the injection interval.
- Corrosive media may react with the injection well components, injection zone formation, or confining strata with very undesirable results. Wastes should be neutralized.
- High iron concentrations may result in fouling when conditions alter the valence state and convert soluble to insoluble species.
- Organic carbon may serve as an energy source for indigenous or injected bacteria, resulting in rapid population growth and subsequent fouling.
- Wastestreams containing organic contaminants above their solubility limits may require pretreatment before injection into a well.
- Site assessment and aquifer characterization are required to determine suitability of site for wastewater injection.
- Extensive assessments must be completed prior to receiving approval from regulatory authority.

A typical application for a nonhazardous Class I underground injection control application would restrict the injection of nonhazardous wastes only such as spent lines, demineralization waters [reverse osmosis (RO) reject waters, etc.] cooling tower blowdown water, nonhazardous waste generated during the closure of the well and associated facilities that are compatible with the permitted wastestreams, injection zone reservoir, and well materials. There is obviously a strict prohibitor for any hazardous wastes, pH ranges less than 5 or greater than 10, and low-level radioactive wastes (NORM). There are also strict operating, monitoring, and testing provisions. Drilling, completion, abandonment, and closure regulations also are included.

Officials agree that waste disposal through properly constructed and operated injection wells is safer and less likely to contaminate surface water or potable groundwater than are landfills and other forms of land treatment. For example, injection of hazardous wastes into aquifers that serve or could serve as groundwater supplies for communities is not allowed.

Class II injection wells also are subject to the permitting process. These wells are used to inject oil and gas wastes, primarily salt water. These injection wells are used primarily for secondary recovery, although some are used to store hydrocarbons for underground storage.

12.4 Membrane Processes

Membrane filtration includes a broad range of separation processes from filtration and ultrafiltration to reverse osmosis. Generally, those processes defined as filtration refer to systems in which discrete holes or pores exist in the filter medium, generally in the order of 10^2 to 10^4 nm or larger. The efficiency of this type of filtration depends entirely on the difference in size between the pore and the particle to be removed.

The various filtration processes relative to molecular size are shown in Fig. 12.4 and Table 12.6. Membrane separation technologies for wastewater treatment are shown in Table 12.7.

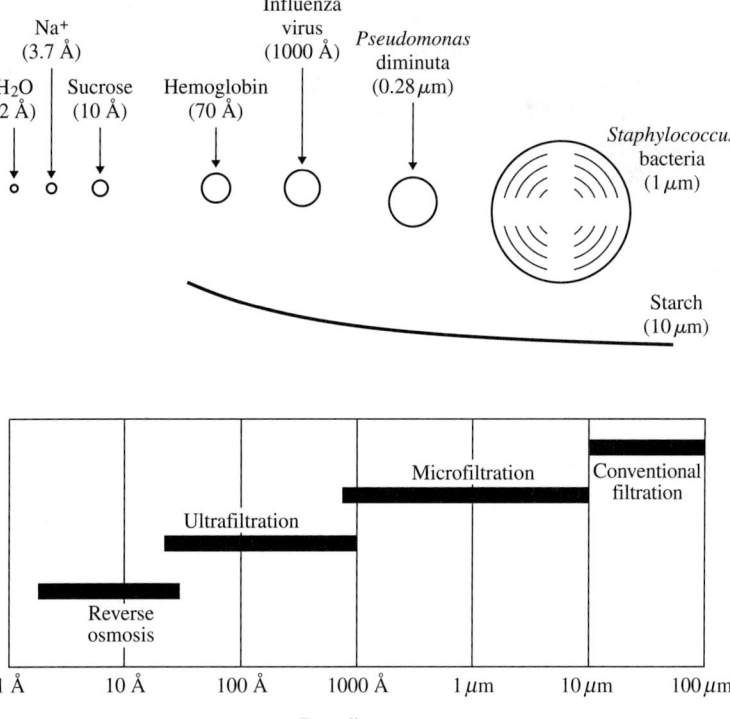

FIGURE 12.4 Membrane processes and pore sizes.

Material to Be Removed	Approximate Size, nm	Process
Ion removal	1–20	Diffusion or reverse osmosis
Removal of organics in true solution	5–200	Diffusion
Removal of organics: subcolloidal—not in true molecular dispersion	200–10,000	Pore flow
Removal of colloidal and particulate matter	75,000	Pore flow

TABLE 12.6 Membrane Processes

Reverse osmosis employs a semipermeable membrane and a pressure differential to drive freshwater to one side of the cell, concentrating salts on the input or rejection side of the cell. In this process, freshwater is literally squeezed out of the feedwater solution.

The reverse osmosis process can be described by considering the normal osmosis process. In osmosis, a salt solution is separated from a pure solvent into a solution of less concentration by a semipermeable membrane. The semipermeable membrane is permeable to the solvent and impermeable to the solute. Such an arrangement is shown in Fig. 12.5. The chemical potential of the pure solvent is greater than that of the solvent in solution, and therefore drives the system to equilibrium. If an imaginary piston applies an increasing pressure on the solution compartment, the solvent flow through the membrane will continue to decrease. When sufficient pressure has been applied to bring about thermodynamic equilibrium, the solvent flow will stop. The pressure developed in achieving equilibrium is the osmotic pressure of the solution, or the difference in the osmotic pressure between the two solutions if a less concentrated salt solution is used instead of pure solvent in the right chamber of the cell.

If a pressure in excess of the osmotic pressure is now applied to the more concentrated solution chamber, pure solvent is caused to flow from this chamber to the pure solvent side of the membrane, leaving a more concentrated solution behind. This phenomenon is the basis of the reverse osmosis process.

The criteria of membrane performance are the degree of impermeability (how well the membrane rejects the flow of the solute) and the degree of permeability (how easily the solvent is allowed to flow through the membrane). Cellulose acetate membranes provide an attractive combination of these criteria. Membrane technology uses cross-flow filtration in which the feed stream flows across (tangential to) the membrane surface, as shown in Fig. 12.6. Cross-flow filtration

Feature	Micro-filtration	Ultra-filtration	Nano-filtration	Reverse Osmosis	Per-vaporation
Suspended solids removal	Excellent	Impractical	Impractical	Impractical	N/A
Dissolved organic removal	N/A	Excellent*	Excellent*	Excellent*	Good‡
VOC removal	N/A	Poor	Fair*	Fair–good*	Excellent
Dissolved inorganic removal	N/A	N/A	Good (function of salt species)	Very good (90–99% removal)	N/A
Osmotic pressure effects	None	Minor	Significant	High	None
Concentration capabilities	Up to 5% total solids	Up to 50% total organics	Up to 15%†	Up to 15%†	N/A
Permeate quality	Excellent	Excellent	Good	Excellent	Excellent
Energy requirements	1–3 bars	3–7 bars	5–10 bars	15–70 bars	<25% of distillation
Capital costs ($/GPD)	0.15–1.5	0.15–1.85	0.15–1.5	0.15–1.5	1.85–4.00
Operating cost ($/1000 L feed rate)	0.15–1.10	0.15–0.80	0.20–0.80	0.25–0.80	0.80–1.30

All technologies are application-specific, and testing is required to develop precise data. Estimating specific capital and operating costs is difficult because some wastestreams may require special materials of construction or additional design considerations as a result of high potential.
*Function of molecular weight.
†Function of osmotic pressure.
‡Function of vapor pressure.
N/A—Not applicable.
GPD—Gallon per day.

TABLE 12.7 Membrane Separation Technologies for Wastewater Treatment[15]

reduces fouling and concentration polarization. Membrane types are shown in Fig. 12.7. Each of these design configurations uses cellulose acetate membranes, except certain of the hollow fine-fiber systems that employ a nylon polymer. The more common membranes are described below:

1. *Tubular.* Manufactured from ceramics, carbon, or any number of porous plastics, these tubes have inside diameters ranging

682 Chapter Twelve

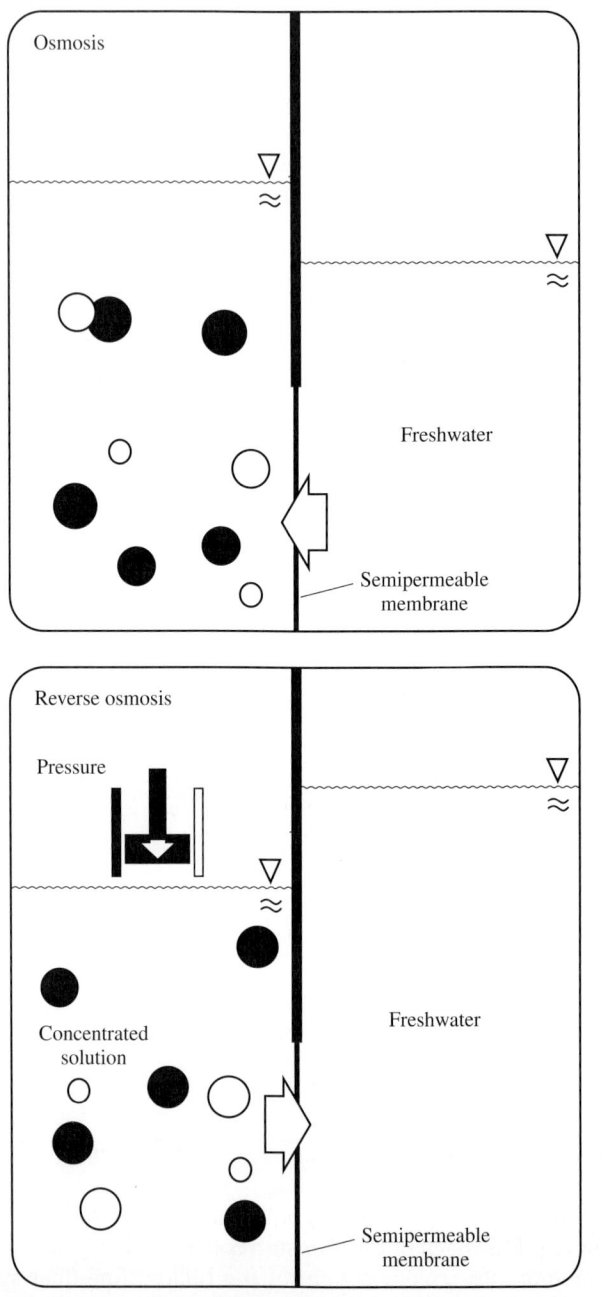

Figure 12.5 Osmosis and reverse osmosis.

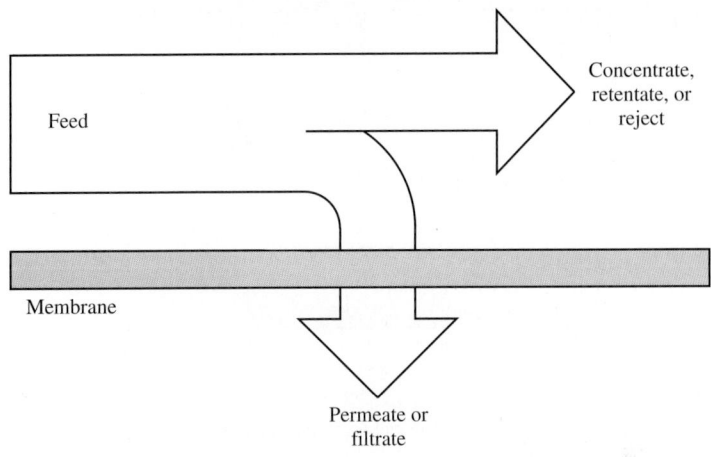

FIGURE 12.6 Cross-flow filtration: The one input feed stream is divided into two output streams—the concentrate and the permeate streams.

from 1/8 in (3.2 mm) up to approximately 1 in (2.54 cm). The membrane is typically coated on the inside of the tube, and the feed solution flows through the interior from one end to the other, with the "permeate" or "filtrate' passing through the wall to be collected on the outside of the tube.

2. *Hollow fiber.* Similar to the tubular elements in design, hollow fibers are generally much smaller in diameter and require rigid support such as is obtained from the "potting" of a bundle inside a cylinder. As with tubular elements, feed flow is usually down the core of the fiber.

3. *Spiral wound.* This device is constructed from an envelope of sheet membrane wound around a permeate tube that is perforated to allow collection of the permeate or filtrate.

4. *Plate and frame.* This device incorporates sheet membrane that is stretched over a frame to separate the layers and facilitate collection of the permeate.

With regard to membrane element configuration, the important physical characteristics of the various membrane element device configurations available today are listed in Table 12.8.

Because of the propensity of suspended or precipitated materials settling out on the membrane surface and plugging the membrane pores, turbulent flow conditions must be maintained (Reynolds numbers in excess of 2000). For high recovery systems, this usually requires recycling a significant percentage of the concentrate back to the feed side of the pump. The addition of this concentrate stream into the feed solution obviously increases the dissolved solids concentration, further increasing osmotic pressure.

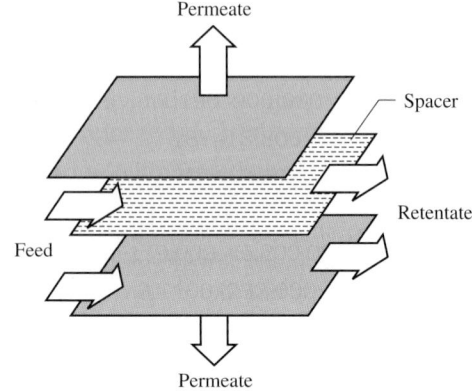

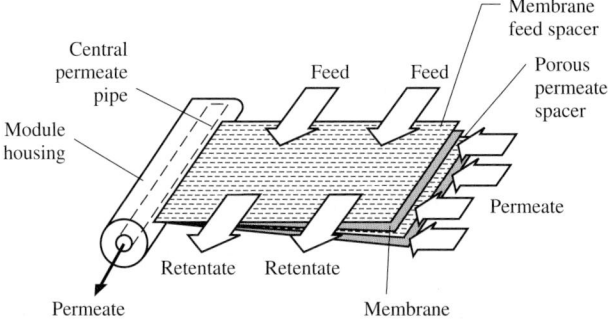

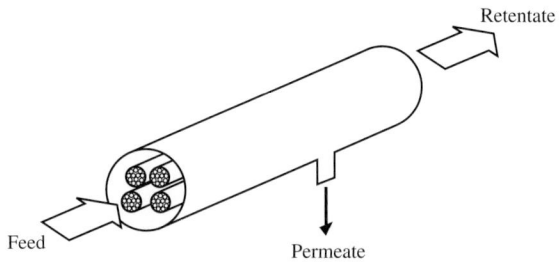

FIGURE 12.7 Membrane shapes.

In order to minimize membrane fouling, pretreatment is required for the removal of suspended matter, bacteria, and precipitatable ions. A typical reverse osmosis process schematic is shown in Fig. 12.8. The design and operating parameters for a reverse osmosis system are summarized from Agardy.[16]

Element Configuration	Packing Density*	Suspended Solids Tolerance
Spiral wound	High	Fair
Tubular	Low	High
Plate and frame	Low	High
Hollow fine fiber	Highest	Poor

*Membrane area per unit volume of space required.

TABLE 12.8 Common Membrane Characteristics

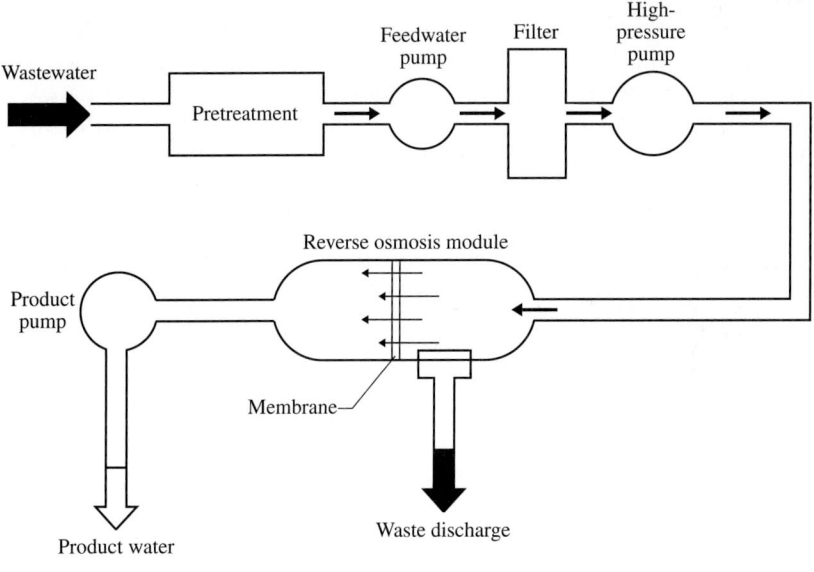

FIGURE 12.8 Basic reverse osmosis process schematic.

Pressure

The water flux is a function of the pressure differential between the applied pressure and the osmotic pressure across the membrane. The higher the applied pressure, the greater the flux. However, the pressure capability of the membrane is limited, so the maximum pressure is generally taken to be 1000 lb/in² (6895 kPa or 68 atm) gauge. Operating experience dictates in the 400 to 600 lb/in² (2758 to 4137 kPa or 27 to 41 atm) gauge range, with 600 lb/in² (4137 kPa or 41 atm) gauge normally being the design pressure.

Temperature

The water flux increases with increasing feedwater temperature. A standard of 70°F (21°C) is generally assumed and temperatures of up to 85°F (29°C) are acceptable. However, temperatures in excess of 85°F (29°C) and up to 100°F (38°C) will accelerate membrane deterioration and cannot be tolerated for long operating periods.

Membrane Packing Density

This is an expression of the unit area of the membrane, which can be placed per unit volume of pressure vessel. The greater this factor the greater will be the overall flow through the system. Typical values range from 50 to 500 ft^2/ft^3 (160 to 1640 m^2/m^3) of pressure vessel.

Flux

Although the flux rates for hollow fibers are from 6.0×10^{-3} to 10.2×10^{-3} m^3/(d · m^2) [0.15 to 0.25 gal/(d · ft^2)] versus 6.1×10^{-1} to 10.2×10^{-1} m^3/(d · m^2) [15 to 25 gal/(d · ft^2)] for sheet systems, the stacking densities for fibers are about 10 times greater, making the sizes of the two systems competitive. This flux tends to decrease with length of run and over a period of 1 or 2 years of operation might be reduced by 10 to 50 percent.

Recovery Factor

This consideration actually represents plant capacity and is generally in the range of 75 to 95 percent, with 80 percent being the practical maximum. At high recovery factors there is a greater concentration in the process water as well as in the brine. At higher concentrations salt precipitation on the membrane increases, causing a reduction in operation efficiency.

Salt Rejection

Salt rejection depends on the type and character of the selected membrane and the salt concentration gradient. Generally, rejection values of 85 to 99.5 percent are obtainable, with 95 percent being commonly used.

Membrane Life

Membrane life can be drastically shortened by undesired constituents in the feed water, such as phenols, bacteria, and fungi as well as high temperatures and high or low pH values. Generally, membranes will last up to 2 years with some loss in flux efficiency.

pH

Membranes consisting of cellulose acetate are subject to hydrolysis at high and low pH values. The optimum pH is approximately 4.7, with operating ranges between 4.5 and 5.5.

Turbidity

Reverse osmosis units can be used to remove turbidity from feedwaters. They operate best if little or no turbidity is applied to the membrane. Generally, it is considered that the turbidity should not exceed one Jackson turbidity unit (JTU) and that the feedwater should not contain particles larger than 25 µm.

Feedwater Stream Velocity

The hydraulics of reverse osmosis systems are such that velocities in the range of 0.04 to 2.5 ft/s (1.2 to 76.2 cm/s) are common. Plate and frame systems operate at the higher velocity, while the hollow fine-fiber units operate at the lower velocities. High velocities and turbulent flow are necessary to minimize concentration polarization at the membrane surface.

Power Utilization

Power requirements are generally associated with the system pumping capacity and operational pressures. Values range from 9 to 17 kW · h/ thousand gal (2.4 to 4.5 kW · h/m^3), with the lower figure taking into account some power recovery from the brine stream.

Pretreatment

The present stage of development of membranes limits their direct application to feedwaters having a TDS not exceeding 10,000 mg/L. Further, the presence of scale-forming constituents, such as calcium carbonate, calcium sulfate, oxides and hydroxides of iron, manganese, and silicon, barium and strontium sulfates, zinc sulfide, and calcium phosphate, must be controlled by pretreatment or they will require subsequent removal from the membrane. These constituents can be controlled by pH adjustment, chemical removal, precipitation, inhibition, and filtration. Organic debris and bacteria can be controlled by filtration, carbon, pretreatment, and chlorination. Oil and grease must also be removed to prevent coating and fouling of the membranes.

Cleaning

Since, under continuous use, membranes will foul, provision must be made for mechanical and/or chemical cleaning. Methods include periodic depressurizations, high-velocity water flushing, flushing with air-water mixtures, backwashing, cleaning with enzyme detergents, ethylene diamine, tetraacetic acid, and sodium perborate. The control of pH during cleaning operations must be maintained to prevent membrane hydrolysis. Approximately 1 to 1.5 percent of the process water goes to waste as a part of the cleaning operation, with the cleaning cycle being every 24 to 48 h.

A summary of operational parameters is shown in Table 12.9.

Parameter	Range	Typical
Pressure, lb/in^2 gauge	400–1000	600
Temperature, °F	60–100	70
Packing density, ft^2/ft^3	50–500	—
Flux, gal/(d · ft^2)	10–80	12–35
Recovery factor, %	75–95	80
Rejection factor, %	85–99.5	95
Membrane life	—	2
pH	3–8	4.5–5.5
Turbidity, JTU	—	1
Feedwater velocity, ft/s	0.04–2.5	—
Power utilization, kW · h/thousand gal	9–17	—

Source: Adapted from Warner, 1965.

TABLE 12.9 Summary of System Operational Parameters for Reverse Osmosis

Applications

Reverse osmosis has been applied to the treatment of plating wastewaters for the removal of cadmium, copper, nickel, and chromium at pressures of 200 to 300 lb/in^2 (1378 to 2067 kPa or 13.6 to 20.4 atm). The concentrated stream is returned to the plating bath and the treated water to the next-to-last rinse tank, as shown in Fig. 12.9.

Pulp mill effluents have been treated by reverse osmosis at a pressure of 600 lb/in^2 (4137 kPa or 41 atm). Wastestreams were

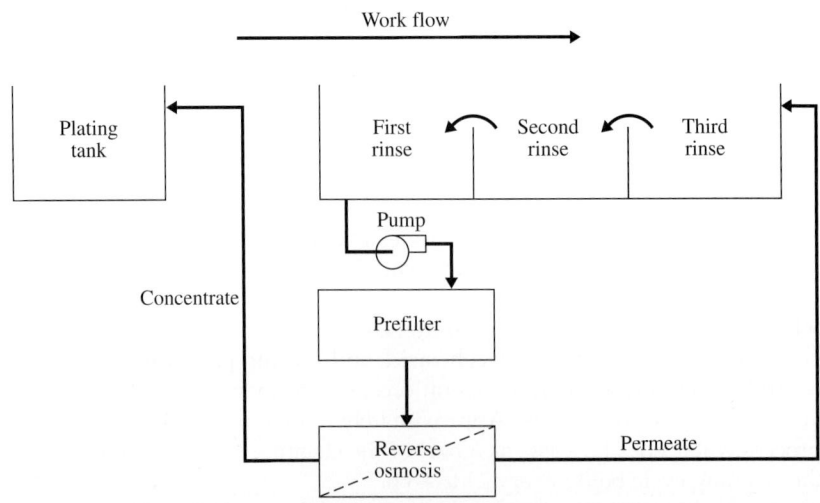

FIGURE 12.9 Treatment of plating wastewaters by reverse osmosis.

concentrated up to 100,000 mg/L total solids. The flux was found to be a function of total solids level and varied from 2 to 15 gal/(d · ft²) [0.08 to 0.61 m³/(d · m²)].[17]

Oily wastes can be treated by ultrafiltration in which the permeate can be recycled as rinsewater and the concentrate can be hauled or incinerated, as shown in Fig. 12.10.

Leachate from an industrial landfill was treated in a two-stage reverse osmosis system as shown in Fig. 12.11.[18] Performance of the

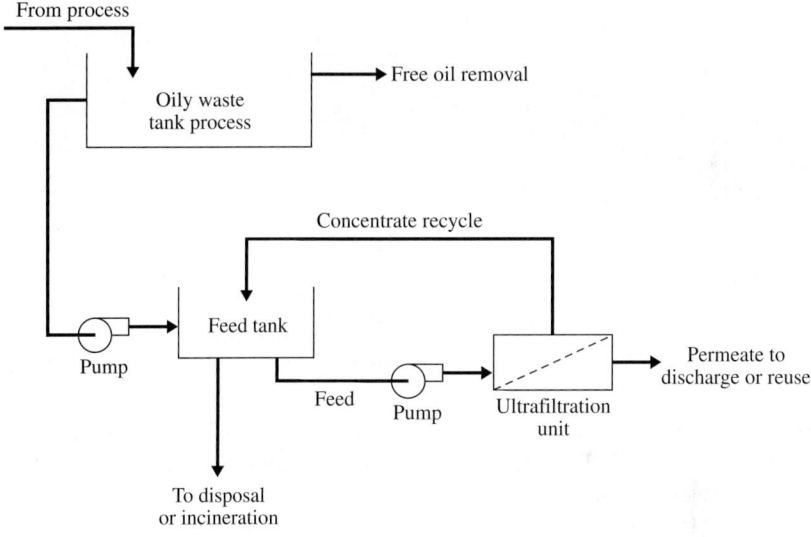

FIGURE 12.10 Treatment of oily wastewaters by ultrafiltration.

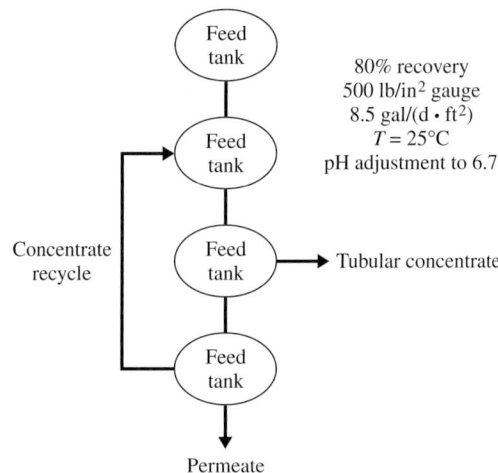

FIGURE 12.11 Two-stage membrane industrial landfill leachate treatment system.

Parameter	Leachate	Permeate
pH	8.2	5.6
COD, mg/L	1948	7
BOD, mg/L	105	2
TKN, mg/L	612	9
Cl, mg/L	2504	33
Zinc, µg/L	630	440
Copper, µg/L	170	45
Lead, µg/L	100	15
Chromium, µg/L	170	60
Nickel, µg/L	150	40
Cadmium, µg/L	1.3	0.7
Arsenic, µg/L	12	4
Mercury, µg/L	0.5	—

TABLE 12.10 Summary of Performance of Organic Chemical Wastewater Treatment by Reverse Osmosis

treatment system is shown in Table 12.10. It was also found that improved operation resulted from biological pretreatment of the wastewater as shown in Fig. 12.12.[19] Membrane treatment of an organic chemicals wastewater is shown in Table 12.11.

An RO design is shown in Example 12.2.

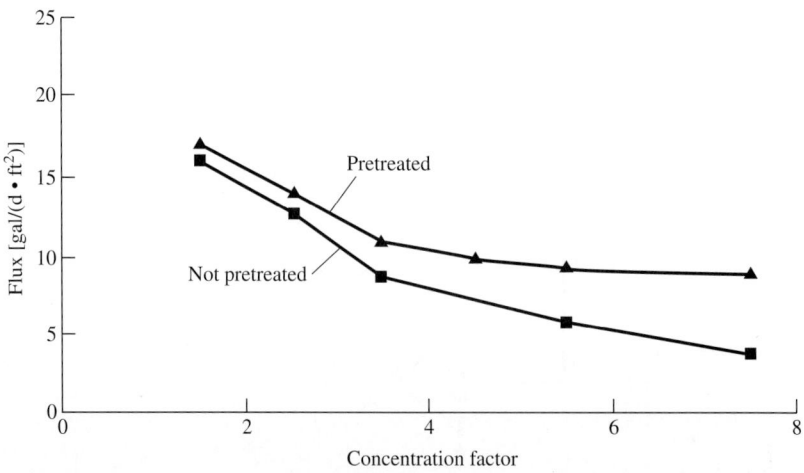

FIGURE 12.12 Effect of biological pretreatment on reverse osmosis performance. (*Adapted from kettern, 1992.*)

Miscellaneous Treatment Processes

Parameter	Unit	Influent	Effluent Nanofiltration	Effluent Reverse Osmosis
Compound A	μg/L	140	ND(1)*	ND(1)
Compound B	μg/L	1.3	ND(1)	ND(1)
Compound C	μg/L	200	11	13
COD	mg/L	12,630	9000	4750
TDS	mg/L	32,950	13,550	1250
C. dubia, LC_{50}	%	1.3	8.8	8.8

*Nonobservable peak at the detection level indicated in parentheses.

TABLE 12.11 Membrane Filtration of a Filtered Industrial Wastewater

Example 12.2. The following information on a wastewater and reverse osmosis system is to be used for renovation. Design an RO system for this wastewater.

Plant design capacity, Q 10 million gal/d (37,850 m³/d)
Recovery factor, R 75%
Salt rejection factor, S 95%
Design pressure   600 lb/in² gauge (40.8 relative atm)
Feedwater temperature   80°F (27°C)
Total dissolved solids   600 mg/L

Solution
System sizing. The wastewater (system feedwater) flow Q_w is

$$Q_w = \frac{Q}{R} = \frac{10 \text{ milliom gal/d}}{0.75} = 13.3 \text{ milion gal/d} \quad (50{,}341 \text{ m}^3/\text{d})$$

The brine flow to be disposed of is then 13.3 − 10, or 3.3 million gal/d (12,491 m³/d).

Assuming a flux of 20 gal/(d · ft²) [0.814 mL/(d · m²)], the membrane area needed, A, is

$$A = \frac{10 \times 10^6 \text{ gal/d}}{20 \text{ gal/(d} \cdot \text{ft}^2)} = 500{,}000 \text{ ft}^2 \quad (46{,}450 \text{ m}^2)$$

Assuming a packing density of 250 ft²/ft³ (820 m²/m³), the total module volume, V, is

$$V = \frac{500{,}000 \text{ ft}^2}{250 \text{ ft}^2/\text{ft}^3} = 2000 \text{ ft}^3 \quad (56.6 \text{ m}^3)$$

Assuming 1 ft³ (0.0283 m³) of volume per module, 2000 modules would be required, and assuming 10 modules per pressure vessel, 200 pressure vessels would be required.

The approximate total dissolved solids content of the product water, TDS_p, is

$$TDS_p = \frac{1}{R}(\text{wastewater TDS})(1-S)$$

$$= \frac{1}{0.75}(600 \text{ mg/L})(1-0.95)$$

$$= 40 \text{ mg/L}$$

Power consumption. The power consumed by the pressurizing pump is

$$\text{Hydraulic hp} = \frac{(\text{wastewater flow in million gal})(\text{pressure in lb/in}^2 \text{ gauge})}{2.74}$$

$$= \frac{(13.3 \text{ million gal})(600 \text{ lb/in}^2 \text{ gauge})}{2.47}$$

$$= 3230 \text{ hp} \quad (2423 \text{ kW})$$

The brake horsepower which must be delivered by the pump motor is

$$\text{Brake hp} = \frac{\text{hydraulic hp}}{\text{pump efficiency}}$$

$$= \frac{3230 \text{ hp}}{0.75}$$

$$= 4000 \text{ hp} \quad (3000 \text{ kW})$$

If provision is desired to operate the system at higher pressures, that is, up to 1000 lb/in gauge (68.05 relative atm), the calculated power can be corrected by multiplying by the ratio of the selected pressure to 600 lb/in² gauge (40.8 relative atm).

Land area required. The land area in acres required for a reverse osmosis system can be estimated by the formula

$$\text{Plant area} = 0.7 + 0.33Q$$

$$= 0.7 + 0.33(10)$$

$$= 4 \text{ acres} \quad (16,300 \text{ m}^2 = 1.6 \text{ ha})$$

Brine disposal. The approximate total dissolved solids content of the rejected brine, TDS_b, is

$$TDS_b = \frac{S}{(1-R)}(\text{wastewater TDS})$$

$$= \frac{0.95}{1-0.75}(600 \text{ mg/L})$$

$$= 2280 \text{ mg/L}$$

The total dry weight of salt to be disposed of daily, W, is

W = 8.34 (2280 mg/L)(3.3 million gal/d) = 62,600 lb/d (28,400 kg/d)

12.5 Membrane Bioreactors

The application of membranes for solids separation has been applied in various manufacturing processes for many years. They were first used for the treatment of wastewater in the early 1970s for solids separation. The membranes were especially useful for the removal of small particulate and were beneficial in reducing total suspended solids and any biological oxygen demanding material in solid form. The first full-scale application of membrane bioreactor (MBR) technology was installed at a General Motors (GM) diesel engineering facility in Mansfield, Ohio, in the early 1990s.[20] Since then this technology has been applied in numerous industrial and, more recently, domestic wastewater applications. Applications of reuse, for example, in the refinery industry include cooling tower makeup and boiler feed water which represent two major uses of water in that industry. The driving forces responsible for industrial reuse include diminishing freshwater supplies, increasingly stringent environmental regulations, and the restriction of access to publicly-owned treatment works for oily wastewater producers.

Conventional secondary treatment systems often suffer from an inability to provide a consistent high quality effluent quality suitable for reuse, especially during flow or load variations typical in industry. In addition, effluents may contain potential toxics such as ammonia and metals and nonreadily biodegradable complex organics which restrict reuse options. These restrictions can be eliminated by the use of membranes in the wastewater treatment flow scheme.

Depending upon the design and application, membranes of various pore sizes may be used as MBRs, for tertiary or RO. The first activated sludge system that incorporated the use of membranes was developed in the research laboratories of Dorr-Oliver, Inc. in the early 1960s.[21] Work continues on the process in order to eliminate the problems encountered, including:

- Limited membrane life
- High energy consumption
- Low membrane flux
- Low permeability

Types of Membranes and Reactor Configuration

The two most commonly used membrane types are (1) microfiltration to remove total suspended solids (TSS), biological oxygen demand (BOD), and other microparticulates and (2) reverse osmosis for dissolved impurities such as salts. When membranes are used as bioreactors, the pore size that is commonly used is between 1 and 4 µm. Ultrafiltration and microfiltration membrane systems operate at low transmembrane pressures compared to nanofiltration or reverse osmosis and thus operate with less energy. The transmembrane pressure is measured as the pressure difference across the membrane required to induce permeate flow through the membrane.

A new generation of MBR technology has overcome many of the issues experienced in the past. These include the development of hollow fiber and flat-plate membranes. The hollow fiber systems were adapted from water supply filtration systems. The flat-plate systems were developed especially to meet the unique characteristics of working in a wastewater environment. Industrial MBRs have been configured with the membrane components being either internal or external to the biological reactor. The GM system discussed earlier used an external membrane configuration. In addition, anaerobic MBRs, first introduced by Dorr-Oliver in the 1980s are being applied for the treatment of industrial wastewater.

The "recirculated MBR" consists of activated sludge mixed liquor being pumped to tubular or flat sheet modules where it flows at high velocities (greater than 2 m/s and often greater than 4 m/s) with consequent high pressure drops and high transmembrane pressure outside of the bioreactor. The second configuration is the submerged membrane module in which hollow fibers or flat-plate membrane panels are immersed in the aeration tank and treated water or permeate is generated by flow through the membrane induced by gravity, a pump, or a small vacuum. The membranes use a transmembrane pressure of 1 to 10 psi. The contaminants in the process wastewater are retained in the process tank by the membrane, providing solids-free effluent.

Benefits of Membranes Compared to Conventional Technology

Although the treatment of wastewater with membrane bioreactors depends upon the same chemical and biological reactions occurring in the activated sludge process, there are numerous advantages that membrane treatment provide. Since the membranes serve as an absolute physical barrier that prevents the passage of particulate from the bioreactor, the MBR is usually operated with much higher mixed liquor suspended solids (MLSS) in the reactor. Systems are designed to operate with a MLSS of between 8000 to 18,000 mg/L for biological treatment of wastewater and 30,000 to 35,000 mg/L when the MBRs are used as membrane biothickeners (MBTs). MBTs can be configured where the thickener liquors have been treated to produce the same quality as the plant effluent, allowing it to be discharged along with the effluent. This eliminates the recycle stream from the thickener, reducing the load on the wastewater system.

The benefits offered by MBRs include[22]

- Provision of an absolutely positive barrier prevents the discharge of particulates.
- The separation of solids by the membrane eliminates the need for clarification.
- The higher MLSS allows for a much smaller footprint than conventional facilities.
- The small membrane pore size prevents the discharge of most pathogens, including all bacteria and most viruses.
- The MBR systems are designed to provide consistent effluent quality, unaffected by variations in flow or concentration of contaminants.
- The higher solids residence time (SRT) in the MBR allows for the development of a population of slower-growing organisms such as nitrifiers.
- Nitrogen and phosphorous can be removed to very low levels by adjusting the process.
- The higher SRT also allows for aerobic digestion of organics, reducing the amount of biosolids generated.
- Extremely small footprint saves space.
- The effluent is suitable for reuse.

Disadvantages include:[23]

- High capital cost
- Limited data on membrane life
- High cost of periodic membrane replacement
- High energy costs
- Need to control membrane fouling

The following comments, however, in this regard are applicable:[24]

- Capital cost of modern systems is in-line with conventional systems when nutrient removal is required.
- Kuboota, flat-plate membranes have been proven to last at least 15 years.
- Energy costs have been significantly reduced and comparable with AWT systems.

Membrane Issues

Presently the market for MBR application is mostly for small wastewater treatment plants where compactness of process and reuse opportunities dominate over conventional processes. However, plants are being built to treat 20 to 40 million gal/d. The small footprint and lack of secondary clarifiers are ideal for expanding large plants where land is not available. As membrane prices decrease, operation at lower fluxes will be developed to simplify scale-up and installation. The membrane flux rate is defined as the mass or volume rate of transfer through the membrane surface, gal/ft^2 · d (L/m^2 · h). It is an important design parameter with lower flux rates expected at high MLSS concentrations. When all factors are considered, concentrations of MLSS in the range of 8000 to 10,000 mg/L seem most cost-effective.[23] Different manufacturers employ different module designs. Optimal pretreatment, design, and operation will be site and waste specific. Hence, treatability studies are recommended, especially for industrial waste applications.

Operational efficiency of MBRs can be significantly reduced by biofouling. Membrane fouling decreases water output and increases maintenance cost. Different fouling mechanisms, such as macromolecular and adsorption, pore plugging, and cake buildup can occur at the membrane surface. The long sludge ages typically employed by these systems have advantages of greatly reduced sludge production, effective nitrification and denitrification, and resistance to flow and load variability. However, microbial viability is reduced owing to the accumulation of dead or inactive microbes, suspended solids, soluble microbial products (SMPs), and extracellular polymeric substances (EPS).

The characteristics of sludge generated by MBRs are different than that of conventional activated sludge systems. Different microbial community and physiological states alter settling and dewatering properties and impact sludge management. Khor et al. reported that EPS distribution in the biofouling layer changed the properties of the submerged tubular ceramic membrane surface, including pore and porosity, and produced a permeate quality from microfiltration to ultrafiltration efficiency.[25] Sun et al. reported that particle size of sludge plays an influential role on specific permeate flux rate. Small colloidal and soluble organic particles deteriorate the permeability of ceramic membranes by directly absorbing into the 0.2 μm pores and contribute to biofilm development. Aeration was found to be a significant factor governing sludge floc size and filtration performance. Particle size distribution and character can be affected by many factors. Some of these include: shear conditions, SRT, temperature, nature of the waste, and nutrient deficiencies.[26]

At present, membranes are cleaned periodically with solutions containing chlorine, acids, and/or bases to maintain and restore membrane flux. The membranes are cleaned in situ, either by dewatering the MBR tank and recovery cleaning the hollow fiber membranes for extended periods of time, or maintenance cleaning membrane plates from the inside out for shorter periods of time. Significant challenges continue including the need to obtain long run times without the need of chemical cleaning and how to reduce energy consumption for membrane operation.

Application

Membrane filtration may be used downstream of a conventional biological process or integrated into the biological process. Either system will produce an effluent quality suitable for direct reuse or for feed to RO system for removal of additional dissolved contaminants. In applications where oily wastewater will be treated for reuse, membrane filtration alone is not sufficient to remove all of the targeted constituents in the wastewater. A biological treatment step is required to remove the soluble organic compounds that are not removed by the membrane filtration process. Therefore, the biological treatment state is critical in the design and operation of all oily wastewater treatment systems.

As noted, the membrane bioreactor process consists of a biological reactor integrated with an ultrafiltration (UF) membrane system that replaces the clarifier of a conventional activated sludge plant providing an ultimate barrier for biomass control. In this configuration solids separation is achieved by means of filtration rather than gravity settling. Thus, effluent quality is independent of the settling characteristics of the sludge. Operation at elevated MLSS levels is therefore possible, thereby reducing treatment process footprint and allowing operation at extended SRTs. The membrane is also capable of separating insoluble solids in the process fluid (bacteria, viruses, colloids, and suspended solids) and higher molecular weight-soluble organics. Lower molecular weight organics are generally readily biodegraded in the aerobic process. The membrane provides a physical barrier to the larger more complex organics. The process has also been reported to produce less sludge. Using a pilot scale submerged bioreactor treating a high strength industrial wastewater at an F/M of 0.11, Sun et al. reported a sludge yield coefficient of 0.115 g VSS/g COD. This is significantly less than the 0.30 to 0.50 g VSS/g COD typical for conventional systems. The observed endogenous decay coefficient, b of 0.024 d^{-1}, is also much lower than the conventional value of 0.06 – 0.20 d^{-1}.[27]

The combination of air scour, back flushing, and maintenance cleaning is not completely effective in controlling membrane fouling and the pressure drop across the membrane will increase with time. At a maximum operating pressure drop of approximately 60 kPa, hollow fiber membranes are removed from the aeration basin for a recovery cleaning.[28] Membrane cassettes are soaked in an external tank with a 1500 to 2000 mg/L sodium hypochlorite solution for about 24 h. Hollow fiber manufacturers currently recommend a combination of frequent clean-in-place recovery cleaning lasting at least 24 h. Flat-plate manufacturers recommend a less frequent, short duration maintenance cleaning.

Full-scale and pilot-plant MBR systems have been operated for effective nutrient removal. These are discussed in Sec. 6.10. The MBR is substituted for conventional biological aeration and sedimentation in Modified Lutzek Ettinger or Bardenpho process, effectively reducing the plant size but achieving the same results.

Case Study A

This case study is taken from a paper by Peeters and Theodoulou.[29] It presents application of hollow fiber ultrafiltration membranes for the treatment of wastewater generated at Marathon Ashland Petroleum (MAP) located in Catlettsburg, Kentucky.

The original wastewater treatment system consisted of an equalization tank, followed by dissolved air flotation. The water contains solids, oil and grease, aromatic hydrocarbons [including benzene, toluene, ethylbenzene, and xylene (BTEX) compounds], metals, BOD, and occasionally arsenic. Due to the nature of the wastewater, MBR treatment was considered as a potential solution to upgrading the treatment facility. In early 2002, a treatability study was performed using reinforced immersed hollow fiber ultrafiltration membranes. Results indicated COD, BOD, and TSS removal, and BTEX compounds and heavy metals removals to acceptable levels. A full-scale membrane bioreactor system was built and commissioned to pretreat the MAP process wastewater and discharges the effluent into the city of Ashland's municipal wastewater treatment system.

Process Overview

Raw water from the plant is pumped to a 150 µm grit removal system for the removal of heavy solids and then through an oil-water separator before entering the bioreactor. A flow schematic is illustrated in Fig. 12.13.

Filtration is achieved by drawing water to the inside of the membrane fibers. Suction is created by permeate pumps. To meet the flow demand of 50,000 m³/d, two membrane cassettes are used. Recirculation

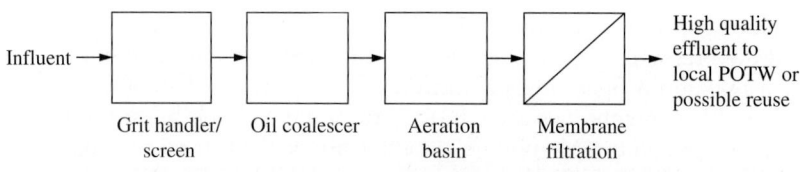

FIGURE 12.13 Process configuration of marathon ashland petroleum.

pumps return the remainder of the flow to the bioreactor to ensure a constant MLSS concentration. Sludge is wasted from the recirculation line to a filter press. Here it is thickened and disposed off site as a nonhazardous solid waste.

Performance of the system during the first year of operation is shown in Table 12.12.[30] As indicated, over 99 percent BOD removal and 95 percent COD removal were achieved, well below permit limits. Typical reduction in BTEX compounds is greater than 98 percent. Oil and grease, which are primarily emulsified oils, are below detectable levels. Free oil is removed upstream of the bioreactor in the oil coalescer.

Case Study B

Brady reported the application of the combination of a hollow fiber UF membrane and a spiral would RO membrane to treat and reuse

Parameter	Influent Avg (mg/L)	Effluent Avg (mg/L)	Permit Limit (mg/L)	Removal (%)
BOD	775	2	250	99.7
COD	1,300	64	658	95.0
Oil and grease (HEM)	165	<5	26	97.0
TSS	66	<7	250	89.4
pH, s.u.	7.7	7.15	6.0–11.0	—
NH_3N	3.3	0.02	20	99.4
Phosphorus	0.7	<0.10	—	85.7
Arsenic	0.061	0.015	0.1	75.4
Cadmium	0.0104	<0.003	0.02	71.1
Chromium	0.1274	<0.002	0.42	98.4
Copper	0.0356	0.011	0.1	69.1
Lead	0.0043	<0.001	0.14	76.7
Mercury	0.0027	<0.0010	0.0013	63.0
Nickel	0.050	0.019	0.58	62.0
Zinc	0.504	0.035	2.74	93.0
Benzene	15.6	<0.01	—	99.9
Toluene	10.5	<0.10	—	99.0
Ethylbenzene	0.61	<0.01	—	98.4
Xylene	3.5	<0.03	—	99.1
Thiocyanate	0.8	0.2	0	75.0
LEL, %	1	<0.1	10	—

TABLE 12.12 System Performance at Marathon Ashland Petroleum During First Year of Operation[30]

the wastewater from a diesel engine manufacturing plant. After 6 years of continuous treatment the UF tubes are consistently producing an average of 7937 gal/d permeate from 8000 gal/d of a heavy oil and grease feed.[31] This is equivalent to a yield of 99.2 percent conversion of the wastewater to permeated from the MBR. As a result, 8000 gal of wastewater is reduced to 62 gal for ultimate disposal. The system operates as a 2-week batch process prior to cleaning.

The permeate has a slight yellow tint (coolant dye) but has <30 mg/L of oil and grease and <0.5 mg/L of Cu and Zn. TSS was reported to be <10 mg/L and turbidity was <2 NTU. Advances in materials of construction, energy reduction, and membrane life have made MBRs a highly viable option for application in the treatment of low-flow, high organic strength industrial wastewaters.

12.6 Granular Media Filtration

Granular media filtration is employed for the removal of suspended solids as a pretreatment for low suspended solids wastewaters, following coagulation in physical-chemical treatment or as a tertiary treatment following a biological wastewater treatment process.

Suspended solids are removed on the surface of a filter by straining, and through the depth of a filter by both straining and adsorption. Adsorption is related to the zeta potential on the suspended solids and the filter media. Particles normally encountered in a wastewater vary in size and particle charge and some will pass the filter continuously. The efficiency of the filtration process is therefore a function of

1. The concentration and characteristics of the solids in suspension
2. The characteristics of the filter medium and other filtration aids
3. The method of filter operation

Granular-medium filters may be either gravity or pressure. Gravity filters may be operated at a constant rate with influent flow control and flow splitting, or at a declining rate with four or more units fed through a common header. To achieve constant flow an artificial head loss (flow regulator) is used. As suspended solids are removed and the head loss increases, the artificial head loss is reduced so the total head loss remains constant. In a declining-rate filter design, the decrease in flow rate through one filter as the head loss increases raises the head and rate through the other filters. A maximum filtration rate of 6 gal/(min · ft^2) [0.24 m^3/(min · m^2)] is used when one unit is out of service. The filter run terminates when the total head loss reaches the available driving force or when excess suspended solids or turbidity appear in the effluent.

Medium size is an important consideration in filter design. The sand size is chosen on the basis that it provides slightly better removal than is required. In dual-media filters, the coal size is selected to provide 75 to

90 percent suspended solids removal across 1.5 to 2.0 ft (0.46 to 0.6 m) of media. For example, if 90 percent suspended solids removal is desired across a filter bed, 68 to 80 percent should be removed through the coal layer and the remaining 10 to 25 percent through the sand layer. If the feed suspended solids particle size is larger than 5 percent of the granular medium particles, mechanical straining will occur.

A 25-µm particle will be mechanically strained by a 0.5-mm filter media. If the feed solids particles have a density of 2 to 3 times that of the suspending medium, then particles as small as 0.5 percent of the filter media particle size can be effectively removed by in-depth granular medium filtration.

Table 12.13 presents available media options. Two general types of monomedia are available, fine and coarse. The fine media are usually found in propriety-type filters such as the automatic backwash filter or the pulse-bed-type filter and rely on a straining mechanism for removal. Frequent backwashes (or pulsing) are required, especially during plant upset and/or high influent turbidity conditions. The coarse monomedia are usually much deeper than larger media requiring scour for cost-effective backwash. Operation of coarse medium filters is characterized by longer filter runs and the ability to respond to plant upset conditions. Dual media and multimedia have traditionally been used in potable water applications and their use has carried over into the tertiary filtration application wastewater treatment.

Filtration rate will affect the buildup of head loss and the effluent quality attainable. The optimum filtration rate is defined as the filtration rate that results in the maximum volume of filtrate per unit filter area while achieving an acceptable effluent quality.

Too high a filtration rate will permit solids to penetrate the coarse media and accumulate on the fine media. Too low a filtration rate is

Type	Material	Size, mm	Depth, in
1. Monomedia			
(a) Fine	Sand	0.35–0.60	10–20
(b) Coarse	Anthracite coal	1.3–1.7	36–60
2. Dual	Sand,	0.45–0.6	10–12
	anthracite coal	1.0–1.1	20–30
3. Multimedia	Garnet,*	0.25–0.4	2–4
	sand,	0.45–0.55	8–12
	anthracite coal	1.0–1.1	18–24

*Other types of material such as metal oxides are also used.
Note: in = 2.54 cm.

TABLE 12.13 Media Options

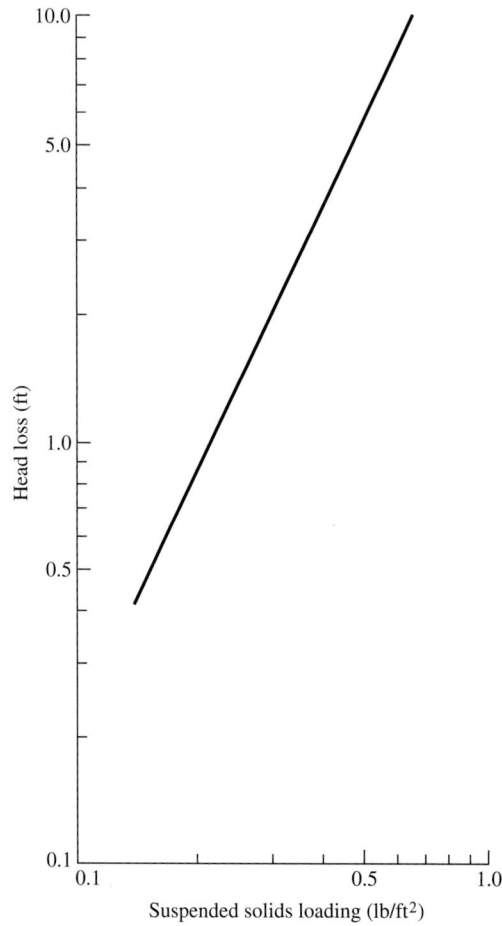

FIGURE 12.14 Head loss versus solids retention in a filter.

insufficient to achieve good solids penetration of the coarse media, resulting in head loss buildup at the top of the coarse media. Filtration rate will also influence effluent quality, depending on the nature of the particles to be removed.

The head loss through the filter is related to the solids loading, as shown in Fig. 12.14:

$$H = aS^n \qquad (12.7)$$

where H = head loss, ft or m
S = solids captured, lb/ft² or kg/m²
a, n = constants

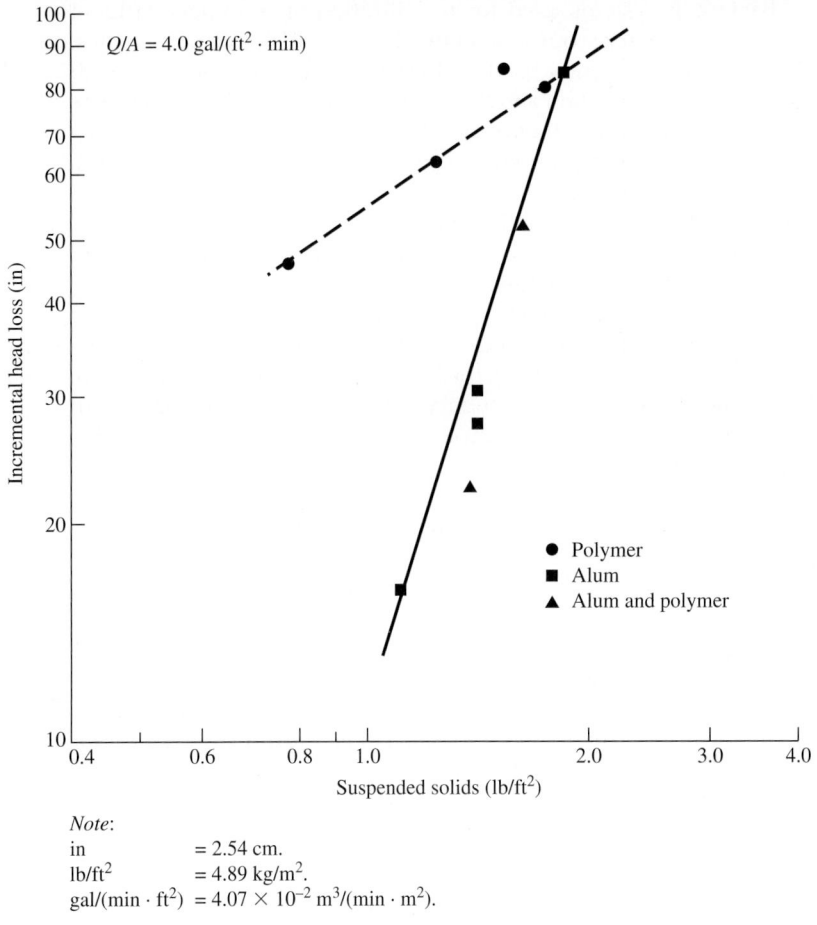

FIGURE 12.15 Head loss as related to solids deposit in a filter.

For a given head loss, the filtration cycle depends on the influent suspended solids and the hydraulic flow rate. The type of coagulant used may also influence the head loss, as shown in Fig. 12.15.

Improved suspended solids removals can be achieved by the addition of coagulants to the wastewater prior to filtration. The use of alum also results in the precipitation and removal of phosphorus through the filter. Flocculation is not needed since the filter serves as a flocculator. Effective mixing is required to disperse the chemicals and initiate the reaction. Since the suspended solids are removed by filtration rather than by sedimentation, 25 to 50 percent less chemicals are required in many cases. For most applications, a maximum of 100 mg/L suspended solids removal is used in order to avoid excessive backwash volumes.

Backwash systems used for the filtration of secondary effluent require an auxiliary scour system to effectively dislodge particulates. The two most frequently used backwash systems are: (1) water with auxiliary surface agitation and (2) water with air scour. Surface systems may be a fixed grid or rotary wash. The filter backwash rate is a function of medium size (both effective size and uniformity coefficient), medium type (specific gravity), and water temperature. Air scour rates are in the range of 1 to 5 standard $ft^3/(min \cdot ft^2)$ [0.305 to 1.53 std $m^3/(min \cdot m^2)$] with auxiliary surface wash rates in the range of 2 to 5 $gal/(min \cdot ft^2)$ [0.081 to 0.204 $m^3/(min \cdot m^2)$].

Granular medium filters can be fully automated by the use of pressure sensors that monitor the head loss across the filter. When the terminal head loss is achieved, the filter will automatically backwash. Turbidimeters serve as a secondary control such that the backwash sequence will be initiated when the turbidity reaches the allowable level in the effluent.

While considerable data are available for the design of filters treating domestic secondary effluents, industrial wastewaters require pilot plant studies to define the type of medium, filter flow rate, coagulant requirements, head loss relationships, and backwash requirements.

There are several types of filters available today. Three of the more common are the dual-media filter consisting of anthrafilt (coal) and sand, the Hydroclear filter, and the continuous backwash filter.

A typical dual-media filter is shown in Fig. 12.16. The Hydroclear filter employs a single sand medium with an air mix for solids suspension and regeneration of the filter surface. Filter operation enables periodic regeneration of the medium surface without backwashing. Typical operating parameters are

Filtration rate	2–5 $gal/(min \cdot ft^2)$ [0.081–0.204 $m^3/(min \cdot m^2)$]
Media size	0.35–0.45 mm sand
Bed depth	10–12 in (25.4–30.4 cm)
Backwash rate	12 $gal/(min \cdot ft^2)$ [0.5 $m^3/(min \cdot m^2)$]
Air mix	0.25 standard $ft^3/(min \cdot ft^2)$ [0.076 std $m^3/(min \, m^2)$]
Terminal head loss	3.5 ft (1.07 m)
Backwash filtrate ratio	0.10

The Dynasand filter (DSF) is a continuous backwash, self-cleaning upflow deep-bed granular medium filter. The filter medium is cleaned continuously by recycling the sand internally through an airlift pipe and sand washer, as shown in Fig. 12.17. The regenerated sand is redistributed on top of the bed, allowing for a continuous uninterrupted flow of filtered water and reject water. Filtration performance is shown in Table 12.14. Filter design is shown in Example 12.3.

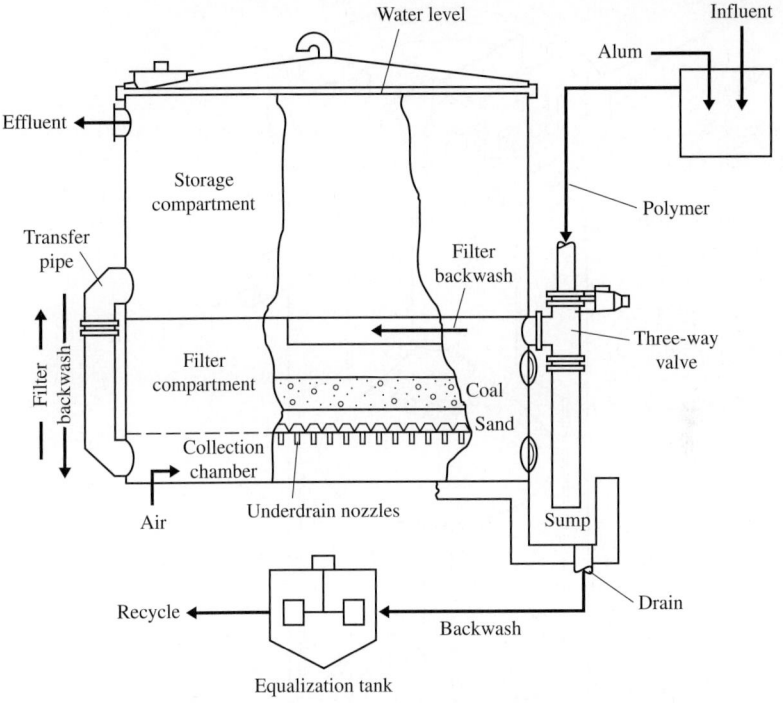

FIGURE 12.16 Typical automatic dual media filter.

Example 12.3. A 1 million gal/d secondary effluent is to be polished by a granular medium filter. If an 8 ft head loss is applied, calculate the filter run time to achieve 80 percent removal from an initial suspended solids of 70 mg/L. The hydraulic loading to the filter is 3.0 gal/(min · ft²).

Solution
The required surface area is

$$A = \frac{1 \times 10^6}{3 \cdot 1440}$$

$$= 230 \text{ ft}^2$$

For a head loss of 8 ft, the suspended solids loading is 0.57 lb/ft² (see Fig. 12.14). The expected filter run is then

$$t = \frac{0.57 \cdot 230}{1 \cdot 0.8 \cdot 70 \cdot 8.34}$$

$$= 0.28 \text{ d or } 6.7 \text{ h}$$

706 Chapter Twelve

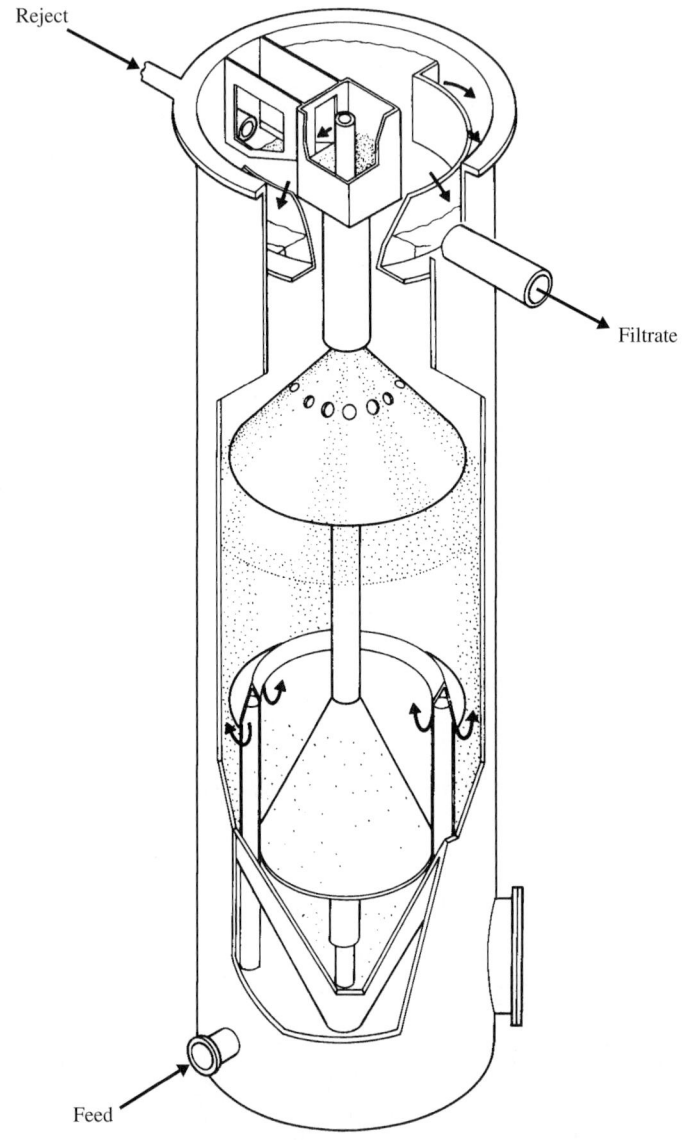

FIGURE 12.17 Dynasand filter (DSF). (*Courtesy of Parkson Corporation*)

12.7 Microscreen

A microscreen is a rotary drum revolving on a horizontal axis covered with stainless steel fabric (Fig. 12.18). The water enters the open end of the drum and is filtered through the fabric, with solids being retained

Filter Type	Wastewater	Filter Depth, ft	Hydraulic Loading, gal/ (min · ft²)	Percent removal		Effluent, mg/L	
				SS	BOD	SS	BOD
Gravity downflow	TF effluent	2–3	3	67	58	—	2.5
Pressure upflow	AS effluent	5	2.2	50	62	7.0	6.4
Dual media	AS effluent	2.5	5.0	74	88	4.6	2.5
Gravity downflow	AS effluent	1.0	5.3	62	78	5	4
Dynasand	Metal finishing	3.3	4–6	90	—	2–5	—
	AS effluent	3.3	3–10	75–90	—	5–10	—
	Oily wastewater	3.3	2–6	80–90*	—	5–10*	—
Hydroclear	Poultry	1	2–5	88	—	19	—
	Oil refinery	1	2–5	68	—	11	—
	Unbleached kraft	1	2–5	74	—	17	—

*Free oil.
Note: ft = 0.305 m.
gal/(min · ft²) = 4.07 × 10⁻² m³/(min · m²).

TABLE 12.14 Filtration Performance

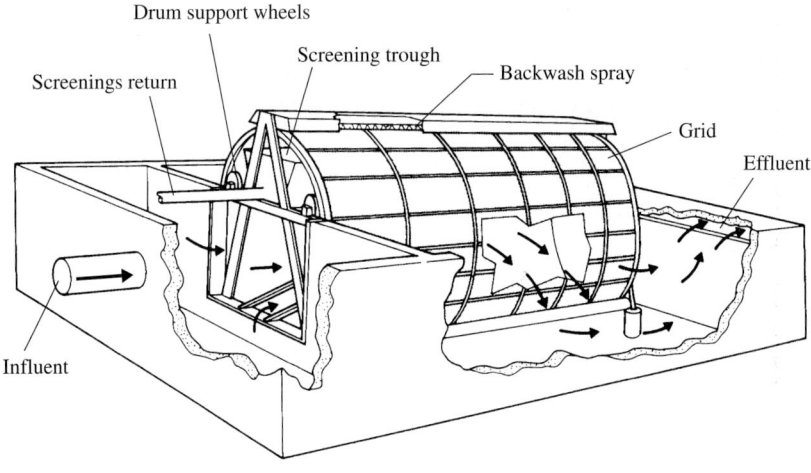

FIGURE 12.18 Microscreen. (*Courtesy of Envirex, Inc.*)

Aperture Screen, μm	Flow Rate, gal/(min · ft²) (submerged)	Percent Removal	
		SS	BOD
35	10.0	50–60	40–50
23	6.7	70–80	60–70

Note: gal/(min · ft²) = 4.07×10^{-2} m³/(min · m²).

TABLE 12.15 Removal Efficiencies

on the inside surface of the fabric. As the drum rotates, the solids are transported and continuously removed at the top of the drum by pumping effluent under pressure through a series of spray nozzles that extend the length of the drum. The head loss is less than 12 to 18 in (30 to 46 cm) of water. The backwash water is 46 percent of the total throughput water. Peripheral drum speeds vary up to 100 ft/min (30.5 m/min) with hydraulic loadings of 2.5 to 10 gal/(min · ft²) [0.1 to 0.4 m³/(min · m²)]. Periodic cleaning of the drum is required for slime control.

For filtration of secondary effluent a maximum solids loading of 0.88 lb/(ft² · d) [4.3 kg/(m² · d)] at a hydraulic loading of 6.6 gal/(min · ft²) [0.27 m³/(min · m²)] has been reported. Lynam et al.[32] reported effluent suspended solids and BOD of 6 to 8 mg/L and 3.5 to 5 mg/L, respectively, with a 20-20 activated sludge effluent of a 23-m microscreen at 3.5 gal/(min · ft²) [0.14 m⁵/(min · m²)]. For design purposes, the removal efficiencies treating a secondary effluent are given in Table 12.15.

The efficiency of the unit is suspended solids sensitive, as indicated by a decrease in the throughput rate from 60 to 13 gal/min (0.227 to 0.049 m³/min) with an increase in influent suspended solids from 25 to 200 mg/L.[33]

References

1. Eckenfelder, W. W., J. P. Lawler, and J. T. Walsch: "Study of Fruit and Vegetable Processing Waste Disposal Methods in the Eastern Region," U.S. Department of Agriculture, Final Report, 1958.
2. Crites, R. W.: *Proceedings of the Industrial Wastes Symposia*, Water Pollution Control Federation, 1982.
3. "Diagnosis and Improvement of Saline and Alkali Solids," U.S. Department of Agriculture, Handbook 60, 1954.
4. Jewell, W. J.: *Limitations of Land Treatment of Wastes in the Vegetable Processing Industries*, Cornell University, Ithaca, N.Y., 1978.
5. Adamczyka, A. F.: *Land as a Waste Management Alternative*, R. C. Loehr, ed., Ann Arbor Science, Ann Arbor, Mich., 1977.

6. Haith, D. A., and D. C. Chapman: "Land Application as a Best Practical Treatment Alternative," in *Land as a Waste Management Alternative*, R. C. Loehr, ed., Ann Arbor Science, Ann Arbor, Mich., 1977.
7. Crites, R.W. et al.: *Proc. 5th National Symposium on Food Processing Wastes*, 1974.
8. Gellman, I., and R. O. Blosser: *Proc. 14th Ind. Waste Conf.*, Purdue University, 1959.
9. Wisneiwski, T. F., A. J. Wiley, and B. F. Lueck: *TAPPI*, vol. 89, pt. 2, p. 65, 1956.
10. Billings, R. M.: *Proc 13th Ind. Waste Conf.*, Purdue University, 1958.
11. Wong, G., and R. Pfarrer: *Proc. 50th Purdue Industrial Waste Conf.*, Ann Arbor Press, Ann Arbor, Mich., 1995.
12. Bausmith, D. S., and R. D. Neufeld: *Proc. WEF*. vol. 3, Dallas, 1996.
13. Warner, D. L.: "Deep Well Injection of Liquid Waste," Environmental Health Series, U.S. Department of Health, Education, and Welfare, Cincinnati, April 1965.
14. Donaldson, E. C.: "Subsurface Disposal of Industrial Wastes in the United States," Bureau of Mines Information Circular 8212, U.S. Department of the Interior, Washington D.C., 1964.
15. Cartwright, W. P.: *Chemical Engineering*, McGraw-Hill, September 1994.
16. Agardy, F. J.: *Membrane Processes, Process Design in Water Quality Engineering*, E. L. Tackston and W. W. Eckenfelder, eds., Jenkins Publishing Co., Austin, Texas, 1972.
17. Okey, R. W.: *Water Quality Improvement by Physical and Chemical Processes*, E. F. Gloyna and W. W. Eckenfelder, eds., University of Texas Press, Austin, Texas, 1970.
18. Logemann, F. P.: *Proc. 11th National Conf. Superfund '90*, Hazardous Materials Control Research Institute, 1990.
19. Kettern, J. T.: *Wat. Sci. Tech.* vol. 26, no. 1–2, p. 137, 1992.
20. Sutton, P. M.: "Membrane Bioreactors for Industrial Wastewater Treatment: Applicability and Selection of Optimal System Configuration," *Proc. Water Environment Federation's Membrane Technology 2008 Specialty Conf.*, Atlanta, GA, CD-ROM, January 27-30, 2008.
21. Smith, C. V., D. O. DiGregorio, and R. M. Talott: "The Use of Ultrafiltration Membranes for Activated Sludge Separation," *Proc. 24th Industrial Waste Conf.*, Purdue University, 1969.
22. Adams, C. E. and Shelby, S. E.: "Comparative Overview of Competitive Activated Sludge Configurations for Industrial Wastewaters," Seminar by ENVIRON International Corp. to AIChE, Baton Kouge, LA, March 14, 2008.
23. Metcalf and Eddy, Inc.: *Wastewater Engineering*, McGraw-Hill Book Company, New York, 2003.
24. Gaines, F. R.: Experience with membrane applications for industrial wastewaters: Personal communication, 2008.
25. Khor, S. L, et al.: "Biofouling Development and Rejection Enhancement in Long SRT MF Membrane Bioreactors," *Proc. Biochem.* 42, pp. 1641–1642, 2007.
26. Sun, D. D., C. T. Hay, and S. L. Khor: "Effects of Hydraulic Retention Time on Behavior of Start-up Submerged Membrane Bioreactor with Prolonged Sludge Retention Time," *Desalination*, 195, pp. 209–225, 2006.
27. Sun, D. D., et al. "Impact of Prolonged Sludge Retention Time on the Performance of a Submerged Membrane Bioreactor," *Desalination*, 208, ElSevier, pp. 101–112, 2007.
28. Fernandez, A., J. Lozier, and G. Daigger: "Investigating Membrane Bioreactor Operation for Domestic Wastewater Treatment: A Case Study," *Municipal Wastewater Treatment Symposium: Membrane Treatment Systems, Proceedings, 73rd Annual Conference*, Water Environment Federation, Anaheim, CA, 2000.
29. Peeters, J. G., and S. L. Theodoulou: "Membrane Technology Treating Oily Wastewater for Reuse."

30. Buckles, J. A., K. Kuljian, and S. Hester: "Full-Scale Treatment of a Petroleum Industry's Wastewater Using an Immersed Membrane Biological Reactor," presented at WEFTEC, October 2-6, 2004.
31. Brady, F. J. "Heavy Industry Plant Wastewater Treatment Recovery and Recycling Using Three Membrane Configurations with Aerobic Treatment—A Case Study," presented at WEFTEC, October 2006.
32. Lynam B. et al.: *J. WPCF*, vol. 41, p. 247, 1969.
33. Carvery, J. J.: *FWPCA, Symposium an Nutrient Removal and Advanced Waste Treatment*, Tampa, Fla., 1968.

CHAPTER 13
Treatment: Oil/Gas Exploration/Production Residuals

13.1 Introduction and Background Information

Introduction

The contamination and control of pollutants resulting from the drilling, production, operation, maintenance, and abandonment of oil and gas production wells is recognized. This century will see increasing exploration and production (E&P) activities as energy demands, both domestically and internationally, become more acute. The high price of oil and gas and the availability of advanced discovery technologies for extracting fossil fuels imply the search for new subterranean oil and gas previously thought unattainable because of economic and technological limitations. This, combined with increasingly restrictive environmental statutes and regulations, leads to the necessity of recognizing and solving environmental safeguards in the E&P industry. The emphasis on E&P pollution control in this chapter is directed toward the onshore sector, particularly with urban E&P as many of the newer oil and gas proven reserves are contiguous to and sometimes within developed urban areas.

Since the first one was drilled in Pennsylvania in 1859, over 1,870,000 oil wells have been drilled in the United States over the years.[1]

Potential oil sources are commonly discovered by the use of gravity meters (devices that measure changes in earth's gravitational field), magnetometers (devices that measure magnetic field changes induced by flowing oil), sniffers (sensitive electronic hydrocarbon detectors), or seismological surveying methods such as 3D Seismic. Seismological methods that measure the reflection of shock waves through various rock layers are the most widely used. Once the most promising sites have been selected, leases have been obtained, and legal issues have been resolved, the land around the site is prepared. Sites may require leveling of the land, the construction of access roads, and the digging of reserve pits to dispose of cuttings, all issues that are associated with potential environmental impacts. Reserve pits in particular, which will contain drilling residues and cuttings, are often lined to protect the surrounding area.

Background Information

Conventional Oil/Gas

Crude oil and gas are recovered from beneath the earth's crust by means of a well. The flow from a well may contain gas, water, sediment, and other impurities in conjunction with crude oil. The flow is piped to separators and treaters, where the natural gas and impurities are removed from the crude oil. The natural gas, when abundant enough, is either separated from the oil and piped to a gas plant for further treatment, or it is flared off (burned). The crude oil is piped to oil battery tanks and then to the pipeline terminals, and subsequently to the refinery. In the absence of a pipeline from the separators, the crude oil is stored and then trucked to the pipeline terminal. The product flow from a well cannot always be easily removed; in some cases the well must be "treated" to enhance oil recovery (i.e., waterflood, CO_2, flood, hydrocarbon [C_5] flood). Field processing also can employ time, gravity, chemical, heat, mechanical or electrical processes, or a combination of these. Figure 13.1 shows a flow diagram of conventional oil/gas recovery from wells.[2]

Conventional Gas

Natural gas may occur in conjunction with crude oil (associated) or by itself (nonassociated). Again, a well is necessary to remove the gas from beneath the earth's crust. Natural gas may be associated with liquid hydrocarbons, hydrogen sulfide, carbon dioxide, water, water vapor, mercaptans, nitrogen, helium, and solids (sediments) as impurities.

The separation of natural gas from these impurities in a separator may combine gravity, time, and mechanical and chemical processes. The gases from a well can be piped directly to a gas plant for processing or may need to be dehydrated at the production site before being piped to the plant. Dehydration is accomplished by several methods, of which glycol dehydration is of significance since glycol absorbs benzene and other organic compounds as well as water from the gas streams. During regeneration of the glycol, it is heated to remove the water. This process also emits organic compounds. Sweetening by removing hydrogen sulfide and other sulfur-containing compounds may also be required. Figure 13.2 shows a flow diagram of a

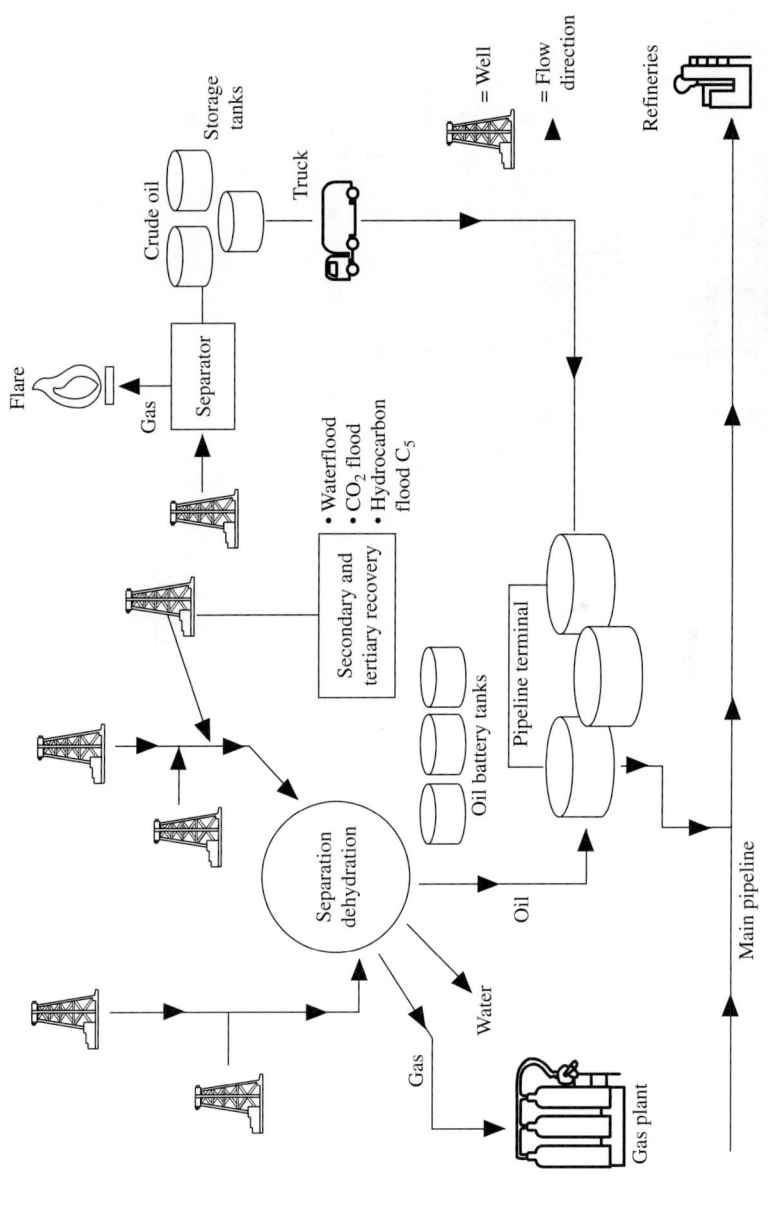

FIGURE 13.1 Flow diagram of conventional oil/gas recovery from wells. *(Adapted from Verma, Johnson, and McLean, 2000.)*

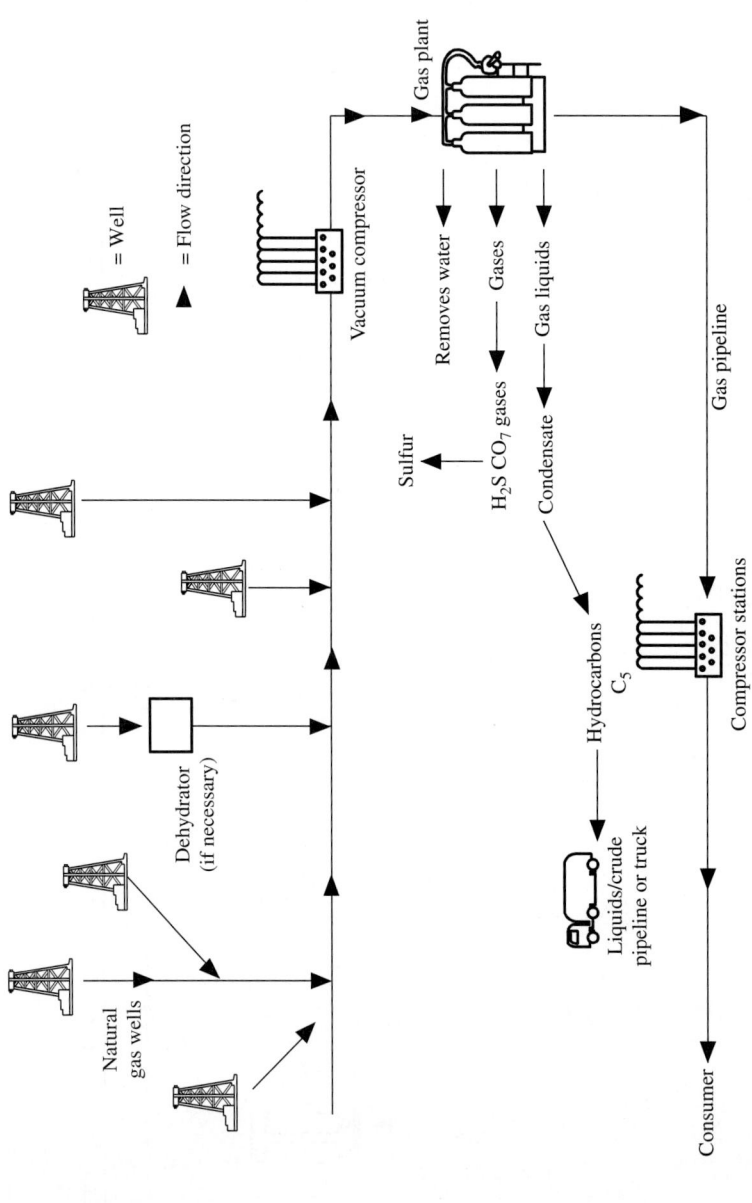

FIGURE 13.2 Flow diagram of conventional gas recovery from wells. *(Adapted from Verma, Johnson, and McLean, 2000.)*

conventional gas operation.[2] The purified natural gas is sent via pipeline to the pipeline station and on to the consumer.

After a site has been prepared, a rig is often set up[3] as seen in Fig. 13.3. Details on rig components are referenced. After a rig is set up, drilling can commence. Mud is circulated through the drill pipe as drilling progresses to float out rock cuttings,[3] as shown in Fig. 13.4. When drilling approaches the oil trap, casing pipe sections are placed in the hole to prevent it from collapsing. Drilling continues in stages until the oil sand at the final depth is reached and confirmed. The well is completed when perforations in the casing are made with the help of explosive charges, and flow control structures are placed in the hole. Acids or fluid containing proponents (i.e., sand, aluminum pellets, and so on) are used to start oil flow of the well. The rig can be removed after flow begins, and the production equipment can then be installed at the site.[3]

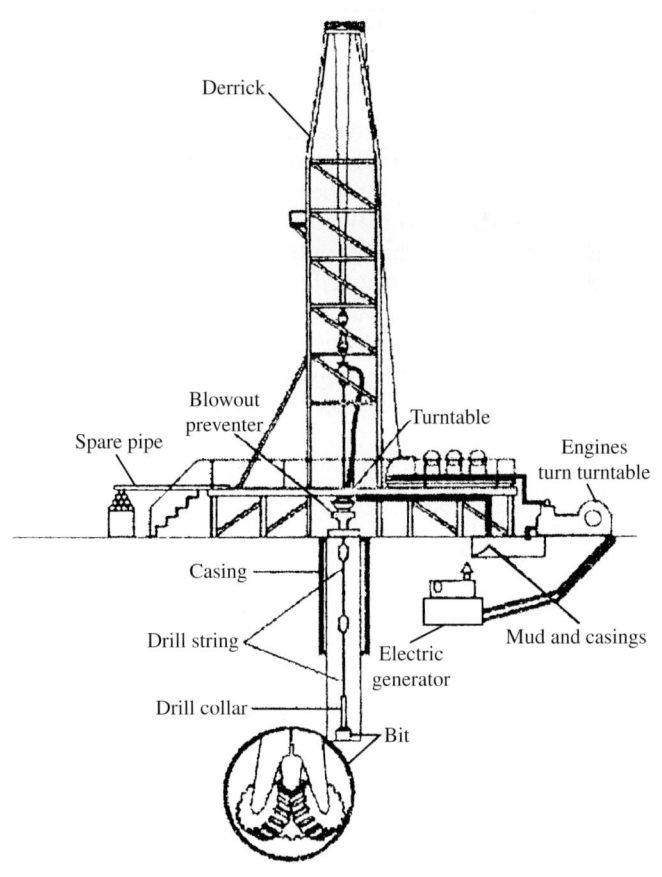

FIGURE 13.3 Oil rig.

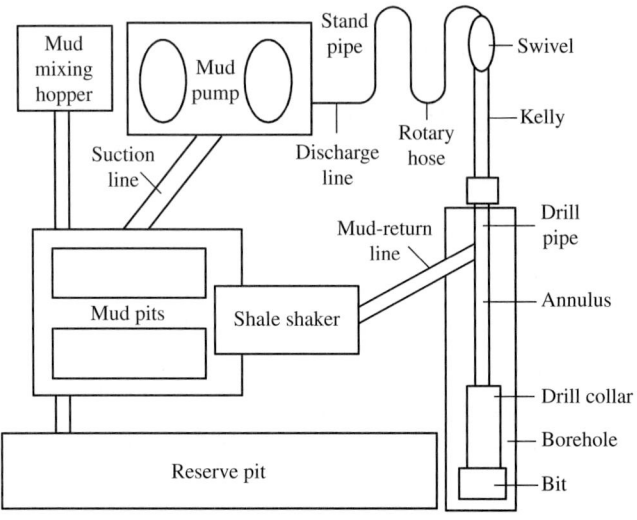

Figure 13.4 Drill mud circulation system. (*Adapted from Freudenrich, 2001.*)

E&P Servicing

Almost all producing wells become deficient sooner or later due to mechanical problems in the well or due to the depletion of the oil or gas reservoir that is being drained. Well-service and workover contractors are employed to make the repairs considered necessary by petroleum engineers. These contractors are organized for the special work needed to return a well to proper production.

Much of the remedial work, though varying from shallow to deep wells, is relatively simple. Nearly every operation can be performed by working the rods or tubing up and down, with a minimum of turning, and there is little or no need to circulate. Other major well operations are:

1. Sand cleanout
2. Liner removal
3. Casing repair
4. Plug-back
5. Squeeze cementing
6. Drilling deeper

These jobs usually require a string of pipe which must be rotated.

Moreover, it will generally be necessary to circulate the well (i.e., to pump fluid to the bottom and back to the surface). The working string may be either the tubing already used in the well or a special string of drill pipe, if hard service is anticipated.

Much work formerly handled by the workover crews is now accomplished by wireline methods—running casing plugs and tubing packers into the hole, changing gas-lift valves, placing cement in a well, determining where pipe is stuck in a well, unscrewing or cutting the pipe, and recompleting the well with a chamber installed for improved lifting ability.

13.2 Regulations

Introduction

The regulatory burden on the oil and gas industry increases the cost of doing business and affects profitability. Because such rules and regulations are frequently amended or reinterpreted, it is difficult to predict the future cost or impact of complying with such laws and regulations.

The Federal Energy Regulatory Commission (FERC) regulates interstate transportation rates and service conditions, which affect the marketing of oil and gas production, as well as the revenues received for sales of such production. Since the mid-1980s, the FERC has issued various orders that have significantly altered the marketing and transportation of oil and gas.

Sales of oil and natural gas liquids are not presently regulated and are made at market prices. The price received from the sale of those products is affected by the cost of transporting the products to market. The FERC has implemented regulations establishing an indexing system for transportation rates for oil pipelines, which generally would index such rate to inflation, subject to certain conditions and limitations.

There are federal statutes that apply to E&P operations, along with state statutes and regulations which may be more stringent than the federal requirements. Each state has its own air and water quality standards, permitting processes, siting criteria, and regulatory agencies dedicated specifically to the E&P industrial sector.

Federal Regulations[4-6]

Federal environmental statutes that might apply to E&P wastes include, but are not limited to: the Resource Conservation and Recovery Act (RCRA), the Safe Drinking Water Act (SDWA), the Clean Water Act (CWA), the Comprehensive Environmental Response, Compensation and Liability Act (CERCLA), the Emergency Planning and Community Right-to-Know Act (EPCRA SARA Title III), the Clean Air Act (CAA), the Toxic Substances Control Act (TSCA), the Oil Pollution Act of 1990 (OPA), the Migratory Bird Treaty Act (MBTA), the Endangered Species Act (ESA), and the Hazardous Materials Transportation Act (HMTA).

State environmental regulations are typically more stringent than federal regulations, and each state has its own air and water quality standards, permitting processes, and siting criteria.

In addition to the above environmental regulations, the Bureau of Land Management (BLM) regulates "onshore leasing, exploration, development, and production of oil and gas on federal lands... and approves and supervises most oil and gas operations on American Indian lands" as codified by 43 CFR Part 3160 (Onshore Oil and Gas Operations). Another Department of the Interior agency, the Minerals Management Service (MMS), is responsible for administration of petroleum resource development outside of state boundaries on the Outer Continental Shelf (OCS) under the Outer Continental Shelf Land Act (OCSLA), as codified by 30 CFR Part 250 (Oil and Gas and Sulfur Operations in the outer Continental Shelf).

Resource Conservation and Recovery Act of 1976

Subtitle C of the Resource Conservation and Recovery Act (RCRA) is intended to regulate and to assure that appropriate safeguards are in place for the generation, storage, treatment, transportation, and disposal of hazardous wastes. It reflects the philosophy that hazardous wastes must be governmentally supervised from cradle to grave. RCRA requires the issuance of a permit before hazardous waste may be treated, stored, or disposed of on a site and regulates these procedures through conditions imposed in the permit process. All information regarding RCRA compliance is publicly available.

The Clean Air Act

The federal Clean Air Act was enacted in 1963 and amended in 1970, 1977, and 1990. It requires the preparation and submission of state implementation plans for the attainment of national ambient air quality standards (NAAQS) by given target dates. The act also requires the state, acting through air districts, to enact regulations sufficient to attain and maintain the federal NAAQS. Due to the 1990 amendments, operating permits (Title V) covering individual facilities and toxic air controls were also mandated on a specified schedule. Title V operating permits require very specific information about exploration and production facilities. The application from the facilities is required to be available to the public. The 1990 Clean Air Act Amendments are in the process of implementation. Currently, most states and USEPA have individual source inventories on an annual basis for criteria pollutants (nitrogen oxides, sulfur oxides, reactive organic compounds, particulates, carbon monoxides, and lead). Air toxics inventories are in various stages of development in individual states due to the Clean Air Act and state regulations.

The Clean Water Act of 1972

The Clean Water Act regulates the discharge of pollutants into surface waters, that is, the National Pollution Discharge Elimination System program (NPDES), management of "nonpoint" sources of waste discharge (i.e., stormwater discharge), and discharges to public sewerage systems [publicly owned pretreatment plants (POTW)]. It was passed in 1972 and has been amended several times. The permits issued under the Clean Water Act limit the composition and volume of a discharge and the concentrations of individual pollutants in it. The discharge requirements are generally based on the quality of the receiving waters. As with the Clean Air Act, the Clean Water Act allows for the delegation of authority for implementation to the states. To date, over 40 states have received the authorization. Under federal or delegated program, all information required for permit application and monitoring for permit compliance is considered public with the exception of confidential business information which may be considered "trade secret." The volume and composition of a discharge are not confidential.

The Safe Drinking Water Act of 1974

The Safe Drinking Water Act regulates the amount of toxic substances in sources of drinking water. It requires the states to establish an underground injection control (UIC) program to prevent endangerment of drinking water supplies including groundwater. Of particular importance to the E&P industry is the control of Class II wells. A *Class II well* is a well that injects fluids either as a means of disposal of produced fluids or as an enhanced recovery fluid. The Class II UIC program dictates well construction standards and mechanical integrity monitoring, as well as the types and composition of fluids that can be injected; all of this information is publicly available.

Class II underground injection wells (Class II wells) are oil- and gas-production–related injection wells used for disposal of produced fluids, for enhanced oil recovery (EOR) projects, or for storage of hydrocarbons which are liquid at standard temperature and pressure. These wells are regulated under the federal Safe Drinking Water Act by the U.S. Environmental Protection agency (EPA) either directly or through states which are granted "primacy" to implement the federal program. EPA plans to issue amended Class II well regulations based on the final recommendations of the Federal Advisory Committee (FAC) formed to review five major Class II well issues: operating, monitoring, and reporting; plugging and abandonment; area of review and corrective action; mechanical integrity testing (MIT); and casing and cementing. An ideal injection well and site[3,4] is illustrated in Fig. 13.5.

720 Chapter Thirteen

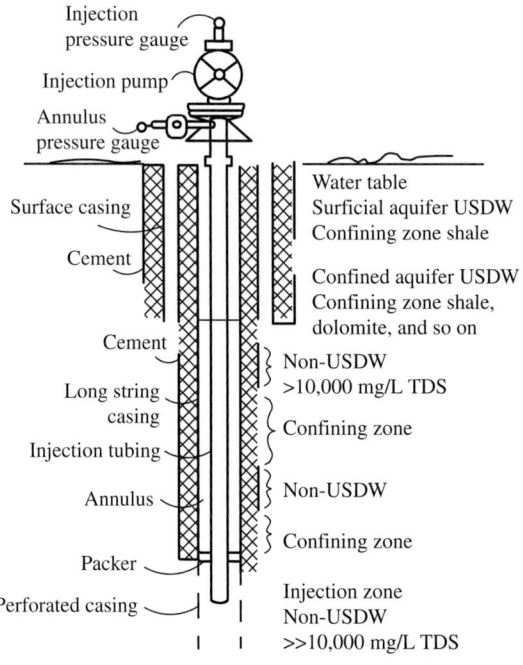

FIGURE 13.5 Ideal injection well and site. (*Adapted from Freudenrich, 2001, and Environmental Protection Agency, Office of Compliance, 2000.*)

Exempt and Nonexempt E&P Wastes[5]

The EPA, following Subtitle C (Hazardous/Nonhazardous) regulations in the RCRA regulations (1980), exempted selective E&P wastes while not exempting others from the hazardous waste definitions.

In its 1988 regulatory determination, EPA published lists of wastes that were determined to be either exempt or nonexempt. These lists are provided as examples of wastes regarded as exempt and nonexempt and should not be considered to be comprehensive. The exempt waste list applies only to those wastes generated by E&P operations. Similar wastes generated by activities other than E&P operations are not covered by the exemption.

As the magnitude of E&P wastes previously described has increased, management of these wastes as to their "hazardous" or "nonhazardous" designations must be clearly understood. All such management and handling of these waste residuals must recognize their impact on "human health and the environment." Understanding the procedures for determining the exempt or nonexempt status is a necessity in the management of E&P activities and proper handling of the associated residuals. The exempt E&P wastes are described in Table 13.1, and the nonexempt wastes[1] are outlined in Table 13.2. This delineation is illustrated diagrammatically in Fig. 13.6. Questionable status wastes are those not specifically listed by EPA as exempt. These include:

Treatment: Oil/Gas Exploration/Production Residuals

Activated charcoal filter media
Basic sediment and water (BS&W)—see Tank bottoms
Caustics, if used as drilling fluid additives or for gas treatment
Condensate
Cooling tower blowdown
Debris, crude oil soaked
Debris, crude oil stained
Deposits removed—from piping and equipment prior to transportation (i.e., pipe scale, hydrocarbon solids, hydrates, and other deposits)
Drilling cuttings/solids
Drilling fluids
Drilling fluids and cuttings from off-shore operations disposed of onshore
Gas dehydration wastes:
    a. Glycol-based compounds
    b. Glycol filters (see process filters), filter media, and backwash
    c. Molecular sieves
Produced sand
Produced water
Produced water constituents removed before disposal (injection or other disposal)
Produced water filters (see Process filters)
Rigwash
Slop oil (waste crude oil from primary field operations and production)
Soils, crude oil-contaminated
Sulfacheck/chemsweet waste
Gas plant sweetening wastes for sulfur removal:
    a. Amines (including amine reclaimer bottoms)
    b. Amine filters (see Process filters), amine filter media, and backwash
    c. Amine sludge, precipitated
    d. Iron sponge (and iron sulfide scale)
    e. Hydrogen sulfide scrubber liquid and sludge
Gases removed from the production stream (i.e., H_2S, CO_2, and VOCs)
Liquid hydrocarbons removed from the production stream but not from oil refining
Liquid and solid wastes generated by crude oil and tank bottom reclaimers
Oil, weathered
Paraffin
Pigging wastes from producer operated gathering lines
Pit sludges and contaminated bottoms from storage or disposal of exempt wastes
Process filters
Tank bottoms and basic sediment and water (BS&W) from: storage facilities that hold product and exempt waste (including accumulated materials such as hydrocarbons, solids, sand, and emulsion from production separators, fluid treating vessels, and production impoundments)
VOCs from exempt wastes in reserve pits or impoundments or production equipment
Well completion, treatment, and stimulation, and packing fluids
Workover wastes (i.e., blowdown, swabbing, and bailing wastes)

TABLE 13.1 Exempt Wastes

Batteries: lead acid
Batteries: nickel-cadmium
Boiler cleaning wastes
Boiler refractory bricks
Caustic or acid cleaners
Chemicals, surplus
Chemicals, unusable (including waste acids)
Compressor oil, filters, and blowdown waste
Debris, lube oil contaminated
Drilling fluids, unused
Drums/containers, containing chemicals
Drums/containers, containing lubricating oil
Drums, empty (and drum rinsate)
Filters, lubrication oil (used)
Gas plant cooling tower cleaning wastes
Hydraulic fluids, used
Incinerator ash
Laboratory wastes
Mercury
Methanol, unused
Oil, equipment lubricating (used)
Paint and paint wastes
Pesticide and herbicide wastes
Pipe dope, unused
Radioactive tracer wastes
Refinery wastes (e.g., unused frac fluids or acids)
Sandblast media
Scrap metal
Soil, chemical-contaminated (including spilled chemicals)
Soil, lube oil-contaminated
Soil, mercury-contaminated
Solvents, spent (including waste solvents)
Thread protectors, pipe dopecontaminated
Vacuum truck rinsate (from tanks containing nonexempt waste)
Waste in transportation pipeline related pits
Well completion, treatment and stimulation fluids, unused

TABLE 13.2 Nonexempt Wastes

- Cement slurry returns from the well and cement cuttings (unused cement slurries would be nonexempt)
- Gas plant sweetening unit catalysts
- Natural gas gathering line hydrotest water
- Produced-water-contaminated soil
- Sulfur recovery unit wastes

Special category wastes are subject to specific state and federal regulations and include naturally occurring radioactive materials

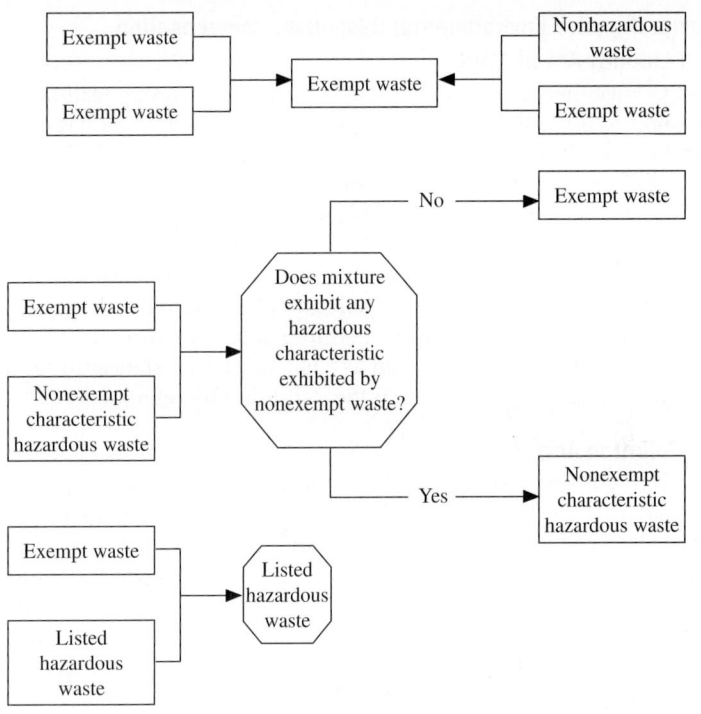

FIGURE 13.6

(NORM) and polychlorinated biphenols (PCBs) and PCB-contaminated soils. Mixing wastes, particularly exempt and nonexempt wastes, creates additional considerations. Determining whether a mixture is an exempt or nonexempt waste requires an understanding of the nature of the wastes prior to mixing and, in some instances, might require a chemical analysis of the mixture. Whenever possible, avoid mixing nonexempt wastes with exempt wastes. If the nonexempt waste is a listed or characteristic hazardous waste, the resulting mixture might become a nonexempt waste and require management under RCRA Subtitle C. Furthermore, mixing a characteristic hazardous waste with nonhazardous or exempt waste for the purpose of rendering the hazardous waste nonhazardous or less hazardous might be considered a treatment process subject to appropriate RCRA Subtitle C hazardous waste regulation and permitting requirements.[7,8]

It should be recognized that specific RCRA statutory and regulatory requirements may apply regardless of the exempt status. E&P activities are subject to enforcement actions under RCRA Section 7003 (eminent hazard) and citizens' suits under Section 7002. States may also bring actions under Section 7002. Some commercial E&P waste facilities have been required to modify their operations by EPA under the provisions of RCRA 7003.

Comprehensive Environmental Response, Compensation, and Liability Act of 1980

The Comprehensive Environmental Response, Compensation, and Liability Act (CERCLA) generally addresses the release of hazardous substances and the procedures, costs, and responsibilities for remediating or removing such substances (cleanup). It authorized USEPA to prepare a list of substances designated as "hazardous" based on the determination that certain substances, when released into the environment, may present a substantial danger to the public health or welfare or to the environment. Petroleum is expressly excluded from the definition of hazardous substances under CERCLA, unless it is otherwise designated as a hazardous substance under one of the other acts incorporated in the definition by reference.

Oil Pollution Act[4-6,9]

The U.S. Oil Pollution Act of 1990 (OPA '90), and similar legislation enacted in Texas, Louisiana, and other coastal states, addressed oil spill prevention and control and significantly expanded liability exposure across all segments of the oil and gas industry. OPA '90 and such similar legislation and related regulations impose a variety of obligations related to the prevention of oil spills and liability for damages resulting from such spills. OPA '90 imposes strict and, with limited exceptions, joint and several liabilities upon each responsible party for oil removal costs and a variety of public and private damages.

Solid Waste Disposal Act

The Solid Waste Disposal Act establishes restrictions on the disposal of solid waste and requires that a notification be placed on a deed or other document normally examined during a title search to alert a potential buyer to the fact that a property has been used to manage hazardous waste. Since most oil field properties may have had releases during their operating lifetimes, this is a source of public information on releases.

Other federal laws that have environmental restrictions but are not extensively discussed in this report are the Federal Water Pollution Control Act, which limits toxic pollutants affecting water resources or the marine environment and the Federal Interstate Land Sales Full Disclosure Act, which requires a statement of record and property report for subdivided lands delineating any unusual conditions relating to safety as well as any information considered necessary for the protection of purchases.

State Regulations

The individual state regulations are paramount in E&P regulatory controls as most states have primacy under the federal statutes, and are directed to state-specific environmental issues. This is underscored by the formation of the Interstate Oil and Gas Compact Commission (IOGCC), which includes five of the more prominent oil- and gas-producing states. In a recent report by the U.S. Department of Energy, the regulatory structure, reporting requirements, and related information in the states of California, Louisiana, New Mexico, Oklahoma, and Texas are documented.[10] Although the Toxic Release Inventory (TRI) Program under Title III of the Federal Superfund Amendments and Reauthorization Act (SARA) did not include the E&P industrial category, the IOGCC compiled a report to EPA concerning new rules that would significantly expand reporting requirements under the TRI program.

These individual activities are regulated under numerous federal environmental statutes previously described, including but not limited to, the federal Clean Air Act (CAA), the Clean Water Act (CWA), the Safe Drinking Water Act (SDWA), the Resource Conservation and Recovery Act (RCRA), the Comprehensive Environmental Response, Compensation, and Liability Act (CERCLA), the Oil Pollution Act (OPA), and the Solid Waste Disposal Act (SWDA). Most of these statutes require subject facilities to monitor and report environmental releases and compliance information to the administering agency. Delegation of authority as well as individual local interests have resulted in many states having similar, or more stringent, environmental laws which regulate air quality, water quality, and waste generation, treatment, storage, and disposal activities of industrial facilities. Generally, these laws also require subject facilities to monitor and report releases to the environment.

Local Regulations

Local regulations, specifically where urban E&P activities are prevalent (such as Southern California, north of the Dallas-Fort Worth complex, and other metropolitan areas contiguous to proven oil and gas reserves), may include a portion or all of the following entities:

- Regional Air and Water Quality Control Board regulations
- Fire Department codes and regulations
- City building codes
- City ordinances
- Local departments of Health Services
- Local department of toxic substance controls
- Municipal public works departments
- Local sanitation districts
- Municipal utility districts
- School districts

Lease Agreements and Miscellaneous Issues

Lease agreements between the land owner and the E&P owner/operators normally stipulate restrictions or ingress and egress, surface restoration, degradation of property, and similar issues. Moreover, claims are sometimes made or threatened against companies engaged in oil and gas exploration and production by owners of surface estates, adjoining properties or others alleging damages resulting from environmental contamination and other incidents of operations. While some jurisdictions limit damages in such cases to the value of land that has been impaired, in other jurisdictions courts have allowed damage claims in excess of land value, including claims for the cost of remediation of contaminated properties.

13.3 E&P-Related Fluid Characterization

The characteristics of E&P-related fluid quality is described herein. Although the characterization of crude oil and natural gas at the wellhead will vary according to location, producing zones, and other variables, the general chemical properties of petroleum crude oils are listed in Table 13.3. American Petroleum Institute (API) gravity expresses the density of liquid petroleum products and is calculated by the following formula:

$$\text{API gravity} = (141.5/\text{specific gravity}) - 131.5 \tag{13.1}$$

Characteristic or Component	Crude Oil		
	Prudhoe Bay	South Louisiana	Kuwait
API gravity (20°C)	27.8	34.5	31.4
Sulfur (wt %)	0.94	0.25	2.44
Nitrogen (wt %)	0.23	0.69	0.14
Nickel (ppm)	10.0	2.2	7.7
Vanadium (ppm)	20.0	1.9	28.0
Naphtha fraction (wt %)	23.2	18.6	22.7
Paraffins	12.5	8.8	16.2
Naphthenes	7.4	7.7	4.1
Aromatics	3.2	2.1	2.4
Benzenes	0.3[d]	0.2	0.1
Toluene	0.6	0.4	0.4
C_8 aromatics	0.5	0.7	0.8
C_9 aromatics	0.06	0.5	0.6
C_{10} aromatics	—	0.2	0.3
C_{11} aromatics	—	0.1	0.1

TABLE 13.3 Physical Characteristics and Chemical Properties of Several Crude Oils

Constituent	Mean Concentration (ppm)
Metals	
Arsenic	1.27
Cadmium	5.90
Chromium	0.63
Lead	0.24
Zinc	15.80
Organics	
Benzo(a)anthracene	1.33
Benzo(a)pyrene	1.38
Total chlorine	152.60

TABLE 13.4 Presence of Some Potentially Hazardous Materials in Crude Oils

No. of Samples	API gravity Range (°C)	Mean	Median	Minimum	Maximum
69 crude oils	8.8–46.4	1,340	780	ND	5,900
14 condensates	45–70.1	10,300	6,400	1,470	35,600

TABLE 13.5 Concentration of Benzene (mg/kg Oil)

The presence of some potentially hazardous materials in crude oils is shown in Table 13.4 and the benzene concentration ranges for crude oils and condensates[5,7,9] is presented in Table 13.5. The typical composition of natural gas is shown in Table 13.6.

There are many other sources of fluids associated with E&P activities, including but not limited to produced water, drilling fluids (muds), workover and completion wastes, dehydration and sweetening residuals, and other miscellaneous wastes and residuals. Although those residuals are discussed more extensively in the following sections, selected characterization of representative produced water quality is present in Table 13.7 and that of drilling fluids (muds) is shown in Table 13.8. Potential pollutant concentrations in treatment, workover, and completion fluids are presented in Table 13.9.

It is interesting to note that there are material safety data sheets (MSDS) for both crude oil and natural gas. The composition as described in these documents is recorded in Table 13.10.

Paraffins			
	Methane	C_1	90.0–95%
	Ethane	C_2	2.0–5%
	Propane	C_3	0.5–3%
	Butane	C_4	0.3–1%
	Pentanes	C_5	0.1–0.5%
	Hexanes	C_6+	0.1–0.5%
Inert gases			
	CO_2, N_2		0.5–2%
BTEX			
	Benzene		60–600 ppm
	Toluene		50–500 ppm
	Ethyl benzene		5–50 ppm
	Xylenes		60–600 ppm
Water			650–1600 ppm
			30–75 lb/mmscf

Source: Adapted from www.naturalgas.org/naturalgas/processing_ng.asp.[11]

TABLE 13.6 Natural Gas Composition

Parameter (mg/L)	Nonassociated Gas
Ph, Units	7.0
Total dissolved solids	20,000–100,000
Total suspended solids	1.0
Chloride	11,000–50,000
Sulfate	0–400
Oil and grease	3–25
Silicon dioxide	
Bicarbonate	
Carbonate	
Fluoride	
Nitrate	
Phenol	0–2
Benzene	1–4
Toluene	0.2–12.3
Ethylbenzene	0–0.3
Xylenes	0.5

TABLE 13.7 Representative Values for Produced Water Quality

Parameter (mg/L)	Nonassociated Gas
Naphthalene	0.03–0.9
Total hydrocarbon	
Aluminum	
Antimony	70
Arsenic	30
Barium	10–100
Cadmium	30
Calcium	
Chromium	20–30
Copper	0–100
Iron	
Lead	100–170
Lithium	
Magnesium	
Manganese	
Mercury	1
Nickel	100
Potassium	
Selenium	60
Silver	10–70
Sodium	
Strontium	
Thallium	90
Vanadium	
Zinc	40–200

Source: Adapted from Veil, 2004.[12]

TABLE 13.7 (*Continued*)

13.4 E&P Treatment Processes, Waste Sources, and Residual Reuse/Disposal

With respect to crude oil, primary field operations include drilling activities, then processing occurring at or near the wellhead through the point where the oil is transferred from an individual field facility or a centrally located facility to a carrier for transport to a refinery.

With respect to natural gas, primary field operations are those activities occurring at or near the wellhead or at the gas plant, but before the point where the gas is transferred from an individual field facility, a centrally located facility, or a gas plant to a carrier for

Parameters	Average	Range
pH	9.57	3.1–12.2
Osmotic Pressure (atm)	76.0	4.3–629
Specific Conductance (μmhos/cm)	4,788.0	383.0–38,600
Pollutants (mg/L)		
Oil & grease	11.9	2.3–38.8
Alkalinity	276.0	18.0–1,594
Bromide	10.2	2.0–56.1
Chloride	1,547.0	12.0–14,700
Phenols	0.288	0.025–0.137
Sulfate	144.0	6.0–785
Surfactants	25.0	1.5–200
Total dissolved solids	3,399.0	386.0–24,882
Total suspended solids	87.0	2.0–395
Aluminum	4,601.0	0.170–16.9
Arsenic	0.032	0.00082–0.117
Barium	2.5	0.078–37.7
Calcium	290.0	8.7–1,900
Copper	0.049	0.012–0.268
Iron	145.0	0.08–3,970
Lead	0.785	0.07–3.46
Lithium	0.46	0.037–2.04
Magnesium	59.0	0.12–1,700
Manganese	2.284	0.01–46.6
Nickel	0.945	0.025–2.4
Silver	0.035	0.035
Sodium	777.0	53.7–5,800
Zinc	0.502	0.014–1.55

TABLE 13.8 Drilling Fluids Characteristics

transport to market. Examples of carriers include trucks, interstate pipelines, and some intrastate pipelines.

Primary field operations include exploration, development, and the primary, secondary, and tertiary production of oil or gas. Crude oil processings, such as water separation, de-emulsifying, degassing,

Treatment: Oil/Gas Exploration/Production Residuals

	Pollutant Concentration (µg/L)	
Pollutant Parameter	Range	Average
Conventionals		
Oil and grease	15,000–722,000	231,688
Total suspended solids	65,500–1,620,000	520,375
Priority pollutant organics		
Benzene	477–2,204	1,341
Ethylbenzene	154–2,144	1,149
Methyl chloride (chloromethane)	0–57	29
Toluene	298–1,484	891
Fluorene	0–123	62
Naphthalene	0–1,050	525
Phenanthrene	0–128	64
Phenol	255–271	263
Priority pollutant metals		
Antimony	0–148	29.6
Arsenic	0–693	166
Beryllium	0–25.1	8.64
Cadmium	7.6–82.3	26.08
Chromium	48–1,320	616.82
Copper	0–1,780	277.20
Lead	0–6,880	1,376
Nickel	0–467	115.52
Selenium	0–139	42.94
Silver	0–8	1.60
Thallium	0–67.3	13.46
Zinc	0–1,330	362.94
Other nonconventionals		
Aluminum	0–13,100	6,408.40
Barium	66.5–3,360	498.10
Boron	4,840–45,200	15,042.0
Calcium	1,070,000–28,000,000	10,284,000.0
Cobalt	0–40.9	8.18
Cyanide	0–52	52.0

Source: Adapted from U.S. EPA, office of water, 1996.

TABLE 13.9 Pollutant Concentrations in Treatment, Workover, and Completion Fluids

Pollutant Parameter	Pollutant Concentration (µg/L)	
	Range	Average
Iron	7,190–906,000	384,412.0
Manganese	187–18,800	5,146.0
Magnesium	10,400–13,500,000	5,052,280.0
Molybdenum	0–167	63.0
Sodium	7,179,000–45,200,000	18,836,000.0
Strontium	21,100–232,000	142,720.0
Sulfur	72,600–646,000	245,300.0
Tin	0–135	27.0
Titanium	0–283	74.58
Vanadium	0–4,850	1,156.0
Yttrium	0–131	41.92
Acetone	908–13,508	7,205.0
Methyl ethyl ketone (2-butanone)	0–115	58.0
m-Xylene	335–3,235	1,785.0
o+p-Xylene	161–1,619	890.0
4-Methyl-2-pentanone	190–5,862	3,028.0
Dibenzofuran	136–138	137.0
Dibenzothiophene	0–222	111.0
n-Decane	0–550	275.0
n-Docosane	237–1,304	771.0
n-Dodecane	0–1,152	576.0
n-Eicosane	0–451	226.0
n-Hexacosane	173–789	481.0
n-Hexadecane	0–808	404.0
n-Tetradecane	513–1,961	1,237.0
p-Cymene	0–144	72.0
Pentamethylbenzene	0–108	54.0
1-Methylfluorene	0–163	82.0
2-Methylnaphthalene	0–1,634	817.0

Source: Adapted from U.S. EPA, office of water, 1996.

TABLE 13.9 Pollutant Concentrations in Treatment, Workover, and Completion Fluids (*Continued*)

Crude Oil
Danger   • Harmful or fatal if swallowed   • Vapor harmful   • Some crude oil may emit hydrogen sulfide (H_2S)   • Prolonged or repeated skin contact may be harmful   • Flammable

Typical Composition
 • Petroleum crude oils (CAS 8002-05-9) are naturally occurring complex mixtures of hydrocarbons containing variable proportions of paraffins, naphthenes, and aromatics and the following:
   1. Small amounts of organic compounds containing sulfur (trace to 8%), nitrogen, and oxygen
   2. Trace quantities of heavy metal such as nickel, vanadium, and lead
   3. Hydrogen sulfide gas (H_2S) may be present in some crude oils and may collect in the headspace of enclosed vessels

Exposure Standard
No Federal OSHA exposure standard of ACGIH TLV has been established for this material.
Natural Gas
Danger   • Extremely flammable
Typical Composition   • Methane (CAS 74-C2-8)   88.0%   • Ethane (CAS 74-84-0)   0.7%   • Propane (CAS 74-98-6)   2.5%   • Butanes (CAS 106-97-8)   0.5%   • Nitrogen (CAS 7727-37-9)   1.5%   • Carbon dioxide (CAS 124-38-9) 0.5%
Exposure Standard
No Federal OSHA exposure standard of ACGIH TLV has been established for this material.

*Prepared according to the OSHA Hazard Communication Standard (29 CFR 1910.1200) (formerly called Material Information Bulletin)

TABLE 13.10 Material Safety Data Sheet for Crude Oil and Natural Gas*

and storage at tank batteries associated with a specific well or wells, are examples of primary field operations. Furthermore, because natural gas often requires processing to remove water, sulfides, and other impurities prior to entering the sale line, gas plants are considered to be part of production operations regardless of their location with respect to the wellhead.

Water Sources Description

A brief description of E&P wastestreams is as follows:[8]

- **Produced water**   Water produced in combination with the production of oil and to a lesser extent, gas. This can constitute 90 percent of the total residuals in drilling and production operations. The flow of produced water can exceed that of oil. These wastes are normally reinjected back into the formation after solids removal (water flooding).

- **Miscellaneous surface water**   Runoff from rainfall or snowmelt which can become contaminated by contact with various oil field contaminants. Control is based on capture of these waters by vertical control and appropriate treatment.

- **Drilling muds**   Proprietary muds inserted down the rotary drill pipes and up through the annular space between the drill pipe and drill hole. The purpose is to cool the drilling action and lift rock chips back to the surface. Moreover, it provides a hydrostatic head in the drill pipe to prevent a rapid release of oil/gas-related subsurface pressure (blowouts). Mud is primarily recirculated, but blowdown mud disposal must be in accordance with the applicable environmental regulations.

- **Drill cuttings**   Rock and soil particles associated in drilling operations and normally disposed in regulated landfills.

- **Workover and completion wastes**   Treated fluids injected in partially open wellheads while tubing strings, valves, packer gaskets, and so on, are undergoing maintenance. These workover and completion fluids normally become contaminated with oil and other hydrocarbons and require oil removal before disposal.

- **Fracturing (frac) sand**   Small aluminum silicon beads used to fracture the formation to allow easier flow of oil/gas from the formation to the production, delivery, and treatment components.

- **Bottom wastes**   Tank, vessel, and pipeline sludges that require treatment and proper disposal.

- **Dehydration and sweetening wastes**   Chemicals are used to dehydrate the oil/gas product (water removal), normally using polyols or glycols to dehydrate, and various other chemicals to sweeten (hydrogen sulfide removal) the oil/gas products to sales quality.

- **Oil debris and filter matter**   Oils from spills, dumps, and spent oil filter media must be handled in accordance with environmental statutes and regulations.

- **Hydrocarbon wastes** Normally termed "dirty diesel" associated primarily with hydrostatic testing of pipelines.
- **Camp wastes** Waste residuals produced at location "camps" such as drum storage, wooden pallets, used pit liners, and other debris.
- **NORM (normally occurring radioactive materials)** Formations-related scales in vessels, piping, and other appurtenances that contain radioactive elements, the dust from which can be a constituent for human-related injury or sickness. There are strict state and federal regulations that establish maximum radiation levels and standards.

E&P Waste Residuals and Treatment Options

The extraction of oil and gas involves a variety of wastestreams and residuals as previously enumerated. These material outputs can be categorized as potential air emissions, process wastewater, and residual wastes. The basic processes are well development (drilling), completion, production, maintenance, and abandoned wells, spills, and blowouts. These waste emissions and residuals are summarized in Table 13.11. The candidate treatment options will depend primarily on location, applicable environmental regulations, and impact on receptors (surface wastes, groundwaters, soil, and so on). Other treatment options include off-site disposal and reuse or reinjection. There are treatment processes common to both oil, associated drilling and production, nonassociated gas, and other treatment systems unique to specific situations. The more common wastestreams and treatment candidates are discussed as follows.

Drilling Fluids (Muds)

Even though drilling muds are exempt from RCRA, environmental concerns and toxicity/storage/discharge requirements are subject to regulation from other statutes such as the Clean Water Act of 1972 and subsequent amendments. For example, NPDES permits are issued in many regions for recirculated and discharged drilling fluids. Toxicity and effluent quality limitations are imposed on these muds. The composition and additive chemicals that constitute these drilling fluids is subject to these regulations. Environmental control and compliance is achieved by concentration of selected chemicals in the base muds and restrictions on additives. Although the "inerts" are required to satisfy the drilling fluid requirements, certain toxicity-based additives are prohibited based on toxicity and/or other NPDES requirements. Moreover, drilling fluid application, storage, and circulation criteria are restricted by several statutes to prevent subsurface contamination.

Process	Air Emissions	Process Wastewater	Residual Wastes Generated
Well development	Fugitive natural gas, other volatile organic compounds (VOCs), polyaromatic hydrocarbons (PAHs), carbon dioxide, carbon monoxide, hydrogen sulfide	Drilling muds, organic acids, alkalis, diesel oil, crankcase oils, acidic stimulation fluids (hydrochloric and hydrofluoric acids)	Drill cuttings (some oil-coated), drilling mud solids, weighting agents dispersants, corrosion inhibitors, surfactants, flocculating agents, concrete, casing, paraffins
Production	Fugitive natural gas, other VOCs, PAHs, carbon dioxide, carbon monoxide, hydrogen sulfide, figitive BTEX (benzene, toluene, ethylbenzene, and xylene) from natural gas conditioning	Produced water possibly containing heavy metals, radionuclides, dissolved solids, oxygen-demanding organic compounds, and high levels of salts, may also contain additives including biocides, lubricants, corrosion inhibitors, wastewater containing glycol, amines, salts, and untreatable emulsions	Produced sand, elemental sulfur, spent catalysts separator sludge, tank bottoms, used filters, sanitary wastes
Maintenance	Volatile cleaning agents, paints, other VOCs, hydrochloric acid gas	Completion fluid, wastewater containing well-cleaning solvents (detergents and degreasers), paint, stimulation agents	Pipe scale, waste paints, paraffins, cement, sand
Abandoned wells, spills, and blowouts	Fugitive natural gas, other VOCs, PAHs, particulate matter, sulfur compounds, carbon dioxide, carbon monoxide	Escaping oil and brine	Contaminated soils, sorbents

TABLE 13.11 Waste Emissions and Residuals

Produced Water

As noted in Sec. 13.3, produced water contains high concentrations of dissolved solids, as well as lesser amounts of heavy metals and various hydrocarbons such as phenols and benzene, toluene, ethylbenzene, and xylenes. The volume of produced water resulting from oil and gas extraction can constitute as much as 90 percent of the wastewater residuals produced. Current disposal practices include pretreatment and reinjection, discharge to surface water, evaporation, and possible land application. A produced water management treatment matrix is illustrated in Fig. 13.7. As noted, this can be an elaborate and comprehensive treatment complex, depending on the applicable environmental regulations that govern the ultimate disposal options. Particular emphasis in unit process evaluation should be placed on sulfide production by sulfate-reducing bacteria and the organic contaminants which may require biological treatment and filtration, as well as inorganic salt reduction options. Any requirement for desalination necessitates significant energy and financial outlays. For example, if a NPDES permit is required for off-site disposal, a typical effluent requirement is shown in Table 13.12

Produced Oil, Associated Gas, and Natural Gas Liquids

Produced oil, before it meets sales quality, most likely requires associated gas separation, possible recovery of natural gas liquids, free water removal through air flotation and slop recovery, free water knockout drums, and coalescers. Each of these processes to upgrade the crude oil to sales quality have the potential of polluting the environment through air emissions, spills, valve, seal and pump releases, leaking storage tanks, and separated water treatment inefficiencies. A flow diagram of an urban drilling and production site is shown in Fig. 13.8.

Associated and Nonassociated Gas Treatment

Associated gas, unprocessed gas that is associated with oil production (casinghead gas generally comes under this category, whether coming from the casing or the tubing), and nonassociated gas (not accompanied by oil) require treatment at or close to the wellhead or at a centralized gas processing plant which receives gas from low pressure gas pipeline gathering systems. A typical flow diagram of a gas processing plant (primarily nonassociated) is illustrated in Fig. 13.9.

Separation[6,11]

The actual processes used to separate gas from oil vary widely, depending on the quality of oil and gas and the separation requirements. Generally, associated gas dissolved in the oil due to formation pressure can be separated rather easily at the surface by decreasing pressure. The most basic conventional separator is a closed tank where the gravity forces separate the heavier oil phases from the

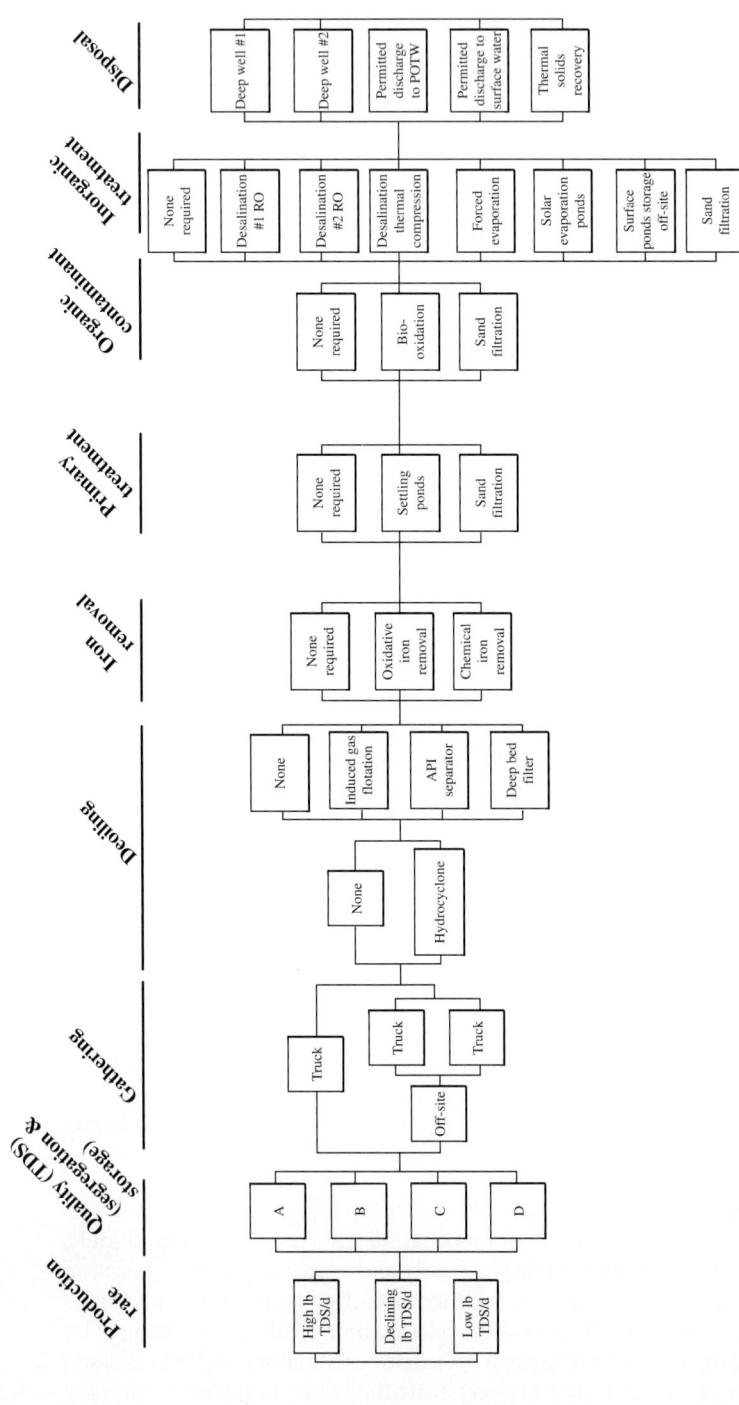

FIGURE 13.7 Produced water management treatment matrix.

Treatment: Oil/Gas Exploration/Production Residuals

Parameter	Average	Daily Maximum	Average	Daily Maximum
pH	6–9	9	6.5–9.0	
BOD (5-d) (mg/L)	30	45		
Conductivity (µmhos/cm)			600	1,200
Total suspended solids (mg/L)			30	60
Total dissolved solids (mg/L)			500	1,000
Total BTEX (µg/L)				100
Total arsenic (µg/L)				50
Total barium (µg/L)				1,000
Total cadmium (µg/L)				10
Total chromium (µg/L)				50
Total mercury (µg/L)				2
Total selenium (µg/L)				5
Benzene (µg/L)				5

*Aquatic toxicity testing is also part of permit requirements.
Source: Adapted from U.S. EPA, office of water, 1996.

TABLE 13.12 NPDES Produced Water Effluent Requirements*

lighter gas phases. In some cases, however, oil/gas separation is more complex.

An example of this type of equipment is the low-temperature separator (LTS). This is most often used for wells producing high-pressure gas along with light crude oil or condensate. These separators use pressure differentials to cool the wet natural gas and separate the oil and condensate. Wet gas enters the separator, being cooled slightly by a heat exchanger. The gas then travels through a high pressure liquid "knockout" vessel, which serves to remove any liquids into a low-temperature separator. The gas then flows into this low-temperature separator through a choke mechanism, which expands the gas as it enters the separator. This rapid expansion of the gas

Chapter Thirteen

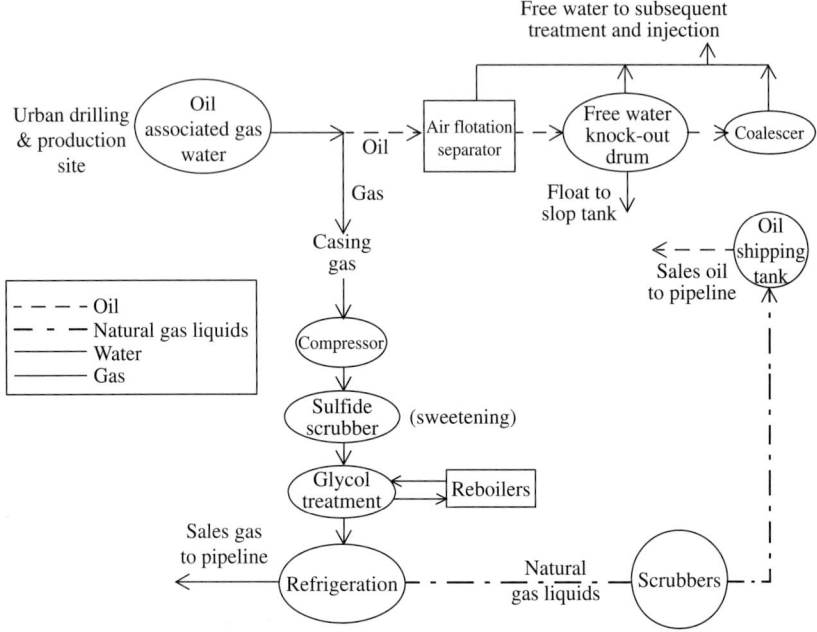

FIGURE 13.8 Generalized flow diagram representing urban drill and production site (oil and associated gas).

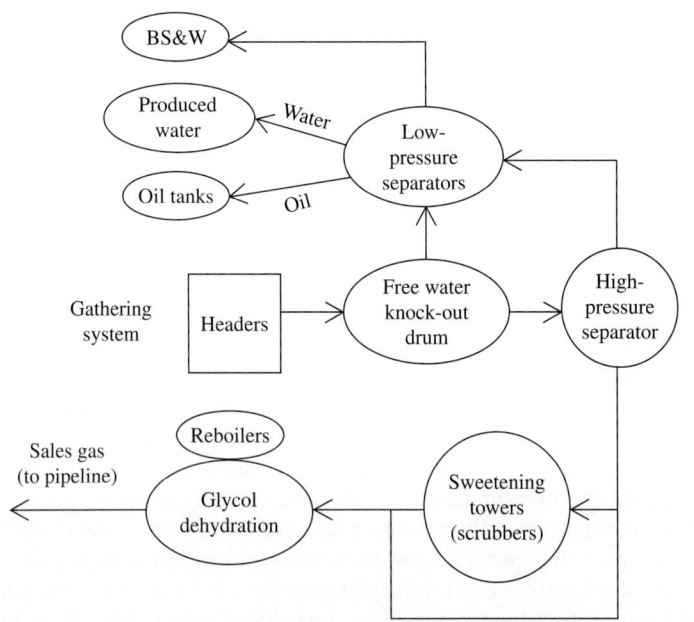

FIGURE 13.9 Simplified flow diagram representing nonurban gas treatment plant.

allows for the lowering of the temperature in the separator. After liquid removal, the dry gas then travels back through the heat exchanger and is warmed by the incoming wet gas. By varying the pressure of the gas in various sections of the separator, it is possible to vary the temperature, which causes the oil and some water to be condensed out of the wet gas stream. This basic pressure-temperature relationship can work in reverse as well to extract gas from a liquid oil stream.

Dehydration
As natural gas is extracted from reservoirs formed over centuries in a marine environment, natural gas at the wellhead is usually produced with some water. A lean gas mix will be saturated with water vapor at a down hole pressure at 1000 psi, for example, which contains 9 lb of water per million standard cubic foot (mmscf). This gas at the surface will be saturated with water vapor at 14.7 psi and contain 400 lb of water vapor per mmscf. Most interstate gas transmission lines require that "dry gas" be delivered to the pipeline, or less than 7 lb of water vapor per mmscf. This requires dehydration at the wellhead or in nearby gas processing plants.

Essentially, glycol dehydration[11] involves using a glycol solution, usually either diethylene glycol (DEG) or triethylene glycol (TEG), which is brought into contact with the wet gas stream in what is called the *contactor*. The glycol solution will absorb water from the wet gas. Once absorbed, the glycol particles become heavier and sink to the bottom of the contactor where they are removed. The natural gas, having been stripped of most of its water content, is then transported out of the dehydrator. The glycol solution, bearing the water stripped from the natural gas, is put through a specialized boiler designed to vaporize only the water out of the solution. While water has a boiling point of 212°F, glycol does not boil until 400°F. This boiling point differential makes it relatively easy to remove water from the glycol solution, allowing it for reuse in the dehydration process.

Glycol dehydration is the most widely used method of removing or absorbing water from the gas stream. This process is illustrated by a flow diagram shown in Fig. 13.10. A more recent innovation in this process has been the addition of flash tank separator-condensers. As well as absorbing water from the wet gas stream, the glycol solution occasionally carries with it small amounts of methane and other compounds found in the wet gas. In the past, this methane was simply vented out of the boiler. In addition to losing a portion of the natural gas that was extracted, this venting contributes to air pollution and potential greenhouse effect. In order to decrease the amount of methane and other compounds that are lost, flash tank separator-condensers work to remove these compounds before the glycol solution reaches the boiler. Essentially, a flash tank separator

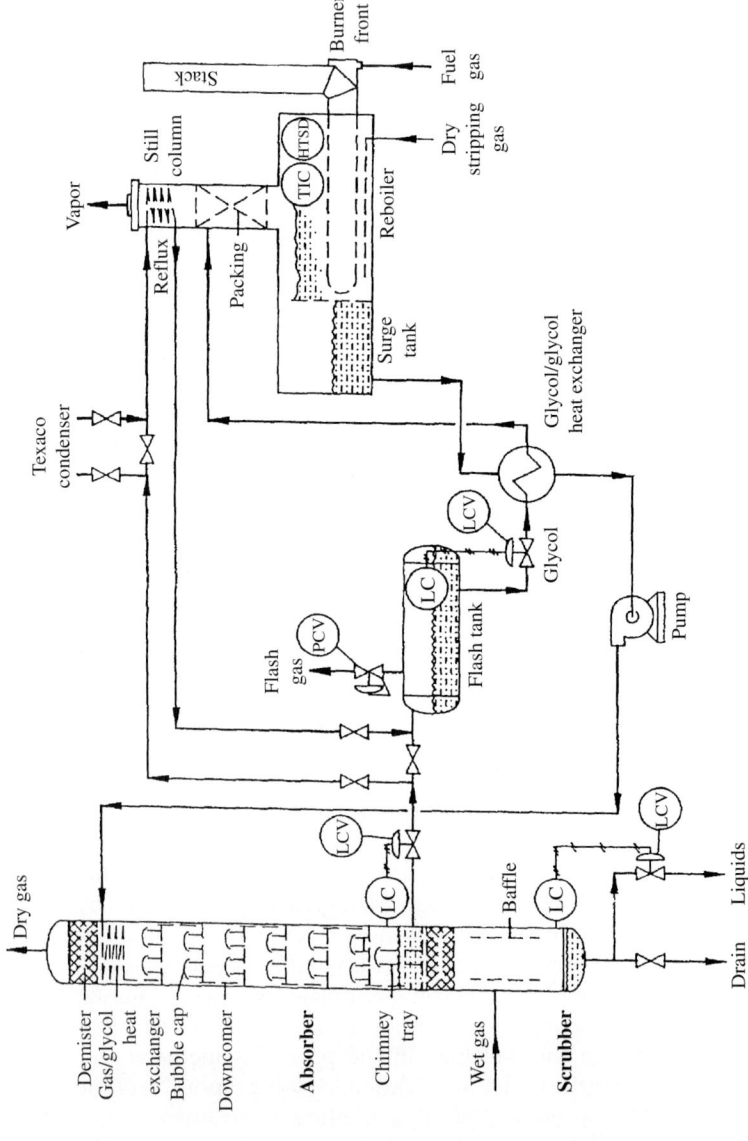

Figure 13.10 Process flow sheet for glycol dehydration. (*Adapted from www.naturalgas.org/naturalgas/processing_ng.asp.*)

consists of a device that reduces the pressure of the glycol solution stream, allowing the methane and other hydrocarbons to vaporize ("flash"). The glycol solution then travels to the boiler that may also be fitted with air or water-cooled condensers, which serve to capture any remaining organic compounds that may remain in the glycol solution.

There are other methods used for gas dehydration, including solid desiccants, dehydrators, refrigeration, and molecular sieve technologies.

Sweetening[11,14]

This process is required in virtually all natural gas processing plants to remove hydrogen sulfide, carbon dioxide, and other constituents. Most wellhead natural gases contain far more corrosive hydrogen sulfide gases than allowed in pipelines. For example, the H_2S content of natural gases sampled throughout the United States contains 2 to 3 percent (20,000 to 30,000 ppm) or higher. As the pipeline sales content must be 0.0015 percent or less (150 ppm as H_2S), a sweetening process is required. The amount of sweetening can be graphically illustrated, as shown in Fig. 13.11.

Several corrosion inhibitors are common in oil field water technology to control hydrogen sulfide, carbon dioxide, and oxygen,

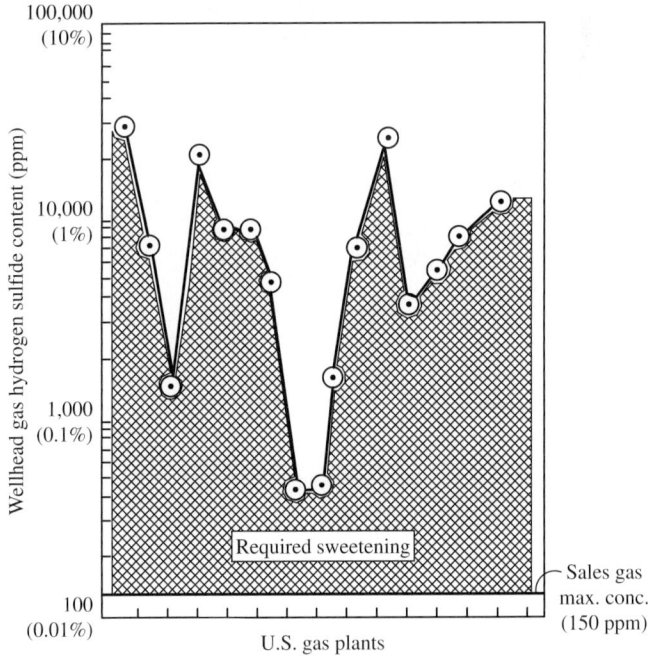

FIGURE 13.11

but the most common sweetening methodology is the amine solution (Giodler) process, specifically:

$$2RN\,H_2 + H_2S \rightarrow (RN\,H_3)_2S \qquad (13.2)$$

where R = amine functional group
       N = nitrogen
       H = hydrogen
       S = sulfur

The sour gas is run through a tower containing the amine solution. This solution has an affinity for absorbing sulfur, somewhat similar to the glycol absorption of water. The two principal amine solutions used are monoethanol amine and diethanolamine. The spent amine solution can be regenerated, removing the absorbed sulfur, and then reused. Although most sour gas sweetening processes involve amine absorption, it is also possible to use solid desiccants, such as iron sponges, to remove the sulfides and carbon dioxide. A recent patented process called the Lo-CAT system has been used for sweetening wellhead casing gas. In this process, hydrogen sulfide is converted to elemental sulfur by applying a chelated iron catalyst according to the following equation:

$$H_2S + 1/2O_2 \xrightarrow{Fe} H_2O + S \qquad (13.3)$$

The sour gas stream is treated in separate pressurized oxidizer and absorber vessels. The primary chemical consumption are the chelated iron and caustic addition to maintain the desired pH.[16]

13.5 Problems

13.1. Given the following characterization data of produced water resulting from E&P extraction activities:

Total Volume Produced	8 bbl/bbl oil
Daily Production	200 bbl/d
TDS	18,000 to 25,000 ppm
BTEX	23 ppm
TSS	150 ppm
Cadmium	56 ppm
Zinc	250 ppm
SO_4	450 ppm

Outline a treatment process flow diagram for the produced water to

(a) Reinject in a Class II disposal well for the purposes of secondary recovery, or
(b) Discharge to a fresh water receiving stream under a NPDES permit, or
(c) Discharge to an ocean outfall under a NPDES permit

13.2. A wellhead nonassociated gas stream has the following characteristics:

Water vapor content (147 psi)
84 lb water per mmscf
1.5 percent sulfides

Describe the potential contaminants or pollution sources at a gas processing plant and the gas treatment required to produce pipeline quality.

References

1. American Petroleum Institute (API): "Guidelines for Commercial Exploration and Production Waste Management Facilities." Exploration and Production Waste Management Facility Guidelines Workgroup, API No. Goooo4, March 2001.
2. Verma, D. K., D. M. Johnson, and J. D. McLean: "Benzene and Total Hydrocarbon Exposures in the Upstream Petroleum Oil and Gas Industry," *Journal AIHA*, 61, March–April, 2000.
3. Freudenrich, C. C.: "How Oil Drilling Works." Available at http://science.howstuffworks.com/oil-drilling.htm. Accessed March 2005. Posted 2001.
4. Environmental Protection Agency (EPA), Office of Compliance: "Profile of the Oil and gas Extraction Industry," EPA/310-R-99-006, 2000.
5. Environmental Protection Agency: "Exemption of Oil and Gas Exploration and Production Wastes from Federal Hazardous Waste Regulations," EPA530-K-01-004, 2002.
6. Sublett, K. L., (ed.): "Environmental Issues and Solutions in Petroleum Exploration and Refining," *Proceedings of the International Petroleum Exploration, Production and Refining Conference*, Houston, Texas, March 1994.
7. Environmental Protection Agency: "Spill Prevention, Control, and Countermeasure: A Facility Owner/Operator's Guide to Oil Pollution Prevention," EPA Office of Emergency Remedial Response, Soil Program Center, 540-K-02-006, 2002b.
8. U.S. Department of Energy: "Risk-Based Decision Making for Assessing Petroleum Impacts at Exploration and Production Sites," 2001.
9. Oil and Gas Accountability Project (OGAP): "The Facts About Oil and Gas Wastes." Available at http://www.ogap.org/waste_products_facts_sheet.htm. Posted September 16, 2003.
10. Interstate Oil & Gas Compact Commission: "Review of Existing Reporting Requirements for Oil and Gas Exploration and Production Operators in Five Key States," Oklahoma City, Oklahoma, December 1996.
11. Internet search, www.naturalgas.org/naturalgas/processing_ng.asp.
12. Veil, J. A., M. Puder, D. Elcock, and R. Redweik, Jr.: "A White Paper Describing Produced Water from Production of Crude Oil, Natural Gas, and Coal Bed Methane." Prepared for U.S. DOE National Technology Laboratory, Contract W-31-109-Eng-38, 2004.
13. U.S. EPA, Office of Water: Development Document for "Final Effluent Limitations and Guidelines for Oil and Gas Extraction Point Source Category," October 1996.
14. Internet search: www.newpointgas.com/amine_treating.php.
15. Environmental Protection Agency (EPA), Office of Solid Wastes, 1987.
16. Gas Technology Products, Merichem Chemicals & Refinery Services LLC. www.gtp-merichem.com.

CHAPTER 14
Chlorinated Compounds, VOCs, and Odor Control

14.1 Introduction

The Industrial Water Quality Control of chlorinated compounds and volatile organic compounds (VOCs) has become an increasingly more complex industry-wide problem over the past four decades. Little, if any, attention, scrutiny, or research was directed toward the environmental health, safety, or impacts of chlorinateds prior to the early 1970s. Many chlorinated compounds which were utilized globally are now restricted or banned entirely, while others are subject to increasingly stringent regulations. The challenges for modern day engineers, environmental scientists, and regulators are severalfold. The health impacts, applicable treatment processes, regulatory issues, and cleanup of historical dump sites are all issues with which problem solvers in the industrial pollution control sector must deal.

The past and present use of chlorinated compounds is ubiquitous, ranging from industrial solvents, fuel and lubricating additives, transformer fluids, rocket propellants, plastic manufacturing to many household uses. Moreover, past disposal practices have resulted in extensive contamination of surficial and subsurface soils, drinking water aquifers, and air quality. It should be recognized, however, that these chlorinated organics have contributed significantly to industrial safety and efficiency as well as pestilence control and agricultural enhancement throughout most of the twentieth century.

A tremendous volume of credible research and literature on this subject has resulted in response to this problem. Control technologies and solutions continue to evolve. This chapter does not attempt to address these chlorinated compound issues or specific VOCs in depth. However, it is designed to highlight and categorize the more environmentally significant chlorinated compounds and address fundamental technical issues such as identification, characterization, and amenability to scientifically based solutions to treat, control, remediate, and/or isolate these contaminants. The evolution of knowledge concerning chlorinated compounds is included to provide a chronological perspective. The five classes of chlorinated compounds included in this chapter are:

1. Polychlorinated biphenyls (PCBs)
2. Chlorinated solvents or intermediates such as perchloroethylene (PCE), trichloroethylene and trichloroethane (TCE and TCA), dichloroethylene and dichloroethane (DCE and DCA), carbon tetrachloride, and vinyl chloride
3. Chlorinated pesticides such as dichlorodiphenyl trichloroethylene (DDT) and associated compounds
4. Perchlorates such as ammonium perchlorate
5. Other miscellaneous chlorinated organics, chlorinated by-products, VOCs, and odor control

Moreover, these and other VOCs and their amenability to stripping are discussed to some extent in Chap. 6. However, the chlorinated organics are discussed in more detail in this chapter along with odor control.

14.2 Polychlorinated Biphenyls

Introduction

PCBs are synthetic chemical compounds consisting of chlorine, carbon, and hydrogen. First synthesized in 1881, PCBs are relatively fire-resistant, very stable, dielectric, and have low volatility at normal temperatures. These and other properties have made them desirable components in a wide range of industrial and consumer products. Some of these same properties make PCBs environmentally hazardous—especially their extreme resistance to chemical and biological breakdown by natural processes in the environment. PCBs are also known by their various brand names which include Aroclor, Pyranol, Interteen, and Hyrol.[1]

Polychlorinated biphenyls (PCBs) are chemical compounds with the empirical formula $C_{12}H_{10-n}Cl_n$ with $n = 1$–10. They generally are a mixture of chlorinated biphenyls congeners. Theoretically, 209 such congeners are possible, but at least 20 congeners have never been identified in commercial products. In addition, PCBs may contain polychlorinated dibenzofurans and chlorinated quaterphenyls as impurities.

Monsanto purchased a chemical plant near Anniston, Alabama, and began commercial production of PCBs in 1929. In 1966, the discovery of PCBs in environmentally related samples raised interest in the analysis and toxicity of these compounds. As Monsanto was the sole manufacturer of PCBs, the company began to focus on the PCB environmental effects in the late 1960s. It began a comprehensive action plan to develop analytical techniques for analyses of PCBs in water and sediments, minimize PCBs in the plant effluent, and commence sampling for PCBs in the receiving ditches and creeks. Following the discovery of PCBs in marine animals from Swedish waters in 1966, and in rice consumed by humans in Japan in 1969, there developed a worldwide health awareness of PCBs in the environment. As PCBs are relatively resistant to biodegradation, their stability and lipophilicity led to an increased concern by public health and environmental regulatory authorities.

In 1971 to 1972, Monsanto voluntarily stopped open-ended uses of PCBs and produced only the lower chlorinated biphenyls, Arochlor 1242 and 1016. (Arochlor designations are four digits, 12 indicating a chlorinated biphenyl and the last two digits indicating the percent chlorine by weight.) During 1972 through 1977, the company ceased PCB production entirely. Between 1930 and 1975, approximately, 1900 million pounds of PCB were produced domestically.[2]

As PCBs have great thermal and electrical stability properties, they had many attractive industrial applications. These PCB mixtures for industrial use were solid, highly viscous, or liquid. They are insoluble in water but soluble in most organic solvents and in vegetable oil. They were used effectively for many years in heat transfer fluids for heat transfer, dielectric fluids in electrical transformers and capacitors, metal cutting hydraulic oils, additives in plastic materials, paint lacquers, varnishes, glass, papers, inks, insecticide formulations, and bactericides, now all banned.

Environmental Impacts

As PCBs were widely used, through the aforementioned applications, they found their way into the environment through many pathways. These include, but are not limited to, the following:

1. Leaking transformers and capacitors
2. Draining and replacement of PCB fluids into transformers
3. Leaks and spills of hydraulic fluids from compressors and other heavy machines
4. Stormwater runoff into storm sewers, contaminated by flow over spills, soil stains, and leaks of PCB-containing materials, the storm flow ending up in stream rivers, bays, and estuaries
5. Point source industrial discharges
6. Pipeline-related discharges, primarily from pigging operations discharged to the environment, attributable

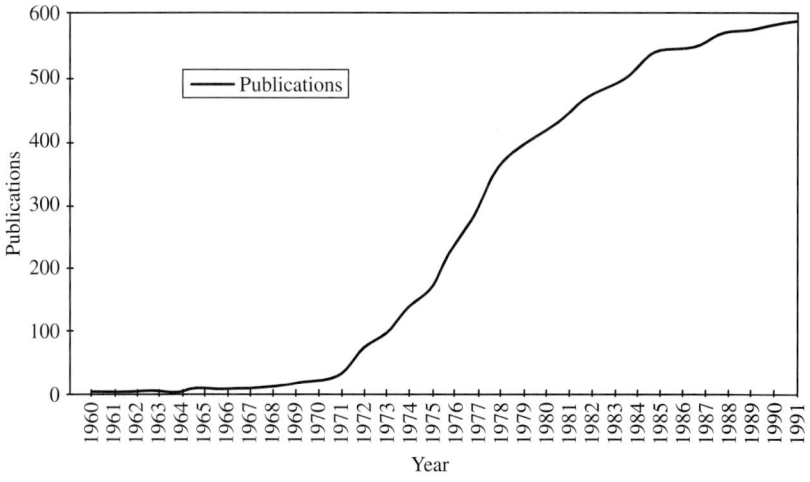

FIGURE 14.1   PCB publication history. (*Adapted from Ford, 1996.*)

to hydraulic fluid leaks in pumping or compressor stations, and

7. Discarded oils, paints, lacquers, transformers, and other sources disposed or stored in dump sites, municipal and industrial landfills, and disposal pits

Virtually all of the PCBs found in the environment today are attributable to early disposal practices with little or no regulation at the state or federal levels, and lack of awareness and enforcement. For example, PCBs are the primary "constituents of concern" and/or remedy driver in many state or federal superfund or RCRA disposal sites today. The number of publications relating to PCBs in environmentally related journals, research reports, and texts indicate the awareness of PCBs over time,[3] as shown in Fig. 14.1

Regulatory History of PCBs

The late 1960s and early 1970s was a period of major transition in the scientific and regulatory community with respect to PCBs. From an analytical perspective, it was not until the 1960s that gas-liquid chromatography with highly sensitive and selective detectors was routinely employed for the measurement of trace quantities of PCBs in the environment.[4] In 1971, several governmental agencies established an interdepartmental task force to better understand the family of compounds known as PCBs.[5] Several of the major conclusions from this study were:

1. PCBs have been used so widely over such a long period of time that they are ubiquitous.

Chlorinated Compounds, VOCs, and Odor Control

2. The major gap in the regulatory system to deal with PCBs is the absence of any broad federal authority to restrict use or distribution of the chemical, to control imports, and to collect certain "types of information."
3. Housekeeping is particularly important in the manufacture, use, and disposal of PCBs.
4. The use of PCBs should not be banned entirely (1972).
5. Most capacitors containing PCBs have been disposed in landfills.
6. More scientific information about PCBs is needed.
7. Current scientific knowledge gained from laboratory animal experiments is often inadequate to allow interpretation of the data in terms of possible effects on man.

As late as 1972, the National Academy of Sciences in considering PCBs for the establishment of water quality criteria stated that "because too little is known about levels of PCBs in waters, the retentive and accumulation in humans, and the effects of very low rates of ingestion, no defensible recommendation can be made at this time."[6]

Throughout the 1970s, there were no drinking water standards for PCBs.[7]

In 1977 a report was published by EPA describing for the first time available management and treatment technology for the purpose of determining toxic pollutant effluents concentrations and daily loads achievable in three industrial categories: PCBs manufacturing, capacitor manufacturing, and transformer manufacturing.[8] In 1978, under the Federal Clean Water Act, a list of priority pollutants for "best available treatment" (BAT) was specified by EPA, which included PCBs 1242, 1254, 1221, 1232, 1248, 1260, and 1015.[9] PCBs were not regulated by either the state or federal governments prior to passage of the Toxic Substances Control Act (TSCA) in 1976, which authorized EPA to regulate PCBs. The regulations enacted in 1978 addressed labeling and disposal, and in 1979 banned manufacturing of PCBs with restrictions on distribution and use. TSCA is unique in that the statute specifically cites PCBs, has no parallel state provisions, and required the phase out of electrical transformers and capacitors containing PCB-contaminated oil.

The chronology of laws and regulations for PCBs with the Federal Register citation is available from 1976 to 2003 on the Internet.[10] A summary of selected citations is presented in Table 14.1.

Treatment Methodologies

Candidate technologies for treating PCBs were first outlined by EPA during the mid-1970s based on extensive research, discussions with treatment equipment supplies, and other sources. They concluded that activated carbon adsorption was the best "current candidate for

Date	Title	Cite
April 1, 1976	PCB Containing Waste; Disposal Procedures	41 FR 14133
February 17, 1978	PCBs; Marking and Disposal; **Final Rule**	43 FR 7150
November 1, 1978	Interim Procedural Rules for Exemptions from PCB Manufacturing Ban	43 FR 50905
January 2, 1979	Policy for Implementation and Enforcement of PCB Ban Rule	44 FR 108
July 9, 1979	Disposal of PCB Contaminated Soil and Debris; **Denial of Citizens' Petition**	44 FR 40132
September 19, 1979	Disposal Requirements; Immediately Effective Amendment to the 5/31/79 Final Rule Comment Period	44 FR 54296
March 28, 1980	Disposal Requirements for PCB Capacitors in Chemical Waste Landfills; **Final Amendment**	45 FR 20473
May 20, 1981	PCBs at Concentrations below 50 ppm; **Court Order**	46 FR 27615
August 25, 1982	PCB Use in Electrical Equipment; **Final Rule**	47 FR 37342
November 17, 1983	TSCA Statement of Policy for Compliance and Enforcement of PCB Storage for Disposal Regulations	48 FR 52304
March 22, 1984	PCBs; Manufacturing, Processing, Distribution in Commerce and Use Prohibitions; Use in Electrical Transformers; **ANPR**	49 FR 11070
April 2, 1987	PCB Spill Cleanup Policy; **Final Rule**	52 FR 10688
July 19, 1988	PCBs in Electrical Transformers; **Final Rule**	53 FR 27323
May 19, 1989	Procedures for Rulemaking under Section 6 of TSCA; **Final Rule**	54 FR 21622
April 2, 1991	PCBs in Natural Gas Pipelines; Availability of Draft Guidance Documents	56 FR 13473
June 8, 1993	Use of Waste Oil	58 FR 32061
June 29, 1998	Disposal of Polychlorinated Biphenyls (PCBs); **Final Rule**	63 FR 35384
April 2, 2001	Reclassification of PCB and PCB-Contaminated Electrical Equipment; **Final Rule**	66 FR 17602

TABLE 14.1 PCB Laws and Regulations, 1976–2003

Carbon	Residual (ppb)		
Dosage (mg/L)	Arochlor 1242	Arochlor 1254	Aldrin
Control	45	49	48
1.0	—	—	—
2.0	7.3	37	26
2.5	—	—	—
5.0	1.6	17	15
10.0	1.1	4.2	12
12.50	—	—	—
25.0	—	1.6	6.3
50.0	—	1.2	4.4

TABLE 14.2 Results of Calgon Corp. Laboratory Isotherm Tests for Carbon Removal of PCBs

successful removal of PCBs from wastewaters."[8] This technology today is still appropriate as it has long been established that activated carbon has an affinity for chlorinated hydrocarbons.[10,11]

For example, results of activated carbon isotherm studies conducted by a carbon manufacturer show excellent removal of PCBs (Arochlor 1242 and Arochlor 1254 as well Aldrin),[8] as shown in Table 14.2 and Fig. 14.2. These are similar isotherm results observed by the author with chlorinated hydrocarbons.[11]

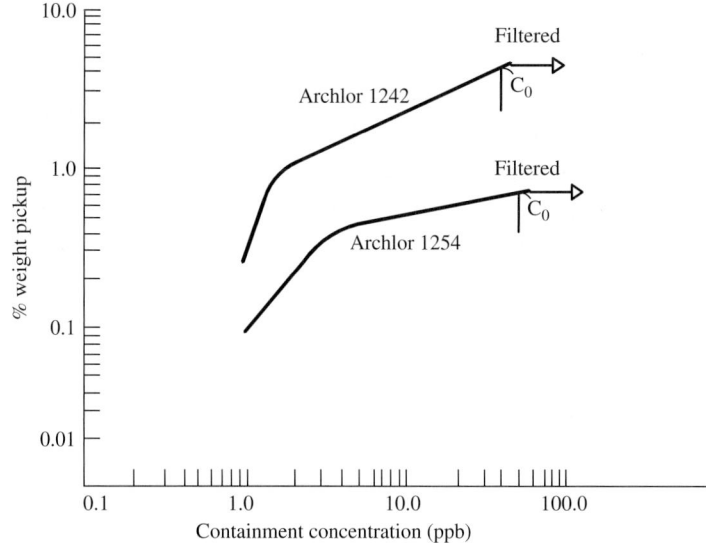

FIGURE 14.2 PCB isotherm results. (*Adapted from U.S. EPA, Office of Toxic Substances, 1976.*)

Other candidate systems proposed by EPA included ultraviolet-assisted ozonation, incineration, and dry carbon filter adsorption for control of air emissions. It should be noted that the desorption of PCBs and associated chlorinated compounds have been studied to check the effects of "reversibility." It appears, based on isotherm comparison, that the more highly chlorinated PCBs are, the more persistent and less likely for future desorption.[12]

The major thrust of PCB removal today, however, is not waste water treatment but removal or containment of PCBs entrained in a soil matrix. This is the result of PCB-laden materials such as capacitors, transformers, oils, paints, and other similar items which ended up in municipal, industrial, and codisposal landfills, many of which are superfund sites today. This is based on two decades or more of disposal site cleanup.

Slurry reactor remediation at a superfund site, using pilot scale reactors, resulted in a moderate reduction of Arochlor 1232. The PCBs were partitioned to the cellular lipids of the bacteria, and additional PCB mass was recovered from the biomass. Therefore, the actual PCB removal potential of the microbial process was a function of the presence of PCBs in both the biotreated residue and the sieved mixed liquor (biomass) fractions. The partitioning of the PCBs was evidenced by the temporal distribution of Archlor 1232 between the biotreated residue and the sieved mixed liquor containing biomass during the experiment. Similar results were obtained with the measurement of the PCB congeners. Congeners with molecular weight greater than tetrachlorobiphenyl appeared to be recalcitrant to microbial attack. The results are shown[13,14] in Table 14.3.

Sample	Sieved Mixed Liquor (mg)	Biotreated Residue (mg)	Total Remaining (mg)	Percent Reduction (%)
Initial waste load	—	—	113	—
Reactor at 90 days				
A*	52	—	52	54
B	25	10	35	69
C	19	12	31	73
Control†	57	9	66	42

*Unsieved or unfiltered mixed liquor.
†Unseed reactor contained indigenous microorganisms.

TABLE 14.3 Mass of PCB-1232 in Bioreactors at 90 Days

A power plant in Austin, Texas, at and before the time of closure, exhibited significant PCB contamination attributable to oil coolers, lubrication oil reservoirs, transformer areas, and turbine generator areas. As the plant was built in the late 1950s, most of the PCB contamination likely occurred in the first 20 to 30 years of operation. The cleanup minimum goal ranged from 10 to 25 mg/kg PCBs. The closure and cleanup activities included:

- Removal of capacitors off-site for incinerator destruction
- Off-site removal of high-level contaminated soils
- In situ remediation of low-level contaminated soils

A similar cleanup of the Pacific Gas & Electric Company power plant near Sacramento, California, was undertaken. Fuel tanks were removed from the site as well as 6200 tons of contaminated soil for off-site incineration, containment, and disposal. Less contaminated soils were capped in accordance with accepted health risk analyses. Other similar sites (including national priority lists [NPLs], NPL sites) have gone or are going through remediation. PCB cleanup standards range from 10 to 25 mg/kg for soil and as low as 0.5 ppb for liquid discharge criteria. Methodologies include, but are not limited to, off-site disposal to hazardous waste disposal facilities, biological reduction using various methods, incineration, containment using RCRA criteria vaults (slurry well, synthetic liners, and so on) and slurry remediation, chemical oxidation, and/or activated carbon for liquid effluents.

14.3 Chlorinated Solvents

Introduction and Historical Perspective

This section addresses the more prominent chlorinated solvents from an environmental perspective, namely PCE, TCE, TCA, DCE, DCA, and CT (carbon tetrachloride). Even though vinyl chloride (VC) is not a solvent but a basic ingredient in plastics manufacturing (polyvinylchloride, PVC, copolymers, and adhesives), it is a daughter product in the biotic reductive dechlorination of PCE, TCE, and DCE and is possibly produced through the abiotic reactions of TCA. VC therefore is included in this section.

These compounds constitute a major environmental problem in the United States today, particularly in the subsurface, including groundwater used for drinking water as well as air quality concerns because of their volatility. As these compounds did not come into focus from an environmental/health perspective until the late 1970s to early 1980s, it is important to understand their chronological history.

Compounds such as TCE, PCE, other chlorinated solvents, and their respective isomers are within a family of compounds which represented a significant breakthrough in workplace safety commencing in the early part of the twentieth century based on their resistance to explosion or flammability. These attributes led to an exponential rise in U.S. production, particularly of TCE and PCE, from the 1930s to the mid-1970s. They were known as the *safety solvents* and have been used for many industrial, commercial, and household applications. Commercial and industrial users considered these chlorinated solvents as excellent metal degreasers, particularly in the vapor phase. The use of TCE for vapor degreasing and other industrial applications was overwhelmingly popular in the United States, particularly during World War II and the following years up through the 1970s. Aside from industrial use, TCE also was commonly used as a general and dermal anesthetic, dry-cleaning solvent, varnish thinner, extraction agent for coffee decaffeination (up to 20 ppm through the mid-1970s), and in various household uses such as septic tank degreasers, furniture polish, and sink cleanser.[15] It was only during the mid- to late 1970s that TCE began to be replaced with TCA, PCE, and miscellaneous nonchlorinated solvents. This, combined with occupational exposure concerns, led to common TCE replacements, including PCE and TCA. It should be noted, however, that today there is still no prohibition on the production of TCE or PCE. The solvent is being produced in the United States today, and is used as a chemical intermediate. EPA has listed TCE as an acceptable substitute for TCA in adhesives, coatings, and other selected applications. There is at least one application today (2008) with which the author is familiar where TCE is still being used as a vapor degreaser and is, however, operated under the stringent restriction of the National Emission Standards for Hazardous Air Pollutants (NESHAPs) regulations.

Environmental scientists and engineers did not address TCE, PCE, and other chlorinated solvents as potential soil and groundwater contaminants until the early 1980s, long after the profession began addressing salts, inorganics, and the few organic parameters previously considered to be important. This was attributable to a number of factors, including the absence of regulatory guidelines and water quality criteria addressing these compounds, the lack of expectation or awareness that land disposal practices could impact the subsurface, a lack of literature in the environmental journals and publications, and the general belief that chlorinated solvents were not a health threat at low levels or through groundwater exposure.

An important reason that managers and engineers had no reason to focus on TCE, PCE, and the other chlorinated solvents as potential surface or subsurface contaminants was that environmental regulations and water quality criteria did not address these materials until the late 1970s and 1980s. Chlorinated solvents, including TCE and PCE, were not singled out as specific "command and control" constituents in state or NPDES permits during the 1960s and 1970s, nor were they included in lists of water quality criteria for drinking water and other beneficial uses. In 1972, the National Academy of Science published a book on water quality criteria which did not include TCE, PCE, and the other chlorinated solvents.[15] In 1976, EPA published *Quality Criteria for Waters* in which chlorinated solvents were still not included.[16] TCE and PCE were not originally included as primary drinking water standards when the Safe Drinking Water Act (SDWA) regulations were first promulgated in 1975. TCE and PCE were not listed as "toxic water pollutants" under Section 307 of the Clean Water Act (CWA) when the first regulations under CWA were promulgated in 1973, and were not listed as "priority pollutants" until 1978.

When articles and other literature concerning subsurface contamination did appear, these references generally focused on a narrow list of just a few specific potential contaminants. Principally these were bacterial contamination emanating from septic tanks or other domestic sources, salt brines, boron, phosphates, nitrogen, cyanides, heavy metals, and several other specific inorganic constituents. There were references to a few specific organic compounds, including surfactants (foaming agents), phenols, pesticides, and petroleum derivatives, but organic compounds were not generally considered potential groundwater contaminants and (with the exception of pesticides) were generally not included for reasons other than specified toxicity. It was not until the mid- to late 1970s that the scientific and technical literature began to describe incidents of subsurface contamination by a broad variety of synthetic organic compounds. Literature describing incidents of contamination by TCE, PCE, and other chlorinated compounds generally did not start appearing until the late 1970s and early 1980s.[17–19] Before 1977, the groundwater literature indicated no reference to TCE or PCE contamination. A 1975 EPA study of TCE, including a literature survey, reported that "no information at all is available about the levels of TCE in the environment in the United States."[15] One group of hydrogeologists noted that very little was known about the environmental impacts of chemical wastes, including chlorinated solvents, until the late 1970s when newly discovered hazardous waste sites attracted attention.[20]

In the late 1970s, there was a prevailing belief that chlorinated solvents would evaporate or be assimilated into the soil and would not have adverse environmental impacts as reflected by the previous discourse on TCE and PCE regulations and publications in the literature. In 1979, an estimated 400,000 gal of septic tank cleaning fluids, most of which contained TCE, benzene, and/or methylene chloride, were used on Long Island alone. In addition to household uses of TCE throughout the 1970s, reliance on evaporation and soil assimilative removal mechanisms of TCE and PCE when applied to the land was underscored by manufacturers' and other industry groups' recommendations at the time. For instance, a 1971 publication by the National Safety Association discussing solvent cleaning of electric motors and machinery discussed waste disposal, and stated that "strong chemical solutions should be preferably disposed of in a dump where they will seep into the earth, and where employees and the public will not be likely to come into contact with them." This type of disposal practice did not contemplate any special measures, beyond the assimilative capacity of soil, to protect groundwater. Recommendations were made by the Manufacturing Chemists Association[21] and various TCE and PCE manufacturers in their material safety data sheets (MSDSs) through at least 1974. It was not until the late 1970s and early 1980s that these disposal recommendations began to change. For example, the MSDSs for PCE and methylene chloride were still recommending placing the solvent on the ground to evaporate as late as 1979. In fact, the 1978 MSDS for PCE contained exactly the same disposal language as the 1974 MSDS.

The use of TCE, PCE, and other chlorinated solvents specifically relating to dry-cleaning establishments have been in place since the 1920s, replacing carbon tetrachloride and petroleum-based solvents because of the relatively low toxicity and flammability of these chlorinated compounds. Prior to 1972, there was little if any environmental concern with PCE and TCE for dry cleaners. It was not until the late 1970s and early 1980s that dry cleaning establishments came under environmental and regulatory scrutiny. The environmental informational assistance and regulatory controls commenced during this post-1972 period through the manufacturers, distributors, trade associations, and state and federal regulatory agencies.

A historical time line for PCE/TCE is presented in Fig. 14.3. When one reviews these time lines, it is the lack of regulation and technical understanding of these chlorinated solvents through periods of extensive use that has resulted in widespread contamination, particularly in the subsurface, caused by these solvents.

It should be noted that the Organic Chemicals, Plastics, and Synthetic Fibers (OCPSF) guidelines established by EPA (40 CFR 414) cover more than 1000 chemical facilities producing over 25,000 end products, such as benzene, toluene, polypropylene, polyvinyl chloride, chlorinated solvents, rubber precursors, rayon, nylon, and polyester. The OCPSF industry is large and diverse, and many plants are highly complex. Some plants produce chemicals in large volumes through continuous chemical processes, while others produce only small volumes of "specialty" chemicals through batch chemical processes.

The OCPSF regulation applies to process wastewater discharges resulting from the manufacture of the products or product groups listed in the rayon fibers, other fibers, thermoplastic resins, thermosetting resins, commodity organic chemicals, bulk organic chemicals, and specialty organic chemicals subcategories.

Treatment Methodologies

The biodegradability of chlorinateds such as chlorinated methanes, chlorinated ethanes, chlorinated ethenes, and other chlorinated compounds have been widely reported.[21] A broad range of chlorinated

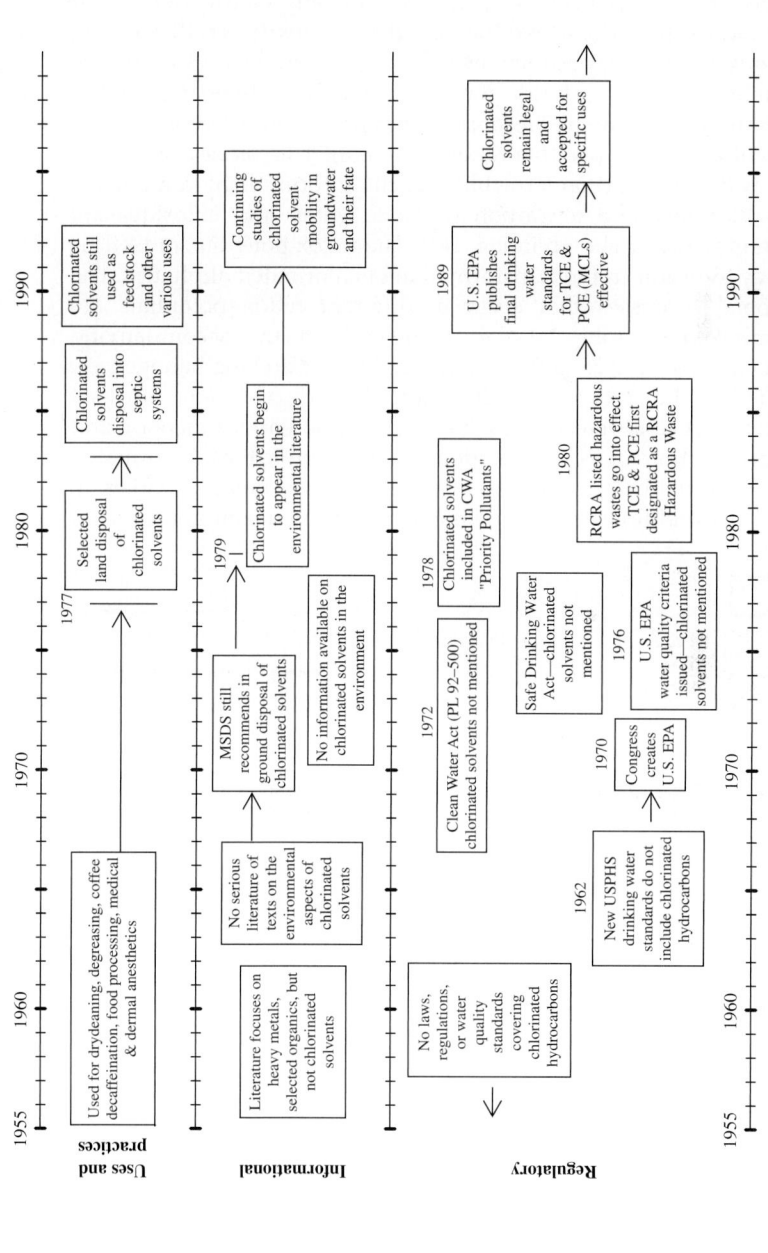

Figure 14.3 PCE/TCE time line.

aliphatic compounds are susceptible to biodegradation, based on the oxidation-reduction potential (ORP or redox potential), cometabolism, and other factors. Under aerobic conditions, the heavily chlorinated compounds such as PCE, TCE, and DCE are only slightly oxidized if at all. However, under anaerobic conditions, reductive dechlorination occurs and daughter products are produced, including dechlorination all the way to vinyl chloride. Vinyl chloride then can be oxidized to ethene and ethane in the presence of ample electron acceptors such as molecular oxygen (positive redox potential). This biotic reaction as well as those of carbon tetrachloride, chloroform, dichloromethane, and chloromethane to carbon dioxide, water, and chlorides are diagrammatically shown in Fig. 14.4. The redox potential, or ORP, is the key indicator in the transformation of chlorinated aliphatics. The microbial processes that occur at different redox potentials are summarized in Table 14.4. It is noted that the accompanying denitrification, iron reduction, sulfate reduction, and methanogenesis results as the redox potential drops. Biodegradation rates of chlorinateds can occur rapidly if microbial populations increase and the combinations of redox conditions are favorable to those chlorinated compounds which are more resistant to aerobic biochemical oxidation. Basically, conditions may allow highly chlorinated compounds to be

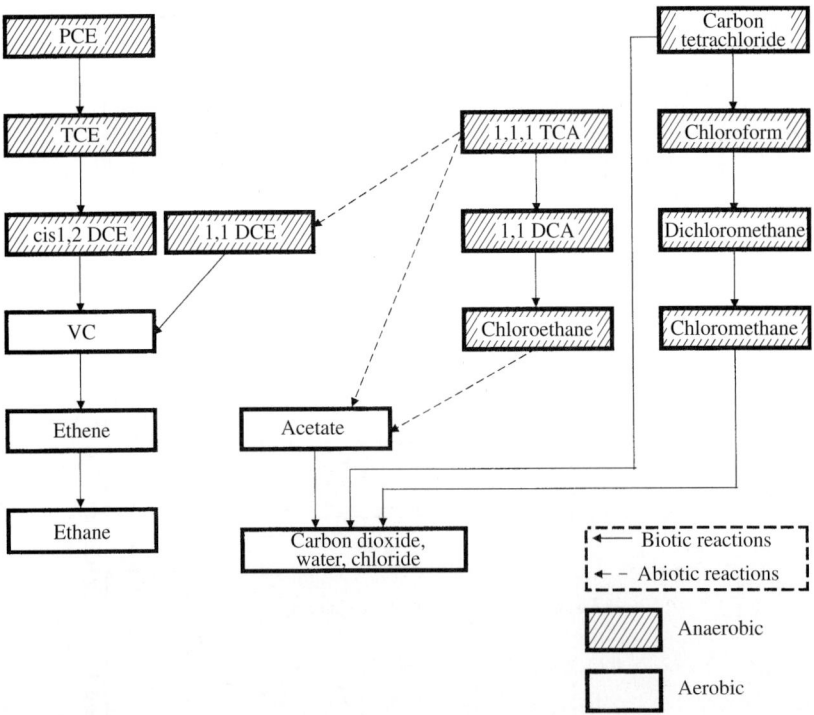

Figure 14.4 Reductive dechlorination of chlorinated compounds.

Chlorinated Compounds, VOCs, and Odor Control

Microbial Process	Electron Acceptor	Products	Eh (mV)
Aerobic respiration	Oxygen	H_2O	+810
Denitrification	Nitrate, nitrite, nitrous oxide	N_2	+50 to –100
Iron reduction	Fe^{3+}	Fe^{2+}	–100
Sulfate reduction	SO_4^{2-}	H_2S	–220
Methanogenesis	CO_2	CH_4	–250

TABLE 14.4 Microbial Processes that Occur at Different Oxidation-Reduction Potentials

partially dehalogenated; if so, the resultant lower chlorinated compounds are more readily oxidized and even mineralized under aerobic conditions.[22]

The *vapor pressure* (the pressure of a compound when its vapor is in equilibrium with its nonvapor phases) relates to the tendency of molecules and atoms to escape from a liquid or a solid. For example, the vapor pressure, which increases with temperature, for volatiles at 25°C is 95.2 mm Hg for benzene and 57.9 mm Hg for TCE, yet only 0.029 mm Hg for acenaphthylene which is much less volatile. More importantly, however, is Henry's constant for a compound (a constant which takes into account the partial pressure of a solute above the solution and the concentration of a solute in solution—provided no chemical reaction takes place between the liquid and the gas). The Henry's constant for benzene is 5.6×10^{-3} atm-m³/mol and 9.1×10^{-2} atm-m³/mol for TCE, while only 10^{-6} atm-m³/mol for a nonvolatile compound like pyrene.

Atmospheric evaporation is a function of vapor pressure, and a factor of chlorinated solvents released to the atmosphere. One predictive equation is:[23]

$$\tau = 12.48 \frac{LP_w C}{EPM} \qquad (14.1)$$

where τ = half-life (time required for compound to evaporate to half its original value) (min)
L = depth of water (m)
P_w = partial pressure of water (mm Hg)
C = concentration of compound in water (i.e., TCE) (mg/L)
E = water evaporation rate (actual rate stipulated at site conditions) (g/m²-d)
P = vapor pressure of compound (mm Hg)
M = molecular weight of compound

Calculations for saturated compound, TCE, in a 1-m-deep pond (25°C) are:

$\tau = (12.48)(1\text{ m})(31\text{ mm Hg})(1200\text{ mg/L})\text{m}^2$

$\times (d/4734\text{ g})(1/59.7\text{ mm Hg})(1/130)(1000\text{ L/m}^3)(g/1000\text{ mg})(1440\text{ min/d})$

$= 18.2\text{ min}$ (14.2)

The calculated half-life for TCE at an initial concentration of 1200 ppm is 18.2 min. The relationship between varying initial concentrations and resultant half-lives is graphically illustrated below (including the remaining half-life concentration). McKays[23] calculated half-lives for other organic compounds as follows.

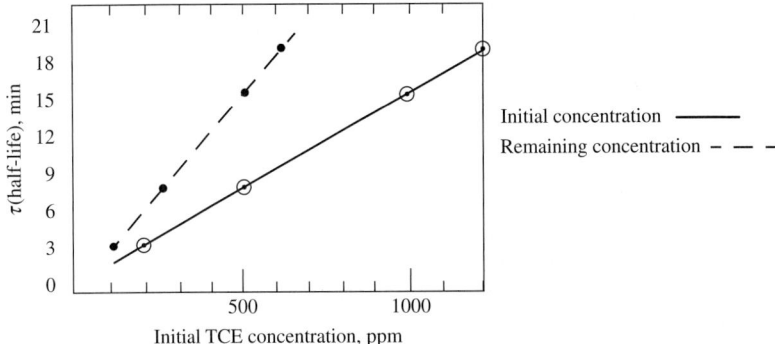

Evaporation parameters and half-life values (τ) for various compounds

Compounds	Vapor pressure (mm Hg)	25°C τ(half-life)
Alkanes		
n-Octane	14.1	3.8 sec
2,2,4-trimethylpentane	49.3	4.1 sec
Aromatics		
Benzene	95.2	37.3 min
Toluene	28.4	30.6 min
o-Xylene	6.6	38.8 min
Cumene	4.6	14.2 min
Napthalene	0.23	2.9 h
Biphenyl	0.057	2.2 h
Pesticides		
DDT	1×10^{-7}	3.7 d
Lindane	9.4×10^{-4}	289 d
Dieldren	1×10^{-7}	723 d
Aldrin	6×10^{-8}	10.1 d
PCBs		
Arochlor 1242	4.06×10^{-4}	5.96 h
Arochlor 1248	4.94×10^{-4}	58.3 min
Arochlor 1254	7.71×10^{-5}	1.2 min
Arochlor 1260	4.05×10^{-5}	28.8 min
Other		
Mercury	1.3×10^{-3}	17.9 min

These half-life approximations, combined with the permeability or seepage rates, can be used to predict the seepage and evaporation fractions of the targeted organic compound. It should be recognized, however, that other variables can affect the actual seepage/evaporation fractions. These variables include but are not limited to biochemical oxidation (or reduction) in the liquid phase, mixing coefficients, and possible sorption in the liquid phase. EPA has developed a model (Water9) which is a wastewater treatment model for estimating air emissions of individual waste constituents in wastewater collection, storage, treatment, and disposal facilities. It includes a database listing many organic compounds and procedure for obtaining reports of constituent fates, including air emissions and treatment effectiveness. The Water9 has been upgraded to include more variables and extend compound properties to more accurately predict VOC emissions. Many, if not most, of these variables are included in Chap. 6.

It should be recognized that in the last 20 years, EPA and state regulatory agencies have increased their focus on air emissions from wastewater collection and treatment systems. This brings into focus the importance for the design engineers to incorporate into their design philosophy an industrial wastewater and collection treatment system that incorporates technologies which minimize VOC emissions into the atmosphere. This includes, for example, covering select tanks with vapor control and recovery facilities, designing aeration systems that supply the required oxygen but minimize air stripping, designing pump and wet well processes with emission control systems, installing scrubber systems as required, and have the automation and monitoring controls necessary to ensure compliance with NESHAPs and other applicable regulations pertaining to air emissions from such treatment plants. For example, one large regional wastewater industrial treatment plant has expended over 30 million dollars solely to meet air emission requirements from selected unit processes.

Granular activated carbon (GAC) has proven extremely effective in treating chlorinated compounds in general and chlorinated solvents in particular. This is based on isotherm tests conducted in the 1980s,[24,25] and subsequently proven throughout the United States, primarily through surface treatment of contaminated groundwater restoration of aquifers used for drinking water. The results based on full-scale case histories are consistent with earlier studies, indicating excellent GAC removal of TCE, TCA, and PCE. Ironically, the higher the percentage of chlorine in the molecule, the apparent higher affinity for being removed by activated carbon.[26]

Subsurface restoration of chlorinated solvent contamination is a major environmental issue resulting from past disposal practices (primarily industrial related) previously discussed. Although the treatment systems are similar to those for surface and point source treatment, the design and infrastructure must be adapted to subsurface restoration projects. Moreover, new factors become increasingly important in predicting the fate and transport of these compounds in the subsurface saturated and unsaturated zones. The fate of these compounds by microorganism transformation takes place by the use of a primary substrate in the more conventional microbial degradation. Cometabolism on the other hand is the beneficial transformation or chlorinateds solvents and other organic compounds by enzymes or cofactors produced by organisms for other purposes.[27-29] Additionally, other compound-related coefficients become important in subsurface restoration:

$$\text{Octanol/water: } K_{ow} = \frac{\text{solute in octanol}}{\text{solute in water}}$$

$$\text{Distribution coefficient: } K_d = \frac{\text{contaminant in soil}}{\text{contaminant in water}}$$

$$\text{Retardation factor: } R = \frac{\text{groundwater velocity}}{\text{solute velocity}}$$

These coefficients assist in the subsurface remediation by predicting the specific compound's affinity for like hydrocarbons (K_{ow}), affinity for the soil matrix as compared to water (K_d), and the mobility as compared to water (R), as shown in Table 14.5.

This information and other coefficients must be utilized in subsurface restoration evaluation projects to predict relative dispersion and movement based on the presence of the cocontaminants for which the target compound has affinity, the quantity that partitions to the soil matrix as compared to groundwater, and the relative velocity of the compound compared to groundwater flow. For example, it is noted the PCBs have a high affinity for any octanol type compounds present, a high affinity for attachment to soil, and are very immobile in groundwater as compared to benzene, PCE, and TCE.[27-29]

There are many enhancements that can expedite and enhance the remediation of chlorinated contamination in the subsurface through in situ remediation. The addition of lactate appeared to enhance the

Coefficient	PCE	TCE	Benzene	PCB
Octonal/water partition: K_{ow}	398	195	135	$1,289 \times 10^3$
Distribution: K_d	2.51	1.23	0.85	8,121
Retardation factor: R	13.54	7.14	5.25	40,604

TABLE 14.5 Coefficients for Selected Compounds

reduction of TCE. Following 8 months of lactate addition, complete dechlorination was occurring within 100 ft of the injection well. Daughter products DCE and VC appeared, accompanied by sulfate reduction and methane production. The ferrous sulfide catalytic process has been used by injecting ferrous sulfide to reduce free sulfides which may be toxic to the dehalogenating bacteria, enhancing the reductive dechlorination process. Oxygen-releasing compounds (ORC) in the subsurface control the redox and enhance aerobic biochemical oxidations. Subsurface hydrogen injection by hydrogen-releasing compounds (HRC) as an electron donor has resulted in enhanced reductive dechlorination. These time release products have been available for several years, and have proven successful in several locations.[30,31] There are countless other methodologies for in situ contamination reduction such as in situ permeable barriers, reactive walls of various varieties, and other enhancement injections. Part of this evolution is the recognition that the "pump and treat" approach (pumping the contaminated groundwater to the surface, treating by stripping, GAC, and so on), then reinjecting or discharging under a NPDES permit, in many cases, is much less cost-effective than in situ enhanced alternatives. However, where large cities depend on contaminated aquifers for their drinking water supply, surface pumping, treatment, and direct input to the water distribution system are the only logical choices. This is particularly true when dense phase nonaqueous liquids (DNAPLs) are present and have the potential to constantly resupply those removed in the pump and treat process.

VC is a key ingredient in the formulation of PVC and a by-product of PCE, TCE, and DCE biodegradation as discussed in this section. The presence of VC in the subsurface can be attributable to surface spills in the VC production facilities and/or the daughter product of PCE, TCE, or DCE transformations. This distinction can take on added importance when allocating cleanup costs to the responsible party involved with production or to other parties which may be using or producing chlorinated solvents with no VC on-site. This can be a function of the subsurface redox potential where the VC is found; that is, high redox potential most likely corresponds to a direct release of VC product, while VC found in an area of low redox potential is most likely a daughter product of the biotransformation of other chlorinated organic solvents.

Vinyl chloride monomer (VCM) traditionally has been produced through the hydrochlorination of acetylene using a mercuric chloride catalyst. Waste products from this process include VCM "heavy ends" (still bottoms) and chloropropene and lesser chlorinated residuals. This production process is no longer used in the United States.

Current VCM production uses ethylene as the primary feedstock which is much cheaper than acetylene. Ethylene is first chlorinated to produce DCA which is then pyrolized to produce VC and hydrogen chloride. This, plus additional ethylene, is oxygenated (oxychlorination reactions). Waste residuals are primarily VCM heavy bottoms, which contain some unreacted VC and various chlorinated by-products. VCM bottoms from either process are listed hazardous wastes under RCRA, and VC is a proven carcinogen. These are the primary reasons that VC may be critical constituent of concern (COC) that drives the remediation remedy for many disposal sites. VC can be oxidized aerobically by acclimated microbes, and can be removed by activated carbon or by more costly means such as incineration with proper emission controls.

14.4 Chlorinated Pesticides

Introduction
In the early twentieth century, agricultural chemicals consisted mainly of insecticides, fungicides, and fertilizers. This has since been expanded to a wide variety of chlorinated and nonchlorinated compounds. The raw materials, primarily synthetic organic pesticides, have been the more prominent environmentally related categories and therefore is the focus of this section.

Regulatory History
There were virtually no pesticide regulations prior to 1910. In that year, federal regulation of pesticide started with the Insecticide Act of 1910 which made it unlawful to manufacture any insecticide or fungicide that was "adulterated or misbranded."[31]

By World War II, synthetic organic pesticides had emerged, demonstrated great effectiveness, and received rapid market acceptance. Organochlorine insecticides, such as aldrin, dieldrin, chlordane, heptachlor, lindane, endrin, toxaphene, and DDT, were highly effective in controlling insects at very low dosages. In 1945, Congress replaced the Insecticide Act with the more comprehensive Federal Insecticide, Fungicide, and Rodenticide Act (FIFRA). FIFRA required that pesticides be registered by the Secretary of Agriculture before their sale or distribution in interstate or foreign commerce.

Under Section 304 of the Clean Water Act, EPA was required to establish "effluent guidelines" for a number of different industrial categories, specifying the effluent limits which must be met by dischargers in each category. Two types of standards were required to be established for each industry: (1) effluent limitations which require the application of the best practicable control technology (BPT) currently available and (2) effluent limitations which require application of the best available technology (BAT).

Effluent limitations reflecting the best practicable technology (BPT) available for the pesticide manufacturing and formulating industrial category were promulgated by EPA in 1978.[32] The effect of the regulations was to require treatment of process wastewater discharged from each of the following point source subcategories: existing manufacturers of organic pesticide chemicals, manufacturers of metallic-organic pesticide chemicals, and formulators and packagers of all pesticide chemicals.

Section 307 of the Clean Water Act required EPA to maintain and publish a list of toxic pollutants, to establish effluent limitations for the best available technology economically achievable (BATEA) for control of such pollutants, and to designate the category or categories of sources to which the effluent standards shall apply. The limitations promulgated must be at a level which provides "an ample margin of safety." The standards (or prohibitions) must be complied with within 1 year after promulgation.

Under the RCRA (1976) regulations, promulgated in 1980, once a waste generator determined that a material was a "solid waste," the next issue is whether it is "hazardous." The subset of solid wastes which are RCRA hazardous wastes are those wastes which are either specifically listed in 40 CFR 261, or else exhibit one of the four "characteristics" (ignitability, corrosivity, reactivity, and "extraction procedure" [EP] toxicity) identified in Part 261.

Both the characteristics and the lists swept many pesticides and pesticide wastes into the RCRA regulatory program. The EP-toxicity characteristic captures numerous pesticides, because it requires analysis of a material to identify whether 14 enumerated constituents are present at specified concentrations. Out of the 14 current constituents, 6 are pesticides (endrin; lindane; methoxychlor; toxaphene; 2,4-D; 2,4,5-TP). Concentrations range from 0.02 (endrin) to 10.0 mg/L (2,4-D), except that barium's limit is 100 mg/L. Eight metals are also included (arsenic, barium, cadmium, chromium, lead, mercury, selenium, and silver).[32]

The general time lines for pesticide regulations are tabulated in Table 14.6 and graphically described in Fig. 14.5.

Pesticide Characterization

The EPA development document for the pesticide chemicals manufacturing sector used three subcategories for the purpose of regulation, namely:

- Organic Pesticide Chemicals Manufacturing
- Metallo-Organic Pesticide Chemicals Manufacturing
- Pesticides Chemicals Formulating and Packaging

The insecticides were overwhelmingly chlorinated organics while the balance of organics were organophosphorus in nature. The metallic-organic pesticides were mercuric and arsenic hydrocarbons, and the third category covered plants which only purchased the

1910	Federal Insecticide Act—regulation was a low priority—primarily directed toward ineffective products and deceptive labeling. No* federal registration requirement or significant safety standards.
1947	Congress passes Federal Insecticide, Fungicide, and Rodenticide Act (FIFRA)—registration required with U.S. Department of Agriculture for interstate commerce and a initial labeling provision. Labeling requirements did not address health issues.
1964	Secretary of Agriculture given power for the first time to cancel labeling registration, if necessary, to "prevent an eminent hazard to the public."
1970	Creation of U.S. EPA. EPA incorporates personnel from Departments of Interior, Agriculture, and other governmental departments. The USDA Pesticide Division is transferred to EPA.
Prior to 1972	Federal government imposed no requirement on the actual use of a pesticide once it had left a manufacturer or distributor if properly labeled.
1972	EPA investigates pesticide literature and the waste disposal practices of the pesticide industry. The Pesticide Manufacturing Industry—Current Waste Treatment and Disposal Practices: Without more information on waste stream quality, toxicity, and flows, it is impossible to isolate areas of the pesticide industry where the problems may be more pressing—EPA and the University of Texas.
Prior to 1975	No drinking water standards on any pesticides in the United States, World Health Organization (WHO), or international community.
1975	Interim drinking water standards established by EPA for endrin, lindane, methyoxychlor, toxaphene, 2,4-D, and 2,4,5-TP (Silex) (not DDT).
1975–1988	FIFRA Amendments.
1978	Development Document for Effluent Limitation Guidelines—EPA
1982	Congress passes Federal Environmental Pesticide Control Act (FEPCA) for the Pesticide Chemicals Manufacturing—Point Source Category.
1982	Proposed Development Document for Effluent Limitations Guidelines and Standards for the Pesticide Point Source Category—EPA.
1984	World Health Organization establishes drinking water standards for selected pesticides, including DDT.
2001	No EPA drinking water standard for DDT.

*Government Affairs Institute, Inc. *Environmental Law Handbook*, 10th Edition (1989).

TABLE **14.6** General Time Lines for Pesticide Regulations

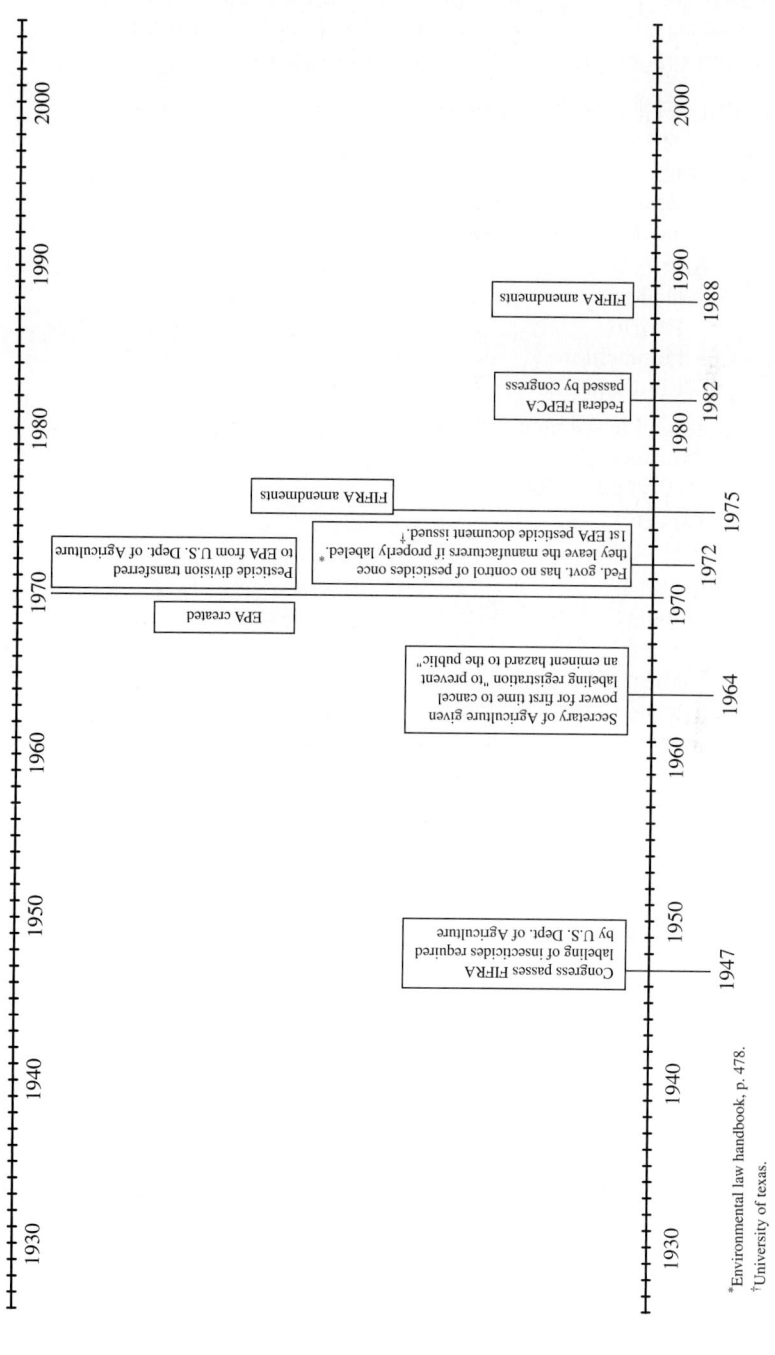

FIGURE 14.5 General time line for U.S. pesticide regulations.

insecticide/pesticide chemicals from the manufacturer, then formulated and packaged these products with a recipe devised by entomologists tailored to the local pestilence and soil conditions. Typical formulation plants included the following formulations:

A. Chlorinated Hydrocarbons
    DDT
    BHC
    Toxaphene
    Chlordane
    Aldrin
        Dieldrin
        Endrin
        Heptachlor
        Rhothane

B. Organic Phosphates
    Parathion
    Methyl parathion
    Vapo-tox
    Systox

C. Fungicides
    Dithane
    Parzate
    Manzate
    Copper
    Sulfur

D. Arsenates
    Calcium arsenate
    Lead arsenate
    Paris-green
    DSMA
    MSMA

E. Carbamates
    Sevin

For example, most of the formulation would contain a 5 to 10 percent DDT component, which were top sellers up through the early 1970s.[33] DDT was one of the most effective and controversial chlorinated hydrocarbons. DDT was the primary pesticide used to control insects in agriculture such as those which carry malaria. It is a white crystalline solid with no odor or taste. Its use in the United States was restricted in 1972 but is still used in other countries and is accepted by the World Health Organization and the U.S. Agency for International Development. (It should be noted that the resurgence of malaria deaths and the outbreak of West Nile virus in the United States has resulted in rethinking the use of DDT.) DDT remains a significant environmental problem in the United States today because of its widespread use by the military and agriculture interests in the 1950s through the 1970s. Residuals persist in such locales today and require cleanup and remediation. The time line for DDT production and application is presented in Table 14.7. Properties of the selected "contaminants of concern" for many remediation projects of such sites today are outlined in Table 14.8.

1939	Dr. Paul Muller discovered the ability of DDT to kill flies, mosquitoes, and Colorado potato beetles.
1943	Geigy Corp. patents DDT in the United States.
1943–1945	DDT used by the U.S. Army to dust soldiers and refugees for lice control.
1948	Dr. Muller wins Nobel Prize for his work on DDT.
1955	Eighth World Health Assembly adopts a Global Malaria Eradication campaign based on the use of DDT to treat malaria and parasite infestation in humans.
1962	Peak U.S. DDT production—176 million pounds.
1967	All developed countries where malaria was endemic and large areas of tropical Asia and Latin America were freed from the risk of infection.
1970	The National Academy of Science concluded that "To only a few chemicals does man owe as great a debt as to DDT. In little more than two decades, DDT has prevented 500 million deaths, due to malaria, that otherwise would have been inevitable."*
1970	EPA formed.
1971–1972	EPA held extensive hearings and administration law Judge Edmund Sweeney concluded that "DDT is not a carcinogenic hazard to man."†
1972–1973	DDT labeling registration cancelled and distribution of and use of DDT for most applications in the United States is prohibited.
1972–present	DDT manufacturing in the United States and/or other countries continues.
2001	No DDT drinking water standards have been established by EPA.

*National Academy of Sciences, Committee on Research in the Life Sciences on the Committee on Science and Public Policy (1970).
†Sweeny, E. M., EPA Hearing Examiners Recommendations and Findings Concerning DDT Hearings, 40CFR164.32, April 25, 1972.

TABLE 14.7 General Time Line for DDT Production and Application

Category	Compound	Primary Residulas	Properties					
			Molecular Weight	Vapor Pressure (mm Hg)	Solubility (ppm)	Partition Coefficient	Toxicity	Biological Degradability
Insecticide	**DDT** $C_{14}H_8Cl_5$ (dichlorodiphenyl trichloroethane)	Unreacted feedstocks Chloral Chlorobenzene Sulfuric Acid	355	1.9×10^{-7}	0.003	263×10^4	48 h LC (Daphnia) 0.36 ppb	Resistant (aerobic), Moderate (anaerobic)
DDT daughter compound	**DDD** (dichlorodiphenyl dichloroethane)	Reaction product Daughter compound of DDT			0.09	1×10^6		Resistant
DDT daughter compound	**DDE** (dichlorodiphenyl dichloroethylene)	Reaction product Daughter compound of DDT	318		0.12	4.5×10^6		Resistant
DDT intermediate	**Chloral** CCl_3CHO (trichloroacetaldehyde)	Unreacted feedstocks Chlorobenzene Sulfuric acid	147		1×10^6	29	1.6 ppm bacteria	Moderate
Insecticide	**Lindane** $C_6H_6Cl_6$ (gamma) (hexachlorocyclohexane)	Unreacted feedstocks and unsaleable isomers (about 85% of product) Hexachlorobenzene 516 ppb	291	9.4×10^{-6}	7.3		48 h LC_{50} (Daphnia) 460 ppb bacteria 5 ppm	Moderate

Category	Name	Byproducts/Feedstocks	MW				Toxicity	Resistance
Lindane by-product	**Hexachlorobenzene** (phenyl chloride)	HCB	285	1.08×10^{-5}	0.006	5.5×10^4	14 d LC (fish) > 0.32 ppb	Moderate
Herbicide	**Arsenates** (monosodiummethanearsenate) $CH_3\,ASO(OH)(Na)$	Unreacted feedstocks Arsenic oxides Other arsenic compounds methylchloride methanol					Bacteria 13–65 ppm	Resistant
Herbicide	**Dacthal** $C_6\,C_4\,(COOCH_3)_2$ (dimethyltetrachloroterephthalate)	Carbon tetrachloride triphenol phosphate Sodium hydroxide Ammonia hydrochloric acid methanol	304	1.25×10^{-5}	0.5	21×10^3	48 h LC_{50} (fish) 700 ppm	Resistant
Fungicide	**Deconil** $C_6Cl_4\,(CN)_2$ (tetrachloroisophthalonitrile)	Unreacted feedstocks Ammonia isophthalonitrile (IPN), Carbon tetrachloride Xylene, hydrochloric acid Sulfuric Acid	266	0.01	100			Resistant

TABLE 14.8 Chemical and Environmental Properties of Contaminants of Concern

Category	Compound	Primary Residulas	Properties					
			Molecular Weight	Vapor Pressure (mm Hg)	Solubility (ppm)	Partition Coefficient	Toxicity	Biological Degradability
Feedstock	Caustic chlorine production (Mercury cell process)	Mercury PCBs (anode preservative) Caustic soda						Resistant
Feedstock	Chlorobenzene		113	8.8	490	219		Degradable
Feedstock	Carbon tetrachloride		154	91.3	800	174		Moderate
Feedstock	Methylene chloride		85	350	13,200	11.7		Degradable
Feedstock	Xylene		106	9	198	374		Degradable

TABLE **14.8** Chemical and Environmental Properties of Contaminants of Concern (*Continued*)

Treatment Methodologies

Extensive research on waste treatment and disposal practices in the pesticide manufacturing industry has been conducted since the 1970s.[34-36] The early candidate treatment methodologies focused on GAC (again, because of the affinity of activated carbon for chlorinated hydrocarbons), hydrolysis, chemical oxidation, and biological oxidation. Activated carbon results indicated 50 to 95 percent removal. Hydrolysis at pH values of 10 to 12 was similarly successful as was chemical oxidation.

The success of biochemical oxidation has been more problematic, depending primarily on the molecular configuration of the chlorinated compound and the ability of the microbial population to properly acclimate. The biodegradation of the chlorophenols and chlorobenzenes has been more successful using high-rate biological systems. Incineration at 1000°C with adequate excess air has been shown to achieve 99 percent destruction of organic pesticide.[36]

14.5 Perchlorates

Introduction

Perchlorates (ClO_4) and perchlorate-containing chemicals were heavily produced beginning in the 1940s as an oxidizer component and primary ingredient in solid propellants for rockets, missiles, and fireworks. Additionally, perchlorates are used as a component of air bag inflators.[36] Because of limited shelf life, perchlorates relating to missile and rocket inventory must be removed and replaced. Therefore, large volumes have been disposed of in Nevada, California, Utah, and other states since the 1950s. Since 1997, perchlorates have been discovered in 35 states' sources of drinking water, mostly in the 4 to 100 ppb range and higher. Time lines for Perchlorate Awareness and Regulation are shown in Table 14.9.

Treatment Technologies

Perchlorates are nonvolatile and highly soluble in water and, therefore, cannot be removed from groundwater by the conventional methods of sedimentation, filtration, and air stripping.

Ion exchange and membrane processes can remove perchlorate from water. Ion exchange is being evaluated in California's San Gabriel Valley for removing approximately 30 to 200 ppb perchlorate from groundwater. Ion exchange resins that can selectively remove perchlorate rather than competing ions (e.g., chloride, sulfate, bicarbonate) that may be present at higher concentrations are required.

Nanofiltration and reverse osmosis will also remove perchlorate, but these technologies are costly. Ion exchange and membrane

1940s	Production of ammonium chlorates begins
1950s–present	Ammonium chlorates and other perchlorate salts used in a wide range of applications, including pyrotechnics, blasting agents, matches, lubricating oils, textile dye finishing, nuclear reactors, electronic tubes, electroplating, automobile air bag inflators, paint and enamel production, pharmaceuticals, and aluminum finishing
1997	Analytical detection method developed to detect quantification level of 4 ppb
1998	EPA placed perchlorate on its Contaminant List for possible regulation
1999	EPA requires drinking water monitoring for perchlorate under the Unregulated Contaminant Monitoring Rule
2001	Publication—Roote, D. "Technology Status Report—Water Remediation Technologies," Analysis Center, U.S. EPA Press Release
2005	EPA established its official dose of perchlorate at 0.0007 mg/kg/d and translated that number to a drinking water equivalent level (DWEL) of 24.5 ppb
2006	State of California proposes a primary drinking water standard (MCL) of 6 ppb
2007	State of California requires this proposed regulation be submitted to the office of Administrative Law
2007	No federal drinking water standard (MCL) yet established for perchlorate

TABLE 14.9   Time Lines for Perchlorate Awareness and Regulation

processes also generate concentrated perchlorate-containing waste brines that may be difficult to dispose. Treatment of the brine may be needed to lower its volume or toxicity before disposal. Ozone-peroxide treatment has minimal effect on perchlorate in water. Ozone-peroxide followed by GAC, however, has been found promising at one site in the San Gabriel Valley, but additional studies are needed to evaluate the long-term effectiveness, reliability, and cost.

Most attention to date has been focused on developing an anaerobic or anoxic biochemical reduction process, whereby microbes convert perchlorate to a less toxic or innocuous form in the absence of molecular oxygen. The Air Force Research Laboratory, Materials and Manufacturing Directorate began developing biochemical reactor systems for treating high-level perchlorate-contaminated wastewater (i.e., 1000 to 10,000 ppm) in the early 1990s. A continuous-stirred-tank-reactor system began treating wastewater from rocket motor production operations in Utah in 1997. Additional pilot tests have been conducted to evaluate removal of low-level perchlorate contamination, and although results appear promising, additional studies are needed to evaluate the cost, reliability, and public acceptance of this technology.[37] In situ removal of perchlorates by bioremediation in soil and groundwater continues to be the subject of extensive study and research.[38–41]

Geosyntec conducted studies for Aerojet General in California where soil and groundwater were contaminated with perchlorate resulting from handling and use of solid rocket motors. Simulated aquifer microorganisms were tested under various control mechanisms such as

1. No added election donor or carbon source
2. Simple electron donor (ethanol)
3. Complex electron donor (molasses)
4. Bioaugmented treatment

This study concluded that rapid biodegradation took place, the main factors being proper redox conditions and electron donor selection.[38] Microbiological studies further investigated and isolated the microorganisms which were responsible for the perchlorate reductions. All of the identified isolates could couple the oxidation of acetate to the reduction of perchlorates. These were motile gram-negative, nonfermentative, facultative anaerobes. In addition to acetates, lactates, ethanol, propionate, and succinate to serve as alternative electron donors, oxygen and nitrate served as alternative electron acceptors.[39]

A major remediation project for the Metropolitan Water District of Southern California was undertaken to demonstrate biochemical reduction technology to reduce chlorate concentrations in the 8000 to 10,000 ppb range to 100 ppb or less. The resultant technology included a fixed film bioreactor using GAC operated as a fluidized bed.[40] Another study on the subject supported the results of perchlorate serving as the electron acceptor for a number of different bacterial strains. Particular emphasis was directed toward the use of hydrogen-oxidizing bacteria to support the microbial reduction of perchlorate.[42]

14.6 Other Miscellaneous Chlorinated Organics

There are many other chlorinated compounds that deserve attention. Those that are often associated with specific industrial production and resultant wastewaters include dioxin pentachlorophenols and hexabetadine. A brief discussion on each of these that are selected by the authors are presented as follows:

- **Dioxins** Dioxin (2,3,7,8-tetrachlorodibenzo-p-dioxin, or TCDD) has been a more recent target compound of EPA and the State regulatory agencies. *Dioxin* is a term used to identify several chemical compounds in closely related families, namely, compounds which are chlorinated singly or any multiple at any of eight positions. Dioxin theoretically includes 75 different species (congeners), each with different physical, biochemical, and toxicological properties. Other related compounds include polychlorinated dibenzofurans (CDF), and the previously discussed polychlorinated biphenyls (PCBs).[43] Toxic characteristics of these chemicals vary significantly from those of TCDD.

  TCDD and related compounds are always generated as unintended by-products of reactions of organic chemicals at high temperatures when chlorine is present. They are not and never have been a desired component of a chemical product. Dioxins have been associated with the pulp and paper industries, PCB-related production, catalyst regeneration process, incineration of chlorinated compounds, and defoliant production. Regulation for TCDD has been evolving for the past 15 to 20 years, with water quality limitations in the part per quadrillion range. Expert opinions still differ on TCDD and its congeners relative to the issues of standards, health effects, and carcinogenicity. Phaseout of compounds which are involved in the potential of TCDD by-products, incineration controls, and activated carbon removal are a few of the control measures available.

- **Pentachlorophenol** Pentachlorophenol, C_6H_5OH (PCP), has wide application in the organic chemicals industry, pesticides production, and the wood products industry. Commercial PCP contains significant quantities of tetrachlorophenol (TCP). It is subject to biodegradation and has a relatively low water solubility and vapor pressure. As of the date of the referenced publication, it is a probable but not confirmed carcinogen.[44]

  As with most organic chlorinated compounds, it is susceptible to activated carbon removal often combined with biological degradation.

- **Hexachlorobutadine** Hexachlorobutadine (C_4Cl_6) is used primarily in rubber products, hydraulic fluids, lubricants, and heat transfer fluids. It has a moderately high vapor pressure. As of the date of this reference it is a probable but not confirmed carcinogen.

14.7 Chlorinated By-Products, Other VOCs, and Odor Control

There are many chlorinated hydrocarbons that are by-products of chlorination, attributed to either the disinfection for domestic water treatment plants or available chlorine oxidants in industrial intake or effluent waters. These include, but are not limited to, the following:

- **Trihalomethanes** Trihalomethanes, first coming into focus in the late 1970s, have been observed when lakes or reservoirs containing naturally occurring organics primarily form decaying vegetation, which react with chlorine, used to treat drinking waters. This "disinfection" by-product (THM) is one of the most commonly detected reactants in POTWs and in drinking water supplies. Since 1993, concentrations of THMs have been detected in the range from 0.04 to 0.1 ppm. They also have appeared in industrial waters from refrigerant and solvent origins. EPA has set a drinking water maximum contaminant level (mcl) at 80 ppb (0.08 ppm) for total trihalomethanes (TTHM). (This has led to alterations in disinfection processes, including step chlorination/dechlorination, and/or substituting chlorinating disinfection with alternative processes such as ozonation.) Moreover, as with most chlorinated organic compounds, removal can be accomplished by several methods, primarily activated carbon.

- **Haloacetic acids** Haloacetic acids (HAA) such as chloroacetic acid, bromoacetic acid, dichloroacetic acid, and other sister compounds may cause a problem, particularly in water treatment plants. An mcl for drinking water limitation was established by EPA in 2004 at a level of 60 ppb (0.06 ppm). Reduction of HAA can be achieved by several methods, namely, pretreatment by chemical oxidation, such as hydrogen peroxide, or removal in the water treatment plant system using activated carbon. For example, one of the authors reduced the HAA to below drinking water standard by using a GAC substitute medium in the conventional filtration plant, adsorbing the HAA compounds. As the concentrations were low and the carbon capacity high, the spent carbon in the filters had to be changed infrequently and the process was cost-effective.

- **Bromochloromethane** This compound may also appear in drinking water or industrial waters. If the concentration is excessive, then reduction can be obtained by activated carbon application or other effective removal processes.

- **Methylene chloride (dichloromethane)** This compound is highly volatile, and used commonly in industry as a solvent, paint stripper, fumigant, and sometimes enclosed in Christmas tree lights. When industrial waters require concentration reduction, for example, to meet NPDES or bioassay requirements, activated carbon and/or biochemical oxidation or stripping/vapor recovery are candidate systems.

- **Methyl chloride (chloromethane)** This compound is often produced synthetically in industrial applications, used primarily as a chemical intermediate or a methylating and chlorinating agent in organic chemical applications. It is also used as an extractant for oils and a propellant in polystyrene foaming production. It is subject to removal by activated carbon, biochemical oxidation, and/or stripping/vapor recovery.

- **Chloroform (trichloromethane)** This compound has had many previous uses, such as an anesthetic and solvent, although today a primary use is in refrigerants. Because of its potential human toxicity potential and propensity for ozone depletion, it is not widely used currently. It is highly volatile, and subject to stripping/vapor recovery as well as activated carbon removal.

There are two major considerations involved with many industrial wastewater treatment facilities, specifically, odor control and VOC stripping removal and control. The two are not mutually exclusive, and VOCs may be associated but masked by odorous gases such as hydrogen sulfide.[45,46] There are, however, some differences in odor and VOC control technologies. As a cross reference, sour water strippers are covered in Chap. 3 (Sec. 3.5) and air stripping of VOCs is cited in Chap. 5 (Sec. 5.3). It is the purpose of this section to complement and add to the previous citations in Chaps. 3 and 5 as well as the first sections of this chapter.

Odor Control[47–49]

Odors can be generated in wastewater treatment facilities when dissolved gases come into contact with air and are stripped out of the liquid by agitation or aeration. The most common odorants are hydrogen sulfide, ammonia, and sulfonated organics such as mercaptans. There are three major issues involved with these compounds, namely noxious odors, corrosion, and safety. Hydrogen sulfide, for example, has an odor threshold at approximately 0.1 to 0.2 ppm, strong health effects in the 10 to 400 ppm range, and possibly fatal at the 500 to 1000 ppm level.

The effects of pH on the hydrogen sulfide concentration is shown in Table 14.10, and the effects of hydrogen sulfide gas on humans is included in Table 14.11.

pH (1)	Percentage of H_2S (2)	Percentage of HS (3)	Solubility (mg/L) (4)
4	99.9	0.1	3,470
5	98.9	1.1	3,510
6	90.1	9.9	3,840
7	47.7	52.3	7,270
7.5	22.5	77.5	15,400
8	8.3	91.7	41,800
8.5	2.80	97.20	124,000
9	0.89	99.11	390,000
10	0.09	99.91	

TABLE 14.10 Percentages of H_2S and HS- and Solubility of Hydrogen Sulfide as a Function of pH

Concentration (in) ppm by Volume	Effect
<0.00021	Olfactory detection threshold
0.00047	Olfactory recognition threshold
0.5–30	Strong odor
10–50	Headache, nausea, and irritation of eye, nose, and throat
50–300	Eye and respiratory injury
300–500	Life threatening (pulmonary edema)
700 or more	Immediate death

TABLE 14.11 Effects of Hydrogen Sulfide Gas on Humans

Sulfides occur primarily when high sulfate effluents are subjected to anoxic or anaerobic conditions such as stagnation in rivers, wet wells in lift stations, anaerobic treatment processes, and septicity in clarifiers, internal piping, and low pH wastestreams. In petroleum refineries, for example, sour crude and crude nitrogen produces ammonia and sulfides through process severity such as desulfurization, denitrification, hydroprocessing, and others. As stripping of sulfides occurs at low pH ranges and ammonia stripping occurs in the higher pH ranges, two-stage stripping at controlled pH levels is normally required.

There are several candidate systems which can be considered for sulfide-related odor control. These include:

- **Chemical scrubber systems** Scrubbers are designed to remove sulfide odors by oxidizing sulfur from its lowest oxidation state (−2) to its highest oxidation state (+6). Scrubbers accomplish this by bringing sulfide-laden air in contact with an aqueous solution of strong oxidizing agents such as an alkaline solution of sodium hypochlorite. The scrubber is a tall cylindrical vessel packed with polymeric packing material. It is a countercurrent operation, with a strong aqueous solution of sodium hypochlorite and sodium hydroxide introduced at the top of the vessel above the packing, with the sulfide gases introduced at the bottom. While flowing up, the gas comes in contact with the liquid solution flowing down by gravity. The mixing action through the interstices of the packing result in the following overall reaction.

$$H_2S + 4NaOCl + 2N_2OH \rightarrow Na_2SO_4 + 4NaCl + 2H_2O \quad (14.3)$$

$$(E° = 587 \text{ mV})$$

Makeup of the alkaline feed is required as the reactions take place, and a blowdown of the reactant salts is required to prevent their accumulation in the system. This requires a continuous flow of makeup water (to prevent fouling of the packing system) to chemical scrubbers, which are commonly used in POTWs, industrial pretreatment systems, and other applications as applicable.

Scrubber systems should be complete with redundancy, pH and ORP probes with feedback controllers to control pH, and ORP of the circulating liquid by modulating the chemical feed metering pumps and gas phase H_2S analyzers.

- **Air/oxygen injection** If an aerobic biological treatment plant is available in the wastewater treatment complex, odor control can be accomplished by directing the gas stream and injecting it directly into the aeration tank. Biochemical oxidation is thereby used to oxidize the sulfur compounds. There are several precautionary factors, however. First, the aeration capacity must be available to satisfy this additional demand; second, the sulfide loading should not be excessive to the point where unoxidized sulfide or ammonia compounds escape the aeration basin; and third, the proliferation of sulfur oxidizing bacteria do not result in sludge bulking. Oxygen injection is an alternative as the oxygen transfer efficiency is higher and can be injected at a controlled optimal rate. It is more expensive, however.
- **Biofilters** are biological systems using biomass on a fixed media. Odorous gases pass through the media where biochemical oxidation occurs. The presence of VOCs may

adversely affect the efficiency of biofilters and should be evaluated before installation.

- **Miscellaneous chemical additives** Ozone is a highly energized form of oxygen which is generated on-site. The odor oxidation reaction is:

$$3H_2S + 4O_3 \rightarrow 3H_2SO_4 \qquad (14.4)$$

The generation of ozone is energy-intensive and not generally cost-effective. Hydrogen peroxide is more intensively used in odor control and is a strong but unstable oxidant. It reacts quickly, but in the absence of reduced chemicals, it decomposes to water and oxygen. It can add dissolved oxygen to an aerobic biological reactor or an oxygen starved wastewater stream as a temporary additive. Potassium permanganate effectively destroys hydrogen sulfide, as shown in the following equation:

$$3H_2S + 8KMnO_4 \rightarrow 3K_2SO_4 + 2H_2O + 2KOH + 8MnO_2 \qquad (14.5)$$

Sodium hypochlorite in a solution form also is effective, although it is potentially corrosive and unstable. If VOCs are present, this approach may not be applicable as there can be an organic chlorination reaction from chloromines or other forms of chlorinated organics. The addition of nitrates can suppress sulfide odors, as shown by the following redox reaction.

$$5HS^- + 8NO_3^- + 3H^+ \rightarrow 4H_2O + 4N_2 + 5SO_2^{-2} \qquad (14.6)$$

It is a safe and effective approach but unreacted NO_3^- in the water could be a drinking water problem as well as a eutrophication causing nutrient. Direct addition of ferrous and ferric salts can be added on a continuous or an emergency basis, particularly as a backup when chemical scrubbers are out of service.

$$FeCl_2 + H_2S \rightarrow FeS + 2HCl \text{ (ferrous salts)}$$

or

$$2Fe^{3+} + 3HS^- \rightarrow 2FeS + S^\circ + H^+ \text{ (ferric salts)} \qquad (14.7)$$

Ferric is normally recommended but the wastewater needs to be sufficiently buffered to offset the added acidity.[1,2]

Odor control systems should be carefully monitored for H_2S and oxidation reduction potential (ORP). It is not uncommon to have property line (fence line) H_2S monitors to confirm and control odor control systems so that off-property odors are not prevalent.

- **Volatile organic compound (VOC) removal and control (odorants or nonodorants)** VOCs may or may not be an odorant, but their presence has other environmental implications. The previously discussed odor control systems may not be applicable in removing VOCs, so VOC recovery and control is basically a separate regulatory and technological issue. Process areas in an industrial complex subject to vapor control and recovery are numerous. For example, there are standards of performance for VOC emissions from industrial wastewater treatment plants such as petroleum refineries under National Emission Standards for Hazardous Air Pollutants (NESHAPs, Petroleum Refinery, 40 CFR 60 Subpart QQQ). In a petroleum refinery, for example, oil separators, slop tanks, storage tanks, equalization basins, and aeration systems possibly require VOC controls, such as condensate return, activated carbon adsorption, and other vapor removal systems.

    Systems include the ventilation and capture of VOCs under fixed or floating roof tanks, covered separators, sumps, and other related process, processing the vapors through activated carbon canisters or routing the VOC gases to flares or incinerators. Some examples include one organic chemical complex which strips and collects VOCs in an upstream production area; so, VOCs are not an issue in the wastewater treatment system. One petroleum refinery has vapor controls on all the potential VOC sources of concern, including the covering of the biological aeration basin.

    A preferred approach on VOC controls, however, is to reduce VOCs upstream to a level that aeration basin covering is not necessary. For example, an older refinery was able to reduce VOCs in a pretreatment "BTEX" biological reactor with minimal turbulence, reducing these compounds (all of which are easily biodegradable) to the level that air stripping in the downstream open aeration system was not an emission problem. However, primary clarifiers, wet wells, equalization basins, knockout drums, open sumps, and even manholes might be subject to installing vapor control and recovery systems, using activated carbon canisters or other removal/recycle/control devices.

As aerobic biological treatment is a critical unit process in industrial treatment plants as well as POTWs receiving significant amounts of industrial wastewaters, the removal of VOCs by air stripping in the aeration basin should be the subject of investigation.

Chlorinated Compounds, VOCs, and Odor Control

The ways to mitigate this potential is prestripping and recovery of VOCs upstream, and/or using upstream aeration systems which can satisfy the dissolved oxygen requirement for biochemical oxidation in the aqueous phase while minimizing turbulence which enhances the stripping process. The maximum possible removal of a compound of a diffused aeration system at equilibrium can be estimated by assuming that the air leaving the process is saturated with the compound in accordance with Henry's law. The following equation can approximate this loss:[50,51]

$$(1 - S_e/S_o) = 1 - [1/1 + (Q_A/Q_L)H_c] \quad (14.8)$$

where S_e = effluent concentration of the compound
S_o = initial concentration of the compound
Q_A/Q_L = air to water volume ratio
H_c = Henry's constant for the compound(s)

$$H_c = \frac{\text{Pollution concentration in the gas phase}}{\text{Pollution concentration in the liquid phase}}$$

An example of this calculation is presented as follows:
Using equation cited above:

$$(1 - S_e/S_o) = 1 - [1/1 + (Q_A/Q_L)H_c] \quad (14.9)$$

where Hc for BTEX approximates 0.24 at 20°C and 0.15 at 10°C
Water flow = 5000 gpm
Air flow = 240 cfm with 2 operating aerators
480 cfm with 4 operating aerators
720 cfm with 6 operating aerators

(2) $Q_A/Q_L = (240 \text{ ft}^3/\text{min})(7.48 \text{ gal/ft}^3) = 1795/5000 = 0.359$

(4) $Q_A/Q_L = (480 \text{ ft}^3/\text{min})(7.48 \text{ gal/ft}^3) = 3590/5000 = 0.718$

(6) $Q_A/Q_L = (720 \text{ ft}^3/\text{min})(7.48 \text{ gal/ft}^3) = 5386/5000 = 1.077$

and

	20°C	10°C
$(1 - S_e/S_o)_2 =$	8.0%	5.1%
$(1 - S_e/S_o)_4 =$	14.7%	9.7%
$(1 - S_e/S_o)_6 =$	20.5%	14.0%

Note: Q_A for surface aerators based on vendor information.

The results of the estimated fraction of BTEX stripped in a mechanically aerated basin for 2, 4, and 6 operating surface aerators at 10°C and 20°C liquid temperature are plotted in the following graph. (As previously noted, the BTEX stripped can be reduced significantly upstream, lowering such emissions from air aerobic basin to acceptable levels without vapor controls, i.e., reduction of the S_o component.)

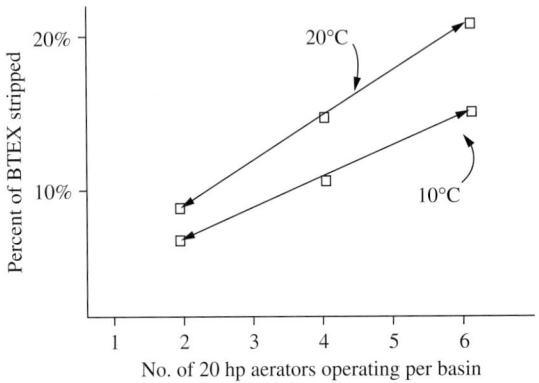

14.8 Problems

14.1. A stripping tower is 30 ft in height. The following design factors are:

L_v (volumetric liquid rate) = 0.07 ft/s
A (cross sectional area) = 160 sq ft
S (stripping factor) = 4.7
A_w (volumetric to air ratio) = 20
K_{La} (transfer coefficient) = 0.015
(See Chap. 5)

Determine the effect of a changing air flow rate on the stripping efficiency of the subject solvent (use 4000 sctm to 16,000 sfm). State your conclusion as to which air flow rate is optimal based on energy efficiency.

14.2. A confined aquifer has the following characteristics:

TCE	138 ppb
PCBs	12 ppm
Benzene	820 ppb
Ammonium perchlorate	920 ppb

It is determined that pump and treat is the only practical remediation candidate as the surface treated water will go into a public drinking water storage and distribution system.

Prepare a conceptual flow diagram for surface treatment which will meet drinking water MCLs. Discuss the rationale for the design.

14.3. An in situ long-term remediation program for a contaminated disposal site has been determined to be the best approach based on the EPA record of decision (ROD). Recommend an in situ solution based on the following data:

Zone of contamination depth   20 to 65 ft
Vadose zone   20 to 32 ft
Saturated zone   32 to 65 ft
Redox potential, vadose zone   +820 mV
Redox potential saturated zone   −210 mV
Contaminants of concern (COCs)
Vinyl chloride
DCE
DCA
Perchlorates

Assume cleanup criteria is 10 times drinking water standards (MCLs).

References

1. Kembrough, R. D.: "Human Health Effects of PCBs and PBBs," *Pharmacological & Toxicological Journal*, 27:87-11, 1987.
2. Ford, D. L.: "A Review of Historical Waste Disposal Practices and Manufacture of Polychlorinated Biphenyls at the Monsanto Plant, Anniston, Alabama, September 1998," February 2003.
3. Ford, D. L.: Search of PCB literature sources, general files, and RREL Database, Version No. S.O., CAS No. 53469-21-9, 1996.
4. Nebeker, A. V., and F. A. Puglisi: "Effect of PCBs on Survival and Reproduction of Daphnia, Gammarus, and Tanytarsus," *Trans. Am. Fish. Soc.*, 1974.
5. Interdepartmental Task Force on PCBs, "Polychlorinated Biphenyls and the Environment," COM-72-10419, Washington D.C., 1972.
6. National Academy of Sciences, National Academy of Engineering, A Report to EPA, *Water Quality Criteria*, 1972.
7. Sanks, R. L.: *Water Treatment Plant Design*, Butterworth Pub. Co., 4th ed., 1982.
8. U.S. EPA, Office of Toxic Substances, *Assessment of Wastewater Management, Treatment Technology and Assigned Costs for Abatement of PCB Concentrations in Industrial Effluents*, February 1976.
9. Federal Register, Sec. 307, 40 CFR 131.36.
10. Internet search: epa.gov/pcb/laws, November 15, 2003.
11. Ford, D. L.: "Optimization of Activated Carbon—Biological Processes in Hazardous Waste Treatment," Hazmat Conference, Orlando, Florida, 1994.
12. Horzempa, L., and D. M. DiToro: "The Extent of Reversibility of Polychlorinated Biphenyl Adsorption," Water Resources, vol. 17, No. 8, 1983.
13. Adams, C., F. Davis, and W. W. Eckenfelder: *Development of Design and Operational Criteria for Wastewater Treatment*, CBI Publishing Co., Boston, 1981.
14. Castaldi, F. J., and D. L. Ford: "Slurry Bioremediation of Petrochemical Waste Sludges," *Applied Bioremediation*, 1993.
15. National Research Council (NRC), *Water Quality Criteria*, Washington, D.C., 1972.
16. U.S. EPA, *Preliminary Study of Selected Potential Environmental Contaminants*, EPA 560/2-75-002, July 1975.
17. Pankow, J. F., S. Fienstra, J. Cherry, and M. C.: *Dense Chlorinated Solvents and Other DNAPLs in Groundwater*, Waterloo Press, 1996.
18. Schaumburg, F. D.: "Banning Trichloroethylene: Responsible Reaction or Overkill," *Environmental Science and Technology*, 1990.
19. Loehr, R. C.: "Development and Assessment of Environmental Quality Standards," *Proceedings of the Seminar on Development and Assessment of Environmental Standards*, American Academy of Environmental Engineers, Annapolis, MD, February 1983.

20. Bedient, P. B., H. S. Fifai, and C. J. Newell: *Groundwater Contamination Transport and Remediation,* Prentice Hall, Englewood Cliffs, N.J., 1994.
21. Field, J. A., and R. Sierra-Alvarez: *Reviews in Environmental Science and Biotechnology,* vol. 3, September 2004.
22. Zehnder, A., and W. Stumm: *Biology of Anaerobic Microorganisms,* John Wiley & Sons, 1988.
23. McKay, D., and A. Walkoff: "Rate of Evaporation of Low-Solubility Contaminants from Water Bodies to Atmosphere," *Environmental Science and Technology,* vol. 7, July 1973.
24. Ford, D. L., and W. W. Eckenfelder: *Design and Economics of Powdered Activated Carbon in the Activated Sludge Process,* Progressive Water Tech., U.K., Pergamon Press, 1980.
25. Ford, D. L.: *The Use of Biological Processes to Remove Residual Materials from CPI Wastewaters,* AICLE Annual Meeting, 1983.
26. U.S. EPA: "Adsorption Capacity for Specific Organic Compounds," EPA-600/8-80-023, April 1980.
27. Norris, D.: *Handbook of Bioremediation,* Robert S. Kerr Environmental Research Laboratory, Lewis Publishers, 1994.
28. Ward, C. H., J. A. Cherry, and M. R. Scalf: *Subsurface Restoration,* Ann Arbor Press, 1997.
29. Brubaker, G., D. L. Ford, and J. Smith: Bioremediation of Organic Constituents in Soil and Groundwater," National Groundwater Association Short Course, Boston, MA., July 1993.
30. U.S. EPA, *Groundwater Currents,* No. 38, December 2000.
31. McKenna, Conner & Cuneo, *Pesticide Regulation Handbook,* Executive Enterprises Publications Co., Inc., 1987.
32. U.S. EPA, *Development Document for Effluent Limitation Guidelines for the Pesticide Chemicals Manufacturing—Point Course Category,* April 1978.
33. Ford, D. L.: Private Communications, Unpublished.
34. U.S. EPA: *The Pesticide Manufacturing Industry—Current Waste Treatment and Disposal Practices,* 12020 FYE 01/72, 1972.
35. National Academy of Science, Committee on Research in the Life Sciences on the Committee on Public Policy, 1970.
36. U.S. EPA and TRW Systems Group: "Assessment of Industrial Hazardous Waste Practices, Organic Chemicals, Pesticides, and Explosives Industry," April 1975.
37. Pontius, F., et al.: "Regulation Perchlorate in Drinking Water," AWWA, Division of Environmental Chemistry Preprints, vol. 39(2), August 1999.
38. Cox, E., et al.: "Regulation Perchlorate in Drinking Water," AWWA, Division of Environmental Chemistry Preprints, vol. 39(2), August 1999.
39. Coates, J., et al.:, "Regulation Perchlorate in Drinking Water," AWWA, Division of Environmental Chemistry Preprints, vol. 39(2), Southern Illinois University, August 1999.
40. Catts, J., et al.: ,"Regulation Perchlorate in Drinking Water," AWWA, Division of Environmental Chemistry Preprints, vol. 39(2), Harding Lawson Association, August 1999.
41. Logan, B., et al., Penn State University, "Regulation Perchlorate in Drinking Water," AWWA, Division of Environmental Chemistry Preprints, vol. 39(2), August 1999.
42. Toxfaqs Fact Sheet, September 1995.
43. Verschueren, K.: *Handbook of Environmental Data on Organic Chemicals,* Von Nostrand Reinhold, 1983.
44. Toxfaqs Fact Sheet, September 2001.
45. Ward. C. H., James Baker Professor, personal communication, Rice University, February 2008.
46. Tischler, L. F.: personal communications, January & February, 2008.
47. American Society of Civil Engineers: "Sulfide in Wastewater Collection and Treatment Systems," ASCE Manuals and Reports on Engineering Practices No. 69.

48. Parson of Puerto Rico, *Conceptual Design for odor Control, Barceloneta Regional Treatment Plant*, July 1999.
49. Water Pollution Control Federation, *Manual of Practice* (OM-4, Operations & Maintenance).
50. Roberts, P., C. Munz, and P. Dandliker: "Modeling Volatile Organic Solute Removal by Surface and Bubble Aeration," Eq. 10, p. 159, *Journal WPCF*, vol. 56, No. 2, February 1984.
51. Patterson, J.: *Industrial Wastewater Treatment Technology*, 2d., 1985.

CHAPTER 15
Waste Minimization and Water Reuse

15.1 Introduction

Industrial waste minimization, reuse, and even the concept of "zero discharge" have been evaluated and implemented with various levels of success over the past four to five decades. There are several drivers which give an industrial complex the incentive to strive toward waste minimization, reuse, and possibly, zero discharge. These are either singularly or in combination, listed as follows:

1. *Regulatory driven.* Discharge permits force effluent treatment levels to quality requirements which open up reuse options.
2. Input water availability is limited/costly which mandates maximum waste minimization/water reuse.
3. Economic cost-effectiveness provides the incentive to reuse water and minimize waste residuals as it is more cost-effective than disposal. (A good example of this is the reuse of Resource Conservation and Recovery Act [RCRA] hazardous wastes back to feedstock/product.)

This chapter discusses some of the more attractive concepts of water recycle and minimization and zero discharge.

15.2 Water Recycle and Reuse

Although water recycle and reuse may not be the primary or only objective in an industrial complex, it is the first step in an overall integrated process of waste minimization and zero discharge.

For purposes of definition, *water reuse* is defined as the utilization of water that has been used previously for another purpose whereas *recycle* is the reuse of the same water one or more times for the same purpose. In-plant reuse and recycling of industrial waters are modes of water conservation which are being pursued aggressively. For example, in as early as 1969 a study found that Texas-based industries reused or recycled their intake waters, on an average, about five times prior to discharge.[1]

By definition, reused or recycled water can be quantified by a recycle ratio defined as:

$$\text{Recycle ratio} = \frac{\text{freshwater intake required with no recycle}}{\text{actual freshwater intake with reuse/recycle}}$$
(in effect, the number of times a gallon of water is recycled)

These ratios will vary from different industrial category, depending on process configurations, treatment levels, and water quality tolerances of the various recycle receivers.[1,2]

Limits of Water Reuse[3]

It should be noted that water reuse in an industrial environment involves taking water back for various uses somewhere in the process. Frequently this use involves evaporation, which means an additional concentration of salts. When used as cooling tower makeup or as scrubber water makeup, various alternatives are available to minimize the effect of these high conductivity waters. These include chemical corrosion inhibitors, use of sidestream softening, installation of different metallurgy, and so on.

However, with the issue of whole effluent toxicity testing on the permits for industrial dischargers, the impact that water reuse has on toxicity must be included in any water reuse plan. Many industrial facilities are mandated to use freshwater organisms, such as *Daphnia* magna, for effluent toxicity testing. Higher salinity levels in wastewater will cause osmotic stress to these organisms; so, there is an antagonistic effect between compounds or metals in the wastewater and the salinity level. Water Quality Criteria reports that the threshold concentration of sodium chloride for immobilization of *Daphnia* magna varies between 2100 and 6100 mg/L. Other salts, such as potassium chloride, may be the critical salt with respect to toxicity compliance.

Increased water reuse causes the salinity level of the wastewater to increase. As previously mentioned, the increase in salinity or other conservative constituents may be the driving force which limits recycle unless membrane or similar processes are incorporated. Moreover, concentration-based limitations in certain states, such as selenium concentration limits for refineries in California, limit water conservation and reuse options.

Reuse or Wastewater Treatment Plant Effluents

The practice of reusing the effluent from industrial, municipal, or combined wastewater treatment plants (WWTPs) has been utilized for years. Reusing effluents from publicly owned treatment plants (POTWs) and even those combined with an industrial component (particularly following the industrial pretreatment standards prior to discharging to a POTW) has an intrinsic advantage because of lower salt content (total dissolved solids [TDS]). However, with better water management and the increasing use of membrane processes such as reverse osmosis, microfiltration, and similar technologies, the reuse of industrial wastewater treatment processes is becoming more viable. The reuse of effluent from WWTPs is possible in a wide range of categories, as shown in Fig. 15.1.[4]

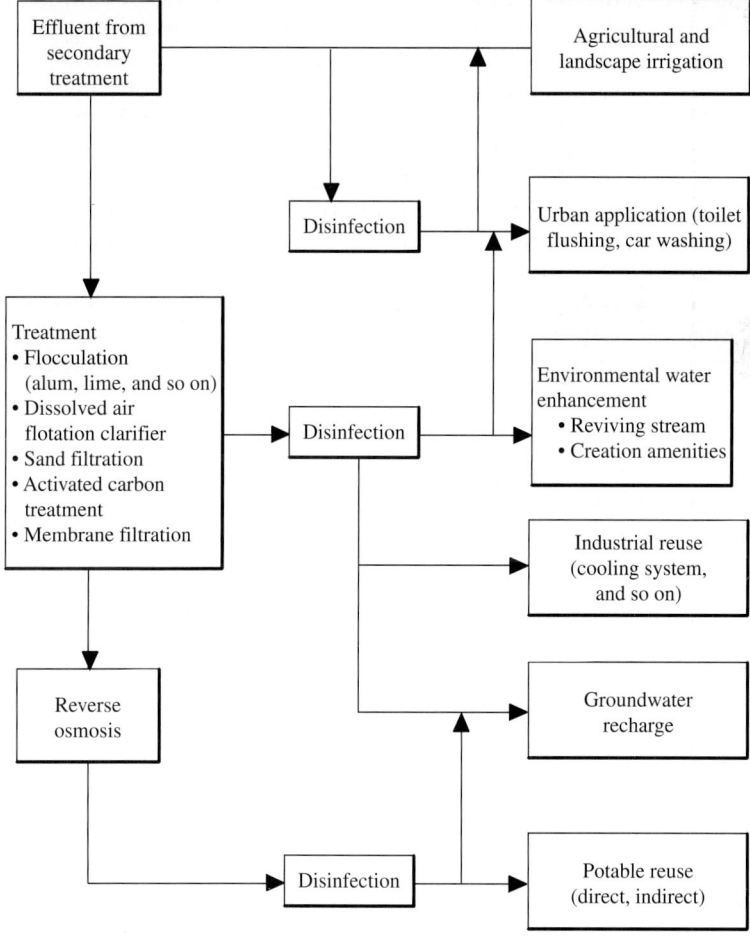

FIGURE 15.1 Wastewater treatment plant effluent reuse.

The Decision-Making Process[5]

The criteria recently developed in analyzing candidate systems and concepts in the recycle/reuse decision process are:

- Is it environmentally superior?
- Is it economically beneficial?

These are logical criteria in the decision-making process, with the *economically beneficial* questions easier to quantify than the *environmentally superior* ones.

The first logical step is to develop a plantwide water balance throughout the industrial complex. The next logical step in the economically beneficial phase of inquiry is to analyze the current wastewater collection system to determine the effectiveness of collection system segregation in terms of such parameters as organics/inorganics, oily/nonoily, high and low dissolved salts, hazardous/nonhazardous, and metallics/nonmetallics. Costing out the necessary collection system modifications—the unit processes required to handle the constituents of concern to satisfy water reuse/recycle quality criteria—and crediting corresponding reduction in raw water use, monitoring costs, and other economic benefits lead one to an economic analysis and definition. A simplified illustration of this concept is graphically presented in Fig. 15.2. A detailed cost analysis of (A), (B), and (A-B) can be developed once the process flow diagram is finalized based on treatability studies, process engineering, equipment selection, and material balances. A cost analysis based on capital costs amortized

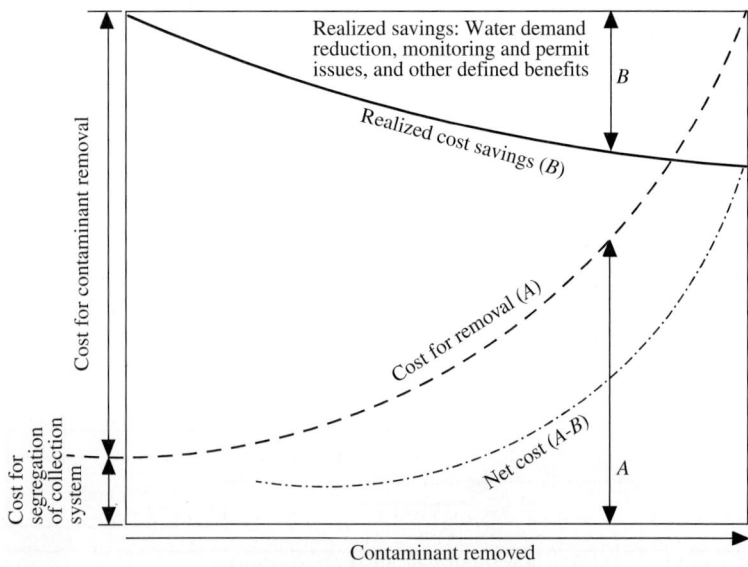

Figure 15.2 Recycle/reuse decision-making concept.

over the life of the system and operating costs, for example, can now give definition to "economically beneficial." There are obviously other less quantifiable components to this analysis, such as litigation exposure, public relations, and regulatory implications, but these can be weighed on a site-specific basis following the cost-estimating process.

The environmentally superior part of this criterion definition is more complex, more subjective to interpretation, and more subservient to statute-driven regulations than are the natural laws or strict economic projections. The biosphere in which people live is a mosaic of an infinite amount of environmental variables within the media of land, air, and water bodies. Ostensibly, the environmentally superior definition has evolved subtly over the second half of the century in a regulatory-driven fashion, namely Water Quality Criteria (1965), Environmental Impacts (1969), Clean Air Requirements (1970), Clean Water Permits (1972), Primary Drinking Water Standards and Underground Injection Control (UIC) (1974), Hazardous Waste Definition and Control (1976, 1984), Site Cleanup (1980), Underground Tank Control (1984), and corresponding legislative amendments and reauthorizations to present.

Case Histories

There are numerous case histories which add pragmatism to the concepts of water recycle and reuse. Some of the earlier pioneers in recycling have been in the petroleum refining sector. A typical refinery flow diagram without recycle is shown in Fig. 15.3.[1,2] Under this configuration, refineries typically would consume in the range of 20 to 45 gal of consumptive water per barrel of crude throughput (once-through water cooling and no coking). The Petroleum Refinery Effluent Guidelines under U.S. Environmental Protection Agency (EPA) set forth the basis for National Pollutant Discharge Elimination System (NPDES) permits for best practicable control technology (BPT) currently available, new source performance standards (NSPS), and best available treatment (BAT).[6] As these limitations are based on pounds of contaminants per day (although some permits also include concentrations limitations), the flow is implicit, calculated as increasing allowed effluent concentrations such as biochemical oxygen demand (BOD) and chemical oxygen demand (COD) at lower effluent flows. Although this is a regulatory incentive to recycle and reduce effluent flow, it also can limit reuse of treated effluent based on salinity buildup and possible effluent toxicity problems as discussed previously in the section "Limits of Water Reuse." For example, some industrial water quality tolerances are listed in Table 15.1, indicating a reduction in water reuse possibilities, primarily limitations on dissolved solid allowables. This mandates the inclusion of some combination of membrane processes, ion exchange, activated carbon, filtration, and so on, before reuse and waste minimization can be maximized.

Typical material outputs from selected petroleum refining processes are listed in Table 15.2. The wastewater flows from various

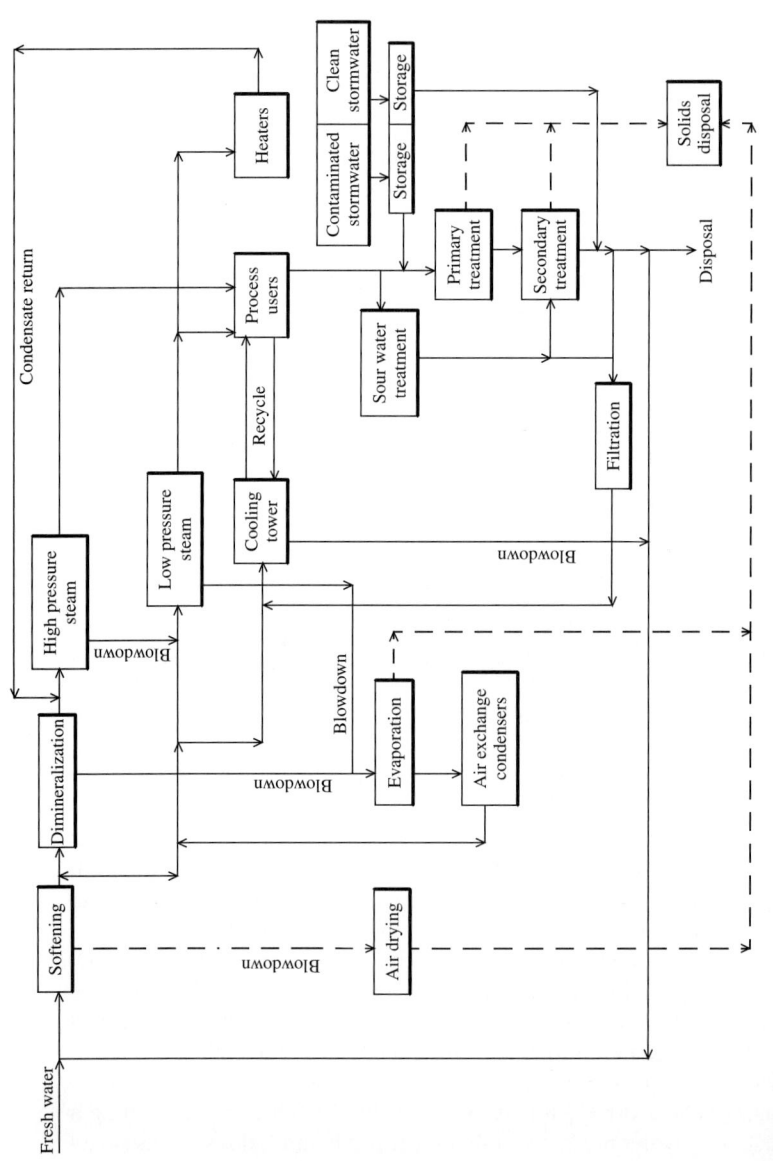

Figure 15.3 Typical petroleum refinery flow diagram.

Water Use	Turbidity	Oxygen Consumed	Dissolved Oxygen	pH	Hardness	Calcium	Chlorides	Dissolved Solids	Suspended Solids	Iron and Manganese	Aluminum Oxide	Silica	Carbonate	Bicarbonate	Hydroxide	Hydrogen Sulfide	Ammonium Nitrogen
Boiler feedwater																	
0–150 psi	20	15	1.4	8.0+	80			500–3000			5.0	40	200	50	50	5.0	
150–250 psi	10	10	0.1	8.4+	40			500–2500			0.5	20	100	30	40	3.0	
250–400 psi	5	4	0	9.0+	10			100–1500			0.05	5	40	5	30	0	
>400 psi	1	3	0	9.6+	2			50			0.01	1	20	0	15	0	
Cooling water	50			6.5–7.5		100	250	1300	3	0.5	5.0			275			1

TABLE 15.1 Industrial Water Quality Tolerances

Process	Air Emissions	Process Waste Water	Residual Wastes Generated
Crude oil desalting	Heater stack gas (CO, SO_x, NO_x, hydrocarbons and particulates), fugitive emissions (hydrocarbons)	Flow = 2.1 gal/bbl Oil, H_2S, NH_3, phenol, high levels of suspended solids, dissolved solids, high BOD, high temperature	Crude oil/desalter sludge (iron rust, clay, sand, water, emulsified oil and wax, metals)
Atmospheric distillation	Heater stack gas (CO, SO_x, NO_x, hydrocarbons and particulates), vents and fugitive emissions (hydrocarbons)	Flow = 26.0 gal/bbl Oil, H_2S, NH_3, suspended solids, chlorides, mercaptans, phenol, elevated pH	Typically, little or no residual waste generated
Vacuum distillation	Steam ejector emissions (hydrocarbons), heater stack gas (CO, SO_x, NO_x, hydrocarbons and particulates), vents and fugitive emissions (hydrocarbons)		
Thermal cracking/ visbreaking	Heater stack gas (CO, SO_x, NO_x, hydrocarbons and particulates), vents and fugitive emissions (hydrocarbons)	Flow = 2.0 gal/bbl Oil, H_2S, NH_3, phenol, suspended solids, high pH_2 BOD_5, COD	Typically, little or no residual waste generated
Coking	Heater stack gas (CO, SO_x, NO_x, hydrocarbons and particulates), vents and fugitive emissions (hydrocarbons), and decoking emissions (hydrocarbons and particulates)	Flow = 1.0 gal/bbl High pH, H_2S, NH_3, suspended solids, COD	Coke dust (carbon particles and hydrocarbons)

Process	Air emissions	Wastewater	Residual wastes
Catalytic cracking	Heater stack gas (CO, SO_x, NO_x, hydrocarbons and particulates), fugitive emissions (hydrocarbons), and catalyst regeneration (CO, NO_x, SO_x, and particulates)	Flow = 15.0 gal/bbl High levels of oil, suspended solids, phenols, cyanides, H_2S, NH_3, high pH, BOD, COD	Spent catalysts (metals from crude oil and hydrocarbons), spent catalyst fines from electrostatic precipitators (aluminum silicate and metals)
Catalytic hydro-cracking	Heater stack gas (CO, SO_x, NO_x, hydrocarbons and particulates), fugitive emissions (hydrocarbons), and catalyst regeneration (CO, NO_x, SO_x, and catalyst dust)	Flow = 2.0 gal/bbl High COD, suspended solids, H_2S, relatively low levels of BOD	Spent catalysts fines (metals from crude oil, and hydrocarbons)
Hydrotreating/hydroprocessing	Heater stack gas (CO, SO_x, NO_x, hydrocarbons and particulates), vents and fugitive emissions (hydrocarbons), and catalyst regeneration (CO, NO_x, SO_x)	Flow = 1.0 gal/bbl H_2S, NH_3, High pH, phenols suspended solids, BOD, COD	Spent catalyst fines (aluminum silicate and metals)
Alkylation	Heater stack gas (CO, SO_x, NO_x, hydrocarbons and particulates), vents and fugitive emissions (hydrocarbons)	Low pH, suspended solids, dissolved solids, COD, H_2S, spent sulfuric acid	Neutralized alkylation sludge (sulfuric acid or calcium fluoride, hydrocarbons)
Isomerization	Heater stack gas (CO, SO_x, NO_x, hydrocarbons and particulates), HCl (potentially in light ends), vents and fugitive emissions (hydrocarbons)	Low pH, chloride salts, caustic wash, relatively low H_2S and NH_3	Calcium chloride sludge from neutralized HCl gas

Source: Assessment of Atmospheric Emissions from Petroleum Refining, Radian Corp., 1980; *Petroleum Refining Hazardous Waste Generation*, U.S. EPA, Office of Solid Waste, 1994.

TABLE 15.2 Typical Material Outputs from Selected Petroleum Refining Processes

Process	Air Emissions	Process Waste Water	Residual Wastes Generated
Polymerization	H_2S from caustic washing	H_2S, NH_3, caustic wash, mercaptans and ammonia, high pH	Spent catalyst containing phosphoric acid
Catalytic reforming	Heater stack gas (CO, SO_x, NO_x, hydrocarbons and particulates), fugitive emissions (hydrocarbons), and catalyst regeneration (CO, NO_x, SO_x)	Flow = 6.0 gal/bbl High levels oil, suspended solids, COD Relatively low H_2S	Spent catalyst fines from electrostatic precipitators (alumina silicate and metals)
Solvent extraction	Fugitive solvents	Oil and solvents	Little or no residual wastes generated
Dewaxing	Fugitive solvents, heaters	Oil and solvents	Little or no residual wastes generated
Propane deasphalting	Heater stack gas (CO, SO_x, NO_x, hydrocarbons and particulates), fugitive propane	Oil and propane	Little or no residual wastes generated
Merox treating	Vents and fugitive emissions (hydrocarbons and disulfides)	Little or no wastewater generated	Spent Merox caustic solution, waste oil-disulfide mixture
Wastewater treatment	Fugitive emissions (H_2S, NH_3, and hydrocarbons)	Not applicable	API separator sludge (phenols, metals and oil), chemical precipitation sludge (chemical coagulants, oil), DAF floats, biological sludges (metals, oil, suspended solids), spent lime

Gas treatment and sulfur recovery	SO_x, NO_x, and H_2S from vent and tail gas emissions	H_2S, NH_3, amines, Stretford solution	Spent catalyst
Blending	Fugitive emissions (hydrocarbons)	Little or no wastewater generated	Little or no residual waste generated
Heat exchanger cleaning	Periodic fugitive emissions (hydrocarbons)	Oily wastewater generated	Heat exchanger sludge (oil, metals, and suspended solids)
Storage tanks	Fugitive emissions (hydrocarbons)	Water drained from tanks contaminated with tank product	Tank bottom sludge (iron rust, clay, sand, water, emulsified oil and wax, metals)
Blowdown and flare	Combustion products (CO, SO_x, NO_x and hydrocarbons) from flares, fugitive emissions	Little or no wastewater generated	Little or no residual waste generated

Source: Assessment of Atmospheric Emissions from Petroleum Refining, Radian Corp., 1980; *Petroleum Refining Hazardous Waste Generation*, U.S. EPA, Office of Solid Waste, 1994.

TABLE 15.2 Typical Material Outputs from Selected Petroleum Refining Processes *(Continued)*

Unit Operation	Percent of Total Wastewater Flow
Desalter	23%
Sour water	10%
Benzene stripper	5%
Process units	23%
Cooling tower blowdown	33%
Boiler blowdown	2%
Miscellaneous sources	4%

TABLE 15.3 Typical Discharge Ratio from Various Sources

sources from one refinery, which are somewhat typical, are shown in Table 15.3.

Therefore, a characterization of these individual flows, the required treatment for each stream, total stream, or some combination thereof is required before developing the unit treatment processes necessary to maximize recycle/reuse.

Benchmarking[7]

Benchmarking is a means of making comparisons in order to measure a refinery's performance against other refineries. Using benchmarking information, a refinery can adopt practices learned from other refineries to strive toward "best in class." A series of tables that highlight benchmarking data gathered from more than 20 refineries worldwide are presented next. Table 15.4 provides benchmarking information developed by Environmental Consultants and Engineers showing typical ranges of volume per barrel or percentage of crude for several key waste categories.

Wastewater that is accounted for in the benchmarking data shown above includes:

- Desalter blowdown
- Process wastewater

Category	Units	Range
Wastewater	Gal/bbl	10–30
Sour water	Gal/bbl	1.1–6
Spent caustic	Gal/1000 bbl	2–12.2
Slops	% of crude	~0.5

TABLE 15.4 Benchmarking Data by Category

Category	Percent of Total Flow
Oily wastewater	10–20
Desalter effluent	10–15
Cooling tower blowdown	10–15
Sour water	5–15
Rain water	5–10
Raw water treatment	5–10
Boiler blowdown	3–10
Tank BS&W	1–10
Ballast water	1–5
Miscellaneous	15–30

TABLE 15.5 Range of Wastewater Flow

- Excess sour water
- Tank bottom sediments and water (BS&W)
- Cooling tower blowdown
- Boiler blowdown
- Rain water

The benchmarking range for sour water is quite variable and depends on the operating complexity of the refinery. Similarly, the spent caustic data varies widely and depends on the level of amine treatment in the refinery.

The ranges of typical wastewater flow refineries are shown in Table 15.5.

Table 15.6 shows the streams with the highest loading of COD in refineries.

A common benchmarking reference used to evaluate refineries is wastewater flow per crude capacity. A comparison of the data available

Category	COD (mg/L)
Tank BS&W	>1000
Desalter effluent	500–1000
Oily wastewater	500–1000
Spent caustic	>20,000
Sour water	200–600

TABLE 15.6 Typical Contaminant Loading

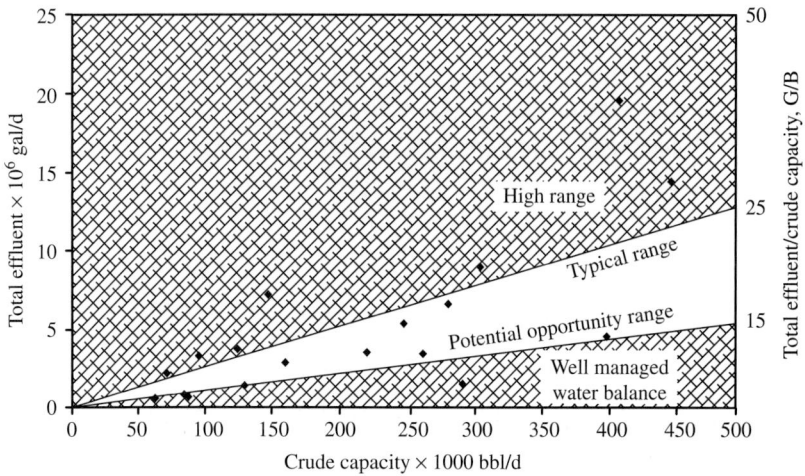

Figure 15.4 Wastewater flow data benchmarking.

for many refineries is provided in Fig. 15.4 to show the range of wastewater flow reduction opportunities available to refineries. Those refineries with well-managed water balance systems have data points below 15 gal of wastewater effluent per barrel of crude capacity. The exact effluent flow per barrel of crude capacity, as previously noted, will depend on reuse limitations based on concentration buildup potentials affected by permit constraints and total maximum discharge limitations (TMDLs).

The most attractive water recycle reuse candidates and source minimization for a petroleum refinery include, but are not limited to, the following:

1. Use of effluent from wastewater treatment plant treating domestic sewage within the industrial complex as cooling tower makeup.

2. Segregate low TDS process streams (such as condensates) as practical in the collection system, store/treat, and use as input to such uses as desalter makeup, cooling tower makeup, boiler makeup, firewater, and/or washdown water. Return condensate from various process units, refineries reduce the makeup water requirements and lower discharges from regeneration systems since condensate does not typically require water treatment. Condensate streams have low TDS and there is a potential for energy recovery.

3. Capture or harvest the maximum amount of rainfall runoff from uncontaminated areas as possible such as parking lots and unused contiguous land. (Storm runoff from process areas which becomes heavily contaminated is usually controlled by curbs and dikes, then sent to the wastewater treatment process.)

Waste Minimization and Water Reuse

This can be an enormous source of freshwater which can easily be used as desalter makeup, firewater, cooling tower makeup, and with minimal treatment, possibly as process water and boiler makeup. This water can be stored in large reservoirs, tanks, or other impoundments and used as required. The storage volume should be as large as possible.

As the accumulation of "clean" storm runoff which is low in TDS is a significant resource for reuse water, a storm runoff management plan can be developed for any industrial site. An excellent application of storm runoff is the collection, storage, and use as a firewater reservoir. Normally, the quality required for firewater is much less rigid than those for other uses. There are three fundamental techniques that form the basis for most storm runoff management systems, namely:

1. Segregation of clean and contaminated areas
2. The availability to store runoff (reservoirs, or other impoundments), and
3. Controlled routing to points of reuse within the industrial complex

A 10-year design storm is typically used as the basis of design. (This is a statistical prediction of a one-time occurrence in 10 years, based on the predicted rainfall at the specific location.) The cumulative runoff for a specific duration and the 10-year–statistics-based design storm for a given location is shown in Fig. 15.5. At various pumpoff rates to

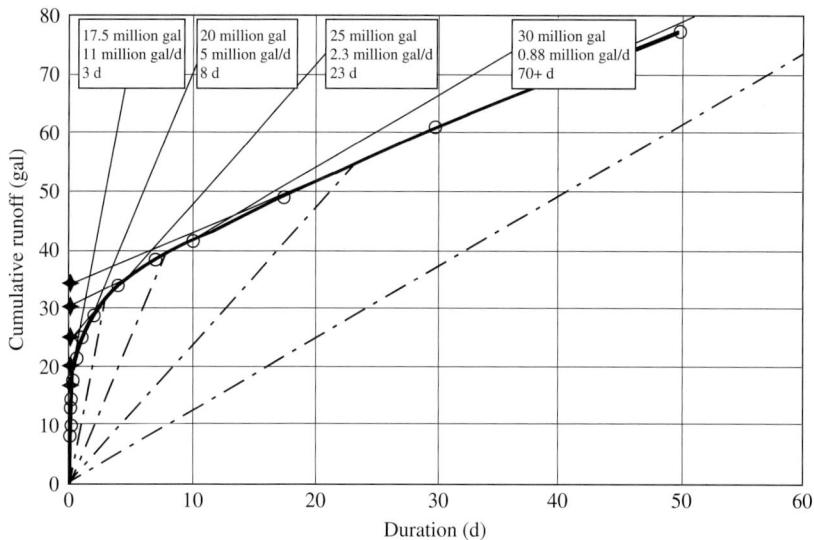

FIGURE 15.5 Development of storage volume and treatment rate tradeoffs for 10-year–statistics-based design storm.

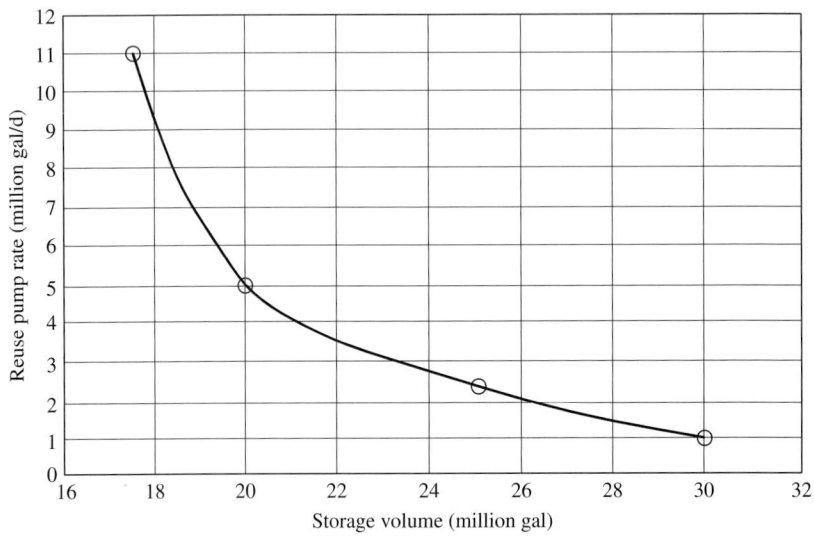

FIGURE 15.6 Projected available reuse water and runoff storage requirements.

selected points of reuse, the difference in the pumpoff rate lines and the accumulation mass curve at the maximum vertical distance is the required storage area. This is plotted in Fig. 15.6, illustrating the reuse pump rate and the corresponding required storage.[8]

A comprehensive report underscored these water reuse and recycle concepts by proposing a material reuse model that identifies lost-optimal reuse scenarios to a water reuse planning program. This model was illustrated by identifying cost and water savings of a large industrial complex in Pasadena, Texas (Bayport Industrial Complex, Gulf Coast Waste Disposal Authority). As a basis for applying this model, the effluent quality and flow rate data were tabulated, as shown in Table 15.7.[9]

15.3 Waste Minimization—RCRA Hazardous Waste Issues

Waste minimization has taken on a new and more important meaning with the promulgation of the RCRA regulations in 1980. RCRA hazardous wastes can be designated by one or more of seven pathways, listed in Table 15.8. As the disposal costs of designated hazardous wastes can be 10 or more times more costly than nonhazardous waste, there is a significant incentive to minimize the production of hazardous wastes within the industrial complex. This waste minimization process can be accomplished in several ways, depending on the production processes and the feasibility of reducing the generation of hazardous residues in a cost-effective manner.

Industry	Flow Rate (1000 gal/d)	TOC (mg/L)	TSS (mg/L)	TDS (mg/L)
1	176	137	126	11,408
2	138	2248	299	776
2	120	440	147	1608
2	91	675	106	556
3	300	22	66	488
4	86	18	72	284
6	993	375	1619	10,220
6	99	215	2657	7176
5	447	347	190	19,312
5	91	3011	244	18,384
5	363	1930	29	4204
5	134	2223	212	41,020
5	947	484	105	904
5	282	431	60	1240
5	149	146	256	740
5	55	1695	795	2324
5	81	3869	257	8960
7	255	46	50	536
7	292	454	99	4148
7	113	202	1895	4236
8	172	83	49	244
3	465	26	34	1468
4	980	22	14	1948
6	932	3218	1287	36,708
5	145	32	29	2388
5	802	1439	75	1336

Legend
1 Pesticides and agricultural chemicals
2 Chemicals and chemical preparations
3 Cyclic organic crudes, intermediates, organic dyes, and pigments
4 Industrial gases
5 Industrial organic gases
6 Industrial inorganic chemicals
7 Plastic mateials, synthetic resins, nonvulcanizable elastomers
8 Synthetic rubbers

TABLE 15.7 Bayport Facilities' Effluent Quality and Flow Rate Data (Provided by the Gulf Coast Waste Disposal Authority)

1. Characteristic: Ignitable
    Corrosive
    Reactive
    EP toxicity (TCLP)
	D001-D017

2. Hazardous waste from nonspecific sources	F001-F028
3. Hazardous waste from specific sources	K001-K136
4. Discarded products, off-spec materials, spill residues, container residuals	P001-P122
5. Discarded commercial, products, off-spec species, intermediates	U001-U228
6. Mixture rule—If listed waste is mixed with other waste, the mixture is hazardous	
7. Derived-from rule—Residue generated from treatment, storage, or disposal of hazardous waste is a hazardous waste	

TABLE 15.8 RCRA Hazardous Waste Designations

The recycling and reuse aspects of RCRA wastes is one of the more attractive approaches of exempting materials from RCRA regulations which would otherwise be considered hazardous wastes. Examples of RCRA exemptions include (40 CFR 261):

1. Using or reusing materials in an industrial process, provided the materials are not to be reclaimed (a material is "reclaimed" if it is processed to recover a useful product or if it is regenerated, like recovery of lead from batteries or regeneration of spent solvents, i.e., the recovery of spent TCE in a still)
2. Using or reusing the material as an effective substitute for commercial products
3. Returning materials to the original process from which they are generated without first being reclaimed (where the material is returned as a substitute for raw material feedstock)

Materials managed by recycling are not classified as "solid wastes," and are therefore out of the scope of RCRA Subtitle C regulations. Exceptions, however, include:

1. Materials used in a manner constituting disposal or used to produce products that are applied to the land
2. Materials burned as fuel for energy recovery
3. Materials that are speculatively accumulated
4. Inherently waste-like materials (those materials include hazardous wastes that are always subject to RCRA regulations)

These regulations and exemptions are complex and not always clear as to which wastes and which activities qualify for recycling exemptions. To reduce uncertainty, EPA has released regulatory determinations that address specific production processes and wastes. It is therefore necessary that industrial personnel or their consultants develop a clear understanding of the regulations so a legal and technically supportable approach can be taken to develop a successful waste minimization program. Generators should always work in concert with the appropriate regulatory authorities to determine if the federal interpretations apply.

Several examples of hazardous waste minimization by substitution treatment or utilization of the recycle exemption include: the recycle of bottoms from vinyl chloride monomer distillation bottoms, rich in chlorinated compounds; reprocessing as a feedstock in the production of associated chlorinated compound products; the reuse of solvents and bathwater in light manufacturing processes; the use of organic solvent substitutes where possible such as detergents; soaps or less toxic acetic acid type of solvents; recover metals from fixing baths using chemical recovery methods, such as electrolytic recovery cells or ion exchange resins; and the use of membrane processes, such as reverse osmosis to reduce saline streams to a specific conductivity of 200 $\mu m/cm$ (or a TDS of approximately 118 ppm) or less and use as cooling tower makeup. The potential for reuse of listed oily wastes from refinery WWTPs-API (American Petroleum Institute) separator bottoms, DAF float, and so on as coker feedstock or coker quench makeup water should be examined.

15.4 Zero Effluent Discharge and Economic Concepts

The natural laws, such as the Second Law of Thermodynamics, state there is no such thing as zero discharge. That, notwithstanding, it is technically possible to achieve "zero effluent discharge." The quest for zero effluent discharge goes a step beyond "the conventional approach of maximum recycle/reuse/waste minimization to reduce industrial residuals released to the environment short of zero effluent discharge."

This is conceptually illustrated in Fig. 15.7.[5,10,11] There is a direct relationship between capital cost of technologically available treatment process and the resulting quality of the treated water. For example, if the NPDES permits allow an effluent quality from secondary treatment alone (advanced biological treatment and possibly post filtration), then cost "A" is incurred. If additional treatment is used, that is, activated carbon, which opens up additional reuse possibilities, then a possible resultant savings can be realized by cost "B." (A typical cost per unit of water treated can be calculated by taking the amortized capital cost over the life of the system plus associated operation and maintenance costs divided by the volume of water

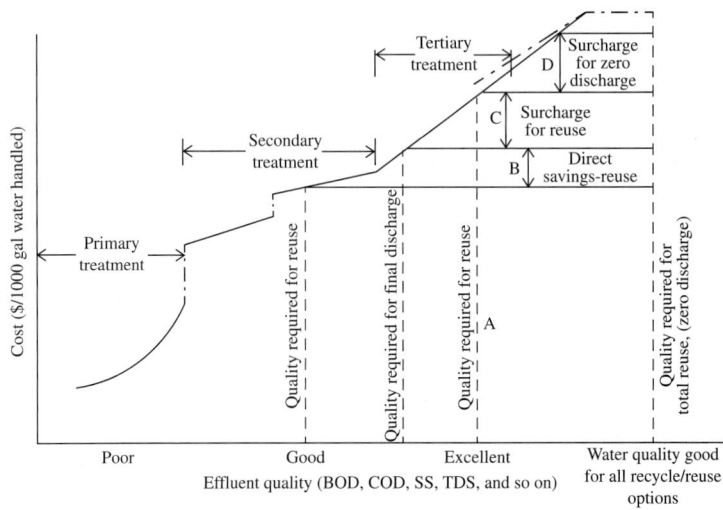

FIGURE 15.7 Cost/effluent quality relationship.

handled.) If additional units are then added, such as membrane processes, ion exchange, and so on, the costs of which exceed the water savings, then the surcharge "C" can be calculated. However, the discharge to the environment, although low in volume, will most likely still be of lesser quality than that originally extracted. If still an additional step is taken to remove the resultant salts, such as nonhazardous deep-well disposal, mechanical vapor recompression evaporation, or solar ponds, and so on, then zero discharge can be achieved at surcharge "D." The comparison of "D" versus "C" can then be used in the decision-making process to approach or achieve "zero effluent discharge." If the latter is truly accomplished, there could be a significant savings bonus both in terms of eliminating monitoring costs relating to compliance and favorable publicity. In the final analysis, the zero effluent discharge goal is basically economic and perception driven rather than a regulatory one.

15.5 Case History

Formosa Plastics, Point Comfort, Texas[12]

Formosa Plastics is a major producer of chlorine, caustics, organic chemicals, and plastics. It is currently a multibillion dollar complex located proximate to Lavaca Bay on the Texas Gulf Coast. It includes two olefins plants, caustic-chlorine ion exchange membrane plant, ethylene dichloride plant, two high-density polyethylene plants, two polypropylene plants, ethylene glycol plant, ethylene dichloride plant, low-density polyethylene plant, cogeneration plants, and

VCM/PVC (vinyl chloride polymer/polyvinyl chloride) plant. The raw waste is supplied through a pipeline from which involves a long-term take or pay contract with Lavaca Navidad River Authority. The plant is located adjacent to the Calhoun County Navigation district bulk liquids port. Plant expansion and water constraints have led Formosa into an aggressive water reuse, waste minimization, zero discharge concept evaluation.

Although zero discharge of pollutants is a goal of the federal Clean Water Act, EPA has never implemented zero discharge requirements for the industrial category, including plastics manufacturers. Instead, EPA has concentrated on reductions of pollutant discharge. In fact, the national discharge standards for the plastics industry do not require the tertiary treatment that is currently being implemented by Formosa. This equipment was put in because of potential problems with water quality in Lavaca Bay, attempting therefore to ensure that there would not be deterioration of the receiving water quality.

History of Zero Discharge Technologies

The history of zero discharge technologies is an evolution from primitive solar evaporation ponds to sophisticated membrane and mechanical evaporation processes. This evolution has produced a great diversity of alternatives from which to achieve zero discharge. The alternatives listed below comprise some of the many "tools" available to the field of water conservation and reuse for zero discharge applications.

1. *Solar ponds.* The beginning of the evolutionary cycle was the solar pond. It became a popular zero discharge method in the 1960s and 1970s as some power plants were required to remove their discharge from a receiving body of water. However, solar ponds are now becoming outdated due to large construction costs, and the inability to recover the evaporated water. These costs are the result of the land area requirements for the ponds. Pond size is governed by the minimal evaporation rates experienced in the winter season. Because of these practical and economic restraints, solar ponds are now being replaced by brine concentrators.

2. *Sidestream softening.* The next step in the evolution of zero discharge was sidestream softening. It is a process in which lime and soda ash are added to precipitate scale-causing materials such as calcium, magnesium, and silica. Sidestream softening is commonly applied to cooling tower blowdown to prevent scale formation in the tower. This allows the cooling water to be concentrated in the cooling tower and thereby reduce or possibly eliminate cooling tower blowdown; in the latter case zero discharge is achieved. One such example is the Laramie River Power Station.

3. *Brine concentrator.* The evolutionary stage after sidestream softening was the brine concentrator. It is the favored zero discharge method for wastewater containing high dissolved solids such as power plant effluent, mainly because of the brine concentrator's ability to control scale. Also, a brine concentrator produces a high quality distillate stream that can be recovered and reused. However, brine concentrators can become very expensive for high flows due to their substantial energy requirements. Many operators look for the recycled distillate to offset the high energy costs. This is common where brine concentration is a major component of the zero discharge system.

4. *Wastewater cooling towers.* Today, wastewater cooling towers are evolving as an accepted zero discharge method. The towers can be used in place of or to supplement brine concentrators, utilizing waste heat as an energy source. The wastewater cooling tower evaporates and therefore concentrates the wastewater similar to a brine concentrator. Wastewater cooling tower zero discharge is accomplished by evaporating a significant portion of the wastewater. This amount of evaporation requires a cooling tower to be operated at high cycles of concentration. Fortunately, sidestream softening can be applied to control potential scaling and corrosion.

5. *Staged high recycle cooling process.* An advance in the evolution of the wastewater cooling tower is the patented staged high recycle cooling process. The process is characterized by two cooling towers in series in which the blowdown from the first is softened and sent to the second cooling tower. This technique allows higher cooling water recycle with fewer chemicals in achieving zero discharge. One such staged high recycle cooling process is in operation at the Signal/Shasta power plant in Northern California.

6. *Reverse osmosis (RO).* An important and oftentimes essential component in the evolution of zero discharge is reverse osmosis. This process uses a semipermeable membrane to remove dissolved solids in an effort to improve water quality for industrial reuse. Reverse osmosis, although energy intensive, is a proven technology capable of producing a high quality permeate stream which can be reused as process water for other applications. The reject stream, approximately 10 to 20 percent of the flow, depending on the initial concentration, is concentrated in salts and a suitable method of disposal must be developed. The quality of membranes, the extended membrane life, and other technological upgrades in RO have resulted in it becoming an increasingly attractive treatment system.

Zero discharge technologies have evolved as such to separate a given wastewater into two main residuals. The residuals are either a high quality permeate/distillate stream or a concentrated brine/reject stream. The clean permeate/distillate stream is recycled. This brine can be sent to a crystallizer which further concentrates the brine until the solids precipitate out of solution and only a dry salt remains. The dry salt may then be disposed of in a secure landfill. As an alternative, deep-well disposal of salt can be utilized. Such injection wells are typically nonhazardous, expediting the permitting process.

Industry Applications of Zero Discharge Technology

The evolution of a new zero discharge technology does not necessarily outdate a prior method. Rather, technologies can be combined in numerous applications to achieve zero discharge. An appropriate combination may create a more efficient zero discharge system. An example of this concept is the operation at Northwest Alloys, Inc. in Addy, Washington. This zero discharge system uses a brine concentrator to reduce the wastewater flow to the solar pond while the concentrator distillate is used as makeup water to the cooling tower. Indeed, both the solar pond and the cooling tower loading conditions are reduced by the installation of the brine concentrator.

The Phelps Dodge Hidalgo Smelter in Playas, New Mexico, also utilizes a combination of technologies to achieve zero discharge. In this application, sidestream softening allows increased cycles in the cooling tower, thereby reducing blowdown. A brine concentrator recycles this blowdown to further reduce discharge to a solar pond.

Another illustration of a combination of zero discharge technologies is the Desalination Plant at Debiensko, Poland. In August 1993, the plant was in the process of constructing an RO unit to reduce the wastewater load to the brine concentrators. This can decrease the flow to the evaporator by as much as 75 percent. Consequently, the size of the evaporator is reduced and the energy needed for evaporation is decreased.

It is apparent from each of these zero discharge examples that the combination of the appropriate technologies will create a synergy that makes zero discharge an achievable and practical goal. Also of note is the wide range of processes and industries that are achieving zero discharge despite adverse conditions. The zero discharge technology is available; the challenge lies in applying these technologies to a specific wastewater and evaluating the cost/benefits of such an endeavor.

Formosa Studies Zero Discharge Options

The initial thrust of the study was to delineate all possible ways to eliminate the plant's wastewater discharge to Lavaca Bay. The following techniques were considered:

- Wastewater disposal into a salt dome
- Deep well injection

- Discharge into a manmade wetland
- Evaporation using brine vapor compression concentration and crystallization of the remaining salt
- Recycle of the wastewater back into the plant

Note that the first three techniques are not technically zero discharge because they merely divert the discharge to another place in the environment. More accurately, they are zero "effluent" discharge into Lavaca Bay.

To decide which of these methods could be used to bring Formosa to zero discharge, the quality of the wastewater needed to be determined, as shown in Table 15.9. Wastewater samples were taken from the four wastestreams exiting the plant. The quality and flows of these wastestreams are reported in this table.

From these data, it was determined that the final discharge (from the total combined streams) was low in organic contaminants and high in dissolved solids, especially salts of Na and Cl. Most of this salt was a waste product of the chlor/alkali plant (IEM). Each day, roughly 80 tons of salt were generated as wastewater brine by the IEM plant. Due to this high concentration, the salt (brine) was considered the main contaminant in the wastewater. Thus, to attain zero discharge, the salt had to be significantly reduced.

Recognizing that the chlor/alkali plant consumed over 3400 tons per day of salt to produce chlorine gas and caustic introduced the possibility of recycling the wastewater brine as a "feedstock" to the chlor/alkali plant. The critical question was: "Could the wastewater

	Organic Stream (ppm)	Demin-Eralizer Stream (ppm)	Cooling Tower Blowdown (ppm)	IEM Stream (ppm)	Total Combined (ppm)
Na^+	3580	485	55	17,180	5356
Cl^-	4570	2230	171	28,920	14,071
Ca^{+2}	0	297	70	1025	259
Mg^{+2}	3	33	10	30	11
HCO_3	1160	0	58	256	609
SO_4^{-2}	821	40	98	7684	1413
TDS (%)	1	0.4	0.05	5.5	1.8
Flow (million gal/d)	3.56	1.22	1.05	1.73	7.56

TABLE 15.9 Significant Wastestream Water Quality

brine be made suitable for makeup to the chlor/alkali plant?" In order for this to happen, two issues had to be addressed:

1. The required makeup for the chlor/alkali plant was at a much higher brine concentration (26 percent) than the wastewater (2 percent).
2. There were potential impurities in the wastewater that had to be removed before it could be reintroduced into the plant.

To raise the brine concentration in the wastewater, either water had to be removed or more salt had to be added to the wastewater. Since the addition of salt to the wastewater did not appear feasible, various technologies were considered to remove water. The two types of technologies considered were:

1. Evaporation of water through a solar pond or wastewater cooling tower
2. Concentration of the brine solution through reverse osmosis or vapor compression

To address the second concern about impurities in the wastewater, Formosa took samples to characterize 16 wastestreams throughout the plant. From the results, Formosa created a water balance to define the entire plant's waste flows and characteristics. Sulfate was singled out as a major potential contaminant for brine reuse due to its potential to cause scaling throughout the plant. It was found that the sulfate could be removed by precipitation with calcium chloride.

Zero Effluent Discharge Technologies Appropriate for Formosa

Several technologies were combined to create possible zero effluent discharge systems for implementation at Formosa. The individual components of these zero discharge systems are described in Fig. 15.8 while Fig. 15.9 lists schematic diagrams of these systems. With the exception of System 1, the no-action alternative, each of the systems meets the goal of zero liquid discharge to Lavaca Bay. Also, each of these alternatives uses reverse osmosis to concentrate the biologically treated wastewater. This wastewater is then combined with the previously concentrated brine (IEM) stream from the chlor/alkali plant for disposal or further treatment. These systems include:

- System 1: No-action, continue existing treatment and discharge (baseline)
- System 2: Reverse osmosis with deep-well disposal
- System 3: Wastewater recycle using reverse osmosis, solar pond evaporation, and sulfate removal

Chapter Fifteen

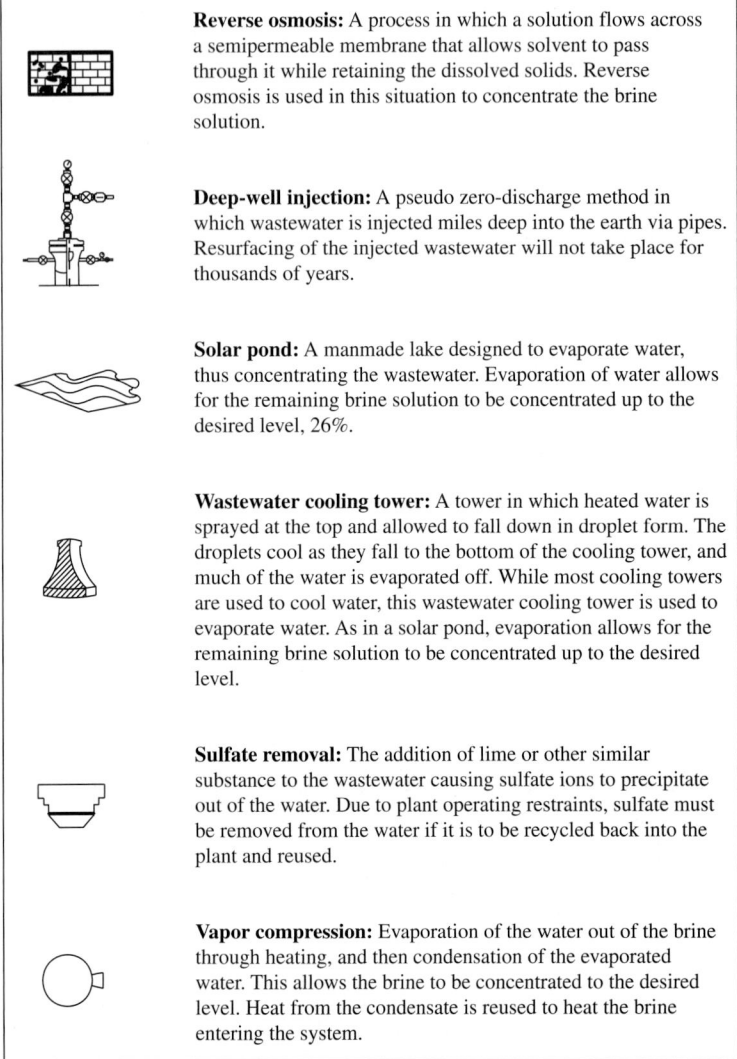

Reverse osmosis: A process in which a solution flows across a semipermeable membrane that allows solvent to pass through it while retaining the dissolved solids. Reverse osmosis is used in this situation to concentrate the brine solution.

Deep-well injection: A pseudo zero-discharge method in which wastewater is injected miles deep into the earth via pipes. Resurfacing of the injected wastewater will not take place for thousands of years.

Solar pond: A manmade lake designed to evaporate water, thus concentrating the wastewater. Evaporation of water allows for the remaining brine solution to be concentrated up to the desired level, 26%.

Wastewater cooling tower: A tower in which heated water is sprayed at the top and allowed to fall down in droplet form. The droplets cool as they fall to the bottom of the cooling tower, and much of the water is evaporated off. While most cooling towers are used to cool water, this wastewater cooling tower is used to evaporate water. As in a solar pond, evaporation allows for the remaining brine solution to be concentrated up to the desired level.

Sulfate removal: The addition of lime or other similar substance to the wastewater causing sulfate ions to precipitate out of the water. Due to plant operating restraints, sulfate must be removed from the water if it is to be recycled back into the plant and reused.

Vapor compression: Evaporation of the water out of the brine through heating, and then condensation of the evaporated water. This allows the brine to be concentrated to the desired level. Heat from the condensate is reused to heat the brine entering the system.

FIGURE 15.8 Zero-discharge candidate technologies.

- System 4: Wastewater recycle using reverse osmosis, wastewater cooling tower, and sulfate removal
- System 4A: Wastewater recycle using reverse osmosis, wastewater cooling tower, vapor compression concentration, and sulfate removal
- System 5: Wastewater recycle using reverse osmosis, vapor compression concentration, and sulfate removal

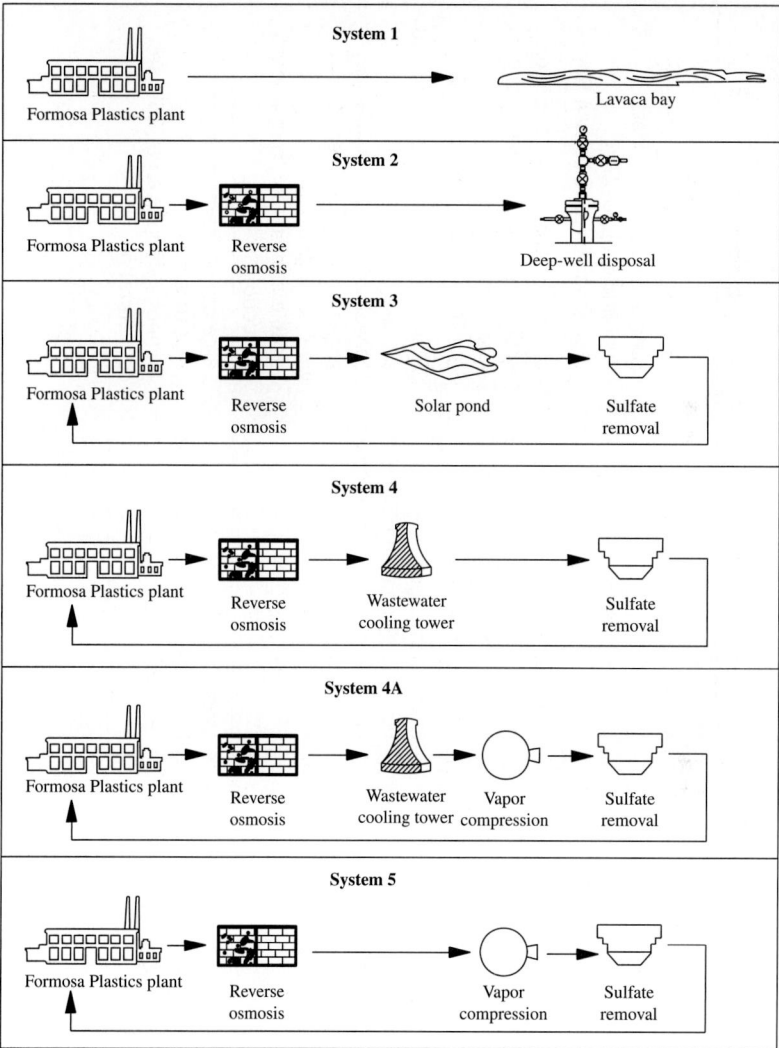

FIGURE 15.9 Zero-discharge system schematics.

Initial Evaluation Results

This study provided the parties involved with technical data and systems for reference in determining possible application of zero discharge at Formosa. These systems are compared in Table 15.10 according to economic benefit and environmental superiority from the standpoint of the authors.[10]

These technologies are all state of the art. However, the wastewater cooling tower presents the highest level of technical uncertainty as in System 4. In order to meet the chlor/alkali plant requirements, the

System No.	Description	Capital $MM	O&M* $/1000 Gal.† Removed from Discharge	Economic and Environmental Issues	
				Advantages	**Disadvantages**
1	No additional action	0‡	$4.15	Meets permit requirements No additional capital costs	Not zero discharge Continuous monitoring costs Continued permitting costs Local public opposition Potential damage to Lavaca Bay
2	Reverse osmosis with deep-well injection	$37.1	$7.12	Low energy cost No additional solids generated or atmospheric emissions Proven technology	Potential negative public opinion Holding capacity or spare wells Water lost to well Significant capital investment
3	Reverse osmosis with solar ponds and sulfate removal	$164.9 ($72.9)§	$18.92	Meets zero discharge Low energy consumption Potential for $CaSO_4$ by-product recovery	Significant land sacrifices to dead salt lake Technology unproven in full-scale Large quantities of chemicals input to system Variable brine quality may impact production
4	Reverse osmosis with wastewater cooling towers with sulfate removal	$29.6	$12.27¶	Meets zero discharge Potential for $CaSO_4$ by-product recovery	Large quantities of chemicals input to system Technology unproven at 26% salt concentration

4A	Reverse osmosis with wastewater cooling towers and Vapor compression concentrator with sulfate removal	$35.5	$11.05¶	Meets zero discharge Wastewater recycled not lost to atmosphere or in wells Proven technology Potential for $CaSO_4$ by-product recovery	High energy requirement Large quantities of chemicals input to system Large quantities of sludge generated requiring disposal
4A	Reverse osmosis with vapor compression concentrate and sulfate removal	$33.2	$13.27	Meets zero discharge Wastewater recycled not lost to atmosphere or in wells Proven technology Potential for $CaSO_4$ by-product recovery	High energy requirement Large quantities of chemicals input to system Large quantities of sludge generated requiring disposal Significant capital investment

*Includes $4.15 for existing treatment system.
†One dollar/1000 gal represents an approximate cost of $700,000/year to Formosa.
‡Initial investment in existing wastewater treatment was approximately $40 million.
§Without liner in ponds.
¶Assuming recoverable "waste" heat is available for heating.

TABLE 15.10 Summary of Costs and Environmental Issues for Viable Zero Discharge Alternatives (1998 Dollars)

water would have to be concentrated to 26 percent brine. To date, wastewater cooling towers have only been operated up to 15 percent brine. System 4A was developed to remove this technical risk by adding vapor compression concentration to the system. The cooling tower would concentrate the brine to 15 percent and the vapor compressor would further concentrate it to 26 percent. Also, the tower requires the water to be heated for evaporation to take place. Thus, to operate the cooling tower, a source of heat had to be found. If waste heat is available, it could be transferred to the wastewater through heat exchangers. A separate loop of heated wastewater could then be circulated in the tower, giving rise to evaporation.

The next step in the study was to cost each system at a planning level, that is approximately 30 percent accuracy. Capital costs were amortized over 11 years. Equipment costs were obtained from vendor estimates. Energy costs were determined to be 3.0 cents/kWh from the cogeneration plant on-site. The results of this cost analysis and associated environmental issues for the systems are presented in Table 15.7. While this concept appears technically feasible, economic considerations, the excellent compliance record of the current wastewater treatment and discharge system, and future water availability all enter into the final decision-making process. A proposed reuse-zero discharge flow diagram is shown in Fig. 15.10.

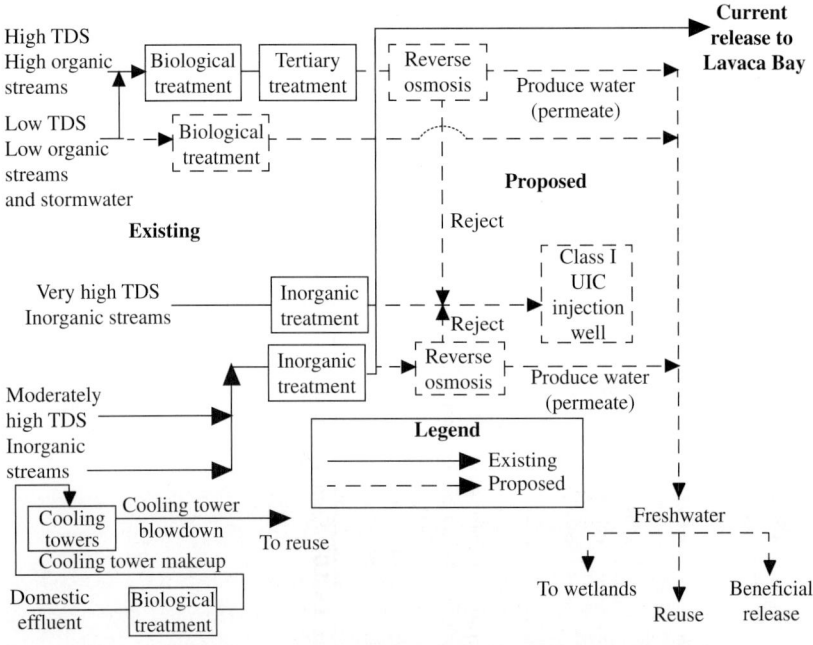

FIGURE 15.10 Proposed reuse-zero discharge flow diagram.

Potential for Mysid Shrimp Toxicity

- Major ion concentrations in deficit or excess cause toxic conditions for mysid shrimp - Pillard, et al. (2000)
- Major ions applicable to FPC are presented below along with the range of concentrations found in the final effluent

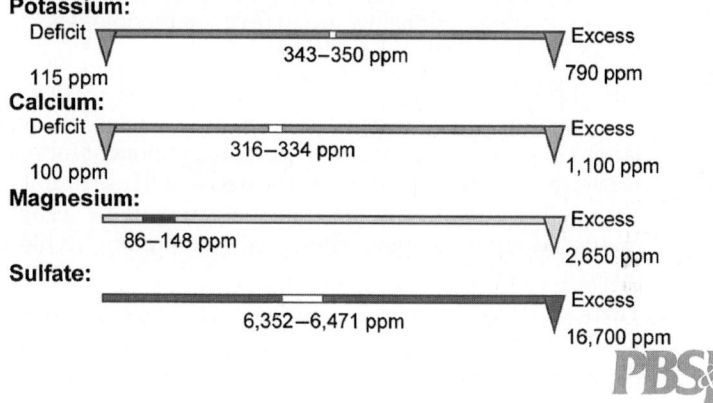

FIGURE 15.11 Effects of potential inorganic toxicity on mysid shrimp.

Recycle Effects on Effluent Toxicity Testing

Before a recycle/reuse program is finalized, one must make sure the test species in the effluent toxicity permitted criteria is not adversely affected, attributable to the change in magnitude and/or constituent ratios in the resultant dissolved solid makeup. A recent study evaluated this specific effect. In the case of Formosa, Mycid shrimp are the test species and major deficit or excess concentration can result in toxicity. Based on the proposed recycle flow mass balances, the range of anticipated cations and anions are predicted in Fig. 15.11. In this case, these concentrations are not expected to adversely affect compliance with the permit bioassay limitations. Such studies should be included in a reuse/recycle program to minimize the chance with problems with toxicity tests required in discharge permits.[13, 14]

15.6 Summary

The concepts of zero discharge, zero effluent discharge, available technologies, and regulatory conflicts are obviously complex. However, these can be summarized as follows:

1. Zero effluent discharge does not conflict with the natural laws (as does "theoretical" zero discharge), and for selected industries and locations, it is a goal which can be both environmentally superior and economically beneficial.

2. There are technologies in place to achieve this goal, but they must be intelligently selected, tested as required, carefully designed, properly costed, and effectively operated.

3. The inclusion of certain residual chemicals can be injected into subsurface formations by UIC Class I or II nonhazardous wells, or disposed using alternative methodologies, thus achieving "zero effluent discharge."

4. The application of deepwells to inject selected wastes into isolated and defined subsurface formations under UIC regulations may be a prudent and important component for many wastewater treatment systems. The decision to use or not use deepwells should be site-specific and based on the technical merits. Alternative approaches may be more applicable for given locations.

5. There needs to be a continuous reevaluation of selected regulations which result in disincentives pursuant to the quest of designing environmentally superior and economically beneficial wastewater treatment systems.

6. Zero effluent discharge potentially can provide an array of additional benefits such as a reduction in raw water demand, possible plant expansion under current permits, less risk for noncompliant discharge events, reduced monitoring costs, and better public perception. Moreover, it improves security of a plant's water supply during drought conditions. This is an environmental enhancement/cost benefit decision.

15.7 Problem

15.1. An industrial complex has a raw water supply with a TDS of 500 ppm. The water demand is 12 million gal/d. The TDS standard for the effluent back to the same water source is 1500 ppm. The estimated flow to the end-of-pipe treatment plant is 4 MGD. The treatment plant influent design characteristics are:

Average design flow	4.0 million gal/d
BODs	500 ppm
COD	3560 ppm
TDS	3150 ppm
Oil and grease	650 ppm
TSS	350 ppm
Benzene	180 ppm
Total chlorinated hydrocarbons	250 ppm

1. Create a detailed process flow diagram (PFD) with the unit processes necessary to meet the NPDES permit:

BODs	50 ppm
COD	700 ppm
TSS	<10 ppm
TDS	1500 ppm
Benzene	<3 ppm
Oil and grease	<10 ppm
Total chlorinated hydrocarbons	20 ppm

2. If effluent flow reductions of 20 to 40 percent are necessary because of water limitations and still meet the effluent criteria, how can this be achieved within the treatment complex? Show the revised PFD to accomplish a recycle range of 20 to 40 percent.

3. Estimate a constituent profile across such unit process for the required effluent parameters. Minimal required unit processes include, but are not limited to:
   - Aerobic biological treatment
   - Activated carbon
   - Effluent polishing
   - Reverse osmosis or equal

References

1. Eller, J., D. Ford, and E. F. Gloyna: "A Review of Water Reuse in Industry," Presented at the AWWA Meeting, San Diego, California, May 19–22, 1969.
2. Carnes, B. A., J. Eller, and D. L. Ford: "Integrated Reuse-Recycle Treatment Processes Applicable to Refinery and Petrochemical Wastewaters," American Society of Mechanical Engineers, Bulletin 72-PID-2, New York, 1973.
3. McIntyre, J. P.: *Industrial Water Reuse and Wastewater Minimization*, BetzDearborn, Inc., Horsham, Pa, 1998.
4. GESAP Sanitation Program, Internet, http://net21.gec.jp/GESAP/themes/themes2.html.
5. Ford, D. L.: "Zero Discharge and Environmental Regulations, The Toxic Release Inventory and Natural Laws," *Environmental Engineer*, vol. 32, no. 4, October 1996.
6. EPA Clean Water Act as Amended, 47FR46446, 1982, as amended, 1985.
7. Venkateh, M., and T. Pellerin: "Source Minimization Techniques and Concepts for Petroleum Refineries," *Environ. Conf.*, ENSR International, Dallas, September 2005.
8. Ford, D. L., and J. M. Eller: "An Evaluation of Storm Runoff Management," unpublished report, 2000.
9. Nobel, C. E., D. Allen, and D. R. Maidment: "A Model for Industrial Water Reuse—A Geographic Information Systems (GIS) Approach to Industrial Ecology," Center for Research in Water Resources, The University of Texas at Austin, 1998.
10. Matson, J., D. Tiffin, J. McLeod, and B. Jordan: "Zero Discharge Technology, a Case Study," *EPA Region III Waste Minimization Pollution Prevention Conference for Hazardous Waste Generators*, Philadelphia, June 3–5, 1996.

11. Ford, D. L.: "A Case History of Environmental Evolution in a Complex Chemical Plant," *DeLange Woodlands Conference*, Rice University, Houston, March 3–5, 1997.
12. Ford, D. L., J. Blackburn, and K. Mounger: "Wilson-Formosa Zero Discharge Agreement," Unpublished, July, 1994.
13. Pillard, D. A., D. L. DuFresne, J. E. Caudle, and J. M. Evans, "Predicting the Toxicity of Major Ions in Seawater to Mysid Shrimp (*Mysidopsis Bahia*), Sheephead Minnow (*Cyprinodon variegatus*), and Inland Silverside Minnow (*Menidia beryllina*)," *Environmental Toxicology and Chemistry*, Vol. 10, *Annual Review,* 2000. Setac Press, Printed in the USA.
14. Jensen, Paul, Horne, J., Lahr, E., Hyak, John, and Ford, D., "How Knowledge Gained in Toxicity Testing Can Help in Water Conservation." Paper presented at SETAC Regional Conference, Houston, Texas, May 16, 2008.

CHAPTER 16
Allocation of Superfund Disposal Site Response Costs

16.1 Introduction and Literature Review

Equitable allocation of environmental cleanup costs between the multiple party users of disposal sites or industrial production areas is one of the more contentious issues currently facing environmental engineers/scientists. These potentially responsible parties (PRPs) obviously have a vested interest in minimizing their allocated costs, so the potential for dispute and subsequent litigation is enormous. This is true not only with Superfund sites where joint PRP defense groups are formed as the process moves through the National Contingency Plan (NCP), but also with State Superfund, Resource Conservation and Recovery Act (RCRA), and voluntary cleanup projects requiring remediation to levels complying with specified numerical or risk-based criteria.

The allocation of the costs can be extremely complex based on the physical, chemical, and biochemical nature of the contaminants and their behavior in the environment. This situation is exacerbated by the paucity of historical information common with many sites, particularly those which were active prior to enactment of the Comprehensive Environmental Response, Compensation, and Liability Act (CERCLA) and RCRA implementing regulations in 1980. Moreover, there are no uniformly accepted guidelines or formulae for allocation calculations.

The intent of this chapter is therefore to outline several approaches the authors have applied in previous allocation projects, and to bring forward the corresponding literature and references which have benefited the quest of developing a format for equitable allocation. One should recognize, however, that each allocation project is unique, and sufficient flexibility is necessary to incorporate basic engineering and scientific principles with site-specific factors.

A cynic might state (with some justification) that if one does the allocation properly, then none of the PRPs are satisfied. The allocator might start with this rather pessimistic approach, but work toward a scientifically and factually based methodology which is both equitable and sufficiently documented to withstand peer review or reversal via negotiation, arbitration, or litigation processes.

There are several literature sources that provide starting points on the allocation process. One of the earliest approaches was developed by Harry LeGrand, who developed a system for estimating contamination potential at disposal sites.[1] He incorporated five site-specific factors, including water table locations, sorption, permeability, gradient, and distance, which translated into a point system to quantify contamination potential. This method was modified by the U.S. Environmental Protection Agency (EPA) in their Surface Impoundment Assessment (SIA) manual and further refined by EPA in the development of a uniform rating system for use in Air Force installation restoration programs.[2] Additional background information and approaches are cited in an unsuccessful amendment to CERCLA outlining an approach to joint and several liability issues known as the *Gore factors*,[3] an expert report on Superfund allocation principles,[4] and experience with the allocation process.[5] The Gore factors include the amount of hazardous substances involved, the degree of hazard, the degree of party involvement, and the degree of care exercised by the respective party.[3] These factors are directionally correct, but obviously oversimplified without some form of technical quantification. The Bernheim expert report partially addressed some of these vagaries by proposing a calculation based on "pure volumetric shares" and "weighted waste strength shares" using toxicity, mobility, and physical state as input numbers.[4]

There is a recent court opinion that partially incorporates the principles outlined by Bernheim.[6,7] The Court's opinion can be summarized as follows:

- It is impossible to calculate with mathematical precision the extent to which volume and toxicity contribute to the extent of response costs.
- Half of the response costs will be allocated to wastestreams which contributed to response costs primarily attributable to volume.
- Half of the response costs will be allocated to high strength wastestreams that contributed to response costs primarily attributable to their toxicity.

The Court further acknowledged that the "toxicity side" of the allocation most likely influenced the site's placement on the National Priority List (NPL) and the EPA's selection of the remedy. To further quantify a scoring approach on the "toxicity" side of the allocation, a firm was retained to develop a scoring format using toxicity, mobility, content (relative concentration), and physical state (liquids, sludges, or solids) as the key components.[7] The authors added persistence factors and interim capital improvements credits to this list when applicable in previously conducted allocation projects as will be discussed subsequently.

These and other allocation methods have been summarized previously by Ford and are presented in Table 16.1.[5]

Allocate by	Applicability
Volume/weight	Most applicable when wastes over time are generally similar in characteristics and at least some historical records are available.
Operating time	Applicable when no major change in mode of operation and/or production occurs throughout the period in question. Can be adjusted if certain "operation eras" can be defined.
Productive history	Applicable if residual generation per unit of production can be estimated.
Degree of hazard or toxicity weighting factors	Difficult to quantify, but may be applicable if key residuals are EPA-listed as hazardous or toxic. Pathways to receptors, including such factors as attenuation, retardation, and/or natural removal/reduction, should be considered.
Source identification and allocation	Applicable when distinct timelines can be defined with respect to production modes, feedstock or product characteristics, use of additives, or other chronological idiosyncrasies.
Remedy driven allocation	Applicable when required remediation methods are attributable to selected constituents and source(s) of such constituents can be reasonably defined. May defer allocation resolution until remedy has been finalized.
Spills, major leaks, cataclysmic events	Applicable when major spills, leaks, or other events are documented and reasonably quantified. Can be put into perspective with residuals generated through normal operations generally in compliance with regulations and standards all the time.
Plume evaluation and proration	Applicable when subsurface plumes can be reasonably defined and attributable to source(s). Less appropriate when there are multiple potential sources and plumes are defined by excessive interpolation.

TABLE 16.1 Cleanup Allocation Method Applicability

16.2 Cost Allocation Principles

The major principles underlying cost allocation are discussed next.

Volume, Weight, Operating Time History

The volume, weight, and operating time history for various owners, operators, or lessors is obviously a starting point. The simplest calculation is based on the site use history (illustrated in Fig. 16.1[a]). If the regulatory climate, production, or use modes and volume/production history are similar throughout the allocation regime, then the use percentage or volume/production percentage (illustrated in Fig. 16.1[b]) can be used directly for cost allocation purposes.

This most likely is an oversimplification primarily because:

- A more restrictive regulatory climate, particularly from 1970 to present, infers increasingly less contamination to the environment over time, that is, older, unregulated "dirty" years versus more recently regulated "cleaner" years.

- Production configuration and levels change throughout the allocation regime, as does the "contamination or residual generation per unit of production."

- Capital expenditures in environmental protection improvements can be made during the allocation regime, and this, combined with the "aging" effect of certain contaminants and contract off-site disposal, may act as credits in the overall allocation process.

A more complex and equitable allocation involves extensive research and document review. The on-site generation of residuals should be approximated over time as a starting point. This is based on production rates, resultant solid and liquid waste generation, and the on-site or off-site receptors of such residuals, either within the industrial complex or off-site disposal areas. These waste depositories are often subdivided into operable units or solid waste management units (SWMUs) for administrative, remedy, and cost allocation purposes.

An example of a stepwise procedure for the data gathering phase for a petroleum refinery site used by three different refining companies is illustrated as follows, and discussed in more detail in Sec. 16.4.

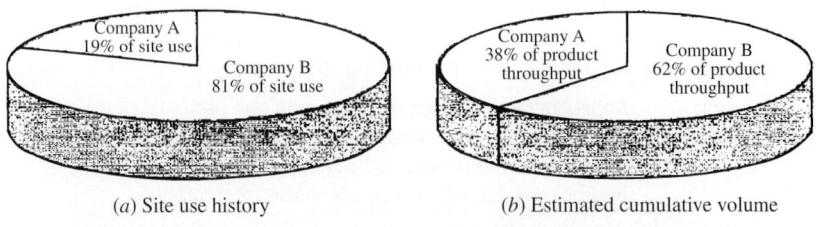

(a) Site use history   (b) Estimated cumulative volume

Figure 16.1 Allocation on basis of ownership or product throughput.

Allocation of Superfund Disposal Site Response Costs   829

- Gather production rates over time, such as the crude throughput rates for the petroleum refinery, shown in Fig. 16.2.
- Estimate effluent flow rates for the same period depicted in Fig. 16.3.

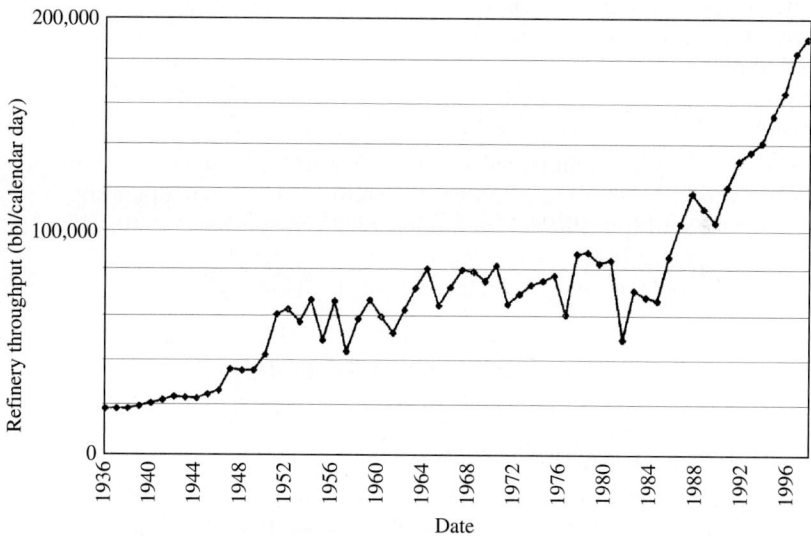

FIGURE 16.2 Refinery throughput.

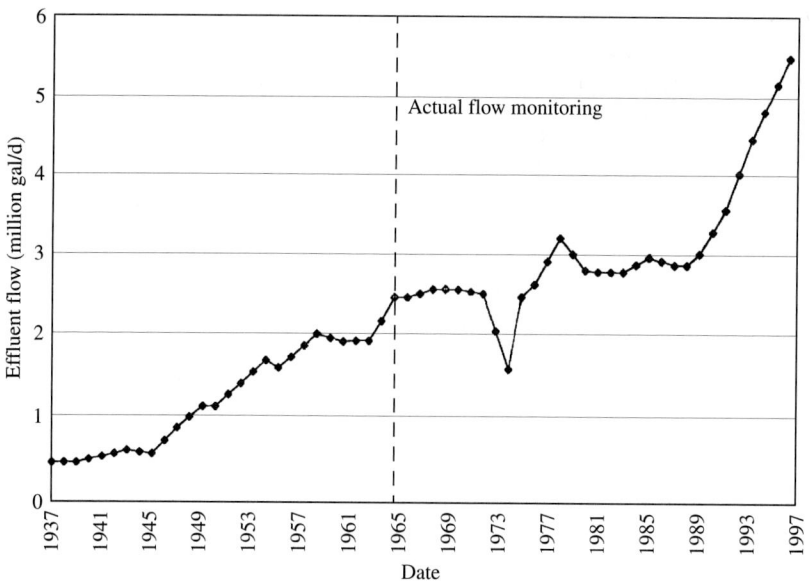

FIGURE 16.3 Estimated and actual refinery wastewater effluent flows.

- Estimate solid wastes generated during the same period, such as effluent flow-related sludges. An example generation estimate is shown in Table 16.2.
- Estimate the portion, if applicable, which was transported off-site to public or commercial waste disposal site (trip tickets, waste manifests, truck drivers' testimonies, and so on) for a given period.
- Attempt to quantify the origins of waste residual hauled off-site as illustrated in Fig. 16.4.

For example, commercial, industrial and municipal disposal facilities will receive many different residuals over their operating lives from multiple PRPs. Evaluation of the respective "contents" from these contributors, therefore, must be conducted.

An example of this "content" is shown in Table 16.3.

16.3 Contaminant Selection, Mobility, Toxicity, Persistence, Content, and Physical State— Multiple Off-Site Contributors

There are a variety of approaches to quantify the response cost drivers such as contaminant selection, mobility, toxicity, persistence, content, and physical state. One approach, previously applied by the authors, starts with the format shown in Table 16.4. This example applies to off-site contributors sending wastestreams to a common disposal site. The methodology that serves as the predicate for the recommended allocation formula is described as follows:

1. *Select the target constituents.* In this example, 12 organic and inorganic constituents were selected for allocation purposes on the basis of potential toxicity (inclusion as hazardous constituents, which have been detected at the targeted site at concentrations that exceed regulatory criteria or action levels). The selected constituents are listed in Table 16.5.

2. *Calculate a combined mobility, toxicity, and persistence factor to each target constituent.* The mobility factor components for the target constituents include water solubility (the ability to move in the liquid phase), adsorption affinity (the effect of sorbing to the solid phase), and volatility (the relative ability to be stripped from solution based on constituent vapor pressure). In this example, adsorption is estimated to be 80 percent and volatility is assumed to be 20 percent as significant as solubility. The weighting factors assigned to each of the 12 constituents and this proration are subjective but are based on personal engineering and scientific experience, the literature, and best professional judgment.

Type of Solid Waste	Generation	Percent of Total	Generation Frequency
API sludge	1,148,500	34%	3 units on 6–8 year cycle
Cooling tower sludge	83,300	2%	On 2–5 year cycle
Tank bottoms	243,963	7%	10 years of product change
Wastewater sludge	1,332,800	39%	Lagoon removal 2–3 year cycle
Spent caustic solution	228,327	7%	Norm neutralized on-site
Spent acid solution	103,168	3%	HCK for HX tubes neutralization on-site
Gasoline steamout condensate	261,000	8%	Not frequently
Phosporic acid catalyst fines	0	0%	Every other month
Sum	3,401,058	100%	

TABLE 16.2 Estimated Refinery Waste Sludge Generation Over a Defined Time Period

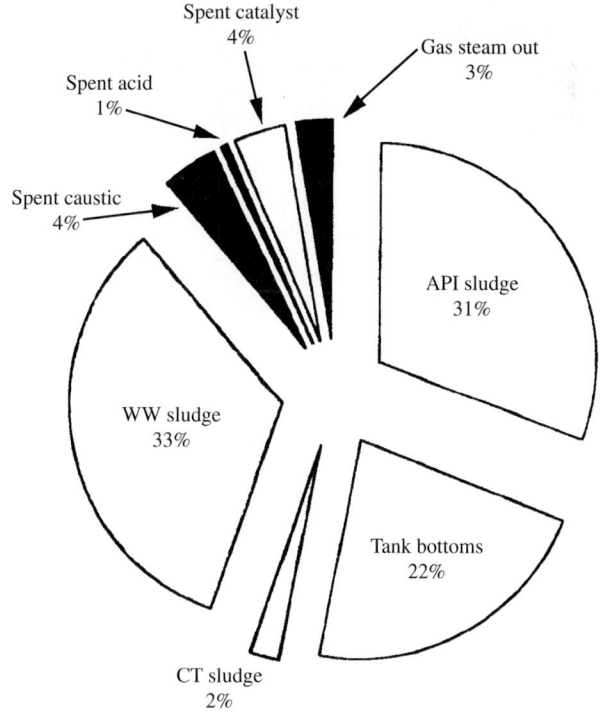

FIGURE 16.4 Estimated waste shipments to an off-site disposal facility.

Party	Contribution of Total lb (%)	(1000 lb) Deposited
PRP A	42.5	1,546
PRP B	1.7	62
PRP C	0.7	24
PRP D	1.1	40
PRP E	16.2	587
PRP F	33.0	1,199
PRP G	2.6	93
PRP H	2.2	81
	100.0	3,634

TABLE 16.3 Mass Sludge Contributions from Multiple PRPs

Waste Source	% Contribution
Spent caustic	4%
Spent acid	1%
Spent catalyst	4%
Gas steam out	3%
CT sludge	2%
WW sludge	33%
API sludge	31%
Tank bottoms	22%

TABLE 16.4 Estimated Waste Shipments to Off-Site Disposal Facility

These numbers obviously can be adjusted or refined as required by the allocator. The combined mobility factor for each target constituent is the sum of solubility weighting factor plus 80 percent of the adsorption factor plus 20 percent of the volatility factor.

3. *Calculate the combined toxicity/mobility/persistence (tox/mob/per) factor for each target constituent.* This factor is calculated by adding the combined mobility factor and the toxicity factor, then multiplying by the persistence factor. The logic of this approach can be evaluated by noting the range of the results shown in Table 16.5. For example, arsenic ranks highest in

		Mobility Factor Components						
Constituent	Solubility Factor	Adsorption Factor	Volatility Factor	Combined Mobility Factor	Tox Factor	Persistence Factor	Combined Tox/Mob/Per Factor
Benzene	8	8	6	15.6	27	0.2	8.52
Chlorobenzene	6	6	6	12	9	0.2	4.2
1,1-Dichloroethane	8	8	8	16	3	0.5	9.5
cis-1,2-Dichloroethene	8	8	8	16	9	0.5	12.5
Ethylbenzene	6	6	4	11.6	1	0.2	2.52
Trichloroethene	8	6	6	14	27	0.5	20.5
Toluene	8	6	6	14	1	0.2	3
1,4-Dichlorobenzene	2	4	2	8	3	0.5	5.5
Arsenic	2	4	2	8	81	1	89
Chromium	2	2	2	6	3	1	9
Lead	1	2	2	5	9	1	14
Mercury	1	1	2	4	9	1	13

Notes: **(a) Solubility factor:** Based on solubility of target compound in water @20°C. The factor ranges from 1 (solubility < 1 mg/L) to 10 (solubility > 10,000 mg/L) generally following order of magnitude delineation.
(b) Adsorption factor: Based on sediment adsorption affinity (SAA) of target compound. The factor ranges from 10 (SAA < 10) to 1 (SAA > 100,000).
(c) Volatility factor: Based on vapor pressure of target compound. The factor ranges from 2 (VP < 1 mm Hg) to 10 (VP > 1,000 mm Hg).
(d) Combined mobility factor: Calculated as follows:
(Toxicity factor) + 0.8 (adsorption factor) + 0.2 (volatility factor).
(e) Toxicity factor: Taken as the reciprocal of the regulatory or action level concentration of the target compound and classified categorically as shown in Column (e), implying a relative "toxicity potential" magnitude.
(f) Persistence factor: Based on resistance to biodegradation. The factor ranges from 1.0 (relatively nondegradable) to 0.2 (relatively degradable).
(g) Combined tox/mob/per factor: Calculated as follows:
([Toxicity factor] + [combined mobility factor]) × (persistence factor).
Source: Adapted from Verschueren, 1983.[8]

TABLE 16.5 Tox/Mob/Per Factor Development

this example because of its exceedance over the regulatory limit (high toxicity factor) and its ability to persist in the environment (resistance to biodegradability) which more than offsets its lower mobility relative to such compounds as benzene, dichloroethane, and dichloroethene.

The weighting factor development diagram is illustrated in Fig. 16.5.

The basis for quantifying each of the factors shown in Table 16.5 (columns [a], [b], [c], [d], and [f]) into a combined tox/mob/per factor [g]) is described as follows:

- The toxicity factor for the grouped constituents are based on the reciprocal of the specified cleanup criteria (groundwater, RCRA TCLP [toxic characteristic leaching procedure], and so on), as shown in Table 16.6.
- The solubility factor is based on the grouped constituent solubility in water, as shown in Table 16.7.
- The adsorption affinity factor combines the log of the solubility and log of partition characteristic of the constituent, then grouped, as shown in Table 16.8.
- The volatility factor development is based on the constituent vapor pressure and grouped, as shown in Table 16.9.
- The persistence factor is based on the relative biodegradability of the specified constituent, ranked based on personal experience and the literature, and grouped from 0.2 for readily biodegradable constituents such as benzene to 1.0 for the more inert constituents such as the heavy metals (lead, arsenic, and so on).
- Once the tox/mob/per factor has been determined, then each compound is multiplied by a "physical state" factor (liquids, sludges, solids, or some combination thereof), then multiplied by the combined "content" of each stream or mass, shown in Table 16.10. The physical state factor depends on the significance of volume or weight applicability for the waste site receptor. This exercise can be repeated for each PRP and the allocation can be calculated for each. The respective allocations depend on the combined remedy-driven characteristics, the physical state, and the mass from each PRP contributor, then dividing that value by the total value from all contributors.

Persistence and the other factors discussed herein can be estimated to some extent based on the applicability of the selected remediation process. An excellent example of this is the use of Table 16.11 as a roadmap for the selection of the aforementioned factors.[9]

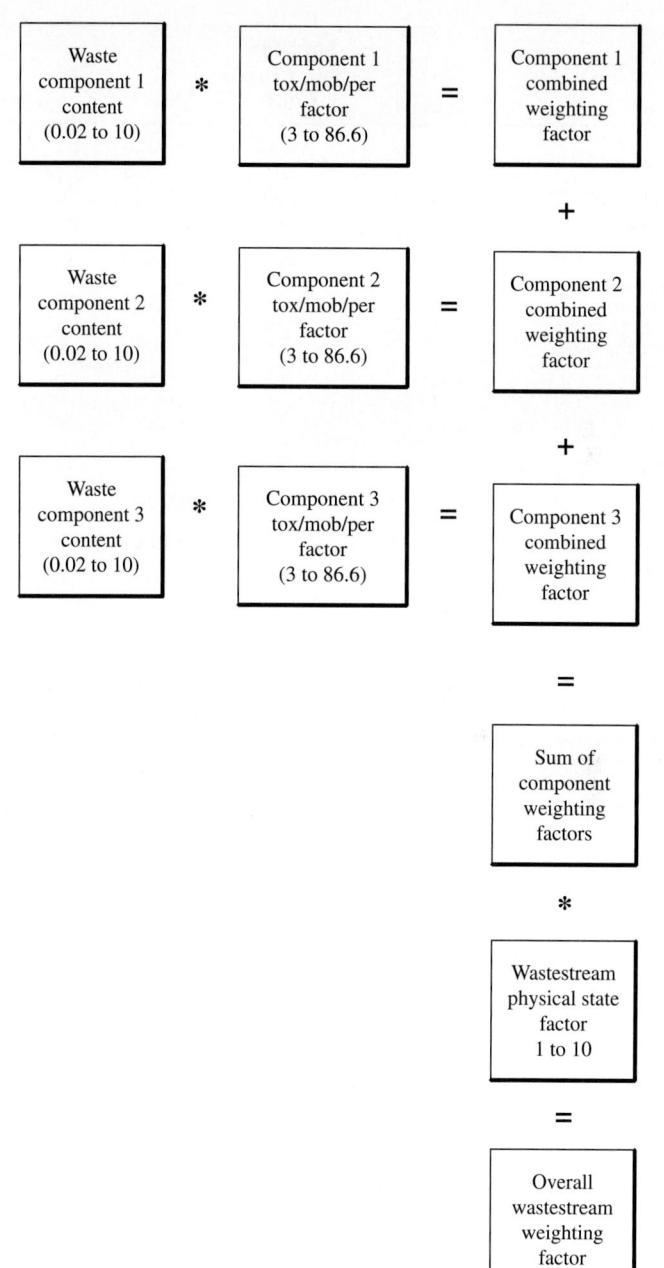

Figure 16.5 Weighting factor development diagram.

Constituent	Com. No.	Specified GW Criteria (μg/L)	1/X	Tox Factor
Arsenic	16	0.02	50.000	81
bis(2-Chloroethyl)ether	13	0.03	33.333	81
Vinyl chloride	10	0.08	12.500	81
Benzene	1	0.2	5.000	27
Trichloroethene	8	1	1.000	27
Methylene chloride	6	2	0.500	9
Mercury	20	2	0.500	9
bis(2-Ethylhexyl)phthalate	14	3	0.333	9
Chlorobenzene	2	5	0.200	9
Lead	19	5	0.200	9
1,2,4-Trichlorobenzene	15	9	0.111	9
cis-1,2-Dichloroethene	4	10	0.100	9
Xylene	11	40	0.025	3
Asbestos	17	N/A	N/A	N/A
1,1-Dichloroethane	3	70	0.014	3
1,4-Dichlorobenzene	12	75	0.013	3
Chromium	18	100	0.010	3
MEK	7	300	0.003	1
Ethylbenzene	5	700	0.001	1
Toluene	9	1000	0.001	1

TABLE 16.6 Toxicity Factor Development

In this approach, the combined tox/mob/per factor calculated for benzene and shown in Table 16.5 is:

Combined mobility factor + toxicity factor × persistence factor = combined tox/mob/per factor

$$[\{(8) + 0.8(8) + 0.2(6)\} + 27](0.2) = 8.52$$

Although columns (a) through (e) are based on constituent properties, the weighting factors involved in calculating the combined factors are more subjective and can be adjusted according to the allocator's own judgment.

Com. No.	Constituent	Solubility in Water (mg/L) @ 20°C	Solubility Factor
7	MEK	26,8000	10
6	Methylene chloride	20,000	10
13	bis(2-Chloroethyl)ether	10,200	10
3	1,1-Dichloroethane	5500	8
4	cis-1,2-Dichloroethene	3500	8
10	Vinyl chloride	2670	8
1	Benzene	1750	8
8	Trichloroethene	1100	8
9	Toluene	1000	8
2	Chlorobenzene	466	6
11	Xylene	175	6
5	Ethylbenzene	120	6
12	1,4-Dichlorobenzene	7.9	2
15	1,2,4-Trichlorobenzene	3	2
16	Arsenic	1.3	2
18	Chromium	1	2
19	Lead	0.9	1
14	bis(2-Ethylhexyl)phthalate	0.285	1
20	Mercury	0.01	1
17	Asbestos	Nil	1

TABLE 16.7 Solubility Factor Development

The columns are then combined according to the weighting factor formula:

$$\text{Combined tox/mob/per factor} = [\{(a) + 0.8(b) + 0.2(c)\} + e](f)$$

16.4 Allocation Methods for Same-Site Multiple Contributors

Whereas the aforementioned scoring format based on weighting factors could logically be applied to disposal sites with multiple off-site

Com. No.	Constituent	Solubility in Water (mg/L) @ 20°C	Log Sol ppb	Estimated* Log P Sed	Estimated† Sediment Adsorption Affinity	Adsorption‡ Affinity Factor
7	MEK	268000	8.4281	0.437	3	10
6	Methylene chloride	20000	7.3010	1.181	15	8
13	bis(2-Chloroethyl)ether	10200	7.0086	1.374	24	8
3	1,1-Dichloroethane	5500	6.7404	1.551	36	8
4	cis-1,2-Dichloroethene	3500	6.5441	1.681	48	8
10	Vinyl chloride	2670	6.4265	1.759	57	8
1	Benzene	1750	6.2430	1.880	76	8
8	Trichloroethene	1100	6.0414	2.013	103	6
9	Toluene	1000	6.0000	2.040	110	6
2	Chlorobenzene	466	5.6684	2.259	181	6
11	Xylene	175	5.2430	2.540	346	6
5	Ethylbenzene	120	5.0792	2.648	444	6
12	1,4-Dichlorobenzene	7.9	3.8976	3.428	2676	4
15	1,2,4-Trichlorobenzene	3	3.4771	3.705	5071	4
16	Arsenic	1.3	3.1139	3.945	8806	4
18	Chromium	1	3.0000	4.020	10471	2
19	Lead	0.9	2.9542	4.050	11225	2
14	bis(2-Ethylhexyl)Phthalate	0.285	2.4548	4.380	23977	2
20	Mercury	0.01	1.0000	5.340	218776	1
17	Asbestos	NIL	N/A	N/A	N/A	1

*Log P sed = 6−0.66(log sol.) based on Verschueren, 1983, p. 82–88.
†Antilog of log P sed.
‡Adsorption (partition) weighting factors grouped so that low adsorption affinity contributes to high mobility factor.

TABLE 16.8 Adsorption Factor Development

Allocation of Superfund Disposal Site Response Costs

Com. No.	Constituent	Vapor Pressure (mm Hg)	Volatility Factor
10	Vinyl chloride	2580	10
6	Methylene chloride	349	8
4	cis-1,2-Dichloroethene	200	8
3	1,1-Dichloroethane	182	8
1	Benzene	95	6
7	MEK	75	6
8	Trichloroethene	60	6
9	Toluene	28	6
2	Chlorobenzene	12	6
5	Ethylbenzene	9.5	4
11	Xylene	6.6	4
13	bis(2-Chloroethyl)ether	0.7	2
12	1,4-Dichlorobenzene	0.6	2
14	bis(2-Ethylhexyl)phthalate	NIL	2
15	1,2,4-Trichlorobenzene	NIL	2
16	Arsenic	NIL	2
17	Asbestos	NIL	2
18	Chromium	NIL	2
19	Lead	NIL	2
20	Mercury	NIL	2

TABLE 16.9 Volatility Factor Development

contributors, different elements of allocation must be considered when there are multiple users over time at the same site. An example of the authors' approach in this regard involves a large petroleum refinery operating from the mid-1930s until today, but owned by three different oil companies throughout the operating regime. Various questions for each ownership period must be answered in this regard, such as:

- Refinery throughput and process configuration
- Estimated residual generation
- Degree of regulatory control
- Capital improvements and degree of environmental oversight

Hazardous Component	Tox/mob/per Factor
Benzene	8.52
Chlorobenzene	4.2
1,1-Dichloroethane	9.5
cis-1,2-Dichloroethene	12.5
Ethylbenzene	2.52
Trichloroethene	20.5
Toluene	3
1,4-Dichlorobenzene	4.3
Arsenic	86.6
Chromium	7
Lead	12
Mercury	11.2

Physical State Factors		
Liquids	=	10
Sludges	=	6
Solids	=	1

			Physical State Factors		Mass of Wastestream Contributed		Relative Wastestream Score
Sum of tox/mob/per factors	182	×	5	×	10 tons	×	9097

TABLE 16.10 Conceptual Wastestream

Remediation Technology	Example of Chemical Property Influencing Remediation Difficulty	Example of Chemicals Easier to Remediate	Example of Chemicals More Difficult to Remediate
Groundwater extraction	Solubility	TBA, benzene	Styrene, naphthalene
Aerobic biodegradation	Degradability with oxygen	Benzene toluene, vinyl chloride TBA	1,1,2-TCA
Soil vapor extraction	Volatility from water phase	Benzene, toluene, vinyl chloride, 1,2-DCA	TBA, styrene, naphthalene
Thermal remediation	Volatility	Benzene, toluene, vinyl chloride	Styrene, naphthalene
Chemical oxidation	Degradability by oxidant (permanganate)	Vinyl chloride, toluene	Benzene, 1,2-DCA
Excavation	None	No difference	No difference

TABLE 16.11 Remediating Technology Factor Selection Guidance

- Off-site disposal
- Aging effects of previous contamination

There were over 15 SWMUs designated in the remediation action plan, necessitating a separate allocation for each SWMU as remediation costs were assigned to each unit. If a SWMU was unique to one company, that is, closed before the ownership change or opened and/or closed during one ownership regime, then a 100 percent allocation could be assigned to that company. Most, however, transcended ownership regimes and had to be allocated accordingly.

In order to develop a rational database, an extensive record search was conducted. This included, but was not limited to, crude throughput information available from the 1930s to present from the *Oil and Gas Journal*, petroleum refinery guidelines developed by EPA, survey data provided by the American Petroleum Institute (API), technical papers and texts, historical drawings, specifications, correspondence, and associated information from the respective owners and regulatory agencies.

A dynamic model was developed to incorporate this information into a basis for estimating production-based residuals deposited in each SWMU over time. The major inputs include:

- The time-dependent residual generation (chemical oxygen demand [COD] or solids load) based on each refinery throughput and residual generation per unit of throughput
- The wastewater flow over time based on the effluent flow per unit of throughput as per EPA guidelines, API survey, or actual records data
- The "aging effect" based on the in situ biological removal of the contaminants over time

The results of one of these applications for a selected SWMU are shown in Fig. 16.6.[10] The refinery production rate (crude throughput) is first plotted, with the corresponding total solid waste generation accumulation similarly shown. The resultant oil waste generation (such as tank farm bottoms, API separator bottoms, and dissolved air flotation float solids) is plotted separately as they were deposited in a separate SWMU. There are several events that affect the waste curves downward, in effect, serving as credits. This occurs when treatment units are installed, reducing or eliminating residuals placed on the site, recovery and reuse or sale of residuals, and/or contract disposal of residuals off-site because of regulatory and/or economic factors. For example, the designation of a target company as a PRP under Superfund at another location provides evidence that residuals were taken off-site and should be excluded from the on-site allocation process.

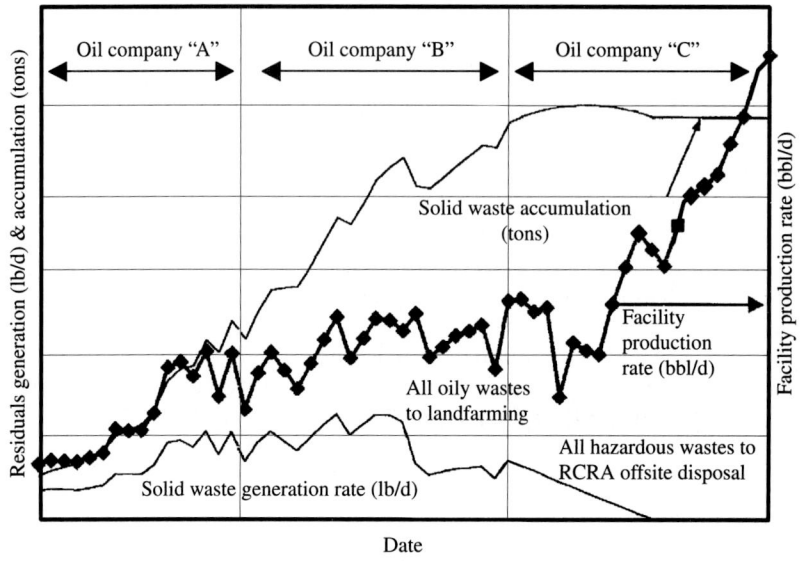

Figure 16.6 Production-based residual estimates.

The biological removal of organics in the subsurface over time is somewhat nebulous, but, in this case, was estimated from an extensive review of literature and the authors' experiences based on the oil fraction in question. The net estimated output of residuals shown in Fig. 16.6 is the primary basis for assignment of allocation. This is calculated by dividing the target contaminant accumulation for each ownership company by the total.

16.5 Summary

In summary, methodologies have been presented that are technically based for consideration in the allocation process. Some of the major underlying allocation principles are highlighted as follows:

- There should be a major effort to quantify the impact of target compounds using sound technological concepts.
- The source of the target compounds and the date(s) of releases to the environment should be researched thoroughly. In the absence of "paper trails," historical information, interviews, trade association data, literature, and other data sources should be applied in reconstructing a most probable scenario from a forensic perspective.

Allocation of Superfund Disposal Site Response Costs

- Weighting factors should be developed which directionally predict relative toxicity levels, mobility, and persistence. These factors can be adjusted by the allocators based on site-specific information.
- Consideration should be given to specific events such as spills, leaks, production levels, off-site disposal, and system upgrades when formulating allocation responsibility.
- Recent characterization data at the site or sites should be thoroughly analyzed and applied retroactively using the scientifically accepted principles of fingerprinting, aging effects, and source identification.

The magnitude of weighting factors and residual generation rates obviously can be adjusted as required, based on the information available. By its inherent nature, the involved entities will understandably dispute their particular allocation. It is with this understanding and in the interest of fairness that the allocator incorporate all of the experience, information, and appropriate engineering and scientific principles into the process with the ultimate objective of obtaining equitability.

16.6 Problem

16.1. Three large chemical companies have joint and several liability at a Superfund site and have been designated as the only PRPs. The constituent remediation "driver" chemicals as determined by EPA are benzene, vinyl chloride, PCBs, and mercury. Based on the Remedial Investigation Feasible Study (RI/FS), the following data have been established.

Quantities Sent (Tons)	PRP A	PRP B	PRP C
Benzene	8620	15,260	890
Vinyl chloride	none	1150	3260
PCB1254	1950	none	560
Hg	5260	5220	none

Solution The EPA cleanup criteria for the groundwater is (ppb):

Benzene	0.2
Vinyl chloride	0.06
PCBs	0.008
Hg	5.0

Using the approach discussed in this chapter, allocate the cleanup costs to each PRP.

844   Chapter Sixteen

The critical constituent characteristics are:

	Toxicity* Factor	Solubility* Factor	Adsorption* Factor	Volatile* Factor	Persistence* Factor	Combined Mobility* Factor
Benzene	5(27)	8	8	6	0.2	
Vinyl chloride	16.7(81)	8	8	10	0.5	
PCB1254	125(81)	1	1	2	1	
Hg	0.2(9)	1	1	2	1	

*See grouping factors, Table 16.6.

Combined factor = [{(solubility factor) + 0.8(adsorption factor) + 0.2(volatile factor)} + toxicity factor] × persistence factor

Benzene [{(8 + 0.8(8) + 0.2(6)} + 27] × 0.2 = 8.5
Vinyl chloride [{8 + 0.8(8) + 0.2(10)} + 81] × 0.5 = 48.7
PCB1254 [{1 + 0.8(1) + 0.2(2)} + 81] × 1.0 = 82.8
Hg [{1 + 0.8(1) + 0.2(2)} + 9] × 1.0 = 10.8

Calculated combined factor × physical state

Benzene	8.5	liquid(10)	=	85
Vinyl chloride	48.7	slurry(9)	=	438
PCB1254	82.8	sludge(3)	=	248
Hg	10.8	sludge(8)	=	86

Summary of PRP scores

PRP A	Combined factor	Physical state	Quantity		Total
Benzene	8.5	10	=	85(8620)	732,700
Vinyl chloride	48.7	9	=	438(0)	0
PCB1254	82.8	3	=	248(1950)	483,600
Hg	10.8	8	=	86(5260)	452,360
					1,668,660

PRP B	Combined factor	Physical state	Quantity		Total
Benzene			85(15,260)	=	1,297,100
Vinyl chloride			438(1150)	=	503,700
PCB1254			248(0)	=	0
Hg			86(5220)	=	448,920
					2,249,720

PRP C	Combined factor	Physical state	Quantity	Total	
Benzene			85(890)	=	75,650
Vinyl chloride			38(3260)	=	1,427,880
PCB1254			248(560)	=	138,880
Hg			86(0)	=	0
					1,642,410
			Total		5,560,000

Allocation

PRP A $\dfrac{1,668,660}{5,560,000}$ 30 percent

PRP B $\dfrac{2,249,720}{5,560,000}$ 40.5 percent

PRP C $\dfrac{1,642,410}{5,560,000}$ 29.5 percent

References

1. LeGrand, H. E.: "System for Evaluation of Contamination Potential of Some Waste Disposal Sites," *J. AWWA*, August 1964.
2. CH2M Hill and Engineering-Science, Inc., *Memo to USAF*, Meeting Summary, June 29, 1981.
3. *United States v. A & F Materials Co., Inc.*, 578 F. Supp. 1248, 1256 (S.D.Ill. 1994).
4. Bernheim, D. D.: Expert report "Superfund Allocation Principles," United States v. Atlas Minerals & Chemicals, Inc., Civil Action No. 91-5118, Eastern District of Pennsylvania, September, 1993.
5. Ford, D. L.: "The Technical & Institutional Implications of Equitable Cleanup Allocation," *Environmental Engineer*, vol. 32, no. 2, April 1996.
6. United States District Court, Eastern District of Pennsylvania, *United States v. Atlas Minerals & Chemicals, Inc.*, Civil Action No. 91-5118, Opinion: C. J. Cahn, August 22, 1995.
7. RT Environmental Services, Inc., "Downey Road Landfill Waste Stream Ranking Report," prepared for Manko, Gold, and Katcher, City Line Ave., Bala Cynwyd, Pa., September 9, 1993.
8. Verschueren, K.: *Handbook of Environmental Data on Organic Chemicals*, Von Nostrand Reinhold Co., New York, 1983.
9. Newell, C. J.: Expert Opinion Regarding Waste Volumes, Waste Constituents and Remediation at the Turtle Bayou Site, Liberty County, Texas (Groundwater Services, Inc.), Houston, Texas, February 7, 2007.
10. Eller, J. M., and D. L. Ford: "The Technical Implications of Equitable Cleanup Application II," *Environmental Engineer*, vol. 34, no. 2, April 1998.

CHAPTER 17
Industrial Pretreatment

17.1 Introduction

The National Industrial Pretreatment Program is designed to reduce the amount of pollutants discharged by industry and other nondomestic wastewater sources into municipal sewer systems and the amount of pollutants released into the environment from publicly owned wastewater treatment works (POTWs). The program is a cooperative effort of federal, state, and local regulatory environmental agencies established to protect water quality. The objectives of the program are to protect the POTWs or municipal wastewater treatment facility from pollutants that may interfere with plant operation or pass through the plant untreated and to improve opportunities for the POTW to reuse treated wastewater and residuals that are generated. The term *pretreatment* refers to pollutant control requirements for nondomestic sources discharging wastewater to sewer systems that are connected to POTWs. Limits on the amount of pollutants allowed to be discharged are established by the U.S. Environmental Protection Agency (EPA), the state, or the local authority. Pretreatment limits may be met by the industry through pollution prevention (e.g., production substitution, recycling and reuse of materials) or treatment of the wastewater.

The National Pretreatment Program's authority comes from Sec. 307 of the Federal Water Pollution Control Act (more commonly referred to as the Clean Water Act [CWA]). The federal government's role in pretreatment began with the passage of the CWA in 1972. The act called for the EPA to develop national pretreatment standards to control industrial discharges into sewerage systems. Certain wastewater discharges are prohibited for all industrial facilities that discharge to POTWs because of the potential hazards these discharges create. Specific prohibited discharges include:

- Pollutants that would create a fire or explosion
- Pollutants that would cause corrosive structural damage to a POTW
- Solid or viscous pollutants that would obstruct flow in a POTW
- Pollutants that result in toxic gases, vapors, and fumes
- Ignitable wastes
- Oil and grease
- Discharges with temperatures above 140°F (40°C) when they reach the treatment plant or hot enough to interfere with the biological processes

The general pretreatment regulations were originally published in 1978 and have been updated periodically; the latest at the time of this writing have been in 1999.[1]

17.2 National Categorical Pretreatment Standards and Local Limits Development

Since the National Pretreatment Program was developed by the EPA through the statutory authority of the CWA in 1972, several amendments and updates have occurred. The 1977 amendments to the CWA required POTWs to ensure compliance with the pretreatment standards by each significant local source introducing pollutants subject to pretreatment standards into a POTW. Based on the 1977 amendments, the EPA developed the General Pretreatment Regulations for Existing and New Sources of Pollution (40 CFR Part 403).

The National Pretreatment Program consists of three types of national pretreatment standards established by regulation that apply to industrial users (IUs). These include prohibited discharges, categorical standards, and local limits. *Prohibited discharges*, comprising general and specific prohibitions, apply to all IUs regardless of the size or type of operation. Categorical standards apply to specific process wastewater discharges from particular industrial categories. Local limits are site-specific limits developed by the POTW to enforce general and specific prohibitions on IUs. The prohibited discharges are basically described in Sec. 17.1.

Categorical standards are uniform, technology-based, and applicable nationwide. Developed by the EPA, these standards apply to specific categories of IUs and limit the discharge of specified toxic and nonconventional pollutants to POTWs. Expressed as numerical limits and management standards, the categorical standards are found at 40 CFR Part 405 through 471. They include specific limitations for 35 industrial sectors.[2,3]

Local limits are developed by POTWs to enforce the specific and general prohibitions, as well as any state and local regulations. The prohibitions and categorical standards are designed to provide a minimum acceptable level of control over IU discharges. They do not, however, take into account site-specific factors at POTWs that may necessitate additional controls. For example, a POTW that discharges into a river designated a "scenic river" under the Wild and Scenic Rivers Act may have extremely stringent discharge limits. To comply with its discharge permit, the POTW may need to exert greater control over IU discharges. This additional control can be obtained by establishing local limits.[2,3]

Categorical standards and local limits are complementary types of pretreatment standards. The former are developed to achieve uniform technology-based water pollution control nationwide for selected pollutants and industries. The latter are intended to prevent site-specific POTW and environmental problems due to nondomestic discharges. As shown in Table 17.1, local limits can be broader in scope and more diverse in form than categorical standards. The development of local limits requires the assessment of local conditions and the judgment of POTW personnel.

EPA's promulgation of categorical standards does not relieve a POTW from its obligation to evaluate the need for and to develop local limits to meet the general and specific prohibitions in the General Pretreatment Regulations. Because specific prohibitions and categorical standards provide only general protection against pass-through and interference, local limits based on POTW-specific conditions may be necessary. Developed in accordance with 40 CFR Part 403.5(c), local limits are pretreatment standards for the purposes of CWA Section 307(d) (see 40 CFR 403.5[d]). Therefore, EPA can take enforcement actions against an IU that violates a local limit. Affected third parties also may sue IUs or POTWs with approved pretreatment programs for violations of local limits under the CWA's citizen suit provisions. A POTW may impose local limits on an IU that are more stringent, or cover more pollutants, than an applicable categorical standard. This may be necessary for the POTW to meet its discharge permit or sludge quality limits. If a local limit is less stringent than an applicable categorical standard, however, the industry to which the local limit applies still must meet the applicable categorical standard.[4-6]

17.3 Pretreatment Compliance Monitoring for Industrial Users

The general pretreatment regulations establish responsibilities of federal, state, and local government, industry, and the public to implement pretreatment standards to control pollutants from the industrial

Characteristic	Categorical Standards	Local Limits
Agency responsible for development	EPA	Control Authority (usually POTW)
Potential sources regulated	Industries specified in the Clean Water Act, or as determined by the EPA	All nondomestic dischargers
Objectives	Uniform national control of nondomestic discharges	Protection of POTW and local environment
Pollutants regulated	Primarily priority pollutants listed under the Clean Water Act Sec. 307 (toxic and nonconventional pollutants only)	Any pollutant that may cause pass-through or interference
Basis	Technology based	Technically based on site-specific factors: • Allowable headworks loadings • Toxicity reduction evaluation • Technology in use • Management practice
Point of application	At the end of the regulated process(es) or in-plant	Depends on development methodology (usually at the point of discharge(s) into the collection system

TABLE 17.1 Comparison of Categorical Pretreatment Standards and Local Limits

users who may pass-through or interfere with POTW treatment processes or which may contaminate sewage sludge. The POTW acts as the control authority for these industrial users and monitors the wastewater they discharge to determine whether they are in compliance with the pretreatment standards.

EPA conducts both inspections and audits of a control authority, normally a POTW, to assess the effectiveness of the pretreatment program. Pretreatment audits are designed as a comprehensive review of all facets of a control authority's pretreatment program. The audit addresses all of the items covered in a pretreatment inspection, but in greater detail.

Pretreatment inspections involve:

- Reviewing the approved program, annual reports, National Pollutant Discharge Elimination System (NPDES) compliance status, previous inspection reports, pretreatment files
- Interviewing officials knowledgeable of the program
- Inspecting various industrial user operations, if appropriate

17.4 POTW Ordinance Guidelines for Industrial Users (EPA Model)

The EPA began developing an ordinance model to assist POTWs as a guidance document as early as the 1980s.[7] The EPA then developed a more universal and updated document through the Office of Wastewater Enforcement and Compliance in 1992.[8] It is intended for use by municipalities operating POTWs that are required to develop pretreatment programs to regulate industrial discharges to their systems. This guidance document for adopting new or revised legal authority to implement and enforce a pretreatment program that fulfills requirements set forth in the Code of Federal Regulations (40 CFR Part 403). The table of contents of this document are included as Table 17.2.

The municipalities may need to establish limits for some or all of the pollutants listed in Table 17.3. Such limits are established to

Table of Contents
Section 1. General Provisions
Section 2. General Sewer Use Requirements
Section 3. Pretreatment of Wastewater
Section 4. Wastewater Discharge Permit Application
Section 5. Wastewater Discharge Permit Issuance Process
Section 6. Reporting Requirements
Section 7. Compliance Monitoring
Section 8. Confidential Information
Section 9. Publication of Users in Significant Noncompliance
Section 10. Administrative Enforcement Remedies
Section 11. Judicial Enforcement Remedies
Section 12. Supplemental Enforcement Action
Section 13. Affirmative Defenses to Discharge Violations
Section 14. Wastewater Treatment Rates
Section 15. Miscellaneous Provisions
Section 16. Effective Date

TABLE 17.2 EPA Model Pretreatment Ordinance

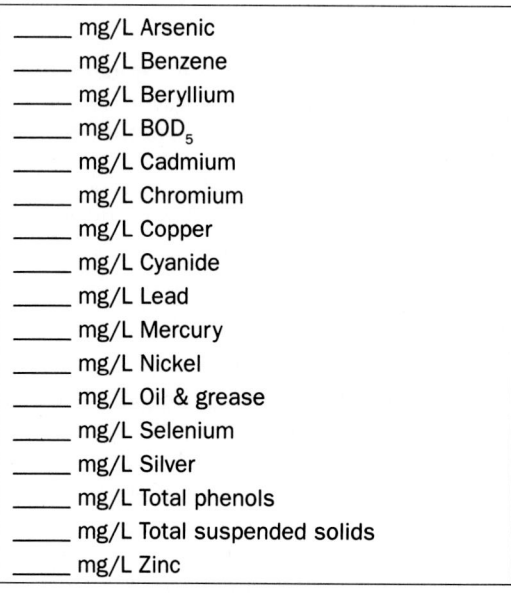

TABLE 17.3 Recommended Instantaneous Maximum Discharge Limits

protect against pass-through and interference, especially assuring the POTW that such contaminant restriction allows the POTW to comply with their own discharge permit.

17.5 User Charge Rates and POTW Cost Recovery

The basis for establishing rates to the industrial user by the POTW is based on several factors. Simply, the POTW needs to recover operational, sampling, maintenance, and replacement (OM&R) costs as well as administrative costs. Reserve funds for POTW employees pension programs for retired employees and employee disability payments, workman's compensation, liability claims, and catastrophic events also are factored in the POTW's "cost for doing business." Funds for major capital expenditures and infrastructure upgrades are normally funded by long-term revenue bonds authorized by the states.

The OM&R costs typically include the following[9]:

- Collection
- Treatment
- Solids processing
- Flood and pollution control
- Solids utilization
- General support

- Annuity and benefit fund
- Reserve claim fund
- Construction and working cash fund

The user charges industry must provide is generally based on the rates clearly outlined in the pretreatment contract between the industries and the POTW. The basis for a monthly charge includes, but is not limited to, flow, 5-d BOD, total suspended solids (TSS), as well as surcharges for other specified constituents. The categorical pretreatment standards for heavy metals and other specified compounds are established by the EPA for the respective industrial categories and subcategories. However, additional requirements can be imposed by the POTW as required. The annual summary of all the industrial charges and payments in effect is the yearly *income statement* for the POTW and sets the basis for the POTW cost recovery budgeting and subsequent rate establishment process. As major metropolitan treatment authorities have thousands of industrial users, this is obviously a complex budgeting endeavor.

The annual net OM&R expenditures by the POTW are analyzed over recent periods and used to project user rates. For example, the user unit costs recently calculated by the Metropolitan Water Reclamation District of Chicago is calculated in Table 17.4.[9]

Total District Loadings for 2005* Volume = 426,690 million gal BOD = 810,199 lb SS = 1,187,924 lb Total OM&R Cost = $247,398,000
Allocation of Cost According to Parameters of Flow, BOD, & SS[†] Flow = 28.4% × $247,398,000 = $70,261,032 BOD = 38.3% × $247,398,000 = $94,753,434 SS = 33.3% × $247,398,000 = $82,383,534
Unit Costs of Treatment Volume = $70,261,032/426,690 million gal = $164.67/million gal BOD = $94,753,434/810,199 lb = $116.95/lb SS = $82,383,534/1,187,924 lb = $69.35/lb

*The 2005 District loadings are used in the calculation of 2007 rates because this is the latest full year's operating data at the time the calculations were made.
(*Source*: R&D Department Water Reclamation Plant 2005 Operating Records.)
[†]Percent distribution of cost-to-load parameters derived from the Maintenance and Operations Memorandum dated June 6, 2006.
Source: Adapted from Metropolitan Water Reclamation District of Greater Chicago, Research and Development Department, Report No. 06-64, 2006.

TABLE 17.4 Unit Cost of Treatment

Annual flow (million gal/d)	426,690
Annual BOD (1000 lb)	810,199
Annual TSS (1000 lb)	1,187,924
Total OM&R costs	$247,398,000
Cost allocation*:	
Flow	28.4%
BOD	38.3%
TSS	33.3%

*Distribution derived by cost-to-load parameters determined by operations department.

17.6 Case Histories

Several examples of pretreatment programs instituted by POTWs in various locations with differing controls and restrictions are presented in this section. These examples are selected to illustrate the various approaches to adopt both the regulations and controls necessary to protect the integrity of the POTW wastewater treatment systems receiving industrial effluents.

City of Chicago, Illinois

The Metropolitan Water Reclamation District of Greater Chicago (MWRDGC) is a sanitary district and all-purpose district chartered in northern Illinois. It is independent of local government with an elected Board of Commission. The district operates the largest treatment plant in the world which is the Stickney Water Reclamation Plant in Stickney, Illinois. The district collects and treats municipal and industrial wastewaters generated by the City of Chicago and 124 suburban communities in Cook County, Illinois. The service area is approximately 872 square miles and serves over 10 million population equivalent, which includes the industrial contribution. There are seven water reclamation plants with a hydraulic capacity of approximately 2 billion gal of water per day. As there are several hundred significant industrial users (SIUs) (over 25,000 gal/d per SIU), pretreatment regulations are comprehensive and restrictive. Each SIU pollutant loads and flows are evaluated by the district, and the pretreatment permits must not only comply with 40 CFR 403 but also meet local limits to protect against pass-through and interference with the district's POTWs and ensure compliance with the POTW effluent permits.

There are complex approaches in allocating costs in an equitable manner by the MWRDGC. The billable flows and loads are allocated to three major categories, namely:

- Residential, small nonresidential commercial
- Large commercial industrial
- Tax exempt and governmental

It is the nature of the collection system to have a significant volume of stormwater (inflow) as well as infiltration (I&I). For example, a calculated dry weather flow and total flow comparison indicated that approximately 26 percent of the annual flow can be attributed to I&I/rain recycle. The allocation of this flow and loading is based on the equalized assessed value (EAV) for the three categories cited above, based on their respective dry weather flows.[9]

Although it is not addressed here, stormwater runoff from land and impervious areas such as paved streets, parking lots, and building rooftops from rainfall and snow events often contain pollutants that can add organic and solids loading to POTWs or separate storm sewers. This applies to direct stormwater discharges and not to combined sewers or dedicated sanitary sewers connected within a POTW collection system.

City of Indianapolis, Indiana

The city of Indianapolis has prepared a decision-making checklist to outline the criteria for evaluating each connection sewer operation (CSO) for its potential impact on the POTW wastewater treatment facility capacity, combined sewer overflows, and receiving streams. Such information can be used to modify any industrial contributor pretreatment permit if required.[10] This document is included as Table 17.5.

City of San Diego, California

The city of San Diego's Water Utilities Department has a wide range of industrial contributors to its POTW wastewater treatment system. As many of its IUs are aircraft and related manufacturers, extensive pretreatment steps are required to protect the integrity of the POTW treatment facilities. These aircraft and space-related manufacturing operations have effluents from processes which use alkaline and acid cleaning baths, anodizing systems, water softening regenerate, reverse osmosis reject, degreasing, and a myriad of other industrial processes. The city's Metropolitan Industrial Waste Program administered by the Water Utilities Department manages and enforces the restrictions imposed on the industrial contributors. Under 40 CFR 403.6(1) metal finishing subcategory, permit limits are imposed as well as additional and more restrictive limitations imposed by the city. Typical pretreatment limitations are shown in Table 17.6.

City of Shreveport, Louisiana

The City of Shreveport's POTW has several major industrial dischargers. As effluent from the POTW and industrial direct dischargers go to the Red River, a low flow river and environmentally sensitive, both

	Factors	Criteria	How Criteria Are Applied
		Location of Discharge and Impact on Receiving Stream	
1	Number of CSOs between discharger and AWT plant*		Minor concern
		2 to 10	Moderate concern
		>10	Major concern
2	Frequency of discharges from affected CSOs*	>40 events/year	Major concern
		4–39 events/year	Moderate concern
		<4 events/year	Minor concern
3	Magnitude of discharges from downstream CSOs (overflow volume million gal/year)*	Do affected CSOs include one or more of the 15 largest overflow points (based upon average annual overflow volume)?	If yes, major concern
		How many million gallons a year are discharged from the affected CSOs?	>100 million gal/year, major concern; 50–99 million gal/year, moderate concern; <50 million gal/year, minor concern
4	Magnitude of pollutant load from CSOs (Toxicity: load/d and concentration)	Likelihood of significant industrial concentration (is industrial concentration >1.1, as determined by the toxics computation worksheet?)	If the answer to any of the three questions on magnitude of pollutant load from CSOs is yes and the overflow frequency from any affected CSOs is >40/year, major concern. If yes and overflow frequency >4 but <39/year, moderate concern. If yes and overflow frequency ≤ 4/year, minor concern.
		Likelihood of significant industrial flow percentage (is the percentage >1.0%?)	
		Do any of the affected CSOs rank in the top 5 as determined by the 2004 analysis of industrial user discharge characteristics?	

5	Stream reach characteristics*	Does affected stream segment flow through areas with opportunities for recreational use?	If yes, moderate to major concern.
		Flow levels in receiving stream	If $7Q_{10}$ <5 cfs, major concern; <40 cfs moderate concern; >41 cfs, minimal concern.
6	Conventional pollutant parameters found in the affected CSOs (BOD, TSS, other)	Would the increased load cause or contribute to NPDES permit violations?	Qualitative analysis of new pollutant load on stream.

*For data related to each stream, see http://www.indygov.org/eGov/City/DPW/Environment/Wastewater/Pretreatment/hom.htrr.

TABLE 17.5 Industrial Pretreatment Permitting Process Decision-Making Factors and Criteria

Constituent	Units	Local Daily	Federal Daily	Federal 30-d Avg.
Acids & alkalies	pH	Range 5–11	Lower 5	
Oil & grease	mg/L	500		
Dissolved sulfides	mg/L	1.0		
Cyanide	mg/L	1.9	1.2	0.65
Antimony	mg/L	2.0		
Arsenic	mg/L	2.0		
Beryllium	mg/L	2.0		
Cadmium	mg/L	1.2	0.69	0.26
Chromium	mg/L	7.0	2.77	1.71
Copper	mg/L	4.5	3.38	2.07
Lead	mg/L	0.6	0.69	0.43
Mercury	mg/L	2.0		
Nickel	mg/L	4.1	3.98	2.38
Selenium	mg/L	2.0		
Silver	mg/L	2.0	0.43	0.24
Thallium	mg/L	2.0		
Zinc	mg/L	4.2	2.61	1.48
Pesticides & PCBs	mg/L	0.04		
Phenolic compounds	mg/L	25.0		
TTOs	mg/L		2.13	
TSS	mg/L	N/A		

TABLE 17.6 City of San Diego Pretreatment Limits Aircraft Manufacturing

the POTW pretreatment and industrial NPDES are quite restrictive. In this setting, many industries require extensive end-of-pipe treatment prior to going either to the Red River directly or to the POTW. In Table 17.7, the 40 CFR Part 437 Subpart D limitations are listed as are the Shreveport discharge limits and the imposed pretreatment limits to selected IUs.

City of Austin, Texas

The city of Austin has an extensive and comprehensive wastewater pretreatment program. As the city has become a major center for

Parameter	40 CFR Part 437 Subpart D Multiple Wastestreams Daily Max/Monthly Avg. (mg/L)	POTW Local Limits (mg/L)	Permit Limit Daily Max/ Monthly Avg. (mg/L)
Antimony	0.249/0.206	0.07	0.07/0.07
Arsenic	0.162/0.104	1.2	0.162/0.104
Cadmium	0.474/0.0962	0.10	0.10/0.0962
Chromium	0.746/0.323	4.70	0.746/0.323
Cobalt	0.192/0.124	N/A	0.192/0.124
Copper	0.500/0.242	3.80	0.500/0.242
Cyanide	N/A	1.50	1.50
Lead	0.350/0.160	1.00	0.350/0.160
Mercury	0.00234/0.000739	0.005	0.00234/0.000739
Molybdenum	N/A	1.50	1.50
Nickel	3.95/1.45	3.60	3.60/1.45
Selenium	N/A	0.14	0.14
Silver	0.120/0.0351	0.10	0.10/0.0351
Tin	0.409/0.120	N/A	0.409/0.120
Titanium	0.0947/0.0618	N/A	0.0947/0.0618
Vanadium	0.218/0.0662	N/A	0.218/0.0662
Zinc	2.87/0.641	3.20	2.87/0.641

TABLE 17.7 City of Shreveport Local Limits and Categorical Standards Comparison

Parameter	40 CFR Part 437 Subpart D Multiple Wastestreams Daily Max/Monthly Avg. (mg/L)	POTW Local Limits (mg/L)	Permit Limit Daily Max/Monthly Avg. (mg/L)
Bis-2-ethylhexylphthalate	0.215/0.101	N/A	0.215/0.101
Carbazole	0.598/0.276	N/A	0.598/0.276
Fluoranthene	0.0537/0.0268	N/A	0.0537/0.0268
n-Decane	0.948/0.437	N/A	0.948/0.437
n-Octadecane	0.589/0.302	N/A	0.589/0.302
o-Cresol	1.92/0.561	N/A	1.92/0.561
p-Cresol	0.698/0.205	N/A	0.698/0.205
2,4,6-Trichlorophenol	0.155/0.106	N/A	0.155/0.106
TTO EPA 624 & 625 and pesticide/PCBs (608)	N/A	2.13	2.13
Chlorides	N/A	N/A	N/A
pH	N/A	6.0–10.5 std. units	6.0–10.5 std. units
BOD	N/A	>250 surcharge	
COD	N/A	reserved	
Oil & grease	N/A	100	100
TSS	N/A	>250 surcharge	

TABLE 17.7 City of Shreveport Local Limits and Categorical Standards Comparison (*Continued*)

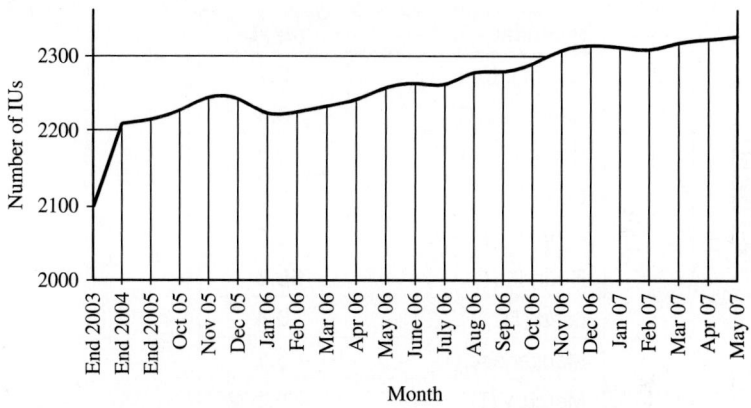

FIGURE 17.1 Permitted industrial users.

high-tech industries such as Dell and Freescale Semiconductors, the industrial users from semiconductor and related industries has grown significantly. For example, the number of industrial users has increased from less than 2000 in 2002 to more than 2400 in 2007, as shown in Fig. 17.1. This, combined with a high population growth, has resulted in increasing loads to the POTW wastewater treatment plants. As the city discharges its effluent to the Colorado river which is a low flow and ecologically sensitive stream, combined with the POTW's stringent effluent standards (10 ppm CBOD, 15 ppm TSS, and 2 ppm ammonia nitrogen), pretreatment from increasing IUs is critical. The state and city pretreatment standards for semiconductor industries (new source, electrical and electronic manufacturing, subject to 40 CFR Parts 401, 403, and 469, Part A, 469.18) are shown in Table 17.8.[11] For combined streams within an industrial use, an alternative limit calculation for concentration limit can be calculated as follows:

$$C_T = \frac{\left(\sum_{i=1}^{N} C_i F_i\right)}{\left(\sum_{i=1}^{N} F_i\right)} \frac{(F_T - F_D)}{(F_T)}$$

where C_T = the alternative concentration limit for the combined wastestream
 C_i = the categorical pretreatment standard concentration limit for a pollutant in the regulated stream i
 F_i = the average daily flow (at least a 30-d average) of stream i to the extent that it is regulated for such pollutant
 F_D = the average daily flow (at least a 30-d average) from such streams as boiler blowdown, cooling streams, and storm water streams

Pollutant	mg/L
Arsenic, total (T)	0.2
Cadmium (T)	0.4
Chromium (T)	2.4
Copper (T)	1.1
Cyanide (T)	1.0
Fluoride (T)	65.0
Lead (T)	0.4
Manganese (T)	6.1
Mercury (T)	0.002
Molybdenum (T)	1.1
Nickel (T)	1.6
Selenium (T)	1.8
Silver (T)	1.0
Zinc (T)	2.3
Total toxic organics end-of-process	1.31
Total toxic organics end-of-pipe	2.0

TABLE 17.8 City of Austin Local Limits

The average daily flow through the combined pretreatment facility F_T includes F_I, F_D, and unregulated streams.

There is a surcharge (S) imposed by the POTW for excess strength wastewater from industrial users in terms of biochemical oxygen demand (BOD) and chemical oxygen demand (COD). Using a standard BOD of 200 ppm, COD of 450 ppm, and suspended solids (SS) of 200 ppm (typical for raw domestic sewage), the surcharge formula is:

$$S = V(8.34)\,[A(BOD - 200) + B(SS - 200)]$$

and

$$S = V(8.34)\,[C(COD - 450) + B(SS - 200)]$$

where S = surcharge in dollars that appears on the users' monthly bills
V = wastewater billed in millions of gallons during the billing period

8.34 = conversion factor (lb per gallon of water)
A = unit charge in dollars per pound of BOD
B = unit charge in dollars per pound of SS
C = unit charge in dollars per pound of COD

Noncompliance of the pretreatment ordinances is issued notice of violation, forcing a corrective plan for voluntary compliance (although not relieved of civil or criminal liability) and enforcement action as required. Continued noncompliance may result in a show cause hearing, cease and desist order, and possible permit cancellation.

The successful history of the city of Austin's pretreatment program has been recognized by both the EPA and the state of Texas. Moreover, the city awards annually those industrial users who have demonstrated outstanding compliance with pretreatment requirements. Over the past 15 years an award ceremony has been held and well attended by the industrial awardees, the city of Austin's officials, the press, and other local dignitaries. This initiative has resulted in many beneficial responses. For example, the industrial user commitment to the environment is honored, water conservation is highlighted, and continual internal industrial control to ensure compliance, and reducing the load to the POTW (such as shown in Fig. 17.2) is underscored.[12]

Barceloneta POTW, Puerto Rico, and Pharmaceutical Pretreatment

The Puerto Rico Aqueduct and Sewage Authority (PRASA) owns and operates the Barceloneta Regional Wastewater Treatment Plant (BRWTP) near Barceloneta, Puerto Rico. It is a POTW but most of the wastewater loading is generated by plants in the pharmaceutical manufacturing industry. Even though it is a classical POTW with industrial contributors under its pretreatment program, it is unique.

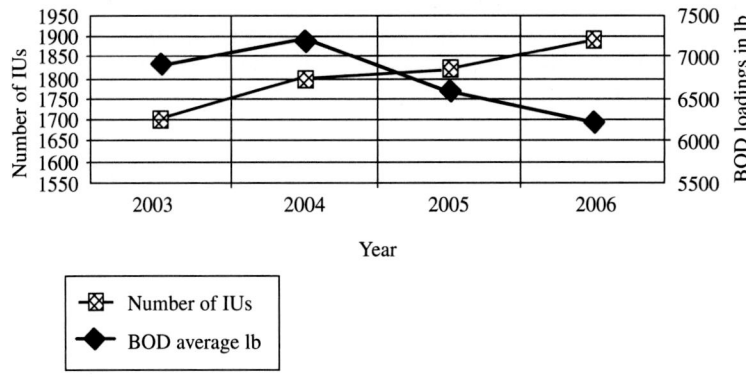

FIGURE 17.2 Average BOD loadings discharged by each IU in lb/year.

A facility agreement was entered in 1978 where an advisory council was created consisting of the pharmaceutical company's representatives and a temporary arrangement was developed to upgrade the then primary treatment plant to a secondary treatment facility, issuing bonds for paying the construction and related costs. Once the plant started operations, the bonds were paid, then ownership was transferred to PRASA, which has owned and operated the upgraded treatment plant since 1982. The EPA promulgated revised effluent guidelines and pretreatment standards for the pharmaceutical industry point source category pretreatment standards in the late 1990s.[13] Since that time, there have been numerous consultant reports addressing operations, notification, odor control, capacity, operator training, and other technical issues.[14–20]

The regulations for the pharmaceutical industry (40 CFR 439) point source categories are delineated into five subparts as follows:

- Subpart A—Fermentation Products Subcategory
- Subpart B—Extraction Products Subcategory
- Subpart C—Chemical Synthesis Subcategory
- Subpart D—Mixing, Compounding, and Formulating Subcategory
- Subpart E—Research Subcategory

Pharmaceutical facilities that discharge effluents into receiving streams or POTWs are required to meet one or more of the effluent limitation guidelines and standards established under the Clean Water Act as set forth in Table 17.9.

As a variety of pharmaceutical companies falling under different subparts are connected to the BRWTP, calculation of the applicable federal pretreatment standards becomes more complex to calculate. An example of such a calculation is included at the end of this chapter.

The BRWTP upgrade in the 1980s was designed for a hydraulic loading of 8.3 million gal/d and an average BOD loading of 88,000 lb/d. This dropped dramatically since 1999 because the contributor pharmaceutical plants have reduced their hydraulic and organic loadings to the BRWTP through recycling, reuse, and expanded pretreatment system upgrades. This reduction has resulted in unused BRWTP capacity now available to other users, and affirms the economic incentives of selected contributor industries to reduce their loadings to the POTW, avoid surcharges, and still have the ability to discharge their reduced streams to the POTW as compared to being a direct discharger.[4]

Example 17.1 Facility B is an existing multiple-subcategory indirect discharging pharmaceutical manufacturing facility which discharges to a municipal POTW. The flow schematic for Facility B shows the flow from each operation and is presented in Fig. 17.3.

Industrial Pretreatment

Program	Type of Discharger	Existing or New Source	Applicable Guidelines and Standards Previously Established	Additional Guidelines and Standards (from 9/21/98 Rule)
NPDES permit program	Direct discharger	Existing source	BCT BPT BAT	BPT BAT
		New source	NSPS	NSPS
National pretreatment program	Indirect discharger	Existing source	PSES	PSES
		New source	PSNS	PSNS

Where PSES are pretreatment standards for existing sources and PSNS are pretreatment standards for new sources.

TABLE 17.9 Effluent Limitations Guidelines and Standards Applicable to Each Program

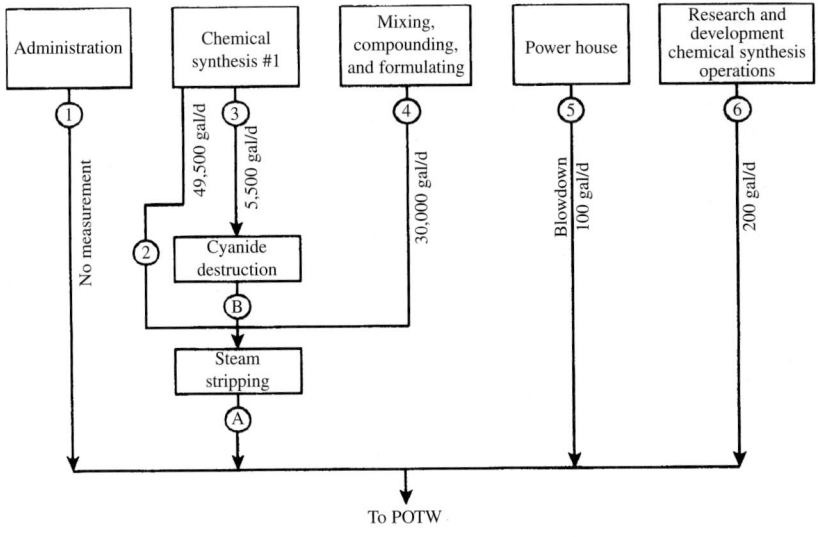

FIGURE 17.3 Flow schematic for facility B.

The information below summarizes the information from the permit application needed to calculate discharge limits for the reissued pretreatment permit.

- What type of discharger is the facility?
  - Indirect.
- Under which subparts do the facility's operations fall?
  - Subparts C, D, and E.
- The facility is subject to which effluent limitations guidelines and standards?
  - Pretreatment Standards for Existing Sources (PSES) (40 CFR Part 439).

Determining Limits for Pollutants Regulated under PSES

PSES has been revised for subparts A, B, C, and D. The final effluent limitation standards are concentration-based and, as such, do not regulate wastewater flow. The limitations apply at the end-of-pipe, except for cyanide. If end-of-pipe measurement is infeasible, control authorities may set a monitoring point at a more suitable location. Compliance monitoring for cyanide should occur in-plant, prior to commingling with noncyanide bearing wastewaters. EPA has regulated 24 priority and nonconventional pollutants (including ammonia, where applicable, and cyanide) for indirect dischargers in subparts A and C. The effluent limitations for subpart A and C operations are presented in Table 17.10. EPA has regulated five priority and nonconventional pollutants for indirect dischargers in subparts B and D. Table 17.11 presents the effluent limitations for subpart B and D operations.[21]

The first step in establishing permit limitations is to determine the types of wastestreams (i.e., regulated process, unregulated process, and dilute). The flow breakdown for Facility B is shown in Table 17.12.

Streams 2, 3, and 4 are regulated process wastestreams because effluent limitations have been established for chemical synthesis operations (subpart C) and mixing, formulating, and compounding operations (subpart D). However, only five pollutants are regulated at subpart D; therefore the facility may have a pollutant regulated in streams 2 and 3 but unregulated in stream 4.

It is assumed that Facility B has provided the permit writer with an accurate characterization of its process wastestreams by means available such as solvent use and disposition data, and chemical analysis of each stream. Permit writers should establish permit limitations and require compliance monitoring for each regulated pollutant generated or used at a pharmaceutical manufacturing facility. Routine compliance monitoring is not required for regulated pollutants not generated or used at a facility. Facilities should make a determination that regulated pollutants are not generated or used based on a review of all raw materials used, and an assessment of all chemical processes used, and consideration of resulting products

Pollutant or Pollutant Property	Maximum for Any One Day (mg/L)	Monthly Average (mg/L)
PSES/PSNS for in-plant monitoring points		
Cyanide*	33.5	9.4
PSES/PSNS for end-of-pipe monitoring points		
Ammonia as N[†]	84.1	29.4
Acetone	20.7	8.2
n-Amyl acetate	20.7	8.2
Benzene	3.0	0.6
n-Butyl acetate	20.7	8.2
Chlorobenzene	3.0	0.7
Chloroform	0.1	0.03
o-Dichlorobenzene	20.7	8.2
1,2-Dichloroethane	20.7	8.2
Diethylamine	255.0	100.0
Ethyl acetate	20.7	8.2
n-Heptane	3.0	0.7
n-Hexane	3.0	0.7
Isobutyraldehyde	20.7	8.2
Isopropyl acetate	20.7	8.2
Isopropyl ether	20.7	8.2
Methylene chloride	3.0	0.7
Methyl formate	20.7	8.2
MIBK	20.7	8.2
Tetrahydrofuran	9.2	3.4
Toluene	0.3	0.2
Triethylamine	255.0	100.0
Xylenes	3.0	0.7

*Cyanide effluent limit established in the 1983 final rule.
[†]Ammonia is only regulated for indirect dischargers that discharge to non-nitrifying POTWs.
Source: Adapted from U.S. EPA, 40 CFR Part 439, 2006.

TABLE 17.10 PSES and PSNS for Subpart A and C Operations

	PSES/PSNS for End-of-Pipe Monitoring Points	
Pollutant or Pollutant Property	Maximum for Any One Day (mg/L)	Monthly Average (mg/L)
Acetone	20.7	8.2
n-Amyl acetate	20.7	8.2
Ethyl acetate	20.7	8.2
Isopropyl acetate	20.7	8.2
Methylene chloride	3.0	0.7

Source: Adapted from U.S. EPA, 40 CFR Part 239, 2006.

TABLE 17.11 PSES and PSNS for Subpart B and D Operations

Wastestream		Flow (gal/d)
1. Administration		No measurement
2. Chemical synthesis	49,500	(Regulated, subpart C)*
3. Cyanide-bearing chemical synthesis	5,500	(Regulated, subpart C)*
4. Mixing/compounding and formulation	30,000	(Regulated, subpart D)*
5. Power house boiler blowdown	100	(Dilute)
6. Research and development chemical synthesis	200	(Unregulated, subpart E)
Total measured wastewater flow	85,300	
Total regulated process flow	85,000	
Total unregulated process flow	200	

*Pollutants regulated at subpart C operations may not be regulated at subpart D operations.

TABLE 17.12 Flow Breakdown for Facility B

Stream	Subpart	Flow (gal/d)	Pollutant
1	N/A	Not measured	No PSES pollutants
2	C	49,500	Acetone, chloroform, toluene
3	C	5,500	Acetone, cyanide
4	D	30,000	Acetone, isopropyl acetate, toluene
5	N/A	100	No PSES pollutants
6	E	200	Chloroform, toluene

TABLE 17.13 Regulated Pollutants Found in the Wastewater of Facility B

and by-products. The determination that a regulated pollutant is not generated or used should be confirmed by annual chemical analyses of wastewater from each monitoring location, and these analyses must be submitted to the permit writer.

Table 17.13 presents a summary of regulated pollutants found in this facility's wastestreams.

Based on the above data, permit limitations would be established for acetone, chloroform, cyanide, isopropyl acetate, and toluene. Acetone and isopropyl acetate are regulated in wastewater discharges from subpart A, B, C, and D operations. Chloroform, cyanide, and toluene are regulated in wastewater discharges from subpart A and C operations only.

Step 1. Determining PSES Maximum Limitations for Any One Day

In this case study, the total flow going to the POTW cannot be measured as the amount of water from the administrative building cannot be determined. Thus, it is not possible to calculate the appropriate concentration of pollutants at the end of pipe. In this case study, the limitations for all pollutants except cyanide would be applied at monitoring point A. Cyanide limitations would apply in-plant at point B prior to any dilution or commingling with noncyanide-bearing wastestreams unless the facility can show that cyanide is detectable at point A.

Concentration-based limits for indirect discharging facilities are listed in Tables 17.10 and 17.11.[21]

In this example, the following maximum for any one day effluent limitations apply:

Acetone	20.7 mg/L (subparts C and D)
Chloroform	0.1 mg/L (subpart C)
Cyanide	33.5 mg/L (subpart C)
Isopropyl acetate	20.7 mg/L (subparts C and D)
Toluene	0.3 mg/L (subpart C)

The concentration-based limit for acetone is 20.7 mg/L for both subpart C and D operations. This limit would be applied at monitoring point A, after the steam-stripping unit operations on streams 2 and 3. Concentration-based limits for chloroform, isopropyl acetate, and toluene would be applied in a similar manner.

Step 2. Determining PSES Monthly Average Limitations

Concentration-based monthly average effluent limitations for each of the pollutants can be calculated in the same manner as the daily maximum effluent limitations. The following monthly average limitations apply for Facility B:

Acetone	8.2 mg/L (subparts C and D)
Chloroform	0.03 mg/L (subpart C)
Cyanide	9.4 mg/L (subpart C)
Isopropyl acetate	8.2 mg/L (subparts C and D)
Toluene	0.2 mg/L (subpart C)

Facility B would show compliance by averaging the daily maximum values in a 30-d period and showing the monthly average concentrations as equal to or less than the numbers above. For this example, Facility B should perform compliance monitoring at point A on Table 17.14 for all regulated pollutants, except cyanide.

Monthly average limitations for cyanide would be calculated using the flow from stream 3 of subpart C operations, as other streams do not contain cyanide. The concentration-based monthly average limitation is 9.4 mg/L. This monthly average limitation is compared

	Effluent Limitations for Point A Monitoring Points		Effluent Limitations for Point B Monitoring Points	
	Max. for Any One Day	Monthly Average	Max. for Any One Day	Monthly Average
Pollutant	(mg/L)	(mg/L)	(mg/L)	(mg/L)
Acetone	20.7	8.2	—	—
Chloroform	0.1	0.03	—	—
Cyanide	—	—	33.5	9.4
Isopropyl acetate	20.7	8.2	—	—
Toluene	0.3	0.2	—	—

TABLE 17.14 Final Limits for Facility B

to the average of daily discharge amounts in a calendar month to determine facility compliance. If only one sample is taken in the calendar month, the sample must meet both the daily maximum limitation and the monthly average limitation.

Determining Compliance Monitoring for PSES Pollutants

Facilities discharging more than one regulated pollutant may request to monitor for a single surrogate pollutant to demonstrate an appropriate degree of control for a specified group of pollutants. For the purpose of identifying surrogates, pollutants have been grouped according to treatability classes. Table 17.15 presents the treatability classes identified for steam stripping.

Strippability Group	Compound	Surrogate (yes/no)
High	Methylene chloride	Yes
	Toluene	Yes
	Chloroform	Yes
	Xylenes	No
	n-Heptane	No
	n-Hexane	No
	Chlorobenzene	No
	Benzene	No
Medium	Acetone	Yes
	Ammonia as N	Yes
	Ethyl acetate	Yes
	Tetrahydrofuran	Yes
	Triethyamine	No
	MIBK	No
	Isopropyl acetate	No
	Diethylamine	No
	1,2-Dichloroethane	No
	n-Amyl acetate	No
	Isopropyl ether	No
	n-Butyl acetate	No
	Methyl formate	No
	Isobutraldehyde	No
	o-Dichlorobenzene	No

Notes: Yes—May be a surrogate pollutant for the group.
No—Should not be used as a surrogate pollutant for the group.
Source: Adapted from U.S. EPA, 40 CFR Part 239, 2006.

TABLE 17.15 Steam Stripping Surrogates for Indirect Dischargers

For this example, the control authority may require compliance at Point A prior to dilution with nonrecess or unregulated process wastewater or may require compliance at the point of discharge to the POTW by using the combined wastestream formula, if the additional dilution or nonregulated flows are known. However, cyanide should be monitored in-plant at point B on Table 17.14 prior to commingling with noncyanide-bearing wastewaters, unless Facility B can show a cyanide value other than nondetect at point A or the discharge point to the POTW.

Since Facility B performs steam stripping wastewater treatment on the subpart C wastewaters, Table 17.15[21] can be used as a guide to determine if surrogate pollutants may be appropriate for compliance monitoring.

In Table 17.15, chloroform and toluene are both classified in the high strippability group, and both are listed as appropriate surrogate pollutants for that group. Acetone and isopropyl acetate are both classified in the medium strippability group, and acetone is listed as an appropriate surrogate pollutant for that group. If the use of surrogates is requested by a facility, control authorities may decide on the use and choice of surrogate pollutants on a facility-by-facility basis.

In this example, the choice of surrogate pollutant for the high strippability group will be based on the pollutant concentrations since two pollutants (chloroform and toluene) are listed as appropriate surrogates. Assuming the average pollutant concentrations are known to be 0.01 mg/L for chloroform and 0.1 mg/L for toluene, the permit writer would choose toluene as the surrogate pollutant. For the medium strippability group, the permit writer can base the choice of surrogate pollutant on the guidance provided in Table 17.15; thus, acetone would be chosen as the surrogate pollutant.

Therefore, Facility B would be required to routinely monitor for toluene and acetone at monitoring point A or the discharge point to the POTW, and for cyanide at monitoring point B, assuming cyanide is not detectable at point A.

Final Limits as They Would Appear in a Permit for Facility B

Table 17.14 presents the final limits as they would appear in a permit for Facility B on a concentration basis. If all cyanide-bearing wastestreams are diverted to a cyanide destruction unit, self-monitoring for cyanide should be conducted after cyanide treatment and before dilution with other streams.

If sufficient flow information is available, the permit writer may determine compliance concentrations at the discharge to the POTW point using the combined wastestream formula (CWF).

The limitations presented in Table 17.14 should have been complied with on or before September 21, 2001.

Pretreatment of Leachate Discharges

The pretreatment of industrial wastes required for discharge to POTWs is not limited to operating industries within the POTW collection systems. For example, there are many situations where leachate collection systems required as part of the remediation process at former disposal sites can use POTWs for discharge, assuming certain quality requirements can be met. An example of a remediation leachate discharge to a POTW is included.[22] The pretreatment quality requirements established by the city of Toledo, Ohio, is included as Table 17.16.

Analyte	Maximum Allowed Concentration (mg/L)
Arsenic, total	0.6
Cadmium, total	0.3
Chromium, hexavalent	0.8
Copper, total	1.0
Mercury, total	1.5
Nickel, total	0.03
Silver, total	2.9
Zinc, total	0.2
Cyanide, total	6.3
Total petroleum hydrocarbons	15.0 average/15.0 grab
Total organic carbon*	200
Total organic halogens†	0.5
Total toxic organics‡	5.0
BTEX	0.5
Benzene	0.05

*The "total organic carbon" standard does not have to be met if either of the "total organic halogens" or the "total toxic organics" concentrations are below their respective standards.
†The "total organic halogens" standard does not have to be met if the "total toxic organics" concentration is below its standard.
‡The "total toxic organics" concentration must be met at all times.
Key: BTEX = benzene, toluene, ethylbenzene, and total xylenes.

TABLE 17.16 Sanitary Sewer Discharges Standards Leachate Extraction System XXKEM Site Toledo, Ohio

References

1. Environmental Health and Safety Online, EHSO, Atlanta, January, 2008 www.ehso.com.
2. U.S. EPA: *Local Limits Development Guidance*, EPA 833-R-04002A, Office of Wastewater Management, Washington D.C., July 2004.
3. U.S. EPA: *Introduction to the National Pretreatment Program*, EPA 833-B98-002, Office of Wastewater Management, Washington D.C., February 1999.
4. Tischler, L. F.: Pretreatment Standards, 2008.
5. U.S. EPA: *Industrial User Permitting Guidance Manual*, EPA 833-B-89-001, September 1989.
6. U.S. EPA: *Guidance Manual for the Use of Production-Based Pretreatment standards and the Combined Wastestream Formula*, EPA 833-B-85-210, September 1985.
7. Patterson Associates, Inc.: "A Model Municipal Pretreatment Ordinance for Existing and New Sources of Pollution," prepared for U.S. EPA Region V, 1980.
8. U.S. EPA, Office of Wastewater Enforcement and Compliance: *EPA Model Pretreatment Ordinance*, June 1992.
9. Metropolitan Water Reclamation District of Greater Chicago, Research and Development Department: "Calculation of 2007 User Charge Rates," Report No. 06-64, 2006. Also, personal communication with Dr. Cecil Lue-Hing, 2008.
10. City of Indianapolis: Correspondence to Industrial Discharges Advisory Committee, c/o Eli Lilly & Co., January 18, 2005.
11. City of Austin: www.ci.austin.tx.us/water/wwwssd_iw_wrppm.htm
12. Bhattarai, R.: Pretreatment Standards, January 11, 2008.
13. Facility Agreement, Selected Pharmaceutical Companies, PRASA and Interim Authorities, May 31, 1978; Amended August 23, 1978.
14. U.S. EPA: *Development Document for Final Effluent Limitations Guidelines and Standards for the Pharmaceutical Manufacturing Point Source Category BPA 821-R-98-005*, 1998.
15. Tischler, L. F. and D. Kocurek: "Evaluation of Nitrification Capability at the BRWTP," prepared for the BRWTP Advisory Council, Barceloneta, Puerto Rico, July 1999.
16. Malcolm Pirnie, Inc.: "Baseline Audit Report of the BRWTP," August 1995.
17. Eckenfelder, Inc.: "Preliminary Evaluation of Rehabilitation and Capacity of the BRWTP," December 1992.
18. Buck, Siefort, and Yost, "BRWTP Detailed Assessment of Plant Rehabilitation Needs," August 1988.
19. Montgomery Watson: "Operations and Maintenance Study, BRWTP," August 1988.
20. Parsons, Engineering-Science, Inc.: "Conceptual Design for Odor Control at the BRWTP," July 1999.
21. U.S. EPA: *Permit Guidance Document: Pharmaceutical Manufacturing Point Source Category* (40 CFR Part 439), January 2006.
22. ENVIRON International Corporation: "Operations and Maintenance Plan, Leachate Collection System, XXKEM Site, Toledo, Ohio," November 25, 1998.

CHAPTER 18
Environmental Economics

PART 1 INDUSTRIAL ENVIRONMENTAL ECONOMICS

18.1 Introduction

Part 1 of Chap. 18 addresses economic and related issues for industry in terms of environmental regulation, compliance, planning, site selection, litigation, and governance. Part 2 of Chap. 18 applies to consulting engineering firms that industry often engages to provide design, operations, process upgrades, permitting, remediation, auditing, and other environmental services.

18.2 Industrial Environmental Economics and Regulatory Compliance Metrics

Many, if not most, of the major industrial environmental projects such as design, operations, process upgrades, remediation, permitting, compliance issues, and others require the services of the consulting community as well as corporate staffs and specialists. The economics, management, and related aspects of environmental consulting engineering firms are addressed in the second part of this chapter. This part focuses on the industrial corporate sector which, as the primary owner and/or permitee, has direct responsibility for environmental control and compliance. Most industries, whether public or privately-held entities, are subject to different sets of economically related issues as compared to consulting firms. Industrial environmental economics balances many direct or indirect corporate priorities, including, but not limited to,

- Capital expenditure and profit
- Cognizance of applicable state, federal and local laws and regulations

- Permit compliance
- Operational, training, and troubleshooting responsibilities
- Preemptive steps to ensure compliance
- Exposure to litigation
- Shared costs for joint and several/retroactive cleanup (remediation) measures based on past industrial activities

Evolution of Environmental Laws and Regulations in the United States

There has been a pronounced evolution in the field of environmental knowledge, disposal practices, and regulatory control in the United States over the past 50 years. Industrial water quality, to a great extent, has evolved and become more stringent as a direct effect of this regulatory control. Industrial environmental awareness, knowledge, capital expenditures, management, and governance can be related directly to the federal statutes and their regulations, particularly since 1970. Basically, prior to the formation of the U.S. Environmental Protection Agency (EPA) in 1970, there were virtually no clearly enforceable statutes at the federal or state levels. Instead, regulatory control was based primarily on "nuisance" or "do not pollute" language which was broad, difficult to interpret, and even more difficult to enforce. This is evidenced by the fact that both municipal and industrial entities were not subject to significant enforcement actions or penalties prior to the latter 1960s and early 1970s.

The regulation of pollution control was primarily vested to the states and state health departments until the early 1960s. During this decade, the states began forming separate pollution control agencies, although enforceable permit systems were either nonexistent or general in nature. The first serious attempt to modify this approach was to require permits through the U.S. Corps of Engineers using the 1899 Rivers and Harbors Act. However, this Act applied only to "navigable waterways," provided no standards for granting or denying permits, and there was little or no basis for enforcement with adequate penalties.

The first significant federal statute to address these issues was the 1972 Clean Water Act (CWA) Amendments, or Public Law 92-500. The CWA initially did not include control of "toxic constituents," and addressed the point source contamination to surface waters. Numerical limitations were based on general organic parameters, and specific constituents were rarely included. Subsurface contamination was ignored in the CWA. The first serious statute addressing the subsurface issue was the Safe Drinking Water Act of 1974 (SDWA), followed by the Resource Conservation and Recovery Act of 1976 (RCRA), the regulations of which were not promulgated until November 1980. The RCRA legislation was the first federal action to designate defined hazardous status to wastes and establish specific criteria for treatment, storage, and disposal of such residuals. The CWA was amended in 1977 to broaden control of designated "toxic" constituents and provide pretreatment standards for industrial dischargers to publicly owned treatment works (POTWs). In 1980, the Comprehensive Environmental Response, Compensation, and Liabilities Act (CERCLA) was passed to identify hazardous disposal sites, evaluate damages, require responsible parties to participate in remedial activities, and establish a claims procedure for restoration. RCRA was significantly amended in 1984 [Hazardous Solid Waste Amendment (HSWA)] to close regulatory loopholes, promote more recycling of hazardous materials, significantly restrict the use of land disposal technologies, and create a comprehensive registration and control program for underground storage tanks (USTs). The CWA was again amended in 1987 to tighten discharge standards beyond technology-based requirements to ensure that water quality standards for toxic residuals were met. A graphical illustration of the federal statute history over the past 50 years is presented in Fig. 18.1.[1] This underscores the increasing compliance costs experienced by both industries and municipalities.

The basic regulatory strategy over the past 50 years has evolved from a disease control and public health perspective in the early part of the twentieth century with little control over environmental impacts, to a general environmental awareness in the 1960s, to a more defined "command and control" strategy with enforceable statutes in the 1970s and 1980s. This transitional strategy continued in the 1990s, namely, a redefinition of "sustainable development," risk assessment, and more intensive linkage of human health to the environment.

Pollution control technologies and practices have evolved over this period, in partial response to and driven by the corresponding statutes and regulations. Moreover, practitioners in the field of

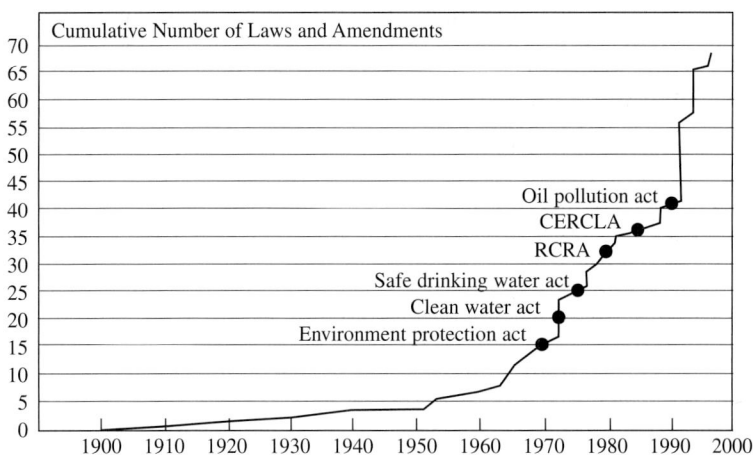

Source: Davis L. Ford & Associates, 2004.

FIGURE **18.1** Federal environmental legislation.

sanitary and environmental engineering assumed a significant assimilative capacity for waste in the subsurface medium in the 1950s, 1960s, and 1970s. As evidenced by the regulatory chronology, priorities for controlling surface water contamination was established in the early 1970s, followed by establishment of universal drinking water standards and subsurface water quality control in the mid-1970s, to defining and controlling hazardous and nonhazardous waste disposal in the early 1980s.

Since the 1990s through 2007 and beyond, the amendments of the aforementioned statutes and new regulations have had a significant impact on the industrial sector:

- More restrictive effluent discharge permit limitations, issued primarily by the states, subject to federal industrial guideline standards
- Extensive spill reporting
- Nonpoint source contamination control
- Recycling issues based on water limitations
- Expanded groundwater contamination control measures
- Inclusion of additional specific compounds in permits
- Expansion of specific regulations of single compounds, numbering less than 200 regulated compounds by 1980 and increasing to over 1000 in the late 1990s

These are but a few of the technical issues industry is facing presently with respect to industrial water quality. The economic implications to industry, resulting from this increasing stringency, are profound, all of which are discussed in the following sections of this chapter:

- Capital expenditures
- Direct and indirect compliance metrics
- Environmentally related litigation exposure
- Joint and several/retroactive cleanup responsibilities
- Corporate governance with respect to industrial water quality

Each has short- and/or long-term economic impacts on industry in the quest for the efficient use of the environmental dollar.

18.3 Capital and Operational Economic Planning

Industry investment in water quality upgrades, plant expansion, and grassroots (new) construction can be measured in different ways. Industrial financial planning and funding for environmental projects, including water quality, must be justified using several parameters.

From an engineering/financial perspective, these might include the following questions:

1. Will a treatment plant upgrade for a given capital outlay meet federal and state standards? At what compliance frequency? What is the cost/risk analysis?
2. What is the upgrade life with respect to establishing the amortization schedule (and corresponding tax implications) and how does this comport with the expected plant life for future expansion planning?
3. For specific process upgrades, what is the payback period? Is this measured in expended capital recovery or risk reduction, or both? Payback period refers to the period of time required for the return on investment to repay the sum of the original investment. For example, a million dollar investment which returns an investment or cost savings of $500,000 a year would have a 2-year payback period. The basic formulas are

Payback period = investment/cash flow (if cash flows are the same duration of the project)

or

Payback period = (last year with a negative cash flow) + (absolute value of net benefits/total cash flow next year) Applicable when benefits change over time.

4. What is the capital/operation and maintenance (O&M) relationship? Has a present worth analysis on the low capital, higher O&M versus higher capital, lower O&M scenario been performed? How does this decision impact cash flow? Have critical philosophical decisions on the level of automation/instrumentation as compared to more labor-intensive options been made?
5. Have operational staffing decisions been made to ensure optimal operational control? Will the operators be dedicated strictly to industrial water quality control or will they have other responsibilities?
6. Has the life of existing permits and possibly more restrictive effluent quality standards for renewed permits been factored into future capital expenditure estimates?
7. Has an analysis been made on capital/operational environmental upgrades regarding what can be expensed as compared to amortized?
8. How much redundancy is incorporated into critical process treatment units, and how much equalization and/or off-site storage is required? What is the economic impact (in terms of

lost or reduced production, and so on) if key environmental control systems are unavailable or bottlenecked?

9. Is there a buffer zone between the complex units and the contiguous areas?

As an example, downstream chemical, petrochemical, and petroleum refinery has invested 15 to 25 percent of their plant capital investment to comply with environmental regulations during the 1990s and through 2001. Between 1992 and 2001, the oil industry spent more than 100 billion dollars to bring existing refineries into compliance with environmental regulations.[1] Grassroots (original) construction has been restricted as it generally is presumed to be less expensive to simply add on the existing refineries. However, changing prices of crude oil and energy demands may change this approach in the next decade. For example, a new 400,000 bbl/d refinery is in the planning and approval stages. If this project comes into fruition, it will be the first grassroots refinery to be built in the United States since 1976.[2]

18.4 Economics of New Facility Siting Analysis and Planning

The authors have been involved in several grassroots industrial plant siting projects over the past decade. These include, but are not limited to, a new Union Carbide plant in the Bay City, Texas, area in the late 1970s; Formosa Plastics, USA, at Point Comfort, Texas, in the 1990s; and Hyperion Energy, South Dakota, in the 2005–2008 era. In the first two examples, multibillion dollar plants were constructed and started up, and are currently operating. The latter is still in the planning stage, planning to begin construction in 2009, depending on future approvals, financing, and other factors.

There are many factors that must be evaluated on finalizing the selection of the industrial site, such as feedstock availability, transportation, market locations, soil structural consideration, existing or proposed pipelines and/or port facilities, ability to secure land and right-of-ways, labor markets, and other economic factors.

Net present value (NPV) or **net present worth (NPW)** is defined as the present value of net cash flows. It is a standard method for using the time value of money to appraise long-term projects. Used for capital budgeting, and widely throughout economics, it measures the excess or shortfall of cash flows, in present value (PV) terms, once financing charges are met.

Each cash inflow/outflow is discounted back to its present value (PV). Then they are summed. Therefore

$$\text{NPV} = \sum_{t=0}^{N} [C_t/(1+r)^t]$$

where t = the time of the cash flow
  N = the total time of the project
  r = the discount rate (the rate of return that could be earned on an investment in the financial markets with similar risk)
  C_t = the net cash flow (the amount of cash) at time t (for educational purposes, C_0 is commonly placed to the left of the sum to emphasize its role as the initial investment)

NPV is an indicator of how much value an investment or project adds to the value of the firm. With a particular project, if C_t is a positive value, the project is in the status of discounted cash inflow in the time of t. If C_t is a negative value, the project is in the status of discounted cash outflow in the time of t.

Internal rate of return (IRR) is a capital budgeting metric used by firms to decide whether they should make investments. It is an indicator of the **efficiency** of an investment, as opposed to net present value (NPV), which indicates value or magnitude.

The IRR is the annualized effective compounded return rate which can be earned on the invested capital, i.e., the yield on the investment.

A project is a good investment proposition if its IRR is greater than the rate of return that could be earned by alternate investments (investing in other projects, buying bonds, even putting the money in a bank account). Thus, the IRR should be compared to any alternate costs of capital including an appropriate risk premium.

Mathematically the IRR is defined as any discount rate that results in a net present value of zero of a series of cash flows.

In general, if the IRR is greater than the project's cost of capital, or *hurdle rate*, the project will add value for the company.

To find the internal rate of return, find the value(s) of r that satisfies the following equation:

$$\text{NPV} = \sum_{t=0}^{N}[C_t/(1+r)^t] = 0$$

(See aforementioned net present value for details on this formula.)

Example. Calculate the internal rate of return for an investment of 100 value in the first year followed by returns over the following 4 years, as shown below:

Year	Cash Flow
0	−100
1	39
2	59
3	55
4	20

Solution We use an iterative solver to determine the value of r that solves the following equation:

$$NPV = -100 + 39/(1+r)^1 + 59/(1+r)^2 + 55/(1+r)^3 + 20/(1+r)^4 = 0$$

The result from this numerical iteration is $r = 28.09\%$.

It should be recognized that IRR is simply a directional investment decision tool. It should not be used to compare projects of different durations or cash flow projections. Moreover, as intermediate cash flows are likely not to be reinvested in the projects, IRR, the actual rate of return will be lower.[7]

One of the major considerations is the environmental setting. Major factors in planning a new industrial facility relating to water quality include, but are not limited to, the following:

1. Available water, firm quantity, and acceptable quality (this can be estimated by historical water demand per unit of production).
2. Proximity of water to the proposed site.
3. Receiving body of water for treated effluent.
4. Quantity and quality of treated water to meet receiving water quality standards and comply with point source permit limitations, including bioassay toxicity.
5. Susceptibility of site to possible subsurface contamination (saturated and unsaturated zones).
6. Effect of nonpoint source runoff to contiguous environment.
7. Impairment of contiguous areas such as noise, lighting, odor, and other facets.
8. Emission rates of air, water, and hazardous wastes and compliance with ambient background quality standards.
9. Unique and restrictive regulations specific to the proposed site (i.e., nonattainment areas, animal and fish habitats, inability to expand, and so on).
10. Total maximum daily loads (TMDLs) is the amount of a particular pollutant that specified receiving bodies of water can "handle" without violating water quality standards. The TMDL process is described briefly as follows:
    - Identify waters that do not meet water quality standards. In this process, the state identifies the particular pollutant(s) causing the water not to meet standards.
    - Prioritize waters that do not meet standards for TMDL development (e.g., waters with high naturally occurring "pollution" will fall to the bottom of the list).
    - Establish TMDLs (set the amount of pollutant that needs to be reduced and assign responsibilities) for priority

waters to meet state water quality standards. A separate TMDL is set to address each pollutant with concentrations over the standards.
- Strategy to reduce pollution and assess progress made during implementation of the strategy. This is when a watershed partnership most likely will want to get involved. If the partnership has already developed a plan of action, it should be shared with the state. In fact, several states have incorporated watershed partnership plans in the state's strategy for specific TMDLs.

The site selection process is enormously complex and requires an expensive investigative phase, including meetings and hearings with stakeholders and regulatory agencies, site and utility procurement, attaining air, water construction and other permits, complying with local building codes, and possible rezoning issues, just to name a few. For example, a list of required permits and approvals for a potential grassroots refinery project before construction can begin includes the following:

- State Construction Storm Water Discharge Permit
- State Wastewater Discharge Permit
- State Water Rights Permit
- State Prevention of Significant Deterioration (PSD Air Quality Permit)
- State Section 106 National Historic Preservation Act (NHPA) consultation
- State and Federal Section 401 Water Quality Certification
- Federal Section 404/10 Permit under the Clean Water Act for Impacts on the Waters of the United States, including wetlands
- RCRA permits as applicable; RCRA permit not required if facility will be generator only
- Federal Section 7 Endangered Species Act (ESA) consultation
- Federal Aviation Administration (FAA) Obstruction and Lighting Clearance
- Other permits and approvals required by the state and federal law

18.5 Environmental Compliance

There is a significant economic impact resulting from efforts to attain and maintain industrial environmental compliance. There are several facets in establishing a base to ensure maximum permit compliance, including the development of a methodology to measure it. Environmental audits, operator training programs, operation manuals, baseline monitoring, spill prevention, and other proactive measures are discussed in this section.

1. *Environmental audits.* Conducted by third-party consultants, environmental audits can be valuable to an industrial entity in many ways. Basically, they are designed to obtain accurate and current information about a specific industrial plant that can quantify its environmental impact as a basis for making management decisions. Such audits also can include safety [Occupational Safety and Health Administration (OSHA) related], mechanical integrity, and other factors which have environmental, health, and safety implications. Moreover, they can be used for accreditation of environmental management systems (EMS) such as those under ISO 14001. ISO is the International Organization for Standardization for a voluntary compliance with ISO-developed environmental standards. Benefits of an EMS based on ISO 14001 include

- Improvements in overall environmental performance and compliance
- Providing a framework for using pollution prevention practices to meet EMS objectives
- Increased efficiency and potential cost savings when managing environmental obligations
- Promoting predictability and consistency in managing environmental obligations
- More effective targeting of scarce environmental management resources
- Enhancing public posture with outside stakeholders

There are various subsets of environmental audits, such as waste disposal, water intake use, management, effluent disposal, compliance, EMS, and environmental "due diligence" (for sale or purchase of an industrial facility). The overall objective of the environmental audits is economically driven, such as reducing potential for fines and liability exposure, ensuring legislative compliance, reducing waste-related costs, reducing energy and water expenditures, and promoting good public relations. Additionally, the control of volatile organic compounds (VOCs) from industrial products and waste residuals and from emissions from the liquid to the gas phase can be reviewed [compliance with the National Emission Standards for Hazardous Air Pollutants (NESHAPs)].

2. *Operator training.* Industrial water quality control can be enhanced through planned and continual training courses for operators who have access to current operating manuals, corporate environmental manuals, material safety data sheets, and troubleshooting guidelines. Operator training programs for industrial wastewater treatment programs must be adapted to site-specific treatment systems,

both in-plant and end-of-pipe processes. Operator training programs can be conducted by outside experts, in-house experts, or both. An example of a 3-d operator training course jointly conducted by an outside consultant and in-house supervisor at a complex chemical plant is outlined in Table 18.1. A typical troubleshooting guide for operators is illustrated in Table 18.2, and a detailed maintenance instruction manual attachment is included in Table 18.3.

It should be noted that extensive monitoring should be incorporated within the treatment system as well as fenceline and buffer zones. The permits have selected monitoring requirements for point source discharges, but profile monitoring throughout the treatment system can be valuable in enhancing process performance and compliance.

3. *Spill prevention.* Spill prevention plays an important role as a proactive compliance measure for industry, which is not only in the interest of the facility, but also a statutory requirement under the CWA. A Spill Prevention Control and Countermeasure (SPCC) Plan

Day One
Introduction
Overview
Pretreatment concepts
Oil removal procedures
Equalization and off-spec storage
Optimization of biological treatment
Introduction to the fluidized bed systems and unit operations
Day Two
Physical/chemical systems
Activated carbon/filtration
Plant start-up and general procedures
Process control
Day Three
Site visit
Sanitary wastes/unit operations
Use of polymers/chemicals
Monitoring safety procedures, standard operating procedures (SOP)

TABLE 18.1 Operator Training Course Industrial Wastewater Treatment System

Operational Problem	Possible Cause	Recommended Corrective Action
High suspended solids concentration in effluent, poor plant efficiency (sludge bulking or gasification)	(1) Excessive organic loading.	(1) Reduce organic loading to system capacity or below by temporary diversion of wastewater.
	(2) Nature of the wastewater results in proliferation of filamentous microorganisms.	(2) Maintain aeration basin contents at neutral pH; add coagulant to basin to enhance solids-liquid separation.
	(3) Predominance of filamentous microorganisms resulting from low N or P concentration in aeration basin.	(3) Add inorganic N or P to system.
	(4) Filamentous microorganisms persist in system.	(4) Subject sludge to prolonged anaerobic holding time if possible, chlorinate, or "purge system of sludge" and reseed.
	(5) Gasified sludge caused by excessive anaerobic sludge holding time in clarifier.	(5) Increase sludge recycle rate and/or increase sludge wastage.
	(6) Excessive sludge in clarifier caused by plugged sludge line.	(6) Increase sludge recycle pumping rate as required to flush lines.
	(7) Excessive sludge in system.	(7) Increase sludge wastage and disposal.
	(8) Excessive oil concentration in aeration basin.	(8) Check primary treatment units for malfunction.

Excessive odors associated with designed aerobic systems (basin dissolved oxygen 0.5 mg/L)	(1) Oxygen deficit resulting from organic overloads.	(1) Increase aeration capacity.
	(2) Localized anaerobic "pockets" due to incomplete mixing.	(2) Increase aeration basin power level or baffle system.
Dissolved oxygen levels in aeration basin exceeding 4–5 mg/L	(1) Reduced organic load, insufficient MLSS concentration; inactive biological population.	(1) Increase MLSS concentration; reseed with active microbial sludge.
Excessive foaming	(1) High concentration of surface active agents in aeration basin, possible surfactant resistance to biological degradation.	(1) Activate foam suppressing spray system and add foam suppressant if necessary.

TABLE 18.2 Biological Treatment Section: Troubleshooting Guide

Chapter Eighteen

Section No.	Description
1	Tables of Contents from installation, operation, and maintenance manuals submitted under the contract for supply of equipment
2	Design Information Fine Bubble Aeration Secondary Clarifiers
3	Design Outline—Fine Bubble Aeration Aeration Equipment Description Blower Sizing and Performance Curve Fine Bubble Aeration Installation, Operation, and Maintenance
4	Clarifier Equipment Specifications Clarifier Installation, Operation, and Maintenance
5	Fluidized Bed Reactor Design Process Description Process Overview Chemistry and Theory Fluidized Bed Mechanics Stoichiometry

6	Fluidized Bed Startup
	Granular Activated Carbon (GAC) Addition
	Reactor Seeding
	Pre-Startup Preparation
	Startup
	Normal Operation: Adjusting Oxygen
	    Adjusting Nutrients
	    Adjusting Biomass Discharge Rate
	    Information Collection and Checks
	Manual Shutdown
	Standby Operation

7	System Interlocks Aeration and Clarifiers Fluidized Bed
8	Fluidized Bed Reactor Maintenance and Troubleshooting
9	Fluidized Bed Reactor—Auxiliary Systems Oxygen Generation System Bulk Carbon Transfer System Fluidization and Flow Distribution System Carbon Wasting System

TABLE 18.3 Operator Maintenance Attachment

Section No.	Description
10	Typical Drawings
	Process Flow Diagrams (PFDs)
	Process and Instrumentation Diagrams (PIDs)
	Instrument Lists
	Isometric Views of Equipment and System Piping

TABLE 18.3 *(Continued)*

is required by regulation contained in 40 CFR 112. The SPCC rule was amended by the U.S. EPA in August 2002, and compliance with the amended rule is set for July 1, 2009. SPCC guidelines are outlined in 40 CFR 112.7(a)(2). The elements of an SPCC Plan include:

1. Personnel listing
2. SPCC for bulk storage containers, secondary containment for bulk storage containers, fluid loading transfer areas, and facility drainage system; for example,
   - Oil/water separators
   - Security and lighting
3. Emergency spill response procedures
4. Spill cleanup mitigation
5. Notification
6. Amendments

A Best Management Plan (BMP) is an EPA requirement which deals with prevention from storm runoff, spills, and improper hazardous waste handling as opposed to that from point sources regulated by federal or state NPDES permits.

Development of a BMP plan requires identification of the hazardous materials located in the plant, a written description of the methods and procedures used to protect against inadvertent releases of these materials, and a detailed analysis of the entire plant as to the probability for such releases and containment/mitigation capabilities. The written plan must cover general practices such as preventive maintenance, visual inspections, personnel training, and spill reporting and investigation, as well as specific containment and/or treatment systems available to prevent or mitigate releases.

Federal spill/release reporting requirements for episodic releases of spills are mandated by the Superfund Amendments and Reauthorization Act of 1986 (SARA). Title III of SARA contains a freestanding act entitled the Emergency Planning and Community Right to Know Act (EPCRA), which creates a framework for local emergency contingency planning and requires owners and operators who handle hazardous

890   Chapter Eighteen

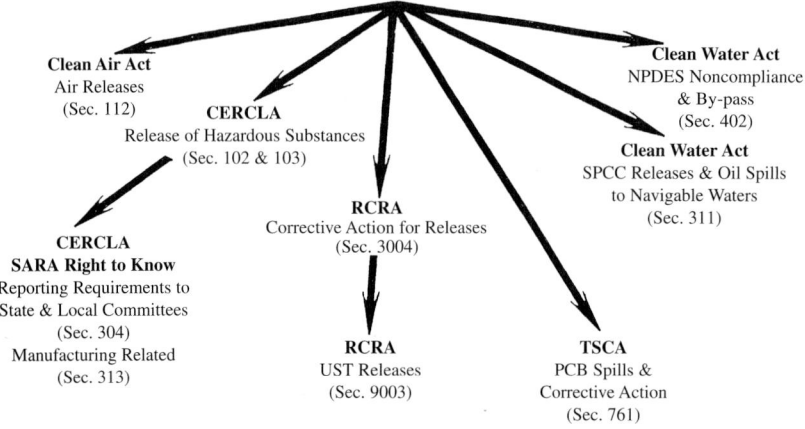

FIGURE 18.2   Federal spill/release reporting requirements.

chemicals to comply with a number of reporting and record keeping requirements. These are grouped into four categories, namely:

- Emergency Planning Reporting
- Community Right to Know Reporting
- Emergency Release Reporting
- Toxic Chemical Release Reporting [Toxics Release Inventory (TRI)]

The federal spill/release reporting requirements are complex, as illustrated in Fig. 18.2.

The TRI is a publicly available database that contains information on toxic releases and other waste management activities reported annually by certain covered industry groups as well as federal facilities. This inventory was established under EPCRA and was expanded by the Pollution Prevention Act of 1990. The 2007 toxic chemical list contains 581 individually listed chemicals and 30 chemical categories. Twenty industrial categories (SIC code 20-39) are included, along with emissions to the air (point source and fugitives) discharges to bodies of water, releases at the facility to land, and disposal to underground injection wells. Off-site transfers are also included. A discussion of TRI with respect to the request for zero discharge and other implication is provided elsewhere.[3]

4. *Compliance documentation.* Compliance documentation can be interpreted as an "environmental report card" and can be expressed in several ways as there is not a standard format. One can develop a mosaic, however, using several metrics (or yardsticks). This documentation should be invaluable to industrial discharges in many ways as it includes increased awareness, action items for improvement,

regulatory impact, significant economic savings through priority environmental expenditure of capital and O&M, and informing stakeholders. Annual recording of reportable spills, employee injuries, discharge trends, and permit exceedances, along with other "yardsticks" are tabulated in Table 18.4, which outlines a proposed matrix for environmental control quantification.

	Year 1	Year 2	Year 3	Year 4	Year 5
Point source NPDES permit discharge (% compliance)	—%	—%	—%	—%	—%
RCRA violations (#): Major	—	—	—	—	—
Minor	—	—	—	—	—
EPCRA violations	—	—	—	—	—
Notice of violations (NOVs) (#)	—	—	—	—	—
Paid fines ($)	—	—	—	—	—
ISO 14,001 (yes, no)					
Environmental-related costs ($)					
Capital	—	—	—	—	—
O&M	—	—	—	—	—
Environmental audits (yes, no)	—	—	—	—	—
Safety (accidents/year)	—	—	—	—	—
Continuing operator training (yes, no)					
Current standard operating procedures (SOP) (yes, no)					
Current material safety data sheets (MSDS) (yes, no)					
Energy efficiency	←		Narrative		→
Employee participation/ incentives	←		Narrative		→
Product environmental impact reduction (est. %)	—%	—%	—%	—%	—%
Runoff management control	←		Narrative		→

TABLE 18.4 Industrial Water Quality Control Environmental "Report Card" Matrix

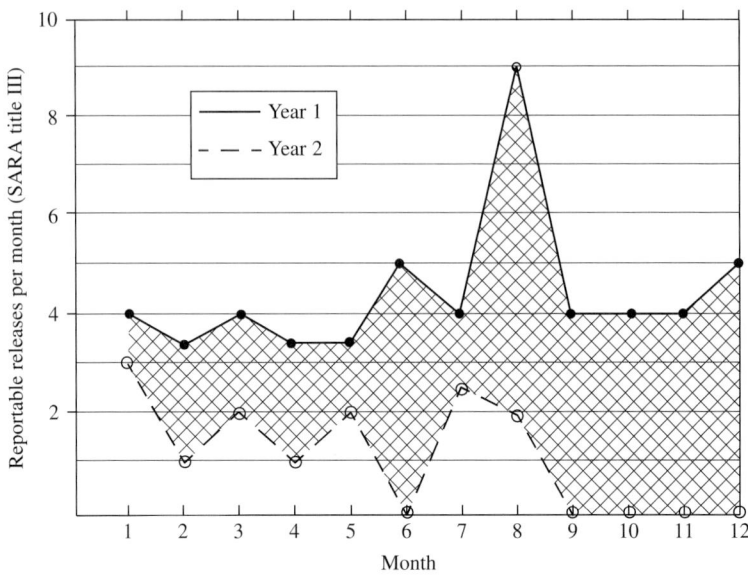

FIGURE 18.3 Spill prevention and control.

Graphically depicted examples are typically used by industries to record environmental metrics and trends. For example, the reduction of reportable spills is shown in Fig. 18.3 for successive years, and illustrates a successful spill control program for the industrial complex. The reduction of point source effluent chlorinated hydrocarbons over three decades is documented in Fig. 18.4, resulting from more stringent permit requirements by the regulatory agencies and matched by capital expenditures necessary to upgrade the industrial wastewater treatment system.

Employee safety metrics based on injuries, near miss rates, and exposure incident rates over a 6-year period for an industrial complex are shown in Fig. 18.5. These data, along with OSHA compliance information, provide a health and safety record over defined periods of time. RCRA compliance is somewhat more difficult to quantify numerically because of the complications in the RCRA-based regulations, ranging from severe violations such as the treatment, storage, and disposal of hazardous waste violations to the more minor violations such as labeling and handling requirements.

The point source discharge (NPDES) record under the CWA is more definitive. For example, a graphical analysis of chemical oxygen demand (COD) for an industrial wastewater discharge can be taken from monthly discharge monitoring reports (DMRs) and plotted over time, as illustrated in Fig. 18.6. The annual outfall excursions (permit exceedances) for both point source wastewater and stormwater

Environmental Economics 893

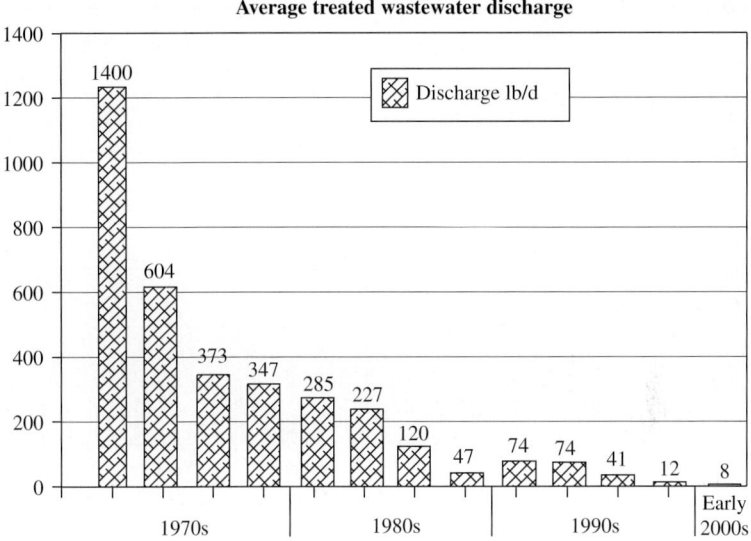

FIGURE 18.4 Effluent reduction—chlorinated hydrocarbons.

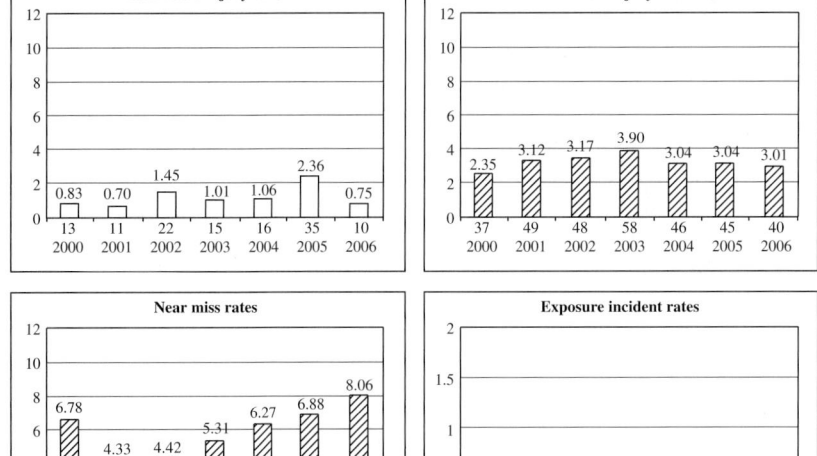

FIGURE 18.5 Injury and exposure incidents.

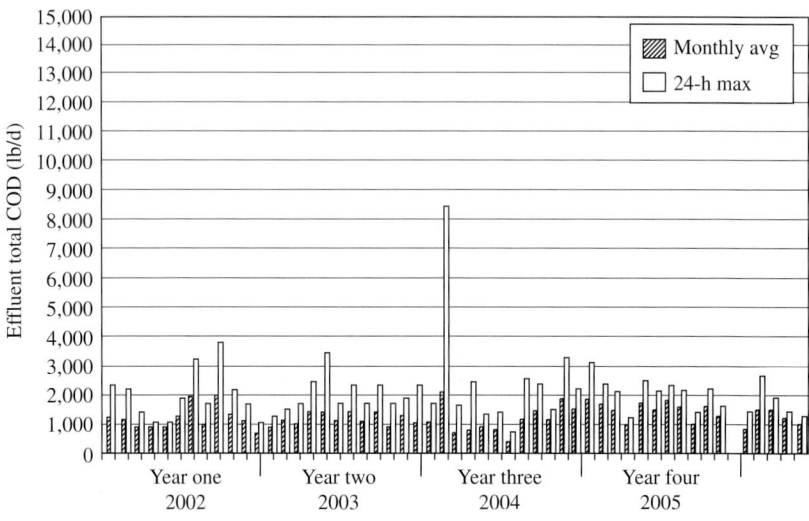

FIGURE 18.6 Effluent COD performance data (2002–2005).

outfalls are recorded in Fig. 18.7. The composition of such exceedances can be delineated as shown in Fig. 18.8. This analysis can be used by the treatment system designers to select unit processes for a treatment system upgrade to improve compliance metrics. Based on the experience of the authors, the federal (or states with permit primacy) NPDES permit compliances for most industries have been in the range of 98 to 99 percent. However, with increasing end-of-pipe technologies, operational experience, and corporate commitment, many annual compliance records recently have approached or even

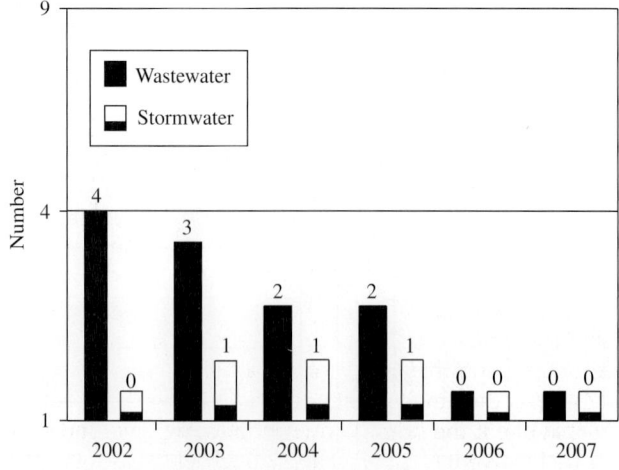

FIGURE 18.7 Annual outfall excursions.

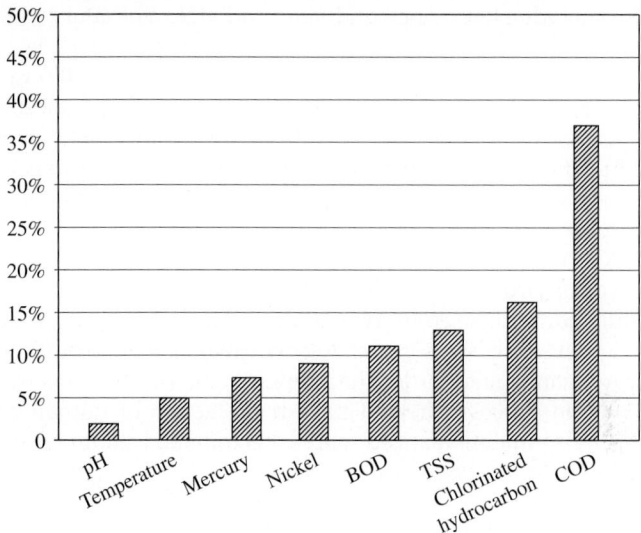

FIGURE 18.8 Composition of specific permit exceedances.

attained 100 percent. Even though the EPA industrial guidelines assumed more of a measure of noncompliance, expectations have been significantly elevated since the 1990s and before.

A chronological paper trail of permits and renewals, installation of treatment process and conveyance upgrades, related capital expenditures, and related correspondence is helpful to industrial permittees to enhance the decision-making process as well as assist in the resolution of legal and regulatory disputes. The recordation of all of these "environmental report card" issues possibly can be used to assess past cost/compliance relationships.

18.6 CERCLA, Superfund, and Joint and Several and Retroactive Economic Exposure

Superfund is a common name for the U.S. environmental law officially known as the Comprehensive Environmental Response, Compensation, and Liability Act (CERCLA). This law and its regulations have a significant impact on industries as "potentially responsible parties" (PRPs). Basically, it opens up a *joint and several, retroactive liability* on removal actions and remediation actions which potentially has pronounced economic consequences on an industrial complex. For example, an industry that sent waste residuals to an off-site landfill or recycler (legal at the time) might have to underwrite the total cleanup costs for that site if the original owners and/or other industrial participants are insolvent (orphan shares). Such sites are not only limited to those on the National Priority List (NPL or federal Superfund sites) but also include state superfund sites and state voluntary cleanup locations.

Many of these cleanup or remediation projects cost 100 million dollars or more, with a possible continuing cleanup responsibility for many years. When two or more PRPs are solvent and share their allotted portion of cleanup costs (plus a share of the orphan costs), then allocation of costs becomes a major factor and often results in legal disputes between the PRPs. (See Chap. 16.)

CERCLA holds past and future landowners fully liable for cleanup of hazardous wastes existing at a site as well as for present and future harm to the environment. Risk Reduction Standards (RRS) are being developed by some states to provide some flexibility with respect to cleanup standards. The intent is to provide a consistent corrective action process directed toward protection of human health and the environment balanced with the economic welfare of the citizens of the state. These standards include:

- State superfund
- Voluntary cleanup program
- Petroleum storage tank
- Industrial and hazardous waste and underground injection control

Cleanup standards can be exceedingly rigorous, driving up PRP costs and having a significant effect on an industry's balance sheet and price per share. Even before any of these corrective action monies are spent, the industrial entity must set aside a contingency account to satisfy the costs of meeting the applicable cleanup standards.

18.7 Environmental Litigation Exposure

Industries always are subject to environment-related lawsuits under a variety of state and federal statutes. Certainly, Superfund liability (release or threat of a release of hazardous wastes), negligence, property diminution, personal injury, status, trespass, property damage, stigma, and a myriad of other civil (and sometimes criminal) allegations end up in lawsuits with the industrial facility generally in a defendant role. Moreover, there are lateral disputes between industries or related parties in the cost allocation of cleanup sites, mergers and acquisitions subject to prior contamination liabilities, and contractual disputes. Any of these can significantly affect the balance sheet of an industrial facility.

1. *Statute-based litigation.* CERCLA, RCRA, CWA, SDWA, and Oil Pollution Act (OPA) liabilities are extensive. For example, present and past landowners are fully liable for cleanup of hazardous

substances. Courts have made clear that present landowners are not excused from liability, even though they had nothing to do with causing the original contamination. Many persons in the refining and production of oil and gas believe that the "petroleum exclusion" under CERCLA and the E&P "associated waste exemption" under RCRA effectively remove oil and gas properties and operations from the hazardous waste designations. Although each law contains special exceptions from its application to the oil and gas industry, the courts have generally held that each statute, and its associated exemption, does not apply to other statutes that could otherwise be used to address the situation. For example, the petroleum exclusion under CERCLA does not limit EPA or third parties from addressing petroleum-related contamination under the OPA, CWA, and RCRA. Similarly, wastes excluded from hazardous waste regulation under RCRA may nevertheless create liability under CERCLA or the CWA or OPA.

2. *Acquisition/merger due diligence.* Another significant litigation exposure is in the acquisition of, or merger with, entities that own industrial interests. There are many good reasons for conducting extensive environmental due diligence related to a planned acquisition or merger, as these issues could translate to long-term exposure to financial or public relations liabilities. If the properties to be acquired have been used to manufacture, store, or use hazardous and toxic chemicals, then it is possible that leaking underground storage tanks, spills and other releases of chemicals into the soil and adjacent waterways, on-site disposal, and contamination of the interior surfaces of buildings and equipment have resulted in conditions that may pose some risk to human health or the environment. Should these conditions ultimately be judged sufficiently hazardous to require some kind of remedial action, then someone has to pay. It is better if the buyer discovers these conditions before the acquisition or merger agreements are executed so that appropriate responsibility for ultimate cost or remedial actions can be allocated between the parties.

Many industries have gone to their insurance companies with claims for contamination and cleanup costs under environmental coverage under comprehensive general liability (CGL) policies. There has been extensive litigation (and case law) between the industrial policy holder and the insurance companies. CGL insurance policies issued after the early 1970s contain some form of a pollution exclusion clause. Although industries that are subject to environmental liability may purchase separate policies to address environmental hazards, most industries have not yet obtained such coverage.

CGL policies issued prior to the early 1970s did not contain a pollution exclusion. However, from the early 1970s through approximately 1986, standard CGL policies contained a "sudden and accidental" pollution exclusion. From approximately 1986 to the present, the standard CGL policy contains a so-called *absolute exclusion* that was intended to eliminate coverage associated with pollution-related events. Some CGL policies exclude from coverage injury or damage arising from "discharge, dispersal, seepage, migration, release, or escape" of pollutants. Courts have construed this language to apply to pollutants which travel "from a contained place to the insured person's surroundings and then cause injury." In contract, injuries caused by irritants that normally are stationary, but can be shifted or moved manually, are not excluded from coverage because they do not cause injury to one of the prescribed methods.

One can see the ambiguities in the CGL policies on "pollution exclusion" and why policyholders in many cases have to take legal action against the insurance companies to let the court decide if remuneration for environmental costs can be realized.

3. *Court decision citations*. Selected court decisions in which the coauthor has participated are cited to illustrate how some of these disputes have been resolved, either through court decisions, arbitration, or settlement.

> Pacific Gas & Electric (PG&E) recently settled a major lawsuit filed by Hinkley, California, residents who claimed personal injury from their proximity to a gas compressor station in a pipeline network that moved natural gas from Texas to the San Francisco Bay Area. Hexavalent chromium from the cooling tower blowdown contaminated the groundwater, commencing in the 1950s until cessation of chrome additives to cooling towers in the 1970s and 1980s. In this litigation (the "Brockovich" case) PG&E paid an initial settlement amount in arbitration proceedings, then paid a second amount to settle the case prior to trial.
>
> Beverly Hills High School (*Lori Lynn Moss et al., Plaintiffs, v. Veneco, Inc., et al.*). Plaintiffs sued Veneco, as well as previous owners/operators of an oil and gas production operation contiguous to Beverly Hills High School, Beverly Hills, California, for medical causation. This case was very high profile and set for trial. The paper trails on this case were extensive due to years of public hearings, monitoring reports, consultants' reports, leases, and royalty issues. Codefendants included a central power plant which supplies heating and air conditioning to Century City, several oil companies, and the City of Beverly Hills and the Beverly Hills Consolidated School District that received royalty payments. Judge Wendell Mortimer, Jr., Los Angeles Superior Court, granted the defendants' selected motions for summary judgment based on the defendants' and applicable agencies' extensive monitoring of benzene and hexavalent chromium and other compounds over the

years which were below control standards and the failure of plaintiffs' epidemiological and toxicological experts to cite authoritative books, peer-reviewed articles, or data to link the chemicals of concern to the alleged diseases. "Defendants' motions for summary judgment shifted the burden to Plaintiffs and Plaintiffs did not meet their burden of proof to show a triable issue of fact." (Judge Mortimer's Granting Motion for Summary Judgment, December 12, 2006.)

Bolinder Real Estate v. United States (2002 WL 732:155—D. Utah). Plaintiffs are the owners of real property in Tooele County, Utah. The real property is located near Tooele Army Depot (TEAD), a U.S. military installation. Plaintiffs brought suit against defendant United States claiming that during operations at TEAD, the defendant discharged trichloroethylene (TCE) into the environment which migrated to and contaminated a well on plaintiffs' property. In their complaint, plaintiffs asserted eight causes of action. Before trial, the court dismissed all claims, except for the state law claims of continuing tort. Trial on this sole issue was held before the court in May 2001.

Before and during trial, there was little or no dispute regarding two facts. First, the well on plaintiffs' property had been contaminated by TCE. Second, discharges from TEAD are the source of the TCE contamination in plaintiffs' well. At trial, the remaining question to be resolved was whether the United States' discharge of TCE was negligent under the relevant standards of care at the time of discharge. To determine this question, then, the court had to determine: (1) What area or areas of TEAD were the sources of the TCE that contaminated the Bolinder well? (2) When did the discharges of TCE from these sources occur? (3) What was the standard of care regarding the discharge of TCE during the relevant time period.?

The final factual questions for resolution focused on the standard of care for the handling and disposal of TCE during the relevant period of time and whether the practices at TEAD violated that standard. The court has found that the TCE that migrated to plaintiffs' property in the mid-1990s was discharged at least 23 years earlier, and most likely before 1960. Therefore, the relevant standards of care were those that existed between 1942, when TEAD was established, and 1974, the time of the last potential TCE discharge that could have migrated to plaintiff's property in the mid-1990s.

The court's findings of fact lead to the conclusion that defendant was not negligent in the discharge of TCE that led to contamination of plaintiffs' property in the mid-1990s, because the discharges did not violate any relevant standards of care at the time those discharges occurred.

The court has also found that defendant's handling and disposal of TCE waste were consistent with waste disposal practices followed by the military and private industry during this period before 1974. Other courts have also recognized that neither industry nor the military were aware of the potential for TCE to contaminate groundwater in hazardous quantities until after the late 1970s and early 1980s.

Western Greenhouses v. United States (878 Supp. 917.927 N.D. Texas—1995). Western Greenhouses brought action against the United States under the Federal Tort Claims Act (FTCA). The plaintiffs sought to recover damages from groundwater contamination (TCE) caused by the activities of the U.S. employees at Reese Air Force Base, Lubbock, Texas. The conclusion by Sam R. Cummings, U.S. District Judge, was that the United States was not negligent and the plaintiffs did not demonstrate that the contamination was the kind of event which would not have ordinarily occurred in the absence of negligence. "In fact, the standard industry practices throughout the period between the 1940s and 1970s unfortunately led to the widespread contamination across the country . . . it was the Plaintiffs own negligence which was responsible for their failure to detect the presence of contamination in close proximity to their property before purchasing the land . . . here, Plaintiffs acted unreasonably in seeking a "dirt cheap" Phase I environmental assessment, when a proper and complete Phase I environmental assessment would have disclosed to the Plaintiffs the true extent of the known environmental contamination at Reese Air Force Base."

The federal court ruling established that the plaintiffs did not carry their burden of prior negligence, nuisance, or trespass under Texas law and the FTCA. Judgment on all claims was entered against the plaintiffs and in favor of the United States.

Snyder v. United States [504 F. Supp 2d 136 (SD MIS? 2007) Camp LeJuene, North Carolina]. The son of a Marine Corps officer and his parents brought suit against the United States under the FTCA, alleging that the toxic chemicals, trichloroethylene (TCE) and tetrachloroethylene (PCE), at Camp LeJuene caused a congenital heart defect.

The defendant's experts correctly pointed out that TCE and PCE were not regulated as toxic pollutants under the Clean Water Act until August 25, 1978, well after the time period relevant to this lawsuit. The EPA did not regulate TCE or PCE as hazardous wastes under the RCRA until November 19, 1980. The EPA did not regulate TCE or PCE as drinking water contaminants under the Safe Water Drinking Act until January 1989, and July 1992, respectively. In short, there were no government regulations which specifically regulated TCE and PCE at the time plaintiffs allege they were exposed to those chemicals.

The court's decision was "after careful consideration of the pleadings filed in this case and the history of the parties' litigation of the issues, it is the opinion of the Court that Plaintiffs' Complaint is barred by the FTCA's discretionary function exception. The Defendant's Motion to Dismiss must therefore be granted."

Lyondell Chemical Company, et al. v. Albermarle Corporation, et al. ("Turtle Bayou," Civil Action No. 1:01 - CV - 890, Eastern District of Texas). This is a classic remediation allocation case between multiple industries with plaintiffs Lyondell Chemical Company and Atlantic Richfield Company, and third party plaintiffs El Paso Tennessee Pipeline Company v. Defendant Albermarle Corp., Ethyl Corporation, Exxon Mobil Corporation, GATX Corporation, Lubrizol Corporation, and PPG Industries Inc. These companies were identified as PRPs for a closed dump site that received hazardous and nonhazardous wastes during the late 1960s and early 1970s. This complex technical and legal dispute came before federal Judge Marcia A. Crone. The reconstruction of historical documents, drivers' testimony, trip tickets, estimated volumes, expert reports and depositions, and production rates for both the active plaintiff and defendant companies as well as other companies which settled out of court, was a complex task. The court retained an expert allocator as an independent assistant to the judge. At the time of this writing, the allocation has not been finalized by the court.

Such allocation proceedings are significant to each of the parties, as the remediation costs have run into millions of dollars, with a corresponding economic impact on those industries with the highest allocation. One can see the complexity and equity issues which are involved in such disputes.

18.8 Industrial Environmental Governance

Environmental governance is relatively a new term, particularly with respect to publicly owned corporations as boards of directors and management teams have been forced to adapt to significant changes attributable to more complex and rigorous standards in business. Similarly, environmental governance has underscored the importance of economics, management, compliance documentation, regulatory relationships, and public relations within these privately or publicly owned entities. The topics presented in this chapter have either a direct or indirect relationship to environmental governance. A brief summary of the more important components of environmental governance for both publicly or privately-owned industrial entities is as follows:

Sarbanes-Oxley Act (SOX) (Publicly Owned Companies)
The Sarbanes-Oxley Act of 2002 established new or enhanced standards for all U.S. public company boards, management, and public accounting firms. Environmental governance, although not directly addressed in this statute, is applicable in some of the 11 titles which describe specific mandates and requirements for financial reporting. Examples are Title III Corporate Responsibility, Title IV Enhanced Financial Disclaimers, Title V Analyst Conflicts of Interest, and Title VIII Corporate and Criminal Fraud Accountability. The most contentious aspect of SOX is Section 404, Assessment of Internal Control. Both management and the external auditor are responsible for performing an assessment in the context of a top-down risk assessment, which requires management to evaluate both the size of its assessment and evidence "based on risk." This could include environmentally-related risks, liabilities, and exposures.

ISO 14001 Certification
The attainment of ISO 14001 certification can enhance environmental governance from small businesses to large manufacturing industries. It is designed to incorporate environmental considerations into manufacturing operations and product quality standards and is the world's most recognized environmental management system (EMS) framework.

Creation of Internal Environmental Cost Accounting System
The economic effectiveness of environmental cost control and accountability can be evaluated using an informal "cost center" for amortized capital and operations and management costs incurred at a central treatment system in a large industrial complex. This approach is being used as an internal accounting system to charge treatment costs back to individual industrial production units. For example, one treatment complex serving multiple production units measures the quality and quantity from the user production units, then allocates a user cost for treatment back to the production unit "profit centers." This not only informs management the costs of treatment per unit of production, but also gives the production "profit centers" an incentive to pretreat, recycle, or minimize their water quantity streams prior to discharging to the centralized treatment system.

Examples of allocating industrial wastewater treatment incurred costs in the central plant back to the users in a chemical complex would take the following form:

- Vinyl chloride monomer   —dollars/month
- High density polyethylene   —dollars/month
- Ethylene dichloride   —dollars/month

- Olefins —dollars/month
- Ethylene glycol —dollars/month
- Utilities, and so on —dollars/month

In a petroleum refinery:

- Tank farms —dollars/month
- Desalter —dollars/month
- Distillation —dollars/month
- Cracking unit —dollars/month
- Coking —dollars/month
- Utilities, and so on —dollars/month
- Sulfur removal —dollars/month
- Catalytic reformer —dollars/month

Annual Reports for Publicly Owned Industrial Complexes

Annual reports include environmental sections, which should be complete and concise relative to the company's environmental standing, compliance record, accomplishments, impairments, pending litigation, and impacts of current and future regulations on the company's environmental costs and their impact on a company's financial condition and operations. Two good examples of such annual report sections of publicly owned corporations are referenced.[4,5] Specific disclosure of any current claims against the company and the estimated costs of their resolutions should be included in this section.

Utilization of Third-Party Consulting Specialists, Advisory Groups, and Regulation Experts

Industrial environmental governance often can be enhanced by selection of environmentally trained board members, use of scientific and technical advisory groups, and retention of regulatory experts. This is common with many industrial and service-related companies.

Organizational Structure and Participatory Policies

A direct and defined line of environmental responsibility needs to be clearly delineated in the company's organizational structure. Moreover, individuals who have specific environmental responsibilities within the management team should be encouraged to join professional environmental organizations, trade associations, publish technical papers, participate in seminars, and develop a professional reputation on a personal basis, all of which enhances the governance capabilities of the employing company.

PART 2 ENVIRONMENTAL ECONOMICS FROM THE CONSULTANT AND CLIENT PERSPECTIVES[6-8]

18.9 Introduction

Environmental projects cost money. They are a cost of doing business for the owner. Oftentimes the owner may engage an environmental consulting firm to execute the project. The project is a source of revenue for the consulting firm. Thus, the two parties can have radically different ways of accessing the economics of an environmental project. This part includes

- Explanation of the business economics of an environmental consulting firm
- Relating this economic model to the pricing and execution of an environmental project
- Enumerating the various ways in which the owner of the project can evaluate its economics (particularly the selection of the economic optimum solution)
- Relating the owner's economic evaluation to the consultant's project budget

An example will make this clearer. Suppose XYZ Chemical Company has a contaminated site in Toxicsville, USA. The chemical plant is closed, but the site is still owned by XYZ. Contamination has been discovered in several monitoring wells sited on the property boundary. XYZ wants to hire an environmental engineering consulting firm to better define the problem, develop several potential solutions, determine their capital and operating costs, pick the most cost-effective solution, negotiate with the relevant environmental regulatory agencies to get the preferred alternative approved, design it, build it, and finally operate it.

This project is an expenditure of operating and capital dollars for XYZ Chemical Company. Their main economic interests are

- Primarily to find the most cost-effective solution to the problem
- Secondarily to have a consulting firm develop the solution using a reasonable expenditure of time and money

From the consulting firm's standpoint the project is an opportunity to earn revenue and profit. This is not to say that it is in the consultant's best interests to maximize their revenue on this project: to do so consistently would lose them business. The consultant is reasonably responsible to help XYZ Chemical attain its two economic goals. This roughly sets the consultant's expected revenue; whatever profit can be attained from the engagement is strictly the purview of the consulting firm. Frankly, XYZ's or the consulting firm's profit margin are not an interest of the opposite party, except that both must be minimally profitable enough to stay in business and work jointly on this environmental problem.

18.10 SUM Concept

Basic Equation

An environmental consultancy is a professional services firm, just as firm's engaged in law, accounting, public relations etc. At the end of the day, the firm, in one form or another, sells the time of its professional staff by the hour. The firm does not have a lot of capital equipment. It doesn't manufacture anything. It gives advice.

There is an engineering analysis that can be done on the income statement for a professional services firm: a material balance on money! The result is an expression which calculates the firm's profit using an equation with three dimensionless parameters:[6]

$$P = 1 - 1/(SUM) \qquad (18.1)$$

where P = profit as a fraction of net revenue
 S = salary to expense ratio = $T_S/(T_S + E)$ = total salaries divided by total salaries plus net overhead expenses
 U = utilization = D_S/T_S = [direct salaries (salaries charged to projects) divided by the firm's payroll], that is, the portion of payroll charged to projects
 M = overall multiplier = N_R/D_S = net revenue divided by direct salaries

It should be noted that direct salaries equals the sum of the salaries (or payroll) charged to billable projects. *Net revenue* is the firm's total revenue after subtracting out the cost of project-related nonpayroll charges, such as the cost of subconsultants, drillers, laboratories, travel, and so on.

A simplified income statement for a consulting firm is shown in Table 18.5.

Note how the net revenue was calculated, by subtracting the *cost* of subcontractors, the *cost* of direct project costs [like travel, fees for environmental data bases, etc., and the *cost* of any overhead recovered from clients (such as for computers)]. Often, consultants will add a handling fee to these costs and charge the total to the client.

Total Revenue	10,000,000	
Cost of subcontractors	1,500,000	
Direct project costs	700,000	
Recovery of overhead Costs on projects	800,000	
Total	3,000,000	
Net Revenue		7,000,000
Cost of project labor (D_s)	2,000,000	
Cost of nonproject labor	1,000,000	
Total Labor		3,000,000
Benefits	900,000	
Rent	250,000	
Communications	200,000	
Advertising	50,000	
Information technology	350,000	
Business travel	400,000	
Facility operations	150,000	
Equipment cost	50,000	
Local taxes and insurance	250,000	
Marketing costs	400,000	
Other costs	100,000	
Bad debt expense	50,000	
Total Overhead Costs	3,150,000	
Recovery of overheads	−800,000	
Net Overhead Costs		2,350,000
Profit		1,650,000

TABLE 18.5 Simplified Income Statement of ABC, Inc., Environmental Consultants

Another way to look at net revenue is that it represents charges for the firm's labor plus any handling fees.

The values for P, S, U, and M derived from the income statement are shown in Table 18.6. Substituting S, U, and M back into the SUM equation:

$$P = 1 - 1/(3.5) \times (0.6666) \times (0.561) = 0.236 \text{ or } 23.6\%$$

Parameter	Calculation	Comments
P (profit as fraction of net revenue)	P = $1,650,000/ $7,000,000 = 0.236 or 23.6%	
S (salary to expense ratio)	S = $3,000,000/ (3,000,000 + 2,350,000) = 0.561 or 56.1%	Note that the cost of recovered overheads was subtracted from total overhead costs to get the net overhead cost.
U (utilization)	U = $2,000,000/ $3,000,000 = 0.6666 or 66.7%	
M (overall multiplier)	M = $7,000,000/ $2,000,000 = 3.5	

TABLE 18.6 Calculation of SUM Parameters from ABC, Inc., Income Statement

The equation and the income statement are consistent with each other.

Parameter Sensitivity

Table 18.7 shows typical ranges of values for S, U, and M for environmental consulting firms. The average profit in the United States for such firms ranges from 7 to 11 percent, depending on the demand for services.

Inspection of the SUM equation suggests that P increases for any increase in S, U, or M. This is another way of saying that profit increases when net overhead costs are reduced, the efficiency of use of personnel on projects is increased, and/or pricing is improved. Sell high, keep overhead costs low, and concentrate on project work.

Figures 18.9 to 18.11 show typical profit sensitivity curves. As a generality, for every 1 percent increase in utilization, there is a 1.5 percent absolute increase in profitability. Similarly, a 1 percent

Parameter	Typical Range	Best in Class
S (salary to expense ratio)	55–62% (0.52–0.62)	72%
U (utilization)	58–62% (0.58–0.62)	78%
M (overall multiplier)	2.7–3.2	3.9
P (profit as fraction of net revenue)	5–15%	35%

TABLE 18.7 Typical Values for SUM Parameters

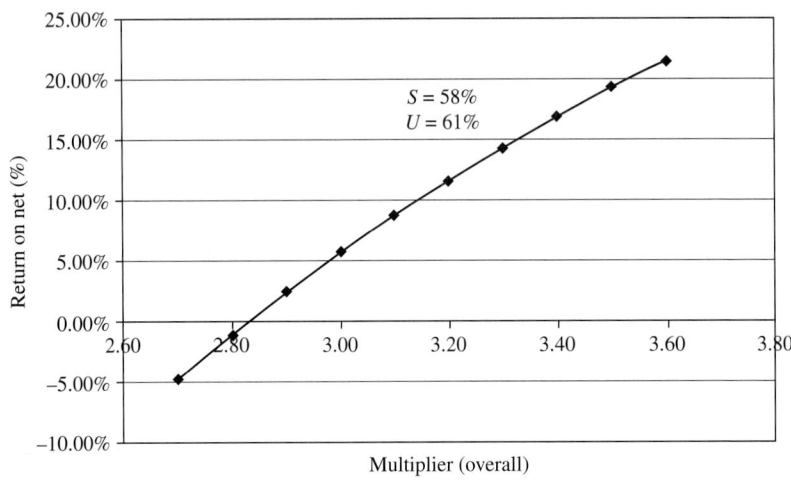

FIGURE 18.9 Sensitivity of P to M.

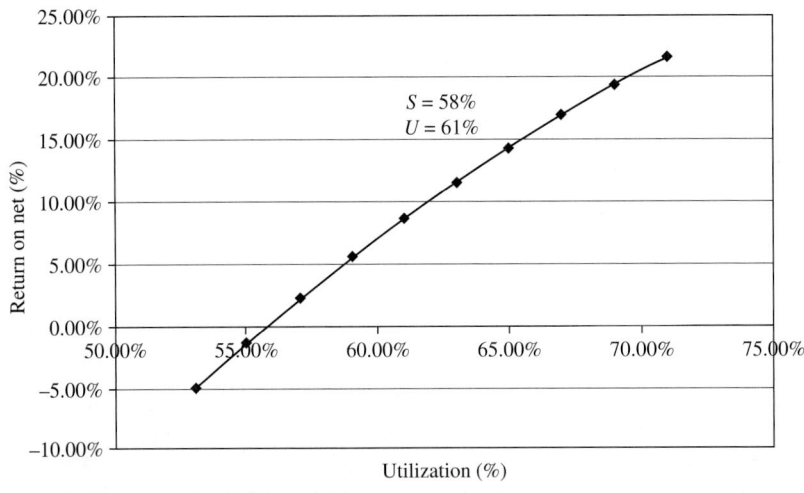

FIGURE 18.10 Sensitivity of P to U.

increase in the salary to expense ratio generates a 1.5 percent absolute increase in profitability. A four basis point (0.04) increase in multiplier yields a 1 percent absolute increase in profitability.

Example 18.1. Take the curves shown in Figs. 18.9 to 18.11. The current operating point is at $S = 0.58$, $U = 0.61$, and $M = 3.1$. At this operating point, what is the exact change in profit per unit change of each SUM parameter?

Solution Use differential calculus:[7]

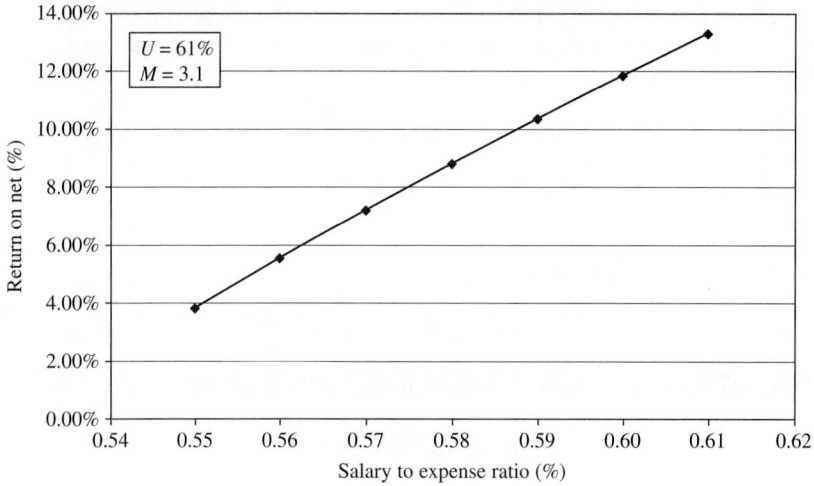

FIGURE 18.11 Sensitivity of P to S.

$$P = 1 - 1/SUM$$

$$dP/dU = [-1/(SM)] \, d \, (1/U)/dU = 1/(SMU^2) \quad (18.1a)$$

Substituting the values for S, U, and M into Eq. (18.1a):

$$dP/dU = 1/[(0.58)(3.1)(0.61)^2] = 1.495$$

That is, every percentage point increase in U increases P about 1.5 percent. By similarity the following equations are defined:[7]

$$dP/dM = 1/(SUM^2) \quad (18.1b)$$

and[7]

$$dP/dS = 1/(UMS^2) \quad (18.1c)$$

Substitutions yield:

$$dP/dM = 1/[(0.58)(0.61)(3.1)^2] = 0.294$$

or profit increases 0.29 percent for every basis point (0.01) increase in multiplier, or 4 basis points will equate to a 1 percent absolute profit increase.

$$dP/dS = 1/[(0.61)(3.1)(0.58)^2] = 1.572$$

or a 1.5 percent absolute increase for each 1 percent absolute increase in salary to expense ratio.

Another important concept in consulting firm economics is the breakeven multiplier. This is often (erroneously) used in analyzing project economics when there is perceived to be fierce price competition.

If a firm's salary to expense ratio and utilization are considered to be constant over a short period of time, and profit is calculated at zero (breakeven), then

$$P = 1 - 1/SUM = 0$$
$$1/SUM = 1$$
$$SUM = 1$$

Defining M at breakeven as M_b,[8]

$$M_b = 1/US \quad (18.2)$$

Thus, for the firm previously shown in Tables 18.5 and 18.6,

$$M_b = 1/(0.61)(0.58) = 2.83$$

18.11 Consulting Internals—Salary to Expense Ratio and Utilization

The private sector client or owner of a project generally does not care about the consulting firm's salary to expense ratio or utilization. This is an internal matter for the consultant. A private sector client does care about the multiplier, because it represents the price it pays for service (hourly charges and handling fees).

Public sector clients usually do care about the firm's salary to expense ratio and utilization. It is in the form of a demonstrated overhead factor to which the firm adds a profit in order to develop a price. This overhead factor allows for overhead costs plus the costs of unbillable time (e.g., marketing, administration). In theory, it is equivalent to the breakeven multiplier, M_b. Taking the definition of breakeven multiplier [Eq. (18.2)] and substituting it into the SUM equation [Eq. (18.1)]:[8]

$$P = 1 - M_b/M \quad (18.3)$$

Control of nonlabor overhead costs is very important. A well structured firm will attempt to control these costs such that they are low enough that the firm can breakeven in a poor year with utilization at only 50 percent, using their normal pricing.

Example 18.2. Again, take the firm with the current operating point at $S = 0.58$, $U = 0.61$, and $M = 3.1$. Does this firm breakeven at 50 percent utilization? If it doesn't, how much must overhead costs be reduced to get the firm to that point?

Solution Calculate the breakeven multiplier at the current salary to expense ratio if utilization is only 50 percent.

$$M_b = 1/US = 1/(0.50 \times 0.58) = 3.45$$

Since this is larger than the actual M of 3.1, the firm is operating at a loss, that is,

$$P = 1 - (3.45/3.1) = -0.113 \text{ or } -11.3 \text{ percent}$$

If the breakeven multiplier was reduced to the current multiplier of 3.1, then the required salary to expense ratio would be [by rearrangement of Eq. (18.2)]:

$$S = 1/(UM_b) = 1/(0.50 \times 3.1) = 0.645$$

Since,

$$S = T_S/(T_S + E)$$

Rearrange and solve for E,

$$E = T_S(1 - S)/S$$

From Table 18.5, T_S (total payroll) is $3,000,000. Therefore at the desired S of 64.5 percent,

$$E = \$3MM(1 - 0.645)/0.645 = \$1,651,000$$

This contrasts with the existing net overhead costs of $2,350,000. Therefore, the net overhead costs must be reduced by $699,000 through a combination of cost cutting and increased recovery of some costs on projects.

The salary to expense ratio at breakeven with 50 percent utilization is known as the *target salary to expense ratio* (S_T). This is a useful concept for building an economically robust consulting firm.

If $P = 0$ (breakeven) and utilization is 50 percent, substitution into the SUM equation yields

$$S_T = 1/(0.5 \times M) \qquad (18.4)$$

The target S is plotted as a function of overall multiplier in Fig. 18.12. Since fringe benefits currently typically run about 31 percent of total payroll, if one has no other overhead costs, then

$$S = T_S/(T_S + E) = T_S/(T_S + 0.31\, T_S) = 0.763 \text{ or } 76.3\%$$

Thus, 76.3 percent is a reasonable estimate of the highest practical S (other overhead costs are exceedingly low or nonexistent, and they are recovered from the client). Experience shows that 65 percent can be done by a large well-run, integrated firm. An average firm is at about 58 percent. The graphical analysis in Fig. 18.12 reaches the interesting conclusion that an economically robust firm (breakeven at

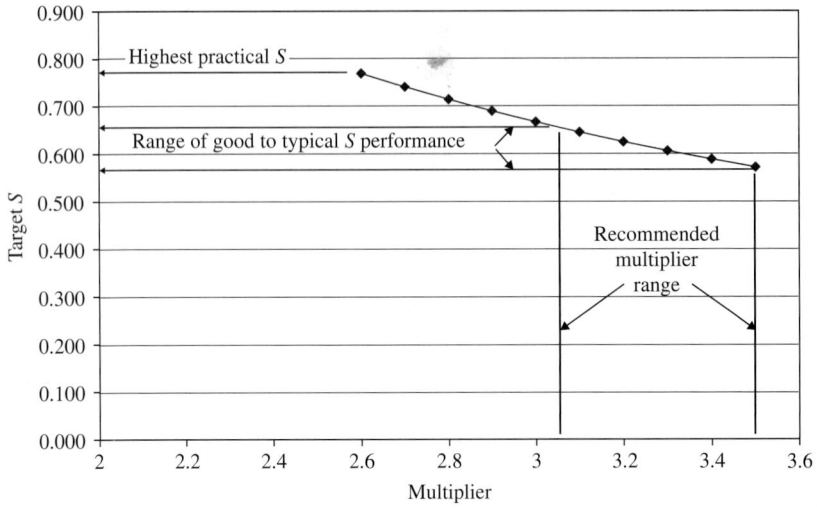

FIGURE 18.12 Target salary to expense ratio (S) versus multiplier.

50 percent utilization) must have an overall multiplier between 3.1 and 3.5.

A concept often used by consulting firms is to calculate net overhead costs as a fraction of payroll. Applying this idea to the salary to expense ratio, one gets

$$E = x T_S$$

where x = net overhead to total salary ratio. Substituting into the definition of S,

$$S = 1/(1 + x) \qquad (18.5)$$

This is shown in Fig. 18.13. In the preferred range of S equal to 58 to 65 percent, net overhead costs as a fraction of total salaries range from 54 to 72 percent.

Finally, what is the sensitivity of salary to expense ratio to net overhead costs themselves? Table 18.8 shows the reductions needed in net overhead costs (assuming an overall payroll of $3,000,000 from Table 18.5), to move each percentage point upward in salary to expense ratios from 58 to 65 percent. This develops the rule of thumb that for every absolute 1 percent increase in S, net overhead costs must be cut 4 percent. This rule of thumb works for this normal operating range because the net overhead costs in this specific table would be proportional to total salaries for firms of other sizes.

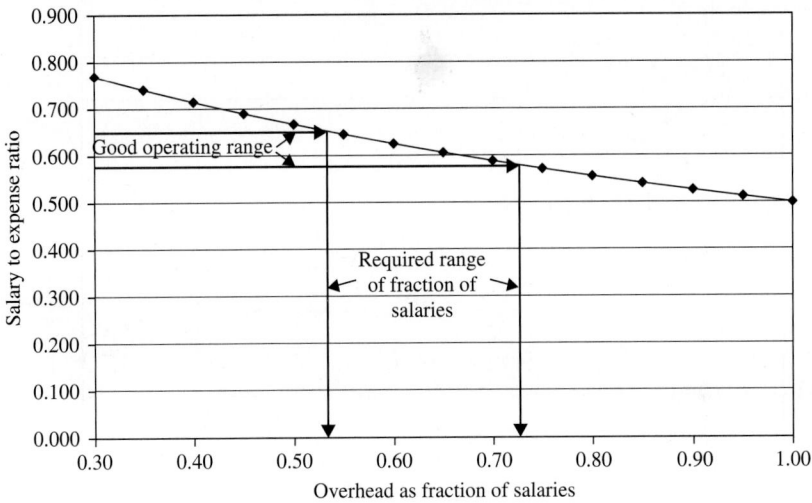

Figure 18.13 S as a function of x.

Salary to Expense Ratio (S)	Net Overhead Cost ($)	Reduction Needed to Improve 1%	Percentage Reduction
58%	2,172,400	—	
59%	2,084,700	87,700	4.05%
60%	2,000,000	84,700	4.06%
61%	1,918,000	82,000	4.10%
62%	1,838,700	79,300	4.13%
63%	1,761,900	76,800	4.17%
64%	1,687,500	74,400	4.22%
65%	1,615,400	72,100	4.27%

Note: Total payroll is $3,000,000.

Table 18.8 Sensitivity of Salary to Expense Ratio to Net Overhead Costs

It is believed that a well-run environmental consultancy should achieve a minimum 20 percent return on net revenue. Figure 18.14 translates this into a parametric plot of minimum required multiplier (M) and utilization (U) for varying salary to expense ratios (S). Any point about its requisite curve is above the minimum 20 percent profit level. If S was on a third (z) axis, the reader can visualize the universe of points above 20 percent as a block with a bowed bottom, which is inclined slightly away from the viewer.

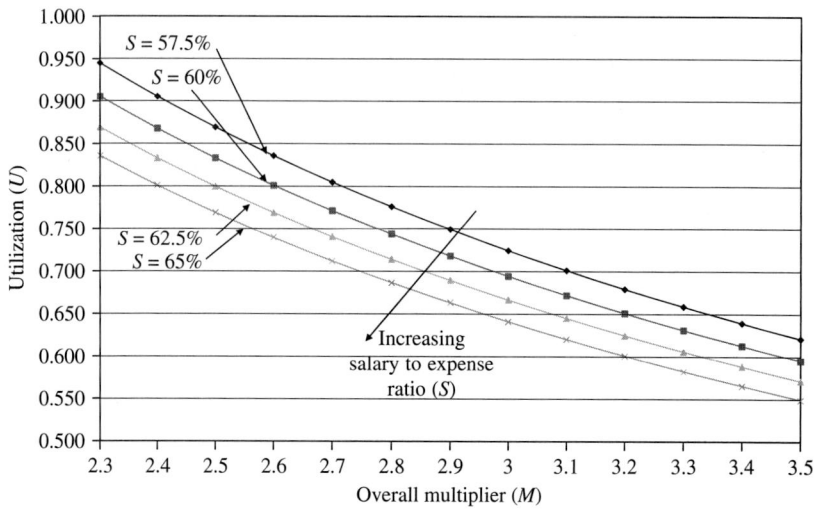

FIGURE 18.14 Required *M* and *U* for 20 percent profitability.

A simplifying concept often used by environmental consulting firms is to track the product of multiplier times utilization (*MU*). This is a surrogate, not an exact replica, of net revenue. Figure 18.15 shows the same data as Fig. 18.14, except *MU* is tracked versus *S* for minimum 20 percent profitability.

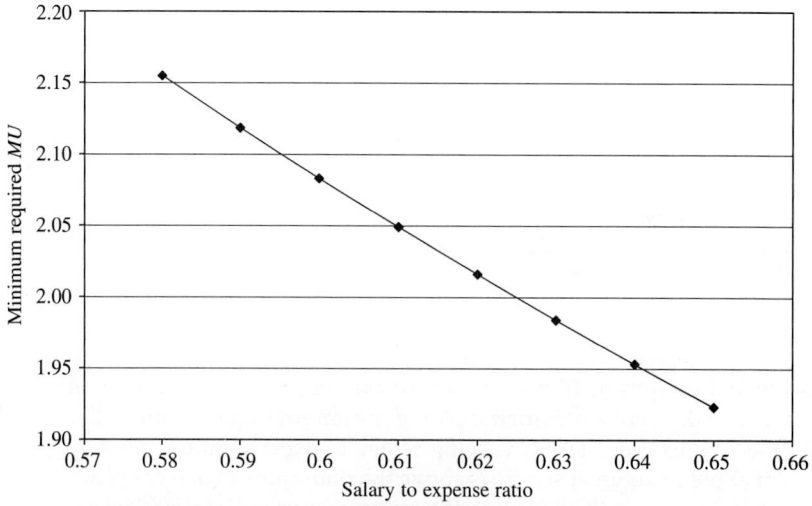

FIGURE 18.15 Minimum *MU* required for 20 percent profitability.

During the daily operation of any reasonably sized consulting firm, the most important economic parameter is utilization, because both multiplier and salary to expense ratio change slowly. The latter is sensitive to cumulative pricing and overhead costs decisions, respectively. Thus, most firms pay heavy attention to utilization.

Example 18.3. What is the utilization for an individual employee with 3 weeks/year of vacation, with an annual hourly breakdown as follows:

Billable work	1722 h
Marketing	50 h
Proposal writing	100 h
Vacation	120 h
Sick time	24 h
Holiday (8 per year)	64 h
Total	2080 h

Assuming this employee is a salaried employee paid $62,400/year, the standard hourly salary rate is $62,400/2080 h (52 weeks times 40 h/week) or $30/h. The direct salary (charged to projects) is 1722 h times $30/h, or $51,660. The employee's utilization is:

$$U = D_S / T_S = \$51,600/\$62,400 \text{ or } 82.8\%$$

The utilization for an individual employee is the ratio of hours charged to projects to hours paid. This is not true for the entire firm, as the following example illustrates.

Example 18.4. What is the utilization for a firm with the following employees and time charges?

Person	Salary ($/h)	Billable Hours	Nonbillable Hours	Standard Hours
Partner	60.00	850	450	1200
Project manager #1	45.00	1000	350	1200
Project manager #2	40.00	950	400	1200
Engineer #1	35.00	1100	200	1200
Engineer #2	30.00	1200	0	1200
Geologist #1	30.00	1050	250	1200
Geologist #2	25.00	600	0	600

Note that these are data for a part of a year, that is, standard hours are 30 weeks times 40 h/week (the standard workweek for this firm).

Geologist #2 is a part-time employee.

This problem is solved by calculating total and direct salaries.

Person	Salary ($/h)	Billable Hours	Nonbillable Hours	Standard Hours	Direct Salary	Total Salary	Individual Utilization
Partner	60.00	850	450	1200	51,000	72,000	70.8%
Project manager #1	45.00	1000	350	1200	45,000	54,000	83.3%
Project manager #2	40.00	950	400	1200	38,000	48,000	79.1%
Engineer #1	35.00	1100	200	1200	38,500	42,000	91.6%
Engineer #2	30.00	1200	0	1200	36,000	36,000	100.0%
Geologist #1	30.00	1050	250	1200	31,500	36,000	87.5%
Geologist #2	25.00	600	0	600	15,000	15,000	100.0%
Total		6750	1650		255,000	303,000	

The firm's utilization will be direct salaries divided by total salaries or 255,000/303,000, 84.1 percent. Notice that this differs from the ratio of billable hours to standard hours (6750/7800 = 86.5 percent), because of the weighting effects of different salaries.

18.12 Consulting Externals: Multiplier and Pricing

While a firm's clients do not care about the firm's internal operating efficiency (S and U), they are often interested in the firm's billing rates and handling fees, and are always interested in the total price for each project (or phase of a project). These elements represent the firm's overall multiplier (M) and net revenue.

The firm's overall multiplier (M) is determined by two components: billing rates actually attained for personnel and handling fees realized on subcontractors. Both items are attained solely from the pricing that the consultant can obtain from its clients.

An example based on Example 18.4 above will illustrate this.

Example 18.5. A consulting firm has been able to bill (and be paid for) the hours shown in the table in Example 18.4. The actual average billing rates shown in the following table were paid by the clients. In addition, $500,000 of subcontractor charges were charged to the clients at an average handling fee (also known as markup) of 10 percent. What is the total revenue? Net revenue? Overall multiplier? Labor multiplier?

Person	Salary ($/h)	Billable Hours	Direct Salary	Labor Rate Attained ($/h)	Individual Labor Multiplier	Total Labor Billings
Partner	60.00	850	51,000	180.00	3.0	153,000
Project manager #1	45.00	1000	45,000	139.50	3.1	139,500
Project manager #2	40.00	950	38,000	128.00	3.2	121,600
Engineer #1	35.00	1100	38,500	112.00	3.2	123,200
Engineer #2	30.00	1200	36,000	95.00	3.2	114,000
Geologist #1	30.00	1050	31,500	95.00	3.2	99,750
Geologist #2	25.00	600	15,000	85.00	3.4	51,000
		Total	255,000		Total	802,050

The total charged for subcontractors is $500,000 plus 10 percent or $550,000.

Thus, the total revenue is $550,000 + $802,050 or $1,352,050.

Net revenue is personnel revenue plus handling fees, that is, $802,050 + $50,000 = $852,050.

The firm's overall multiplier is net revenue divided by direct salaries, $852,050/255,000 = 3.34.

The firm's labor multiplier is labor billings divided by direct salaries, $802,050/255,000 = 3.15.

Note that if the firm had charged its clients for expenses (travel, communications, and so on) to the extent that the charges exceeded the cost to the consultant, the difference would be an additional handling fee that increases net revenue and M.

There are many ways to package the price of a project to a client. Most are either lump sum or time and materials or some variant on these two themes. In a lump sum project, the firm presents a fixed price to the client based on a firm definition of the scope of services. As long as the scope doesn't change, the firm simply charges the stated price with no breakdown of individual time charges or expenses. The price is fixed.

In a time and materials project, an estimate is given to the client for the assumed scope of work. This estimate includes all labor charges and expenses (including subcontractors). The firm then charges the client for the actual time and material costs incurred. The invoice provides a breakdown of all hourly time charges and billable expenses. The final total charge to the client may be less than, equal to, or greater than the original estimate. Thus, the price can vary.

The mode of pricing is either determined by the client's preference, or if the client has none, by the consultant. Most consultants will (wisely) not perform a project for a lump sum if the scope of the project is ill defined.

The following example basically shows how a price is determined for a project.

Example 18.6. This is a simple project consisting of a meeting to review the client's compliance with a RCRA permit and a short letter to document the findings.

The estimate is started by breaking the project into steps that are simple enough for an experienced project manager to determine a reasonable level of effort for each task. Thus,

Project Task	Partner	Project Engineer	Word Processing
Hourly billing rate	180.00	95.00	45.00
Review permit and data	8	8	
Interview key personnel	8		
Review meeting	4	4	
Draft letter report	6	8	4
Finalize letter report	3	4	4
Total hours	29	24	8
Total charge	$5,220	$2,280	$720

Total personnel charges are $8,220.

There are no subcontractors required for accomplishing this project, but there are expenses. Estimate:

Item	Charge Rate	Units	Total
Computer usage	$10/h	45 h	$450.00
Telephone, fax, Internet	3% of personnel charges	8,220	247.00
Copying	$0.10/page	500 pages	50.00
		Total	$747.00

The client wants a lump sum price.
The price is calculated as follows:

Personnel charges	$8,220
Expenses	747
Preliminary total	$8,967
Add contingency at 15 percent	$1,345
Total lump sum price	$10,312 say, $10,300

A contingency should always be added: generally 15 percent for lump sum, 10 percent for time and material. Leaving the contingency out ensures that the project has about a 50 percent probability of going over budget.

18.13 Project Economics

The SUM equation can be used on individual projects to calculate their expected or actual profitability. This can be done as long as the firm's overall utilization and salary to expense ratio are used in tandem with the actual multiplier from the project. To state the obvious: in the long run, the sum of all project profits will equal the firm's profit.

Equation (18.2) is very useful for calculating the expected (before execution) and actual (after execution) profit P_p of a project, if the project overall multiplier (M_p) is substituted for M, that is,

$$P_p = 1 - M_b/M_p \qquad (18.6)$$

Remembering that

$$M_b = 1/US$$

The calculation of project profit is illustrated in the following example.

Example 18.7. Take the priced project in Example 18.6 and calculate the project profit (P_p), assuming that the firm has a U of 70 percent and an S of 60 percent. Also assume that this project requires a subcontractor for which a 10 percent handling fee is attached.

Solution The new project price is:

Personnel charges	$8,220
Expenses	747
Preliminary total	$8,967
Add contingency at 15 percent	$1,345
Total lump sum price	$10,312 say, $10,300
Subcontractor	$5,000
Add 15 percent contingency	$750
Handling fee at 10 percent	$575
Subtotal	$6,325
New price	$16,600

The project multiplier is equal to project net revenue divided by project direct salaries. In this case, net revenue is:

Total price	$16,600
Minus expenses	−$860 ($747 plus 15 percent)
Minus cost of subcontractor	−$5,750
Net revenue	$9,990

The direct salaries used in this cost estimate are:

	Partner	Project Engineer	Word Processing
Hourly salary rate	60.00	30.00	15.00
Total hours	29	24	8
Total direct salary	$1,740	$720	$120

The total is (after adding the 15 percent contingency), $2,967. Thus,

$$M_p = \$9,990/\$2,967 = 3.367$$
$$M_b = 1/US = 1/(0.60 \times 0.70) = 2.381$$

and

$$P_p = 1 - M_b/M_p = 1 - (2.381/3.367) = 29.3\% \text{ profit}$$

After the project is executed, a similar calculation can be done using actual direct salaries expended, the actual cost of the subcontractor, and the actual cost of expenses to arrive at a project profit.

References

1. Ford, D. L.: Report, U.S. Department of Justice, "Review of Disposal Practices," Camp Lejeune, North Carolina, 2006.
2. Hyperion Energy, LLC: Dallas, Texas, Web site, 2008.
3. Ford, D. L.: "Zero discharge and environmental regulation, the toxic release inventory, and natural laws," *Environmental Engineer*, vol. 32, no. 4, October 1996.
4. Clayton Williams Energy, Inc.: CWEI, (NASDQ) Annual Report, 2006.
5. TECO Energy: (NYSE) Annual Report, 2005.
6. Flynn, B. P.: "Maximizing engineering firm profits," *Profit Fundamentals*, vol. 1, Pine Tree Press, 2001.
7. Internal Communication, MRE, LLC.
8. Flynn, B. P.: "Project profitability and pricing," MRE Associates, LLc, Copyright 2005.

Index

Note: Page numbers referencing figures are italicized and followed by an "f"; page numbers referencing tables are italicized and followed by a "t".

A

absolute exclusion, 898
absorbable organic halides (AOX), 281
acceptable degradation, 577
acclimation
  of activated sludge to specific organics, *234f*
  for degradation of benzidine, *234f*
acid
  lime-waste titration curve for, *91f*
  reagents, *90t*
  wastes, 83–86
  weight of, 85
acidic neutralizing agents, 86–87
acidic sulfide oxidation, 596
acids, anaerobic fermentation and, 499
acquisition/merger due diligence, 897
activated alumina contact beds, 167
activated carbons
  adsorption
    amenability of organic compounds to, *528t*
    overview, *59t*
    for PCBs, 751, 753
    system design, 536–549
  carbon regeneration, 534–536
  continuous carbon filters, 532–534
  GAC small column tests, 550–553
  laboratory evaluation of adsorption, 532
  overview, 530–531
  performance of activated carbon systems, 553–554
activated silica, 144
activated sludge, 56, *207t*, 208
activated sludge processes
  ammonia inhibition in, *354f*
  batch activated sludge, 445–448
  Biohoch process, 451–453
  bioinhibition of
    fed batch reactor (FBR), 322–323
    glucose inhibition test, 323–325
    OECD method 209, 321–322
    overview, 318–321
  complete mix activated sludge, 436–437
  deep-shaft activated sludge, 451
  effluent suspended solids control, 465–474
  extended aeration, 437–439
  final clarification, 454–462
  flocculation and hydraulic problems, 462
  integrated fixed film activated sludge, 453
  municipal activated sludge plants, 462–465
  oxidation ditch systems, 439–440
  oxygen activated sludge, 448–451
  plug flow activated sludge, 435–436
  SBR, 440–445
  thermophilic aerobic activated sludge, 454

924 Index

acute toxicity, *393t–394t*
  of refinery effluent to stickleback, *33f*
  of selected compounds, *33f*
  of six species to refinery effluent, *33f*
ADF (airport deicing fluid), 672
ADI-BVF process, 497
ADI-BVF reactor, *498f*
adsorbability, *527t*
adsorption
  adsorption factor development, *838t*
  isotherms, *555t*
  overview, *59t*, *62t*, 525
  PACT process, 554–561
  problems, 562
  properties of activated carbon
    adsorption system design, 536–549
    carbon regeneration, 534–536
    continuous carbon filters, 532–534
    GAC small column tests, 550–553
    laboratory evaluation of adsorption, 532
    overview, 530–531
    performance of activated carbon systems, 553–554
  removing heavy metals with, 156
  theory of, 525–530
advisory groups, environmental governance, 903
aerated lagoons
  aerobic lagoons, 415–418
  defined, 405
  facultative lagoons, 418–420
  overview, 414–415
  systems, 422–433
  temperature effects in, 420–422
  treatment method, 56
aerated ponds, COD removal, *425f*
aeration
  air stripping of VOCs
    overview, 211
    packed towers, 213–221
    size of towers, 216–220
    steam stripping, 220–221
  equalization, 69
  equipment
    diffused aeration, 197–201
    measurement of oxygen transfer efficiency, 205–210
    measuring techniques, 210–211
    overview, 195–197
    surface-aeration, 202–205
    turbine aeration, 201–202
  mechanism of oxygen transfer, 179–195
  overview, 179
  problems with, 221–223
aerobic biological oxidation
  bioinhibition of activated sludge process
    fed batch reactor (FBR), 322–323
    glucose inhibition test, 323–325
    OECD method 209, 321–322
    overview, 318–321
  biooxidation
    mathematical relationships of organic removal, 256–278
    nutrient requirements, 253–256
    overview, 233–237
    oxygen requirements, 246–252
    sludge yield and oxygen utilization, 237–246
    specific organic compounds, 278–284
  denitrification
    design procedure, 372–375
    overview, 358–368
    systems, 368–369
  effect of temperature
    effect of pH, 293–294
    overview, 284–293
    toxicity, 294–299
  laboratory and pilot plant procedures for development of process design criteria
    reactor operation, 388–389
    reduction of aquatic toxicity, 392–396
    volatile organic carbon, 389–392
    wastewater characterization, 386–388
  nitrification
    batch activated sludge (BAS), 355–357
    design procedure, 369–372
    fed batch reactor (FBR) nitrification test, 357–358
    of high-strength wastewaters, 342–343

Index   925

inhibition of, 343–355
kinetics, 335–342
overview, 334–335
systems, 368–369
organics removal mechanisms
biodegradation, 232–233
overview, 225–227
sorbability, 229–231
sorption, 227–228
stripping, 228–229
overview, 225
phosphorus removal
biological, 380–381
chemical, 375–380
design considerations, 383–385
GAOs, 383
MBRs, 385–386
mechanism for, 381–383
problems, 396–399
sludge quality
biological selectors, 308
design of aerobic selectors, 308–315
filamentous bulking control, 307–308
overview, 299–307
soluble microbial product (SMP) formation, 316–318
stripping of VOCs
emissions treatment, 332–333
overview, 326–332
aerobic digestion, 609–616, 658
aerobic lagoons, 414, 415–418
aerobic selectors, design of, 308–315
Aerojet General, 777
Aeromonas Punctata, 382
agents, neutralizing, 86–87
air bubbles, 189–190
air emissions, 763
air flotation treatment, $129f$
Air Force Research Laboratory, Materials and Manufacturing Directorate, 776–777
air injection, 782
Air National Emission Standards for Hazardous Air Pollutants, 6
air solubility, 120–123
air standards, 6
air stripping
defined, 38

problem, 223
of VOCs
overview, 211
packed towers, 213–221
size of towers, 216–220
steam stripping, 220–221
aircraft manufacturing, 855, $858t$
airport deicing fluid (ADF), 672
air/solids (A/S) ratio, 122, 126, $622f$
aliphatics, 25
alkaline chlorination, $595f$, 600
alkaline peroxidation, 587
alkaline reagents, $90t$
alkaline sulfide oxidation, 596
alkaline wastes, 83, 86–87
alkalinity
anaerobic process, $508f$
direct precipitation and, 380
alum, $145t$
alum salts, 468
aluminum hydroxide, 143
aluminum sulfate, 143
American Petroleum Institute (API) gravity, 726
amine solution process, 744
ammonia
anaerobic process, 511
heavy metals removal, 155–156, $158f$
inhibition, in activated sludge process, $354f$
oxidation, effect of pH on, $341f$
removal, 572
removal, SRT and, $336f$
stripping, 119
ammonia nitrogen, $339f$
Anabaena, 405
Anacystic, 405
anaerobic contact process, 495
anaerobic fermentation, 499–505
anaerobic filter reactors, 495
anaerobic ponds, 404
anaerobic toxicity assay (ATA), 513
anaerobic treatment processes
biodegradation of organic compounds, 505–507
factors affecting process operation, 507–511
laboratory evaluation of, 511–517
mechanism of anaerobic fermentation, 499–505

Index

anaerobic treatment processes (*Cont.*):
  overview, 494–495
  process alternatives, 495–499
anaerobic waste treatment, *57t, 496f*
anaerobiosis, filamentous organisms and, 312
anion exchangers, 572–575
anionic ion exchange, 565–567
anionic polymers, 144
annual outfall excursions, 894
antagonistic ions, anaerobic process, 511
AOX (absorbable organic halides), 281
API (American Petroleum Institute) gravity, 726
API oil separators, 111, *112f*
applications
  chemical oxidation, 581–582
  lagoons, 405–413
  membrane bioreactors, 697–698
  membrane processes, 688–693
  for source treatment of toxic persistent wastewaters, *54f*
  trickling filtration, 482–485
aquatic toxicity, 5, 392–396
Archlor 1232, 754
aromatic compound oxidation, 585, *591t, 597f*
aromatics, 25
arsenates, 770
arsenic, 157–158, *175t*
arsenic removal, 572
A/S (air/solids) ratio, 122, 126, *622f*
associated gas, 737
associated waste exemption, 897
ATA (anaerobic toxicity assay), 615
ATAD (autothermal aerobic digestion), 615
atmospheric evaporation, 761
audits, pretreatment, 850
Austin, Texas, 755, 858–863
autothermal aerobic digestion (ATAD), 615

B

backwash systems, 704
baffles, 202
ball-bearing manufacturing wastes, 148, 150
Barceloneta POTW, Puerto Rico
  determining compliance monitoring for PSES pollutants, 871–872
  determining limits for pollutants regulated under PSES, 866–871
  final limits as they would appear in permit for facility B, 872–873
  overview, 863–866
barium, 159, *175t*
barium sulfate, 159
BAS (batch activated sludge), *260f*, 355–357, 445–448
basic neutralizing agents, 87
basic wastes, 86–87
basket centrifuge, 626–627
batch activated sludge (BAS), *260f*, 355–357, 445–448
batch denitrification test, 363
batch flux curves, *619f–620f*
batch leaching tests, 609
batch operation, variation in flow from, *8f*
batch oxidation, *261f, 610f*
batch settling data, *620f*
batch settling, of activated sludge, 454
batch systems, *535f*
batch treatment of chromium wastes, 161–162
BATEA (best available technology economically achievable), 767
BAT-equivalent treatment performance, 175–176
Bayport Industrial Complex, *807t*
bed depth service time (BDST) approach, 546–549
belt filter presses, 645–647
benchmarking, 802–806
benthal stabilization, 651
benzene, 223, *726t*
  in activated sludge reactors, *231t*
  stripping, *328f*
benzidine, *234f*
Bernard-Englande equation, 636
Bernheim expert report, 826
best available technology economically achievable (BATEA), 767
Best Management Plan (BMP), 889
best practicable control (BPT) technology, 766
bio-acclimation, *21t*
bioassays, 33
biochemical methane potential (BMP), 511

Index

biochemical oxidation of pesticides, 775
biochemical oxygen demand (BOD)
  BOD/TKN ratio, *338f*
  characteristics of effluents from anaerobic treatment of wastewaters, *507t*
  COD and SMP relationships for industrial wastewaters, *27t–28t*
  COD ratio, *592f*
  curves, *21f*
  loadings discharged by Austin industrial users, *863f*
  organic content of wastes, 18, 26
  probability analysis, *74f*
  probability of occurrence in raw waste, *15f*
  rate constants at 20°C, *20t*
  reactions occurring in bottle, *20f*
  relationship between COD for chemical refinery wastewater, *22f*
  removal, *434f, 492f*
  statistical correlation of data, *17t*
  values for wastewaters, *501f–502f*
  variations in, *11f*
  wastewaters, 7
biochemical reactor systems, 776–777
biochemical reduction processes, 776–777
biodegradability
  biotoxicity data and, *345t–352t*
  of chlorinated solvents, 758, 760
  test, 37
  wastewater, 54
biodegradation, 232–233
  of organic compounds under anaerobic conditions, 505–507
  stripping and, *327f*
biofilters, 782–783
Biohoch process, 451–453
bioinhibition, of activated sludge process
  fed batch reactor (FBR), 322–323
  glucose inhibition test, 323–325
  OECD method 209, 321–322
  overview, 318–321
biological nitrogen conversion process, *334f*
biological oxidation process, *188f*
  trace nutrient requirements, *253t*
biological oxidation rate constant K, effect of temperature on, *285f*
biological phosphorus removal, 380–381
biological pretreatment, *690f*
biological reactor, with alternative PAC addition, *390f*
biological selectors, 308
biological wastewater treatment
  activated sludge processes, 434–474
    batch activated sludge, 445–448
    Biohoch process, 451–453
    complete mix activated sludge, 436–437
    deep-shaft activated sludge, 451
    effluent suspended solids control, 465–474
    extended aeration, 437–439
    final clarification, 454–462
    flocculation and hydraulic problems, 462
    integrated fixed film activated sludge, 453
    municipal activated sludge plants, 462–465
    oxidation ditch systems, 439–440
    oxygen activated sludge, 448–451
    plug flow activated sludge, 435–436
    SBR, 440–445
    thermophilic aerobic activated sludge, 454
  aerated lagoons
    aerated lagoon systems, 422–433
    aerobic lagoons, 415–418
    facultative lagoons, 418–420
    overview, 414–415
    temperature effects in aerated lagoons, 420–422
  anaerobic treatment processes
    biodegradation of organic compounds, 505–507
    factors affecting process operation, 507–511
    laboratory evaluation of, 511–517
    mechanism of anaerobic fermentation, 499–505

928 Index

biological wastewater treatment,
 anaerobic treatment processes
 (*Cont.*):
  overview, 494–495
  process alternatives, 495–499
 lagoons and stabilization basins
  aerated lagoons, 405
  anaerobic ponds, 404
  facultative ponds, 404
  lagoon applications, 405–413
  overview, 403–404
 overview, 403
 problems, 518–521
 rotating biological contactors,
  488–494
 trickling filtration
  applications, 482–485
  effect of temperature, 482
  oxygen transfer and utilization,
   479–482
  tertiary nitrification, 485–488
  theory, 474–479
biooxidation
 mathematical relationships of
  organic removal, 256–278
 nutrient requirements, 253–256
 overview, 233–237
 oxygen requirements, 246–252
 reactions occurring during, *237f*
 sludge yield and oxygen utilization,
  237–246
 specific organic compounds,
  278–284
bioreactors. *See* membrane bioreactors
bioscrubbers, 332
biosorption
 batch activated sludge and, *260f*
 relationship for soluble degradable
  wastewaters, 235
biotoxicity data, biodegradability and,
 345t–352t
bleached sulfite mill wastewater, *288f*
BLM (Bureau of Land Management),
 718
BMP (Best Management Plan), 889
BMP (biochemical methane potential),
 889
BOD. *See* biochemical oxygen demand
Bohart-Adams equation, 541–546
Bolinder Real Estate v. United States, 899
bottom wastes, 734

BPT (best practicable control)
 technology, 766
breakpoints, 533
breakthrough curves, 536–538, *547f*,
 562
breakthrough process, 568
brewery wastewater treatment, *452t*
brine concentrators, 812
brominated flame retardants, 5
bromochloromethane, 779
brush aerators, *189t*, 204
BTEX biological reactors, 784–786
bubble diffusers, 184, 186–187, *189t*
bubble-aeration data, *191f*
bubbles, air, 189–190
Büchner funnel test, 631–632
Bureau of Land Management (BLM),
 718
by-products, chlorinated, 779–786

C

cadmium, 159, *160t*, *175t*
calcium carbonate, *88t–89t*
calcium hydroxide, *88t–89t*
calcium oxide, *88t–89t*
Calgon Corp. Laboratory isotherm
 tests, *753t*
California, 855
camp wastes, 735
capillary suction time (CST) test,
 634–635
capital and operational economic
 planning, 878–880
carbamates, 770
carbon adsorption, activated, *528t*
carbon capacity, *535f*, 536
carbon dosages, 558, *559f*
carbon regeneration, 534–536
carbons, activated
 adsorption system design, 536–549
 carbon regeneration, 534–536
 continuous carbon filters, 532–534
 GAC small column tests, 550–553
 laboratory evaluation of adsorption,
  532
 overview, 530–531
 performance of activated carbon
  systems, 553–554
carbon-treated effluent, *34f*
case studies. *See also* Barceloneta
 POTW, Puerto Rico

industrial pretreatment
  Austin, Texas, 858–863
  Chicago, Illinois, 854–855
  Indianapolis, Indiana, 855
  pretreatment of leachate
    discharges, 873
  San Diego, California, 855
  Shreveport, Louisiana, 855–858
membrane bioreactors
  A, 698–699
  B, 699–700
water recycle and reuse, 795–802
zero effluent discharge at Formosa
    Plastics
  history of technologies, 811–813
  industry applications of
    technology, 813
  initial evaluation results, 817–821
  overview, 810–811
  recycle effects on effluent toxicity
    testing, 821
  studies of zero discharge options,
    813–815
  technologies appropriate for,
    815–817
casings, disposal well, 675
catalyzed hydrogen peroxide, 587–590
categorical pretreatment standards,
    848–849, 850t, 859t–860t
cation exchange capacity (CEC), 653
cation exchange resins, 569f
cation exchangers, 572–575
cation valence, 141
cationic ion exchange, 565–567
cationic polyelectrolytes, 468
cationic polymers, 144, 638f
CEC (cation exchange capacity), 653
cell synthesis relationship, for soluble
    pharmaceutical wastewater, 244
center sludge, 109
center-feed circular clarifiers, 107
centrifugation, 635–638, 639f
CERCLA (Comprehensive
    Environmental Response,
    Compensation, and Liability Act),
    723–724, 877, 895–896
CFR (Code of the Federal Register), 6
CGL (comprehensive general liability)
    policies, 897
charge rates, user, for industrial
    pretreatment, 852–854

chemical coagulant applications, 145t
chemical desalting, 117
chemical flocculation, 127f
chemical industry waste reduction
    techniques, 45
chemical inhibition, to anaerobic
    process, 509t–510t
chemical oxidation
  applicability, 581–582
  chlorine, 590–595
  hydrogen peroxide, 586–590
  hydrothermal processes, 596–600
  overview, 58t, 62t, 577–578
  ozone, 582–586
  potassium permanganate, 596
  problem, 600
  stoichiometry, 578–581
chemical oxygen demand (COD), 892
  BOD and SMP relationships for
    industrial wastewaters, 27t–28t
  COD/TOC ratio, 581, 582f, 583f
  organic content of wastes, 18
  reduction by ozonation, 585, 586f
  relationship between BOD for
    chemical refinery wastewater,
    22f
  removal
    aerated ponds, 425f
    from brewery wastewater, 417f
    relative to initial COD, 514f
    stoichiometry, 580–581
  and TSS composition for influent
    and effluent, 29f
  values
    of activated sludge during an SBR
      cycle of reactor for pulp and
      paper mill wastewater, 441f
    for wastewaters, 501f–502f
chemical phosphorus removal,
    375–380
chemical precipitation. *See*
    precipitation
chemical scrubber systems, 782
chemical treatment of paperboard
    waste, 149t
chemical wastewaters, 52, 442t
chemicals, neutralization, 88t–89t
Chicago, Illinois, 854–855
Chlamydomonas, 405
chlor/alkali plants, 814–815
Chlorella, 405

chlorinated benzenes, *559f*
chlorinated compounds
  by-products, 779–786
  miscellaneous organics, 777–778
  odor control, 780–786
  overview, 747–748
  perchlorates, 775–777
  pesticides
    overview, 766
    pesticide characterization, 767–774
    regulatory history, 766–767
    treatment methodologies, 775
  polychlorinated biphenyls
    environmental impacts, 749–750
    overview, 748–749
    regulatory history of, 750–751
    treatment methodologies, 751–755
  problems, 786–787
  solvents
    historical perspective, 755–758
    overview, 755–766
    treatment methodologies, 758–766
chlorinated hydrocarbons, 770, 893
chlorine, 314, 590–595
chloroform, 780
chloromethane, 780
chlorophenols, 281
chromate removal, 572–573
chrome content fluctuations, 162
chromium, 156, 159–166, *176t*, 576
chronic bioassays, 32
circular clarifiers, 107, 109
clarification, 106, 454–462
clarifier design and operation diagram, *459f*
clarifiers, 107–110, *124f*, *148f*
Class II wells, 719
clays, *145t*
Clean Air Act, 718
Clean Water Act (CWA), 2, 719, 766–767, 847–848, 876
cleaning
  allocation method applicability, *827t*
  membrane bioreactors, 696–698
  membranes, 687–688
  PCB cleanup, 755
client perspective, industrial environmental economics
  multiplier and pricing, 916–918
  overview, 904–905
  project economics, 918–920
  salary to expense ratio and utilization, 910–916
  SUM concept, 905–910
coagulant aids, 144–145
coagulants, 143–144, 630–631, *632f*
coagulation
  coagulant aids, 144–145
  equipment, 147–148
  of industrial wastes, 148–151
  laboratory control of, 146–147
  mechanism of, 141–142
  overview, *62t*, 137–139
  properties of coagulants, 143–144
  of textile wastewaters, *152t*
  zeta potential, 139–141
coarse-bubble diffusers, 198–200, *206t*
cobalt, 205–206
COCs (contaminants of concern), *772t–774t*
COD. *See* chemical oxygen demand
Code of the Federal Register (CFR), 6
coefficients, compound-related, 764
coke plant wastewaters, *558t*
colloids, 137–139
color removal, *153t*, *559f*, 590, *592f*
column systems, *535f*
column tests, GAC small, 550–553
combined carbon oxidation–nitrification, *486f*
combustion, self-sustaining, 658
cometabolism, 764
competitive adsorption, 530
complete mix activated sludge, 436–437
complete mix performance, versus plug-flow, *265t*
completion wastes, 734
compliance, 849–851, 883–895
composite reaction rate coefficients, *303t*
compound-related coefficients, 764
compounds, chlorinated. *See* chlorinated compounds
Comprehensive Environmental Response, Compensation, and Liability Act (CERCLA), 724–725, 877, 895–896
comprehensive general liability (CGL) policies, 897
compressibility, sludge, 630

Index 931

concentration limit equation, 861–862
concentration-based effluent limits, 867t–868t, 869–870
conceptual wastestream, *840t*
connection sewer operation (CSO), 855
constant effluent flow systems, 76–77
constant outflow equalization basins, *72f–73f*
constant volume equalization basins, 69, *70f*, 73, *80t*
constituents, 830, 832
consultant perspective, industrial environmental economics
  multiplier and pricing, 916–918
  overview, 904–905
  project economics, 918–920
  salary to expense ratio and utilization, 910–916
  SUM concept, 905–910
consulting specialists, third-party, 903
contaminant loading, *803t*
contaminants
  allocation of Superfund disposal site response costs, 830–837
  for TCLP, *607t–608t*
contaminants of concern (COCs), *772t–774t*
content in allocation of Superfund disposal site response costs, 830–837
continuous carbon column breakthrough curves, 536–538
continuous carbon filters, 532–534
continuous countercurrent solid bowl conveyor discharge centrifuge, *635f*
continuous counterflow carbon column design, 539
continuous treatment of chromium wastes, 162, *163f*
continuous-flow sequentially aerated activated sludge, *444f*
conventional bleaching, versus oxygen bleaching, *283t*
conventional oil/gas, 712–716
conventional pollutants, AOX and, 281
cooling towers, 812, 820
copper, 166, *175t–176t*, *298f*
coprecipitation, 156
corrugated plate separators (CPS), 111, *116f*

cost analysis
  Formosa plant, 820
  for recycle/reuse decision process, 794–795
cost centers, 902
cost recovery, POTW, 851–852
cost/effluent quality relationship, *810f*
costs of zero discharge alternatives, *818t–819t*
court decision citations, 898
CPS (corrugated plate separators), 111, *116f*
critical depth, 541
cross-flow filtration, 680–681, *683f*
crude oils, *726t–727t*, 729–730
CSO (connection sewer operation), 855
CST (capillary suction time) test, 634–635
cumulative pollutant loading rates, *654t*
CWA (Clean Water Act), 2, 719, 766–767, 847–848, 876
cyanide
  alkaline peroxidation of, 587
  cadmium removal, 159
  heavy metals removal, 155
  limits on, 866–869
  oxidation of with chlorine, 591, 593–595
  relative rate of nitrification as function of, *354f*
cycle time equation, 641

D

DAF (dilution attenuation factor), *128f–129f*, 608
DAF (dissolved air flotation), 124, 621
dairy wastewater
  bench-scale aerobic aerated lagoon test results, *418f*
  treatment of by intermittent activated sludge system, *442t*
damage claims, 726
Daphnia magna, 792
DCP (dichlorophenol), 280
DDT (dichlorodiphenyl trichloroethylene), 770–771
decanting devices, 444
dechlorination, 760

decision-making process, water recycle and reuse, 794–795
declining-rate filter design, 700
decomposition, ozone, 583
deep-shaft activated sludge, 451
deep-shaft process, brewery wastewater treatment, *452t*
deep-well disposal, 672–679, 822
defusing, 577
degradable fraction, *238f*, *390f*
degradable VSS, oxidation of, 241
degradation of oxidation products, 577
dehydration, 712, 734, 741–743
deinking mill wastewater, *310f*
denitrification, 614
 design procedure, 372–375
 industrial wastes or waste by-products for, *365t*
 overview, 358–368
 oxidation ditch with, *439f*
 SBR, 440
 systems, 368–369
depth determination, limestone bed, *84f*
Desalination Plant, Debiensko, Poland, 813
desalting, oil, 117–118
design
 aeration system, 221
 diffused aeration, 200–201
 equalization basin, *82t*
 neutralization system, *92f*
desorption, 211, 754
detention time, *103f*, *375f*
detoxification, of plastics additives wastewater, *321f*
dewatering, *605f*, 606, *624t*, 649–650. *See also* vacuum filtration
diatomaceous earth, 641
dichlorodiphenyl trichloroethylene (DDT), 770–771
dichloromethane, 779–780
dichlorophenol (DCP), 280
diesel engine manufacturing plants, 699–700
diffused aeration systems, 189, *192f*, 197–201, 785
diffusers, bubble, 184
diffusion, 179–180
dilute black liquor, treatment of on plastic packing, *478f*

dilution attenuation factor (DAF), *128f–129f*, 608
dioxins, 778
direct precipitation, 380
discharge. *See also* zero effluent discharge
 pretreatment of leachate, 873
 problems, 131–132
 recommended limits, *852t*
 typical ratio of, *802t*
discharge monitoring reports (DMRs), 892
discrete particles, *95f*
discrete settling, 94–98
disk centrifuges, 626
disk-nozzle separators, 626
dispersed suspended solids (DSS), 462, 466–467
dispersion tests, 110, *111f*
disposal. *See also* sludge; Superfund disposal site response costs
 of chlorinated solvents, 757–758
 of PCBs, 750
 pollution, 40–41
dissolved air flotation (DAF), 124, 621
dissolved oxygen
 effect on filamentous overgrowth, 301
 effect on nitrification rate at 20°C, *340f*
 relationship between F/M and, *304f*
distillery waste, anaerobic treatment of, 506
DMRs (discharge monitoring reports), 892
domestic wastewater, rotating biological contactors and, 489
dosages, carbon, 558, *559f*
downflow carbon column design, 538
drill cuttings, 734
drill mud circulation systems, *716f*
drilling muds, *730t*, 734–736
drilling, oil, 715
dry cleaning establishments, 758
dry time, 640
DSF (Dynasand filter), 704, *706f*
DSS (dispersed suspended solids), 462, 466–467
dual-media filters, 700–701, *705f*
dyes, *598t*
Dynasand filter (DSF), 704, *706f*

E

E&P. *See* exploration and production (E&P) pollution control
EBCT (empty bed contact time), 541, 550
EBPR (excess biological phosphorus removal), *381f*
economic concepts of zero effluent discharge, 809–810
economical limestone bed depth, 84–85
EDCs (endocrine disrupting chemicals), 5
EDTA (ethylenediaminetetraacetic acid), 38
efficacy of oxidation, 581–582
efficiencies
  aerator, *206t*
  oil separation unit, *114t*
effluent COD performance data, 894
effluent quality, effect of industrial wastewaters on, 463
effluent reduction, chlorinated hydrocarbons, 893
effluent suspended solids (TSS), *29f*, *103f*, 465–474, *469f*
effluent total COD (TCOD$_e$), 26
effluent total phosphorus concentration, *379f*
effluent toxicity, 392
effluents. *See also* zero effluent discharge
  BOD, *68f*
  COD and TSS composition for, *29f*
  concentrations, 76
  effects of recycle on toxicity testing, 821
  flows, *829f*
  limitations, 766–767, *865t*, 869–870
  mass rate peaking factor variation, *79f*
  paper mill, *153t*
  toxicity
    identification of effluent fractionation, 38
    overview, 31–35
    source analysis and sorting, 39
    testing, 821
  wastewater treatment plant, 793
EGSB (expanded granular sludge bed) process, 497, *499f*

electrical desalting, 117
electrochemical properties of colloids, 137
electrodialysis, *60t*
electrodialysis reverse, *60t*
electrokinetic coagulation, 141
electrolytic aluminum hydroxide, *143f*
electrolytic treatment of zinc cyanide, *174t*
electrostatic desalting, *118f*
Emergency Planning and Community Right to Know Act (EPCRA), 889–890
emerging pollutants, 5
emissions, VOC, 332–333
empty bed contact time (EBCT), 541, 550
EMS (environmental management systems), 884
emulsified oils, 116, 148
endocrine disrupting chemicals (EDCs), 5
endogenous coefficient, *390f*
environmental audits, 884
environmental compliance, 883–895
environmental governance
  annual reports for publicly-owned industrial complexes, 903
  internal environmental cost accounting system, 902–903
  ISO 14001 certification, 902
  organizational structure and participatory policies, 903
  overview, 901
  Sarbanes-Oxley Act (SOX) (publicly-owned companies), 902
  utilization of third party consulting specialists, advisory groups, and regulation experts, 903
environmental impact of PCBs, 749–750
environmental issues for zero discharge alternatives, *818t–819t*
environmental laws and regulations in United States, 876–878
environmental management systems (EMS), 884
Environmental Protection Agency (EPA), 876
  POTW ordinance guidelines, 851–852
  toxic chemicals, 2

Index

environmental regulations, 718
EPCRA (Emergency Planning and Community Right to Know Act), 889–890
equalization, 67–82
equalization tanks, 131–132
equilibrium biological solids concentration, 415
equipment
  coagulation, 147–148
  diffused aeration, 197–201
  measurement of oxygen transfer efficiency, 205–210
  measuring, 210–211
  surface-aeration, 202–205
  turbine aeration, 201–202
ethylbenzene, activated sludge reactors, 231t
ethylenediaminetetraacetic acid (EDTA), 38
Euglena, 405
excess biological phosphorus removal (EBPR), *381f*
exchange, ion. *See* ion exchange
exempt E&P wastes, 720–724
expanded granular sludge bed (EGSB) process, 497, *499f*
experimental procedures, ion exchange, 570–572
exploration and production (E&P) pollution control
  background information
    conventional gas, 712–716
    conventional oil/gas, 712
    E&P servicing, 716–717
  fluid characterization, 726–729
  overview, 711–712
  primary field operations, 729–730, 733
  problems, 744–745
  regulations
    exempt and nonexempt wastes, 720–724
    federal regulations, 717–720
    lease agreements and miscellaneous issues, 725–726
    local regulations, 725
    overview, 717
    state regulations, 724–725
  waste residuals and treatment options
    associated and nonassociated gas treatment, 737
    dehydration, 741–743
    drilling muds, 735–736
    produced oil, associated gas, and natural gas liquids, 737
    produced water, 737
    separation, 737–741
    sweetening, 743–744
  water sources description, 734–735
extended aeration, activated sludge processes, 437–439

F

facultative lagoons, 414, 418–420
facultative ponds, 404, 405, *406f*
FBR. *See* fed batch reactor
Fe dose, versus soluble P residual curve, *378f*
fed batch reactor (FBR), 54, 281
  nitrification test, 357–358
  overview, 322–323
Federal Energy Regulatory Commission (FERC), 717
federal environmental legislation, *877f*
Federal Industry Point Source Category Limits, 6–7
Federal Insecticide, Fungicide, and Rodenticide Act (FIFRA), 766
Federal Interstate Land Sales Full Disclosure Act, 724
federal regulations for E&P pollution, 717–720
federal spill/release reporting requirements, *890f*
Federal Tort Claims Act (FTCA), 900
Federal Water Pollution Control Act, 724
feed temperature, 215t
feedwater stream velocity, membrane processes, 687
FERC (Federal Energy Regulatory Commission), 717
ferric, 783
ferric chloride coagulation, of activated sludge effluent, 471t
ferric iron, 167–169
ferric salts, 144
ferrous iron, 167–169, 587–588, 783

Index

ferrous sulfate, 159–160
fertilizer wastewater, alkalinity utilization in treatment of, *344f*
Ficks law, 179
field operations, E&P, 729–730, 733
FIFRA (Federal Insecticide, Fungicide, and Rodenticide Act), 766
filament types, found in industrial wastewaters, *302t*
filamentous bulking control, 307–308
filamentous organisms, anaerobiosis and, 312
filamentous overgrowth, 301
films, stagnant, 180
filter belts, 646
filter loading equation, 640
filter matter, 734
filter presses, 660
filter slime, mean retention time and, 476
filterability, sludge, 628, 630
filters, continuous carbon, 532–534
filtration
    granular media, 700–706
    membrane processes, 679–680
    oil, 112, 116
    performance, *707t*
    rates of, 701–702
    sludge handling and disposal
        pressure, 642–645
        vacuum, 638–642
    toxicity identification of effluent fractionation, 37
final clarification, activated sludge processes, 454–462
final clarifier relationships, *457f*
fine bubble diffusers, *467f*
fine-bubble diffusers, 197–198, *199f*, *206t*
fixed-bed exchangers, 568
flash tank separator-condensers, 741, 743
flat-plate membrane systems, 694
flexible sheath tubes, *199t*
floating media, 453
floc load relationship, *311f*
Floc Shear test results, pulp and paper mill wastewaters, 466
flocculated sludge, 106, *108f*, 650
flocculated suspended solids (FSS) test, 462
flocculation
    activated sludge processes, 462
    chemical, *127f*
    colloids, 141
flocculents
    in centrifugation, 636
    settling, 98–106
flooding drop curves, packed tower, *217f*
flotation
    air solubility and release, 120–123
    overview, 120
    system design, 123–131
    thickening through dissolved air, 621–624
flotation units, clarifier, *124f*
flow diagrams
    desalter-oil recovery-wastewater treatment, *118f*
    flotation units, *622f*
    gas recovery from wells, *713–714f*
    glycol dehydration, *742f*
    limestone neutralization, *84f*
    nonurban gas treatment plant, *740f*
    oil recovery from wells, *713f*
    for petroleum refineries, *796f*
    reuse-zero discharge, *820f*
    urban drill and production site, *740f*
flow equalization, 69–70
flow, overland, 667
flow schematic, BRWTP, *865t*
flowsheets, process selection, *55f*
flow-through lagoons, 404–405
flue gases, 86
fluid characterization, E&P-related, 726–729
fluidized beds, 658
fluidized-bed reactor (FBR) wastewater, 495–497
fluorides, 166–167, *168t*, *176t*
flux
    in membrane processes, 686
    rate of membranes, 696
    solids, 106–107, 616–617
    water, 685
food/microorganisms (F/M) ratio, 437
food-processing wastewaters, *673t–674t*
form time, 639

936 Index

Formosa Plastics case history
　history of technologies, 811–813
　industry applications of technology, 813
　initial evaluation results, 817–821
　overview, 810–811
　recycle effects on effluent toxicity testing, 821
　studies of zero discharge options, 813–815
　technologies appropriate for, 815–817
formulation of adsorption, 526–530
fortuitous metabolism, 505
fouling, membrane, 696
fractionation, *36f*
fracturing sand, 734
Freundlich isotherm, 526, *529t*, *533f–534f*
FSS (flocculated suspended solids) test, 462
FTCA (Federal Tort Claims Act), 900
fungicides, 770
fusing, 577

G

GAC. *See* granular activated carbon
GAOs (glycogen accumulating organisms), 383
gas
　E&P waste residual treatment, 737
　gas films, 180
　gas flow, *192f*
　natural, *728t*, 729–730, 733
　processing plants, 737, *740f*
　production, cumulative
　　for inhibitory wastewater, *515f*
　　for nontoxic wastewaters, *514f*
Geosyntec conducted studies, 777
Giodler process, 744
glucose inhibition test, 323–325
glycogen accumulating organisms (GAOs), 383
glycol dehydration, 712, 741, 743
goldmine tailings wastewater, *490t*
Gore factors, 826
governance, environmental. *See* environmental governance
granular activated carbon (GAC)
　column design, *539f*
　columns schematic, *536f*
　process flowsheet, *540f*
　small column tests, 550–553
　TOC and toxicity reduction by columns, *556f*
　in treating chlorinated compounds, 763
granular carbons, 530
granular media filtration, 700–706
granular-activated carbon (GAC), 318
grapefruit processing wastewater, *309t*
grasses, 664–665
grassroots (original) construction, 880
gravity belt thickeners, 625–626, *627f*
gravity filters, 700
gravity thickening, 616–621
Great Lakes Initiative, 7
green algae, 405

H

HAA (haloacetic acids), 779
half-lives for organic compounds, 761–763
half-reactions, oxidant, 578–579
haloacetic acids (HAA), 779
halogenated aliphatics, *589f*
handling. *See* sludge
Hazardous Solid Waste Amendment (HSWA), 677, 877
hazardous waste
　injection well storage of, 677–679
　leaching tests, 606
　MEP for, 609
　RCRA regulations, 117, 718
　SPLP for, 608–609
　TCLP for, 606–608
　WET for, 609
head loss, 702–703
hearth furnaces, 658
heavy metals removal
　on activated carbon, *554t*
　arsenic, 157–158
　barium, 159
　cadmium, 159
　carbon dosages for, *559f*
　chromium, 159–166
　copper, 166
　effluent levels achievable in, *175t*
　fluorides, 166–167
　iron, 167–169

Index

lead, 169
manganese, 169
mercury, 169–170
from municipal sewage, *297f*
nickel, 170–171
overview, 151–157
selenium, 171
silver, 171–172
zinc, 172
heavy-metal selective chelating resins, 566
Henry's law, 181, 212–215, 761
hexachlorobutadiene, 778
hexavalent chromium removal, *164t, 574t*
high-BOD wastewaters, 489
high-effluent suspended solids levels, 466
high-rate trickling filters, *483t, 484f*
high-speed surface aerators, *203f–204f, 205t*
high-valence cations, 141
holding ponds, *81f*
hollow fiber membranes, 683, 694
horsepower guide, activated sludge process mixing, *207t*
hot wastewaters, 290
HRC (hydrogen releasing compounds), 765
HRT (hydraulic retention time), 414
HSWA (Hazardous Solid Waste Amendment), 677, 877
hurdle rate, 881
hydraulic loading
 oxygen transfer rate coefficient and, *481f*
 rate determination, *130f*
hydraulic presses, 645
hydraulic problems, activated sludge processes, 462
hydraulic retention time (HRT), 414
hydrocarbons, *533f–534f*, 735
hydrochloric acid, *88t–89t*
Hydroclear filter, 704
hydrogen peroxide, 314
 chemical oxidation, 586–590
 odor control, 783
 oxidation precipitation system, 159
hydrogen releasing compounds (HRC), 765

hydrogen sulfide, 5, 780–781
hydrolysis, hydrothermal, 597, 599
hydrolytic microorganisms, 499
hydrophilic colloids, 137
hydrophobic colloids, 137
hydrothermal hydrolysis, 597, 599
hydrothermal processes for chemical oxidation, 596–600
hydroxide, *155f*
hydroxide precipitation treatment, *160t, 167t, 173t*
hyperfiltration, *60t*
Hyperion Energy, 880

I

identification matrix, physical-chemical treatment, *58t–61t*
Illinois, 854–855
impellers, 201–202
impounding-adsorption lagoon, 403
incineration of sludge, 606, 657–658
income statements, 853
Indianapolis, Indiana, 855
indirect dischargers, *871t*
induced-air flotation system, 124, *125f*
industrial environmental economics
 capital and operational economic planning, 878–880
 CERCLA, joint and several, and retroactive economic exposure, 895–896
 consultant and client perspectives multiplier and pricing, 916–918
 overview, 904–905
 project economics, 918–920
 salary to expense ratio and utilization, 910–916
 SUM concept, 905–910
environmental compliance, 883–895
environmental governance
 annual reports for publicly-owned industrial complexes, 903
 internal environmental cost accounting system, 902–903
 ISO 14001 certification, 902
 organizational structure and participatory policies, 903
 overview, 901
 Sarbanes-Oxley Act (SOX) 2002, 902

industrial environmental economics, environmental governance (*Cont.*):
  third party consulting specialists, advisory groups, and regulation experts, 903
  litigation exposure, 896–901
  new facility siting analysis and planning, 880–883
  regulatory compliance metrics and environmental laws and regulations in United States, 876–878
  overview, 875–876
industrial pretreatment. *See also* Barceloneta POTW, Puerto Rico
  case histories
    Austin, Texas, 858–863
    Chicago, Illinois, 854–855
    Indianapolis, Indiana, 855
    pretreatment of leachate discharges, 873
    San Diego, California, 855
    Shreveport, Louisiana, 855–858
  compliance monitoring, 849–851
  local limits development, 848–849
  national categorical pretreatment standards, 848–849
  overview, 847–848
  permitting process, *856–857t*
  POTW ordinance guidelines, 851–852
  user charge rates and POTW cost recovery, 852–854
industrial treatment facilities, equalization for, 68–69
industrial users, Austin, Texas, 861, *863f*
industrial wastewaters, *438t*
  adsorption isotherms results on, *555t*
  alternative treatment technologies, *50f*
  coagulation of, 148–151, *150t*
  discharge to municipal plants, *71f*
  effluent toxicity
    overview, 31–35
    source analysis and sorting, 38–39
    toxicity identification of effluent fractionation, 35–38
  estimating organic content, 18–31
  industrial waste survey, 8–18
  in-plant waste control and water reuse, 40–45
  membrane filtration of, 691
  overview, 1–2
  oxygen demand and organic carbon of, *23t*
  physical-chemical treatment, *58–61t*
  problems, 46–47
  quality tolerances, *797t*
  regulations
    air, 6
    liquid, 6–7
    overview, 6–7
  reuse and recycling of, 791–792
  sources and characteristics of, 7–8
  spray irrigation of, *672f*
  treatment system, operator training, *885t*
  undesirable characteristics, 2–5
infiltration, rapid, 666, *667t*
influents
  BOD, *68f*
  COD and TSS composition for, *29f*
inhibitors, 65–66
inhibitory wastewater, cumulative gas production for, *515f*
injection wells, 653–654, 675, 678–679, *720f*
injury and exposure incidents, *893f*
in-plant treatment, 51
in-plant waste control, 40–45
inspections, pretreatment, 851
instantaneous maximum discharge limits, *852t*
integrated fixed film activated sludge, 453
integrated systems, *50f*
internal environmental cost accounting system, 902–903
internal rate of return (IRR), 881
International Organization for Standardization (ISO), 884
Interstate Oil and Gas Compact Commission (IOGCC), 724–725
iodine number, 530
ion exchange
  in fluoride precipitation, 167
  in hexavalent chromium removal, *164t*
  overview, 565
  for perchlorate removal, 775–776

in physical-chemical waste
    treatment, 59t–60t, 62t
plating waste treatment, 572–576
problem, 576
resin, 60t
theory of
    experimental procedure, 570–572
    overview, 565–570
toxicity identification of effluent
    fractionation, 36
IQ toxic units, 35f
iron, 167–169, 176t
iron salts, 468
IRR (internal rate of return), 881
irrigation
    compared to other treatment
        systems, 667t
    design of systems, 668–671
    overview, 663–666
ISO 14001 certification, 902
isoelectric point, 141
isotherms
    adsorption, 555t
    tests for PCBs, 753t

J

jar test analysis, 146–147

K

kinetic relationships, in aerobic
    lagoons, 416
krypton stripping, 211

L

laboratory control of coagulation,
    146–147
laboratory evaluation
    of adsorption, 532
    of zone settling, 106–107
laboratory flotation cells, 125–126
laboratory ion exchange column
    assembly, 570–572
laboratory procedures, specific
    resistance, 631–633
laboratory settling studies, 98–99, 101
lactate addition, 764–765
lagoons
    aerated
        aerobic, 415–418
        facultative, 418–420

overview, 414–415
systems, 422–433
temperature effects in, 420–422
treatment method, 56
applications, 405–413
stabilization basins and
    aerated lagoons, 405
    anaerobic ponds, 404
    facultative ponds, 404
    lagoon applications, 405–413
    overview, 403–404
systems, 422–433
treatment method, 56, 650–653
Land Ban Provisions, HSWA, 677
land disposal of sludges, 606, 650–657
land incorporation systems, 655–657
land treatment
    design of irrigation systems,
        668–671
    irrigation, 663–666
    overland flow, 667
    overview, 663
    performance, 671–672
    rapid infiltration, 666
    waste characteristics, 667–668
landfills, PCBs in, 754
Langmuir equation, 527, 530
large-bubble diffusers, 198–200
latex manufacturing wastes, 150t, 151
laundromat wastes, 150t, 151
laws, PCB, 752t
Lazer Zee meter, 139, 140f
LCA (life cycle assessment), 45
leachate discharges, pretreatment of,
    873
leachate treatment system, 689–690
leaching tests to characterize residuals,
    605–609
lead, 169, 176t
lease agreements, E&P pollution,
    725–726
LeGrand, Harry, 826
life cycle assessment (LCA), 45
life, membrane, 686
liftable sludge, 649f
lignite carbon, 530
lime
    anaerobic process, 507–508
    in coagulant applications, 144, 145t
    in fluoride precipitation, 166
    in nickel precipitation, 171t

lime (*Cont.*):
  requirements, 133
  slaking, 86
  slurries, 86
limestone beds, 83–86
lime-waste titration curve, *91f*
limiting flux, 617, *619f*
liquid depth, aeration system, 185–186
liquid films, 180–181
liquid regulations, 6–7
liquid temperature, 285, *288f*
liquid-film coefficient, 180, 184
litigation exposure, 896–901
load balancing analysis, *81f*
loading rates
  hydraulic, *130f*
  limestone bed, *84f*
  soil, 668
local limits
  Austin, Texas, 861–862
  comparison of categorical pretreatment standards, *850t*
  defined, 849
  Shreveport, Louisiana, *859t–860t*
Local Pretreatment Limits, 7
local regulations, E&P pollution, 725
Lo-CAT system, 744
Louisiana, 855–858
low-speed surface aerators, *203f–204f, 205t*
low-strength wastewaters, 489
low-temperature separators (LTSs), 739, 741
Lyondell Chemical Company, et al. v. Albermarle Corporation, et al, 901

M

macroreticular resins, 571
MACT (Maximum Achievable Control Technology), 6
management programs, 52
manganese, 169
Marathon Ashland Petroleum (MAP), 698–699
mass loading, 616–617
mass sludge contributions, *832t*
mass transfer
  aeration equipment
    diffused, 197–201
    measurement of oxygen transfer efficiency, 205–210
    other measuring techniques, 210–211
    overview, 195–197
    surface-aeration, 202–205
    turbine aeration, 201–202
  air stripping of VOCs
    overview, 211
    packed towers, 213–221
    size of towers, 216–220
    steam stripping, 220–221
  mechanism of oxygen transfer, 179–195
  overview, 179
  problems, 221–223
mass-transfer coefficients, 212, 216–217
mass-transfer process, 179–180
material balance at corn plant, *14f*
material outputs, petroleum refinery, *798t–801t*
material safety data sheets (MSDS), 727, *733t*
Maximum Achievable Control Technology (MACT), 6
maximum limitations, PSES, 869–870
MBRs. *See* membrane bioreactors
MBTs (membrane biothickeners), 694
mean retention time, filter slime and, 476
mean saturation values, aeration tank, 184
measured COD and BOD_5, 25
measuring equipment, 210–211
meat waste, 408
media
  floating, 453
  height, 214
  trickling filter, *476t*
membrane bioreactors (MBRs), 385–386
  application, 697–698
  benefits of compared to conventional technology, 694–695
  case study A, 698–699
  case study B, 699–700
  issues, 695–697
  overview, 693
  reactor configuration, 693–694
  types of membranes, 693–694
  with ZeeWeed membranes, *450f*
membrane biothickeners (MBTs), 694

Index

membrane diffusers, 197–198
membrane filter press, *643f*
membrane filtration units, *450f*
membrane flux rate, 696
membrane processes
  applications, 688–693
  cleaning, 687–688
  feedwater stream velocity, 687
  flux, 686
  membrane life, 686
  membrane packing density, 686
  overview, *59t–60t*, 679–685
  for perchlorate removal, 775–776
  pH, 686
  power utilization, 687
  pressure, 685
  pretreatment, 687
  recovery factor, 686
  salt rejection, 686
  temperature, 686
  turbidity, 687
MEP (multiple extraction procedure), 609
mercury, 169–170, *175t–176t*
mesophilic regime, 284
metal chelation, 38
metal ion precipitation, *157t*
methane fermentation, 503
methanogens, 503
Methanosarcina, 503
Methanothrix, 503
methyl chloride, 780
methylene chloride, 779–780
Metropolitan Industrial Waste Program of San Diego, 855
Metropolitan Water District of Southern California, 777
Metropolitan Water Reclamation District of Greater Chicago (MWRDGC), *853t*, 854–855
microfiltration membrane systems, 693–694
microorganism metabolism, 66–67
microscreens, 706–708
Microtox, 34
mineralization, 577
Minerals Management Service (MMS), 718
minimization of waste. *See also* water recycle and reuse; zero effluent discharge
  overview, 791
  problem, 822–823
  RCRA hazardous waste issues, 806–809
miscellaneous surface water, 734
mixed liquor suspended solids (MLSS), 436, *456f*, 694
mixed liquor TDS, *299f*, *469t*
mixed liquor temperature, *292f*
mixed liquor volatile suspended solids (MLVSS), *242f*
mixed reactors, 611
mixing methods, 69
MLSS (mixed liquor suspended solids), 436, *456f*, 694
MLVSS (mixed liquor volatile suspended solids), *242f*
MMS (Minerals Management Service), 718
mobility in allocation of Superfund disposal site response costs, 830–837
Model Pretreatment Ordnance, 851
molasses number, 530
molecular structure, *527t*
molecular weight
  classification, 37
  distribution, biological effluents, *317t*
monitoring industrial pretreatment compliance, 849–851
monomedia filters, 701
monovalent ions, 297
Monsanto company, 748–749
monthly average limitations, PSES, 870–871
moving supports, 494
MSDS (material safety data sheets), 727, *733t*
MU (multiplier times utilization), 914
muds, drilling, *730t*, 734–736
multicomponent substrate removal, *259f*
multicomponent wastewaters, *538f*
multimedia, 701
multiple contributors, Superfund disposal site
  off-site, 830–837
  same-site, 837–842
multiple extraction procedure (MEP), 609

942　Index

multiple stages BOD removal comparison, *273f*
multiple unit carbon column design, 539
multiple zero-order concept, 270
multiple-hearth furnaces, 658, *659f*
multiplier and pricing, 916–918
multiplier times utilization (MU), 914
multistage neutralization process, *92f*
multistage operation
　pilot plant results for, *272f*
　pulp and paper mill, *423f*
municipal activated sludge plants, 462–465
MWRDGC (Metropolitan Water Reclamation District of Greater Chicago), *853t*, 854–855
Mycid shrimp, 821

N

NAAQS (national ambient air quality standards), 718
national ambient air quality standards (NAAQS), 718
national categorical pretreatment standards, 848–849
National Emission Standards for Hazardous Air Pollutants (NESHAP), 6, 884
National Industrial Pretreatment Program, 847–848. *See also* industrial pretreatment
National Pollution Discharge Elimination System program (NPDES), *739t*
National Priority List (NPL), 895–896
natural gas, *728t*, 729–730, 733, 737
NESHAP (National Emission Standards for Hazardous Air Pollutants), 6, 884
net present value (NPV), 880
net present worth (NPW), 880
net revenue, 905
neutralization
　chemicals for, *88t–89t*
　control of process, 87–93
　neutralizing tank power requirements, *93f*
　overview, 83
　system, 87
　types of processes, 83–87
NH_3 (unionized ammonia), 353
nickel, 170–171, *175t–176t*
nickel cyanide, 170
nickel hydroxide, 170
nitrate reduction recycle ratio, *375f*
nitrates, 783
nitrification
　in aerobic digestion, 611, 614
　ammonia and nitrite inhibition to, *355f*
　BAS, 355–357
　of coke plant wastewaters, *558t*
　design procedure, 369–372
　FBR nitrification test, 357–358
　of high-strength wastewaters, 342–343
　inhibition of, 343–355
　kinetics, 335–342
　overview, 334–335
　oxidation ditch with, *439f*
　pH and, 294
　rate determination, test procedure for, *356f*
　systems, 368–369
　trickling filters, 485
nitrifying bacteria, *336t*
Nitrobacter, 334, 335, 353, 355
nitrogen, 2, *255f*, *335f*, 655, 668–670
nitrogenous organics, 25
Nitrosomonas, 334, 335, 353
Nocardia foams, 314
nonassociated gas, 737, 745
noncompliance of pretreatment ordinances, 863
nondegradable TOC, *316f*
nonexempt E&P wastes, 720–724
nonionic polymers, 144
non-steady-state processes, 205–209, 532–533
nontoxic wastewaters, 52, *514f*
normally occurring radioactive materials (NORM), 735
Northwest Alloys, Inc., 813
NPDES (National Pollution Discharge Elimination System program), *739t*
NPDES (point source discharge) record, 892

NPL (National Priority List), 895–896
NPV (net present value), 880
NPW (net present worth), 880
nutrient requirements, biooxidation, 253–256

O

objectives of residuals management, 603
Occupational Safety and Health Administration (OSHA), 6, 884
OCPSF (Organic Chemicals, Plastics, and Synthetic Fibers) regulations, 758
odor control, 410, 424, 780–786
OECD method 209, 321–322
off-gas analysis, 210
off-gas treatment, air stripper, $220f$
off-site disposal, $831f$, $832t$
off-site multiple contributors, Superfund, 830–837
oil
 conventional, 712–716
 debris, 734
 emulsified, 148
 processing in petroleum refineries, 116–118
 produced, 737
 removal of, $128f$
 separation, 110–116
 as waste constituent, 5
Oil Pollution Act (OPA), 724, 896–897
oil-water separators, $113f$–$115f$
oily wastes, $129f$, 657, 689
OM&R (operational, sampling, maintenance, and replacement) costs, 852–853
OPA (Oil Pollution Act), 724, 896–897
operating characteristics, sour water strippers, 120
operating time history, 828–830
operational, sampling, maintenance, and replacement (OM&R) costs, 852–853
operator maintenance attachment, $888t$–$889t$
operator training, 884–885
optical monitoring applications, $82t$
ORC (oxygen-releasing compounds), 765

ordinance guidelines, POTW, 851–852
organic carbon
 of industrial wastewaters, $23t$
 removal, $559f$
Organic Chemicals, Plastics, and Synthetic Fibers (OCPSF) regulations, 758
organic chemicals wastewater
 Relationship between denitrification rate and temperature for, $361f$
 toxicity reduction, 396
 treatment of in Biohoch Reactor, $453t$
organic compounds, $528t$
 biodegradation of under anaerobic conditions, 505–507
 biooxidation, 278–284
organic constituents of wastewater, 18–31, $29f$
organic phosphates, 770
organic priority pollutants, EPA list of, $3t$–$5t$
organic removal, mathematical relationships of, 256–278
organics removal mechanisms
 biodegradation, 232–233
 overview, 225–227
 sorbability, 229–231
 sorption, 227–228
 stripping, 228–229
organizational structure, participatory policies and, 903
organochlorine insecticides, 766
original construction, 880
ORP (oxidation-reduction potential), 760, $761t$
orthokinetic coagulation, 141
ortho-phosphate, $376f$
Oscillatoria, 405
OSHA (Occupational Safety and Health Administration), 6, 884
osmosis, 680, $682f$
OTEs (oxygen transfer efficiencies), $198f$–$200f$, 201, 205–210
OUR (oxygen utilization rate), 322
overflow operating lines, 455, $461f$
overflow rates, $102t$, $103f$
overhead costs, 910, $913t$
overland flow, 667, $675t$
ownership, $828f$
oxidant reduction, 37–38

oxidation. *See also* chemical oxidation
  of cellular organics, 609
  of degradable VSS, 241
  ditch systems, 439–440
  manganese, 169
  ponds, 653
  processes, 58
oxidation-reduction potential (ORP), 760, 761t
OXY-DEP process, 450f
oxygen
  demand of industrial wastewaters, 23t
  injection, 782
  requirements for biooxidation, 246–252
  saturation, 184
  solubility of, 181–184
  transfer, 179–195, 204
  transfer and utilization, trickling filtration, 479–482
  update rate methodology, 247f
  utilization and sludge yield, 237–246
  utilization coefficients for food processing wastewater, 247f
  utilization of coefficients for a variety of wastes, 247t–248t
oxygen activated sludge, 448–451
oxygen penetration, facultative pond, 405, 406f
oxygen transfer efficiencies (OTEs), 198f–200f, 201, 205–210
oxygen utilization rate (OUR), 322
oxygen-releasing compounds (ORC), 765
oxygen-uptake rate, 238
ozonation, 584–585
ozone, 582–586, 783
ozone-peroxide treatment, 776

P

PAC (powdered activated carbon), 296t, 318, 345
Pacific Gas & Electric (PG&E), 755, 898
packed towers, 211–221
packing density, membrane, 686
packing media volume, 216–217
packing-house waste, 408
PACT process, 51, 554–561
PAO (phosphate accumulating organisms), 381

paperboard wastes, 148, 149t
parameter correlation plots, 391f
partitioning, 227
PCBs. *See* polychlorinated biphenyls
PCE (perchloroethylene), 756–758, 759f
peaking factor (PF), 76
pentachlorophenol, 778
perchlorates, 775–777
perchloroethylene (PCE), 756–758, 759f
performance
  of activated carbon systems, 553–554
  DAF, 128f
  land treatment, 671–672
perikinetic coagulation, 141
peripheral-feed circular clarifiers, 107
perlite, 641
permeability, soil, 665–666
permit exceedances, 895
permit limitations for pollutants, 866–867
permitted industrial users, Austin, Texas, 861f
permitting process, industrial pretreatment, 856t–857t
persistence in allocation of Superfund disposal site response costs, 830–837
persistent organic pollutants (POPs), 5
pesticides, chlorinated
  characterization of, 767–774
  overview, 766
  regulatory history, 766–767
  treatment methodologies, 775
petrochemical industry waste reduction techniques, 45
petrochemical plant influent wastewater, 277t
petroleum exclusion, 897
petroleum hydrocarbons, 672
petroleum refineries
  material outputs from processes, 798t–801t
  oil processing in, 116–118
  wastewater from, 276t, 554t, 795, 804–805
PF (peaking factor), 76
PG&E (Pacific Gas & Electric), 755, 898
pH, 293–294
  in biological systems, 83
  control of, 87
  effect on ammonia oxidation, 341f

membrane processes, 686
versus ortho-phosphate for lime precipitation, *376f*
plot for electrolytic aluminum hydroxide, *143f*
in spray irrigation systems, 667–668
pharmaceutical pretreatment
determining compliance monitoring for PSES pollutants, 871–872
determining limits for pollutants regulated under PSES, 866–871
final limits as they would appear in permit for facility B, 872–873
overview, 863–866
Pharmaceuticals and Personal Care Products (PPCPs), 5
PHB (polyβ-hydroxybutyrate), 381
Phelps Dodge Hidalgo Smelter, 813
phenol, 2, 585
phenol number, 530
phoredox flow sheets, *381f*
Phormidium, 405
phosphate accumulating organisms (PAO), 381
phosphorus, 2
precipitation, 378
removal, *379f*
biological phosphorus removal, 380–381
chemical phosphorus removal, 375–380
design considerations, 383–385
GAOs, 383
MBRs, 385–386
mechanism for, 381–383
SBR, 440
phostrip process, for phosphorus removal, *380f*
photosynthesis, oxygen production and, 405
phthalate esters, 5
physical state in allocation of Superfund disposal site response costs, 830–837
physical-chemical waste treatment methods of, *62t*
screening and identification matrix for, *58t–61t*
pilot belt filter testing, *647t*
plant waste flow, *16f*
plastic-packed filters, *474f*, 482

plastics additives wastewater
activated sludge inhibition from, *320f*
detoxification of, *321f*
plate and frame membranes, 683, *684f*
plate and frame pressure filters, 642–645
plate separators, 111
plating wastewaters, 161–162, 572–576, *688f*
plow-type mechanism, 109
plug-flow, *262f*, *265t*, 435–436
Point Comfort, Texas. *See* Formosa Plastics case history
point source discharge (NPDES) record, 892
pollutants
concentrations of, 66, *731t–732t*
final limits, 872–873
loading rates of, *654t–655t*
PSES
determining compliance monitoring for, 871–872
determining limits for, 866–871
pollution. *See also* exploration and production (E&P) pollution control
cost-effective control of, *43t*
reduction of, 40
Pollution Prevention Act of 1990, 44, 890
pollution prevention information clearinghouse (PPIC), 44
polyβ-hydroxybutyrate (PHB), 381
polychlorinated biphenyls (PCBs)
environmental impacts, 749–750
overview, 748–749
regulatory history of, 750–751
treatment methodologies, 751–755
polyelectrolytes, 144, 630–631, 636
polymers, 144, 151, *639f*, 648
polyvalent complexes, 568
POPs (persistent organic pollutants), 5
pore sizes, *679f*
porous media, 197
potassium permanganate, 596, 783
potentially responsible parties (PRPs), 895
POTWs. *See* publicly owned wastewater treatment works

Index

powdered activated carbon (PAC), 345
powdered-activated carbon (PAC), 318
power levels, aeration equipment, 205f
power plants, PCB cleanup in, 755
power utilization in membrane processes, 687
powered activated carbon (PAC), 296t
PPCPs (Pharmaceuticals and Personal Care Products), 5
PPIC (pollution prevention information clearinghouse), 44
precipitation
  for heavy metals removal
    arsenic, 157
    barium, 159
    cadmium, 159
    chromium, 159–166
    copper, 166
    enhanced removal of soluble metals by, 158t
    fluorides, 166–167
    iron, 167–169
    lead, 169
    manganese, 169
    mercury, 169–170
    nickel, 170–171
    overview, 151, 154–157
    selenium, 171
    silver, 171–172
    zinc, 172, 173t
  in physical-chemical waste treatment, 59t, 62t
precoating, 641
preneutralization, 83
present value (PV) terms, 880
press belts, 646
pressure drop curves, packed tower, 217f
pressure filtration, 642–645
pressure in membrane processes, 685
pretreatment. *See also* case studies
  for disposal wells, 675–676
  equalization, 67–82
  flotation
    air solubility and release, 120–123
    overview, 120
    system design, 123–131
  industrial
    compliance monitoring, 849–851
    local limits development, 848–849
    national categorical pretreatment standards, 848–849
    overview, 847–848
    user charge rates and cost recovery, 852–854
  limits on aircraft manufacturing, 858t
  membrane processes, 687
  neutralization
    control of process, 87–93
    overview, 83
    systems, 87
    types of processes, 83–87
  oil processing in petroleum refineries, 116–118
  oil separation, 110–116
  overview, 65–67
  problems, 131–134
  sedimentation
    calculation of solids flux, 106–107
    clarifiers, 107–110
    discrete settling, 94–98
    flocculent settling, 98–106
    laboratory evaluation of zone settling, 106–107
    overview, 94
    zone settling, 106
  sour water strippers, 119–120
Pretreatment Standards for Existing Sources (PSES)
  determining compliance monitoring for pollutants, 871–872
  determining limits for pollutants regulated under, 866–871
pricing and multiplier, 916–918
primary degradation, 577
primary field operations, E&P, 729–730, 733
primary treatment, 49. *See also* pretreatment
priority pollutants, 2, 226t
private sector clients, 910
process design criteria, laboratory and pilot plant procedures for development of
  reactor operation, 388–389

reduction of aquatic toxicity, 392–396
volatile organic carbon, 389–392
wastewater characterization, 386–388
produced oil, 737
produced water, *728t–729t*, 734, 737, *738f*, 744
product throughput, *828f*
production-based residual estimates, *842f*
profit centers, 902
prohibited discharges, 847–848
project economics, 918–920
properties of activated carbon
 adsorption system design, 536–549
 carbon regeneration, 534–536
 continuous carbon filters, 532–534
 GAC small column tests, 550–553
 laboratory evaluation of adsorption, 532
 overview, 530–531
 performance of activated carbon systems, 553–554
PRPs (potentially responsible parties), 895
PSES. *See* Pretreatment Standards for Existing Sources
psi (ψ) potential, 138
psychrophilic regime, 284
public sector clients, 910
publicly owned wastewater treatment works (POTWs), 877. *See also* Barceloneta POTW, Puerto Rico
 case histories
  Austin, Texas, 858–863
  Chicago, Illinois, 854–855
  Indianapolis, Indiana, 855
  pretreatment of leachate discharges, 873
  San Diego, California, 855
  Shreveport, Louisiana, 855–858
 compliance monitoring, 849–851
 local limits development, 848–849
 national categorical pretreatment standards, 848–849
 ordinance guidelines, 851–852
 overview, 847–848
 user charge rates and cost recovery, 852–854
publicly-owned industrial complexes, annual reports for, 903

Puerto Rico. *See* Barceloneta POTW, Puerto Rico
pulp and paper industry
 Floc Shear test results, 466
 waste stabilization pond performance, 407
pulp and paper-mill wastes, *104t–105t, 107f, 153t, 310f, 650f*
pulp and recycle mills, *651f*
pulp mill effluents, 688–689
PV (present value) terms, 880

Q

quality, attainable, *63t*

R

radioactive tracer technique, 211
range of wastewater flow, *803t*
rapid infiltration, 666, *667t*
rapid small-scale column test (RSSCT), 550–553
raw waste, *15f*
rbCOD (readily biodegradable carbon measured as COD), 381
RCRA (Resource Conservation and Recovery Act), 117, 677, 718, 767, 806–809, 876
Re (Reynolds numbers), 550
reaction rate coefficient
 K, 287
 for pulp and paper mills, *275t*
 two-stage operation on, *274f*
 for wastewaters, *265t*
reactivity, regions of, 584
reactor clarifiers, 109, *148f*
reactor operation, 388–389
readily biodegradable carbon measured as COD (rbCOD), 381
recirculated MBRs, 694
recirculation
 flotation system, *123f*
 pollution, 40
recovery factor, membrane processes, 686
rectangular clarifiers, 107–109
recycle ratio, 792
recycling. *See* water recycle and reuse
Red River, Louisiana, 855, 858
redox potential of chlorinated solvents, 760

reductive dechlorination, *760f*
reed canary grass, 664
refineries
    effluent flows of wastewaters, *829f*
    throughput, *829f*
    waste sludge generation, *831f*
refractory substances, 5, 25
regeneration
    carbon, 534–536, *537f*
    ion exchange resin cycles, 568, *569f*
regional initiatives, 7
regions of ozone reactivity, 584
regulation experts, environmental governance, 903
regulations
    E&P pollution
        exempt and nonexempt wastes, 721–724
        federal regulations, 717–720
        lease agreements and miscellaneous issues, 721–726
        local regulations, 725
        overview, 717
        state regulations, 724–725
    PCB, 750–751, *752t*
    for perchlorates, *776t*
    pesticide, 766–767, *768t*, *769f*
    for pharmaceutical industry, 864
regulatory compliance metrics, 875–878
relative biodegradability, *264t*, *279f*
release, air, 120–123
remediating technologies, *840t*
removal efficiencies, microscreen, *708t*
removal of suspended solids, 98–101
residual generation rates, 843
residuals, 603–604. *See also* waste
resin adsorption, 38, *59t*
resin utilization, 568
resins, ion exchange, 565–566, *569f*
Resource Conservation and Recovery Act (RCRA), 117, 677, 718, 767, 806–809, 876
retroactive economic exposure, 895–896
reuse, 603. *See also* water recycle and reuse
reuse-zero discharge flow diagram, *820f*

reverse osmosis (RO)
    comparison with osmosis, 680, *682f*
    effect of biological pretreatment on, *690f*
    overview, *59t*
    process schematic, *685f*
    system operational parameters for, *688t*
    zero discharge technologies, 812
    of zinc wastewaters, *174t*
Reynolds numbers (Re), 550
rigs, oil, 715
rim-flow circular clarifiers, 107
Ringlace ropelike media, 453
Risk Reduction Standards (RRS), 896
Rivers and Harbors Act 1899, 876
RO. *See* reverse osmosis
rotary drums, 625, 639, 706, 708
rotary-hoe mechanisms, 109
rotating biological contactors, 488–494
RRS (Risk Reduction Standards), 896
RSSCT (rapid small-scale column test), 550–553

S

Sacramento, California, 755
Safe Drinking Water Act of 1974 (SDWA), 719–720, 876
safety solvents, 756
salary to expense ratio and utilization, 910–916
salinity, 668, 792
salt rejection, 686
same-site multiple contributors, Superfund, 837–842
San Diego, California, 855, *858t*
San Gabriel Valley, California, 775–776
sand bed drying, 648–649
sandstone, *676f*
sandy soils, 666
sanitary sewer discharges standards, *873t*
SARA (Superfund Amendments and Reauthorization Act of 1986), 889
Sarbanes-Oxley Act (SOX) 2002, 902
saturation, oxygen, 184
SBR (sequencing batch reactor), 440–445

Index **949**

schematic diagram, air stripping system, *219f*
scour, 97–98, 204
screening and identification matrix, physical-chemical treatment, *58–61t*
screening laboratory procedures, *53f*
screw presses, 647–648, *651f*
scrubber systems, 782
SCWO (supercritical water oxidation), 58, 599
SDWA (Safe Drinking Water Act of 1974), 719–720, 876
secondary clarifier effluents, *585t*
secondary treatment, 49, 51
sedimentation
  clarifiers, 107–110
  discrete settling, 94–98
  flocculent settling, 98–106
  laboratory evaluation of zone settling and calculation of solids flux, 106–107
  overview, 94
  zone settling, 106
segregation, pollution, 41
selector flow sheet, *312f*
selenium, 171, *175t*, 572
self-sustaining combustion, 658
separation, *36f*, 737–741
sequencing batch reactor (SBR), 440–445
servicing, E&P, 716–717
settling
  data, batch, *620f*
  discrete, 94–98
  flocculent, 98–106
  sludge, *618f*
  studies, 98–99, 101
  tanks, *96f*, 101–103
  zone, 106–107
settling basin, 422
settling flux curve, 457
sewage, 253
Shreveport, Louisiana, 855–858, *859t–860t*
SIA (Surface Impoundment Assessment) manual, 826
sidestream softening, 811
signification biological nitrification, SBR, 440
silica, activated, 144

silver, 171–172, *176t*
silver cyanide, 172
silver sulfide, 172
simultaneous precipitation, 380
single component wastewaters, *538f*
single-sludge system, 368
single-stage operation, pilot plant results for, *272f*
single-stage operation, pulp and paper mill, *423f*
single-style BOD removal comparison, *273f*
siting analysis and planning, new facility, 880–883
slaking, lime, 86
slime biomass, rotating biological contactors and, 489
slopes, 667
sludge. *See also* specific resistance
  age of, 238
  characteristics at 35°C and 43°C, *291f*
  effect of age on PAC process, 557–558
  flocculated, 106, *108f*
  handling and disposal
    aerobic digestion, 609–616
    basket centrifuge, 626–627
    belt filter press, 645–647
    centrifugation, 635–638
    characteristics of sludges for disposal, 605–609
    disk centrifuge, 626
    factors affecting dewatering performance, 649–650
    flotation thickening, 621–624
    gravity belt thickener, 625–626
    gravity thickening, 616–621
    incineration, 657–658
    land disposal of sludges, 650–657
    overview, 603–605
    pressure filtration, 642–645
    problems, 658–660
    rotary drum screen, 625
    sand bed drying, 648–649
    screw press, 647–648
    vacuum filtration, 638–642
  handling, effect of industrial wastewaters on, 463
  in MBRs, 696

sludge (*Cont.*):
  quality of
    biological selectors, 308
    design of aerobic selectors, 308–315
    effect of industrial wastewaters on, 463
    filamentous bulking control, 307–308
    overview, 299–307
    yield, 237–246, 359t
sludge volume index (SVI), 455, *456f*
sludge-blanket units, 147–148
slurry reactor remediation of PCBs, 754
SMA (specific methanogenic activity), 514
small column tests, GAC, 550–553
SMP (soluble microbial products), 24, *27t–28t*, 232, 316–318, 506
Snyder v. United States, 900
sodium bicarbonate, 508
sodium carbonate, *88t–89t*
sodium hydroxide, *88t–89t*
sodium hypochlorite, 783
sodium meta-bisulfite, 159–161
sodium nitrate addition, pond treatment, *410f*
soils, 665, 668, 754
solar ponds, 811
solid bowl decanters, 635
Solid Waste Disposal Act, 724
solid waste management units (SWMUs), 828, 841
solids
  flux, 106–107, 616–617
  handling, 603–604
  recovery, *638f*
  removal in spray irrigation systems, 667
  retention, *702f*
solids flux rate, effect of temperature on, *292f*
solids retention time (SRT), 385
  ammonia removal and, *336f*
  relationship between degradable fraction and, *238f*
solubility
  air, 120–123
  oxygen, 181–184
  solubility factor development, *837t*
soluble degradable wastewaters, biosorption relationship for, 235
soluble metal removal, *158t*
soluble microbial products (SMP), 24, *27t–28t*, 232, 316–318, 506
soluble organics, 2
soluble P residual curve, versus Fe dose, *378f*
soluble pharmaceutical wastewater, 244
solvents
  chlorinated
    historical perspective, 755–758
    overview, 755–766
    treatment methodologies, 758–766
  extraction, 38
SOR (standard oxygen rating), 195
sorbability, 229–231
sorption, 227–228
SOUR (specific oxygen uptake rate), 246, 249–250
sour water strippers, 119–120
source management and control, *42t–43t*
SOX (Sarbanes-Oxley Act) 2002, 902
soybean wastewater, *263f*
SPCC (Spill Prevention Control and Countermeasure) Plan, 885
specific methanogenic activity (SMA), 514
specific oxygen uptake rate (SOUR), 246, 249–250
specific resistance
  capillary suction time test, 634–635
  laboratory procedures, 631–633
  overview, 628–631, *649f*
spill basins, 80–82
spill diversion control, *424f*
spill prevention, 885, *892f*
Spill Prevention Control and Countermeasure (SPCC) Plan, 885
spiral wound membranes, 683, *684f*
SPLP (synthetic precipitation leaching procedure), 608–609
sponges, 453
spray irrigation systems, *57t*, 663–665, *673t–674t*
sprinklers, 663–664, 667
SRT. *See* solids retention time
SS. *See* suspended solids

Index

stabilization basins, lagoons and aerated lagoons, 405
  anaerobic ponds, 404
  facultative ponds, 404
  lagoon applications, 405–413
  overview, 403–404
staged high recycle cooling process, 812
Stamford baffle, *468f*
standard acute toxic units, *35f*
standard oxygen rating (SOR), 195
standards, national categorical pretreatment, 848–849
state points, 455, *457f*
state regulations
  E&P pollution, 724–725
  environmental, 718
state water quality standards, 7
static aerators, *189t*, 199
static supports, 494
statute-based litigation, 896
steam stripping, 58, 220–221, *222t*
steam stripping surrogates, *871t*
Stickney Water Reclamation Plant, 854–855
stimulation, well, 677
stirrers, 106
Stockholm Convention Treaty, 5
stoichiometry, 578–581
Stoke's law, 94
storm runoff, 805–806, 855
strippability, versus sorbability, *231f*
stripping, 58, 216, 228–229. *See also* air stripping
  of VOCs
    emissions treatment, 332–333
    overview, 326–332
strong-acid cation resins, 566
strong-base anion resins, 566
substitution, pollution reduction through, 41
substrates, zero-order removal rates for, 257
subsurface contamination, 757–758
subsurface restoration, 763–765
subsurface waste-disposal systems, *677f*
sulfide, *155f*, 511
sulfide oxidation, *587t*, 596
sulfur dioxide, 160–161
sulfuric acid, *88t–89t*

SUM concept, 905–910
SUM Parameters values, *907t*
supercritical water oxidation (SCWO), 58, 599
Superfund. *See* Comprehensive Environmental Response, Compensation, and Liability Act
Superfund Amendments and Reauthorization Act of 1986 (SARA), 889
Superfund disposal site response costs
  cost allocation principles, 828–830
  literature review, 825–827
  multiple off-site contributors, 830–837
  overview, 825–827
  problem, 843–845
  same-site multiple contributors, 837–842
  summary, 842–843
surcharge formula, 862–863
surface active agents, *187f*
surface aerators, *189t*, *206t*
surface application of sludge, 653
Surface Impoundment Assessment (SIA) manual, 826
surface-aeration equipment, 202–205
surrogate pollutants, 871–872
suspended growth systems, *560t–561t*
suspended solids (SS)
  defined, 2
  probability of occurrence in raw waste, *15f*
  removal of, 98–101, *102t*, 703
  variations in, *11f*
suspended solids control, coagulant addition for, *470f*
SVI (sludge volume index), 455, *456f*
Swanwick method, 649
sweetening, 734, 743–744
SWMUs (solid waste management units), 828, 841
synthetic precipitation leaching procedure (SPLP), 608–609
synthetic rubber wastes, 151

T

tanks
  aeration, *190f*, 197
  settling, *96f*, 101–103
tannery wastewaters, 151, *154t*
target constituents, 830, 832

target salary to expense ratio, 911, 912
TCA (trichloroethane), 756–758
TCE (trichloroethylene), 756–758, 759f, 899
TCLP (toxicity characteristic leaching procedure), 606–608
$TCOD_e$ (effluent total COD), 26
TDS. *See* total dissolved solids
TEAD (Tooele Army Depot), 899
temperature
   in aeration systems, 184, *185f*
   of aerobic digesters, 615–616
   effect of industrial wastewaters on, 463
   effect on 30-d average performance of pond treating pulp and paper mill effluent, *409f*
   effect on aerobic biological oxidation
      overview, 284–293
      pH, 293–294
      toxicity, 294–299
   effect on aerobic digestion, 611–612
   effect on denitrification rate, *361f*
   effect on Henry's law constant, 214
   effect on trickling filtration, 482
   effects in aerated lagoons, 420–422
   feed, *215t*
   membrane processes, 686
terminal settling velocity, 94–96
tertiary nitrification, 485–488
tertiary-treatment processes, 51
Texas, 858–863. *See also* Formosa Plastics case history
textile wastewaters, *152t*
theoretical oxygen demand (THOD), 24
theory of adsorption, 525–530
theory of ion exchange
   experimental procedure, 570–572
   overview, 565–570
thermal regeneration, 534–535
thermophilic aerobic activated sludge, 454
thermophilic regime, 284
thickeners, 658
thickening underflow, 106
THOD (theoretical oxygen demand), 24
three-stage oxygen system, *449f*
timelines
   DDT, *771t*

perchlorate awareness and regulation, *776t*
pesticide regulations, *768t*
TMDLs (total maximum daily loads), 882
TOC (total organic carbon), *533f–534f*
TOD. *See* total oxygen demand
Toledo, Ohio, *873t*
toluene
   activated sludge reactors, *231t*
   stripping, *328f*
Tooele Army Depot (TEAD), 899
total dissolved solids (TDS), *182t–183t, 188f*
   effect on activated sludge treatment of agricultural chemicals wastewater, *470t*
   effect on effluent BOD, *298f*
   mixed liquor TDS, *299f*
total maximum daily loads (TMDLs), 882
total organic carbon (TOC)
   adsorption, *533f–534f*
   COD/TOC ratio, 581, *582f, 583f*
   defined, 18
   overview, 23–24, 26
   toxicity reduction by granular carbon columns, *556f*
total oxygen demand (TOD), 18
total petroleum hydrocarbons, 672
tower diameter, 217–218
towers, packed, 213–221
Toxic Release Inventory (TRI) Program, 725
Toxic Substances Control Act (TSCA), 751
toxic wastewater
   oxidation, 581
   source treatment of, *54f*
toxicants, 65
toxicity. *See also* effluents
   aerobic biological oxidation, 294–299
   allocation of Superfund disposal site response costs, 830–837
   aquatic, reduction of, 392–396
   and COD correlation, *39f*
   factor development, *836t*
   before and after oxidation, *598t*
   reduction by granular carbon columns, *556f*
   testing, 792

toxicity characteristic leaching
procedure (TCLP), 606–608
Toxics Release Inventory (TRI), 890
transfer coefficient, *185f*
treatability classes, 871
treatment. *See also* pretreatment
  chlorinated pesticides, 775
  chlorinated solvents, 758–766
  deep-well disposal, 672–679
  E&P waste residuals
    associated and nonassociated gas, 737
    dehydration, 741–743
    drilling muds, 735–736
    produced oil, associated gas, and natural gas liquids, 737
    produced water, 737
    separation, 737–741
    sweetening, 743–744
  granular media filtration, 700–706
  for high strength and toxic industrial wastewater, 52
  ion exchange resin, 568, *569f*
  land treatment
    design of irrigation systems, 668–671
    irrigation, 663–666
    overland flow, 667
    overview, 663
    performance, 671–672
    rapid infiltration, 666
    waste characteristics, 667–668
  membrane bioreactors
    application, 697–698
    benefits of compared to conventional technology, 694–695
    case study A, 698–699
    case study B, 699–700
    issues, 695–697
    overview, 693
    reactor configuration, 693–694
    types of membranes, 693–694
  membrane processes
    applications, 688–693
    cleaning, 687–688
    feedwater stream velocity, 687
    flux, 686
    membrane life, 686
    membrane packing density, 686
    overview, 679–685
    pH, 686
    power utilization, 687
    pressure, 685
    pretreatment, 687
    recovery factor, 686
    salt rejection, 686
    temperature, 686
    turbidity, 687
  microscreen, 706–708
  overview, 663
  perchlorate, 775–777
  polychlorinated biphenyls, 751–755
TRI (Toxic Release Inventory), 725, 890
trichloroethane (TCA), 756–758
trichloroethylene (TCE), 756–758, *759f*, 899
trichloromethane, 780
trickling filter treatment method, 56
trickling filters
  performance, *480t*
  tertiary nitrification through, *487f*
trickling filtration
  applications, 482–485
  effect of temperature, 482
  oxygen transfer and utilization, 479–482
  tertiary nitrification, 485–488
  theory, 474–479
trihalomethanes, 779
trivalent chromium treatment, *164t*
TSCA (Toxic Substances Control Act), 751
TSS (effluent suspended solids), *29f*, *103f*, 465–474, *469f*
tube clarifiers, 110
tubes, flexible sheath, *199t*
tubular membranes, 681, 683, *684f*
turbidity, 2, 687
turbine aerators, *189t*, 201–202, *206t*
turbulence
  effect on effluent suspended solids when fine bubble diffusers are used, *467f*
  oxygen transfer, 186, *187f*
  settling tank, 96–97
turbulent mixing, 185
two-film transfer concept, 180
two-stage activated sludge system, 271
two-stage operation, on reaction rate coefficient, *274f*

U

UASB (upflow anaerobic sludge blanket) process, 497
UF (ultrafiltration) membranes, 697
UIC (underground injection control) program, 719
ultimate degradation, 577
ultrafiltration, 60t, 689
ultrafiltration (UF) membranes, 697
ultraviolet (UV) radiation, 585–586, 588–589
unacceptable degradation, 577
underflow
  operating lines, solids flux curve and, 461f
  sludge, 616–617
  thickening, 106
underground injection control (UIC) program, 719
underground storage tanks (USTs), 877
UNEP (United Nations Environment Programme), 46
Union Carbide Plant (Bay City, Texas), 880
unionized ammonia (NH_3), 353
unit cost of treatment, 853t
United Nations Environment Programme (UNEP), 46
upflow anaerobic sludge blanket (UASB) process, 497
upflow carbon column design, 539
upflow units, 116
upflow-downflow carbon column design, 539
user charge rates, industrial pretreatment, 852–854
USTs (underground storage tanks), 877
UV (ultraviolet) radiation, 585–586, 588–589

V

vacuum filtration, 638–642, 660
vapor pressure, 761
variability, DAF performance, 128f
variable-volume equalization basins, 69, 70f, 78t–80t
variation in flow
  from batch operation, 8f
  for representative industrial wastes, 10t
  tomato waste, 9f
VC (vinyl chloride), 755, 765–766
vegetable-processing wastes, 151
velocity, settling, 94–96
VFA (volatile fatty acids), 508
vinyl chloride (VC), 755, 765–766
vinyl chloride monomer (VCM), 765–766
viscous bulking, 314
VOCs. *See* volatile organic compounds
volatile fatty acids (VFA), 508
volatile materials, 5
volatile organic carbon, 389–392
volatile organic compounds (VOCs), 884. *See also* chlorinated compounds
  removal of, 784–785
  stripping of, 228
    emissions treatment, 332–333
    overview, 211, 326–332
    packed towers, 213–221
    size of towers, 216–220
    steam stripping, 220–221
volatile suspended solids (VSS), 232–233, 241
volatility factor development, 839t
volume in cost allocation principles, 828–830
volumetric capacity, stabilization basins, 403–404
VSS (volatile suspended solids), 232–233, 241

W

WAO (wet air oxidation), 58, 558, 597, 599–600
waste. *See also* sludge; water recycle and reuse; zero effluent discharge
  acid, 83–86
  aeration system, 186
  alkaline, 86–87
  conceptual wastestream, 840t
  flow of
    diagram at corn plant, 14f
    in partially filled sewers, 13f
  injection wells, 676f
  land treatment, 667–668

Index

minimization
  in-plant waste control and water reuse, 40–45
  overview, 791
  problem, 822–823
  RCRA hazardous waste issues, 806
off-site disposal of, *831f, 832t*
plating, 572–576
residuals, E&P
  associated and nonassociated gas treatment, 737
  dehydration, 741–743
  drilling muds, 735–736
  E&P, 734–735
  produced oil, associated gas, and natural gas liquids, 737
  produced water, 737
  separation, 737–741
  sweetening, 743–744
waste extraction test (WET), 609
waste stabilization pond, facultative, *404f*
wastewater
  aerobic biological oxidation, 386–388
  composition, effect on filamentous overgrowth, 301
  treatment. *See also* industrial wastewaters
    biological waste treatment, *57t*
    maximum quality attainable from waste treatment processes, *63t*
    overview, 49–52
    physical-chemical waste treatment, *62t*
    process selection, 52–56
    regulations, 6–7
    screening and identification matrix for physical-chemical treatment, *58t–61t*
wastewater cooling towers, 812
wastewater flow refineries, *803t*
wastewater treatment plant (WTTP) effluents, 793
water conservation, *44f*
water flux, 685
water, produced, 737
water quality, produced, *728t–729t*
water recycle and reuse. *See also* zero effluent discharge
  benchmarking, 802–806

case histories, 795–802
decision-making process, 794–795
effects on effluent toxicity testing, 821
limits of, 792–793
overview, 791–792
problem, 822–823
of wastewater treatment plant effluents, 793
water sources, E&P, 734–735
Water Utilities Department of San Diego, 855
Water9 model, 763
water-soluble compound removal, *215t*
weak-acid cation resins, 566
weak-base anion resins, 566
weight
  acid, 85
  cost allocation principles, 828–830
  losses of carbon, 535
weighting aids, *145t*
weighting factors, *835f*, 843
weirs, 110
well-head pressure, 676
wells
  Class II, 719
  gas, 712
  oil, 712
  stimulation of, 677
Western Greenhouses v. United States, 900
WET (waste extraction test), 609
wet air oxidation (WAO), 58, 558, 597, 599–600
wireline methods, 717
woodlands, 665
workover wastes, 734
WTTP (wastewater treatment plant) effluents, 793

X

xylene, *231t*

Z

ZeeWeed membranes, *450f*
zero dissolved oxygen, 195
zero effluent discharge
  economic concepts, 809–810

zero effluent discharge (*Cont.*):
  Formosa Plastics case history
    history of technologies, 811–813
    industry applications of technology, 813
    initial evaluation results, 817–821
    overview, 810–811
    recycle effects on effluent toxicity testing, 821
    studies of zero discharge options, 813–815
    technologies appropriate for, 815–817
    problem, 822–823
    summary of concepts, 821–822
zero-order removal rates, for specific substrates, 257
zeta potential (ζ)
  in control of coagulants, 146
  defined, 138
  overview, 139–141
zinc, 172, *173t–176t*
zinc cyanide, *174t*
zone settling, 106–107
zone settling velocities (ZSVs), 455